The Organic Chemistry of
Enzyme-Catalyzed Reactions

The Organic Chemistry of Enzyme-Catalyzed Reactions

RICHARD B. SILVERMAN

*Department of Chemistry and Department of
Biochemistry, Molecular Biology, and Cell Biology
Northwestern University, Evanston, Illinois*

ACADEMIC PRESS

An Elsevier Science Imprint

San Diego   San Francisco   New York   Boston   London   Sydney   Tokyo

Copyright © 2002, 2000 by ELSEVIER SCIENCE

Academic Press
Harcourt Place, 32 Jamestown Road, London NW1 7BY, UK
http://www.academicpress.com

Academic Press
An Elsevier Science Imprint
525 B Street, Suite 1900, San Diego, California 92101-4495, USA
http://www.academicpress.com

ISBN 0-12-643731-9

Library of Congress Control Number: 2001098273

A catalogue record for this book is available from the British Library

Typeset by G&S Typesetters, Austin, Texas, USA
Printed and bound in Great Britain by MPG Books Ltd, Bodmin, Cornwall

02 03 04 05 06 07 MP 9 8 7 6 5 4 3 2 1

To Barbara, Matt, Mar, and Phil, for their love, their laughter, and for being a constant source of pride, joy, and admiration.

Contents

3   *Reduction and Oxidation*

4   *Monooxygenation*

5   *Dioxygenation*

6   *Substitutions*

7   *Carboxylations*

8   *Decarboxylation*

9   *Isomerizations*

10   *Eliminations and Additions*

11   *Aldol and Claisen Reactions and Retroreactions*

12 *Formylations, Hydroxymethylations, and Methylations*

13 *Rearrangements*

Preface

This is *not* your standard enzymology text. It actually serves two functions: it is an enzymology text for organic chemists and an organic chemistry text for enzymologists. It does not follow the usual traditions of biochemistry texts, which discuss such topics in enzymology as metabolic pathways, biosynthesis, protein synthesis and structure, and regulatory mechanisms. Instead, it seeks to give organic chemists an appreciation that enzymology is simply a biological application of physical organic chemistry, and to teach biochemists to view enzymology from the perspective of organic chemical mechanisms.

The text is organized according to organic reaction types so that the reader learns to think this way when looking at unknown enzyme systems. Each chapter represents a particular class of organic reactions catalyzed by different classes of enzymes, rather than the typical approach of standard biochemistry texts in which classes of enzymes are discussed and the reactions that they catalyze are mentioned as part of the characterization of the enzyme. This text also emphasizes the design of experiments to test enzyme mechanisms, so that the reader becomes familiar with approaches taken to elucidate new enzyme mechanisms. No attempt has been made to discuss each reference in detail; some experiments are cited and conclusions from these experiments are made. If more detail is desired, the original reference should be consulted.

There is no way that a text designed for a one-semester or even a one-year course could cover all of the enzymes that have been reported in the literature. In fact, that is not the purpose of this text, nor is it important to do so. I believe that what is most important is to be able to recognize and categorize an enzymatic reaction, to associate that reaction with a particular class of enzymes, and then to design experiments to test hypotheses regarding the mechanism for that enzyme-catalyzed reaction. Consequently, only representative enzymatic examples of each of the various reaction types are described here. Therefore, some of your favorite enzymes may not be included in this text, simply because I chose a different enzyme as an example for that particular reaction mechanism. This approach allows the instructor to add other enzymes to the discussion of a particular reaction mechanism. If more in-depth knowledge of a particular enzyme system is desired, or if a full-year course is offered, then the literature references cited can be assigned for critical analysis. I have taken examples from both the current literature and the older literature so that readers can appreciate that clever experiments have been carried out for many years.

I must thank a succession of teachers for my excitement about enzyme-catalyzed reactions. As an organic chemistry graduate student at Harvard, I started with no interest at all in enzymes because I had the impression that they were magic boxes that somehow catalyzed reactions. My only passion was the synthesis of natural products. My mentor, David Dolphin, who had other interests as well, asked me, as a side project, to synthesize a deuterated compound that his "collaborator" needed for studying the mechanism of an enzyme-catalyzed reaction. Having no interest in enzymes, I synthesized the compound without asking its utility in this mechanistic study (something I now tell my students *never* to do). Not long after I began working on my main project, the synthesis of the antitumor antibiotic camptothecin, its first total synthesis was reported. By this time, David had cunningly convinced me that, because I had already synthesized the desired deuterated molecule for his collaborator, it would be easy to attach it to a cobalt complex, which his collaborator would then use in his mechanistic study. So, while working hard on the synthesis of camptothecin, I learned about making cobalt complexes and attached the ligand. It soon became apparent that camptothecin was the focus of no less than six other research groups, because all these groups published syntheses of this molecule by my second year in graduate school! Having no desire to be the seventh (or tenth?) person to synthesize camptothecin, I finally asked David what this cobalt complex was for and found out that it was to carry out a model study of a potential mechanism for the coenzyme B_{12}-dependent rearrangements. Although I had resisted the temptation to become interested in enzymology, my curiosity was piqued. It was not difficult to convince David that this new project sounded interesting, so he agreed to let me work on this project for my Ph.D. thesis. (Was there ever really a collaborator, or was this my introduction to the psychology of assistant professors?)

In my second year, I sat in on a general biochemistry course, which corroborated my suspicions that enzymes were black boxes, and I realized that organic chemists needed to enter this field to clarify the "mysteries" of enzymology. The fog about enzymes began to lift in my third year, when I was fortunate to sit in on a unique course in enzymology taught by a relatively young (and getting younger every year, from my perspective) visiting professor. It was in this class that I was shown the connection between the black box of enzyme-catalyzed reactions and organic chemical mechanisms. The excitement of the subject, the clarity of the exposition, and the wit of the professor changed my opinions about the science of enzymology and changed the direction of my career goals and my research interests. Thanks, Jeremy Knowles.

Not long after I finished this course, another great stroke of fate occurred; Bob Abeles gave a colloquium at Harvard. It was this colloquium, and my two-year postdoctoral stint in his lab, that demonstrated the applications of the concepts in Jeremy's course and the value of organic chemistry to the study of enzyme-catalyzed reactions.

I am very grateful to those who unselfishly agreed to act as reviewers of this text. I selected four scientists, whom I considered to be the real experts in the general areas discussed, for each of the chapters; the editor's assistant, Linda Klinger (née McAleer), tried to get two of these to read each chapter. She was successful in all but three of the chapters, for which only one reviewer participated. Many thanks go to Vern Schramm (for two chapters), Dick Schowen, Frank Raushel, Ted Widlanski, Ben Liu, Paul Ortiz de Montellano, Paul Fitzpatrick, John Lipscomb, Mark Nelson, Richard Armstrong, Steve Withers, Ron Kluger, Marion O'Leary, George Kenyon, Ralph Pollack, Al Mildvan, Chris Whitman, Dennis Flint, Rob Phillips, Eileen Jaffe, Rowena Matthews, Jim Coward, Perry Frey, Bob Abeles, and John Blanchard. Your efforts are much appreciated.

As some of you may recognize, Chris Walsh's textbook *Enzymatic Reaction Mechanisms* (W. H. Freeman: San Francisco, 1979), played an important part in shaping my approach to presenting the intricacies of enzyme-catalyzed reactions. I thank Chris for getting the study of modern mechanistic enzymology off to a great pedagogical start.

For those of you who believe that a textbook should be a formal piece of writing, I apologize for the informality throughout this text; I wanted this book to be read as though I was talking to you about enzyme mechanisms.

As I find mistakes in the text, I will post them on my website, located at http://www.chem.northwestern.edu/~agman/index.htm. When you find errors, please notify me at Agman@chem.northwestern.edu. Among all of us, maybe we'll get this thing right.

Richard B. Silverman

About the Author

Professor Richard B. Silverman received his B.S. degree in chemistry from The Pennsylvania State University, his M.A. and Ph.D. degrees in organic chemistry from Harvard University, and he carried out postdoctoral research in enzymology under the guidance of Professor Robert H. Abeles at Brandeis University. He has been on the faculty of Northwestern University in the Department of Chemistry since 1976 and also in the Department of Biochemistry, Molecular Biology, and Cell Biology since 1986. Professor Silverman is a member of the Northwestern University Institute for Neuroscience, the Lurie Cancer Center, the Center for Biotechnology, and the Drug Discovery Program.

He was named a DuPont Young Faculty Fellow (1976), an Alfred P. Sloan Research Fellow (1981), a NIH Research Career Development Awardee (1982), a Fellow of the American Institute of Chemists (1985), and a Fellow of the American Association for the Advancement of Science (1990). In addition to having been chosen for the Northwestern University Faculty Honor Roll seven times, he was honored with the 1999 E. LeRoy Hall Award for Teaching Excellence, the 2000 Northwestern Alumni Association Award for Teaching Excellence, and in 2001 was named Charles Deering McCormick Professor of Teaching Excellence. He is a member of the editorial boards of the *Journal of Medicinal Chemistry, Archives of Biochemistry and Biophysics,* the *Journal of Enzyme Inhibition,* and *Archiv der Pharmazie-Pharmaceutical and Medicinal Chemistry* and has co-chaired the Gordon Research Conference on Enzymes, Coenzymes, and Metabolic Pathways (1994). He has given numerous two- and three-day short courses at meetings and at companies on drug design and drug action and on enzyme mechanisms and inhibition.

Professor Silverman is the author or co-author of over 190 research publications in enzymology, medicinal chemistry, and organic chemistry and is the holder of 19 patents. He also has written two other textbooks: *Mechanism-Based Enzyme Inactivation: Chemistry and Enzymology* and *The Organic Chemistry of Drug Design and Drug Action.*

CHAPTER 1

Enzymes as Catalysts

I. WHAT ARE ENZYMES, AND HOW DO THEY WORK?

A. Historical

Segel[1] has given a fascinating historical perspective on the discovery of enzymes; some of the notable events are mentioned here. One of the earliest observations of enzyme activity was reported in 1783 by Spallanzani, who noted that the gastric juice of hawks liquefied meat. Although the digestive effects were not ascribed to enzymes per se, Spallanzani recognized that something in the hawk juice was capable of converting solid meat into a liquid. Over the next 50 years many other observations suggested the existence of enzymes, but the first "isolation" of an enzyme is credited to Payen and Persoz. In 1833 they added ethanol to an aqueous extract of malt and obtained a heat-labile precipitate that was utilized to hydrolyze starch to soluble sugar. The substance in this precipitate, which they called *diastase,* is now known as amylase. Schwann "isolated" the first enzyme from an animal source, pepsin, in 1834 by acid extraction of animal stomach wall. Berthelot obtained an alcohol precipitate from yeast in 1860, which converted sucrose to glucose and fructose; he concluded that there were many such ferments in yeast. In 1878 Kühne coined the name *enzyme,* which means "in yeast" to denote these ferments. It was Duclaux who proposed in 1898 that all enzymes should have the suffix "ase" so that a substance would be recognized as an enzyme from the name.

Enzymes are, in general, natural proteins that catalyze chemical reactions; RNA also can catalyze reactions.[1a] The first enzyme recognized as a protein was jack bean urease, which was crystallized in 1926 by Sumner[2] and was shown to catalyze the hydrolysis of urea to CO_2 and NH_3. It took almost 70 years more, however, before Andrew Karplus obtained its crystal structure (for the enzyme from *Klebsiella aerogenes*).[3] As it turns out, urease is one of the few nickel-containing enzymes now known. By the 1950s hundreds of enzymes had been discovered, and many were purified to homogeneity and crystallized. In 1960 Hirs, Moore, and Stein[4] were the first to sequence an enzyme, namely, ribonuclease A, having only 124 amino acids (molecular weight 13,680). This was an elegant piece of work, and William H. Stein and Stanford Moore shared the Nobel Prize in chemistry in 1972 for the methodology of protein sequencing that was developed to determine the ribonuclease A sequence. Ribonuclease A also was the target of the first chemical synthesis of an enzyme; two research groups independently reported its synthesis in 1969.[5]

Enzymes can have molecular weights of several thousand to several million, yet catalyze transformations on molecules as small as carbon dioxide or nitrogen. Carbonic anhydrase from human erythrocytes, for example, has a molecular weight of about 31,000 and each enzyme molecule can catalyze the hydration of 1,400,000 molecules of CO_2 to H_2CO_3 per second! This is almost 10^8 times faster than the uncatalyzed reaction.

In general, enzymes function by lowering transition-state energies and energetic intermediates and by raising the ground-state energy. The transition state for an enzyme-catalyzed reaction, as in the case of a chemical reaction, is a high-energy state having a lifetime of about 10^{-13} sec, the time for one bond vibration.[6] No spectroscopic method available can detect a transition-state structure.

At least 21 different hypotheses for how enzymes catalyze reactions have been proposed.[7] The one common link between all these proposals, however, is that an enzyme-catalyzed reaction always is initiated by the formation of an *enzyme–substrate (or E·S) complex,* in a small cavity called the *active site,* where the catalysis takes place. The concept of an enzyme–substrate complex was originally proposed independently in 1902 by Brown[8] and Henri;[9] this idea extends the 1894 *lock-and-key hypothesis* in which Fischer[10] proposed that an enzyme is the lock into which the substrate (the key) fits. This interaction of the enzyme and substrate would account for the high degree of specificity of enzymes, but the lock-and-key hypothesis does not rationalize certain observed phenomena. For example, compounds whose structures are related to that of the substrate, but have *less* bulky substituents, often fail to be substrates, even though they should have fit into the enzyme. Some compounds with *more* bulky substituents are observed to bind *more* tightly to the enzyme than does the substrate. If the lock-and-key hypothesis were correct, one would think a more bulky compound would not fit into the lock. Some enzymes that catalyze reactions between two substrates do not bind one substrate until the other one is already bound to the enzyme. These curiosities led Koshland[11] in 1958 to propose the *induced-fit hypothesis,* namely, that when a substrate begins to bind to an enzyme, interactions of various groups on the substrate with particular enzyme functional groups are initiated, and these mutual interactions induce a *conformational change* in the enzyme. This results in a change of the enzyme from a low catalytic form to a high catalytic form by destabilizing the enzyme and/or by inducing proper alignment of the groups involved in catalysis. The conformational change could serve as a basis for substrate specificity. Compounds resembling the substrate except with smaller or larger substituents may bind to the enzyme but may not induce the conformational change necessary for catalysis. Also, different substrates may induce nonidentical forms of the activated enzyme. On the basis of *site-directed mutagenesis* studies (site-directed mutagenesis means that an amino acid residue in the enzyme is genetically changed to a different amino acid), Post and Ray[12] showed that a unique form of an enzyme is not required for efficient catalysis of a reaction.

In the case of bimolecular systems, the binding of the first substrate may induce the conformational change that exposes the binding site for the second substrate, and, consequently, this would account for an enzyme-catalyzed reaction that only occurs when the substrates bind in a particular order. Unlike the lock-and-key hypothesis, which implies a rigid active site, the induced-fit hypothesis requires a flexible active site to accommodate different binding modes and conformational changes in the enzyme. Actually, Pauling[13] stated the concept of a flexible active site earlier, hypothesizing that an enzyme is a flexible template that is most complementary to substrates at the transition state rather than at the ground state. This flexible model is consistent with many observations regarding enzyme action.

In 1930 Haldane[14] suggested that an enzyme–substrate (E·S) complex requires additional activation energy prior to enzyme catalysis, and this energy may be derived from substrate strain energy on the enzyme. Transition-state theory, developed by Eyring,[15] is the basis for the mentioned hypothesis of Pauling. According to this hypothesis, the substrate does not bind most effectively in the E·S complex; as the reaction proceeds, the enzyme conforms to the transition-state structure, leading to the tightest interactions (increased binding energy) with the transition-state structure.[16] This increased binding, known as *transition-state stabilization,* results in rate enhancement. Schowen has suggested[17] that all the mentioned 21 hypotheses of enzyme catalysis (as well as other correct hypotheses) are just alternative expressions of transition-state stabilization.

The E·S complex forms by the binding of the substrate to the *active site.* Only a dozen or so amino acid residues may make up the active site, and of these only two or three may be involved directly in substrate binding and/or catalysis. Because all the catalysis takes place in the active site of the enzyme, you may wonder why it is necessary for enzymes to be so large. There are several hypotheses regarding the function of the remainder of the enzyme. One suggestion[18] is that the most effective binding of the substrate to the enzyme (the largest binding energy) results from close packing of the atoms within the protein; possibly, the remainder of the enzyme outside the active site is required to maintain the integrity of the active site for effective catalysis. The protein may also serve the function of channeling the substrate into the active site. Storm and Koshland[19] suggested that the active site aligns the orbitals of substrates and catalytic groups on the enzyme optimally for conversion to the transition-state structure. This hypothesis is termed *orbital steering.* Evidence to support the concept of orbital steering was obtained by structural modification studies with isocitrate dehydrogenase.[20] Small modifications were made to the structures of the *cofactors* (organic molecules or metal ions required for catalysis) for this enzyme, which led to a slight misalignment of the bound cofactors. Because the substrate must react with one of the cofactors during catalysis, misalignment of the cofactor would translate into a perturbed reaction trajectory that should affect the catalytic power of the enzyme. In fact, the reaction with the modified cofactors resulted in large decreases in the reaction rate (factors of one-thousandth to one–hundred-thousandth the rate) with only small changes in the orientation of the substrates, as evidenced by X-ray crystallographic analyses of the active isocitrate dehydrogenase complexes. It appears, then, that small changes in the reaction trajectory, by misalignment of the reacting orbitals, can result in a major change in catalysis.

Enzyme catalysis is characterized by two features: *specificity* and *rate acceleration*. The active site contains moieties, namely, amino acid residues and, in the case of some enzymes, cofactors, that are responsible for these properties of an enzyme. As mentioned, a *cofactor*, also called a *coenzyme,* is an organic molecule or metal ion that binds to the active site, in some cases covalently and in others noncovalently, and is essential for the catalytic action of those enzymes that require cofactors. We will discuss various cofactors throughout the text for those enzyme-catalyzed reactions that require one or more cofactors.

B. Specificity of Enzyme-Catalyzed Reactions

1. Enzyme Kinetics (Definitions only—See Appendix I for Derivations)

Two types of specificity of enzymes must be considered: specificity of binding and specificity of reaction. As mentioned, enzyme catalysis is initiated by a prior interaction between the enzyme and the substrate, known as the *E·S complex* or *Michaelis complex* (Scheme 1.1). The driving force for the interactions of substrates with enzymes is the low-energy state of the E·S complex resulting from the covalent and noncovalent interactions (discussed later). The term k_1, sometimes referred to as k_{on}, is the rate constant for formation of the E·S complex, which depends on the concentrations of the substrate and enzyme; k_{-1}, also called k_{off}, is the rate constant for the breakdown of the complex, which depends on the concentration of the E·S complex and other forces (by the way, Cleland has proposed the use of odd-numbered subscripts for forward rate constants and even-numbered subscripts for reverse rate constants to avoid typos that omit minus signs; this seems quite sensible, but I have not adopted this usage here). The stability of the E·S complex is related to the affinity of the substrate for the enzyme, which is measured by its K_s, the dissociation constant for the E·S complex. When $k_2 \ll k_{-1}$, we refer to the term k_2

$$E + S \underset{k_{-1}}{\overset{k_1}{\rightleftharpoons}} E \cdot S \overset{k_2}{\rightleftharpoons} E \cdot P \rightleftharpoons E + P$$

$$K_s = \frac{k_{-1}}{k_1}$$

SCHEME 1.1   Generalized enzyme-catalyzed reaction.

TABLE 1.1   Examples of Turnover Numbers[a]

Enzyme	Turnover number k_{cat} (s^{-1})
Papain	10
Carboxypeptidase	10^2
Acetylcholinesterase	10^3
Kinases	10^3
Dehydrogenases	10^3
Aminotransferases	10^3
Carbonic anhydrase	10^6
Superoxide dismutase	10^6
Catalase	10^7

[a] Data from Eigen, M.; Hammes, G. G. *Adv. Enzymol.* **1963**, *25*, 1.

as the k_{cat} (the catalytic rate constant) and the dissociation constant K_s is called the K_m (the Michaelis–Menten constant). The k_{cat} represents the maximum number of substrate molecules converted to product molecules per active site per unit of time, that is, the number of times the enzyme "turns over" the substrate to the product per unit of time (called the *turnover number*). Typical values for a k_{cat} are on the order of 10^3 s^{-1} (about 1000 molecules of substrate are converted to product every second!). One of the most efficient enzymes is Δ^5-3-ketosteroid isomerase from *Pseudomonas testosteroni*,[21] having a turnover number of 10^6 s^{-1}. Table 1.1 gives turnover numbers for several enzymes and classes of enzymes. Note, however, that because there are two other important steps to enzyme catalysis, namely substrate binding and product release, high turnover numbers are only useful if these two physical steps occur at faster rates. As we will see, this is not always the case. Once an enzyme has reached "perfection" in efficiency—that is, the k_{cat}/K_m (see later definition) is diffusion controlled (about 10^9 M^{-1} s^{-1} for an enzyme-catalyzed reaction)[22]—the rate-determining step can be release of product! This occurrence makes kinetic studies of individual steps during catalysis very difficult, if not impossible. An example of this phenomenon is the enzyme triosephosphate isomerase.[23]

So how does an enzyme release the product so efficiently? For catalytically efficient enzymes, the Michaelis complex forms with a diffusional rate constant. To obtain the maximal catalytic rate, transition-state formation, product formation, and product release all must occur within this time frame. An enzyme binds the transition state structure about 10^{12} times more tightly than it binds the substrate or products. Following bond breaking (or making) at the transition state, the interactions that match in the transition-state stabilizing complex are no longer present in the products complex, and therefore the products are bound poorly, resulting in their expulsion from the active site. Even more significant than the loss of binding interactions, there can be a change in the electronic distribution as bonds are broken and made, which can generate a repulsive interaction between the products or with groups at the active site, leading to the opening of the active site and the expulsion of products.[24] A demonstration of this phenomenon can be found in a study of the transition-state structure for nucleoside hydrolase.[25] In this study, kinetic isotope effects are utilized to develop a geometric model of the transition state, and the molecular electrostatic potential surface of this geometric model is then determined by molecular orbital calculations. The electrostatic potential surfaces of the enzyme-bound substrate and products were found to differ considerably from those of the transition-state structure. The enzyme-bound products contained adjacent areas of negative charge; this electrostatic repulsion presumably destroys the affinity of the products for the active site, resulting in their expulsion.

The K_m is the concentration of substrate that produces half the maximum rate of which the enzyme is capable. Remember that this is related to a *dissociation* constant, so the smaller the K_m, the stronger the interaction (the tighter the binding) between E and S, and, therefore, the higher the concentration of the E·S complex. Another term utilized quite often is the k_{cat}/K_m, called the *specificity constant*. The k_{cat}/K_m is utilized to rank an enzyme according to how good it is with different substrates. It provides information about how fast the reaction of a given substrate bound to the enzyme (k_{cat}) would be and what concentration of the substrate would be required to reach the maximum rate. The upper limit for the k_{cat}/K_m is the rate of diffusion, which is when a reaction occurs with every collision between molecules. Because of this upper limit to k_{cat}/K_m (10^9 $M^{-1}s^{-1}$), there is a price to pay for a very "fast" enzyme (one with a large k_{cat}), namely, that its K_m also will have to be high. For example, a substrate for an enzyme with a k_{cat} of 10^7 s^{-1} cannot have a K_m lower than about 10 mM, which is very high (poorly stabilized E·S complex), but a substrate for an enzyme with a k_{cat} of 10^3 s^{-1}, can have a K_m of about 1 μM.

There is a more complete, but still simplified, somewhat mathematical treatment of the kinetics of enzyme-catalyzed reactions in Appendix I. Because I am determined to keep this text on an intuitive level, which is the way organic chemists think, I have concealed the mathematical garbage in this appendix, so you can read about it there, if you wish.

The E·S complex results from interactions of the substrate with various amino acid side chains in the active site, mostly via noncovalent interactions, including ionic (electrostatic) interactions, ion–dipole interactions, dipole–dipole interactions, hydrogen bonding, charge transfer, hydrophobic interactions, and van der Waals interactions. In some cases, however, covalent interactions also occur. Examples of these noncovalent interactions are shown in Figure 1.1. Weak interactions, such as noncovalent interactions, usually are possible only when molecular surfaces are close and complementary, that is, when bond strength depends on distance. Because the spontaneous formation of a bond between atoms occurs with a decrease in free energy, $\Delta G°$ is a negative value. The change in free energy is related to the binding equilibrium constant (K_{eq}) by the equation.

$$\Delta G° = -RT \ln K_{eq} \qquad (1.1)$$

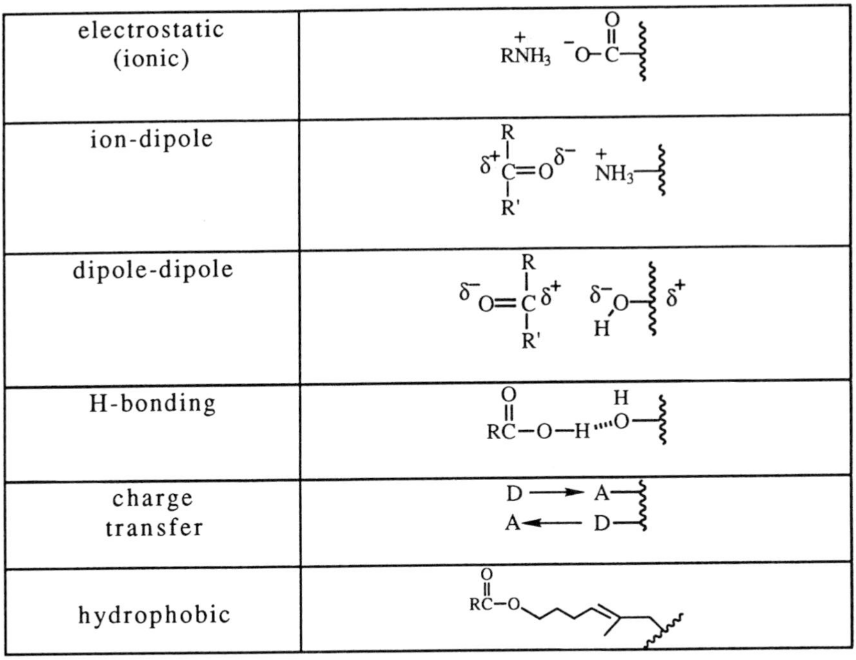

electrostatic (ionic)	RNH_3^+ $^-O-C-$ (with O double bonded)
ion-dipole	$\overset{+}{\delta}C=O\overset{-}{\delta}$ $\overset{+}{NH_3}-$
dipole-dipole	$\overset{-}{\delta}O=C\overset{+}{\delta}$ $\overset{-}{\delta}O-\overset{+}{\delta}$ H
H-bonding	$RC-O-H\cdots O-$
charge transfer	$D \longrightarrow A-$ $A \longleftarrow D-$
hydrophobic	$RC-O$

FIGURE 1.1   Noncovalent interactions.

Therefore, at physiological temperature (37 °C), changes in the free energy of -2 to -3 kcal/mol can have a major effect on the establishment of good secondary interactions. In fact, a decrease in $\Delta G°$ of -2.7 kcal/mol changes the binding equilibrium constant from 1 to 100. If the K_{eq} were only 0.01 (1% of the equilibrium mixture in the form of the enzyme–substrate complex), then a $\Delta G°$ of interaction of -5.45 kcal/mol would shift the binding equilibrium constant to 100 (99% in the form of the enzyme–substrate complex).

2. Specific Forces Involved in Enzyme–Substrate Complex Formation

In general, the bonds formed between a substrate and an enzyme are weak noncovalent interactions; consequently the effects produced are reversible, which is very important for product release. In the following subsections the various types of possible enzyme–substrate interactions are discussed briefly.

a. Covalent Bond

The *covalent bond* is the strongest bond, generally worth anywhere from -40 to -110 kcal/mol in stability. All the enzymes that utilize pyridoxal 5'-phosphate as a cofactor form a covalent E·S complex (see Chapter 8, Section V.A), namely, an imine between the amino group of the amino acid substrate and the aldehyde

group of pyridoxal 5′-phosphate; many other enzymes also utilize covalent catalysis (see Section II.B) to accelerate the rate of the reaction.

b. Ionic (or Electrostatic) Interactions

At *physiological pH* (generally taken to mean pH 7.4, the pH of human blood)
basic groups such as the amino side chains of arginine, lysine and, to a much lesser
extent, histidine are protonated and, therefore, provide a cationic environment.
Acidic groups, such as the carboxylic acid side chains of aspartic acid and glutamic
acid, are deprotonated to give anionic groups. Substrate and enzyme groups will be
mutually attracted provided they have opposite charges. This *ionic interaction* can be
effective at distances farther than those required for other types of interactions, and
they can persist longer. A simple ionic interaction can provide a $\Delta G°$ up to about
-5 kcal/mol, which declines by the square of the distance between the charges. If
this interaction is reinforced by other simultaneous interactions, the ionic interaction becomes stronger and persists longer. Acetylcholine is utilized as an example of
a molecule that can participate in an ionic interaction, which is shown in Figure 1.2.

c. Ion–Dipole and Dipole–Dipole Interactions

As a result of the greater electronegativity of atoms such as oxygen, nitrogen,
sulfur, and halogens relative to that of carbon, C–X bonds in substrates and enzymes, where X is an electronegative atom, will have an asymmetric distribution
of electrons; this produces electronic dipoles. These dipoles in a substrate can be
attracted by ions (*ion–dipole interaction*) or by other dipoles (*dipole–dipole interaction*)
in the active site of the enzyme, provided charges of opposite sign are properly
aligned. Because the charge of a dipole is less than that of an ion, a dipole–dipole
interaction is weaker than an ion–dipole interaction. In Figure 1.2, acetylcholine
also is utilized to demonstrate these interactions.

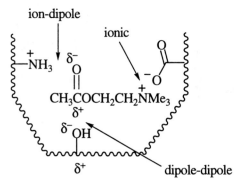

FIGURE 1.2   Examples of ionic, ion–dipole, and dipole–dipole interactions. The wavy line represents the enzyme active site.

d. Hydrogen Bonds

Hydrogen bonds are a type of dipole–dipole interaction formed between the proton of a group X–H, where X is an electronegative atom, and other electronegative atoms (Y) containing a pair of nonbonded electrons. The only significant hydrogen bonds occur in molecules where Y is N, O, or F. X removes electron density from the hydrogen so it has a partial positive charge, which is strongly attracted to nonbonded electrons of Y. This interaction will be denoted as a dashed line, — X — H---Y —, to indicate that an interaction between H and Y occurs.

The hydrogen bond is unique to hydrogen because it is the only atom that can carry a positive charge at physiological pH while remaining covalently bonded in a molecule and that also is small enough to allow close approach of a second electronegative atom. The strength of the hydrogen bond is related to the Hammett σ constants.[26] Hydrogen bonding can be quite important for biological activity; hydrogen bonds are essential for maintaining the structural integrity of α-helix and β-sheet conformations of peptides and proteins (Figure 1.3). Krause and co-workers[27] demonstrated the importance of hydrogen bonding to the integrity of proteins by comparing the hydrogen bonds, salt links, buried surface area, packing density, surface-to-volume ratio, and stabilization of α-helices and β-turns

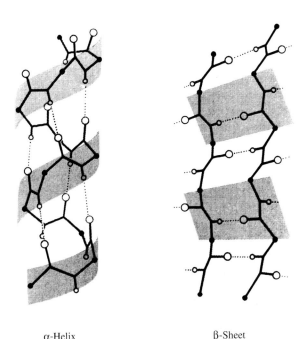

α-Helix   β-Sheet

FIGURE 1.3   Hydrogen bonding in the secondary structure of proteins: α-helix and β-sheet. [From Zubay, G. *Biochemistry*, 4th ed., p. 86. Wm. C. Brown Publishers, Dubuque, IA. Copyright © 1998. Reproduced by permission of the McGraw-Hill Companies.]

from the X-ray crystal structures of the enzyme D-glyceraldehyde-3-phosphate dehydrogenase from four different sources: the hyperthermophile *Thermotoga maritima,* the extreme thermophile *Thermus aquaticus,* the moderate thermophile *Bacillus stearothermophilus,* and the psychrophilic lobster *Homarus americanus,* which grow at temperatures of 80, 70, 58, and 20 °C, respectively. A clear correlation was found between the number of hydrogen bonds, particularly hydrogen bonds to charged amino acids, and thermostability; the thermophilic enzymes have hundreds more hydrogen bonds than does the psychrophilic one. The $\Delta G°$ for hydrogen bonding usually is in the range of -1 to -3 kcal/mol.

e. Charge Transfer Complexes

When a molecule (or group) that is a good electron donor comes into contact with a molecule (or group) that is a good electron acceptor, the donor may transfer some of its charge to the acceptor. This forms a *charge transfer complex,* which, in effect, is a molecular dipole–dipole interaction. The potential energy of this interaction is proportional to the difference between the ionization potential of the donor and the electron affinity of the acceptor.

Donor groups contain π-electrons, such as alkenes, alkynes, and aromatic moieties with electron-donating substituents, or groups that contain a pair of nonbonded electrons, such as oxygen, nitrogen, and sulfur moieties. Acceptor groups contain electron-deficient π orbitals, such as alkenes, alkynes, and aromatic moieties having electron-withdrawing substituents, and weakly acidic protons. There are groups on receptors that can act as electron donors, such as the aromatic ring of tyrosine or the carboxylate group of aspartate, or act as electron acceptors, such as cysteine, or act as electron donors and acceptors, such as histidine, tryptophan, and asparagine.

f. Hydrophobic Interactions

In the presence of a nonpolar molecule or region of a molecule, the surrounding water molecules orient themselves and, therefore, are in a higher energy state than when only other water molecules are present. When two nonpolar groups, such as a lipophilic group on a substrate and a nonpolar active-site group on the enzyme, each surrounded by ordered water molecules, approach each other, the water molecules around one group become disordered in an attempt to associate with the water molecules of the approaching group. This increase in entropy, therefore, results in a decrease in the free energy ($\Delta G° = \Delta H° - T\Delta S°$) that stabilizes the enzyme–substrate complex. This stabilization is known as a *hydrophobic interaction.* Consequently, this is not an attractive force of two nonpolar groups "dissolving" in one another, but rather is the decreased free energy of the nonpolar group because of the increased entropy of the surrounding water molecules. Jencks[28] has suggested that hydrophobic forces may be the most important single factor responsible for noncovalent intermolecular interactions in aqueous solution. Hildebrand,[29] in contrast, is convinced that hydrophobic effects do not exist. Every methylene–methylene interaction (which actually may be a van der Waals interaction; see later) liberates about 0.5 kcal/mol of free energy.

g. van der Waals Forces

Atoms in nonpolar molecules may have a temporary nonsymmetrical distribution of electron density that results in the generation of a temporary dipole. As atoms from different molecules (such as an enzyme and a substrate) approach each other, the temporary dipoles of one molecule induce opposite dipoles in the approaching molecule. Consequently an intermolecular attraction, known as *van der Waals forces,* results. These weak universal forces become significant only when there is a very close surface contact of the atoms; however, when there is molecular complementarity, numerous atomic interactions (each contributing about -0.5 kcal/mol to the $\Delta G°$) result, which can add up to a significant overall enzyme–substrate binding component.

h. Conclusion

Because noncovalent interactions are generally weak, the involvement of multiple types of interactions is critical. To a first approximation, enthalpy terms will be additive. Once the first interaction has taken place, translational entropy (the energy associated with the freedom of molecules to move around) is lost. This results in a much lower entropy loss in the formation of the second interaction. The effect of this cooperativity is that several rather weak interactions may combine to produce a strong interaction. Because several different types of interactions are involved, selectivity in enzyme–substrate interactions can result. As indicated earlier, maximum binding interactions at the active site occur at the transition state of the reaction. It is therefore important that an enzyme does not bind to the ground state or to intermediate states excessively, which would increase the free-energy difference between the ground state or intermediate state and the transition state. The binding interactions between the substrate and the active site of the enzyme set up the substrate for the reaction that the enzyme catalyzes.

3. Binding Specificity

At one end of the spectrum, binding specificity can be absolute, that is, essentially only one substrate forms an E·S complex with a particular enzyme, which then leads to product formation. Examples of enzymes possessing such a property include L-aspartase,[30] glutamate mutase,[31] and 2-methyleneglutarate mutase.[32] At the other end of the spectrum, binding specificity can be very broad, in which case many molecules of related structure can bind and be converted to product, such as alkaline phosphatase,[33] alcohol dehydrogenase,[34] and the family of enzymes known

$$\text{Enz}_L + (R,S) \longrightarrow \text{Enz}_L \cdot R + \text{Enz}_L \cdot S$$

SCHEME 1.2   Resolution of a racemic mixture.

as cytochrome P450,[35] which protects us from the small-molecule toxins we eat and breathe. Specificity may involve E·S complex formation with only one enantiomer of a racemate or E·S complex formation with both enantiomers, but only one is converted to product. Enzymes can accomplish this enantiomeric specificity because they are chiral molecules (mammalian enzymes consist of only L-amino acids); interactions of an enzyme with a racemic mixture therefore result in the formation of two diastereomeric complexes (Scheme 1.2). Diastereomers have different energies (stabilities and reactivities) and different properties. This is analogous to the principle behind the resolution of racemic mixtures with chiral reagents; two diastereomers are produced that can be separated by physical means (such as distillation or chromatography) because they have different properties. When an enzyme is exposed to a racemic mixture of a substrate, the binding energy for E·S complex formation with one enantiomer may be much higher than that with the other enantiomer either because of differential binding interactions as noted earlier or for steric reasons. For example (Figure 1.4), after the ammonium and carboxylate substituents of phenylalanine have interacted with active-site groups, a third substituent at the stereogenic center (the benzyl group) has two possible orientations; in the case of the S-isomer, there is a binding pocket (Figure 1.4A), but the benzyl group of the R-isomer (Figure 1.4B) points in the other direction toward the leucine side chain and causes steric hindrance (i.e., the R-isomer does not bind to the active site). If the binding energies for the two complexes are significantly different, then only one E·S complex may form (as would be the case in Figure 1.4). Alternatively, both E·S complexes may form, but for steric or electronic reasons only one E·S complex may lead to product formation. The enantiomer that forms the E·S complex that is not turned over is said to undergo *nonproductive binding* to the enzyme. Enzymes also can demonstrate complete stereospecificity with geometric isomers, because these are diastereomers already.

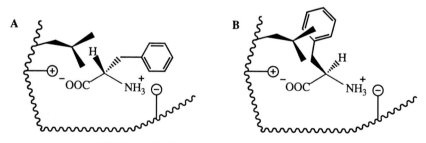

FIGURE 1.4   Basis for enantioselectivity in enzymes.

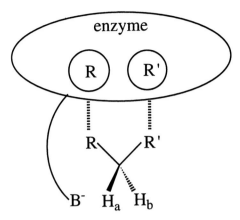

FIGURE 1.5   Enzyme specificity for chemically identical protons. R and R′ on the enzyme are groups that interact specifically with R and R′, respectively, on the substrate.

4. Reaction Specificity

Reaction specificity also arises from constituents of the active site, namely, specific acid, base, and nucleophilic functional groups of amino acids and from cofactors. Unlike reactions in solution, enzymes can show specificity for chemically identical protons (Figure 1.5). If there are specific binding sites for R and R′ at the active site of the enzyme, and a base (B⁻) of an amino acid side chain is juxtaposed so that it can only reach proton H_a, then abstraction of H_a will occur stereospecifically, even though in a nonenzymatic reaction H_a and H_b would be chemically equivalent and, therefore, would have equal probability to be abstracted. The approach taken by synthetic chemists in designing chiral reagents for stereospecific reactions is modeled after this. The chirality of the enzyme should determine the chirality of the reaction. An exquisite example of this principle was provided by Kent and co-workers,[36] who chemically synthesized the D- and L-forms of HIV-1 protease (each enzyme consisted of either all D- or all L- amino acids, respectively). Only peptides made of D-amino acids are hydrolyzed by the D-enzyme, and only the L-amino acid peptides are cleaved by the L-enzyme.

C. Rate Acceleration

In general, catalysts stabilize the transition state relative to the ground state, and this decrease in activation energy is responsible for the rate acceleration that results (Figure 1.6A). Jencks proposed that the fundamental feature distinguishing enzymes from simple chemical catalysts is the ability of enzymes to utilize binding interactions away from the site of catalysis.[37] These binding interactions facilitate reactions by positioning substrates with respect to one another and with respect

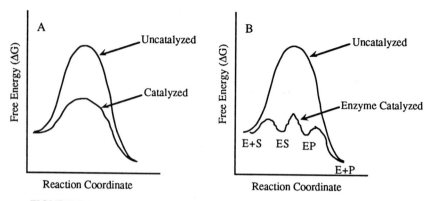

FIGURE 1.6   Effect of (A) a chemical catalyst and (B) an enzyme on activation energy.

to the catalytic groups at the active site. The importance of binding interactions to catalysis was demonstrated with a group I RNA enzyme.[38]

Because an enzyme has numerous opportunities to invoke catalysis—for example, by stabilization of the transition states (thereby lowering the transition-state energy), by destabilization of the E·S complex (thereby raising the ground-state energy), by destabilization of intermediates, and during product release—multiple steps, each having small activation energies, may be involved (Figure 1.6B). As a result of these multiple catalytic steps, rate accelerations of 10^{10}–10^{14} over the corresponding nonenzymatic reactions are common. (Table 1.2 gives some rate acceleration values as high as 10^{17}!) Wolfenden hypothesized that the rate acceleration produced by an enzyme is proportional to the affinity of the enzyme for the transition-state structure of the bound substrate;[39] the reaction rate is proportional to the amount of substrate that is in the transition-state complex. The trick that the enzyme must perform is to be able to bind tightly only to the unstable transition-state structure (with a lifetime of one bond vibration) and not to either the substrate or the products. A conformational change in the protein structure plays an important role in this operation.

Enzyme catalysis does not alter the equilibrium of a reversible reaction. If an enzyme accelerates the rate of the forward reaction, it must accelerate the rate of the corresponding backward reaction by the same amount; its effect is to accelerate the attainment of the equilibrium, but not the relative concentrations of substrates and products at equilibrium.

Knowles and co-workers[40] have suggested that to maximize catalytic efficiency, enzymes have evolved to produce a leveling effect, resulting in approximately equal energies for all the ground states and the transition states bound to the enzyme. The value for the "internal equilibrium constant," K_{int}, the equilibrium constant between the bound substrates and the products for an enzyme that operates near equilibrium, is generally near unity. Therefore, the concentrations of substrate-containing complexes (E·S) and product-containing complexes (E·P) are often

TABLE 1.2   Examples of Enzymatic Rate Acceleration

Enzyme	Nonenzymatic rate k_{non} (s^{-1})	Enzymatic rate k_{cat} (s^{-1})	Rate acceleration k_{cat}/k_{non}
Cyclophilin[a]	2.8×10^{-2}	1.3×10^4	4.6×10^5
Carbonic anhydrase[a]	1.3×10^{-1}	10^6	7.7×10^6
Chorismate mutase[a]	2.6×10^{-5}	50	1.9×10^6
Chymotrypsin[b]	4×10^{-9}	4×10^{-2}	10^7
Triosephosphate Isomerase[b]	6×10^{-7}	2×10^3	3×10^9
Fumarase[b]	2×10^{-8}	2×10^3	10^{11}
Ketosteroid Isomerase[a]	1.7×10^{-7}	6.6×10^4	3.9×10^{11}
Carboxypeptidase A[a]	3×10^{-9}	578	1.9×10^{11}
Adenosine Deaminase[a]	1.8×10^{-10}	370	2.1×10^{12}
Urease[b]	3×10^{-10}	3×10^4	10^{14}
Alkaline Phosphatase[b]	10^{-15}	10^2	10^{17}
Orotidine 5′-Phosphate Decarboxylase[a]	2.8×10^{-16}	39	1.4×10^{17}

[a] Taken from Radzicka, A.; Wolfenden, R. *Science* **1995**, *267*, 90.
[b] Taken from Horton, H. R.; Moran, L. A.; Ochs, R. S.; Rawn, J. D.; Scrimgeour, K. G. *Principles of Biochemistry,* Neil Patterson: Englewood Cliffs, NJ, 1993.

equal at steady state, and all rate constants are approximately equal. Pettersson,[41] however, has reasoned that this assumption—that enzyme reactions exhibit an intrinsic constraint during evolution in the form of a linear free-energy relationship between certain rate and equilibrium constants in the reaction mechanism—is too restrictive to have a bearing on the actual biological problem of enzyme catalytic optimization. Constraints relating to the evolutionary development of the enzyme are suggested to be more important than constraints in the internal properties of the enzyme. According to Pettersson,[42] the enzymes that exhibit K_{int} values close to unity are those which catalyze reactions with equilibrium constants close to unity. Experiments by Sinnott and co-workers[43] with higher evolved forms of β-galactosidases indicate that changes in the free-energy profile of the catalyzed reaction, except for that of the rate-determining transition state, are random and that there are large alterations in the transition-state structure with the evolutionary forms, providing support for Pettersson's proposal.

II. MECHANISMS OF ENZYME CATALYSIS

Once the substrate binds to the active site of the enzyme via the interactions just noted, the enzymes can utilize a variety of mechanisms to catalyze the conversion of the substrate to product. The most common mechanisms[44–46] are approximation, covalent catalysis, general acid/base catalysis, electrostatic catalysis, desolvation, and strain or distortion. All of these act by stabilizing the transition-state energy or destabilizing the ground state (which is generally not as important as transition-state stabilization).

A. Approximation

Approximation is rate enhancement by proximity; that is, the enzyme serves as a template to bind the substrates so that they are close to each other in the reaction center. This results in a loss of rotational and translational entropies of the substrate on binding to the enzyme; however, this entropic loss is offset by a favorable binding energy of the substrate, which provides the driving force for catalysis. Furthermore, because the catalytic groups are now an integral part of the same molecule, the reaction of enzyme-bound substrates becomes first order rather than second order when these compounds are free in solution. Jencks[37] suggests that in addition to lowering the degree of rotation and the translational entropy, the concept of *intrinsic binding energy,* which results from favorable noncovalent interactions with the substrate at the site of catalysis, is largely responsible for the remarkable specificity and high rates of enzymatic reactions. Holding the reaction centers in close proximity and in the correct geometry for reaction is equivalent to increasing the concentration of the reacting groups. This phenomenon can be exemplified with nonenzymatic model studies. For example, consider the second-order reaction of acetate with an aryl acetate (Scheme 1.3). If the rate constant k for this reaction is set equal to $1.0 \ M^{-1} s^{-1}$, and then the effect of decreasing rotational and translational entropy is determined by measuring the corresponding first-order rate constants for related molecules that can undergo the corresponding intramolecular reactions, it is apparent from Table 1.3 that forcing the reacting groups to be closer to each other increases the reaction rate.[47,48] Thirty-six years after the original experimental study by Bruice and Pandit of the effect of restricted rotation on rate acceleration, a theoretical investigation by Lightstone and Bruice using MM3 calculations showed that when the nucleophile and electrophile are closely arranged, and the van der Waals surfaces are properly juxtaposed (called the *near-attack conformation*), the activation energy is lowered as a result of a decrease in the enthalpy of the reaction (ΔH°), and the rate of the reaction really should increase, thereby supporting the earlier experimental observations.[49] Interestingly, there is no correla-

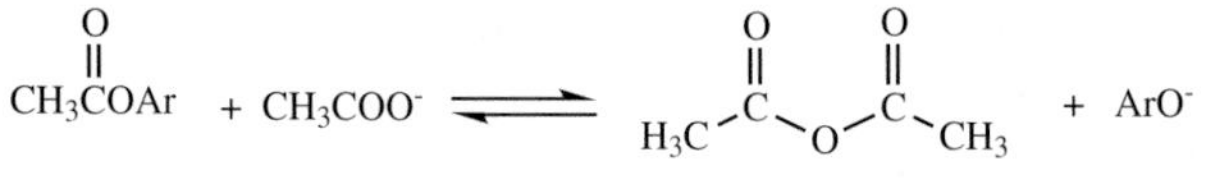

SCHEME 1.3   Second-order reaction of acetate with aryl acetate.

TABLE 1.3   Effect of Approximation on Reaction Rates

		Relative rate (k_{rel})	Effective molarity (EM)
		$1\ M^{-1}\ s^{-1}$	
	Decreasing rotational and translational entropy	$220\ s^{-1}$	$220\ M$
		$5.1 \times 10^4\ s^{-1}$	$5.1 \times 10^4\ M$
		$2.3 \times 10^6\ s^{-1}$	$2.3 \times 10^6\ M$
		$1.2 \times 10^7\ s^{-1}$	$1.2 \times 10^7\ M$

Source: (a) Bruice, T. C.; Pandit, U. K. *J. Am. Chem. Soc.* **1960**, *82*, 5858. (b) *Ibid., Proc. Natl. Acad. Sci. USA* **1960**, *46*, 402.

tion of the rate constants with the transition-state structure[50] or with entropy, only with the ground-state conformations; this effect appears to be entirely enthalpic.

Although first- and second-order rate constants cannot be compared directly, the efficiency of an intramolecular reaction can be defined in terms of its *effective molarity* (EM),[51] also called the *effective concentration,* the concentration of the reactant (or catalytic group) required to cause the intermolecular reaction to proceed at the observed rate of the intramolecular reaction. The EM is calculated by dividing the first-order rate constant for the intramolecular reaction by the second-order rate constant for the corresponding intermolecular reaction (see Table 1.3). This indicates that acetate ion would have to be at a concentration of, for example,

220 M (220 s^{-1}/1 M^{-1} s^{-1}) for the intermolecular reaction of acetate and aryl acetate to proceed at a rate comparable to that of the glutarate monoester reaction. Of course, 220 M acetate ion is an imaginary number (pure water is only 55 M), so the effect of decreasing the enthalpy is quite significant. Effective molarities for a wide range of intramolecular reactions have been measured, and the conclusion is that the efficiency of intramolecular catalysis varies with structure and can be as high as 10^{16} M for reactive systems. Therefore, holding groups proximal to each other, particularly when the reacting moieties in an enzyme−substrate complex are aligned correctly for reaction, can be important contributors to catalysis.

B. Covalent Catalysis

Some enzymes can utilize nucleophilic amino acid side chains, such as acidic groups (aspartate or glutamate carboxylates), neutral groups (serine hydroxyl or cysteine thiol), or basic groups (lysine amino, arginine guanidino, or histidine imidazolyl) or cofactors in the active site to form covalent bonds to the substrate; in some cases, a second substrate then can react with this enzyme−substrate intermediate to generate the product. This is known as *nucleophilic catalysis* (Scheme 1.4 shows active-site amino acid side-chain catalysis), a subclass of *covalent catalysis* that involves covalent bond formation as a result of attack by an enzyme nucleophile at an electrophilic site on the substrate. For example, if Y in Scheme 1.4 is an amino acid or peptide and Z$^-$ is a hydroxide ion, then the enzyme would be a peptidase (or protease). For nucleophilic catalysis to be most effective, Y should be converted into a better leaving group than X, and the covalent intermediate (**1.1,** Scheme 1.4) should be more reactive than the substrate.

Nucleophilic catalysis is the enzymatic analogue of anchimeric assistance by neighboring groups in organic reaction mechanisms. *Anchimeric assistance* is the process by which a neighboring functional group assists in the expulsion of a leaving group by intermediate covalent bond formation.[52] This results in accelerated reaction rates. Scheme 1.5 shows how a neighboring sulfur atom makes the displacement of a β-chlorine a much more facile reaction than it would be without the sulfur atom. If the sulfur atom were part of an active-site nucleophile, such as a

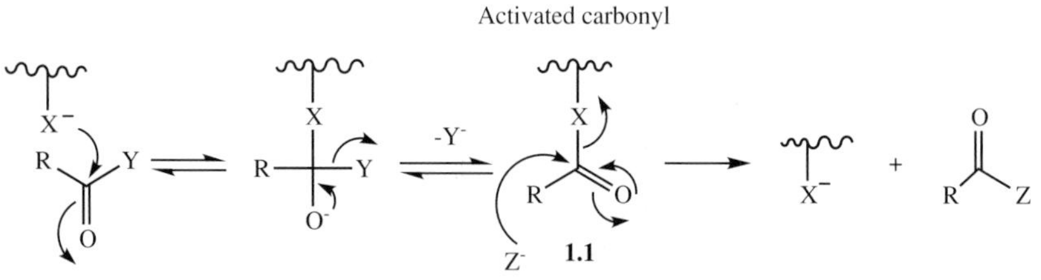

SCHEME 1.4   Nucleophilic catalysis.

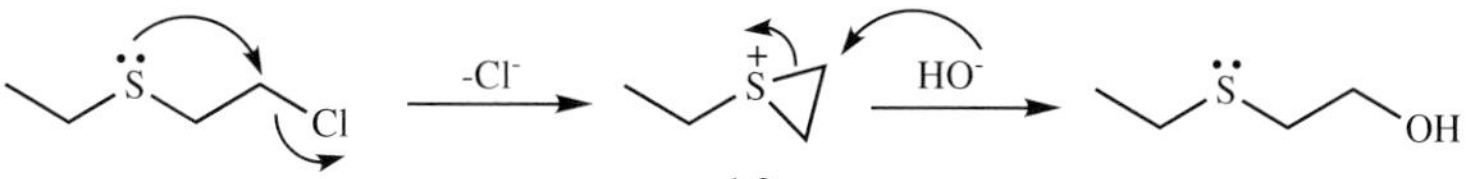

1.2

SCHEME 1.5   Anchimeric assistance by a neighboring group.

methionine residue, the C−Cl bond were part of a substrate, and HO⁻ were generated by enzyme-catalyzed deprotonation of water, this would represent covalent catalysis in an enzyme-catalyzed reaction, where the covalent adduct is **1.2** (Scheme 1.5).

The most common active-site nucleophiles are the thiol group of cysteine, the hydroxyl group of serine, the imidazole of histidine, the amino group of lysine, and the carboxylate group of aspartate or glutamate. These active-site nucleophiles are generally activated by deprotonation, often by a neighboring histidine imidazole or by a water molecule that is deprotonated in a general base reaction (see Section II. C). Therefore, if the substrate in Scheme 1.4 is a peptide (R = $NH_2CH(R')$; Y = amino acid or peptide), then a peptidase would convert the relatively unreactive amide linkage to a covalent intermediate (**1.1**) having a much more reactive ester linkage (if the nucleophile X were the serine hydroxyl group) or thioester linkage (if X were a cysteine thiolate), either of which could be rapidly hydrolyzed ($Z^- = HO^-$). The principal catalytic advantage of using an active-site residue instead of water directly is that the former leads to a unimolecular reaction (because the substrate is bound to the enzyme, attack by the serine residue is equivalent to an intramolecular reaction), which is entropically favored over the bimolecular reaction with water. Also, alkoxides (ionized serine) and thiolates (ionized cysteine) are better nucleophiles than hydroxide ion. An alternative, simpler, rationalization for covalent catalysis may relate to the binding interactions of the enzyme. Some enzymes may not have a sufficient amount of noncovalent binding interactions and may need covalent catalysis to assist in binding.

One of the classic model studies showing the relevance of covalent catalysis in the hydrolysis of an ester was done by Bender and Neveu[53] in which ^{18}O-labeled acetate was added to the aqueous medium during hydrolysis of aryl benzoates (Scheme 1.6). The observation of ^{18}O-labeled benzoic acid as the product supported the importance of covalent catalysis in this reaction.

Typical enzymatic reactions in which nucleophilic catalysis is important include many of the proteolytic enzymes, for example, the serine proteases (proteases that

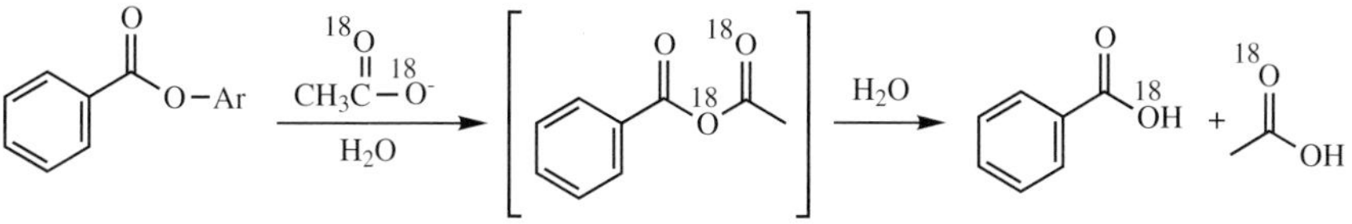

SCHEME 1.6   Early evidence to support covalent catalysis.

use a serine residue at the active site as the nucleophile) such as elastase (degrades elastin, a connective tissue prevalent in the lung) or plasmin (lyses blood clots), and the cysteine proteases (that utilize an active–site cysteine residue as the nucleophile) such as papain (found in papaya fruit and utilized in digestion).

C. General Acid/Base Catalysis

In any reaction in which proton transfer occurs, *general acid catalysis* and/or *general base catalysis* can be an important mechanism for specificity and rate enhancement. As an example of general acid/base (and covalent) catalysis, consider the enzyme *α-chymotrypsin,* a serine protease, which means that it utilizes an active-site serine residue in a covalent catalytic cleavage of peptide bonds. Because the nucleophilic group of serine is hydroxyl, it should be a poor nucleophile. However, aspartic acid and histidine residues nearby have been implicated in the conversion of the serine to an alkoxide by a mechanism called the *charge relay system* by Blow and co-workers,[54] the discoverers of the existence of the hydrogen-bonding network involving Asp-102, His-57, and Ser-195 (Scheme 1.7). This *catalytic triad,* which is easily visible in the 1.68 Å resolution X-ray crystal structure of α-chymotrypsin[55] (Figure 1.7), involves the aspartate carboxylate (pK_a of the acid is 3.9 in solution) removing a proton from the histidine imidazole (pK_a 6.1 in solution), which in turn removes a proton from the serine hydroxyl group (pK_a 14 in solution). Any respectable organic chemist would find that suggestion absurd; how can a base such as aspartate, whose conjugate acid is 2 pK_a units lower than that of the histidine imidazole, remove the imidazole proton efficiently, and, then, how can this imidazole remove the proton from the hydroxyl group of serine, which is 8 pK_a units higher than the (protonated) imidazole of histidine? The equilibrium is only 1% in favor of the first proton transfer, and the equilibrium for the second proton transfer favors the backward direction by a factor of 10^8! One explanation could be that the

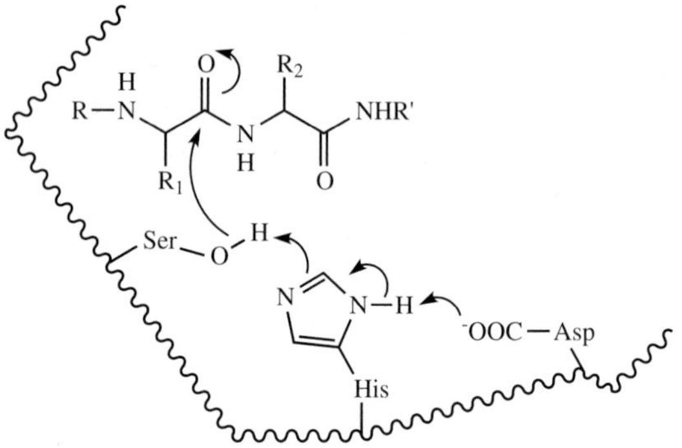

SCHEME 1.7   Charge relay system for activation of an active-site serine residue in α-chymotrypsin.

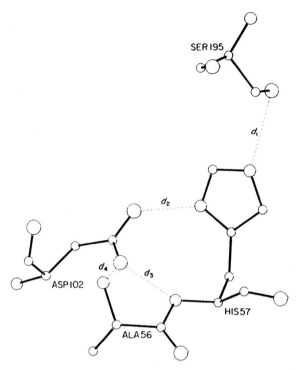

FIGURE 1.7   The catalytic triad of α-chymotrypsin. The distances are as follows: d_1 = 2.82 Å; d_2 = 2.61 Å; d_3 = 2.76 Å. [From Tsukada, H.; Blow, D. M. *J. Mol. Biol.* **1985**, *184*, 703. Copyright © 1985. Reproduced by permission of Academic Press Ltd., London.]

pK_a values of these acids and bases at the active site differ from those in solution. Furthermore, because these groups are held close together at the active site, as the proton is beginning to be removed from the serine hydroxyl group, the charge density proceeds to the next step (attack of the alkoxide at the peptide carbonyl), thereby driving the equilibrium forward. This is the beauty of enzyme-catalyzed reactions; the approximation of the groups and the fluidity of the active-site residues working in concert permit reactions to occur that would be nearly impossible in solution. As an example of this phenomenon, consider the 10^{17} rate enhancement by alkaline phosphatase (see Table 1.2). What this means is that alkaline phosphatase can accomplish in one second the product formation that would take 3,170,979,200 years without this enzyme. Because this amount of time approaches the age of Earth, it is clear that the evolution of enzymes was a necessary precursor to life.

It is very important to appreciate the fact that the pK_a values of amino acid side-chain groups within the active site of enzymes can be quite different from those in solution. This is partly the result of the low polarity inside of proteins. On the basis of molecular dynamics simulations of several proteins in water, Simonson and Brooks calculated that the interiors of these proteins have dielectric constants of about $2-3$,[56] which is comparable to the dielectric constant of benzene (2.28) or p-dioxane (2.21). This is quite different from the dielectric constant of about 78.5 for water as a result of the strong dipole moment of the O$-$H bonds. If a carboxylic acid is in a nonpolar region, its pK_a will rise because the anionic form will be destabilized. Glutamate-35 in the lysozyme$-$glycolchitin complex has a pK_a of 8.2;[57] the pK_a of glutamate in solution is about 4.5. If the carboxylate ion forms a salt bridge, it will be stabilized, and this will lower the pK_a. If several acid groups are near an essential active-site carboxylic acid, the anionic form will be stabilized, and its pK_a also will be lowered.

Likewise, a basic group buried in a nonpolar microenvironment will have a lower pK_a because protonation will be disfavored (to avoid polar cationic character). If the protonated base is in a salt bridge, it will be stabilized (deprotonation inhibited), and the pK_a will rise. The pK_a of a base will fall if it is adjacent to other bases. For example, Westheimer and co-workers found that the ϵ-amino group of the active-site lysine (now known to be Lys-115) in acetoacetate decarboxylase from *Clostridium acetobutylicum* has a pK_a of 5.9,[58] which is 4.5 pK_a units lower than the pK_a of lysine in solution. These researchers suggested that the pK_a of Lys-115 was low because of its proximity to Lys-116. Therefore, the removal of seemingly higher pK_a protons from substrates by active-site bases and protonation by relatively weakly acidic groups may not be as unreasonable as would appear if solution chemistry alone were considered.

There are two kinds of acid/base catalysis. If catalysis occurs by a hydronium (H_3O^+) or hydroxide (HO^-) ion and is determined only by the pH, not the buffer concentration, it is referred to as *specific acid* or *specific base catalysis,* respectively. As an example of how specific acid/base catalysis works, consider the hydrolysis of ethyl acetate (Scheme 1.8). At neutral pH this reaction is exceedingly slow, because both the nucleophile (H_2O) and the electrophile (the carbonyl of ethyl acetate) are unreactive. The reaction rate could be accelerated, however, if the reactivity of either the nucleophile or the electrophile could be enhanced. An increase in the pH increases the concentration of hydroxide ion, which is a much better nucleophile than is water, and in fact the rate of hydrolysis at higher pH increases (Scheme 1.9). Likewise, a decrease in the pH increases the concentration of the hydronium ion, which can protonate the ester carbonyl, thereby increasing its electrophilicity, and this also increases the hydrolysis rate (Scheme 1.10). That being the case, then the

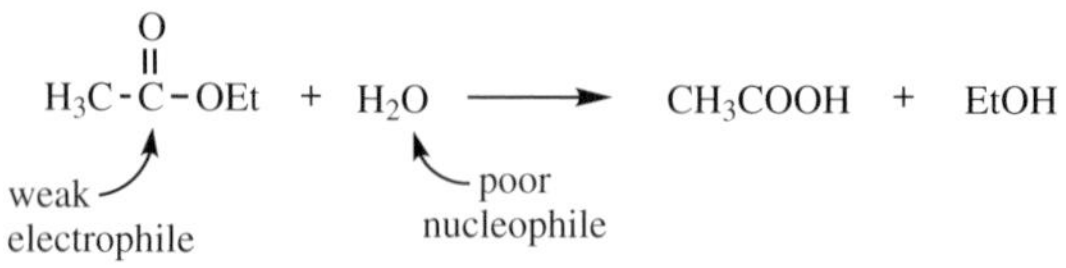

$$H_3C-\overset{\overset{\displaystyle O}{\|}}{C}-OEt \ + \ H_2O \ \longrightarrow \ CH_3COOH \ + \ EtOH$$

weak electrophile   poor nucleophile

SCHEME 1.8   Hydrolysis of ethyl acetate.

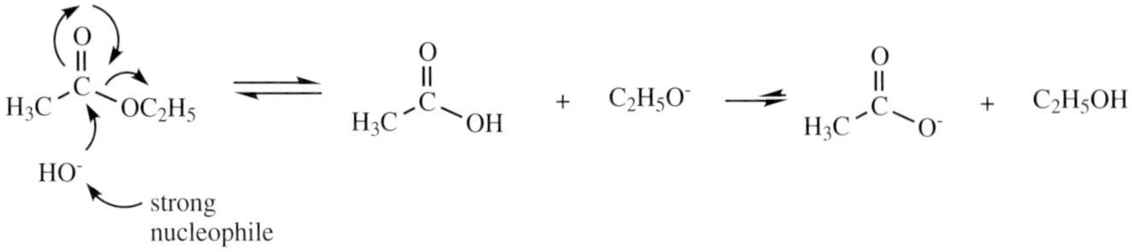

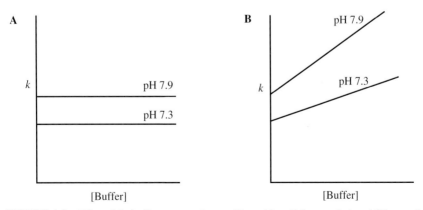

SCHEME 1.9   Alkaline hydrolysis of ethyl acetate.

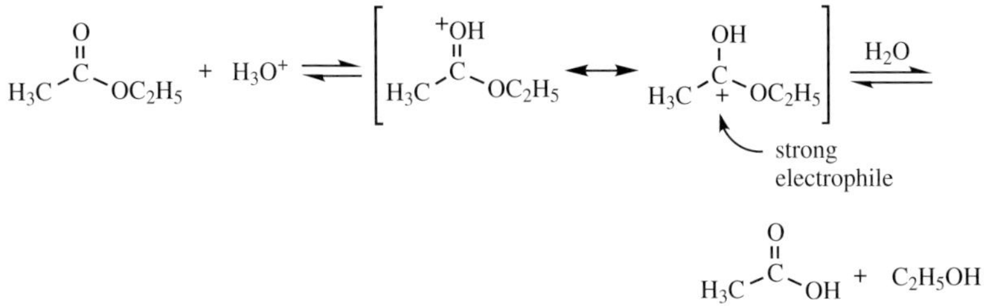

SCHEME 1.10   Acid hydrolysis of ethyl acetate.

hydrolysis rate should double if base *and* acid are added together, right? Of course not. Adding an acid to a base only leads to neutralization and loss of any catalytic effect.

General acid/base catalysis, in contrast, occurs when the reaction rate increases with increasing buffer concentration at a constant pH and ionic strength and shows a larger increase with a buffer that contains a larger acid or base component (Figure 1.8). Because the hydronium or hydroxide ion concentration is not increasing

A

k

pH 7.9

pH 7.3

[Buffer]

B

pH 7.9

k

pH 7.3

[Buffer]

FIGURE 1.8   Effect of the buffer concentration on (A) specific acid/base catalysis and (B) general acid/base catalysis.

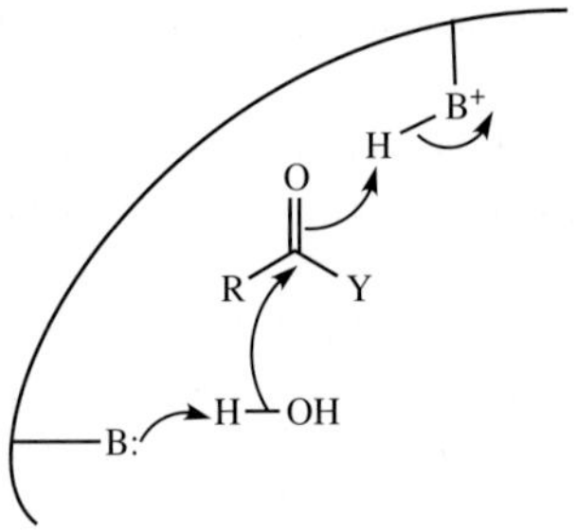

SCHEME 1.11   Simultaneous acid and base enzyme catalysis.

(the pH is constant), the buffer must be catalyzing the reaction. This is *general acid catalysis* (if acids other than hydronium ion accelerate the reaction rate) or *general base catalysis* (if bases other than hydroxide ion accelerate the rate). A classic example of general base catalysis is the imidazole buffer-catalyzed hydrolysis of acetylimidazole.[59] Imidazole in the free base form acts as the catalyst to remove the proton from water, generating a localized concentration of hydroxide that reacts with the acetylimidazole.

Unlike reactions in solution, however, an enzyme *can* utilize acid and base catalysis simultaneously (Scheme 1.11) for even greater catalysis. The protonated base in Scheme 1.11 is an acidic amino acid side chain and the free base is a basic residue.

In solution, the pK_a values of α-protons of aldehydes, ketones, and thioesters are about 16–20, and the pK_a values of α-protons of carboxylate ions are >32. So how can an active-site base, whose conjugate acid has a pK_a of ≤7, abstract those protons? Gerlt and co-workers[60] point out the importance of simultaneous acid and base catalysis for the rationalization of how enzymes, such as mandelate racemase, are capable of deprotonating weak carbon acids (protons α to carbonyls), such as in mandelic acid (**1.3**), to make enols (Scheme 1.12). On the basis of estimates of pK_a

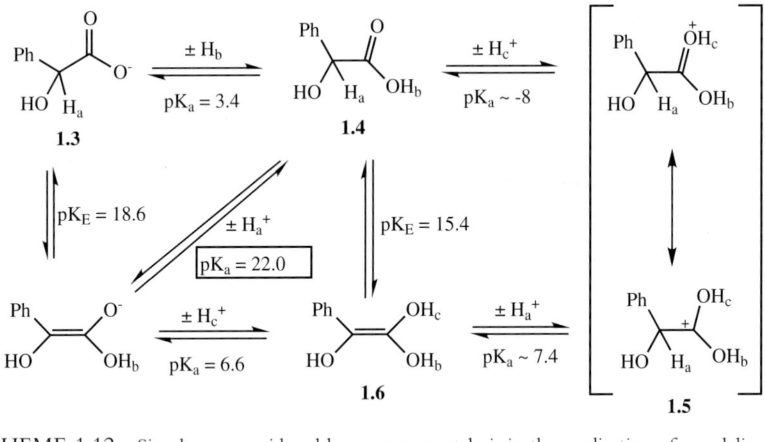

SCHEME 1.12   Simultaneous acid and base enzyme catalysis in the enolization of mandelic acid.

and pK_E values, concerted general acid/general base catalysis was shown to provide a low-energy pathway consistent with the observed enzymatic rates. The K_E relates the concentrations of keto and enol tautomers of a carbon acid; pK_E is either the difference between the pK_a values of the α-proton (H_a) of the keto tautomer (**1.4**) and the hydroxyl group of the enol tautomer (**1.6**) or the difference between the pK_a values of the α-proton (H_a) and the proton bound to the carbonyl group of the protonated carbonyl acid (H_c, **1.5**). Assuming that the pK_a of the proton bound to the carbonyl group of protonated mandelic acid (**1.5**) is about -8, then the pK_a of the α-proton (H_a, **1.5**) is estimated to be about 7.4 (a decrease of 15 pK_a units by protonation of the carbonyl!). This pK_a is in the range of pK_a values of active-site bases, so they would be capable of deprotonation.

Further support for this hypothesis is a theoretical study carried out by Gerlt and Gassman[61] of free energies for the mechanism of enzyme-catalyzed β-elimination reactions. Their findings suggest that the lowest energy pathway is a stepwise general-acid/general-base-catalyzed formation of an enol intermediate followed by 1,4-(E2-like)-elimination from the enol (Scheme 1.13), *not* an E1cB mechanism via an enolate (Scheme 1.14).

This mechanism also can rationalize the observed stereochemistry of enzyme-catalyzed β-elimination reactions. First, consider β-elimination reactions of aldehydes, ketones, and thioesters (Scheme 1.13A). The α-protons in these substrates,

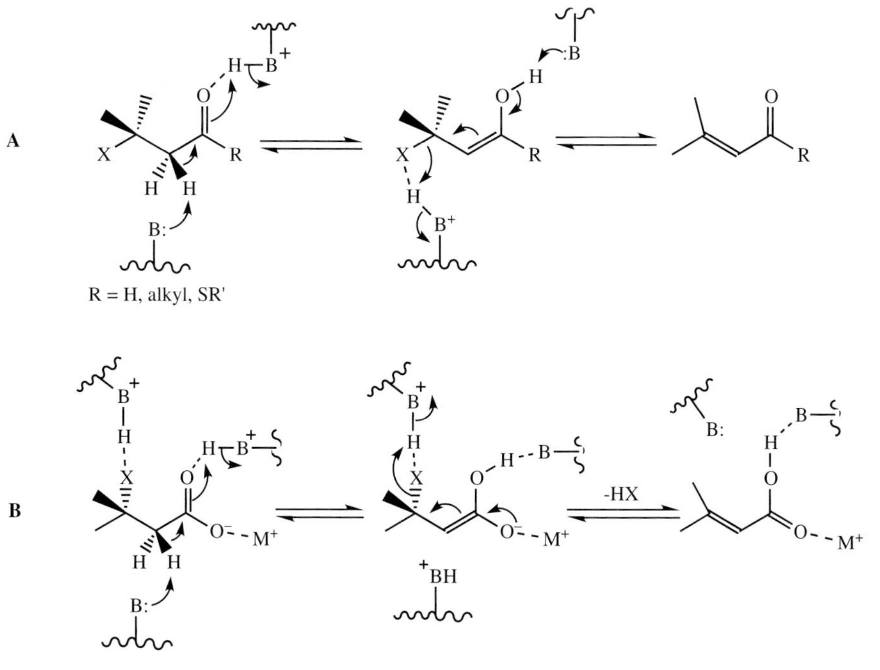

SCHEME 1.13   Simultaneous acid and base enzyme catalysis in the 1,4-elimination of β-substituted (A) aldehydes, ketones, thioesters and (B) carboxylic acids.

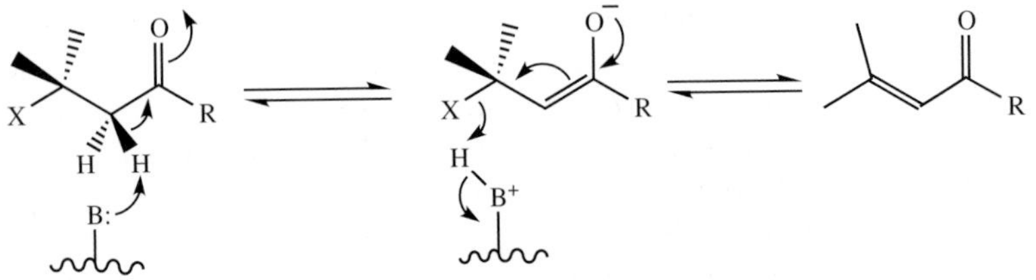

SCHEME 1.14   Base-catalyzed 1,4-elimination of β-substituted carbonyl compounds via an enolate intermediate (ElcB mechanism).

once protonation of the carbonyl occurs, are quite acidic, so the active-site base need not be strongly basic to remove them. Therefore the conjugate acid of this active-site base should be sufficiently acidic (remember, the weaker a base, the stronger its conjugate acid) to catalyze the elimination of the β-leaving group by proton donation. For the same active-site residue that removes the α-proton to then be able to donate a proton to the β-leaving group requires that the α-proton and the β-leaving group be oriented *gauche* (adjacent) to one another and in close proximity to the general base residue. Amazingly, all enzyme-catalyzed β-elimination reactions of aldehydes, ketones, and thioesters proceed with *syn* stereochemistry,[62] as predicted from Scheme 1.13A. This mechanism also predicts that only one base is involved in the proton transfer and elimination.

In the case of carboxylate substrates (Scheme 1.13B), either a metal ion or cationic group would neutralize the carboxylate charge at the active site. In contrast to aldehyde, ketone, and thioester substrates, α-protons of carboxylates are not very acidic. Therefore, it would require a stronger active-site base to remove the α-proton, which means that its conjugate acid, formed after removal of the α-proton, may be too weak to catalyze the elimination of the β-leaving group. That being the case, a second, more acidic, group may be required to protonate the leaving group. To avoid proton transfer from the second acidic group to the base that removes the α-proton and to have the most efficient means of catalysis, these two groups should be oriented on opposite sides of the substrate, and the α-proton and β-leaving group should be oriented antiperiplanar. This leads to an *anti*-elimination. Believe it or not, every enzyme-catalyzed β-elimination reaction with a carboxylic acid substrate, except one,[63] proceeds by an *anti*-elimination! In the case of the one exception, muconate lactonizing enzyme, the leaving group is a carboxylate, a very good leaving group; possibly it does not require acid catalysis to depart, so one base is sufficient for catalysis. In general, though, Scheme 1.13B predicts that a two-base mechanism is required for proton transfer and elimination of carboxylates with *anti* stereochemistry.

This general acid/base enzyme-catalyzed enolization reaction was further analyzed by the Marcus formalism, which partitions the activation energy barrier ($\Delta G^{\ddagger}$) into a thermodynamic component (ΔG°) and an intrinsic kinetic component ($\Delta G_{int}^{\ddagger}$); $\Delta G_{int}^{\ddagger}$ is the activation energy for the reaction in the absence of a thermodynamic barrier.[64] On this basis Gerlt and Gassman proposed that in the enzyme active site both $\Delta G°$ and $\Delta G_{int}^{\ddagger}$ are reduced from those in nonenzymatic reactions because the transition states for enzymatic reactions are late (i.e., the transition states resemble the enol tautomer, which lowers the pK_a of the proton that is abstracted). The reduction in $\Delta G_{int}^{\ddagger}$ could be achieved by a general acid group positioned adjacent to the carbonyl group of the substrate, which could neutralize the development of a negative charge on the carbonyl oxygen as the α-proton is abstracted. They also propose that the reduction in $\Delta G°$ is accomplished by stabilization of the enolic intermediate via formation of a short, strong hydrogen bond between the anionic conjugate base of the acidic catalyst and the hydroxyl group of the enol tautomer. Schowen[65] was the first to propose that these strong hydrogen bonds may exist within the protected interior of proteins. Cleland and Kreevoy,[66] who referred to these short (<2.5 Å) and very strong hydrogen bonds as *low-barrier hydrogen bonds,* suggest that they are involved in transition-state or enzyme-intermediate complex formation and are very important in enzyme catalysis. They suggest that these bonds could contribute to the tighter binding of substrates to enzymes *at the transition state of the reaction* and estimate that they could supply $10-20$ kcal/mol of stabilization energy for utilize toward catalysis. Others, however, have estimated the additional stabilization of low-barrier hydrogen bonds to be much lower. On the basis of a comparison of enzyme-inhibitor complexes, a value of about 2 kcal/mol was calculated.[67] This value is more in line with values for conventional hydrogen bonds obtained through site-directed mutagenesis[68] or in unnatural amino acid replacement experiments.[69] A value of $4-5$ kcal/mol was estimated for a low-barrier hydrogen bond in dimethyl sulfoxide (DMSO) and $2.7-6$ kcal/mol in tetrahydrofuran (THF).[70] By proton and deuterium nuclear magnetic resonance (NMR) spectroscopy, a low-barrier hydrogen bond in benzene was estimated at 2.4 kcal/mol and in methylene chloride 1.4 kcal/mol.[71]

Some of the physicochemical parameters utilized to identify low-barrier hydrogen bonds are (1) extremely low-field ^1H NMR chemical shifts ($16-20$ ppm); (2) deuterium isotope effects on the low-field NMR resonances; and (3) deuterium isotope effects on infrared (IR) and Raman frequencies.[72] Because these low-barrier hydrogen bonds form only when the pK_a values of the atoms sharing the hydrogen are very similar (preferably identical), a weak hydrogen bond in an enzyme−substrate complex could become a strong hydrogen bond if the pK_a's become matched at the transition state. This might occur if the donor−acceptor distance becomes reduced and any competing water is squeezed out by the tight fit of the transition state in the enzyme and/or if the proton affinities of the two heteroatoms bridged by the hydrogen bond are brought to near equality. When the electronegativities and degree of ionization of atoms X and Y become equal, the barrier between the two hydrogen positions becomes low enough that the zero-point energy level is at or above the energy barrier for transfer of the proton between the two electronegative atoms, and the proton is then shared equally between the two groups, i.e., $-X\cdots H\cdots Y-$. For interaction with two oxygen atoms,

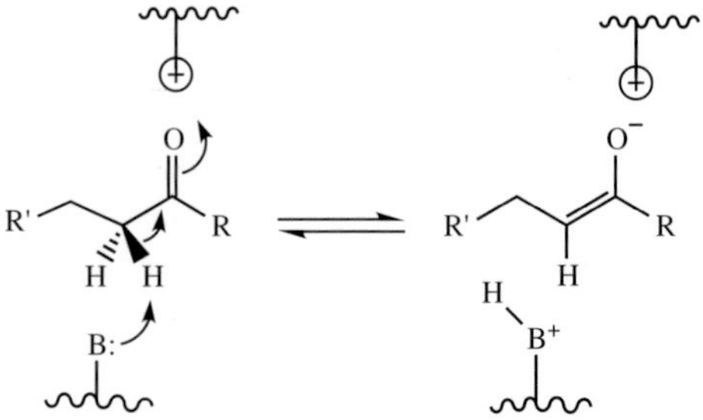

SCHEME 1.15   Electrostatic enzyme catalysis in enolization.

this corresponds to about 2.5 Å. The tighter binding of the low-barrier hydrogen bond stabilizes the transition state (lowers the activation energy) and would be sufficient to facilitate enolization of carbonyl groups.

Guthrie and Kluger,[73] however, argue that there is no need to rationalize enolization in terms of simultaneous protonation of the carbonyl by hydrogen bonding and deprotonation of the α-carbon (see Scheme 1.13). They suggest that the negative charge on an enolate can be stabilized by electrostatic interactions, which they point out can be large in media of low polarity (Scheme 1.15). A theoretical investigation on the energetics of hydrogen bonds as a function of the change in the pK_a found a linear correlation between the increase in hydrogen bond energy and the decrease in the change in pK_a, as expected from simple electrostatic effects.[74] No additional contribution to the hydrogen bond at $\Delta pK_a = 0$ was found, which was utilized as evidence to argue against low-barrier hydrogen bonds and to favor electrostatic stabilization. Another study on the enolization of cationic ketones indicated that electrostatic, through-space, stabilization of enolate ions, and their corresponding transition states, can translate into more than a factor of 10^4 toward the rate of enolization.[75] This indicates that cationic groups at the active site (e.g., protonated lysines, arginines, or histidines) could be very important residues for catalyzing enolization reactions. However, metal ions and low-barrier hydrogen bonds may be equally important alternatives for enolization of carbonyl groups having seemingly high pK_a α-protons.

The debate about the existence and importance of low-barrier hydrogen bonds to enzyme catalysis is far from being over. If this kind of thing intrigues you (maybe as much as the classical versus nonclassical carbocation debate in organic chemistry of the 1960's and 1970's), there are many papers out there now (and many more, I'm sure, coming in the future) that will give you hours of reading pleasure.[76]

D. Electrostatic Catalysis

An enzyme catalyzes a reaction by stabilizing the transition state and by destabilizing the ground state. Stabilizing the transition state may involve the presence of an

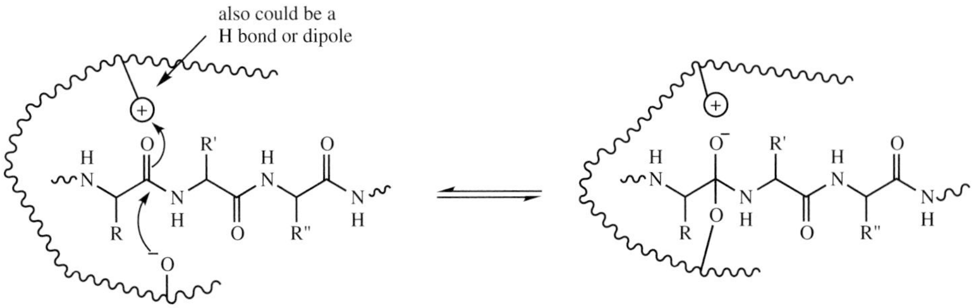

SCHEME 1.16   Electrostatic stabilization of the transition state.

ionic charge or partial ionic charge at the active site to interact with an opposite charge developing on the substrate at the transition state of the reaction (Scheme 1.16). In the case of the tetrahedral intermediate shown in Scheme 1.16, the site in the enzyme that leads to this stabilization is referred to as the *oxyanion hole*.[77] Foje and Hanzlik[78] demonstrated the importance of the oxyanion hole of the cysteine protease papain by comparing the enzyme-catalyzed hydrolysis of a series of peptidyl amides to the corresponding peptidyl thioamides. Whereas all of the amides were very good substrates, none of the thioamides was hydrolyzed, although they were competitive inhibitors. This difference in the two series of compounds can be rationalized by the differences in hydrogen-bonding properties and electrostatic potentials of oxygen versus those of sulfur, and suggests that the oxyanion hole plays a significant role in catalysis. Free-energy calculations on serine protease active sites indicate that the primary mechanism for rate acceleration is derived from electrostatic stabilization in the transition state.[79]

Electrostatic interactions may not be as pronounced as is shown in Scheme 1.16; instead of a full positive charge on the enzyme, there may be one or more local dipoles having partial positive charges directed at the incipient transition-state anion or a protonated group available for hydrogen bonding. In the case of the serine protease subtilisin, it has been suggested that the lowering of the free energy of the activated complex is the result of hydrogen bonding of the developing oxyanion with protein residues.[80] When the suspected active-site proton donor was replaced by a leucine residue using site-directed mutagenesis, the k_{cat} greatly diminished, but the K_m remained the same, indicating the importance of hydrogen bonding to catalysis. Furthermore, a *mutant* (a form of the enzyme that has one or more of its amino acid residues changed, generally by site-directed mutagenesis) of the protease subtilisin, in which all three of the catalytic triad residues (serine-221, histidine-64, and aspartate-32) were replaced by alanine residues by site-directed mutagenesis, still was able to hydrolyze amides 10^3 times faster than the uncatalyzed hydrolysis rate, albeit 2×10^6 times slower than the *wild-type enzyme* (i.e., the unmutated form).[81] This suggests that factors other than nucleophilic and general base catalysis must also be important.

For future reference, the nomenclature for mutants utilizes the one-letter amino acid code of the wild-type residue followed by its position in the enzyme, then the one-letter code of the amino acid change. Therefore, the preceding three mutants for the catalytic triad would be denoted S221A, H64A, and D32A to denote a mutation of serine-221 to alanine, histidine-64 to alanine, and aspartate-32 to alanine, respectively.

E. Desolvation

Ground-state destabilization could occur by desolvation, that is, removal of water molecules from charged groups at the active site on substrate binding. This exposes the substrate to a lower dielectric constant (possibly hydrophobic) environment, which would destabilize a charged group on the substrate. Desolvation also could expose a water-bonded charged group at the active site so it can more effectively participate by electrostatic catalysis in stabilization of a charge generated at the transition state. Because these electrostatic interactions are much stronger in low-dielectric media than in water, the charged groups within the low-dielectric environment more strongly stabilize developing charges at the transition state of the reaction.

F. Strain or Distortion

In organic chemistry, *strain* and *distortion* play an important role in the reactivity of molecules. The much higher reactivity of epoxides relative to other ethers demonstrates this phenomenon. Cyclic phosphate ester hydrolysis is another example. Considerable ring strain in **1.7** (Scheme 1.17) is released on alkaline hydrolysis; the rate of hydrolysis of **1.7** is 10^8 times greater than that for the corresponding acyclic phosphodiester **1.8**.[82] Therefore, if strain or distortion could be induced during enzyme catalysis, then the enzymatic reaction rate would be enhanced. This effect could be induced either in the enzyme, thereby converting it to a high-activity state, or in the substrate, thereby raising the ground-state energy (destabilization) of the substrate and making it more reactive. In Section I.A the induced-fit hypothesis of Koshland[11] was mentioned. This hypothesis suggests that the enzyme

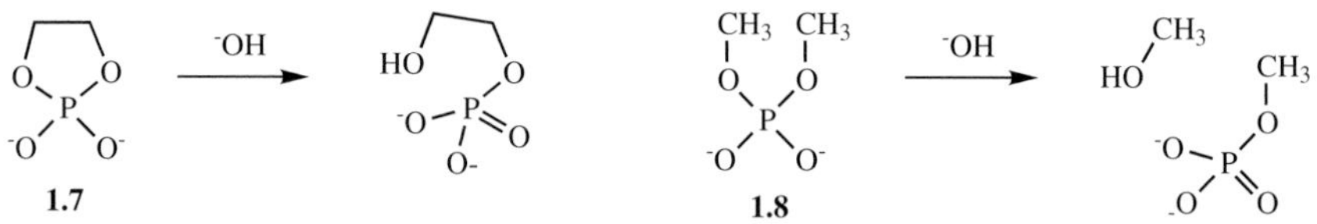

SCHEME 1.17   Alkaline hydrolysis of phosphodiesters.

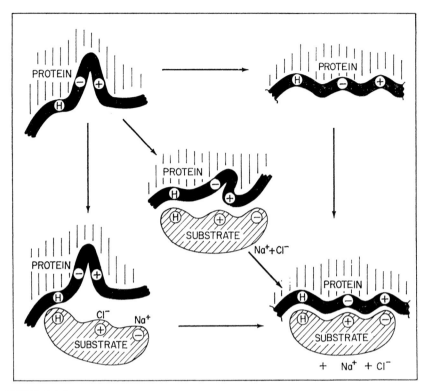

FIGURE 1.9   Schematic of the induced-fit hypothesis. [With permission, from *Annual Review of Biochemistry,* Vol. 37, © 1968 by Annual Reviews, www.annualreviews.org]

need not necessarily exist in the appropriate conformation required to bind the substrate. As the substrate approaches the enzyme, various groups on the substrate interact with particular active-site functional groups, and this mutual interaction induces a conformational change in the active site of the enzyme (Figure 1.9). This can result in a change of the enzyme from a low-catalytic form to a high-catalytic form by destabilization of the enzyme (strain or distortion) or by inducing proper alignment of active-site groups involved in catalysis (orbital steering), which could be responsible for the initiation of catalysis. Inspection of substrate binding sites in protein crystallographic data bases indicates that most enzymes have at least a portion of the active site in a structure that is complementary to the substrate to permit binding on the first collisional event. The conformational change need not occur only in the enzyme; the substrate also could undergo deformation, which would lead to strain (destabilization; higher ground-state energy) in the substrate. The enzyme was suggested to be elastic and could return to its original conformation after the product was released. This rationalizes how high-energy conformations of substrates are able to bind to enzymes.

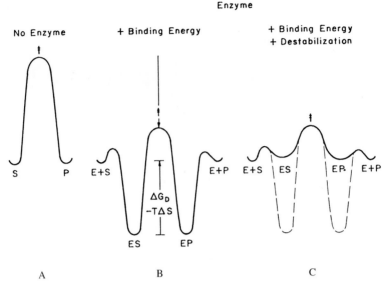

FIGURE 1.10   Energetic effect of enzyme catalysis. [Reproduced with permission from Jencks, W. P. *Cold Spring Harbor Symp. Quant. Biol.* **1987,** *52,* 65. Copyright © 1987 Cold Spring Harbor Laboratory Press.]

According to Jencks,[83] strain or distortion of the bound substrate is essential for catalysis. Because ground-state stabilization of the substrate occurs concomitant with transition-state stabilization, the $\Delta G^{\ddagger}$ is no different from that of the uncatalyzed reaction, only displaced downward (Figure 1.10B). To lower the $\Delta G^{\ddagger}$ for the catalytic reaction, the E·S complex must be destabilized by strain, desolvation, or loss of entropy on binding, thereby raising the $\Delta G^{\ddagger}$ of the E−S and E−P complexes (Figure 1.10C). As the reaction proceeds, the $\Delta G^{\ddagger}$ can be lowered by release of strain energy or by other mechanisms described earlier.

Experimental evidence for strain in a biological catalyst comes from an X-ray crystallographic and mutant study using a catalytic antibody (see Chapter 2, Section I.B.3) that has the ability to catalyze the insertion of a metal ion into a porphyrin ring (**1.9**),[84] a key step in the biosynthesis of the cofactor heme (see Chapter 4, Section IV). Nonenzymatically, *N*-alkylporphyrins are metallated at rates orders of magnitude faster than those of nonalkylated porphyrins,[85] which is believed to be the result of distortion of the planar porphyrin ring toward the transition-state geometry because of the alkyl substituent. The catalytic antibody was raised based on this same principle, and the crystal structure at 2.4 Å resolution of this antibody with an *N*-alkylporphyrin bound shows the porphyrin ring distortion. Furthermore, these distortions are the same as those induced by *N*-methylation of the free porphyrin.

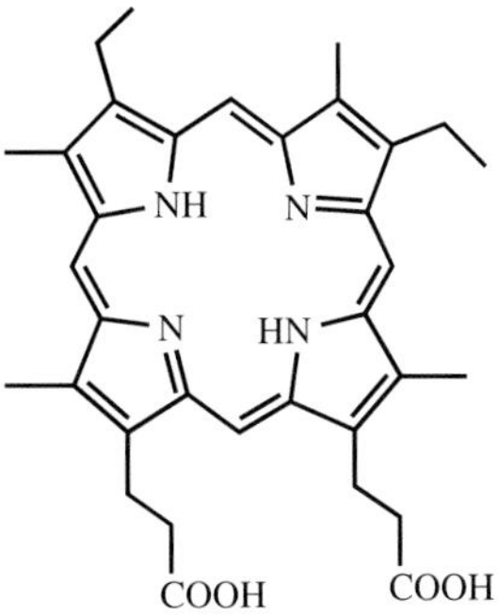

1.9

III. ENZYME CATALYSIS IN ORGANIC MEDIA

The enzymatic reactions we will be discussing throughout the text are, for the most part, *in vitro* reactions in buffer near pH 7. Klibanov[86] and others,[87] however, have shown that enzymes also function in organic solvents with a trace of buffer and also at high temperature. Less than a monolayer of water is required for catalysis.[88] Enzymes that are active in organic solvents include a variety of lipases, which, in buffer, catalyze the hydrolysis of glycerides (**1.10**). Porcine pancreatic lipase, however, also catalyzes the transesterification between tributyryl glycerol and various alcohols. In water or buffer at 100 °C, the lipase loses activity instantaneously. However, when the mixture is heated to 100 °C in tributyryl glycerol–heptanol containing 0.015% water, the $t_{1/2} = 12$ h.[89] Also, the enzyme catalyzes the transesterification at 100 °C five times faster than at 20 °C (no *non*enzymatic reaction occurs at 100 °C).

Three other lipases were shown to catalyze several different reactions (transesterification, esterification, aminolyses, acyl exchange, thiotransesterification, and oximolysis) in various solvents (hexane, acetone, THF, ether, pyridine, toluene) that

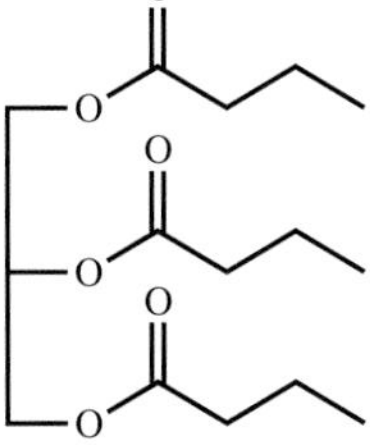

1.10

they could not catalyze in aqueous medium. The catalytic efficiencies of these enzymes were similar to those in water.[90] The effectiveness in organic solvents appears to result from the structural rigidity of proteins in organic solvents leading to high kinetic barriers to denaturation.

Organic solvents do not appear to have much effect on enzyme structure. The crystal structure of the serine protease subtilisin Carlsberg in anhydrous acetonitrile at 2.3 Å resolution is essentially identical to that in water.[91] The hydrogen-bonding system of the catalytic triad remains intact. Most of the enzyme-bound water molecules (99 of 119) are not displaced by acetonitrile. Twelve acetonitrile molecules are bound to the enzyme, four of which have displaced water molecules, and the other eight are bound to where there were no water molecules previously. However, the enzyme had to be crosslinked with glutaraldehyde to prevent fracturing of the crystals so that the crystal structure could be obtained. To be certain that the glutaraldehyde crosslinking did not affect the structure in acetonitrile, the crystal structure of subtilisin Carlsberg crosslinked with glutaraldehyde also was solved in aqueous solution at 2.3 Å resolution.[92] This structure is virtually identical to the structure in anhydrous acetonitrile, although the latter structure was significantly more rigid than in water.

A similar X-ray comparison was made for the structure of γ-chymotrypsin in water and in n-hexane at 2.2 Å resolution; in this case, however, significant changes were found between the two structures in the side chains of both polar and neutral residues, particularly in the vicinity of the bound hexane molecules.[93] Although the details of the structure in hexane are affected, the overall structure is not dramatically altered. The reactivity of the crystalline enzyme in hexane is similar to that of the enzyme in water in the hydrolysis direction, but only in the absence of water can the enzyme catalyze the reaction in the synthetic direction as well.

Because enzymes appear to be capable of catalyzing reactions in organic solvents that are different from those in water, including changes in substrate specificity and enantioselectivity,[59–63] the use of enzymes in organic solvents is becoming increasingly more important in synthetic organic chemistry.[94]

IV. ENZYME NOMENCLATURE

Before we get started with enzyme reactions, let's take a quick look at enzyme nomenclature. As noted earlier, most enzymes are given the suffix -*ase* to distinguish them as enzymes. Furthermore, enzymes are principally classified and named according to the reaction they catalyze: peptidases hydrolyze peptide bonds, esterases hydrolyze esters, oxidases oxidize compounds, and so forth. The Enzyme Commission of the International Union of Biochemistry and Molecular Biology (IUBMB) is the official organization charged with the responsibility of naming enzymes and classifying them.[95] This group decided that there should be two nomenclatures for enzymes, one systematic and one trivial. An Enzyme Commission (EC) number is assigned to every enzyme according to the reaction it catalyzes, so it is possible to determine from the number to what general family the enzyme belongs. The EC number contains four elements separated by periods. The first number shows to which of the six classes the enzyme belongs. Class 1 is the oxidoreductases, which are enzymes that catalyze oxidation/reduction reactions; class 2 comprises transferases, enzymes that transfer a group from one compound to another; class 3 is hydrolases, which catalyze the hydrolytic cleavage of C–O, C–N, C–C, and some other bonds such as phosphoric anhydrides; class 4, the lyases, are enzymes that cleave C–C, C–O, C–N, and other bonds by elimination, leaving double bonds or rings, or conversely adding groups to double bonds; the class 5 enzymes are isomerases, which catalyze geometric or structural changes within one molecule; and class 6 is the ligases, enzymes that catalyze the joining together of two molecules coupled with the hydrolysis of a diphosphate bond in ATP or a similar triphosphate. The second digit in the EC number code indicates a subclass: For class 1 it denotes the group in the hydrogen donor that undergoes oxidation, for class 2 it is the group transferred, in class 3 it indicates the nature of the bond hydrolyzed, for class 4 the second figure indicates the bond broken, for class 5 it is the type of isomerism, and for class 6 it indicates the bond formed. The third number is called the *sub-subclass,* which characterizes the enzyme further, and the fourth number is the serial number, which specifies the particular enzyme (or group of enzymes with the same catalytic property). For example, the esterase acetylcholinesterase is EC 3.1.1.7; it is in class 3 (hydrolases), subclass 1 (acting on ester bonds), sub-subclass 1 (carboxylic ester hydrolysis), and serial number 7.

V. EPILOGUE

This chapter lays the foundation for the discussions of enzyme-catalyzed reactions in subsequent chapters. You have seen the forces involved in the formation of the E·S complex and the clever use of physical organic chemistry that enzymes employ in converting the E·S complex to the E·P complex. We now turn our attention to applying these basic organic principles to the study of mechanisms of enzyme-catalyzed reactions. Each chapter represents a particular class of organic reactions that different classes of enzymes catalyze. This is generally backward from the approach of standard biochemistry texts, in which classes of enzymes are discussed and the reaction that they catalyze are mentioned as part of the characterization of the enzyme. The focus here is on the reaction and its possible mechanisms and on how the enzyme utilizes the "tricks" discussed in this chapter to lower the energy of the reactions.

NOTE THAT PROBLEMS AND SOLUTIONS RELEVANT TO EACH CHAPTER CAN BE FOUND IN APPENDIX II.

REFERENCES

1. Segal, I. H. *Enzyme Kinetics. Behavior and Analysis of Rapid Equilibrium and Steady-State Enzyme Systems,* Wiley: New York, 1975.

1a. (a) Lorsch, J. R.; Szostak, J. W. *Acc. Chem. Res.* **1996,** *29,* 103. (b) Pan, T. *Curr. Opin. Chem. Biol.* **1997,** *1,* 17. (c) Joyce, G. F. *Proc. Natl. Acad. Sci. USA* **1998,** *95,* 5845.

2. Sumner, J. B. *J. Biol. Chem.* **1926,** *69,* 435.

3. (a) Jabri, E.; Carr, M. B.; Hausinger, R. P.; Karplus, P. A. *Science* **1995,** *268,* 998. (b) Jabri, E.; Karplus, P. A. *Biochemistry* **1996,** *35,* 10616.

4. Hirs, C. H. W.; Moore, S.; Stein, W. H. *J. Biol. Chem.* **1960,** *235,* 633.

5. (a) Gutte, B.; Merrifield, R. *J. Am. Chem. Soc.* **1969,** *91,* 501. (b) Hirschmann, R.; Nutt, R.; Veber, D.; Vitali, A.; Varga, S.; Jacob, T.; Holly, F.; Denkewalter, R. *J. Am. Chem. Soc.* **1969,** *91,* 507.

6. Schramm, V. L. *Annu. Rev. Biochem.* **1998,** *67,* 693.

7. Page, M. I. In *Enzyme Mechanisms,* Page, M. I.; Williams, A., Eds., Royal Society of Chemistry: London, 1987, p. 1.

8. Brown, A. J. *Trans. Chem. Soc. (London)* **1902,** *81,* 373.

9. Henri, V. *Acad. Sci., Paris* **1902,** *135,* 916.

10. Fischer, E. *Berichte* **1894,** *27,* 2985.

11. (a) Koshland, D. E., Jr. *Proc. Natl. Acad. Sci. USA* **1958,** *44,* 98. (b) Koshland, D. E.; Neet, K. E. *Annu. Rev. Biochem.* **1968,** *37,* 359.

12. Post, C. B.; Ray, W. J., Jr. *Biochemistry* **1995,** *34,* 15881.

13. (a) Pauling, L. *Chem. Eng. News* **1946,** *24,* 1375. (b) Pauling, L. *Am Sci.* **1948,** *36,* 51.

14. Haldane, J. B. S. *Enzymes,* Longmans and Green: London, 1930 (reprinted by MIT Press: Cambridge, 1965).

15. Eyring, H. *J. Chem. Phys.* **1935,** *3,* 107.

16. Hackney, D. D. *The Enzymes,* 3rd ed., vol. 19, Sigman, D. S.; Boyer, P. D., Ed., Academic Press: San Diego, 1990, pp. 1–37.

17. Schowen, R. L. In *Transition States of Biochemical Processes,* Gandour, R. D.; Schowen, R. L., Eds., Plenum: New York, 1978, p. 77.

18. Richards, F. M. *Annu. Rev. Biophys. Bioengg.* **1977,** *6,* 151.

19. (a) Storm, D. R.; Koshland, Jr., D. E. *Proc. Natl. Acad. Sci. USA* **1970,** *66,* 445. (b) Ibid., *J. Am. Chem. Soc.* **1972,** *94,* 5805.

20. Mesecar, A. D.; Stoddard, B. L.; Koshland, D. E., Jr. *Science* **1997,** *277,* 202.

21. Talalay, P.; Benson, A. In *The Enzymes,* 3rd ed., vol. 6, Boyer, P., Ed., Academic Press: New York, 1972, p. 591.

22. Fersht, A. R. *Enzyme Structure and Mechanism,* 2nd ed., W. H. Freeman: New York, 1985, pp. 147–148.

23. Backlow, S. C.; Raines, R. T.; Lim, W. A.; Zamore, P. D.; Knowles, J. R., *Biochemistry* **1988,** *27,* 1158.

24. Schramm, V. L. *Annu. Rev. Biochem.* **1998,** *67,* 693.

25. Horenstein, B. A.; Schramm, V. L. *Biochemistry* **1993,** *32,* 7089.

26. Jencks, W. P. *Catalysis in Chemistry and Enzymology,* McGraw-Hill: New York, 1969, p. 340.

27. Tanner, J. J.; Hecht, R. M.; Krause, K. L. *Biochemistry* **1996,** *35,* 2597.

28. Jencks, W. P. *Catalysis in Chemistry and Enzymology,* McGraw-Hill: New York, 1969, p. 393.

29. Hildebrand, J. H. *Proc. Natl. Acad. Sci. USA* **1979,** *76,* 194.

30. (a) Karsten, W. E.; Gates, R. B.; Viola, R. E., *Biochemistry* **1986,** *25,* 1299. (b) Falzone, C. J.; Karsten, W. E.; Conley, J. D.; Viola, R. E. *Biochemistry* **1988,** *27,* 9089.

31. Switzer, R. L. In $B_{12},$ vol. 2, Dolphin, D., Ed., Wiley: New York, 1982, p. 289.

32. Michel, C.; Albracht, S. P. J.; Buckel, W. *Eur. J. Biochem.* **1992,** *205,* 767.

33. Coleman, J. E.; Gettins, P. *Adv. Enzymol. Rel. Areas Mol. Biol.* **1983,** *55,* 381.

34. (a) Edenberg, H. J.; Bosron, W. F. In *Comprehensive Toxicology,* vol. 3, Guengerich, F. P., Ed., Perga-

mon, Elsevier Science: New York, 1997, pp. 119–131. (b) Green, D. W.; Sun, H.-W.; Plapp, B. V. *J. Biol. Chem.* **1993**, *268,* 7792.

35. Guengerich, F. P., Ed. *Mammalian Cytochromes P-450,* CRC Press: Boca Raton, 1987.
36. Milton, R. C. deL.; Milton, S. C. F.; Kent, S. B. H. *Science* **1992**, *256,* 1445.
37. Jencks, W. P. *Adv. Enzymol.* **1975**, *43,* 219.
38. Narlikar, G. J.; Hershlag, D. *Biochemistry* **1998**, *37,* 9902.
39. Wolfenden, R. *Acc. Chem. Res.* **1972**, *5,* 10.
40. Burbaum, J. J.; Raines, R. T.; Albery, W. J.; Knowles, J. R. *Biochemistry* **1989**, *28,* 9293.
41. Pettersson, G. *Eur. J. Biochem.* **1992**, *206,* 289.
42. Pettersson, G. *Eur. J. Biochem.* **1991**, *195,* 663.
43. Krishnan, S.; Hall, B. G.; Sinnott, M. L. *Biochem. J.* **1995**, *312,* 971.
44. Jencks, W. P. *Catalysis in Chemistry and Enzymology,* McGraw-Hill, New York, 1969, chaps. 1, 2, 3, 5.
45. Jencks, W. P. *Adv. Enzymol.* **1975**, *43,* 219.
46. Wolfenden, R.; Frick, L. In *Enzyme Mechanisms,* M. I. Page; A. Williams, Eds., Roy. Soc. Chem.: London, 1987, p. 97.
47. Bruice, T. C.; Pandit, U. K. *J. Am. Chem. Soc.* **1960**, *82,* 5858.
48. Bruice, T. C.; Pandit, U. K. *Proc. Natl. Acad. Sci. USA* **1960**, *46,* 402.
49. (a) Lightstone, F. C.; Bruice, T. C. *J. Am. Chem. Soc.* **1996**, *118,* 2595. (b) Bruice, T. C.; Lightstone, F. C. *Acc. Chem. Res.* **1999**, *32,* 127.
50. Lightstone, F. C.; Bruice, T. C. *J. Am. Chem. Soc.* **1997**, *119,* 9103.
51. Kirby, A. J. *Adv. Phys. Org. Chem.* **1980**, *17,* 183.
52. March, J. *Advanced Organic Chemistry,* 3rd ed., Wiley: New York, 1985, p. 268.
53. Bender, M. L.; Neveu, M. C. *J. Am. Chem. Soc.* **1958**, *80,* 5388.
54. Blow, D. M.; Birktoft, J.; Hartley, B. S. *Nature* **1969**, *221,* 337.
55. Tsukada, H.; Blow, D. M. *J. Mol. Biol.* **1985**, *184,* 703.
56. Simonson, T.; Brooks, C. L., III. *J. Am. Chem. Soc.* **1996**, *118,* 8452.
57. Parsons, S. M.; Raftery, M. A. *Biochemistry* **1972**, *11,* 1623, 1630, 1633.
58. (a) Kokesh, F. C.; Westheimer, F. H. *J. Am. Chem. Soc.* **1971**, *93,* 7270. (b) Frey, P. A.; Kokesh, F. C.; Westheimer, F. H. *J. Am. Chem. Soc.* **1971**, *93,* 7266. (c) Schmidt, D. E., Jr.; Westheimer, F. H. *Biochemisty* **1971**, *10,* 1249.
59. Jencks, W. P.; Carriuolo, J. *J. Biol. Chem.* **1959**, *234,* 1272, 1280.
60. Gerlt, J. A.; Kozarich, J. W.; Kenyon, G. L.; Gassman, P. G. *J. Am. Chem. Soc.* **1991**, *113,* 9667.
61. Gerlt, J. A.; Gassman, P. G. *J. Am. Chem. Soc.* **1992**, *114,* 5928.
62. (a) Schwab, J. M.; Klassen, J. B.; Habib, A. *J. Chem. Soc. Chem. Commun.* **1986**, 357. (b) Hanson, K. R.; Rose, I. A. *Acc. Chem. Res.* **1975**, *8,* 1.
63. Chari, R. M. J.; Whitman, C. P.; Kozarich, J. W.; Ngai, K. L.; Ornston, L. N. *J. Am. Chem. Soc.* **1987**, *109,* 5514.
64. (a) Gerlt, J. A.; Gassman P. G. *J. Am. Chem. Soc.* **1993**, *115,* 11552. (b) Gerlt, J. A.; Gassman P. G. *Biochemistry* **1993**, *32,* 11943.
65. Schowen, R. L. In *Mechanistic Principles of Enzyme Activity,* Liebman, J. F.; Greenburg, A., Eds., VCH: New York, 1988, p. 119.
66. (a) Cleland, W. W. *Biochemistry* **1992**, *31,* 317. (b) Cleland, W. W.; Kreevoy, M. M. *Science,* **1994**, *264,* 1887. (c) Cleland, W. W.; Kreevoy, M. M. *Science,* **1995**, *269,* 104. (d) Frey, P. A. *Science,* **1995**, *269,* 104.
67. Usher, K. C.; Remington, S. J.; Martin, D. P.; Drueckhammer, D. G. *Biochemistry* **1994**, *33,* 7753.
68. Fersht, A. R. *Trends Biochem. Sci.* **1987**, *12,* 301.
69. Thorson, J. S.; Chapman, E.; Schultz, P. G. *J. Am. Chem. Soc.* **1995**, *117,* 9361.
70. Shan, S.; Loh, S.; Herschlag, D. *Science* **1996**, *272,* 97.
71. Kato, Y.; Toledo, L. M.; Rebek, J., Jr. *J. Am. Chem. Soc.* **1996**, *118,* 8575.
72. Hibbert, F.; Emsley, J. *Adv. Phys. Org. Chem.* **1990**, *26,* 255.

73. Guthrie, J. P.; Kluger, R. *J. Am. Chem. Soc.* **1993,** *115,* 11569.

74. Shan, S.-O.; Loh, S.; Herschlag, D. *Science* **1996,** *272,* 97.

75. Tobin, J. B.; Frey, P. A. *J. Am. Chem. Soc.* **1996,** *118,* 12253.

76. For other references on this debate, see the following: **low-barrier H-bonds:** Cleland, W. W.; Frey, P. A.; Gerlt, J. A. *J. Biol. Chem.* **1998,** *273,* 25529; Cleland, W. W.; Kreevoy, M. M. *Science,* **1995,** *269,* 104; Frey, P. A. *Science,* **1995,** *269,* 104; Frey, P. A.; Whitt, S. A.; Tobin, J. B. *Science* **1994,** *264,* 1927; Ash, E. L.; Sudmeier, J. L.; De Fabo, E. C.; Bachovchin, W. W. *Science* **1997,** *278,* 1128; Hur, O.; Lejà, C.; Dunn, M. F. *Biochemistry* **1996,** *35,* 7378; Schiøtt, B.; Iversen, B. B.; Madsen, G. K. H.; Larsen, F. K.; Bruice, T. C. *Proc. Natl. Acad. Sci. USA* **1998,** *95,* 12799. **electrostatic interactions:** Warshel, A. *J. Biol. Chem.* **1998,** *273,* 27035; Jackson, S. E.; Fersht, A. R. *Biochemistry* **1993,** *32,* 13909; Warshel, A; Papazyan, A.; Kollman, P. A., *Science,* **1995,** *269,* 102; Alagona, G.; Ghio, C.; Kollman, P. A., *J. Am. Chem. Soc.* **1995,** *117,* 9855.

77. Kraut, J. *Annu. Rev. Biochem.* **1977,** *46,* 331.

78. Foje, K. L.; Hanzlik, R. P. *Biochim. Biophys. Acta,* **1994,** *1201,* 447.

79. Warshel, A.; Naray-Szabo, G.; Sussman, F.; Hwang, J.-K. *Biochemistry* **1989,** *28,* 3629.

80. Bryan, P.; Pantoliano, M. W.; Quill, S. G.; Hsaio, H.-Y.; Poulos, T. *Proc. Natl. Acad. Sci. USA* **1986,** *83,* 3743.

81. Carter, P.; Wells, J. A. *Nature* **1988,** *332,* 564.

82. Covitz, T.; Westheimer, F. H. *J. Am. Chem. Soc.* **1963,** *85,* 1773.

83. Jencks, W. P. *Cold Spring Harbor Symp. Quant. Biol.* **1987,** *52,* 65.

84. Romesberg, F. E.; Santarsiero, B. D.; Spiller, B.; Yin, J.; Barnes, D.; Schultz, P. G.; Stevens, R. C. *Biochemistry* **1998,** *37,* 14404.

85. Bain-Ackerman, M. J.; Lavallee, D. K. *Inorg. Chem.* **1979,** *18,* 3358.

86. (a) Wescott, C. R.; Klibanov, A. M. *Biochim. Biophys. Acta* **1994,** *1206,* 1. (b) Klibanov, A. M. *Trends Biochem. Sci.* **1989,** *14,* 141. (c) Klibanov, A. M. *Chemtech* **1986,** *16,* 354. (d) Klibanov, A. M. *Acc. Chem. Res.* **1990,** *23,* 114.

87. (a) Gupta, M. N. *Eur. J. Biochem.* **1992,** *203,* 25. (b) Dordick, J. S. *Enzyme Microb. Technol.* **1989,** *11,* 194. (c) Halling, P. J. *Enzyme Microb. Technol.* **1994,** *16,* 178.

88. Zaks, A.; Klibanov, A. M. *J. Biol. Chem.* **1988,** *263,* 3194.

89. Zaks, A.; Klibanov, A. M. *Science* **1984,** *224,* 1249.

90. Zaks, A.; Klibanov, A. M. *Proc. Natl. Acad. Sci. (USA)* **1985,** *82,* 3192.

91. Fitzpatrick, P. A.; Steinmetz, A. C. U.; Ringe, D.; Klibanov, A. M. *Proc. Natl. Acad. Sci. USA,* **1993,** *90,* 8653.

92. Fitzpatrick, P. A.; Ringe, D.; Klibanov, A. M. *Biochem. Biophys. Res. Commun.,* **1994,** *198,* 675.

93. Yennawar, N. H.; Yennawar, H. P.; Farber, G. K. *Biochemistry* **1994,** *33,* 7326.

94. Wong, C.-H.; Whitesides, G. M. *Enzymes in Synthetic Organic Chemistry,* Elsevier Science: Oxford, England, 1994.

95. The book of nomenclature is called *Enzyme Nomenclature,* Academic Press: San Diego, **1992.**

CHAPTER 2

Group Transfer Reactions: Hydrolysis, Amination, Phosphorylation

I. HYDROLYSIS REACTIONS

A. Amide Hydrolysis: Peptidases

Peptidases (sometimes referred to as *proteases* when protein hydrolysis is involved) are in a family of enzymes whose function is to catalyze the hydrolysis of peptide bonds. The hydrolysis of a peptide bond is not an easy reaction. The nonenzymatic rate of hydrolysis at pH 9 and 25 °C is 8.2×10^{-11} s^{-1}.[1] The overall reaction catalyzed by peptidases is shown in Scheme 2.1. Note the special nomenclature used for

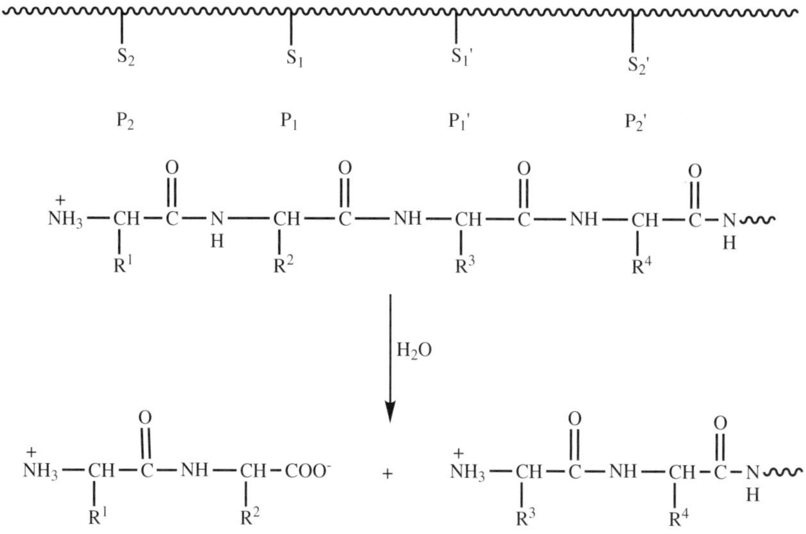

SCHEME 2.1   Reaction catalyzed by peptidases.

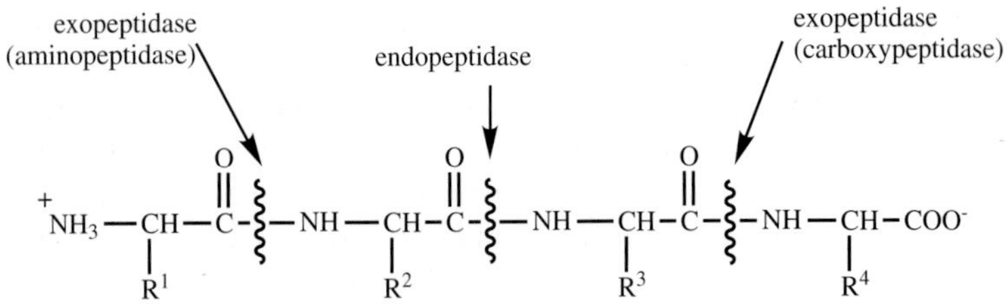

FIGURE 2.1   Classifications of peptidases

peptide (protein) hydrolysis: The amino acid at the C-terminus of the amide bond that is hydrolyzed is termed the P_1 amino acid, the next one away from the bond being hydrolyzed is P_2, and so on. The amino acid at the N-terminus of the newly formed peptide is the P_1' amino acid, next is P_2', and so on. Likewise, the amino acid residue on the enzyme with which the P_1 amino acid interacts is the S_1 site, the residue with which the P_1' amino acid interacts is the S_1' site, and so forth. Some peptidases hydrolyze the carboxy terminus (*carboxypeptidases*), some hydrolyze the amino terminus (*aminopeptidases*), and some hydrolyze interior peptide bonds, depending on what the amino acid side-chain group (*R*) is. The ones that cleave the interior peptide bonds are called *endopeptidases,* and the ones that cleave the terminal bonds are called *exopeptidases* (Figure 2.1).

1. Endopeptidases

a. Types

As a representative example of endopeptidases, consider the digestive enzyme α-chymotrypsin from mammalian pancreas. Note that everything you just learned in Chapter 1 about nomenclature can be trashed! The enzyme name does not give you the slightest clue to what reaction it catalyzes, and does not even end in the suffix -*ase!* Chymotrypsin gets away with having that name because it has been known for a very long time by this name, and the Enzyme Commission went with tradition over regulation and approved the common name. This endopeptidase regiospecifically hydrolyzes peptide bonds of the aromatic amino acids; that is, P_1 is Phe, Tyr, and Trp (Figure 2.1, $R_2 = CH_2Ph$, CH_2Ph-4-OH, and CH_2-indolyl). Another endopeptidase, trypsin, regiospecifically cleaves at Arg and Lys residues (P_1 is Arg or Lys). How can an enzyme be so selective for which peptide bond it hydrolyzes? The active site of the peptidase has amino acid residues that are complementary to the side chains of the amino acid where it cleaves, which holds the appropriate peptide bond in the correct juxtaposition for hydrolysis; hydrophobic interactions of several amino acids with the lysine side chain methylenes and an electrostatic interaction of Asp-189 with the ϵ-ammonium ion of lysine can be seen in the X-ray crystal structure of a trypsin inhibitor bound to trypsin.[2] Although physiologically these enzymes are peptidases, they also can catalyze the hydrolysis of esters, thio-esters, anhydrides, and acid chlorides.

You may be wondering how an enzyme whose function is to catalyze the hydrolysis of particular proteins manages not to hydrolyze other proteins that it is not supposed to hydrolyze or to be hydrolyzed by itself. The reason is that α-chymotrypsin and trypsin are synthesized in the pancreas in the form of inactive *proenzymes* (also called *zymogens*). When they are needed, each zymogen is secreted into the intestines and is converted into the active enzyme by a protease-catalyzed hydrolysis of a terminal peptide of the zymogen.

b. Proposed Mechanism for α-Chymotrypsin

α-Chymotrypsin is an excellent enzyme for introducing mechanistic enzymology because it has been studied by many techniques over an extended period of time. It is the prototype for a variety of classes of enzymes, not only for proteases. Because of the complexity of the mechanism, it may be clearer to see the proposed mechanism at the outset, so the significance of the experimental design can be followed and the results understood more readily. As indicated in Chapter 1, α-chymotrypsin uses covalent catalysis in its mechanism of hydrolyzing proteins and peptides. A *charge relay system,* involving a *catalytic triad* of Asp-102, His-57, and Ser-195, is considered important to the catalysis; this is a common mechanism for many proteases (Scheme 2.2). Some investigators believe that the interaction between Asp-102 and His-57 involves a low-barrier hydrogen bond (see Chapter 1, Section II.C),[3] whereas others favor an electrostatic interaction between the aspartate anion and the protonated histidine cation.[4]

(1) Evidence for an Acyl Intermediate   The mechanism shown in Scheme 2.2 is an example of covalent catalysis (acylation) followed by general acid/base catalysis (deacylation); electrostatic catalysis is important for the lowering of the transition-state energy. No covalent intermediate with good substrates was observed, so a typical approach taken in enzymology to gain information about short-lived intermediates (if spectroscopic techniques are not available or are not adequate) is to use an alternate, poor substrate. This may change the rate-determining step and allow the buildup of an intermediate that could not previously be observed. However, it may also change the mechanism completely. Nonetheless, this is a valid and commonly used approach in mechanistic enzymology.

The substrate chosen to study this mechanism looks nothing at all like a normal substrate; in fact, it is not even a peptide.[5] Generally, proteases also can catalyze the hydrolysis of esters, so a reactive ester, a *para*-nitrophenyl acetate (**2.1**, Figure 2.2), was selected as the alternative substrate. The principle behind this idea is that the rate-determining step in peptide hydrolysis is probably the acylation step because of the stability of peptide bonds; as soon as the acyl intermediate forms, it rapidly

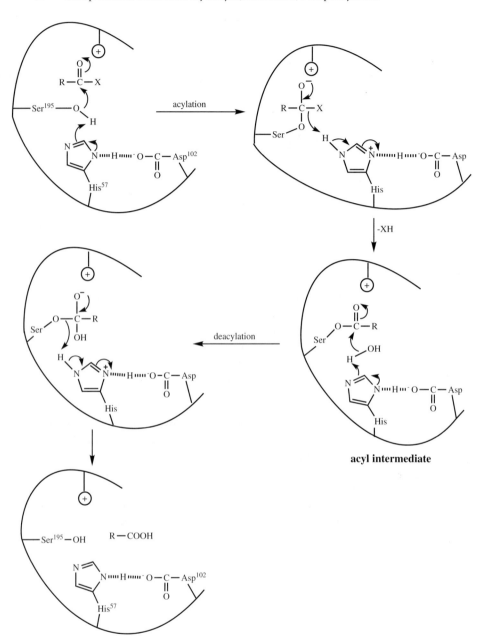

SCHEME 2.2   Mechanism for α-chymotrypsin showing the catalytic triad.

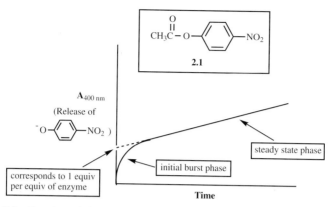

FIGURE 2.2   Reaction of chymotrypsin with *p*-nitrophenyl acetate: demonstration of an initial burst.

undergoes deacylation, and this intermediate is not detected. However, because *para*-nitrophenyl esters are very reactive, the rate of the acylation step should increase, thereby making deacylation the rate-determining step and allowing for a buildup of the acyl intermediate. But **2.1** does not have the hydrophobic and acyl-amino substituents and is thus a poor substrate. The kinetics of the enzyme reaction with **2.1** as substrate are shown in Figure 2.2. The *para*-nitrophenolate anion released by hydrolysis is bright yellow and can easily be detected spectrophotometrically (at 400 nm). The release of *para*-nitrophenolate anion shows an *initial burst*, a fast initial phase, followed by a linear steady-state phase. This is typical of a multistep reaction in which the first step is fast relative to the rest of the steps (Scheme 2.3). If the steady-state phase is extrapolated back to zero time, it should correspond to an amount of product equivalent to the amount of enzyme used in the experiment. This is because every enzyme molecule reacts rapidly with a substrate molecule to release an equivalent amount of product molecules (P_1) prior to the hydrolysis of the acyl intermediate. Note, however, that this experiment is successful only when a large amount of enzyme is used, which then generates enough *para*-nitrophenolate to detect; with a good substrate, much smaller amounts of

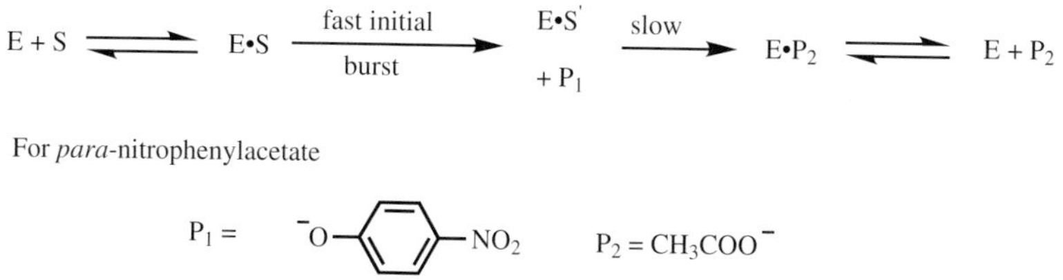

$$E + S \rightleftharpoons E{\cdot}S \xrightarrow[\text{burst}]{\text{fast initial}} \begin{array}{c} E{\cdot}S' \\ + P_1 \end{array} \xrightarrow{\text{slow}} E{\cdot}P_2 \rightleftharpoons E + P_2$$

For *para*-nitrophenylacetate

$$P_1 = \quad {}^{-}O{-}\!\!\!\bigcirc\!\!\!{-}NO_2 \qquad P_2 = CH_3COO^{-}$$

SCHEME 2.3   Typical enzyme reaction in which the first step is fast.

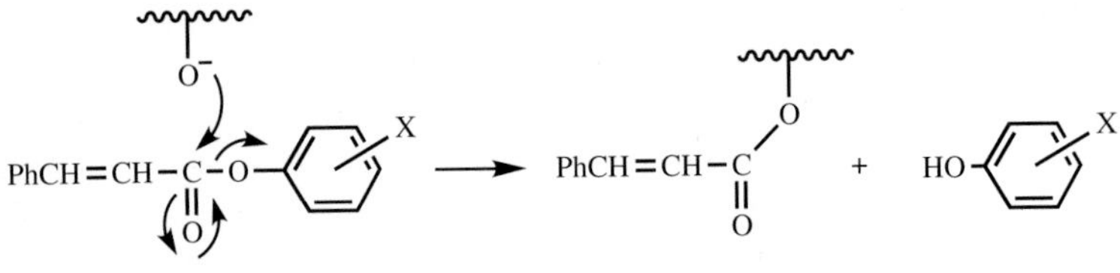

SCHEME 2.4   Reaction of α-chymotrypsin with aryl cinnamate esters.

enzyme are used, and the initial burst is too small to be detected. Once a mechanistic pathway has been supported with a poor substrate, it is important to determine whether the same mechanism is relevant with good substrates.

To show the formation of an acyl intermediate, a series of aryl cinnamate esters (**2.2**, Scheme 2.4) was tested as substrates.[6] Regardless of the structure or position of X, the same first-order steady-state phase was observed; when hydrolysis was carried out nonenzymatically, the rates of the different analogues were very different. This suggests that in the enzymatic reaction a *common intermediate* (**2.3**), the same intermediate produced by all the substrates, is formed in an initial fast step followed by rate-determining product formation (deacylation).

To demonstrate that covalent catalysis occurred and that an acyl intermediate formed, a radioactively labeled substrate, *para*-nitrophenyl [2-^{14}C]acetate (**2.4**, Scheme 2.5), was used. Isotopic labeling is a common and very important tool in mechanistic enzymology. Because of the minute amounts of enzyme that generally are used in enzymological studies, it is necessary to use radioactive isotopes, such as ^{3}H, ^{14}C, ^{32}P, or ^{35}S, all β-emitters, to gain sufficient sensitivity for detection. Because each of these radioisotopes emits β particles of different energies, it also is possible to perform double labeling experiments (^{3}H and ^{14}C are commonly used together) and monitor radioactivity separately in a scintillation counter. In experiments that use a radiolabeled compound, only trace amounts of radioactivity are actually incorporated into the compound, not only because of the expense of radio-

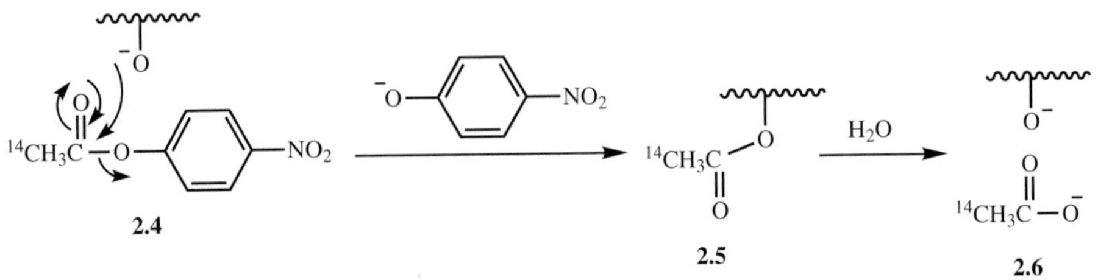

SCHEME 2.5   Formation of an acyl intermediate in the reaction catalyzed by α-chymotrypsin.

active compounds but also because higher concentrations of radiation can decompose molecules. Therefore, maybe only one in a billion or fewer of the molecules will actually contain a radioactive atom; the rest are nonradioactive. All of the same molecules, labeled or not, will react at virtually the same rate, with the exception of the ones that contain tritium, which may exhibit an isotope effect if the $C-^3H$ bond is broken in the rate-determining step and, consequently, will lead to fewer tritium-containing molecules reacting than unlabeled ones. Therefore, the same products will be produced by all of the molecules, labeled or not; however, if the $C-^3H$ bond is broken during the reaction, tritium will be released from the molecule as ^3H_2O. The labeled compounds also will comigrate by chromatography in the same way as the unlabeled compounds and can be detected easily by scintillation counting (unless the tritium was abstracted from the compound).

When chymotrypsin was treated with **2.4** at pH 5 (well below the pH optimum of the enzyme of about 8) to slow down all of the steps in the reaction, inactive and radiolabeled ($[^{14}C]$acetylated) chymotrypsin (**2.5**, Scheme 2.5) was produced. When the pH was raised to 8 so that the enzyme could be fully functional, the radioactivity was released from the enzyme as $CH_3^{14}COO^-$ (**2.6**), and the enzyme became fully active again. The rate at which **2.6** was formed from the inactive, radiolabeled enzyme was comparable to the rate at which **2.6** was formed when **2.4** was used as the substrate under normal conditions, indicating that the intermediate (**2.5**) is *kinetically competent* (i.e., it breaks down at a rate that is consistent with its formation during the enzyme-catalyzed reaction).[7] An experiment that is done to demonstrate formation of a covalent intermediate is to incubate the enzyme with the substrate under nonoptimal conditions so that the reaction stops after the covalent catalytic step, and then to gel filter. *Gel filtration* (or size-exclusion chromatography) uses a column packed with porous beads (generally polyacrylamide or agarose) having a specific pore size. If the pore size is chosen so that the enzyme cannot fit into the pore, but small molecules can, then the small molecules will have farther to travel to get through the column and therefore will be retained longer on the column than the enzyme molecules, which will pass around the beads and come out in the void volume. This is demonstrated in Figure 2.3. The large peak of radioactivity at the later fractions is the excess radiolabeled substrate; the radioactive peak in the earlier fractions comigrates with the protein peak (absorbance at 280 nm). Comigration of radioactivity with protein supports a covalent adduct.

Other experiments were carried out to support the formation of acetylchymotrypsin (**2.5**, Scheme 2.5). One experiment involved the treatment of the acyl intermediate with hydroxylamine (NH_2OH); the enzyme became active, and the product released was identified as *N*-acetylhydroxylamine (**2.7**, Scheme 2.6).[8] Another experiment was performed to show that the rate of base hydrolysis of acetylchymotrypsin *denatured* (i.e., when the protein is unfolded) by the denaturing agent 8 M urea was identical to the rate of base hydrolysis in 8 M urea with a model compound, *O*-acetylserinamide.[9] Further evidence for an acyl intermediate

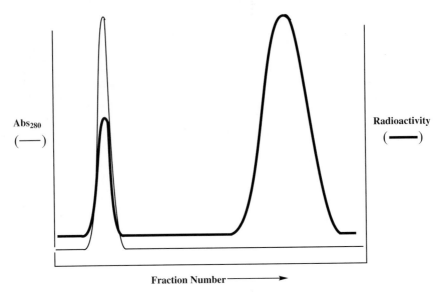

FIGURE 2.3   Gel filtration of radiolabeled enzyme.

involving a serine residue is the reaction of diisopropylphosphorofluoridate (**2.8,** Scheme 2.7) with chymotrypsin.[10] Compound **2.8** is very electrophilic and highly toxic (phosphofluoridates have been used as chemical warfare agents). After binding to the active site this compound undergoes a rapid reaction with the enzyme and inactivates it (**2.9**). *Inactivation* or *irreversible inhibition* is a process in which a compound blocks enzyme activity, usually by covalent modification of the enzyme. This kind of inactivator, which is a reactive compound that undergoes a covalent

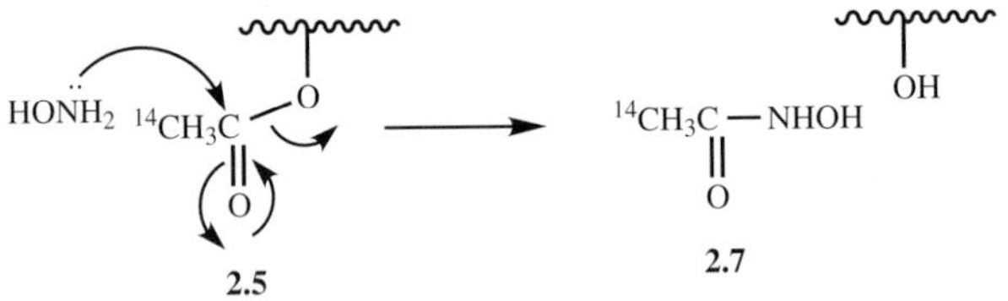

SCHEME 2.6   Reactivation of acetylchymotrypsin by hydroxylamine.

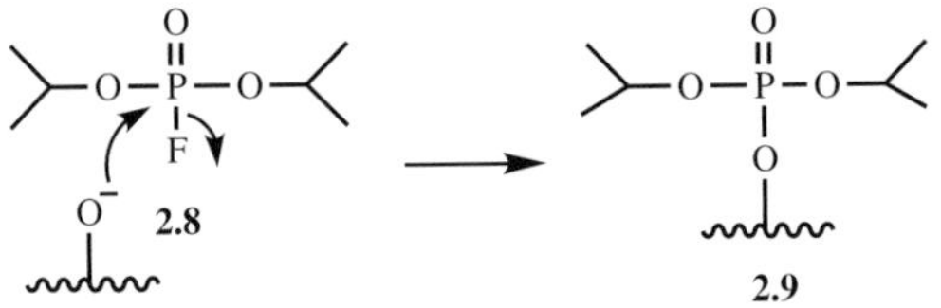

SCHEME 2.7   Reaction of α-chymotrypsin with an organophosphofluoridate affinity labeling agent.

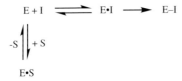

SCHEME 2.8   Kinetics of affinity labeling of enzymes.

reaction with an essential active-site residue (usually acylation or alkylation), is called an *affinity labeling agent*. Once it attaches to the enzyme, it blocks the active site, preventing the substrate from binding and thereby irreversibly preventing catalysis. The kinetics of an affinity labeling agent (see Appendix I, Section II.B.1) are similar to Michaelis–Menten kinetics for substrate turnover (Scheme 2.8). Initially, there is a fast reversible formation of an enzyme–inhibitor (E·I) complex, followed by a covalent reaction to give E–I. As long as an inhibitor is bound to the active site, the substrate cannot get in, and the enzyme cannot catalyze a reaction. When attachment occurs at the active site, the inactivator competes with the substrate for active-site binding.

Irreversible inhibitors, such as affinity labeling agents, exhibit time-dependent inhibition. The initial formation of the E·I complex is generally rapid; the reaction that ensues after E·I complex formation is rate limiting and, therefore, is time dependent. Because the binding occurs at the active site, it is protected by substrates (the rate of inactivation in the presence of a substrate is slower than that in its absence, because the substrate competes with the inactivator for the active site; when the substrate is in the active site, the inactivator cannot be).

The plot in Figure 2.4 depicts the inactivation of an enzyme by a radioactively

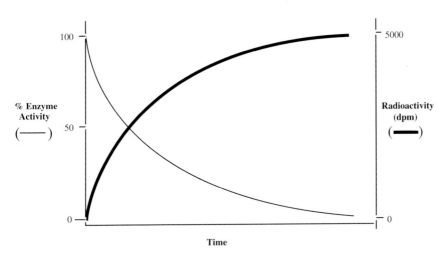

FIGURE 2.4   Correlation between loss of enzyme activity and incorporation of radioactivity during enzyme inactivation.

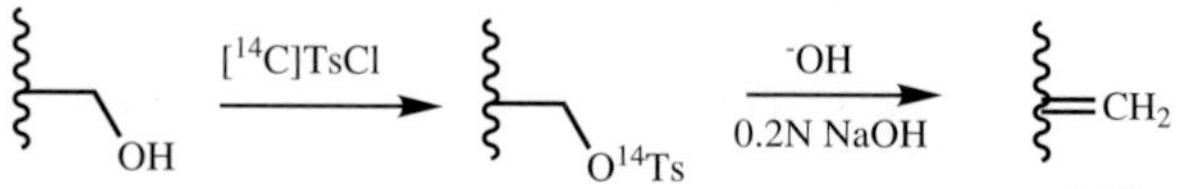

SCHEME 2.9   Chemical modification of Ser-195 in α-chymotrypsin.

labeled inactivator. Note the time-dependent loss of enzyme activity and the concomitant time-dependent increase in the incorporation of radioactivity; at 50% loss of enzyme activity, 50% of the maximum amount of radioactivity has been incorporated into the enzyme. At the end of inactivation a comparison of the amount of enzyme to the amount of radioactivity incorporated often shows a 1:1 correlation. When this kind of experiment was carried out with chymotrypsin and [^{32}P]diisopropylphosphorofluoridate, stoichiometric incorporation of ^{32}P was observed. The radiolabel was stable to strong acid hydrolysis (6 N HCl at 110 °C for 24 h, conditions that hydrolyze all of the peptide bonds). The radioactive compound isolated after acid hydrolysis was shown to be [^{32}P]phosphoserine,[11] the hydrolysis product of **2.9** (Scheme 2.7). Peptidase-catalyzed hydrolysis of **2.9** gave one radiolabeled peptide, which was shown to contain modified Ser-195.[12]

Another clever experiment to show the importance of the active-site serine-195 in the reaction (prior to the advent of site-directed mutagenesis experiments) was the chemical conversion of Ser-195 to a dehydroalanyl residue (**2.10**, Scheme 2.9), which makes chymotrypsin inactive.[13] The base conditions used to cause elimination of [^{14}C]tosylate do not inactivate chymotrypsin. The more "high-tech" approach for determining the importance of active-site residues is the use of site-specific mutagenesis,[14] in which suspected key residues are mutated to other amino acid residues by genetic engineering techniques, and the effect of these mutations on catalysis is determined.[15]

(2) Evidence for Histidine Participation   An approach similar to that used for the identification of the active-site serine was used to show that a histidine is at the active site. α-Chloroketones are alkyl halides that are activated toward nucleophilic addition reactions (e.g., S_N2 reactions). Two arguments, a ground-state argument and a transition-state argument, are generally given for the reactivity of α-chloroketones. The electron–withdrawing inductive effect of the carbonyl on the α-carbon raises the ground-state energy and activates that site for nucleophilic substitution of the halogen. In the transition state the inductive effect of the carbonyl stabilizes the "pentavalent" carbon, which now has additional electron density from the nucleophile. As a mimic of N-tosyl-L-phenylalanine methyl ester (**2.11**), a substrate for α-chymotrypsin,[16] N-tosyl-L-phenylalanine chloromethyl ketone (**2.12**, abbreviated TPCK), was synthesized and shown to be an irreversible inactivator of α-chymotrypsin.[17] Only the L-isomer of **2.12** inactivates α-chymotrypsin, although the D-isomer reversibly binds. This suggests that the active-site nucleophile being labeled is in the appropriate orientation to react with the α-chloromethyl ketone moiety only when it is in the L-geometry. As expected for an affinity labeling agent, substrate protects the enzyme from inactivation. When [^{14}C]TPCK is used as the inactivator, time-dependent inactivation occurs, and when all the enzyme activity is lost, one equivalent of radioactivity is found to be covalently attached to the enzyme. Graphs such as the ones shown in Figures 2.3 and 2.4 can be constructed for this inactivation. No radiolabeling occurs if the enzyme is denatured prior to treatment with **2.12** or if the zymogen of α-chymotrypsin, chymotrypsinogen, is used. Therefore, this is not just a random alkylation of the enzyme, but rather a specific reaction with an active-site residue. Acid hydrolysis of the inactivated enzyme shows that one histidine residue is missing.[18] There are two histidines in α-chymotrypsin (His-40 and His-57). Pepsin-catalyzed hydrolysis of the radiolabeled enzyme followed by separation of the peptides gives only one radiolabeled peptide. From the knowledge of the primary sequence of the enzyme and Edman degradation of the peptide, it was established that His-57 is modified. These results might suggest that the active-site nucleophile is His-57, rather than Ser-195, so we need to put these inactivation experiments in perspective. The only conclusion that can be made from an inactivation experiment is that an active-site nucleophile that may (or may not) be involved in catalysis is present. However, because nucleophiles generally can act as bases, it is not clear if the moiety labeled acts as a nucleophile or as a base during turnover of substrates. In this case, His-57 acts as a base, not as a nucleophile, during catalysis.

It has long been thought that the mechanism for the reaction of nucleophiles with α-chloromethyl ketones was that of a S$_N$2 reaction. However, a clever study by Abeles and co-workers[19] of the reaction of an α-chloromethyl ketone with chymotrypsin indicated otherwise. The reaction of chymotrypsin with (S)-N-acyl-L-alanyl-L-phenylalanine α-chloroethyl ketone (**2.13**), a peptide chloromethyl

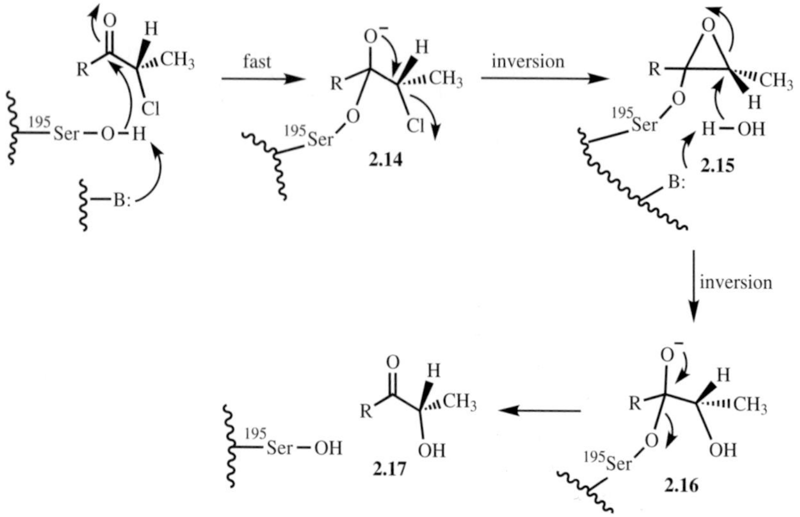

SCHEME 2.10   Double inversion mechanism for inactivation of serine proteases by α-chloromethyl ketones.

ketone, resulted in burst kinetics with release of Cl$^-$. The stereochemistry of the reaction could be determined because the α-chloroethyl ketone has a stereogenic center, and a side reaction that leads to the corresponding α-hydroxyethyl ketone occurs during inactivation (2.17, Scheme 2.10). The stereogenic center in the metabolite was determined to have the (S)-configuration. Consequently, there is retention of configuration, which negates a S$_N$2 mechanism, unless it involves two S$_N$2 reactions (double inversion gives retention). A mechanism that is consistent with a double inversion is depicted in Scheme 2.10. Attack of the active-site serine on the ketone carbonyl (similar to attack at the carbonyl of substrates) gives 2.14. The alkoxide ion could undergo rapid epoxide formation with ejection of the chloride ion, a process that gives inversion of stereochemistry (2.15). Nucleophilic opening of the epoxide ring with water (a second inversion step) leads to 2.16, which would rapidly decompose to 2.17, the product obtained with retention of configuration.

Further support for the mechanism of inactivation of chymotrypsin by α-chloromethyl ketones was obtained by Abeles, Ringe, and co-workers.[20] Let's consider three different mechanisms (Scheme 2.11). Mechanism 1 is a direct S$_N$2 reaction of His-57 with the α-carbon, resulting in 2.18. The second mechanism involves initial attack of Ser-195 on the carbonyl, followed by S$_N$2 attack of His-57 on the α-carbon, but the final adduct also is 2.18. Mechanism 3 is the same as the mechanism shown in Scheme 2.10 for hydrolysis, except His-57 substitutes for water; the adduct (2.19) retains the stereochemistry of the starting material because of a double inversion. An approach similar to that taken for determining the hydrolysis mechanism would differentiate mechanism 3 from 1 and 2. However,

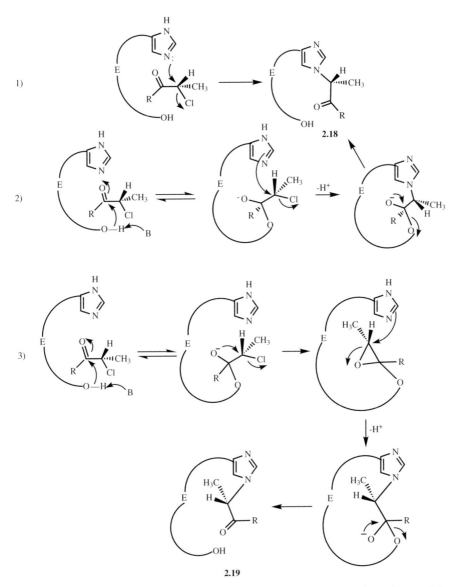

SCHEME 2.11   Three possible mechanisms for inactivation of α-chymotrypsin by α-chloromethyl ketones.

2.20

this problem is much more complicated because we need to elucidate the stereo-chemistry of the inactivator still attached to the enzyme! The problem, however, can be solved with the use of X-ray crystallography. Using (2*S*)-*N*-acetyl-L-alanyl-L-phenylalanyl α-chloroethane (**2.20**) as the inactivator, the crystal structure of the covalent adduct with γ-chymotrypsin clearly shows that His-57 is alkylated, and the stereochemistry of the inactivator is retained, again supporting the double displace-ment mechanism (mechanism 3). This may be a general mechanism for α-chloro-methyl ketone inactivators.

As discussed in Chapter 1 (Section II.C), pK_a values inside an active site can be very different from those in solution. Factors other than those discussed earlier as-sist in these proton transfers. For example, an NMR study using the peptidyl tri-fluoromethyl ketones *N*-acetyl-L-Leu-DL-Phe-CF_3 and *N*-acetyl-DL-Phe-CF_3 as inhibitors showed that the protons bridging His-57 and Asp-102 form low-barrier hydrogen bonds[21] (see Chapter 1, Section II.C, again; obviously, Frey is one of the believers in low-barrier hydrogen bonds). Without the Leu residue in the inhibi-tor, the low-barrier hydrogen bonds are weakened, decreasing the basicity of the His-57/Asp-102 dyad. On the basis of the structure of chymotrypsin and these re-sults, it was suggested that the formation of low-barrier hydrogen bonds during catalysis occurs from a substrate-induced conformational change, which leads to steric compression between His-57 and Asp-102. The low-barrier hydrogen bond may stabilize the tetrahedral intermediate through relief of steric strain between these residues. Substrate-induced steric compression increases the basicity of $N^{\epsilon 2}$ in His-57, making it a more effective base for abstracting a proton from Ser-195. The energy for conformational compression would be supplied by the binding of remote portions of the substrate to the active site.

(3) Evidence for the Deacylation Mechanism (GAB Catalysis)   Hy-drolysis of the acyl intermediate is required to produce the product and regenerate the enzyme. The nonenzymatic hydrolysis of a model compound (**2.21**) was shown

2.21

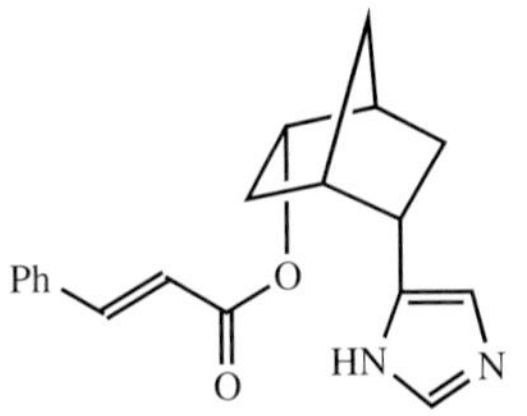

2.22

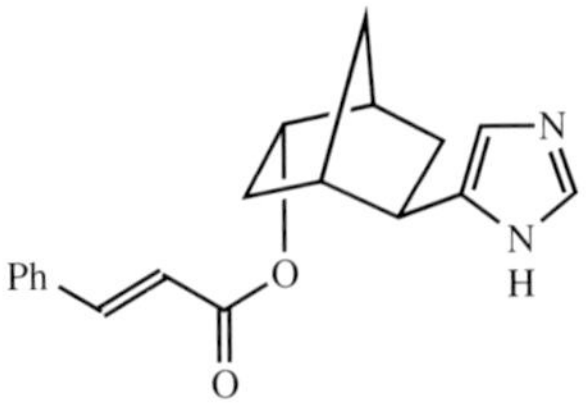

2.23

to be subject to general base catalysis by imidazole.[22] The enzyme reaction exhibits a *solvent deuterium isotope effect* of 2–3 when carried out in deuterated buffer.[23] This indicates that deprotonation of a water molecule is a partially rate-determining step and suggests that general base catalysis is important.

To provide chemical support for the enzymatic observations, a model study for deacylation was carried out by Bender and co-workers.[24] Norbornane was used as the "artificial enzyme" template to which were substituted an imidazolyl group and a cinnamoyloxy group in *cis- (endo, endo;* **2.22**) and *trans- (endo, exo;* **2.23**) configurations. The imidazolyl group is a model for His-57 at the active site, and the cinnamoyl ester mimics the Ser-195–bound acyl intermediate. Because the imidazole in **2.23** is *trans* to the ester carbonyl, it is not geometrically possible for the imidazole to participate in an intramolecular reaction. In fact, only the imidazole group in **2.22**, and not the one in **2.23**, participates in the hydrolysis of the cinnamate ester, as shown by the dependence of the ionization of the imidazole group on the rate for **2.22**, but not for **2.23**. The reaction also shows a k_{H_2O} / k_{D_2O} of 3 as observed for the enzymatic reaction. Addition of benzoate ion (to model the effect of Asp-102 at the active site) produces a 2500-fold increase in the hydrolysis rate. This rate is 6.6×10^{-4} times the enzymatic rate of deacylation of *trans*-cinnamoyl-α-chymotrypsin, which is not so bad for a simple chemical model. Of course, this also is a stoichiometric reaction, not a catalytic reaction.

To improve on this model, Bender and co-workers[25] then synthesized a model with an intramolecular benzoate group (**2.24**) to more closely mimic the deacylation reaction (Scheme 2.12). This model mimics the deacylation reaction of *trans*-cinnamoyl-α-chymotrypsin at one-eighteenth the rate (Table 2.1).

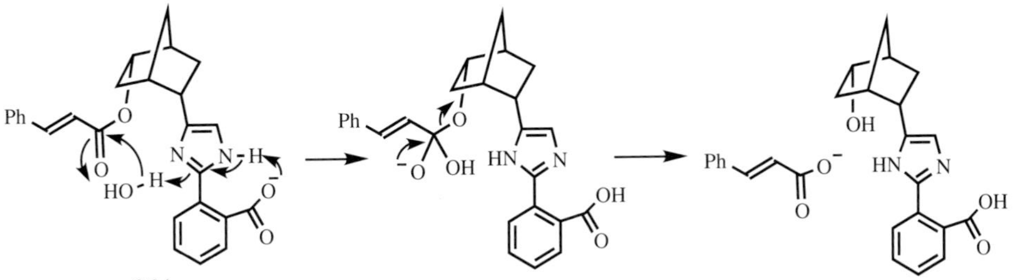

2.24

2.25

SCHEME 2.12   Chemical model for the deacylation step in α-chymotrypsin.

TABLE 2.1   Rate of Deacylation of Model Compounds
Compared to Cinnamoyl-α-chymotrypsin

Compound	Relative rate (k_{rel})
Ph—...—O—chymotrypsin	1.0
2.22	2.6×10^{-7}
2.22 plus benzoate ion	6.6×10^{-4}
2.24	5.6×10^{-2}

Chymotrypsin not only is an excellent peptidase in aqueous buffers but also catalyzes peptide hydrolysis in *n*-octane.[26] However, because hydrophobic interactions in the active site owe their existence to water as the reaction medium (remember, a hydrophobic effect results from the entropy increase of surrounding water molecules), substrate preferences in the organic medium are different from those in water. For example, *N*-acetyl-L-histidine methyl ester, which is only 0.5% as reactive as *N*-acetyl-L-phenylalanine ethyl ester in aqueous buffer, is 20 times more reactive in *n*-octane.

c. Other Endopeptidases

α-Chymotrypsin is just one example in the family of *serine proteases,* which are characterized by having a *conserved* (i.e., all members of the family have it) serine residue at the active site for the purpose of catalyzing the covalent catalytic mechanism. Other active-site residues serve the purpose of permitting binding of certain residues for selective cleavage. Trypsin, mentioned in Section I.A.1.a as another example of an endopeptidase, also is a serine peptidase, but cleaves at lysine and arginine residues. To determine the importance of the catalytic triad of rat trypsin to catalysis, each of the catalytic triad residues was altered by site-directed mutagenesis. Mutants D102N and H57K catalyze hydrolysis of peptide bonds at 0.003% the rate of the wild type.[27] Therefore, the catalytic triad is not an absolute requirement for peptide bond cleavage, but it is very important to catalysis.

Here is an interesting phenomenon. Wu and Hilvert[28] were able to convert subtilisin, another serine protease, into an acyltransferase, an enzyme that takes an acyl group from one molecule and transfers it to a second molecule, by converting the active-site serine OH group into a SeH group. This was accomplished by selective reaction of the hydroxyl group with phenylmethanesulfonyl fluoride and then treatment of this sulfonylated enzyme with excess hydrogen selenide to give the selenol. Selenolsubtilisin reacts with cinnamoylimidazole (**2.26**, Scheme 2.13) to give Se-cinnamoylselenolsubtilisin (**2.27**), which can be isolated; reaction of **2.27** with amines yields the corresponding amides (**2.28**).

Other classes of peptidases have different conserved active-site residues. For ex-

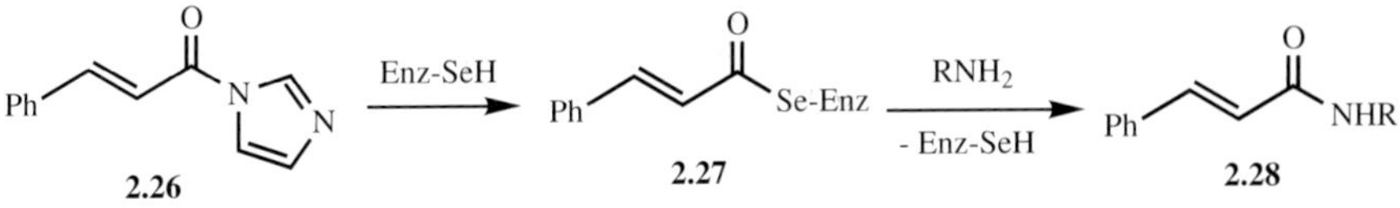

SCHEME 2.13   Reaction of selenosubtilisin with cinnamoylimidazole and amines

ample, the digestive enzyme papain (that is why papaya, from which the enzyme gets its name, is good for digestion) is a member of the family of *cysteine proteases,* which have a conserved cysteine residue that is involved in covalent catalysis. HIV protease, a retrovirus that catalyzes the cleavage of encapsulated viral proteins into the functional enzymes and structural proteins of the virion core of human immunodeficiency virus, thereby completing the viral replication cycle, is a member of the *aspartate protease* family.[29] Inhibition of HIV-1 protease during its maturation within infected human T-lymphocytes makes the resulting virions noninfectious and replication incompetent.[30] Many specific inhibitors of HIV-1 protease have been discovered,[31] and the use of HIV-1 protease inhibitors in treating AIDS has had a major impact on the treatment of this dreaded disease.[32] The enzyme is a *homodimer;* that is, the active enzyme consists of two identical polypeptide subunits, each of which contains one of the two active-site aspartyl residues. The active site is located at the bottom of a large deep cleft into which the substrates bind and that is partially covered by two highly mobile flaps. On the basis of crystallographic data[33] and a variety of isotope effect studies,[34] a mechanism for HIV-1 protease has been suggested (Scheme 2.14).

2. Exopeptidases

a. Aminopeptidases (N-Terminal Exopeptidases)

Aminopeptidases play critical roles in protein maturation, protein digestion, regulation of peptide hormone levels, and plasmid stabilization. Altered aminopeptidase activity has been associated with aging, cancers, cataracts, cystic fibrosis, and leukemia. Most of the aminopeptidases require Zn^{2+} for activity (Zn^{2+} is the cofactor), but some use Co^{2+} or no metal ion. The ones that need metal ions for activity are in the family of peptidases known as *metallopeptidases.* On the basis of an X-ray structure of bovine lens aminopeptidase and with the inhibitor bestatin, $(2S,3R)$-3-amino-2-hydroxy-4-phenylbutanoyl-L-leucine (**2.29**), bound,[35] the function of the Zn^{2+} appears to be coordination to the scissile carbonyl to activate it for general base-catalyzed reaction with water.[36]

b. Carboxypeptidases (Carboxyl-Terminus Exopeptidases)

Carboxypeptidase A from bovine pancreas, which catalyzes the hydrolysis of the carboxyl-terminal peptide bond of proteins that contain a hydrophobic side chain

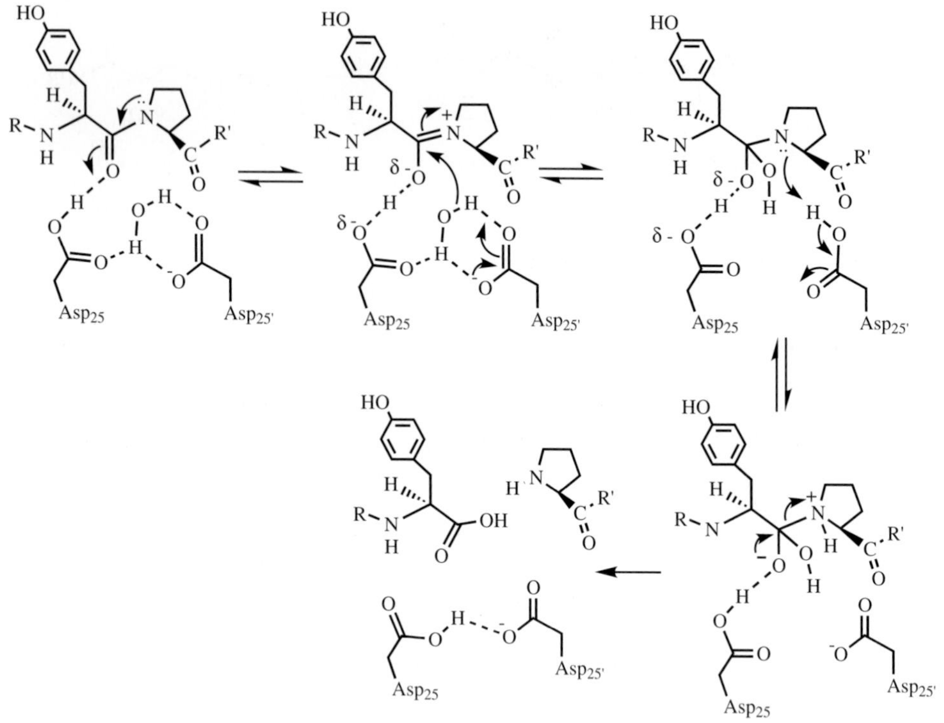

SCHEME 2.14   Proposed mechanism for HIV-1 protease.

on the C-terminal residue, was crystallized in 1937. It was shown to contain one Zn^{2+} ion per active site as the cofactor.[37] The affinity labeling agent **2.30** caused irreversible inactivation and implicated Glu-270 as an important active-site residue.[38]

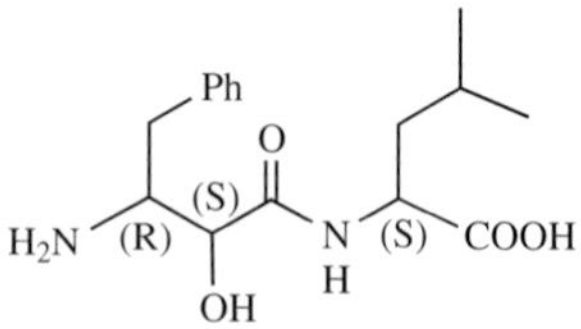

2.29

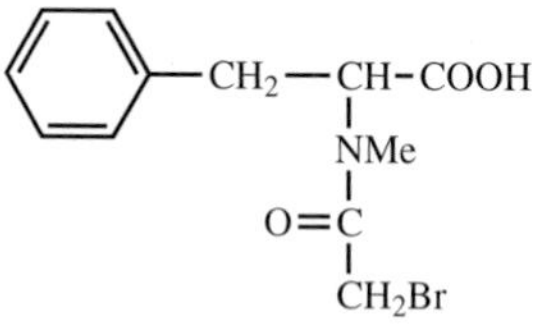

2.30

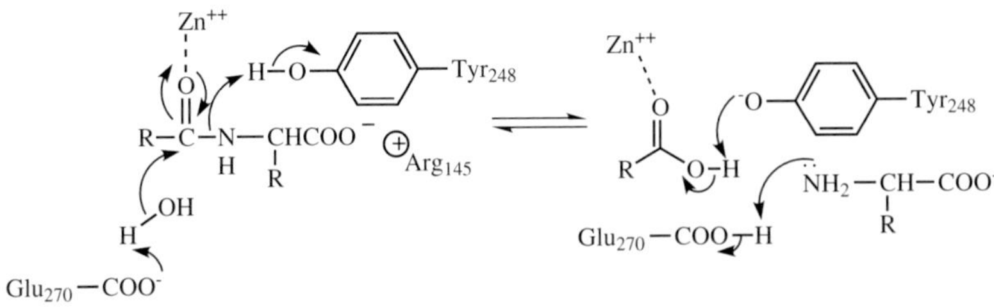

SCHEME 2.15   General base catalytic mechanism for carboxypeptidase A.

Other modification studies using iodine for phenolic iodination demonstrated the presence of an essential tyrosine (Tyr-248).[39] X-ray crystallographic studies also showed that Arg-145 is in the active site, presumably to bind the carboxyl terminus of the substrate and orient it for catalysis.[40] Two possible generalized mechanisms can be drawn for this reaction, a general base mechanism (also known as a *promoted-water mechanism*) (Scheme 2.15) and a nucleophilic mechanism (Scheme 2.16).[41] There is evidence to support both possibilities. The function of the Zn^{2+} in both mechanisms, as drawn, is to coordinate to the scissile carbonyl and make it more electrophilic. The failure to detect an anhydride intermediate[42] or to trap it with nucleophiles[43] when a peptide substrate was used provides evidence against the nucleophilic mechanism (Scheme 2.16). Resonance Raman spectroscopy at low temperatures shows no accumulation of an enzyme-bound mixed anhydride,[44] providing further evidence against the nucleophilic mechanism. However, Kuo and Makinen[45] think that the failure of Carey and co-workers to detect an anhydride intermediate could have resulted from inadequate mixing of the viscous solvents. Using a cinnamate ester substrate at $-40\,^{\circ}C$, Kuo and Makinen observed biphasic kinetics, which they believe indicates that a covalent enzyme intermediate (from the nucleophilic mechanism) accumulates, although Christianson and Lipscomb[46] contend that the "intermediate" may have been an enzyme–substrate or enzyme–product complex. In support of the studies of Kuo and Makinen, single crystals of carboxypeptidase A were soaked with a chromophoric ester substrate and Britt and

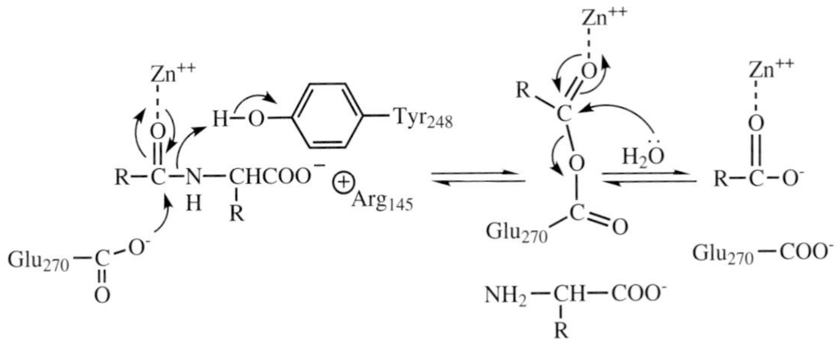

SCHEME 2.16   Nucleophilic mechanism for carboxypeptidase A.

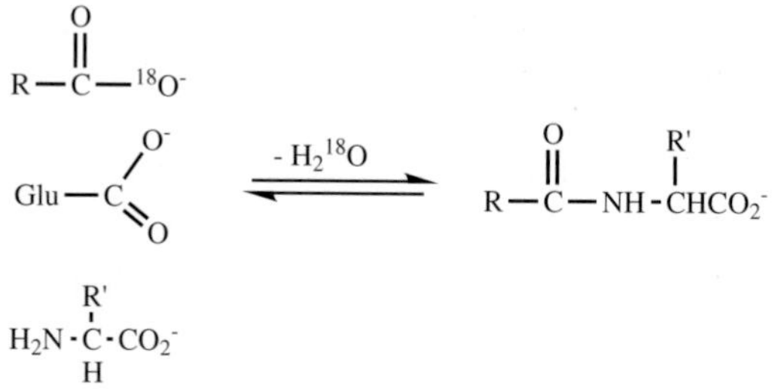

SCHEME 2.17   Reverse general base catalytic reaction of carboxypeptidase A in the presence of $H_2{}^{18}O$.

Peticolas[47] observed the solid-state hydrolysis with resonance-enhanced Raman spectroscopy of the interior of the crystal. Weak, but reproducible bands were observed in the $1720-1800$ cm^{-1} region, supporting the formation of an anhydride intermediate.

Another set of experiments, using ^{18}O labeling, however, supports a general base mechanism.[48] Consider the outcomes of the general base (Scheme 2.17) and nucleophilic (Scheme 2.18) mechanisms with ^{18}O incorporated into the carboxyl terminus of a product, running the reaction in reverse. What? Run the reaction in reverse! Most enzyme-catalyzed reactions are fully reversible, so it does not matter if you start from the substrate or the product. According to the *principle of microscopic reversibility,* for any reversible reaction, the mechanism in the reverse direction must be identical to that in the forward direction (except backward). Therefore, any mechanistic information you obtain by carrying out a reaction in the reverse direction is directly relevant to the mechanism in the forward direction. This gives you two options when studying enzymatic reactions; you can design experiments for whichever direction you wish.

The general base mechanism (Scheme 2.17) requires both products and the active-site glutamate residue to react in one step, with the release of $H_2{}^{18}O$. In the nucleophilic mechanism (Scheme 2.18) $H_2{}^{18}O$ is released in the absence of the second product (the amino acid), which reacts in a subsequent step with the glutamate

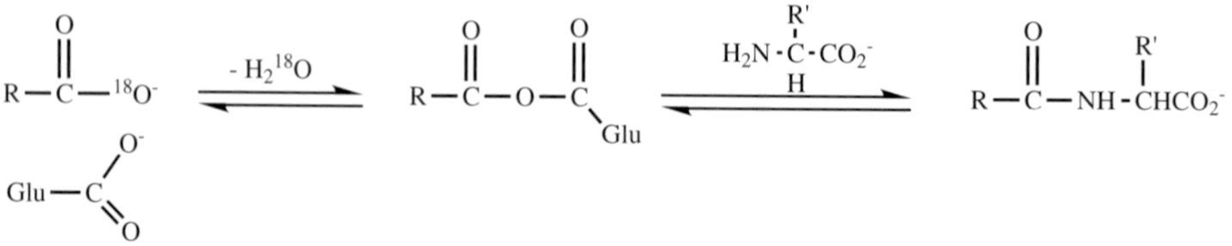

SCHEME 2.18   Reverse nucleophilic catalytic reaction of carboxypeptidase A in the presence of $H_2{}^{18}O$.

anhydride. Therefore, if $H_2{}^{18}O$ is released in the absence of the amino acid, then the nucleophilic mechanism is reasonable, but if it requires the amino acid for $H_2{}^{18}O$ to be released, then this supports the general base mechanism. Release of $H_2{}^{18}O$ requires the presence of an amino acid, lending support to the general base mechanism. There appears to be evidence in favor of both mechanisms, however; the experiments that support the nucleophilic mechanism were carried out with ester substrates and those that favor a general base mechanism were done with peptide substrates. It is conceivable that different mechanisms are operative depending on the type of substrate involved.

The role of the Tyr-198 and Tyr-248 in the catalytic mechanism of carboxypeptidase A was probed by site-specific mutagenesis. Conversion of the Tyr-198 codon (TAT) to the Phe codon (TTT) in the cDNAs for carboxypeptidase A was accomplished by oligonucleotide-directed mutagenesis.[49] The same was done to Tyr-248. The catalytic activities of these mutants and of the double mutant [Y198F, Y248F] are not abolished.[50] Depending on the substrate used, anywhere from 5 to 66% of the k_{cat} of WT (wild type) was observed; K_m values of the mutants and the WT are comparable. Therefore, Tyr-198 and Tyr-248 are *not* involved in general acid catalysis. They may be involved in stabilizing the rate-determining transition-state complex.

Lipscomb and co-workers[51] suggest a function for the Zn^{+2} ion completely different from that shown in Schemes 2.15 and 2.16, a modified promoted-water mechanism, based on the X-ray crystal structure of a tight-binding ketonic substrate analogue bound to carboxypeptidase A. The bound substrate is in the hydrate form; in solution this compound exists in a hydrated form to an extent of less than 0.2%, so the enzyme must have performed a hydration reaction on the ketone carbonyl to give the tetrahedral intermediate. This structure, then, is likely to be important to the mechanism. Further evidence against a nucleophilic mechanism is that a Glu-270 tetrahedral intermediate is not observed in the crystal structure. Note that in this mechanism (Scheme 2.19), the Zn^{2+} ion is not coordinated to the carbonyl of the substrate, but is coordinated to a water molecule, thereby increasing the

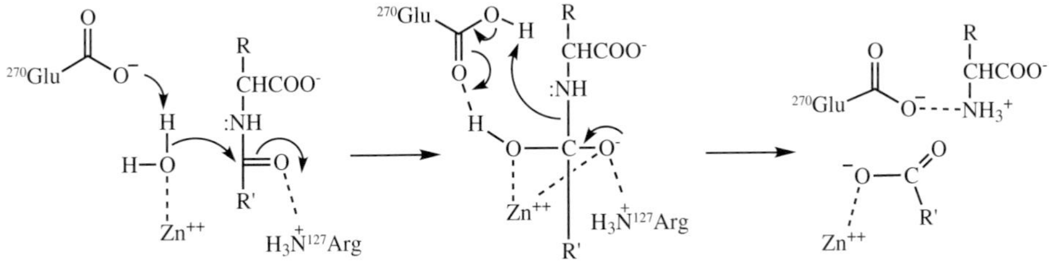

tetrahedral intermediate

SCHEME 2.19   Alternative mechanism for carboxypeptidase A on the basis of the X-ray structure with a ketone bound.

acidity of the water proton. Actually, coordination of the Zn^{2+} ion to the scissile carbonyl would raise the pK_a of the zinc-bound water molecule, possibly sufficiently to hinder nucleophilic attack. Model studies have shown that zinc ion increases the nucleophilicity of a water molecule for both proteolytic and esterolytic reactions.[52] Groves and Olson[53] found that a zinc-coordinated water can have a pK_a as low as 7, in which case, it would be hydroxide-like even at neutral pH. This mechanism, then, is very appealing for carboxypeptidase A. Differences observed between proteolysis and esterolysis may be the result of different roles for zinc ion and Arg-127 rather than of different substrate binding modes.[54] For example, peptides bind to the *apoenzyme* (i.e., the protein without its cofactor) but esters do not, as evidenced kinetically[55] as well as crystallographically[56]; the peptides interact with Arg-127. Also, an arginine residue is required for proteolysis, but not for esterolysis.[57] The mechanism in Scheme 2.19 is consistent with these results.

B. Ester Hydrolysis: Esterases and Lipases

1. Structure and Activity

Esterases, enzymes that catalyze the hydrolysis of esters, also catalyze serine-mediated acyl-enzyme formation, releasing the ester alcohol; enzymatic hydrolysis produces the carboxylic acid (Scheme 2.20). The activities of these enzymes for hydrolysis of esters are generally much greater than those of peptidases for esters. *Lipases* are enzymes that catalyze the hydrolysis of lipid esters (triacylglycerol hydrolases).

Acetylcholinesterase,[58] an enzyme found in nervous tissue of all species of animals, catalyzes the rapid hydrolysis of the excitatory neurotransmitter acetylcholine, thereby terminating impulse transmission at cholinergic synapses. Because of the need for rapid termination of neurotransmission, the rate of acetylcholinesterase approaches that of a diffusion-controlled reaction.[59] The active site of the enzyme was long believed to contain two important binding sites, the "anionic" site that binds the quaternary ammonium cation of acetylcholine and the "esteratic" site where the catalytic hydrolysis of the acetyl group occurs (Scheme 2.21).[60] The esteratic site was believed to resemble the active site of other serine hydrolases.[61] In 1991, however, Sussman et al.[62] reported the X-ray crystal structure (to 2.8 Å reso-

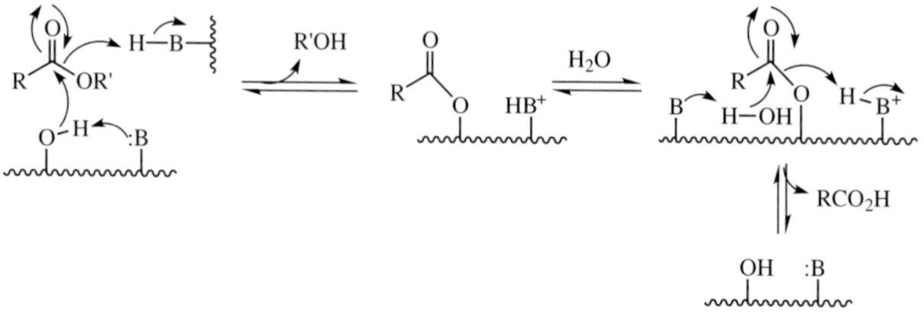

SCHEME 2.20   Esterase mechanism.

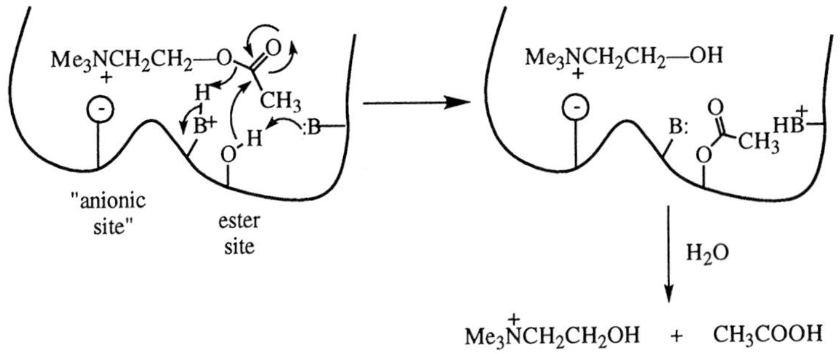

SCHEME 2.21   Mechanism for acetylcholinesterase.

lution) of acetylcholinesterase from *Torpedo californica,* which revealed a number of surprises. First, the active site is unusual, relative to other serine hydrolases, in that the catalytic triad contains a glutamate in place of the usual aspartate residue (Glu–His–Ser). Also, the relationship of this triad to the rest of the enzyme is approximately the mirror image of triads in serine proteases. Probably the most unexpected finding, however, is that the "anionic" site does not contain an anionic group! The active site is near the bottom of a deep and narrow gorge that reaches halfway into the protein. It was assumed that a residue such as Asp or Glu would be present in the active site to bind to the quaternary ammonium ion of acetylcholine. Instead, it was found that a cluster of 14 aromatic residues lines the gorge to the active site, and the "anionic" site is really the π-electron density of some of these aromatic residues! The stabilization of cations by aromatic π-electrons has much support from physical organic studies, principally of Dougherty and co-workers.[62a] Another unusual aspect of acetylcholinesterase is that its inhibition can have either a fatal effect or a medically beneficial result, depending on whether the inhibitor is irreversible or reversible, respectively.[63] The organophosphorus nerve poisons, such as sarin, phosphorylate the active-site serine of acetylcholinesterase irreversibly, leading to uncontrollable cholinergic activity, which results first in weakness and muscle twitching and then in complete paralysis and muscle atrophy. Reversible inhibitors of acetylcholinesterase, however, result in a mild enhancement of cholinergic activity by facilitating transmission of impulses across neuromuscular junctions, which is useful in the treatment of the neuromuscular disease myasthenia gravis (with neostigmine)[64] and by increasing acetylcholine levels in the central cholinergic synapses involved in the memory circuit, which explains its benefit in the treatment of Alzheimer's disease (with E2020).[65]

2. Stereoselectivity of Esterases / Lipases

Many esterases and lipases are used as chiral catalysts in organic synthesis.[66] For an enantioselective catalyst to be synthetically useful, it must accept a broad range of substrate structures and also retain a high degree of enantioselectivity for each. Lipases are particularly useful in this regard. These catalysts not only are relatively

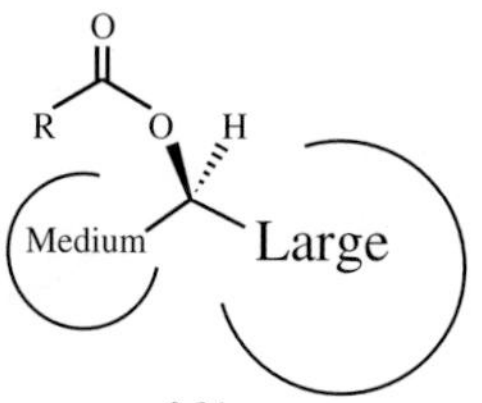

2.31

inexpensive, commercially available, and stable, but also can catalyze the enantio-selective hydrolysis and transesterification of a variety of natural and unnatural esters. Empirical rules for enantioselectivity, based on common structural features among well-resolved substrates, have been devised.[67] For example, the rate of hydrolysis of esters of chiral secondary alcohols (or the reverse reaction) by a variety of lipases and esterases is faster when the enantiomer of the secondary alcohol ester has the larger substituent on the right and the medium group on the left when the ester group is pointing forward (**2.31**); an example is shown in Scheme 2.22.[68] The alcohol product can easily be separated from the unreacted enantiomer. By comparing crystal structures of enantiomeric transition-state analogues (see Section I.B.3) bound to *Candida rugosa* lipase, researchers determined that the group on the slower-reacting enantiomer disrupts the hydrogen bond from the histidine of the catalytic triad to the alcohol oxygen atom, which probably accounts for the decrease in rate of hydrolysis of that enantiomer.

3. Nonnatural Esterases

Met-myoglobin is a nonenzymatic protein that contains an organic cofactor, called a *heme* (see Chapter 3, Section V), that is used to bind to and transport oxygen. Removing the heme from the active site of sperm whale Met-myoglobin gives a new apoprotein with a deep hydrophobic cavity that can serve as a binding pocket for hydrophobic substrates.[69] Furthermore, within this cavity are two histidine residues that could act as general acid/base catalysts for hydrolysis reactions. In fact, this apoprotein has esterase activity and can catalyze the hydrolysis of *para*-nitrophenyl esters at a rate that is 1900 times greater than that in buffer. Because of the catalytic activity of this unnatural enzyme, it is referred to as a *semisynthetic enzyme*.

Another nonnatural approach to the hydrolysis of esters has been the use of *catalytic antibodies* (also called "*abzymes*"). Antibodies are large proteins consisting of two identical heavy chains and two identical light chains roughly in the shape of a Y.

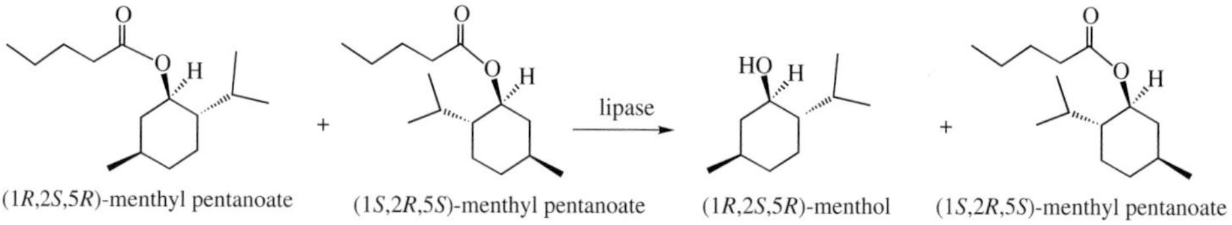

(1R,2S,5R)-menthyl pentanoate   (1S,2R,5S)-menthyl pentanoate   (1R,2S,5R)-menthol   (1S,2R,5S)-menthyl pentanoate

SCHEME 2.22   An example of the enantioselectivity of lipases/esterases.

The two arms of the Y (the *Fab fragments*) contain identical antigen-binding sites, and the stem is joined to the Fabs by a flexible hinge. These proteins, which are scavengers of *macromolecular xenobiotics* (large foreign species) that enter an organism, such as bacteria and viruses, protect that organism from external harm. They function by forming very tight complexes with the macromolecular xenobiotic; the formation of this complex triggers a cascade of events that leads to degradation of the xenobiotic. Jencks was the first to propose that it should be possible for an antibody to be catalytic and to act as if it were an enzyme.[70] The basis for how catalytic antibodies work is the same as that for enzymes, namely, that the rate of a reaction is accelerated by stabilization of the transition state. The enzyme achieves this rate enhancement by changing its conformation so that the strongest interactions occur between the substrate and the enzyme active site *at the transition state* of the reaction. Bernhard and Orgel[71] theorized that inhibitor molecules resembling the transition-state species would be bound to the enzyme much more tightly than the substrate would be bound. Therefore, a potent enzyme inhibitor would be a stable compound whose structure resembles that of the substrate *at a postulated transition state* (or transient intermediate) of the reaction rather than that at the ground state. A compound of this type binds much more tightly to the enzyme and is called a *transition-state analogue inhibitor.* Jencks[72] was the first to suggest the existence of transition-state analogue inhibitors and cited several possible literature examples; Wolfenden[73] and Lienhard[74] developed the concept further. Values for dissociation constants (K_i) of 10^{-14}–10^{-23} M for enzyme–transition-state complexes may not be unreasonable given the normal range of 10^{-3}–10^{-6} M for dissociation constants of enzyme–substrate complexes (K_m).[75]

The following is a brief, general approach to raising catalytic antibodies. First, a transition-state analogue that mimics the transition state of the desired reaction is synthesized. This compound is called a *hapten*. Next, the hapten is attached to a carrier molecule that is capable of eliciting an antibody response (antibodies do not recognize small molecules, so it is necessary to attach the hapten to a macromolecule that the antibody recognizes as a xenobiotic). The hapten attached to the macromolecule is an *antigen*. *Monoclonal antibodies* (antibodies that bind to a specific region of a molecule) that bind very tightly to the transition-state analogue part of the antigen are isolated.[76] Then these monoclonal antibodies are tested as catalysts for the desired reaction. For example, Schultz and co-workers[77] synthesized hapten **2.32** (X = OH), attached it to a protein, and used that antigen (**2.32**,

2.32

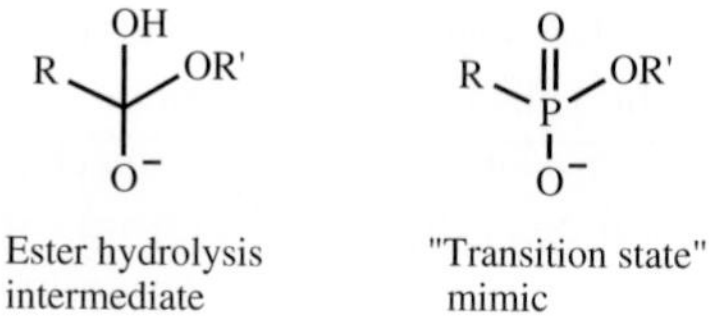

Ester hydrolysis intermediate   "Transition state" mimic

FIGURE 2.5   Comparison of an ester hydrolysis tetrahedral intermediate and a phosphonate "transition state" mimic.

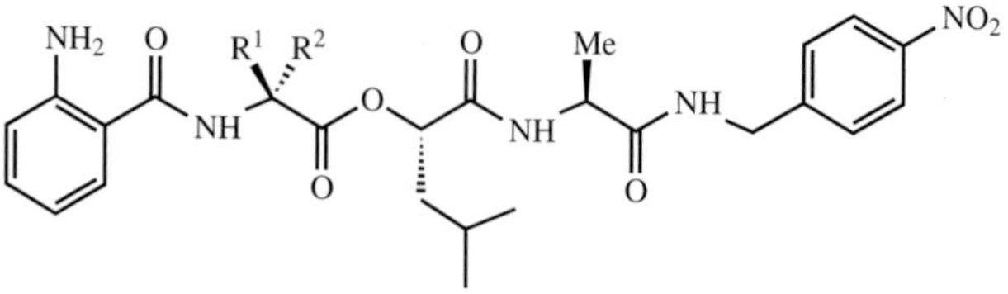

2.33

X = protein) to raise monoclonal antibodies that catalyze the stereospecific hydrolysis of alkyl esters. The phosphonate group of **2.32** mimics the tetrahedral intermediate of ester hydrolysis (Figure 2.5) and selects for antibodies whose binding sites have appropriate amino acids to stabilize this anionic transition state, such as the cation residues arginine and lysine. One monoclonal antibody that was isolated catalyzed the stereospecific hydrolysis of epimer **2.33** (R^1 = PhCH$_2$, R^2 = H) and another catalyzed the hydrolysis of epimer **2.33** (R^1 = H, R^2 = PhCH$_2$). Catalytic antibodies that have stereoselective lipase activity also have been raised.[78] If you want to read more about catalytic antibodies in general, there have been numerous reviews.[79] Throughout this book I will make reference to catalytic antibodies that catalyze interesting reactions, particularly reactions for which no enzyme has yet been found to catalyze them.

II. AMINATIONS

A. Glutamine-Dependent Enzymes

1. Types of Reactions

The types of reactions catalyzed by glutamine-dependent enzymes are summarized in Table 2.2. Many of these reactions are important in the biosynthesis of pyridine nucleotides, purines, pyrimidines, glucosamine, and asparagine.

TABLE 2.2   Types of Reactions Catalyzed by Glutamine-
Dependent Enzymes

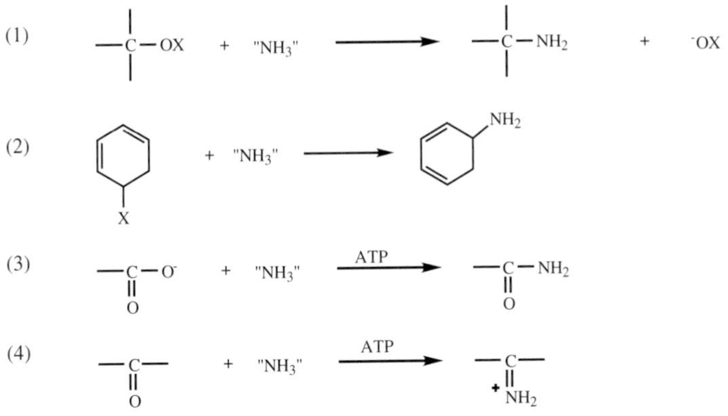

2. Glutamine as a Source of Ammonia

a. Glutaminase Activity (Generation of NH_3)

Ammonia is a toxic and reactive species to cells, yet it is an important molecule for enzyme-catalyzed amination reactions. To avoid the toxicity of this compound, controlled amounts of ammonia are generated from glutamine at the desired site of action. Under these conditions, glutamine is a nontoxic, unreactive, and neutral source of ammonia. A covalent catalytic mechanism for generation of ammonia from glutamine is shown in Scheme 2.23. This converts the nonreactive amide NH_2 group of glutamine into ammonia via a mechanism that resembles that for

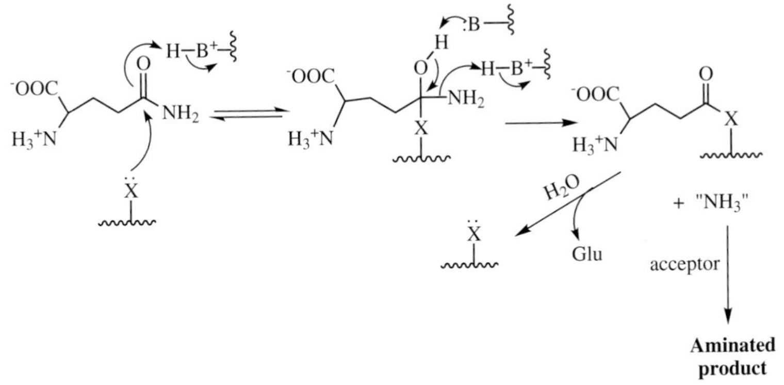

SCHEME 2.23   A covalent catalytic mechanism for the "glutaminase" activity of glutamine-dependent enzymes.

peptidases (see Section I.A) and esterases (I.B.1). Some enzymes catalyze a related reaction, the transfer of the glutamyl group to acceptors other than ammonia, such as to other amino acids or peptides. These enzymes are called *γ-glutamyltransferases.*

Why does NH_4^+ not act as a substrate for these enzymes, if "NH_3" is being generated, and at physiological pH NH_3 would be rapidly converted into NH_4^+ anyway? In fact, NH_3 *is* a substrate. In general, *both exogenous ammonia and glutamine can serve as the amino donor.* However, the K_m value for ammonium ion is generally 100–1000 times larger than that for glutamine, which means that 100–1000 times more ammonium ions are required to reach the same rate of the lower amount of glutamine. This allows the enzyme to regulate when a particular reaction will occur and to avoid the need for high concentrations of ammonia in the bloodstream.

What is the evidence that covalent catalysis is important, and what residue is involved? Some of the experiments described for peptidases have been used to support a covalent catalytic mechanism with glutamine-dependent enzymes. For example, if the substrate reaction is carried out in the presence of hydroxylamine, γ-glutamyl hydroxamate (**2.35**, Scheme 2.24) is produced, suggesting the intermediacy of an activated γ-glutamyl enzyme complex (**2.34**). The principal support for covalent catalysis, however, comes from inactivator studies. The α-chloromethyl ketone of asparagine, namely, L-2-amino-4-oxo-5-chloropentanoic acid (**2.36**, Figure 2.6), is a time-dependent irreversible inactivator of L-asparagine synthetase.[80] Compound **2.36**, an analogue of asparagine rather than of glutamine, was used because it has the methylene of the α-chloromethyl ketone group in line with the scissile carbonyl of glutamine. However, as we discussed in Section I.A.1.b.2, the inactivation mechanism for α-chloromethyl ketones appears to involve attack at the

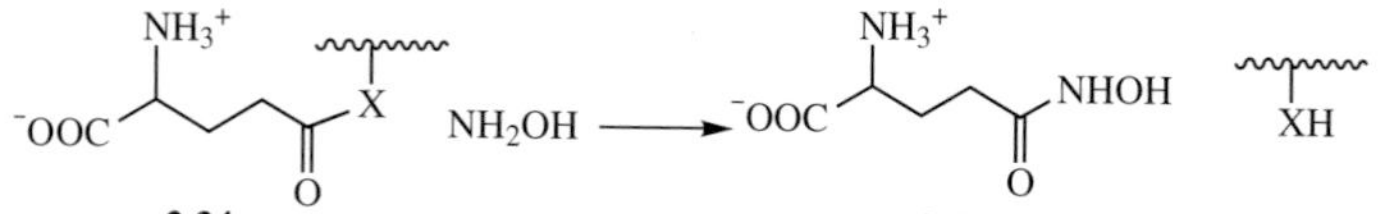

SCHEME 2.24   Evidence for a γ-glutamylenzyme intermediate in glutamine-dependent enzymes.

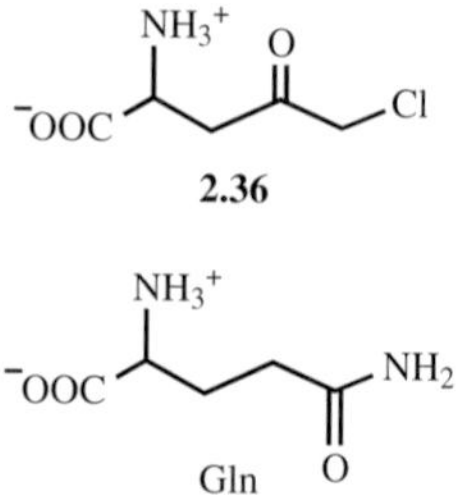

FIGURE 2.6   Comparison of the structure of the α-chloromethyl ketone of asparagine with the structure of glutamine.

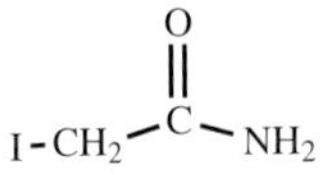

2.37 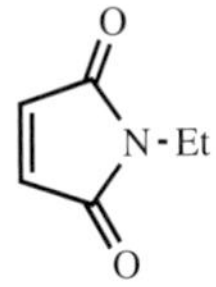   2.38

carbonyl, not at the methylene of the α-chloromethyl ketone. This suggests that the α-chloromethyl ketone of glutamine would have been a better choice for an inactivator. Also, inactivators that are highly selective for cysteine residues in enzymes, such as iodoacetamide (2.37) and N-ethylmaleimide (2.38), are irreversible inactivators.[81] After the enzymes are inactivated by the affinity labeling agents, amidotransferase activity using glutamine is blocked (i.e., the glutaminase activity is lost); however, amine transfer can still occur with high concentrations of exogenous NH_4^+. This suggests that there are two binding sites, one for glutamine and a separate one for the NH_3 that is generated by the glutaminase activity of the enzyme.

Another type of inactivation is termed *mechanism-based enzyme inactivation*.[82] A mechanism-based inactivator (sometimes, much to my dismay, called a suicide substrate) is an unreactive compound that has a structural similarity to a substrate or product for an enzyme. Once at the active site of the enzyme, it is converted into a species that generally forms a covalent bond to the enzyme, producing inactivation. Mechanism-based enzyme inactivators are very useful for the study of enzyme mechanisms because they are really nothing more than substrates for the enzymes. They happened to be converted into products (or intermediates) that are generally reactive, and these products inactivate the enzyme. But the mechanisms by which these compounds are converted into the activated form proceed, at least initially, by the normal catalytic mechanism. Therefore, any information gleaned from the inactivation mechanism is directly related to the catalytic mechanism. Because the inactivator must be converted into the activated form, the kinetic scheme for mechanism-based inactivation (Scheme 2.25; see also Appendix I, Section II.B.2) is different from that for affinity labeling inactivation (Scheme 2.8). As in the case of affinity labeling agents, the first step is formation of the noncovalent E·I complex. At this point the enzyme treats the inhibitor as a substrate and catalyzes a reaction on it, converting it to a modified (activated) form (E·I'), which is still a

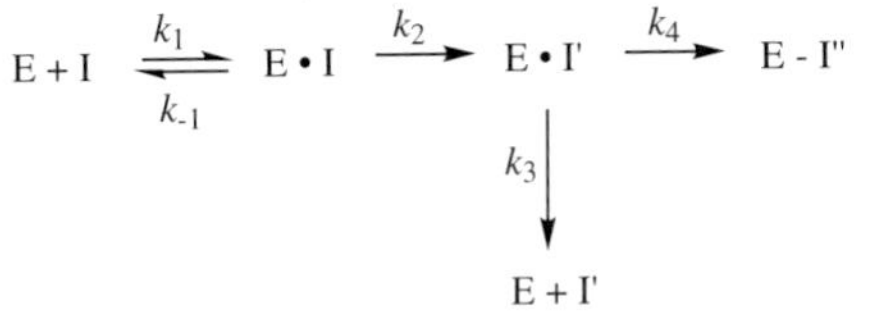

SCHEME 2.25   Kinetics for mechanism-based inactivation.

noncovalent complex. This complex can have two fates: either I′ is recognized as the product and is released from the active site (k_3) or it reacts with the enzyme, generally forming a covalent bond and inactivating it (k_4). The ratio of release of I′ as a product per inactivation event, called the *partition ratio,* is the ratio of the two rate constants (k_3/k_4). Because I′ could be a reactive species, its release could mean the release of a potential affinity labeling agent, which could cause toxic effects. Therefore, the ideal case is when I′ is not released and only inactivation occurs; that is, the partition ratio is zero.

Consider two compounds that could be defined as mechanism-based inactivators that use the simplest of catalytic mechanisms to initiate their reactivity, namely, protonation. The inactivators are 6-diazo-5-oxo-L-norleucine (**2.39**, also called DON)[83] and L-azaserine (**2.40**),[84] both natural products isolated from *Streptomyces*. Because of the structural similarity of these two compounds to glutamine, they inactivate many glutamine-dependent enzymes. Two inactivation mechanisms can be drawn (Scheme 2.26, pathways a and b), based on what we learned from α-chloromethyl ketone affinity labeling agents (Scheme 2.11). The diazoketone inactivators are relatively stable as long as they are kept from proton sources. All of the inactivation mechanisms are initiated by enzyme-catalyzed protonation of the diazo ketone moiety to give the highly reactive E·I′ product, the α-diazonium methyl ketone **2.41**. This product complex, which is more reactive than an α-chloromethyl ketone, then can undergo nucleophilic attack at either the α-carbon (pathway a) to give inactivated enzyme (**2.42**) or attack at the ketone carbonyl (pathway b) to give intermediate **2.43,** which can decompose by two different pathways. Pathway c mimics the mechanism of inactivation of peptidases by α-chloromethyl ketones (see Scheme 2.11), via epoxide **2.44** to inactivated enzyme (**2.45**). Alternatively (pathway d), the tetrahedral intermediate could break down to produce the acyl enzyme (**2.46**) and release diazomethane (**2.47**). When 6-^{14}C-labeled **2.39** was used as the inactivator of glutaminase A, and the reaction was carried out in the presence of benzoic acid, a compound that rapidly reacts with diazomethane, [^{14}C]methyl benzoate and [^{14}C]methyl alcohol were isolated as products, supporting pathway b/d.[85] In addition to these products being formed, the enzyme was irreversibly inactivated with stoichiometric ^{14}C labeling of the active site. Loss of enzyme activity correlated with the incorporation of radioactivity into the enzyme (see Figure 2.4).[86] The partition ratio for the formation of products per inactivation event was found to be 70. To determine if inactivation results from reaction with **2.39** or from the diazomethane released, a double labeling experiment was carried out. The enzyme was inactivated with **2.39** labeled with ^{14}C

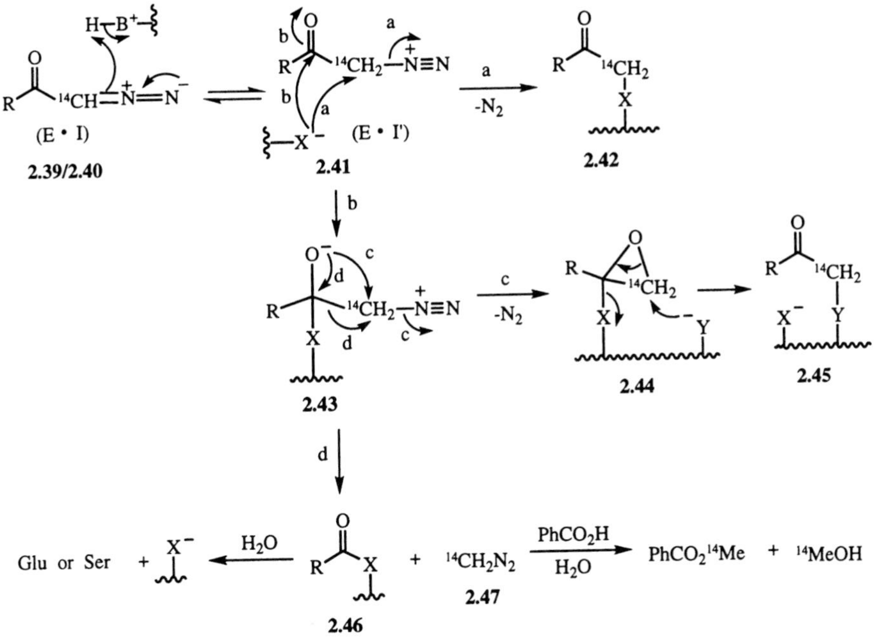

SCHEME 2.26   Mechanisms for inactivation of glutamine-dependent enzymes by α-diazoketones.

in the diazomethyl group and with ^3H in the amino acid backbone (R group) and was tryptic digested, and the peptides were then separated. Because the ratio of $^{14}C/^3H$ remained constant throughout the purification, it can be concluded that **2.39** (via either pathway a or b/c, but not b/d) is responsible for inactivation (if diazomethane were responsible, only ^{14}C would have been bound; the ^3H would have been released in the glutamate). For the glutamine-dependent enzymes, glutamine phosphoribosyl diphosphate amidotransferase,[87] cytidine triphosphate synthetase,[88] anthranilate 5-phosphoribosyl diphosphate phosphoribosyltransferase,[89] anthranilate synthase,[90] and 5-phosphoribosyl diphosphate amidotransferase,[91] inactivation by **2.39** has been used to identify an active-site cysteine residue. This provides evidence for the involvement of a cysteine residue in the catalytic mechanism; it is believed that the X in Scheme 2.23 is this cysteine residue. However, there is some evidence with γ-glutamyl transpeptidase from both rat[92] and human[93] kidney that an active-site cysteine is not involved, indicating that the mechanism in Scheme 2.23, where X = cysteine, is not universal.

Note that even in the absence of the ammonia acceptor, there is a slight amount of hydrolysis of glutamine (glutaminase activity), but this occurs only to a small extent (about 0.5% of the rate in the presence of the acceptor). Presumably, the presence of the acceptor induces a conformational change in the enzyme that sets up the glutaminase activity.

b. Reactions with Nascent Ammonia (Acceptor Reactions)

Once the ammonia is generated from L-glutamine, it is used for the incorporation of amino groups into various molecules. In the X-ray crystal structure of the amidotransferase carbamoyl phosphate synthetase, a protein tunnel can be seen that is believed to be the passageway for the ammonia from the L-glutamine to the acceptor molecule (carboxyphosphate).[94]

Most of the glutamine-dependent reactions require a cosubstrate, namely, adenosine triphosphate (ATP), but some do not. An example of an acceptor reaction that does not need ATP is the reaction catalyzed by 5-phosphoribosyl-1-diphosphate amidotransferase (Scheme 2.27).[95] The ammonia in this reaction is derived from L-glutamine as described, and the acceptor is 5-phosphoribosyl-1-diphosphate (**2.48**). Because of inversion of stereochemistry at the reaction site, a S_N2 mechanism can be proposed.

The reaction catalyzed by 5-phosphoribosyl-1-diphosphate amidotransferase does not require ATP because the diphosphate is an excellent leaving group and there is no need to activate it. ATP provides energy to those molecules that are not sufficiently reactive to undergo chemical reactions. Consider the nonenzymatic reaction of a carboxylic acid with ammonia. If you mix benzoic acid with a base such as ammonia, what are you going to get? You will not get an amide; you get the salt, ammonium benzoate (Scheme 2.28). What if you wanted to get the amide; how would you do it? There are numerous ways to make amides, but in general, the carboxylic acid must first be activated, because once the proton is removed, the carboxylate usually does not react with nucleophiles. Typically, dehydrating agents such as thionyl chloride or acetic anhydride are used first to activate the carboxylic acid to an acid chloride (**2.49,** Scheme 2.29) or anhydride (**2.50**), respectively. Acid chlorides and anhydrides are very reactive toward nucleophiles, so ammonia can displace the good leaving groups chloride ion and acetate ion, respectively, and produce amides. That, in fact, is what enzymes do also, except because thionyl chloride

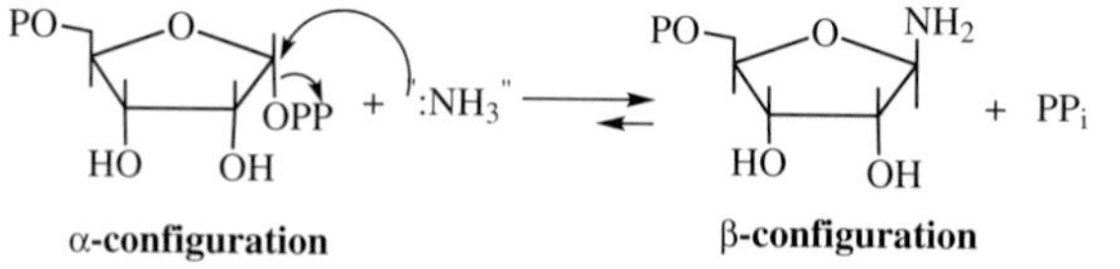

α-configuration
2.48   **β-configuration**

SCHEME 2.27   Amination reaction catalyzed by glutamine phosphoribosyldiphosphate amidotransferase.

$$PhCO_2H \quad + \quad NH_3 \quad \longrightarrow \quad PhCO_2^- \; NH_4^+$$

SCHEME 2.28   Reaction of ammonia with benzoic acid.

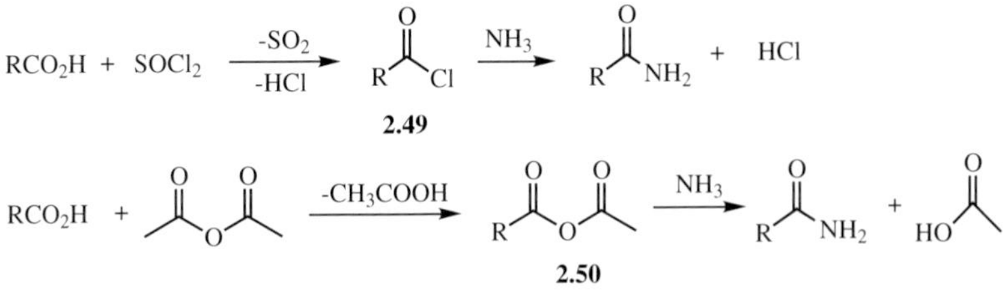

SCHEME 2.29   Activation of carboxylic acids with thionyl chloride and acetic anhydride.

would be a bit harsh on our tissues, ATP is the substitute. Think of ATP, then, as the endogenous form of thionyl chloride or acetic anhydride.

The structure of ATP is shown in Figure 2.7. Nucleophiles can attack at four electrophilic sites, at the γ-phosphate, the β-phosphate, the α-phosphate, and the 5'-methylene group. Attack at the γ- and α-positions is the most common because those reactions are thermodynamically favored. Attack at the 5'-methylene is rare, but does occur, and we will consider this mode of attack in later chapters. For the most part, after the reaction, ATP is converted to ADP (adenosine diphosphate) + P_i (inorganic phosphate) or to AMP (adenosine monophosphate) + PP_i (inorganic diphosphate) (Figure 2.8). To achieve selectivity (each ATP-dependent enzyme catalyzes reactions at only one of these sites), ATP-dependent reactions require Mg^{2+} ion. The Mg^{2+} ion complexes to the appropriate phosphoester oxygens to direct nucleophilic attack, presumably by aligning the site of attack with the nucleophile and by making the appropriate phosphorus atom more electrophilic.

First, consider a reaction that converts the ATP into AMP and PP_i, namely, the conversion of aspartate to asparagine, catalyzed by asparagine synthetase (Scheme 2.30). Because tumor cells have a high requirement for asparagine, this enzyme is

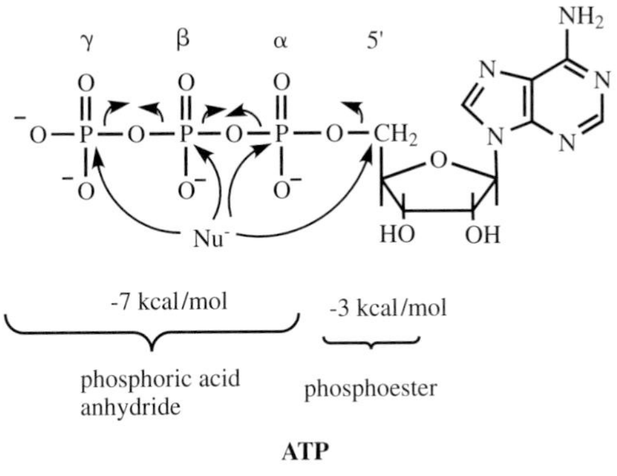

FIGURE 2.7   Electrophilic sites on ATP.

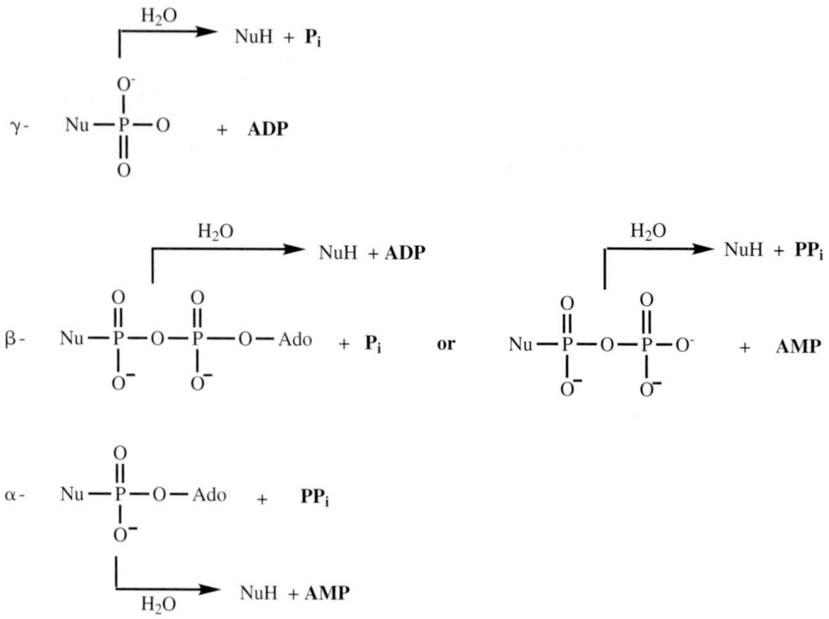

FIGURE 2.8   Products of reaction of nucleophiles at the α-, β-, and γ-positions of ATP.

important for the life of a tumor cell. Note that this enzyme is called a *synthetase,* whereas other enzymes that catalyze the synthesis of molecules are called *synthases.* The difference in nomenclature arises from the requirement for ATP. An enzyme that catalyzes the synthesis of a product that needs the utilization of ATP is called a *synthetase;* if no ATP is used, then the enzyme is called a *synthase.* As discussed earlier, a carboxylic acid in nonenzymatic chemistry must be activated to undergo reaction with a nucleophile; likewise, aspartate must be activated to react enzymatically with the nascent ammonia generated from glutamine (see Scheme 2.23). ATP activates the carboxylic acid functionality of aspartate (Scheme 2.31). However, as shown in Figure 2.8, AMP and PP$_i$ can be formed by nucleophilic attack at either the α-position or the β-position of ATP. That being the case, how can you tell at which position the initial attack of aspartate occurs? Isotopic labeling of the carboxylate gives the answer (Scheme 2.32). If the carboxylate of aspartate is synthesized with an ^{18}O label, and the isolated AMP contains the ^{18}O, attack was at the

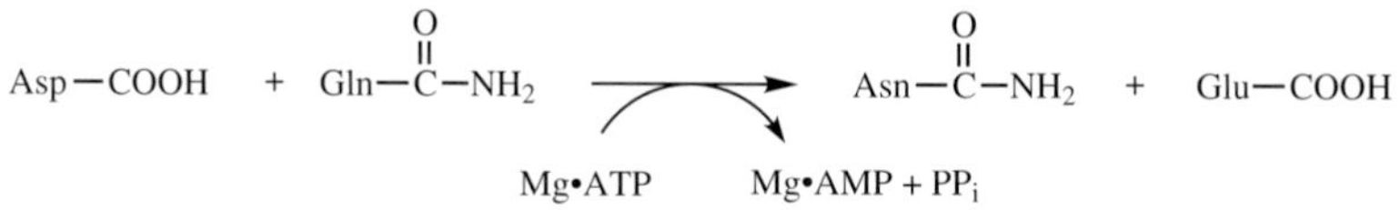

SCHEME 2.30   Reaction catalyzed by asparagine synthetase.

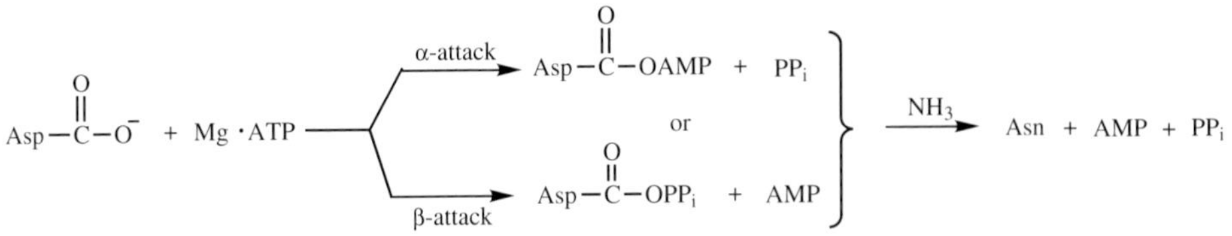

SCHEME 2.31   Activation of aspartate by ATP followed by reaction with ammonia generated from glutamine.

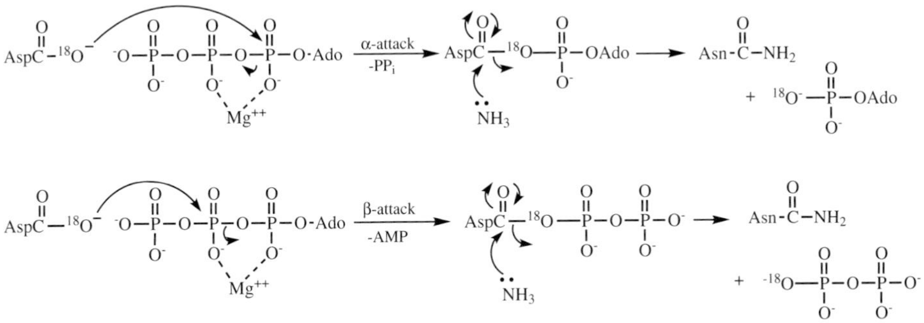

SCHEME 2.32   Use of ^{18}O-labeled aspartate to differentiate attack at the α- or β-positions of ATP.

α-position; if the PP$_i$ contains the ^{18}O, then β-attack occurred. The products can be isolated by HPLC (high-performance liquid chromatography), and the incorporation of the ^{18}O can be determined by mass spectrometry. Because ^{18}O is isotopic, but not radioactive, a high concentration (90% or more) of the label can be incorporated into the compound. In the case of asparagine synthetase, incorporation of the ^{18}O into aspartate gives [^{18}O]AMP, indicating that the initial attack of aspartate is at the α-position.[96]

Next, consider a reaction that converts the ATP into ADP and P$_i$, namely, formylglycinamide ribonucleotide (FGAR) aminotransferase, an enzyme that is important in the biosynthesis of purines.[97] The reaction that the enzyme catalyzes is shown in Scheme 2.33. The amide functionality of FGAR (2.51) is not very reactive toward

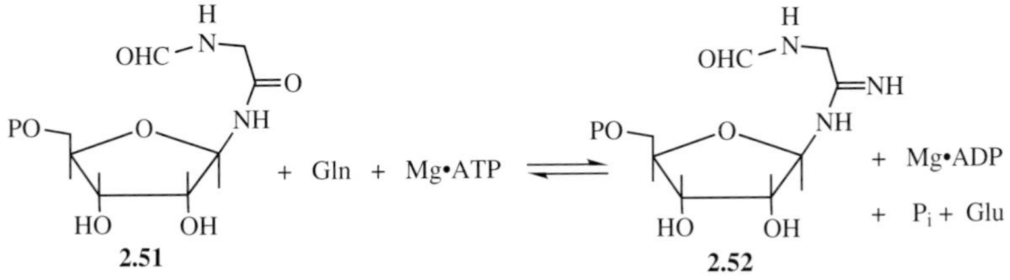

SCHEME 2.33   Reaction catalyzed by formylglycinamide ribonucleotide (FGAR) aminotransferase.

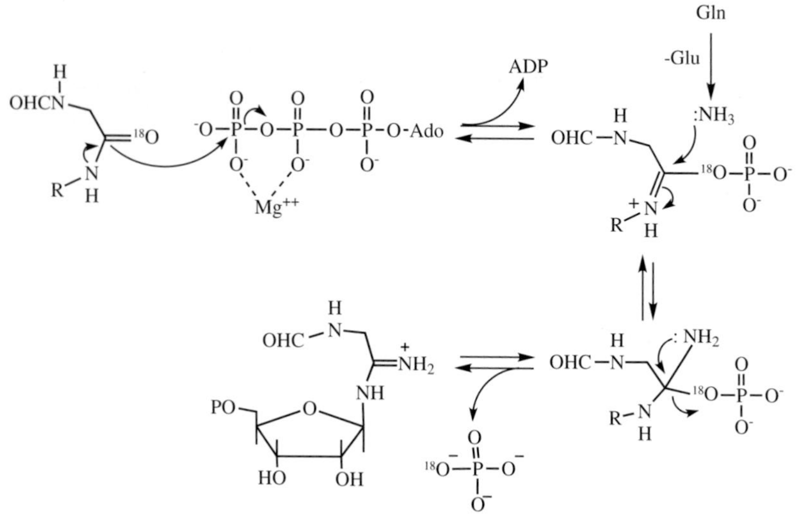

SCHEME 2.34   Use of ^{18}O-labeled FGAR to differentiate attack at the α- or β-positions of ATP.

NH_3. Therefore, it needs ATP for activation. If the amide O is labeled with ^{18}O, the label ends up completely in the P_i. This indicates that the initial attack of the amide oxygen of **2.51** occurs at the γ-position of ATP (Scheme 2.34). Further evidence for a FGAR–phosphate intermediate can be obtained by carrying out a *partial exchange reaction*. This is a common approach in enzymology to gain insight into intermediates. When there are multiple steps in a reversible reaction with multiple substrates, and it is possible to omit one of the substrates and still have a partial reaction occur, then carrying out a partial exchange reaction with an isotopically labeled intermediate can be fruitful for the elucidation of the mechanism. If all of the substrates are present, the complete reaction will occur; when the products form, the reverse reaction will take place, and the isotopic label may be scrambled, giving no useful mechanistic information. This can be avoided by omission of one of the substrates so only a partial reaction proceeds. For example, what if we incubated FGAR aminotransferase with FGAR, ATP, and ^{32}P-labeled ADP *in the absence of glutamine*? According to the generalized reaction (Scheme 2.34), the glutamine is not needed until after the substrate is activated by ATP, so the first partial reaction of FGAR with ATP should proceed without the glutamine to give either FGAR–P_i + ADP or FGAR–ADP + P_i depending on whether attack of FGAR occurs at the γ- or β-position of ATP, respectively. In the absence of glutamine, no ammonia is generated to react with the activated FGAR, so all that can happen is the reverse reaction to FGAR and ATP. After incubation for a period of time, look for isotope incorporation into the ATP (HPLC separation of the ADP and ATP and scintillation analysis). Because only one in maybe 10^9 or 10^{10} molecules of the AD^{32}P actually has radioactive ^{32}P incorporated, there will be a slow, time-dependent increase in ^{32}P incorporation into the ATP (i.e., an increase in the *specific*

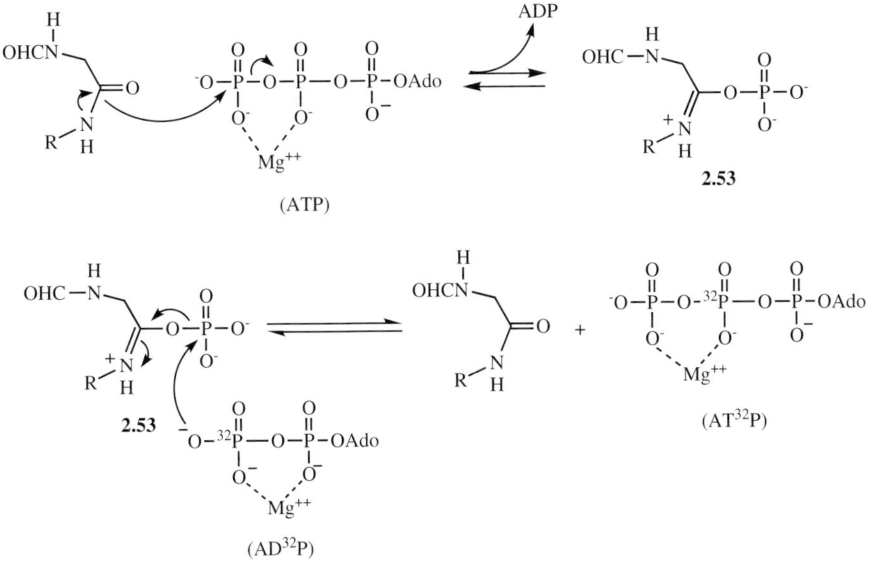

SCHEME 2.35   Use of AD^{32}P in a partial reaction to test for reversibility of FGAR aminotransferase and test whether ADP or P$_i$ is released during the reaction.

radioactivity of the ATP as the ADP exchanges with AD^{32}P and the reaction reverses). This will only occur if ADP is an intermediate and if the reaction is reversible (Scheme 2.35). Because ADP does not react with ATP nonenzymatically, incorporation of ^{32}P from ADP into the ATP cannot occur without the enzyme. ATP is needed to initiate the enzyme-catalyzed reaction to give FGAR–P$_i$ (**2.53**) + ADP. The ADP is released from the enzyme where it equilibrates with the AD^{32}P, which can bind to the enzyme containing the **2.53**; the reverse reaction produces AT^{32}P + FGAR. If FGAR–ADP + P$_i$ were formed, then no ^{32}P could be incorporated into the ATP; in this case ^{32}P$_i$ would give a partial exchange reaction with ATP, producing AT^{32}P (Scheme 2.36). If neither experiment leads to incorporation of ^{32}P into the ATP, it does not necessarily mean that neither intermediate (FGAR–P$_i$ nor

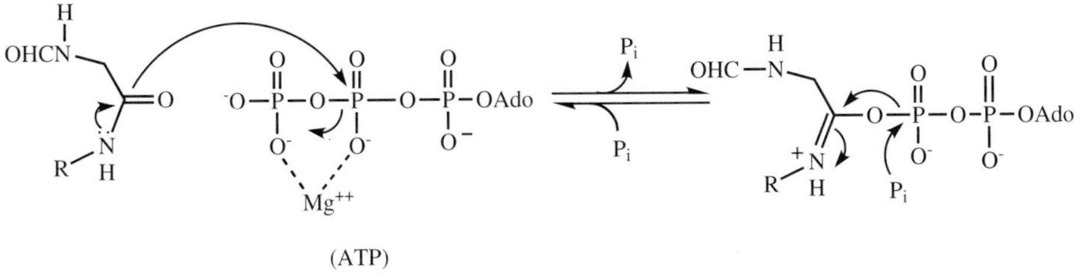

SCHEME 2.36   Outcome if FGAR aminotransferase proceeded by formation of ADP phosphate ester.

FGAR–ADP) is formed. Other causes could obscure this result. First, we assumed that the enzyme followed an *ordered mechanism* (i.e., a specific order to the binding of the substrates for the enzyme reaction to proceed), and that the first partial reaction, FGAR and ATP, could proceed in the absence of glutamine. That need not be the case. Maybe the enzyme needs the glutamine to be bound before activation of the FGAR, so that as soon as the activated FGAR is formed it has the ammonia generated from glutamine available for reaction. To ensure this condition, the binding of glutamine may cause a conformational change in the enzyme that sets up the binding site for FGAR and ATP; until glutamine binds, then, the enzyme cannot bind the other substrates appropriately, and the reaction cannot proceed. Another possibility for the potential failure to observe a partial exchange reaction is that the ADP generated in the first partial reaction may bind very tightly to the enzyme, so its dissociation and exchange with AD^{32}P do not occur (or is too slow to detect). A way of detecting slow dissociation of the intermediate ADP has been reported.[98] It also should be noted that it is not a good idea to start with AT^{32}P and look for AD^{32}P because there may be an unrelated enzyme activity, namely, an ATPase reaction (hydrolysis of ATP), that would generate AD^{32}P, but this would be irrelevant to the FGAR aminotransferase mechanism.

B. Aspartic Acid as a Source of Ammonia

A few enzyme reactions that catalyze aminations use the ammonium group of aspartate as the source of ammonia. For example, in the biosynthesis of L-arginine, L-citrulline (**2.54**) reacts with L-aspartate and Mg·ATP in a reaction catalyzed by L-argininosuccinate synthetase to give L-argininosuccinate (**2.55**), Mg·AMP, and PP$_i$ (Scheme 2.37).[99] L-Argininosuccinate is then converted to L-arginine (**2.56**) and fumarate (**2.57**) by L-argininosuccinate lyase. If ^{18}O is incorporated into the urea oxygen of citrulline, it ends up in the Mg·AMP, indicating that the initial attack of citrulline occurs at the α-position of ATP.[100] Affinity labeling agents were used to show that cysteine and arginine residues are essential for the activity of L-argininosuccinate synthetase[101] and that a lysine residue (Lys-51) is important to the catalytic action of bovine liver L-argininosuccinate lyase.[102] Given that the carboxylate groups of fumarate are *trans*, the β-elimination of arginine from **2.55** must occur with *anti* geometry (the proton removed and the arginine eliminated are *trans*).

III. PHOSPHORYLATIONS: TRANSFERS OF PHOSPHATE AND PHOSPHATE ESTERS TO WATER OR OTHER ACCEPTORS

Three principal classes of enzymes catalyze phosphoryl group transfer reactions: phosphatases, phosphodiesterases, and kinases (Figure 2.9).

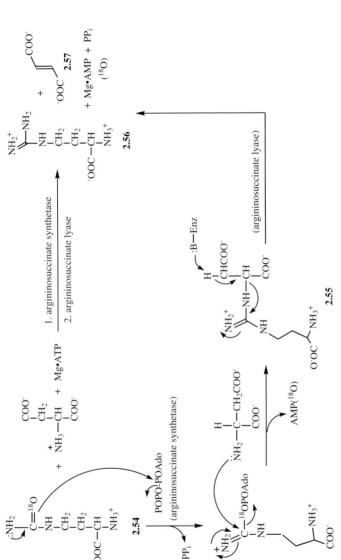

SCHEME 2.37 Mechanisms for the reactions of argininosuccinate synthetase, an aspartate–dependent enzyme, and argininosuccinate lyase. ATP is abbreviated as POPOPOAdo.

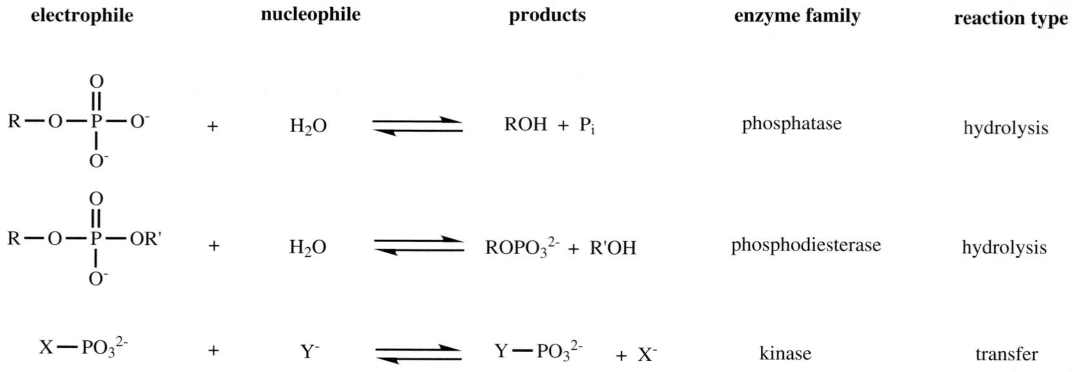

electrophile	nucleophile	products	enzyme family	reaction type
$R-O-\overset{\displaystyle O}{\underset{\displaystyle O^-}{\overset{\|}{P}}}-O^-$ +	H_2O	$\rightleftharpoons$ $ROH + P_i$	phosphatase	hydrolysis
$R-O-\overset{\displaystyle O}{\underset{\displaystyle O^-}{\overset{\|}{P}}}-OR'$ +	H_2O	$\rightleftharpoons$ $ROPO_3^{2-} + R'OH$	phosphodiesterase	hydrolysis
$X-PO_3^{2-}$ +	Y^-	$\rightleftharpoons$ $Y-PO_3^{2-} + X^-$	kinase	transfer

FIGURE 2.9 Comparison of the reactions of a phosphatase, a phosphodiesterase, and a kinase.

A. Phosphatases

Phosphatases catalyze the hydrolysis of phosphates (transfer of the phosphoryl group to water).[102a] This is the phosphate equivalent of a peptidase for peptides. We will consider three general mechanisms for these enzyme-catalyzed reactions (Scheme 2.38). Mechanism A is a general acid/base-catalyzed reaction that can occur either associatively (A1) or dissociatively (via a metaphosphate-like intermediate, A2). Mechanism B is a covalent catalytic mechanism, again, either associative (B1) or dissociative (B2). Mechanism C is a S_N2 reaction. All these mechanisms use the phosphate as the electrophile and water as the nucleophile.

How would you test which mechanism was correct for a particular enzyme? Mechanism C could be differentiated from mechanisms A and B by incubation with $H_2{}^{18}O$. If the ^{18}O is found in the organic part (the ROH) and not in the released P_i, then that would support mechanism C, but if the ^{18}O were found in the P_i then it would exclude mechanism C. Associative and dissociative mechanisms can be differentiated by studies of secondary kinetic isotope effects. The bond order to the equatorial phosphoryl oxygen atoms changes for associative and dissociative pathways in the transition state. Substitution of the phosphate oxygen atoms with ^{18}O will give a slower reaction in an associative mechanism (lower bond order; normal secondary isotope effect), but a faster reaction in a dissociative mechanism (higher bond order; inverse secondary isotope effect).[103] An associative mechanism is expected to give inversion of stereochemistry about the phosphorus atom, but this may or may not occur with a dissociative mechanism.

Let's consider glucose-6-phosphatase, the enzyme that catalyzes the hydrolysis of glucose 6-phosphate (**2.58**) to glucose (**2.59**) (Scheme 2.39) as an example to differentiate the various mechanisms. First, $H_2{}^{18}O$ was shown to cleave the P−O bond (reaction at the phosphorus), not the C−O bond, which excludes mechanism C.[104] Mechanisms A and B were differentiated by the following experiments. The reversibility of the reaction and evidence for a phosphoenzyme were demonstrated by an exchange reaction. When glucose-6-phosphatase was incubated with glucose-6-phosphate and [^{14}C]glucose, [^{14}C]glucose-6-phosphate was isolated; no incorporation of ^{32}P into glucose-6-phosphate occurred when $^{32}P_i$ was used.[105] These experiments support a mechanism in which a phosphoenzyme is formed reversibly with release of glucose, followed by *irreversible* hydrolysis to P_i (see Scheme 2.38, mechanism B1 or B2). A covalent catalytic mechanism was more convincingly supported by incubation of ^{32}P-labeled glucose-6-phosphate with glucose-6-phosphatase followed by a phenol quench. This results in the *denaturation* (unfolding of the tertiary structure) of the enzyme. The denatured enzyme was tryptic digested and ^{32}P was found bound to a peptide, supporting a covalent attachment of the phosphate to the enzyme (mechanism B1 or B2 in Scheme 2.38). The ^{32}P-labeled peptides were saponified with potassium hydroxide, and N-3-[^{32}P]phosphorylhistidine was isolated, thereby revealing the covalent catalytic group as a histidine residue.[106] When the same experiment was done with ^{14}C in the sugar

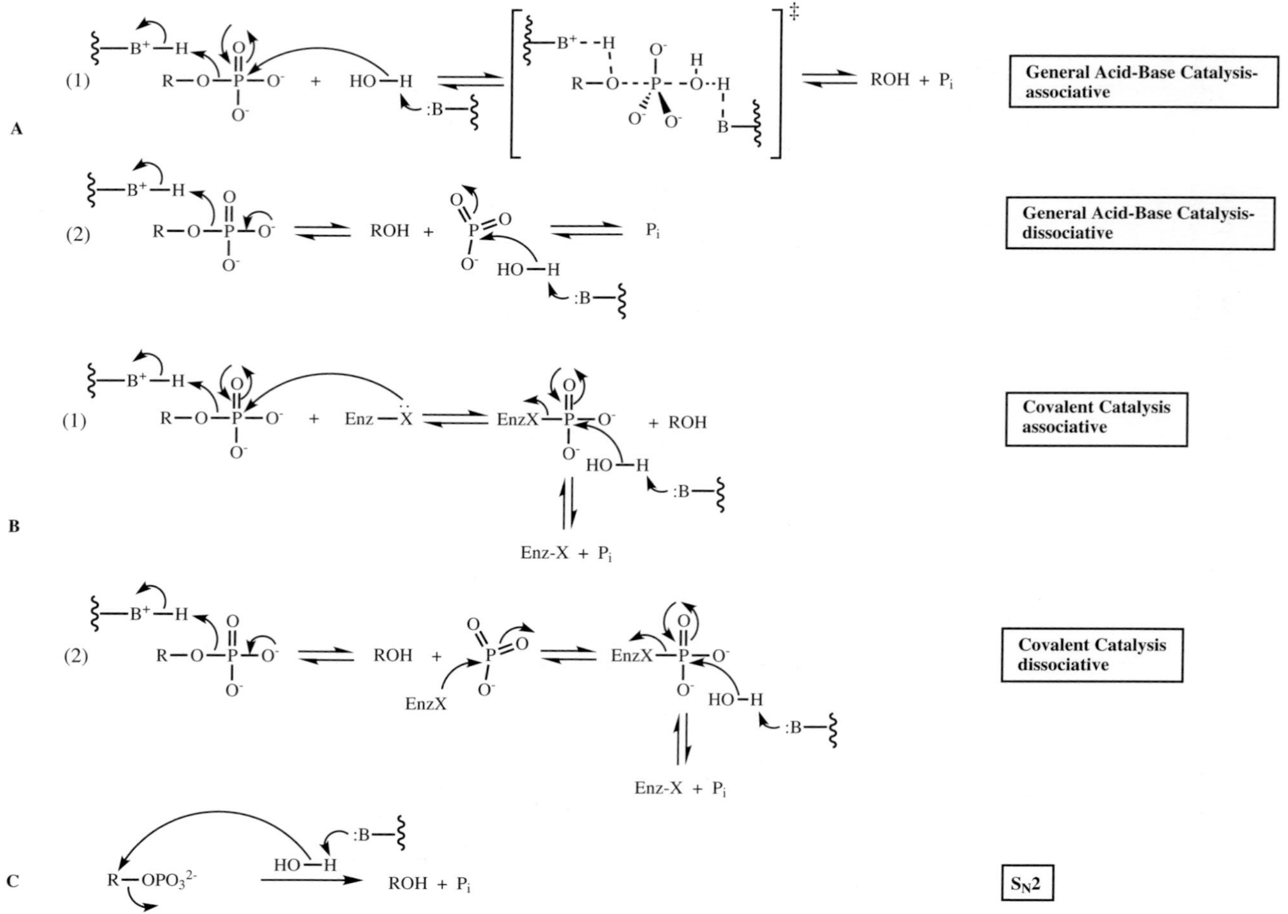

SCHEME 2.38 Three general mechanisms for phosphatases.

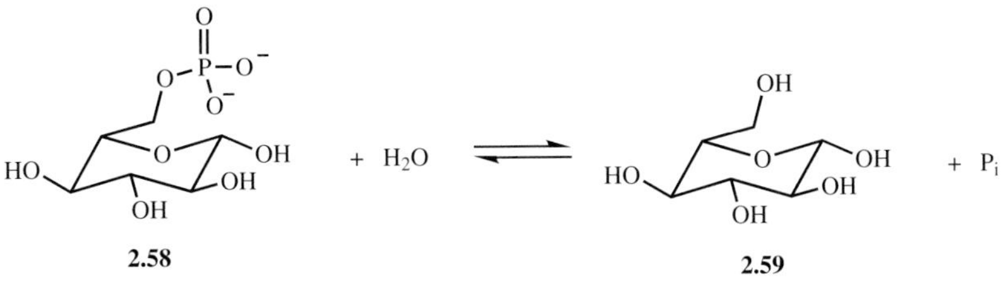

SCHEME 2.39 Reaction catalyzed by glucose-6-phosphatase.

ring of glucose-6-phosphate instead of ^{32}P in the phosphate group, no ^{14}C was incorporated into protein, consistent with transfer of the phosphoryl group to an enzyme nucleophile. Further evidence for the importance of an active-site histidine residue is the observation that glucose-6-phosphatase is inactivated by photooxidation in the presence of the triplet sensitizers methylene blue or rose bengal.[107] Under these conditions photoxidation destroys the imidazole ring of histidines, but the enzyme can be protected from inactivation by incubation with the competitive inhibitor P_i, which binds to the active site and prevents the activated oxygen from approaching the histidine residue.

When similar experiments are done with alkaline phosphatase (except for quenching below pH 5.5), an enzyme that catalyzes the nonspecific hydrolyses of a wide range of phosphate monoesters, a serine residue (now known to be Ser-102) is phosphorylated,[108] suggesting that it acts by a mechanism similar to that of serine peptidases (again, mechanism B1 or B2, Scheme 2.38). Consistent with a covalent mechanism is the finding that the reaction proceeds with retention of configuration around the phosphorus atom, indicating two S_N2 inversions (first by Ser-102, second by water).[109] Other experiments that could be tried to support a covalent catalytic mechanism would be those used to characterize chymotrypsin as a peptidase (see Section I.A.1). A series of phosphates, $ROPO_3^{2-}$, could be synthesized with different R groups and the rate of P_i formation monitored. In fact, that was done, and the rate was shown to be independent of the structure of R (from R = CH_3 to a protein), supporting a two-step mechanism in which either dephosphorylation (the phosphate analogue of the deacylation step in chymotrypsin) or product (P_i) release is the rate-determining step.[110] A variety of NMR spectroscopic studies support product release as the rate-determining step.[111] Or you might want to try trapping the phosphate with nucleophiles other than water, such as hydroxylamine. You could synthesize *para*-nitrophenyl phosphate (**2.60**) and carry out an initial burst experiment, as was done for chymotrypsin (see I.A.1.b.1).

Examples of phosphatases for which associative and dissociative mechanisms have

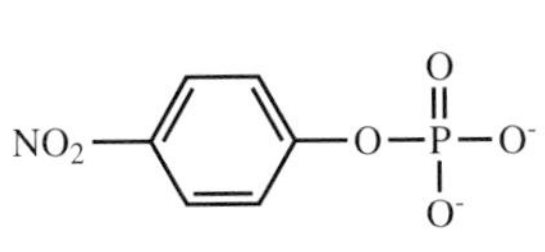

2.60

been differentiated are the tyrosine phosphatase[112] and serine/threonine phospha-tase[113] *superfamilies*. But first let's discuss the term *superfamily*. In general, superfami-lies consist of enzymes with different overall structures (<50% sequence identity), but with common active-site structural scaffolds. Some superfamilies catalyze quite distinct overall chemical reactions, although a common mechanistic strategy for each of these reactions is involved.[114] Other superfamilies share a common struc-tural motif and similar enzymatic activity, but have varying substrate specificities.[115] The former type of superfamilies would include the enolase superfamily,[116] the members of which catalyze at least 11 different chemical reactions, including race-mization, epimerization, and *syn*- and *anti*-β-elimination reactions involving wa-ter, ammonia, and a carboxylate as the leaving group. Despite the very broad dif-ferences in substrate structures and chemical reactions, all the reactions in this superfamily are initiated by a common metal-assisted, general base-catalyzed ab-straction of the α-proton of a carboxylic acid to generate a stabilized enolate anion intermediate, from which all the reactions in this superfamily proceed (Scheme 2.40). The other type of superfamily includes the serine protease superfamily, the members of which all catalyze the hydrolysis of various proteins and peptides using an active-site serine residue in a covalent catalytic reaction.

The tyrosine phosphatase and serine/threonine phosphatase superfamilies of en-zymes catalyze the hydrolysis of protein tyrosine phosphate and protein serine/threonine phosphate residues, respectively. Reversible protein phosphorylation leads to conformational changes in the proteins that trigger important signal trans-duction mechanisms involved in cell growth, proliferation, and differentiation.[117] A significant difference in the structure and mechanism of these two superfamilies is that the protein tyrosine phosphatases do not require metal ions or other cofactors for catalysis, whereas the protein serine/threonine phosphatases are *binuclear metal-lohydrolases* (require two metal ions for catalysis).

All members of the tyrosine phosphatase superfamily share a common active-site motif consisting of a cysteine residue and an arginine residue separated by five resi-dues (CX_5R), which is where the phosphate oxyanions bind. There also is a con-served aspartic acid residue in each member of this superfamily. A number of stud-ies indicate that mechanism B2 (Scheme 2.38), where X = Cys,[118] is the most reasonable; both steps are in-line displacements that are highly dissociative, similar to that in solution, with P—O bond breaking occurring faster than bond making.[119]

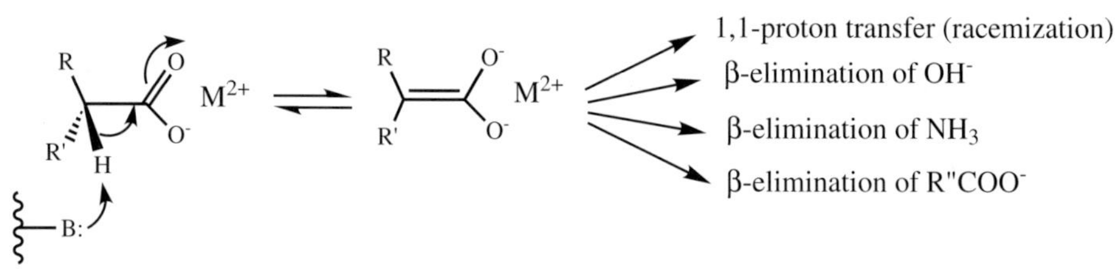

SCHEME 2.40   Common partial reaction catalyzed by the members of the enolase superfamily.

The conserved arginine residue at the active site helps to stabilize the transition state by forming three hydrogen bonds to two of the substrate oxygens. The conserved aspartic acid residue acts as a general acid in the first step by donation of a proton to the leaving group (RO$^-$); it also may be the general base that removes the proton from water in the second step. Typically in the thiol phosphatases, a histidine residue precedes the cysteine residue and a serine or threonine residue follows the arginine residue; these histidine and serine/threonine residues appear to be important in lowering the pK_a of the active-site cysteine residue (by electrostatic and polar interactions). With a lower pK_a, the cysteine can be maintained in an anionic state, so it acts as a better nucleophile in the first step; a lower pK_a also means that it will be a better leaving group for the second step of the reaction.[120] In the case of the phosphatase from *Yersinia*, for example, the pK_a of the cysteine residue is lowered by more than 3 pK_a units to 4.7.[121]

An example of a tyrosine phosphatase is VH1 (*vaccinia* virus H1 gene-related), which actually has dual specificity; that is, it hydrolyzes phosphoserine and phosphothreonine residues as well as phosphotyrosine residues, and is called a *dual-specific phosphatase*.[122] Dephosphorylation of the tyrosine residues is an event that is required for the cell to enter mitosis. The k_{cat}/K_m values rise as a function of pH with a slope of 2 to a peak at about pH 5.6; then they decrease with a slope of -1, indicating that two groups must be unprotonated and one group must be protonated for activity (all with pK_a values of about 5.6).[123] The k_{cat}/K_m values for the D92N mutant reach a peak at about pH 5.6, but then remain constant at higher pH values, suggesting that Asp-92 is the protonated group. Asp-92 is conserved throughout the entire family of dual-specific phosphatases, supporting its importance in catalysis. The pK_a of Cys-124 was determined to be 5.6 by measuring the pseudo-first-order rate constant for inactivation by the affinity labeling agent iodoacetic acid as a function of pH. This cysteine residue was designated as the unprotonated group (the cysteine anion) important to covalent catalysis. With the protein tyrosine phosphatase from leukocyte (antigen-related), the covalent intermediate was detected by rapid chemical quench flow experiments; following quenching with 0.2 N NaOH in D$_2$O, the covalent intermediate was characterized by ^{31}P NMR spectroscopy as a phosphorylated cysteine.[124] A covalent adduct also was demonstrated using ^{32}P-radiolabeled substrate. The mechanism proposed is shown in Scheme 2.41.

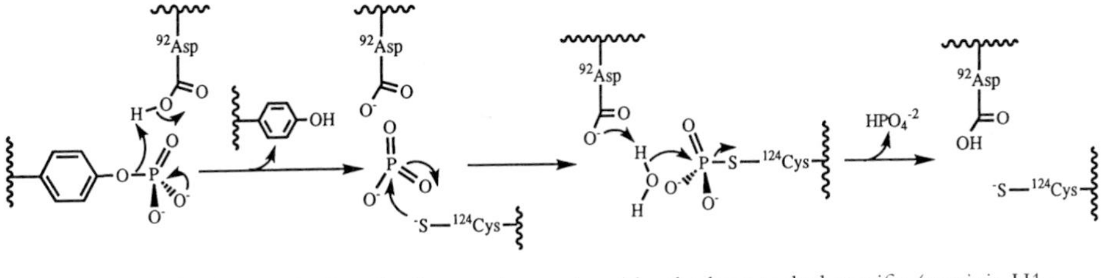

SCHEME 2.41   Mechanism for the reaction catalyzed by the human dual-specific (vaccinia H1-related) protein tyrosine phosphatase.

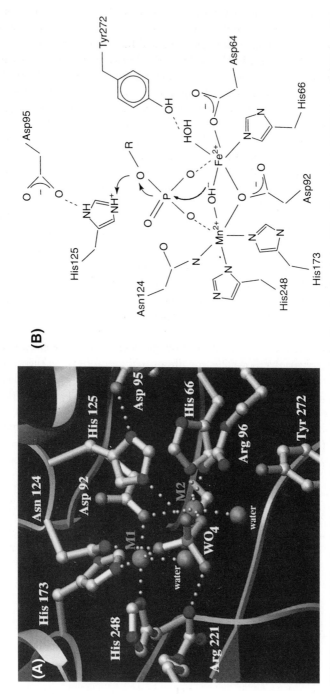

FIGURE 2.10 (a) Molecular modeled view of the active site of protein serine/threonine phosphatase PP1 with tungstate ion (WO_4) bound; (b) Schematic of the catalytic mechanism based on the crystal structure and kinetic studies. [Reprinted from Trends in Biochemical Sciences, **21**, D. Barford, Molecular mechanisms of the protein serine/threonine phosphatases, 407–412. Copyright © 1996, with permission from Elsevier Science.]

The serine/threonine phosphatases require two metal ions for activity; for example, protein phosphatase 1 contains Mn^{2+} and Fe^{2+}, whereas protein phosphatase 2B (also called *calcineurin*) [125] contains Zn^{2+} and Fe^{2+}. The X-ray crystal structures at 2.1 Å resolution of PP1 [126] and of PP2B [127] reveal the active-site catalytic domains and suggest a Lewis acid-catalyzed associative mechanism (Scheme 2.38, A1). Although the stereochemistry of dephosphorylation of these enzymes is not known, the reaction catalyzed by purple acid phosphatase, which has an active-site structure similar to that of the serine/threonine phosphatases,[128] proceeds with inversion of configuration at phosphorus, consistent with a single-step displacement.[129] The metal ions are bound to the phosphate oxyanions, a fact that also supports an associative mechanism. An associative mechanism proceeds via a pentacoordinate intermediate/transition state with three equatorial phosphoryl oxyanions (see Scheme 2.38, A1), but a dissociative mechanism gives a metaphosphate intermediate with reduced oxyanion character (one oxyanion). Therefore, the metal ions stabilize the ground-state structure of a dissociative mechanism (two oxyanions) better than the metaphosphate intermediate (one oxyanion), whereas in an associative mechanism the metal ions would stabilize the transition state (three oxyanions) to a larger extent than the ground state (two oxyanions); this favors an associative mechanism. Also, because of the high reactivity of a metaphosphate intermediate, activation of the nucleophile (water) is not necessary, but in an associative mechanism this activation is important; the crystal structures show that the nucleophilic water molecule is activated by metal ion binding (Figure 2.10).[130] As Lewis acids, the metal ions enhance both the nucleophilicity of the water (by making it more acidic, thereby favoring the generation of hydroxide ion) and the electrophilicity of the phosphorus atom (by neutralization of the phosphate oxyanion). In PP1 the side-chain His-125 probably donates a proton to the leaving group oxygen of the serine or threonine residue; mutation of His-125 results in loss of catalytic activity.[131]

B. Phosphodiesterases

As an example of a phosphodiesterase, consider ribonuclease A (RNase), the enzyme that catalyzes the hydrolysis of RNA. The affinity label, 1–fluoro-2,4–dinitrobenzene (**2.61**), attaches to an active-site lysine (Lys-41), and iodoacetamide labels one of two different histidine residues, His-12 or His-119, but not both in any one enzyme molecule.

The classical general acid/base mechanism for ribonuclease A, based on these studies and the isolation of the cyclic phosphate ester **2.62,** is shown in Scheme 2.42;

2.61

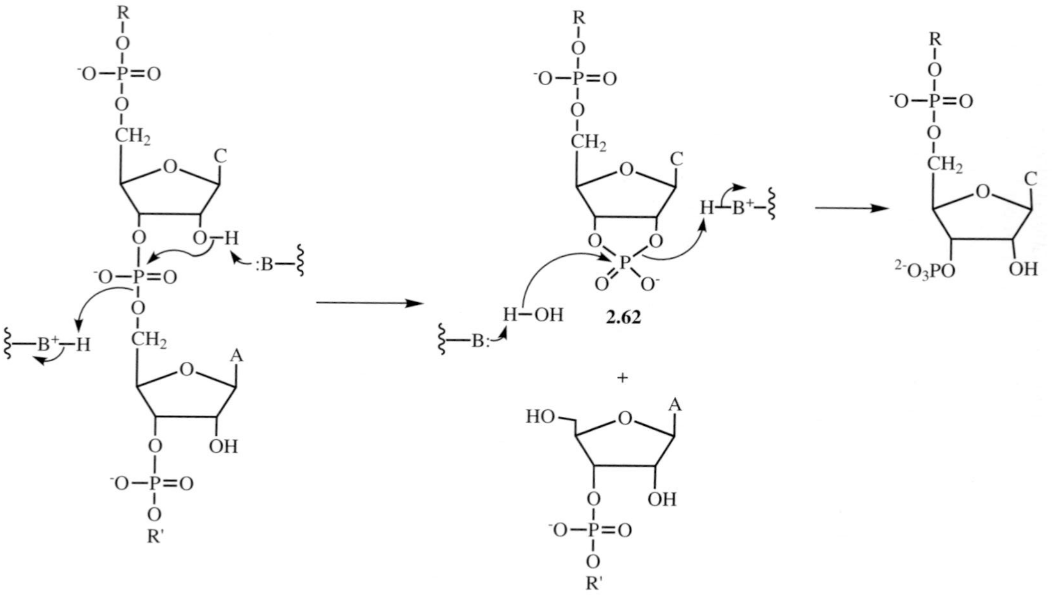

SCHEME 2.42   General acid/base-catalyzed mechanism for ribonuclease A.

the two active-site bases involved are now known to be His-119 (the proton donor in the first step) and His-12 (the proton abstractor in the first step). Hydrolysis of **2.62** gives the 3'-phosphate exclusively, not the 2'-phosphate.

An alternative mechanism, however, has been proposed (Scheme 2.43).[132] Protonation of one of the nonbridging phosphoryl oxygens (**2.63**) renders the transition state more like a phosphate triester (**2.64**), which is 10^3–10^5 times more reactive than the corresponding phosphate diester.[133] To differentiate these mechanisms, rate effects as a result of substitution of a nonbridging phosphoryl oxygen by sulfur (called a *thio-effect*) for enzymatic and nonenzymatic reactions of RNA were compared.[134] The thio-effect of substitution by sulfur slows down the reaction. A small thio-effect would be inconsistent with a triester-like mechanism and would support the classical mechanism. The thio-effect on k_{cat} for opening of the 2', 3'-cyclic phosphate of uridine by RNase A is only 5 for both stereoisomers (replacement of one of the nonbridging oxygens with sulfur makes the phosphorus a stereogenic center; replacement of the other gives the stereoisomer), which is within the range observed for nonenzymatic reactions of phosphate diesters, but smaller than observed for phosphate triesters. Another approach to differentiating the two mechanisms was by considering the effect of proton affinities as a result of substitution of sulfur for oxygen. In the triester-like mechanism, one of the phosphoryl oxygens is protonated in the transition state. Because sulfur has a much lower proton affinity than oxygen, a large thio-effect would be predicted for at least one of the epimeric thio-isomers in the triester-like mechanism. Again, it was found that both thio-isomers of a cyclic phosphate ester (cUMP) gave small thio-effects on k_{cat} with RNase, suggesting that neither phosphoryl oxygen is protonated in the transition state. Both of these analyses suggest that RNase A–catalyzed reactions are consistent with the classical mechanism (Scheme 2.42), not with the phosphate triester mechanism (Scheme 2.43). The small thio-effects also argue against the importance of low-barrier hydrogen bonds (see Chapter 1, Section II.C) in this reaction because these hydrogen bonds would be strongly affected by sulfur substitution.

Another test of the mechanism of ribonuclease A was the use of the internal competition method, which measures the isotope effect on V_{max}/K_m for the reaction.[135] A pseudo-substrate (**2.65**) was synthesized with ^{18}O at the bridge position or nonbridge positions of the phosphodiester bond and with ^{15}N in the nitro group of the benzyl ring.[136] This compound was mixed with the corresponding compound having ^{16}O at the bridge position or nonbridge positions and ^{14}N in the nitro group. Normal secondary isotope effects were observed, arguing against a phosphorane intermediate and in favor of a concerted mechanism with a transition state having slightly associative character (again, Scheme 2.42).

C. Kinases

True kinases transfer the γ-phosphoryl group of nucleoside triphosphates (e.g., ATP, GTP, UTP, CTP) to an acceptor. Therefore, the step in glutamine-dependent

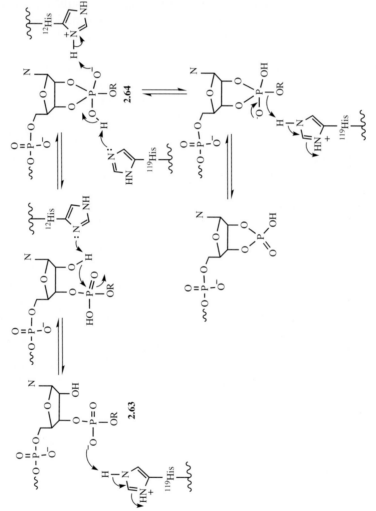

SCHEME 2.43　Alternative mechanism for ribonuclease A via a phosphate triester intermediate.

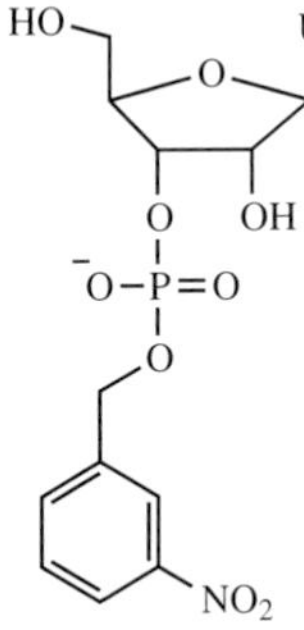

2.65

enzyme reactions that use ATP to activate the substrate (see Section II.A.2.b) has kinase activity, but it is not a kinase, because that is just one step in the overall mechanism for that reaction. Kinases are now generalized to reactions at the α-, β-, or γ-position of nucleoside triphosphates.

An example of a true kinase is pyruvate kinase, which catalyzes the transfer of the γ-phosphate of ATP to pyruvate (**2.66**, Scheme 2.44), giving an important metabolite in enzymology, phosphoenolpyruvate (**2.68**, PEP). There is no evidence for a covalent catalytic mechanism via the phosphoenzyme. In the presence of an analogue of ATP that cannot transfer a phosphate (actually, P_i was used) in ^3H_2O, a tritium atom was incorporated into pyruvate, whereas in the absence of the ATP mimic no tritium was incorporated.[137] This supports a conformational change in the enzyme on binding of the ATP (or mimic) to induce the deprotonation. Enolpyruvate (**2.67**) was shown to be an intermediate in the pyruvate kinase reaction by chemical trapping with bromine under acid conditions.[138] It also was synthesized and was shown to have sufficient stability ($t_{1/2}$ = 1.8 min in 10 mM phosphate buffer, pD 7.0 at 20 °C or 7 min at no buffer concentration) to allow future studies of its intermediacy in other enzyme reactions.[139]

Another important example of an enzyme with kinase activity is acetyl-CoA synthetase (Enzyme Commission name: acetate-CoA ligase), which catalyzes the synthesis of acetyl-CoA from acetate, ATP, and coenzyme A (**2.69**, Scheme 2.45).

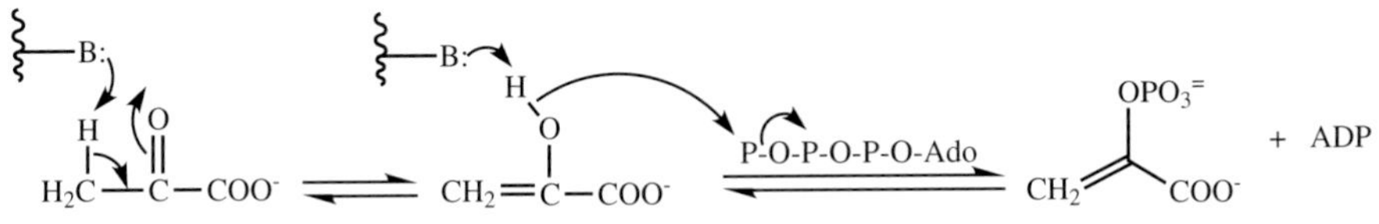

2.66   **2.67**   **2.68**

SCHEME 2.44   Mechanism for pyruvate kinase (ATP is abbreviated as POPOPOAdo).

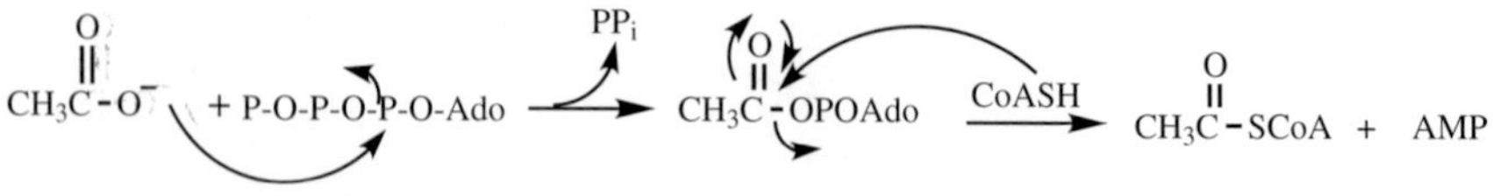

SCHEME 2.45   Mechanism for acetyl-CoA synthetase (ATP is abbreviated as POPOPOAdo).

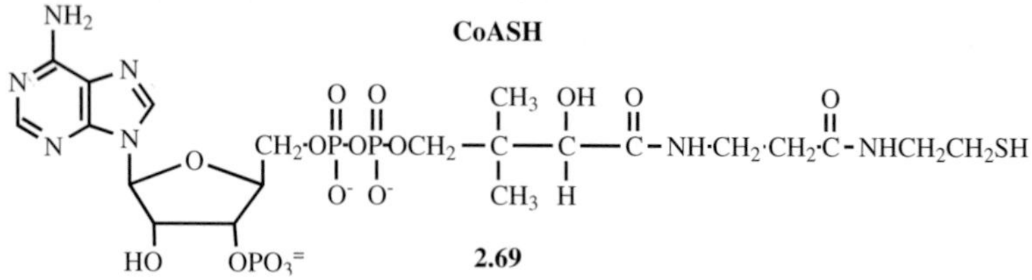

The name coenzyme A is actually a misnomer. Coenzymes for mammalian enzymes are generally made from vitamins, but coenzyme A is unrelated to vitamin A. As a reagent, coenzyme A is abbreviated CoASH to emphasize the important structural feature of this molecule, namely, the thiol group. It functions as a potent nucleophile and in the formation of CoA thioesters, which not only are relatively reactive species, but also have considerable size and functionality for easy transport throughout the cell. This is an important property for acetyl CoA, because it is used as the prime source for transferring the acetyl group to other molecules in the cell.

NOTE THAT PROBLEMS AND SOLUTIONS RELEVANT TO EACH CHAPTER CAN BE FOUND IN APPENDIX II.

REFERENCES

1. Bryant, R. A. R.; Hansen, D. E. *J. Am. Chem. Soc.* **1996**, *118*, 5498.
2. Bode, W.; Schwager, P. *J. Mol. Biol.* **1975**, *98*, 693.
3. Frey, P. A.; Whitt, S. A.; Tobin, J. B. *Science* **1994**, *264*, 1927 (but also read the rebuttal: Ash, E. L.; Sudmeier, J. L.; De Fabo, E. C.; Bachovchin, W. W. *Science* **1997**, *278*, 1128).
4. Warshel, A.; Naray-Szabo, G.; Sussman, F.; Hwang, J.-K. *Biochemistry* **1989**, *28*, 3629.
5. (a) Hartley, B. S.; Kilby, B. A. *Biochem J.* **1954**, *56*, 288; (b) Gutfreund, H,; Sturtevant, J. M. *Proc. Natl. Acad. Sci. USA* **1956**, *42*, 719.
6. (a) Bender, M. L.; Zerner, B. *J. Am. Chem. Soc.* **1961**, *83*, 2391. (b) Bender, M. L.; Zerner, B. *J. Am. Chem. Soc.* **1962**, *84*, 2550. (c) Zerner, B.; Bond, R. P. M.; Bender, M. L. *J. Am. Chem. Soc.* **1964**, *86*, 3674.
7. Cleland, W. W. *Biochemistry* **1990**, *29*, 3194.
8. Epand, R. M.; Wilson, I. B. *J. Biol. Chem.* **1963**, *238*, 1718.
9. Anderson, B. R.; Cordes, E. H.; Jencks, W. P. *J. Biol. Chem.* **1961**, *236*, 455.
10. Balls, A. K.; Jansen, E. F. *Adv. Enzymol.* **1952**, *13*, 321.
11. Schaffer, N. K.; May, S. C.; Summerson, W. H. *J. Biol. Chem.* **1953**, *202*, 67.
12. Hartley, B. S. *Nature* **1964**, *201*, 1284.
13. Strumeyer, D. H.; White, W. N.; Koshland, D. E., Jr. *Proc. Natl. Acad. Sci., USA* **1963**, *50*, 931.

14. (a) *Directed Mutagenesis: A Practical Approach,* McPherson, M. J., Ed., IRL Press: Oxford, 1991. (b) *Protein Engineering: A Practical Approach,* Rees, A. R.; Sternberg, M. J. E.; Wetzel, R., Ed., IRL Press: Oxford, 1992.
15. Plapp, B. V. *Methods Enzymol.* **1995,** *249,* 91.
16. Schoellmann, G.; Shaw, E. *Biochem. Biophys. Res. Commun.* **1962,** *7,* 36.
17. (a) Ong, E. B.; Shaw, E.; Schoellmann, G. *J. Biol. Chem.* **1965,** *240,* 694. (b) Shaw, E. In *The Enzymes,* 3rd edit.; vol. 1, Boyer, P., Ed., Academic Press: New York, 1970, p. 91.
18. Shaw, E.; Schoellmann, G. *Biochemistry* **1963,** *2,* 252.
19. Prorok, M.; Albeck, A.; Foxman, B. M.; Abeles, R. H. *Biochemistry* **1994,** *33,* 9784.
20. Kreutter, K.; Steinmetz, A. C. U.; Liang, T.-C.; Prorok, M.; Abeles, R. H.; Ringe, D. *Biochemistry* **1994,** *33,* 13792.
21. Cassidy, C. S.; Lin, J.; Frey, P. A. *Biochemistry* **1997,** *36,* 4576.
22. Anderson, B. R.; Cordes, E. H.; Jencks, W. P. *J. Biol. Chem.* **1961,** *236,* 455.
23. Bender, M. L.; Clement, G. E.; Kezdy, F. J.; Heck, H. D'A. *J. Am. Chem. Soc.* **1964,** *86,* 3680.
24. Komiyama, M.; Roesel, T. R.; Bender, M. L. *Proc. Natl. Acad. Sci. USA* **1977,** *74,* 23.
25. Mallick, I. M.; D'Souza, V. T.; Yamaguchi, M.; Lee, J.; Chalabi, P.; Gadwood, R. C.; Bender, M. L. *J. Am. Chem. Soc.* **1984,** *106,* 7252.
26. Zaks, A.; Klibanov, A. M. *J. Am. Chem. Soc.* **1986,** *108,* 2767.
27. Corey, D. R.; Craik, C. S. *J. Am. Chem. Soc.* **1992,** *114,* 1784.
28. Wu, Z.-P.; Hilvert, D. *J. Am. Chem. Soc.* **1989,** *111,* 4513.
29. Pearl, L. H., Ed. *Retroviral Proteases: Control of Maturation and Morphogenesis,* Stockton Press: New York; 1990.
30. Meek, T. D.; Lambert, D. M.; Dreyer, G. B.; Carr, T. J.; Tomaszek, Jr., T. A.; Moore, M. L.; Strickler, J. E.; Debouck, C.; Hyland, L. J.; Matthews, T. J.; Metcalf, B. W.; Pettyway, S. R. *Nature (London)* **1990,** *343,* 90.
31. Meek, T. D. *J. Enzyme Inhib.* **1992,** *6,* 65.
32. (a) Perelson, A. S.; Neumann, A. U.; Markowitz, M.; Leonard, J. M.; Ho, D. D. *Science* **1966,** *271,* 1582. (b) Kempf, D. J.; Marsh, K. C.; Denissen, J. F.; McDonald, E., et al. *Proc. Natl. Acad. Sci. USA* **1995,** *92,* 2484. (c) Vacca, J. P.; Dorsey, B. D.; Schleif, W. A.; Levin, R. B., et al. *Proc. Natl. Acad. Sci. USA* **1994,** *91,* 4096.
33. (a) Miller, M.; Schneider, J.; Sathyanarayana, B. K.; Toth, M. V.; Marshall, G. R.; Clawson, L.; Selk, L.; Kent, S. B. H.; Wlodawer, A. *Science* **1989,** *246,* 1149. (b) Wlodawer, A.; Miller, M.; Jaskolski, M.; Sathyanarayana, B. K.; Baldwin, E.; Weber, I. T.; Selk, L. M.; Clawson, L.; Schneider, J.; Kent, S. B. H. *Science* **1989,** *245,* 616.
34. Rodriguez, E. J.; Angeles, R. S.; Meek, T. D., *Biochemistry* **1993,** *32,* 12380.
35. Burley, S. K.; David, P. R.; Sweet, R. M.; Taylor, A.; Lipscomb, W. N. *J. Mol. Biol.* **1992,** *224,* 113.
36. (a) Taylor, A. *TIBS,* **1993,** *18,* 167; (b) Taylor, A. *FASEB J.* **1993,** *7,* 290.
37. Lipscomb, W. N.; Reeke, G. N.; Harsuck, J. A.; Quiocho, F. A.; Bethge, P. H. *Phil. Trans. Roy. Soc. Ser. B* **1970,** *257,* 177.
38. Hass, G. M.; Neurath, H. *Biochemistry* **1970,** *10,* 3535.
39. Simpson, R. T.; Vallee, B. L. *Biochemistry* **1966,** *5,* 1760.
40. (a) Lipscomb, W. N. *Proc. Natl. Acad. Sci. USA* **1967,** *58,* 2220; (b) Lipscomb, W. N. *Chem. Soc. Rev.* **1972,** *1,* 319.
41. Breslow, R.; Wernick, D. *Proc. Natl. Acad. Sci. USA* **1977,** *74,* 1303.
42. Makinen, M. W.; Yamamura, K.; Kaiser, E. T. *Proc. Natl. Acad. Sci. USA* **1976,** *73,* 3882.
43. Sander, M. E.; Witzel, H. *Biochem. Biophys. Res. Commun.* **1985,** *132,* 681.
44. Hoffman, S. J.; Chu, S. S.-T.; Lee, H.; Kaiser, E. T.; Carey, P. R. *J. Am. Chem. Soc.* **1983,** *105,* 6971.
45. Kuo, L. C.; Makinen, M. W. *J. Am. Chem. Soc.* **1985,** *107,* 5255.
46. Christianson, D. W.; Lipscomb, W. N. *Acc. Chem. Res.* **1989,** *22,* 62.

47. Britt, B. M.; Peticolas, W. L. *J. Am. Chem. Soc.* **1992**, *114*, 5295.

48. Breslow, R.; Wernick, D. *J. Am. Chem. Soc.* **1976**, *98*, 259.

49. (a) Craik, C. S. *BioTechniques* **1985**, *3*, 12, (b) Gardell, S. J.; Craik, C. S.; Hilvert, D.; Urdea, M. S.; Rutter, W. J. *Nature* **1985**, *317*, 551.

50. Gardell, S. J.; Hilvert, D.; Barnett, J.; Kaiser, E. T.; Rutter, W. J. *J. Biol. Chem.* **1987**, *262*, 576.

51. (a) Christianson, D. W.; David, P. R.; Lipscomb, W. N. *Proc. Natl. Acad. Sci. USA* **1987**, *84*, 1512. (b) Christianson, D. W.; Lipscomb, W. N. *Acc. Chem. Res.* **1989**, *22*, 62.

52. Fife, T. H.; Przystas, T. J. *J. Am. Chem. Soc.* **1986**, *108*, 4631.

53. Groves, J. T.; Olson, J. R. *Inorg. Chem.* **1985**, *24*, 2715.

54. (a) Vallee, B. L.; Galdes, A. *Adv. Enzymol. Relat. Areas Mol. Biol.* **1984**, *56*, 283.

55. Auld, D. S.; Holmquist, B. *Biochemistry* **1974**, *13*, 4355.

56. Rees, D. C.; Lipscomb, W. N. *Proc. Natl. Acad. Sci. USA* **1983**, *80*, 7151.

57. Riordan, J. F. *Biochemistry* **1973**, *12*, 3915.

58. Quinn, D. M. *Chem. Rev.* **1987**, *87*, 955.

59. (a) Hasinoff, B. B. *Biochim. Biophys. Acta* **1982**, *704*, 52. (b) Bazelyansky, M.; Robey, C.; Kirsch, J. F. *Biochemistry* **1985**, *25*, 125.

60. Nachmansohn, D.; Wilson, I. B.; *Adv. Enzymol.* **1951**, *12*, 259.

61. Roseberry, T. L. *Adv. Enzymol.* **1975**, *43*, 103.

62. Sussman, J. L.; Harel, M.; Frolow, F.; Oefner, C.; Goldman, A.; Toker, L.; Silman, I. *Science*, **1991**, *253*, 872.

62a. Dougherty, D. A. *Chem. Rev.* **1997**, *97*, 1303.

63. Silverman, R. B. *The Organic Chemistry of Drug Design and Drug Action*, Academic Press: San Diego, 1992, p. 382.

64. Taylor, P. In *Goodman and Gilman's The Pharmacological Basis of Therapeutics*, 7th ed., Gilman, A. G.; Goodman, L. S.; Rall, T. W.; Murad, F., Eds., Macmillan: New York, 1985, p. 110.

65. Kawakami, Y.; Inoue, A.; Kawai, T.; Wakita, M.; Sugimoto, H.; Hopfinger, A. J. *Bioorg. Med. Chem.* **1996**, *4*, 1429.

66. Wong, C.-H.; Whitesides, G. M. *Enzymes in Synthetic Organic Chemistry*, Pergamon Press: Oxford, 1994.

67. (a) Kazlauskas, R. J.; Weissfloch, A. N. E.; Rappaport, A. T.; Cuccia, L. A. *J. Org. Chem.* **1991**, *56*, 2656. (b) Kim, M. J.; Choi, Y. K. *J. Org. Chem.* **1992**, *57*, 1605. (c) Feorgens, U.; Schneider, M. P. *J. Chem. Soc. Chem Commun.* **1991**, 1066.

68. Cygler, M.; Grochulski, P.; Kazlauskas, R. J.; Schrag, J. D.; Bouthillier, F.; Rubin, B.; Serreqi, A. N.; Gupta, A. K. *J. Am. Chem. Soc.* **1994**, *116*, 3180.

69. Zemal, H. *J. Am. Chem. Soc.* **1987**, *109*, 1875.

70. W. P. Jencks, *Catalysis in Chemistry and Enzymology*, McGraw-Hill: New York, **1969**, p. 288.

71. Bernhard, S. A.; Orgel, L. E. *Science* **1959**, *130*, 625.

72. Jencks, W. P. In *Current Aspects of Biochemical Energetics*; Kennedy, E. P., Ed., Academic: New York, 1966, p. 273.

73. (a) Wolfenden, R. *Annu. Rev. Biophys. Bioengg.* **1976**, *5*, 271; (b) ibid., *Nature* **1969**, *223*, 704; (c) ibid., *Meth. Enzymol.* **1977**, *46*, 15.

74. (a) Lienhard, G. E. *Science* **1973**, *180*, 149; (b) ibid., *Annu. Repts. Med. Chem.* **1972**, *7*, 249.

75. Schramm, V. L. *Annu. Rev. Biochem.* **1998**, *67*, 693.

76. Goding, J. W. *Monoclonal Antibodies: Principles and Practice*, Academic Press: New York, 1986.

77. Pollack, S. J.; Hsiun, P.; Schultz, P. G. *J. Am. Chem. Soc.* **1989**, *111*, 5961.

78. Benkovic, S. J.; Lerner, R. A. *Science* **1989**, *244*, 437.

79. (a) Schultz, P. G. *Acc. Chem. Res.* **1989**, *22*, 287; (b) Schultz, P. G. *Angew. Chem. Int. Ed. Engl.* **1989**, *28*, 1283; (c) Lerner, R. A.; Benkovic, S.J.; Schultz, P.G. *Science* **1991**, *252*, 659; (d) Benkovic, S. J. *Annu. Rev. Biochem.* **1992**, *61*, 29; (e) Schultz, P. G.; Lerner, R. A. *Acc. Chem. Res.* **1993**, *26*, 391; (f) Stewart, J. D.; Liotta, L. J.; Benkovic, S. J. *Acc. Chem. Res.* **1993**, *26*, 396; (g) Hilvert, D. *Acc. Chem. Res.* **1993**, *26*, 552; (h) Schultz, P. G.; Lerner, R. A. *Science* **1995**, *269*,

1835.

80. Jayaram, H. N.; Cooney, D. A.; Milman, H. A.; Homan, E. R.; Rosenbluth, R. J. *Biochem. Pharmacol.* **1976**, *25*, 1571.

81. Lundblad, R. L. *Techniques in Protein Modification;* CRC Press: Boca Raton, FL, 1994.

82. (a) Silverman, R. B. *Mechanism-Based Enzyme Inactivation: Chemistry and Enzymology,* vols. 1, 2, CRC Press: Boca Raton, FL, 1988. (b) Silverman, R. B. *Methods Enzymol.* **1995**, *249*, 240.

83. Dion, H. W.; Fusari, S. A.; Jakubowski, Z. L.; Zora, J. G.; Bartz, Q. R. *J. Am. Chem. Soc.* **1956**, *78*, 3075.

84. Fusari, S. A.; Haskell, T. H.; Frohardt, R. P.; Bartz, Q. R. *J. Am. Chem. Soc.* **1954**, *76*, 2878.

85. Hartmann, S. C.; McGrath, T. F. *J. Biol. Chem.* **1973**, *248*, 8506.

86. Hartmann, S. C. *J. Biol. Chem.* **1968**, *243*, 853.

87. (a) Tso, J. Y.; Hermodson, M. A.; Zalkin, H. *J. Biol. Chem.* **1982**, *257*, 3532. (b) Vollmer, S. J.; Switzer, R. L.; Hermodson, M. A.; Bower, S. G.; Zalkin, H. *J. Biol. Chem.* **1983**, *258*, 10582.

88. Levitzki, A.; Stallcup, W. B.; Koshland, D. E., Jr. *Biochemistry* **1971**, *10*, 3371.

89. Nagano, H.; Zalkin, H.; Henderson, E. J. *J. Biol. Chem.* **1970**, *245*, 3810.

90. Zalkin, H.; Hwang, L. H. *J. Biol. Chem.* **1971**, *246*, 6899.

91. Hartmann, S. C. *J. Biol Chem.* **1963**, *238*, 3036.

92. Inoue, M.; Horiuchi, S.; Morino, Y. *Eur. J. Biochem.* **1979**, *99*, 169.

93. Tate, S. S.; Ross, M. E. *J. Biol. Chem.* **1977**, *252*, 6042.

94. (a) Thoden, J. B.; Holden, H. M.; Wesenberg, G.; Raushel, F. M.; Rayment, I. *Biochemistry* **1997**, *36*, 6305. (b) Raushel, F. M.; Thoden, J. B.; Holden, H. M. *Biochemistry* **1999**, *38*, 7891.

95. Rudolph, J.; Stubbe, J. *Biochemistry,* **1995**, *34*, 2241.

96. Horowitz, B.; Meister, A. M. *J. Biol. Chem.* **1972**, *247*, 6708.

97. Buchanan, J. W. *Adv. Enzymol.* **1973**, *39*, 91.

98. Middlefort, C.; Rose, I. *J. Biol. Chem.* **1976**, *251*, 5881.

99. (a) Ratner, S. *The Enzymes,* 3rd ed.; vol. 7, Boyer, P., ed.; Academic Press: New York, **1973**, p. 168. (b) Ratner, S. *Adv. Enzymol. Rel. Areas Mol. Biol.* **1973**, *39*,1.

100. Rochovansky, O.; Ratner, S. *J. Biol. Chem.* **1961**, *236*, 2254.

101. Kumar, S.; Leannane, J.; Ratner, S. *Proc. Natl. Acad. Sci. USA* **1985**, *82*, 6745.

102. Lusty, C. J.; Ratner, S. *Proc. Natl. Acad. Sci. USA* **1987**, *84*, 3176.

102a. Widlanski, T.; Taylor, W. *Comprehensive Nat. Prod. Chem.* **1999**, *5*, 139.

103. (a) Cleland, W. W. *Bioorg. Chem.* **1987**, *15*, 283. (b) Cleland, W. W. *FASEB J.* **1990**, *4*, 2899. (c) Cleland, W. W.; Hengge, A. C. *FASEB J.* **1995**, *9*, 1585.

104. Hass, L. F.; Boyer, P. D.; Reynard, A. M. *J. Biol. Chem.* **1961**, *236*, 2284.

105. Hass, L. F.; Byrne, W. L. *J. Am. Chem. Soc.* **1960**, *82*, 947.

106. (a) Feldman, F.; Butler, L. G. *Biochem. Biophys. Res. Commun.* **1969**, *36*, 119. (b) Feldman, F.; Butler, L. G. *Biochim. Biophys. Acta* **1972**, *268*, 698.

107. Feldman, F.; Butler, L. G. *Biochim. Biophys. Acta* **1972**, *268*, 690.

108. (a) Schwartz, J. H.; Lipmann, F. *Proc. Natl. Acad. Sci. USA* **1961**, *47*, 1996. (b) Engström, L. *Biochim. Biophys. Acta* **1962**, *56*, 606. (c) Fernley, H. N.; Walker, P. G. *Nature* **1966**, *212*, 1435.

109. Jones, S. R.; Kindman, L. A.; Knowles, J. R. *Nature* **1978**, *275*, 564.

110. Han, R.; Coleman, J. E. *Biochemistry* **1995**, *34*, 4238.

111. Coleman, J. E. *Annu. Rev. Biophys. Biomol. Struct.* **1992**, *21*, 441.

112. (a) Fauman, E. B.; Saper, M. A. *TIBS* **1996**, *21*, 413. (b) Walton, K. M.; Dixon, J. E. *Annu. Rev. Biochem.* **1993**, *62*, 101. (c) Zhang, Z.-Y. *Crit. Rev. Biochem. Mol. Biol.* **1998**, *33*, 1.

113. Barford, D. *TIBS* **1996**, *21*, 407.

114. Babbitt, P. C.; Gerlt, J. A. *J. Biol. Chem.* **1997**, *272*, 30591.

115. Perona, J. J.; Craik, C. S. *J. Biol. Chem.* **1997**, *272*, 29987.

116. Babbitt, P. C.; Hasson, M. S.; Wedekind, J. E.; Palmer, D. R. J.; Barrett, W. C.; Reed, G. H.; Rayment, I.; Ringe, D.; Kenyon, G. L.; Gerlt, J. A. *Biochemistry* **1996**, *35*, 16489.

117. Cohen, P. *Annu. Rev. Biochem.* **1989**, *58*, 453.

118. Guan, K. L.; Dixon, J. E. *J. Biol. Chem.* **1991**, *266*, 17026.
119. Hengge, A. C.; Sowa, G. A.; Wu, L.; Zhang, Z.-Y. *Biochemistry* **1995**, *34*, 13982.
120. Denu, J. M.; Dixon, J. E. *Proc. Natl. Acad. Sci. USA* **1995**, *92*, 5910.
121. Zhang, Z.-Y.; Dixon, J. E. *Biochemistry* **1993**, *32*, 9340.
122. Guan, K.; Broyles, S. S.; Dixon, J. E. *Nature* **1991**, *350*, 359.
123. Denu, J. M.; Zhou, G.; Guo, Y.; Dixon, J. E. *Biochemistry* **1995**, *34*, 3396.
124. Cho, H.; Krishnaraj, R.; Kitas, E.; Bannwarth, W.; Walsh, C. T.; Anderson, K. S. *J. Am. Chem. Soc.* **1992**, *114*, 7296.
125. (a) Klee, C. B.; Draetta, G. F.; Hubbard, M. J. *Adv. Enzymol.* **1988**, *61*, 149. (b) King, M. M.; Huang, C. Y. *J. Biol. Chem.* **1984**, *259*, 8847.
126. (a) Goldberg, J.; Huang, H.-B.; Kwon, Y.-G.; Greengard, P.; Nairn, A. C.; Kuriyan, J. *Nature* **1995**, *376*, 745. (b) Egloff, M.-P.; Cohen, P. T. W.; Reinemer, P.; Barford, D. *J. Mol. Biol.* **1995**, *254*, 942.
127. (a) Kissinger, C. R.; Parge, H. E.; Knighton, D. R.; Lewis, C. T.; Pelletier, L. A.; Tempczyk, A.; Kalish, V. J.; Tucker, K. D.; Showalter, R. E.; Moomaw, E. W.; Gastinel, L. N.; Habuka, N.; Chen, X.; Maldonado, F.; Barker, J. E.; Bacquet, R.; Villafrance, E. *Nature* **1995**, *378*, 641. (b) Griffith, J.P.; Kim, J.L.; Kim, E.E.; Sintchak, M.D.; Thomson, J.A.; Fitzgibbon, M.J.; Fleming, M.A.; Caron, P.R.; Hsiao, K.; Navia, M.A. *Cell* **1995**, *82*, 507.
128. Strater, N.; Klabunde, T.; Tucker, P. *Science* **1995**, *268*, 1489.
129. Mueller, E. G.; Crowder, M. W.; Averill, B. A.; Knowles, J. R. *J. Am. Chem. Soc.* **1993**, *115*, 2974.
130. (a) Egloff, M.-P.; Cohen, P. T. W.; Reinemer, P.; Barford, D. *J. Mol. Biol.* **1995**, *254*, 942. (b) Barford, D. *TIBS* **1996**, *21*, 407.
131. Zhang, L.; Zhang, Z.; Long, F.; Lee, E. Y. C. *Biochemistry* **1996**, *35*, 1606.
132. (a) Breslow, R.; Xu, R., *Proc. Natl. Acad. Sci. USA* **1993**, *90*, 1201. (b) ibid., *J. Am. Chem. Soc.,* **1993**, *115*, 10705.
133. Chandler, A. J.; Hollfelder, F.; Kirby, A. J.; O'Carroll, F.; Stromberg, R. *J. Chem. Soc. Perkin Trans. 2* **1994**, 327.
134. Herschlag, D. *J. Am. Chem. Soc.* **1994**, *116*, 11631.
135. Cleland, W. W. *CRC Critical Rev. Biochem.* **1982**, *13*, 385.
136. Sowa, G. A.; Hengge, A. C.; Cleland, W. W. *J. Am. Chem. Soc.* **1997**, *119*, 2319.
137. Rose, I. A. *J. Biol. Chem.* **1960**, *235*, 1170.
138. Seeholzer, S. J.; Jaworowski, A.; Rose, I. A. *Biochemistry,* **1991**, *30*, 727.
139. Peliska, J. A.; O'Leary, M. H. *J. Am. Chem. Soc.* **1991**, *113*, 1841.

CHAPTER 3

Reduction and Oxidation

I. GENERAL

Most organic substrates undergo one-electron or two-electron redox reactions, but some undergo four-electron reactions. The most common electron acceptor in biological systems is molecular oxygen, undergoing a two-electron reduction to hydrogen peroxide or four-electron reduction to water. Removal or addition of electrons from or to a substrate requires an electron acceptor or donor, respectively, and there are no obvious redox groups in the side chains of amino acids. Consequently, redox enzymes require either an additional organic molecule or a transition metal, known as *coenzymes or cofactors,* at the active site to assist in electron transfer.

II. REDOX WITHOUT A COENZYME

Before we discuss redox coenzymes, let's take a look at an enzyme that catalyzes an internal redox reaction, one in which part of the substrate is oxidized and part is reduced, and therefore there is no net change in oxidation state of the substrate. Consequently, this enzyme does not require a redox cofactor. The enzyme is glyoxalase, and it catalyzes the conversion of methylglyoxal (pyruvaldehyde, **3.1**) to lactate (**3.2,** Scheme 3.1). For many years this enzyme was considered an unequivocal example of the Cannizzaro reaction, a base-catalyzed conversion of an aldehyde to a carboxylic acid and an alcohol, via a direct hydride transfer (Scheme 3.2).

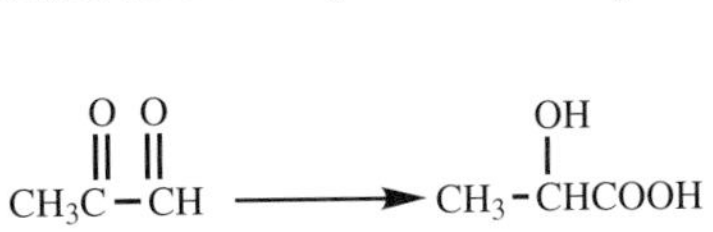

$$\underset{\textbf{3.1}}{\underset{\displaystyle CH_3C-CH}{\overset{\displaystyle \overset{O}{\|}\ \overset{O}{\|}}{}}} \longrightarrow \underset{\textbf{3.2}}{\underset{\displaystyle CH_3-CHCOOH}{\overset{\displaystyle \overset{OH}{|}}{}}}$$

SCHEME 3.1 Reaction catalyzed by glyoxalase.

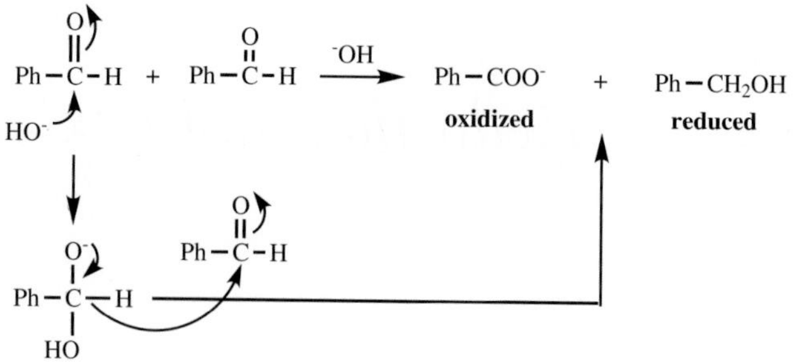

SCHEME 3.2   Cannizzaro reaction mechanism.

Hydride transfer reactions proceed without exchange of protons with solvent (i.e., no deuterium incorporation when carried out in D_2O). However, if in an enzymatic reaction there is no solvent proton exchange, that does not necessarily mean a hydride transfer mechanism has occurred. An active site may be very hydrophobic and contain little or no water. In this case, proton transfers can occur without exchange by solvent because solvent water molecules are excluded.[1]

Two enzymes actually function in tandem to convert methylglyoxal to lactate: glyoxalase I and glyoxalase II. The overall reaction, which requires a second substrate, glutathione (GSH, **3.3**), converts methylglyoxal to the glutathione thioester of lactate (**3.4**, Scheme 3.3) by glyoxalase I (this is the internal redox reaction), and then **3.4** is converted to lactate by glyoxalase II in a hydrolysis reaction.

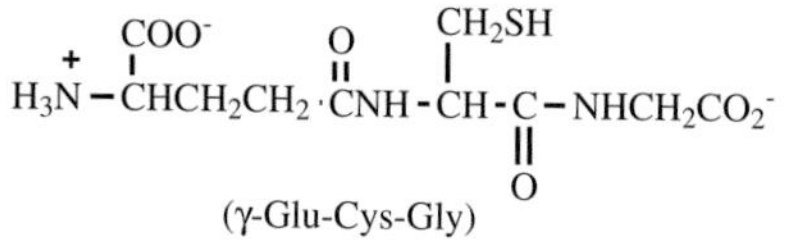

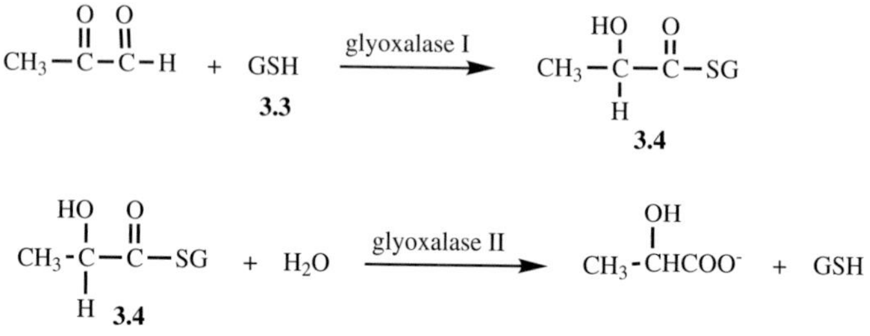

SCHEME 3.3   Reactions catalyzed by glyoxalase I and glyoxalase II.

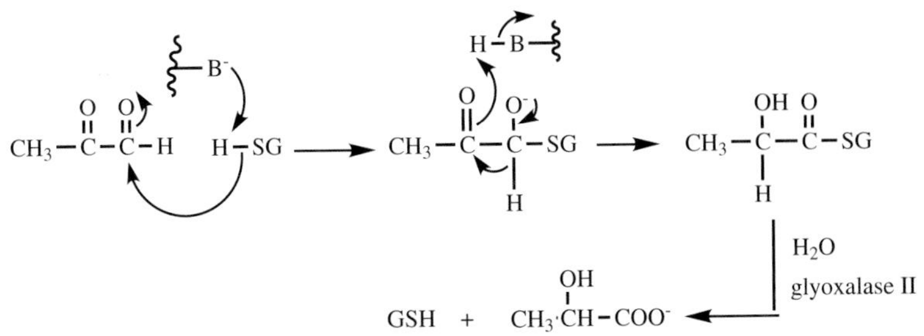

SCHEME 3.4   Hydride mechanism for glyoxalase.

Consider two mechanisms, a hydride mechanism (Scheme 3.4)[2] and a proton transfer (*cis*-enediol [**3.5**]) mechanism (Scheme 3.5).[3] Evidence for a hydride mechanism is that when the enzyme reaction is run in ^3H_2O, the lactate formed contains less than 4% of the expected tritium. As noted, this does not necessarily eliminate a proton transfer mechanism, which would be possible if no solvent protons were exchanged at the active site. An NMR experiment provided evidence for a proton transfer mechanism. The enzyme reaction was followed by NMR at 25 °C in ^2H_2O, and 15% deuterium was incorporated into the product; at 35 °C, 22% deuterium was incorporated. At higher temperatures, there is more motion of the enzyme molecules and therefore more opportunities for solvent to get into the active site and exchange with the active-site bases, leading to more deuterium incorporation. Increased temperature should have no effect on a hydride mechanism.

Another clever experiment that supports an enediol (proton transfer) mechanism takes advantage of the facts that the van der Waals radius of fluorine (1.35 Å) is close in size to that of hydrogen (1.2 Å), and fluorine also acts as a leaving group.

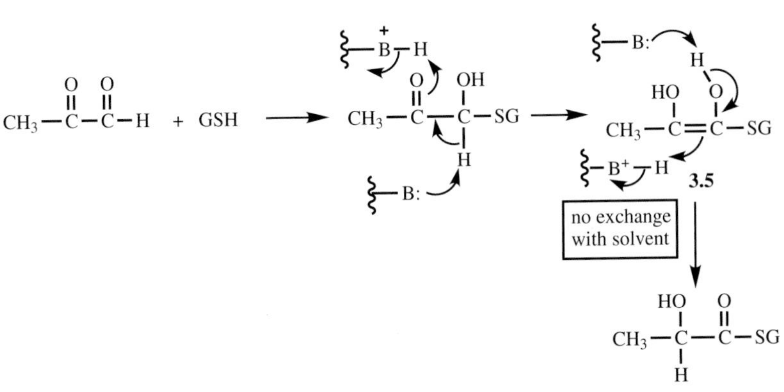

SCHEME 3.5   Enediol mechanism for glyoxalase.

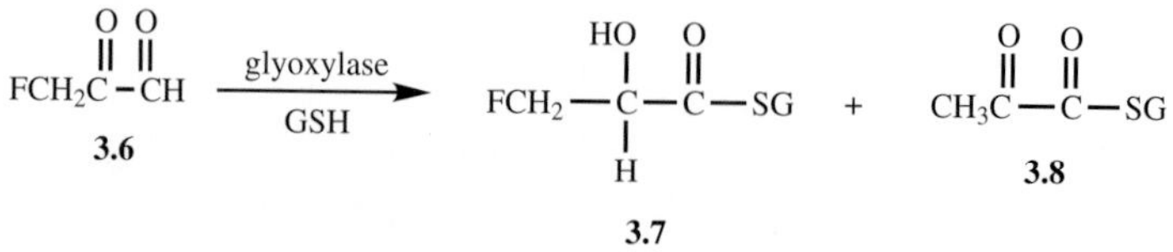

SCHEME 3.6   Reaction of glyoxalase with fluoromethylglyoxal.

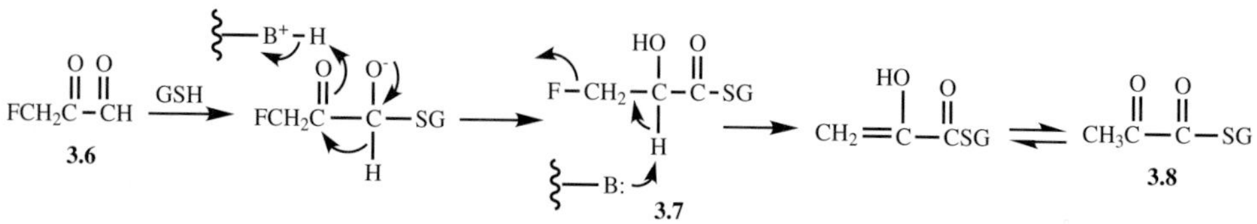

SCHEME 3.7   Hydride mechanism for the reaction of glyoxalase with fluoromethylglyoxal.

Glyoxylase I was incubated with fluoromethylglyoxal (**3.6**, Scheme 3.6), and two products were observed: the fluorolactate thioester **3.7** and the pyruvate thioester **3.8**. Given that with **3.6** as the substrate, **3.7** and **3.8** are the products, let's reconsider the hydride mechanism (Scheme 3.7) and proton transfer mechanism (Scheme 3.8). These two mechanisms can be differentiated by using a deuterated fluoromethylglyoxal analogue (**3.9**). Both reactions should be slower, because there should be a deuterium isotope effect in the first step regardless of whether it is hydride transfer or proton transfer. However, the second step determines the product ratio. In the hydride mechanism (Scheme 3.9), there should be a deuterium isotope effect on fluoride elimination, making it less favorable than when there is

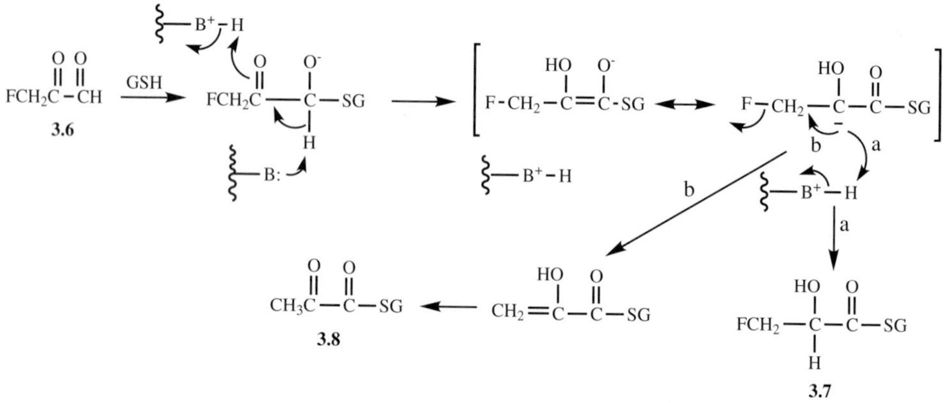

SCHEME 3.8   Enediol mechanism for the reaction of glyoxalase with fluoromethylglyoxal.

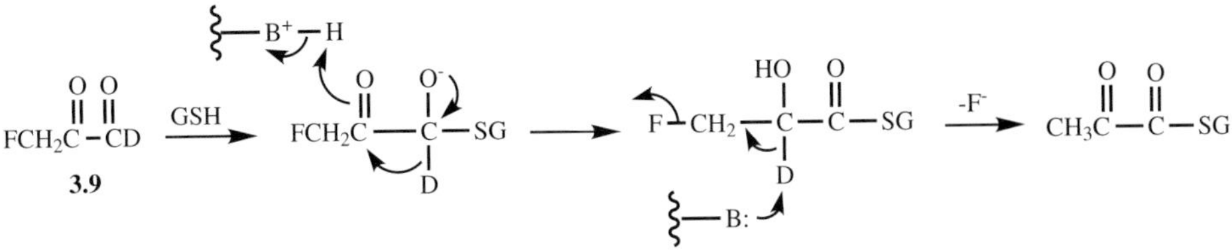

SCHEME 3.9   Hydride mechanism for the reaction of glyoxalase with deuterated fluoromethylglyoxal.

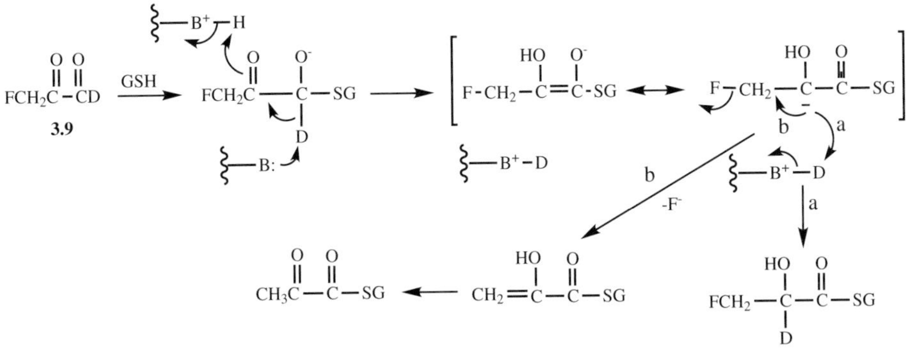

SCHEME 3.10   Enediol mechanism for the reaction of glyoxalase with deuterated fluoromethylglyoxal.

no deuterium present. However, in the proton transfer mechanism (Scheme 3.10) the isotope effect is on reprotonation, so fluoride ion elimination becomes more favorable than in the absence of deuterium. Therefore, if the percentage of fluoride ion elimination increases when the deuterated substrate is used, the proton transfer mechanism is supported. The experiment was carried out with glyoxalase from yeast, rat, and mouse, and the results are given in Table 3.1.[4] In every case the

TABLE 3.1   Comparison of Fluoride Ion Elimination with Fluoromethyl Glyoxal and [1-²H]Fluoromethyl Glyoxal

Source	% Fluoride Ion Elimination	
	FCH_2C-CH	FCH_2C-CD
Yeast	32.2 ± 0.2	40.7 ± 0.2
Rat	7.7 ± 0.1	13.3 ± 0.9
Mouse	26.4 ± 1.0	34.8 ± 0.5
Yeast/D_2O	33.8 ± 0.2	39.1 ± 0.4

deuterated substrate increased fluoride ion elimination, thus supporting a proton transfer mechanism. The experiment done with the yeast enzyme in D_2O gave the same results as in H_2O, as expected for an enzyme in which there is little or no exchange of solvent protons with protons at the active site of the enzyme.

III. REDOX REACTIONS THAT REQUIRE COENZYMES

A. Nicotinamide Coenzymes (Pyridine Nucleotides)

1. General

The *pyridine nucleotide coenzymes* include *nicotinamide adenine dinucleotide* (NAD$^+$, **3.10a**), *nicotinamide adenine dinucleotide phosphate* (NADP$^+$, **3.10b**), reduced nicotinamide adenine dinucleotide (NADH, **3.11a**), and reduced nicotinamide adenine dinucleotide phosphate (NADPH, **3.11b**).[5] Although the nucleotide part is very important for binding to the active site, the chemistry of these coenzymes derives from the pyridine part of the coenzyme; consequently, these oxidation and reduction coenzymes will be abbreviated as shown in **3.12** and **3.13,** respectively. Also, there is no difference in function between NAD$^+$ and NADP$^+$, but some enzymes prefer one coenzyme form and other enzymes prefer the other coenzyme form; generally, a particular enzyme will not be active with both coenzymes. In the older literature, NAD$^+$ and NADP$^+$ are referred to as DPN (diphosphopyridine nucleotide) and TPN (triphosphopyridine nucleotide), respectively.

The pyridine nucleotide coenzymes are bound to the active site of the enzymes by noncovalent interactions. An enzyme without its coenzyme is known as an *apo-*

3.10   a, R' = H   **3.11**
                           b, R' = PO$_3^=$

3.12   **3.13**

enzyme; with the coenzyme, it is the *holoenzyme.* For enzymes with noncovalently bound coenzymes, the coenzyme often can be removed by dialysis or gel filtration to give the apoenzyme. The process of reinserting the coenzyme (or a modified coenzyme) into the apoenzyme to give the holoenzyme (or a modified holoenzyme) is known as *reconstitution.*

Coenzymes typically are derived from *vitamins* (compounds essential to our health, but not biosynthesized; therefore, we must get them from our diet). The pyridine nucleotide coenzymes are derived from nicotinic acid (*vitamin B$_3$,* also known as *niacin* so it is not confused with nicotine) (**3.14**, Scheme 3.11), which is converted into NAD$^+$ with some enzyme-catalyzed reactions that we have discussed in Chapter 2.[6] Nicotinate phosphoribosyltransferase catalyzes the conversion of niacin and 5′-phosphoribosyl-1-pyrophosphate (PRPP, **3.15**) to nicotinate mononucleotide (**3.16**) (this is an S$_N$2-type reaction; see Chapter 6, Section II). Nicotinate adenine dinucleotide (**3.17**) comes from the reaction of **3.16** with ATP (see Chapter 2, Section II.A.2.b), catalyzed by nicotinate–nucleotide adenylyltransferase. NAD$^+$ synthetase catalyzes the conversion of **3.17** to NAD$^+$ (**3.18**) with glutamine and ATP (see Chapter 2, Section II.A.2.b). Pellagra is a nutritional disease caused by a deficiency of niacin.

The oxidation potential for the NAD$^+$/NADH couple is -0.32 V, which allows it to catalyze the reactions shown in Figure 3.1.

2. Mechanism

The pyridine nucleotide-dependent alcohol dehydrogenases constitute a family of enzymes that catalyzes the reversible oxidation of alcohols to the corresponding aldehydes (Scheme 3.12). As we will see, the evidence supports a hydride mechanism; unlike solution chemistry, the hydride only adds to the 4-position of the pyridinium ring in the enzymatic reaction.

If the reaction is run in ^{3}H$_2$O, *no* ^{3}H is found in the NADH produced. This suggests a hydride mechanism (or a proton transfer mechanism with no exchange of solvent). Conversely, if [1-^{2}H]- or [1-^{3}H]alcohol is used in the enzyme reaction in H$_2$O, the deuterium or tritium is found exclusively in the NADH; none is found in the H$_2$O (Scheme 3.13). The reverse also holds; label in NADH is transferred to the substrate with no incorporation of solvent. This specific reversible transfer of label from the substrate to the coenzyme is true of all NAD$^+$/NADH-dependent enzymes. Although this transfer does not require a hydride mechanism, it is strong evidence in favor of such a mechanism. Therefore, think of NAD(P)H as endogenous sodium borohydride.

To probe for radicals in NAD$^+$/NADH-dependent enzyme reactions, researchers used a cyclopropylcarbinyl radical rearrangement approach. A cyclopropylcarbinyl radical (**3.19**, Scheme 3.14) decomposes to the corresponding butenyl radical (**3.20**) at a rate of 10^8 s^{-1},[7] so you might expect that if a cyclopropylcarbinyl radical were to form in the enzyme-catalyzed reaction, the cyclopropyl group would

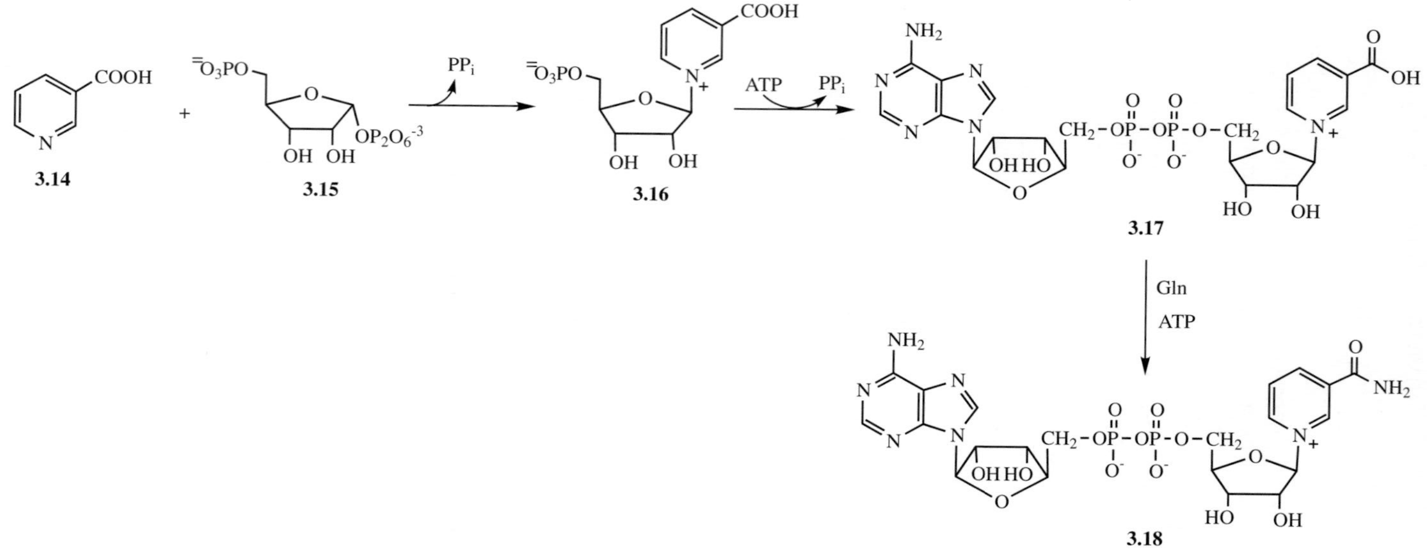

SCHEME 3.11   Biosynthesis of nicotinamide adenine dinucleotide.

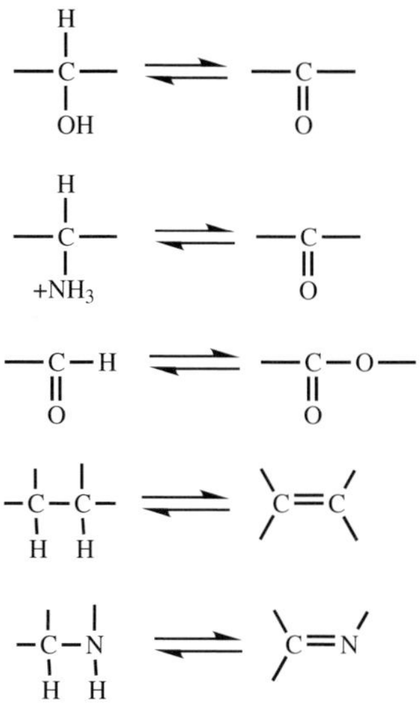

FIGURE 3.1 Reactions catalyzed by pyridine nucleotide-containing enzymes.

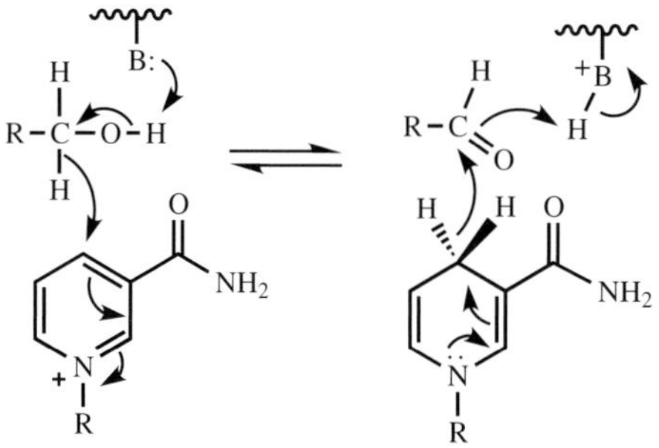

SCHEME 3.12 Reaction catalyzed by alcohol dehydrogenases.

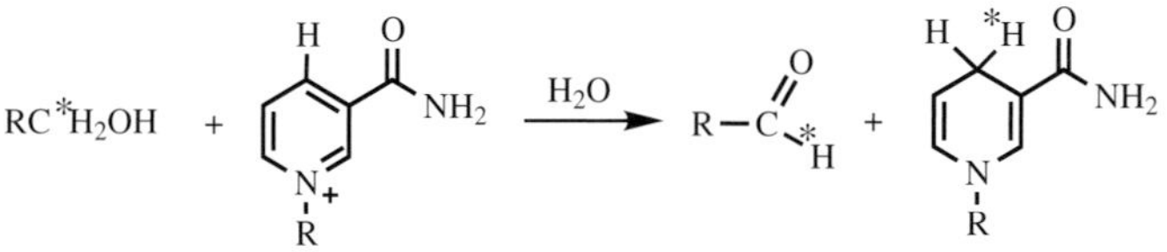

SCHEME 3.13 Reaction catalyzed by alcohol dehydrogenases using labeled alcohol.

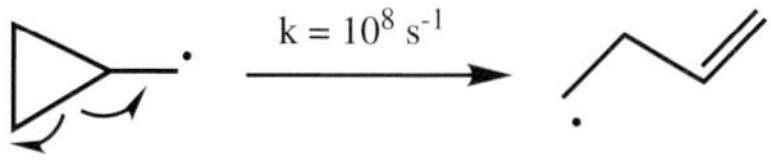

3.19 **3.20**

SCHEME 3.14 Cyclopropylcarbinyl radical rearrangement.

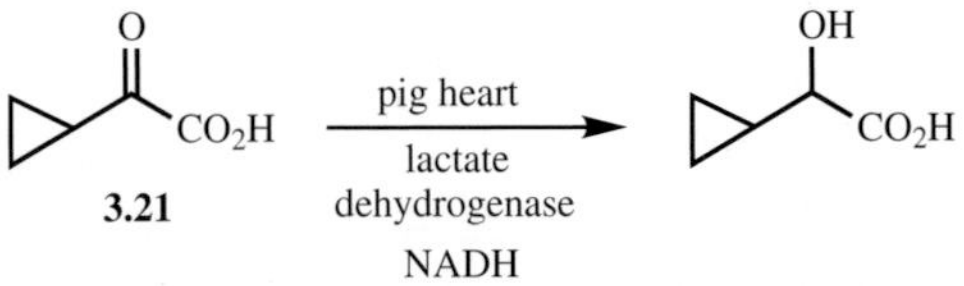

SCHEME 3.15   Test for the formation of a radical intermediate with lactate dehydrogenase.

not remain intact. The cyclopropyl analogue of pyruvate (**3.21**, Scheme 3.15) was used as a substrate to probe the formation of a radical intermediate, and no cyclopropyl ring cleavage product was observed.[8] As a chemical model for this reaction, the corresponding ester was subjected to conditions known to generate the cyclopropylcarbinyl radical (Scheme 3.16), and in fact complete cyclopropyl ring cleavage resulted. This cleavage is further evidence for a hydride mechanism in the enzyme-catalyzed reaction. Later, though, researchers found that the particular cyclopropylcarbinyl radical that was studied in those experiments is unusually stable; in fact, the corresponding O-silyl analogue (**3.22**) produced a cyclopropylcarbinyl radical that was seen in the EPR spectrum up to 408 K (Scheme 3.17)![9]

Another test for radical intermediates involved the use of α-halo ketones. The nonenzymatic reaction of α-chloroacetophenone (**3.23**, Scheme 3.18) with NADH gave the radical reduction product, namely, acetophenone (**3.24**).[10] In the presence of horse liver alcohol dehydrogenase (HLADH), however, only a trace of acetophenone was formed; the major products from a series of α-haloacetophenones were optically active 1-phenyl-2-haloethanol (**3.25**, Scheme 3.19), suggesting an enzyme-catalyzed hydride reduction of the ketone. When this study was extended

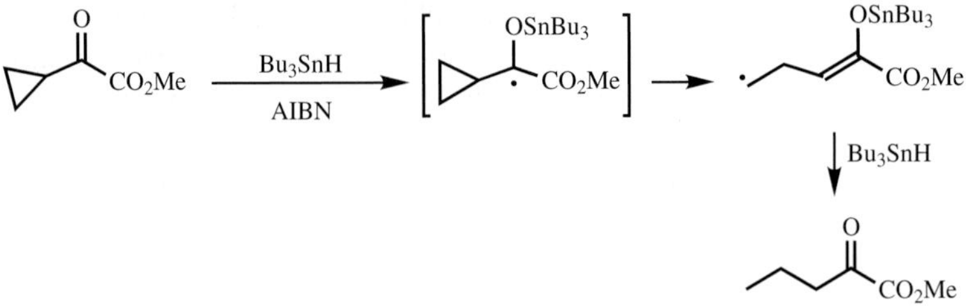

SCHEME 3.16   Chemical model for the potential formation of a cyclopropylcarbinyl radical during the lactate dehydrogenase-catalyzed reaction.

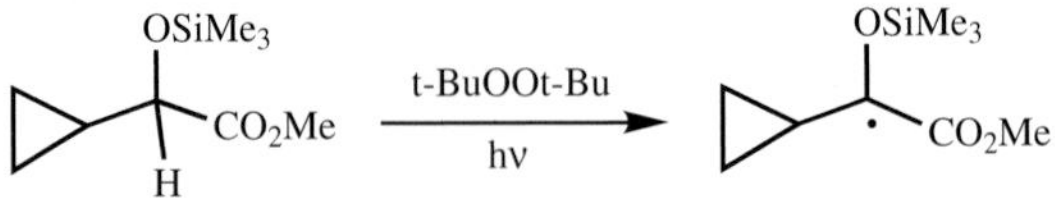

SCHEME 3.17   Formation of a stable cyclopropylcarbinyl radical.

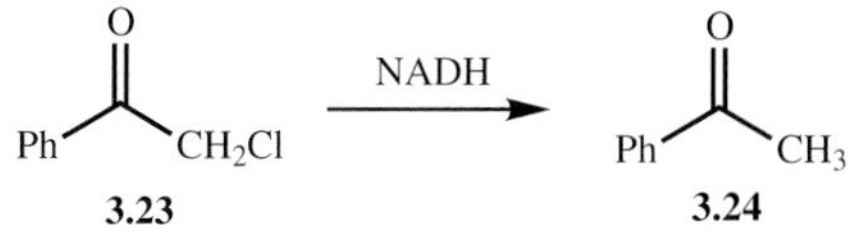

SCHEME 3.18   Nonenzymatic reduction of α-chloroacetophenone.

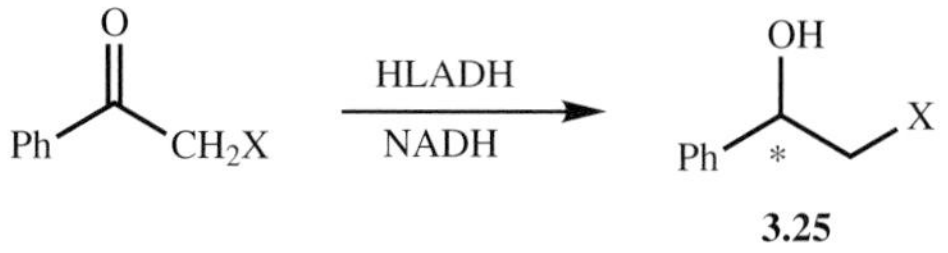

SCHEME 3.19   Horse liver alcohol dehydrogenase-catalyzed reduction of α-haloacetophenones.

to the entire α-haloacetophenone series (X = F, Cl, Br, I) with baker's yeast alcohol dehydrogenase, surprising results were observed.[11] Whereas the fluoro, chloro, and bromo analogues gave only chiral 1-phenyl-2-haloethanols (confirming Tanner and Stein's earlier work), the iodo analogue gave a 2:1 mixture of acetophenone and chiral 1-phenylethanol (both are reduction products). When the radical scavenger m-dinitrobenzene was added, the acetophenone formation was inhibited, and some 1-phenyl-2-iodoethanol (**3.25,** X = I) was produced. The conclusion is that electron transfer (giving the radical intermediate) appears to compete with hydride transfer. If the electron potential of the substrate is low enough, the free-radical chain process becomes favored.

As we will see, this is going to be a general conclusion throughout the redox enzyme section: Coenzymes can undergo more than one kind of chemistry in solution. Although a particular mechanism will be catalyzed by an enzyme with its normal substrate, given an appropriately modified substrate, other mechanisms may emerge. Unfortunately, because of the efficiency of enzyme catalysis, it is often not possible to use natural substrates for detecting intermediates in enzyme-catalyzed reactions. Therefore, enzyme mechanisms typically are tested with modified (nonnatural) substrates to probe intermediate formation. It can always be said that any result obtained with a modified substrate may not be relevant to the "natural" enzyme mechanism, but this approach is the best we have and is often the only approach possible. The same can also be said about nonenzymatic mechanism studies, but the enzyme mechanism community appears to get more criticism than the physical organic chemical community for this same practice. No one experiment should be used to define a mechanism; a variety of different kinds of experiments must be carried out before conclusions about the most likely mechanism can be made. Of course, *no* mechanism is ever *proved;* at best, you can say, given the current data, that the most likely mechanism is X. As soon as you publish the "definitive" mechanism for an enzyme, you challenge half a dozen other enzymologists to prove you wrong!

3. Stereochemistry

As discussed in Chapter 1 (Section I.B.3), because enzymes are chiral (made of chiral amino acids), stereospecific reactions are typical for enzyme-catalyzed reactions. Before we discuss the stereochemistry of the pyridine nucleotide coenzymes, I think it will be worthwhile to discuss briefly some basic stereochemistry nomenclature. If you know the rules for naming stereogenic sp^3 centers (the *R*- and *S*-rules), about *prochirality,* and the rules for naming sp^2 carbon (carbonyl and methylene) chirality, then skip the next three paragraphs.

The rules for the Cahn–Ingold–Prelog or *R,S* nomenclature system are enumerated briefly here: [12]

1. Assign priorities to substituents attached to the stereogenic center in order of decreasing atomic number: $O > N > C > H$.
2. Put the lowest-priority group in back of the tetrahedron.
3. Arrange the other three substituents in decreasing priority. If the descending order follows a clockwise direction, then the stereogenic center has *R* chirality; if it follows a counterclockwise direction, then it is *S*. As an example, the alanine isomer in Figure 3.2 has the *S* chirality: N has the highest priority, then C, then H. Because there are two C substituents, we look to the next atom out; the carboxylate has an oxygen, but methyl has only hydrogens, so the carboxylate is second highest priority and methyl third. The lowest-priority atom, H, is pointing toward the back of the tetrahedron. The direction of the highest three substituents (N, COO^-, and CH_3) is counterclockwise. Therefore, the stereogenic center has *S*-chirality.

A concept that is generally not covered in a basic organic chemistry course is *prochirality.* An atom is prochiral if by changing one of its substituents, it changes from achiral to chiral (Figure 3.3). In changing substituent a to b, priorities must remain in the order $d > c > b > a$. If the stereogenic center generated by that change has *R* chirality, then the "a" group that was changed to "b" is called the *pro-R* group; if *S,* then it is the *pro-S* group. When two hydrogen atoms are involved, it is best to make one a deuterium to determine the prochirality.

A stereochemical nomenclature that has probably escaped all students but those who took an advanced organic class is sp^2 carbon chirality. Here we will discuss carbonyl and methylene chirality; in Chapter 7 (Section II.B.1) we will discuss rules for defining the chirality of higher-substituted alkenes.

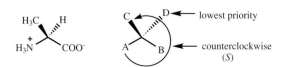

FIGURE 3.2   Determination of the chirality of an isomer of alanine.

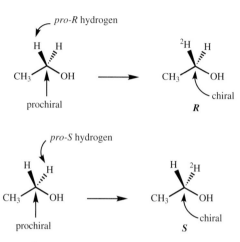

FIGURE 3.3   Determination of prochirality.

1. Determine the priorities of the three substituents attached to the sp² carbon in descending order according to the *R,S* rules.

2. If the priority sequence of the attached groups is clockwise looking down from top, then the top is the *re* face; if it is counterclockwise, then it is the *si* face (Figure 3.4). In Figure 3.4 the alkene has the stereochemistry shown because the CH₂ has higher priority than the CH₃ group. A double bond means that you count, as the third substituent, the group attached to the other end of the double bond. Therefore, CH₂ also has a C connected to it, and CCH₂ has higher priority than CH₃.

The reason we went into this discussion of stereochemistry is that pyridine nucleotide-dependent enzymes are stereospecific. The following yeast alcohol

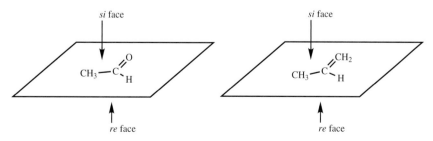

FIGURE 3.4   Determination of carbonyl and alkene chirality.

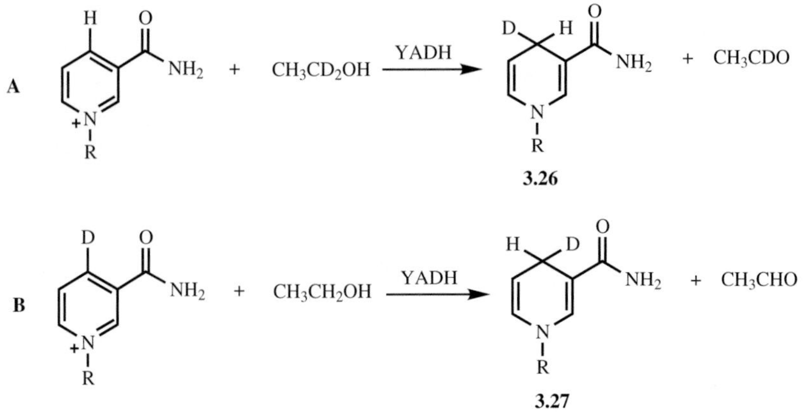

SCHEME 3.20   Reaction of yeast alcohol dehydrogenase with (**A**) $[1,1-^2H_2]$ethanol and NAD$^+$ and (**B**) ethanol and $[4-^2H]$NAD$^+$.

dehydrogenase (YADH)-catalyzed reactions are representative of their stereospecificity.[13] Two different specifically deuterated NADH coenzymes were prepared enzymatically in two ways, one from $[1,1-^2H_2]$ethanol and NAD$^+$ (**A**, Scheme 3.20), giving deuterated NADH **3.26**, and the other from $[4-^2H]$NAD$^+$ and ethanol (**B**), giving deuterated NADH **3.27**. If *apo*-YADH is reconstituted with **3.26**, and incubated with acetaldehyde (Scheme 3.21A), the deuterium is lost from the NAD$^+$ and is found in the ethanol (**3.28**) produced. If **3.27** is reconstituted into *apo*-YADH and incubated with acetaldehyde, the deuterium remains in the NAD$^+$ and is not transferred to the ethanol produced (**B**). Furthermore, if the deuterated ethanol (**3.28**) produced in **A** is incubated with YADH in the NAD$^+$ form (**C**), the acetaldehyde generated contains no deuterium; all the deuterium is in the NADH, which can be shown to be identical with **3.26**. Therefore, the hydride is added to and removed from NAD$^+$–NADH on only one side of the molecule with absolute stereospecificity (**3.29**). The *pro-R* hydrogen is on the *re* face, and the *pro-S* hydrogen is on the *si* face (you also may see in the literature the *re* face referred to as the A side and the *si* face called the B side). Likewise, there is absolute stereospecificity for the α-H of the substrate alcohol. For yeast alcohol dehydrogenase, the active hydrogen is the *pro-R* hydrogen.

Although the stereospecificity is absolute for all the nicotinamide-dependent enzymes, they do not all remove the same hydrogen. For example (Scheme 3.22), if $[1,1-^2H_2]$ethanol is incubated with YADH to give **3.26** (reaction **A**), then the enzyme is denatured, and the isolated **3.26** is reconstituted into *apo*-glyceraldehyde 3-phosphate dehydrogenase (G3PDH), which is run in the reverse direction (**B**), then all the deuterium remains in the NAD$^+$; none is in the glyceraldehyde 3-phosphate (**3.30**) produced. This enzyme uses the *pro-S* hydrogen of NADH stereospecifically.

The transition state for hydride transfer involves a boatlike conformation.[14] Enzymes that catalyze hydride transfer from the A-side (*re* face) are known to bind the

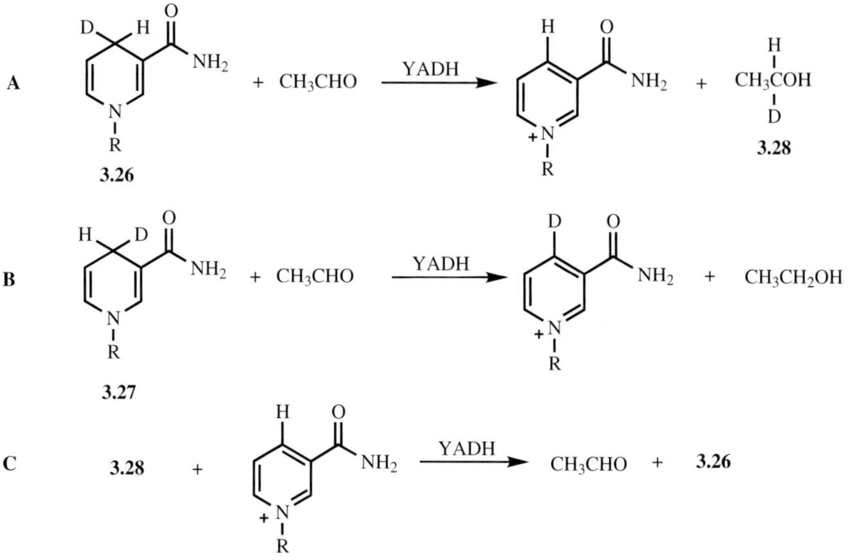

SCHEME 3.21   Reaction of yeast alcohol dehydrogenase with (**A**) $[4\text{-}^2H]NAD^2H$ prepared in Scheme 3.20A; (**B**) reaction of yeast alcohol dehydrogenase with $[4\text{-}^2H]NAD^2H$ prepared in Scheme 3.20B; (**C**) reaction of yeast alcohol dehydrogenase with **3.28** and NAD^+.

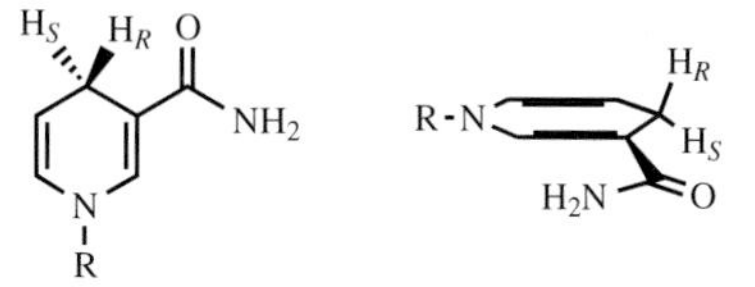

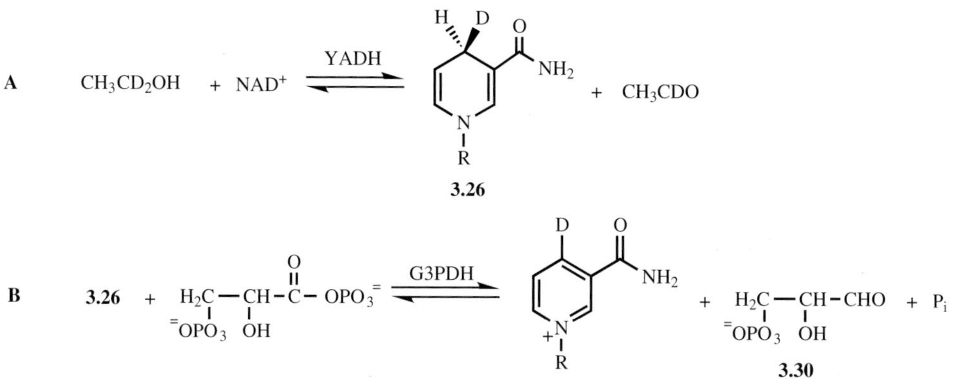

SCHEME 3.22   (**A**) Reaction of yeast alcohol dehydrogenase with $[1,1\text{-}^2H_2]$ethanol and NAD^+; (**B**) reaction of glyceraldehyde-3-phosphate dehydrogenase with the cofactor produced in A and glycerate 1,3-diphosphate.

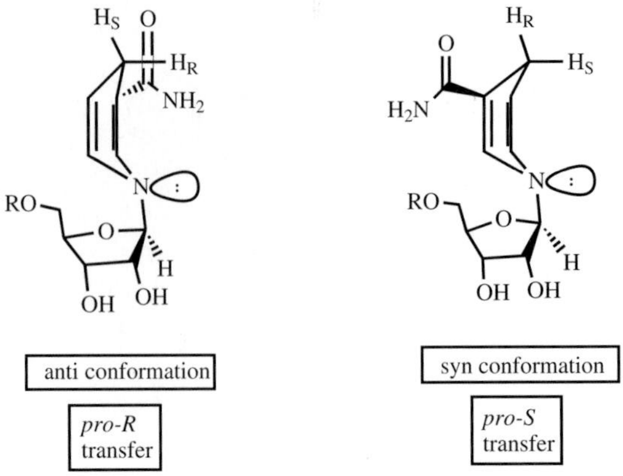

anti conformation

pro-R
transfer

syn conformation

pro-S
transfer

FIGURE 3.5   *Anti-* and *syn-*conformations of NADH.

cofactor with the carboxamide of the NADH *anti* to the ribose, whereas B-specific (*si*-face) enzymes bind the *syn* conformation (Figure 3.5). Transfer of the axial C-4 hydrogen is assisted by a *syn*-axial lone pair of the N-1 nitrogen. On the basis of a molecular dynamics study, Young and Post showed that the NADH coenzyme undergoes thermal conformational fluctuations on binding to the enzyme lactate dehydrogenase.[15] As a result of the correlation between the NADH torsion angle (from either *anti* or *syn* conformations) and the stereospecificity of hydride trans-fer, a catalytic function of the enzyme could be to drive the appropriate NADH conformer to the transition state for hydride transfer, thereby reducing the entropic barrier to activation (Figure 3.6). The $\Delta G°$ of interconversion of the two con-formers was calculated to be less than 1 kcal/mol, so this transformation is easy for the enzyme to catalyze.

Several groups have proposed hypotheses to account for the difference in stereo-specificity of the nicotinamide-dependent enzymes, but none seems to fully account for the results. The simplest, and probably most accepted hypothesis, is that early dehydrogenases merely carried out a reaction with either *pro-R* or *pro-S* stereo-specificity initially, and then all enzymes that evolved from these primordial enzymes retained this original stereospecificity.[16]

Another hypothesis, proposed by Benner and co-workers, is that thermodynami-cally unstable (i.e., more reactive) carbonyls are reduced with the *pro-R* hydro-gen of NADH, whereas thermodynamically stable carbonyls are reduced with the *pro-S* hydrogen.[17] To be valid, the stereoelectronic arguments used for this hypothe-sis must meet three criteria: (1) the enzyme must catalyze the interconversion of "simple" unconjugated carbonyls with the corresponding alcohol; (2) the "natural" substrate must be well defined; and (3) the equilibrium constant for the overall re-action (Scheme 3.23) must lie at least one log unit away from the position of the "break" in the values of $-\log K_{eq}$ between the reactions catalyzed by the *pro-R* and

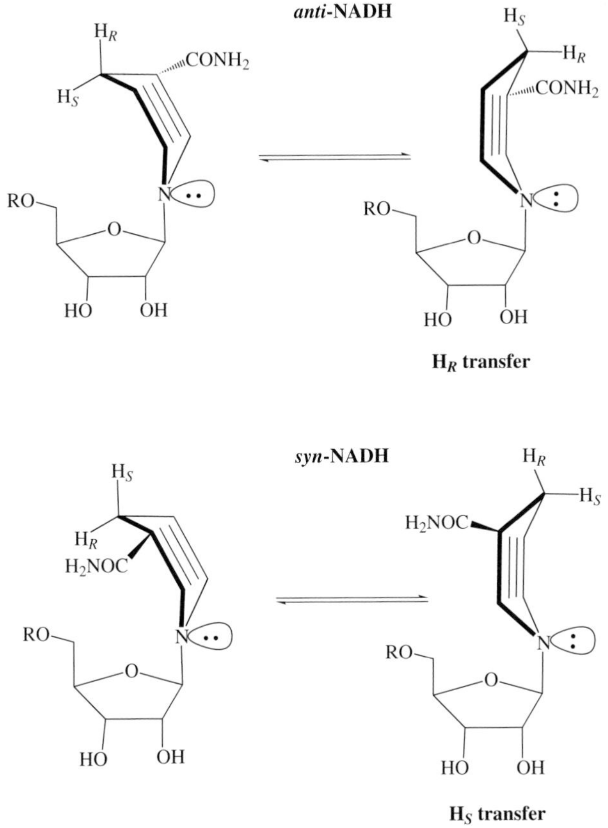

FIGURE 3.6   Boat-boat equilibria of NADH.

$$\begin{array}{c}\diagup\\C=O\\\diagup\end{array} + \; NADH \;\; \rightleftharpoons \;\; \begin{array}{c}\diagup\\CHOH\\\diagup\end{array} + \; NAD^+$$

$$K_{eq} = [C=O][NADH][H^+] \, / \, [C\text{-}OH][NAD^+]$$

SCHEME 3.23   Equilibrium of a carbonyl compound and NADH with the corresponding alcohol and NAD$^+$.

pro-S specific enzymes. For the reactions studied, the $-\log K_{eq}$ values were greater than 11.1 for the pro-R specific enzymes and less than 11.2 for pro-S specific enzymes. Other assumptions necessary to validate this hypothesis are that (1) the pro-R hydrogen is transferred from NAD(P)H in an anti conformation, and the pro-S hydrogen is transferred in a syn conformation; (2) anti–NADH is a weaker reducing agent than syn–NADH; (3) the favored conformation is the boat form for overlap of the nitrogen lone pair with the anomeric carbon antibonding orbital; (4) the axial hydrogen is the one transferred from the NADH; and (5) optimal enzymes bind substrates to match the free energies of the bound intermediates (*anti* and *syn* forms have different energies).

Oppenheimer[18] and You[19] claimed to have found numerous "exceptions" to Benner's rules. However, Benner then counterclaimed that theirs were not proper exceptions.[20]

Reaction trajectories were calculated from the X-ray structure of the nicotinamide-dependent enzyme lactate dehydrogenase using both *anti* and *syn* conformations of NADH. The difference in the AM1 potential energies of the two transition states from the two conformers is less than 1 kcal/mol; that is, the *anti* and *syn* forms are comparable in their reducing ability. This would not be sufficient to account for Benner's stereoelectronic hypothesis.

From molecular dynamics simulation experiments by Bruice and co-workers using $CHARM_m$ and AM1 calculations, a very late transition state is proposed for general acid proton transfer to the substrate carbonyl oxygen and a midway transition state for hydride transfer to the substrate carbonyl carbon (Figure 3.7).[21] The conclusion from this work is that preference for transfer of the *pro-R* or *pro-S* hydrogen of NADH is determined by the shielding of one side of the dihydropyridine ring by the enzyme, which blocks access of the substrate to, and dynamic deformation at, the unblocked side of the dihydropyridine ring.

Another hypothesis regarding the specificity of hydrogen transfer is that the zinc-containing enzymes with a higher subunit molecular weight are *pro-R* specific, and the ones without zinc and having a lower subunit molecular weight are *pro-S* specific.[22]

Of course, none of these hypotheses should be accepted fully. There may be a combination of explanations for the stereospecificity of nicotinamide-dependent enzymes, or maybe it is just serendipitous! In the case of a crotonyl CoA reductase from *Streptomyces collinus,* Reynolds and co-workers showed that the *pro-4S* hydrogen of NADPH is transferred to the *re* face of the β-carbon of crotonyl CoA.[23] The other four known enoyl thioester reductases transfer the *pro-4S* hydrogen of

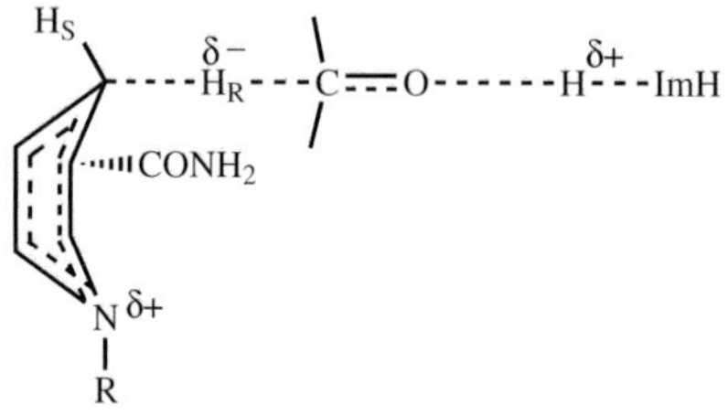

FIGURE 3.7   Late transition state proposed for the reactions catalyzed by NADH-dependent enzymes.

NADPH to the *si* face of their respective substrates. The amino acid sequences of the other four enzymes are unrelated to that of crotonyl CoA reductase, which suggests that the stereochemical course of, at least, an enoyl thioester reductase exhibits no reactivity advantage; it merely reflects the ancestral lineage of the enzyme.

Klibanov[24] has shown that horse liver alcohol dehydrogenase also is active in organic solvents. The enzyme containing NADH in buffer can be mixed with glass powder and allowed to air dry, which forms a glass-adsorbed enzyme that is capable of catalyzing carbonyl reduction in buffer-saturated ethyl acetate containing a little ethanol to regenerate the cofactor (the ethanol converts the NAD^+ that is formed back to NADH to continue the catalysis).

4. Oxidation of Amino Acids to α-Keto Acids

The family of $NAD(P)^+$–amino acid dehydrogenases catalyze the reversible oxidative deamination of various amino acids to the corresponding keto acid and ammonia.[25] The proposed mechanism for one member of the family, glutamate dehydrogenase, whose crystal structure is known,[26] is shown in Scheme 3.24.

5. Oxidation of Aldehydes to Carboxylic Acids

Aldehyde dehydrogenases typically contain an active-site cysteine, which is generally thought to serve two functions: (1) to form a thiohemiacetal that anchors the

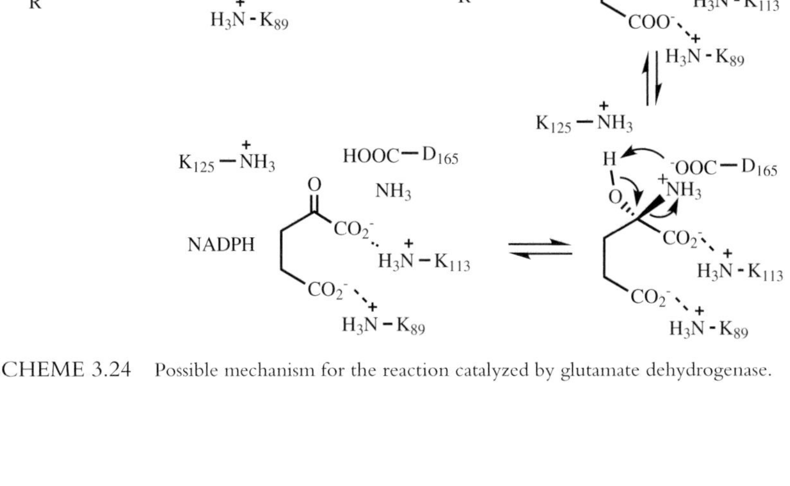

SCHEME 3.24   Possible mechanism for the reaction catalyzed by glutamate dehydrogenase.

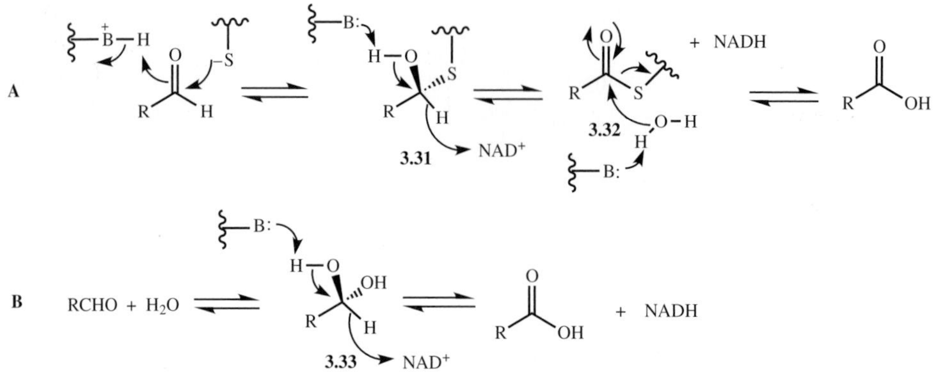

SCHEME 3.25   (A) Covalent catalytic mechanism for the oxidation of aldehydes by aldehyde dehydrogenases; (B) noncovalent catalytic mechanism for the oxidation of aldehydes by aldehyde dehydrogenases.

substrate and (2) to activate the C-1 carbon for hydride transfer. Two mechanisms are reasonable (Scheme 3.25). Mechanism A involves covalent catalysis by the cysteine residue, oxidation of the α-hydroxy sulfide (3.31), followed by hydrolysis of the enzyme-bound thioester (3.32) to the carboxylic acid. Another mechanism can be drawn that does not involve covalent catalysis by an active-site cysteine residue (mechanism B). Instead, the addition of water gives the corresponding hydrate (3.33), which is oxidized by the NAD^+. In this case, an active-site cysteine could act as the base.

Some dehydrogenases convert alcohols to carboxylic acids, presumably via the aldehyde (or hydrate). The mechanism for histidinol dehydrogenase from *Salmonella typhimurium,* a four-electron oxidoreductase that converts the amino alcohol to the amino acid (histidine) at a single active site, presumably via the amino aldehyde, was demonstrated by site-directed mutagenesis.[27] The four active-site cysteines in histidinol dehydrogenase were mutated (C116S, C116A, C153S, and C153A). The active site contains two Zn^{2+}/dimer, presumably held by two or more of the cysteines. All of the mutants exhibited normal k_{cat} and K_m values. Therefore, the oxidation mechanism does not involve covalent catalysis by the active-site cysteine, and the cysteines do not serve a function in the enzyme's catalysis.

Some aldehyde dehydrogenases appear to use a covalent catalytic mechanism. *trans*-4-(N,N-Dimethylamino)cinnamaldehyde (3.34) and *trans*-4-(N,N-

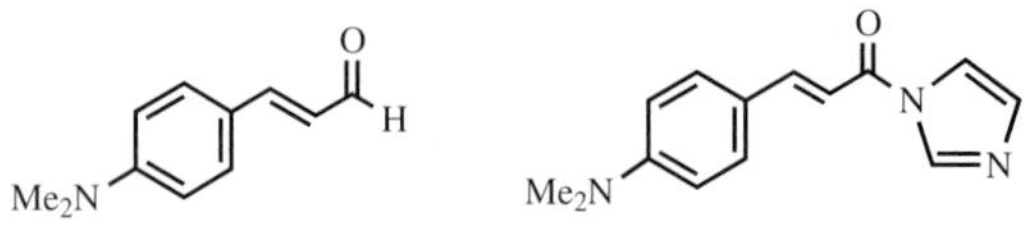

3.34   3.35

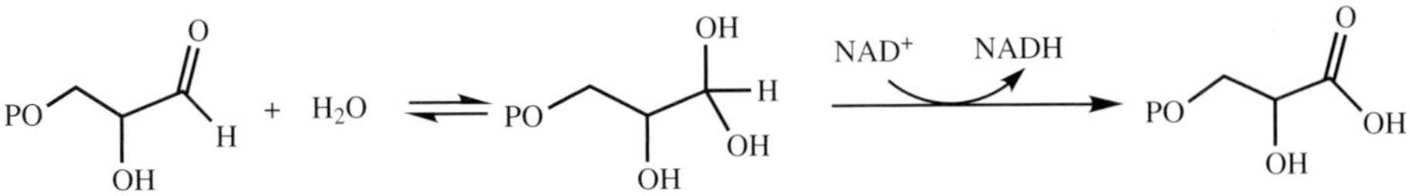

SCHEME 3.26   Mechanism for oxidation of glyceraldehyde 3-phosphate by the C149A mutant of glyceraldehyde-3-phosphate dehydrogenase.

dimethylamino)cinnamoylimidazole (**3.35**) were shown to be substrates for human aldehyde dehydrogenase that form chromophoric covalent intermediates.[28] Using ^{3}H-labeled **3.34** and **3.35,** followed by Pronase (a mixture of proteases) digestion and purification of the radioactive species, researchers showed that the labeled amino acid was a cysteine residue. Tryptic digestion of the tritiated enzyme, followed by HPLC separation of the peptides, gave a tritium-labeled peptide (residues 273–307). Cysteine-302 was identified as the labeled amino acid. This cysteine is the only cysteine conserved in all aldehyde dehydrogenases that have been sequenced and suggests that Cys-302 may be involved in covalent catalysis with aldehyde substrates in general.

The active-site Cys-149 of glyceraldehyde-3-phosphate dehydrogenase, an enzyme mentioned in Section III.A.3 (Scheme 3.22B), was mutated (C149A).[29] The mutant still catalyzed a redox reaction, but without the phosphorylating activity. This result suggests a mechanism that does not involve the active-site cysteine residue, probably oxidation from the hydrate (Scheme 3.26), is more reasonable.

6. Oxidation of Deoxypurines to Purines

Similar to the oxidation of an aldehyde to a carboxylic acid, a nitrogen analogue, a deoxypurine, such as inosine 5′-monophosphate (**3.36**) is oxidized to the purine, xanthosine 5′-monophosphate (**3.37**), in a NAD^+-dependent reaction catalyzed by inosine-5′-monophosphate dehydrogenase (Scheme 3.27). This is the rate-limiting step in guanine nucleotide biosynthesis. Evidence for the presence of an active-site nucleophile in the mechanism comes from a study with the antiviral agent 5-ethynyl-1-β-D-ribofuranosylimidazole-4-carboxamide 5′-monophosphate (**3.38**), an affinity labeling agent.[30] The inactivator was shown to form a covalent bond to Cys-305, suggesting that this cysteine may be involved in the oxidation process, supporting a mechanism similar to mechanism A in Scheme 3.25 for the NAD^+-dependent oxidation of aldehydes to carboxylic acids.

7. An Unusual Mechanism of Action of NAD^+

Up to this point, we have discussed NAD^+ as an important coenzyme for enzyme-catalyzed oxidation reactions in which NAD^+ appears to act as an electrophile for the acceptance of a hydride ion from the substrate. NAD^+, however, also has been

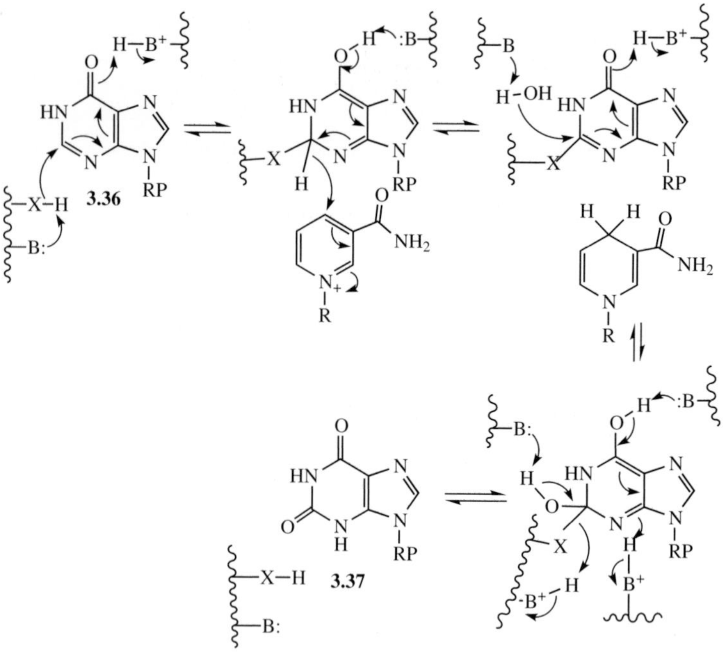

SCHEME 3.27   Mechanism for the oxidation of inosine 5'-monophosphate by inosine 5'-monophosphate dehydrogenase.

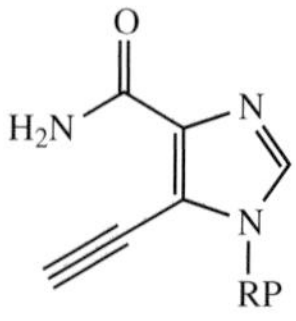

3.38

proposed as a cofactor for a nonredox reaction, catalyzed by urocanase, which converts urocanic acid (3.39) to imidazolonepropionic acid (3.40); Scheme 3.28 shows the outcome of the reaction carried out in D_2O. When reduced urocanic acid labeled with ^{13}C at the 2-position (3.41) was used by Rétey and co-workers as the substrate (actually, it can be considered an inhibitor) with urocanase that was reconstituted with [4-^{13}C]NAD$^+$ (3.42), a stable adduct between these two was isolated, which was further stabilized by oxidation with phenazine methosulfate.[31] The structure was determined by NMR spectroscopy to be 3.43. On the basis of this structure, Rétey and co-workers proposed the mechanism in Scheme 3.29. The inhibitor−NAD$^+$ adduct was observed directly without oxidation using difference

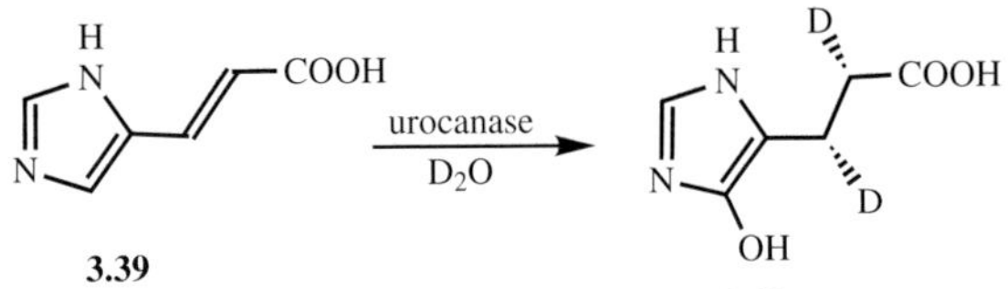

SCHEME 3.28   Reaction catalyzed by urocanase.

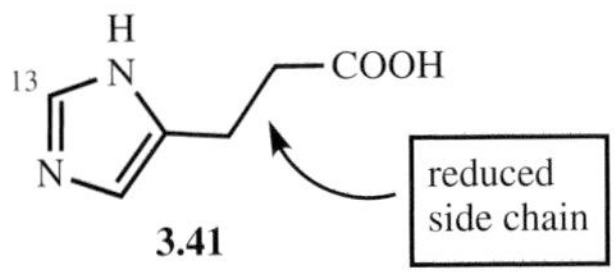

^{13}C NMR spectroscopy.[32] Position 4 of the NAD$^+$ (**3.44**) and position 5' of the reduced substrate (**3.45**) were labeled with ^{13}C. In the ^{13}C NMR difference spectrum (urocanase with NAD$^+$ after addition of **3.45** minus the same before addition of **3.45**), two ^{13}C NMR signals were observed, one at $\delta = 117$ and one at 87.5 ppm, for the free and bound imidazolepropionic acid, respectively. The large upfield shift of the C-5' signal of bound **3.45** indicates that the imidazole ring in the enzyme-bound adduct is not aromatic, suggesting that **3.46** is the structure of the adduct. This would account for the urocanase-catalyzed exchange of the C-5' proton of imidazolepropionic acid.

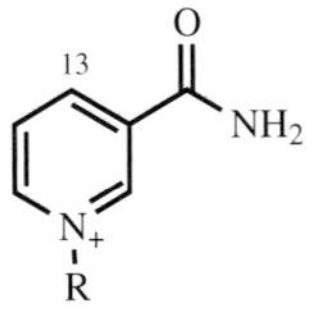

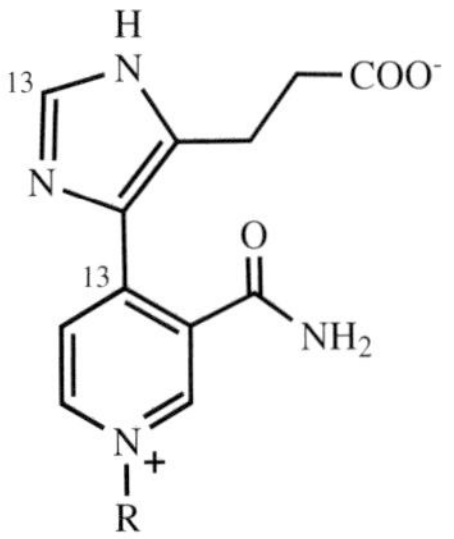

The possibility that the C-4 hydrogen atom of the NAD$^+$ undergoes a 1,5-sigmatropic shift after adduct formation was investigated by deuterium labeling (Scheme 3.30).[33] No incorporation of solvent deuterium from D$_2$O into C-4 of the cofactor was detected, which rules out a 1,5-sigmatropic rearrangement (**3.47**).

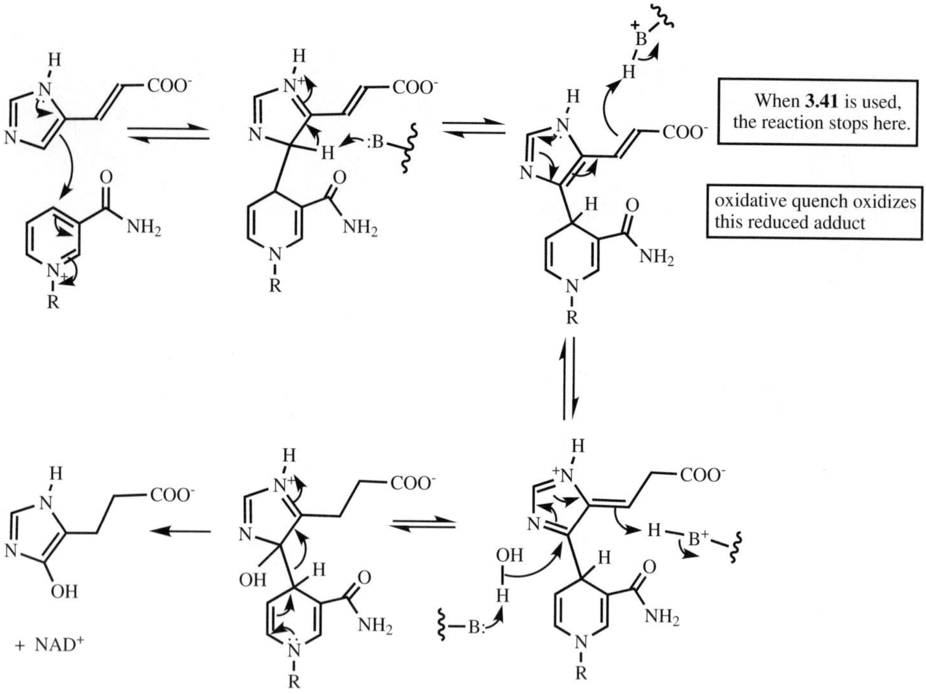

When **3.41** is used, the reaction stops here.

oxidative quench oxidizes this reduced adduct

+ NAD$^+$

SCHEME 3.29   Mechanism proposed for urocanase.

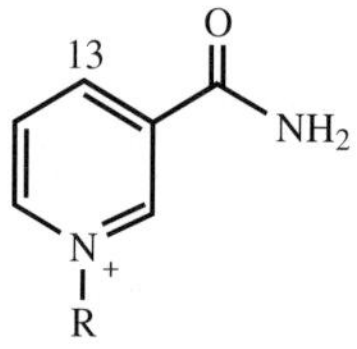

3.44

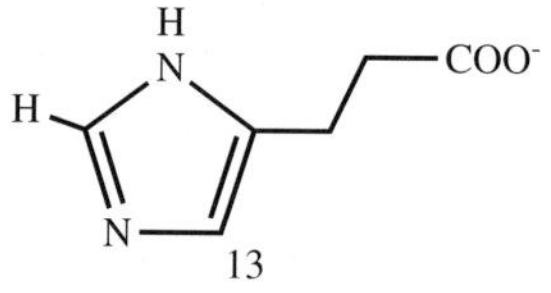

3.45

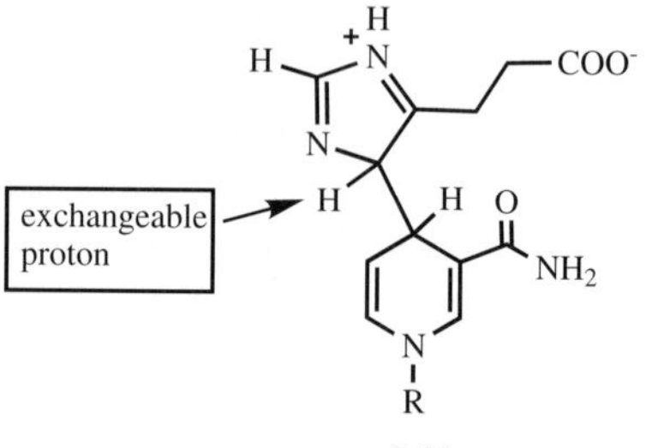

exchangeable proton

3.46

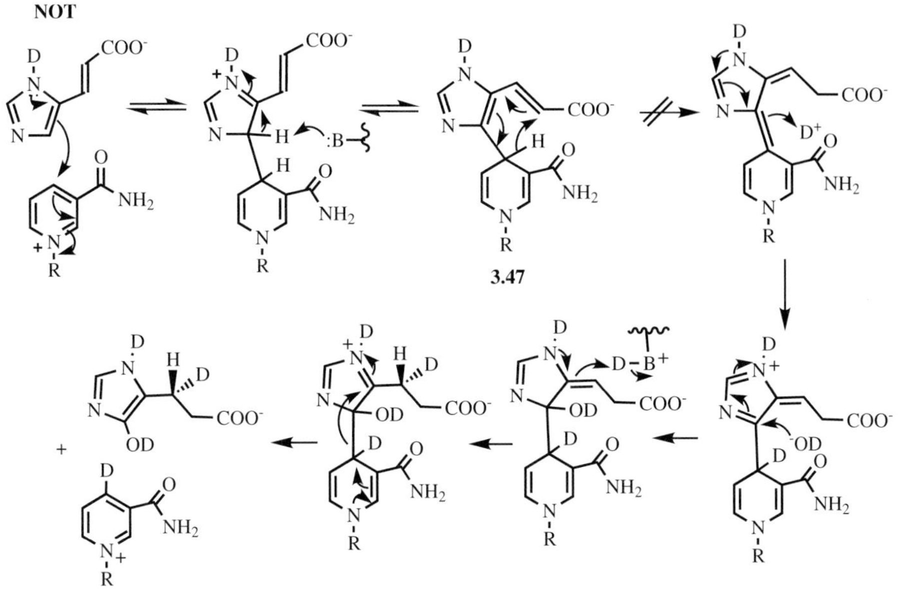

NOT

SCHEME 3.30   Excluded 1,5-sigmatropic rearrangement mechanism for urocanase.

B. Flavin Coenzymes

1. General

As in the case of the pyridine nucleotide coenzymes (see Section III.A), flavin coenzymes exist in several different forms.[34] All of the forms are derived from riboflavin (vitamin B_2, **3.48,** Scheme 3.31), which is enzymatically converted to two other forms, flavin mononucleotide (FMN, **3.49**) and flavin adenine dinucleotide (FAD, **3.50**). Both **3.49** and **3.50** appear functionally equivalent, but some enzymes use one form, some the other, and some use one of each. Because flavin is a very good electron sink, it can accept electrons. In fact, it can accept one electron to give a *flavin semiquinone* (**3.51,** Scheme 3.32) or two electrons to give *reduced flavin* (**3.52**). Note in the projection drawings in Scheme 3.32 that oxidized flavin is planar, but reduced flavin is bent along the $N^5–N^{10}$ axis. This may be important in the control of the redox state in the enzyme. The flavin semiquinone can be detected by EPR spectroscopy in some cases. Although most flavin-dependent enzymes (also called *flavoenzymes*) bind the flavin with noncovalent interactions, some enzymes have covalently bound flavins, which is a posttranslational modification of the protein, in which the flavin is attached at its 8α-position or 6-position (see **3.48** for numbering) to either an active-site histidine[35] or cysteine[36] residue.

The main oxidation reactions that flavin-dependent enzymes catalyze are shown in Figure 3.8. Flavin oxidation potentials generally are in the range of -100 to -200 mV, but they can be -500 mV (flavodoxins).

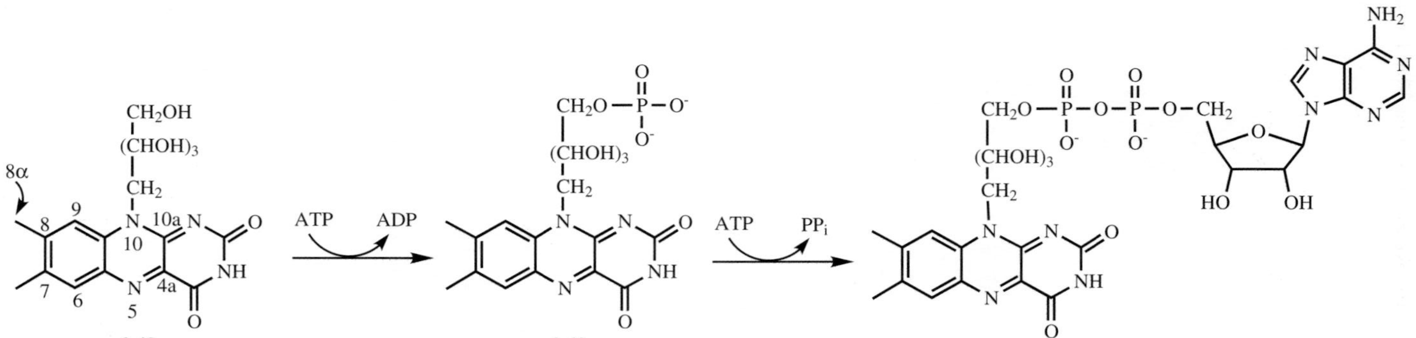

SCHEME 3.31 Biosynthetic conversion of riboflavin to FMN and FAD.

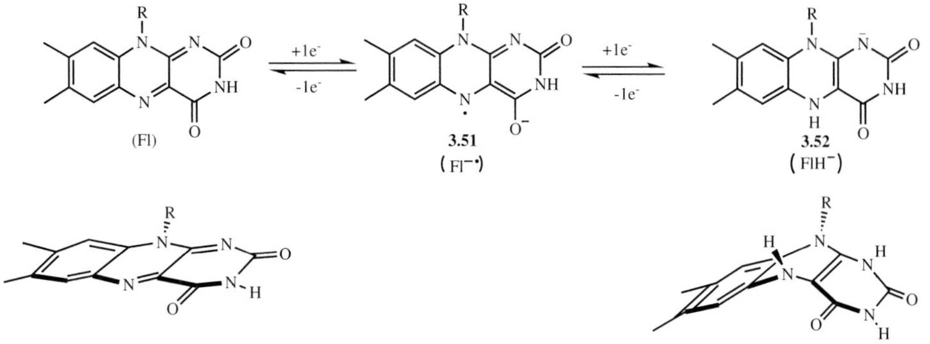

SCHEME 3.32   Interconversion of the three oxidation states of flavins.

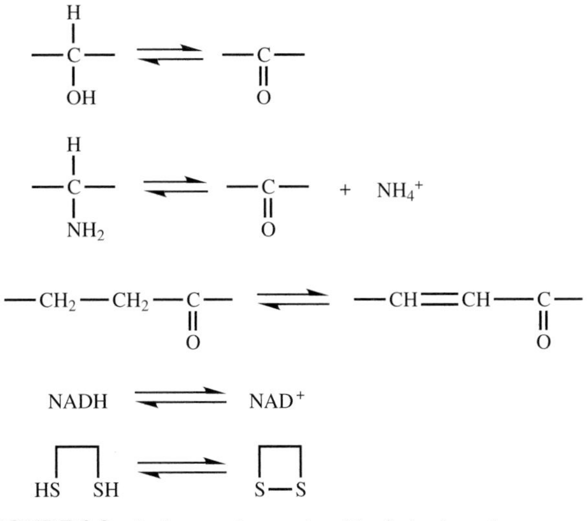

FIGURE 3.8   Redox reactions catalyzed by flavin-dependent enzymes.

2. Oxidases versus Dehydrogenases

Some flavin–dependent enzymes are called *oxidases* and others *dehydrogenases*. The distinction between these names indicates the way in which the reduced form of the coenzyme is reoxidized so the catalytic cycle can continue. Oxidases use molecular oxygen to oxidize the coenzyme with concomitant formation of hydrogen peroxide; Scheme 3.33 shows possible mechanisms for this oxidation. Pathway a shows the reaction with singlet oxygen to give a flavin hydroperoxide (**3.53**), which will only occur if there is a mechanism for spin inversion from the normal triplet oxygen, such as with a metal ion. Loss of hydrogen peroxide gives oxidized flavin. Pathway b depicts the analogous reaction with triplet oxygen, leading first

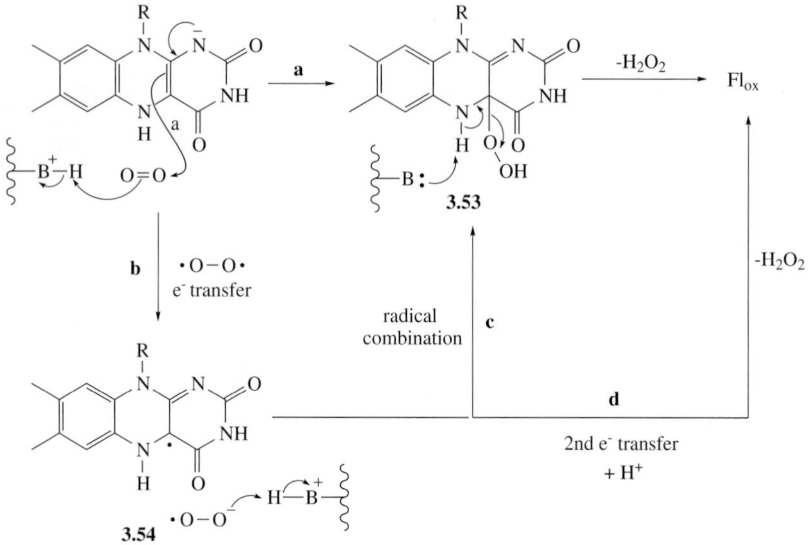

SCHEME 3.33   Mechanisms for an oxidase-catalyzed oxidation of reduced flavin to oxidized flavin.

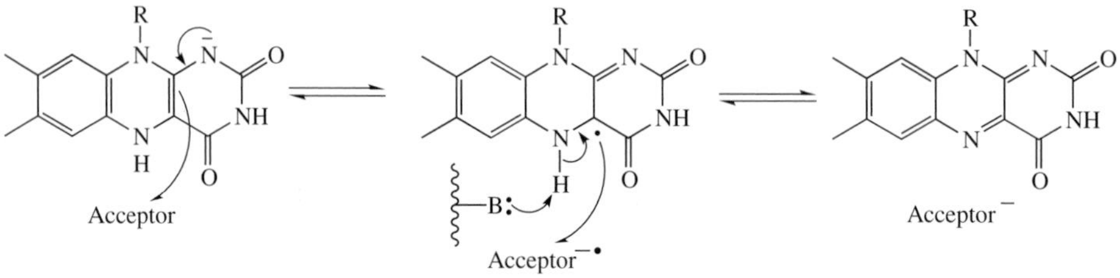

SCHEME 3.34   Mechanisms for a dehydrogenase-catalyzed oxidation of reduced flavin to oxidized flavin.

to the caged radical pair **3.54** by electron transfer, which, after spin inversion, can either undergo radical combination via pathway c to give **3.53**[37] or second electron transfer (pathway d) to go directly to the oxidized flavin. Scheme 3.34 gives a possible mechanism for dehydrogenases, which use electron transfer proteins, such as ubiquinone and cytochrome b_5 to accept electrons from the reduced flavin. Two one-electron transfers generally occur, leading to oxidized flavin. Often the reoxidation step can be observed by EPR spectroscopy.

3. Mechanisms

The total enzyme reaction occurs in at least two steps (Scheme 3.35): The first oxidizes the substrate with concomitant reduction of the flavin, and the second returns

$$\text{Substrate} \quad + \quad \text{Enzyme-Fl}_{ox} \longrightarrow \underset{\text{(product)}}{\text{Oxidized substrate}} \quad + \quad \text{Enzyme-FlH}^-$$

$$\text{Enzyme-FlH}^- \quad + \quad \underset{(O_2)}{\text{Acceptor}} \longrightarrow \text{Enzyme-Fl}_{ox} \quad + \quad \underset{(H_2O_2)}{\text{Reduced acceptor}}$$

SCHEME 3.35   Overall reaction of flavoenzymes.

the reduced flavin to the oxidized form. If the oxidized substrate is released from the enzyme prior to the binding of the acceptor, then this is known as a Ping Pong Bi–Bi kinetic mechanism; see Appendix I, Section VI.A.4.

Three types of mechanisms can be considered, one involving a carbanion intermediate, a radical intermediate (s), and a hydride intermediate. As we will see, there is no definitive mechanism for flavin-dependent enzymes; each of these mechanisms may be applicable to different flavoenzymes and/or different substrates.

a. Two-Electron Mechanism (Carbanion)

Evidence for a carbanion mechanism has been provided for the flavoenzyme D–amino acid oxidase (DAAO), which catalyzes the oxidation of D–amino acids to α-keto acids and ammonia (the second reaction in Figure 3.8). An early approach taken to gain evidence for the type of intermediate that is involved in this enzyme-catalyzed reaction was the use of Hammett plots. Because this is a common approach in mechanistic enzymology, I think we should take a slight digression for those unclear about what a Hammett plot is and what it represents. Those of you who have had a course in physical organic chemistry and know about Hammett relationships can jump ahead a few paragraphs or take a short refresher course right here.

In 1940 L. P. Hammett showed that the electronic effects (both the inductive and resonance effects) of a set of substituents on different organic reactions are similar.[38] He postulated that if values could be assigned to the electronic effects of substituents in a standard organic reaction, these same values could be used to estimate rates in a new organic reaction. Furthermore, evidence for the types of intermediates associated with the reaction could be determined. Hammett chose reactions of benzoic acids as the standard system.

Consider the reaction shown in Scheme 3.36. Intuitively, it seems reasonable that as X becomes electron withdrawing (relative to H), the equilibrium constant (K_a) should increase (the reaction to the right should become favored) because X is inductively pulling electron density from the carboxylic acid group, making it

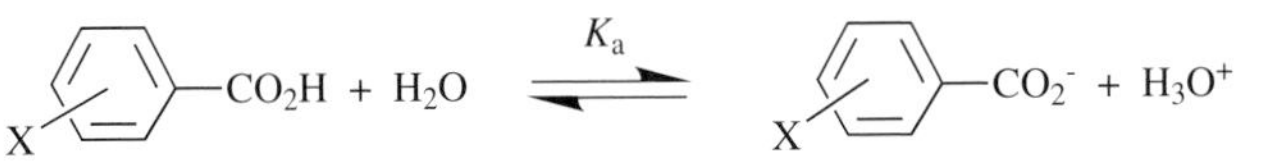

SCHEME 3.36   Ionization of substituted benzoic acids.

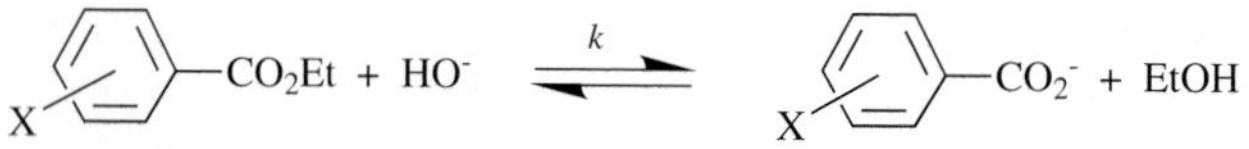

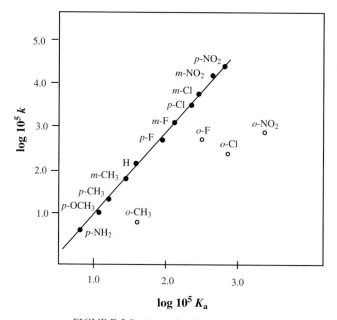

SCHEME 3.37   Reaction of hydroxide ion with ethyl-substituted benzoates.

more acidic (a ground-state argument); it also is stabilizing the incipient negative charge on the carboxylate group in the transition state (a transition-state argument). Conversely, when X is electron donating, the equilibrium constant should decrease because additional electron density is being pushed into an electron–rich carboxylate, thereby destabilizing it. A similar relationship should exist for a rate constant (k) where charge develops in the transition state. Hammett chose the reaction shown in Scheme 3.37 as the standard system. If K_a is measured from Scheme 3.36 and k from Scheme 3.37 for a series of substituents X, and the data are expressed in a double logarithm plot (Figure 3.9), then a straight line can be drawn through most of the data points. This is known as a *linear free-energy relationship*. When X is a *meta-* or *para*-substituent, then virtually all the points fall on a straight line; the *ortho*-substituent points are badly scattered. The Hammett relationship does not hold for *ortho*-substituents because of steric interactions and polar effects. The linear correlation for the *meta-* and *para*-substituents is observed for rate or equilibrium constants for a wide variety of organic reactions. The straight line can be expressed by

$$\log k = \rho \log K + C, \tag{3.1}$$

FIGURE 3.9   Example of a Hammett plot.

where the two variables are log k and log K. The slope of the line is ρ, and the intercept is C. When there is no substituent, that is, when X $=$ H, the equation

$$\log k_o = \rho \log K_o + C \tag{3.2}$$

holds. Subtraction of Eq. (3.2) from Eq. (3.1) gives

$$\log k/k_o = \rho \log K/K_o, \tag{3.3}$$

where k and K are the rate and equilibrium constants, respectively, for compounds with a substituent X and k_o and K_o are the same for the parent compound (X $=$ H). If log K/K_o is defined as σ, then Eq. (3.3) reduces to

$$\log k/k_o = \rho\sigma, \tag{3.4}$$

known as the *Hammett equation*. The *electronic parameter,* σ, depends on the electronic properties and position of the substituent on the ring and is therefore also called the *substituent constant*. The more electron-withdrawing a substituent, the more positive its σ value (relative to H, which is set at 0.0); conversely, the more electron donating a substituent, the more negative its σ value. The *meta* σ constants (σ_{meta}) result from inductive effects, but the *para* σ constants (σ_{para}) correspond to the net inductive and resonance effects, because substituents in the *para*-position of an aromatic ring can contribute both to inductive and to resonance effects, whereas those in the *meta*-position can contribute only to induction (you cannot draw resonance structures with *meta*-substituents). Therefore, σ_{meta} and σ_{para} values for the same substituent, generally, are not the same.

The ρ values (the slope) depend on the particular type of reaction and the reaction conditions (e.g., temperature and solvent) and, therefore, are called *reaction constants*. The importance of ρ is that it is a measure of the sensitivity of the reaction to the electronic effects of the *meta-* and *para*-substituents. A large ρ, either positive or negative, indicates great sensitivity to substituent effects (i.e., a charged transition state). Reactions that have carbocation-like transition states are favored by electron-donating substituents and have negative ρ values; reactions that are aided by electron-withdrawing substituents (e.g., reactions with carbanionic transition states) have positive ρ values.

Now we are ready to see how this applies to the study of the mechanism of D-amino acid oxidase (DAAO). A series of substituted D-phenylglycines (**3.55**) was tested as substrates for DAAO, and it was found that as X becomes more electron donating, the V_{max} decreases.[39] The ρ value was determined to be $+5.44$, a rather large ρ, and its positive slope suggests that carbanionic character develops (electron

3.55   **3.56**

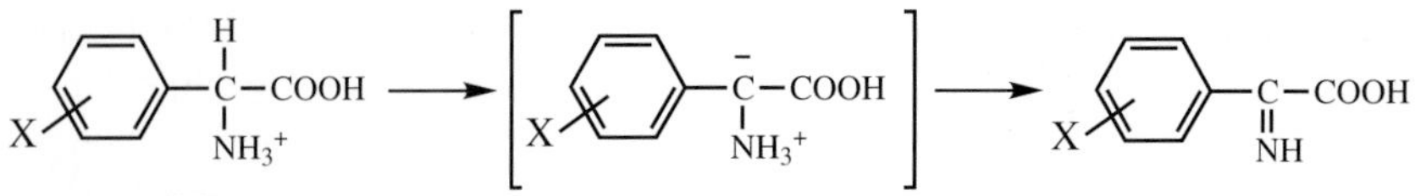

SCHEME 3.38   Proposed intermediate in the D-amino acid oxidase-catalyzed oxidation of substituted phenylglycines.

donation destabilizes the negative character). In accord with this notion, the ρ value for a series of substituted D-phenylalanines (**3.56**) with DAAO is only +0.73, as expected, because of the farther distance of the inductive effect and the inability for a carbanion generated at the α-carbon to be involved in a resonance effect with a phenyl substituent group. These Hammett results are consistent with a carbanion intermediate, such as that shown in Scheme 3.38.

Further evidence for a carbanion intermediate was obtained from the reaction of DAAO with β-D-haloalanines. Incubation of DAAO with β-D-chloroalanine (**3.57**, Scheme 3.39) produced two products, chloropyruvate (**3.59**) and pyruvate (**3.60**), depending on the conditions.[40] When the enzyme reaction was carried out in an atmosphere of 100% O_2, only **3.59** was produced, but when the reaction was done in an atmosphere of 100% N_2, exclusively **3.60** was observed. In air (20% O_2) a 60:40 mixture of **3.59:3.60** was obtained. The total amount of product, however, was the same, regardless of the oxygen content of the atmosphere. The nitrogen atmosphere reaction gives the same amount of product as the one in an oxygen

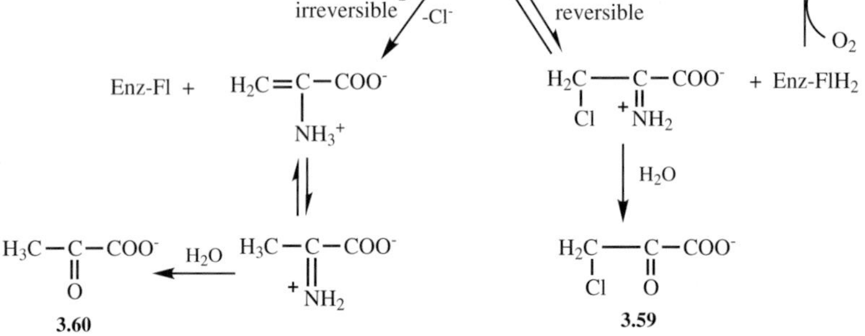

SCHEME 3.39   D-Amino acid oxidase-catalyzed oxidation of β-chloroalanine under oxygen and under nitrogen.

atmosphere, which suggests that the flavin is not reduced; if it were, then, because there is no oxygen to reoxidize it, only one turnover would be observed (which would be too little to detect without a radioactively labeled substrate). Actually, pyruvate is in the same oxidation state as β-D-chloroalanine, so there is no need for the flavin to change oxidation states. But chloropyruvate is one oxidation state higher than β-D-chloroalanine, so the flavin must be involved. So how do you rationalize these results? All conditions give the same amount of product(s), which suggests a common intermediate that can partition to give each product. If the pathway in 100% O_2 is reversible, but oxidation of the reduced flavin strongly displaces the equilibrium in favor of product formation, then only chloropyruvate would form. In 100% N_2 no oxidation of reduced flavin can occur, so the carbanion intermediate (**3.58**) undergoes elimination of chloride ion, leading exclusively to pyruvate (**3.60**). In air the amount of oxygen is not sufficient to drive the equilibrium totally to the oxidation product. If this explanation is correct, then an intermediate with a better leaving group than chloride ion should produce more pyruvate, even in the presence of O_2, and one with a poorer leaving group should give more halopyruvate. In fact, β-D-bromoalanine gives only elimination to pyruvate under all conditions, and β-D-fluoroalanine only undergoes oxidation to fluoropyruvate (except that in 100% N_2 there is no reaction).[41]

Up to this point we have not mentioned how the flavin becomes involved, yet no reaction occurs with these substrates without it. Covalent and noncovalent mechanisms can be drawn. On the basis of molecular orbital calculations, there are two electrophilic sites on oxidized flavin, N^5 and C^{4a} (see **3.48** in Scheme 3.31 for numbering), suggesting that nucleophilic addition could occur at either site. The evidence against addition to C^{4a} is that no C^{4a} substrate adducts have been detected with oxidases, although model studies indicate that thiols readily attack at this position.[42] Furthermore, reconstitution of the enzyme with N^5-ethyl flavin, which contains a reactive C^{4a} iminium, does not lead to formation of a covalent adduct. This should accelerate the rate of nucleophilic attack, which has been demonstrated in chemical model reactions (Scheme 3.40).[43]

There is some evidence for attack at the N^5 position of flavin. If the enzyme AMP–sulfate reductase, which catalyzes the first step in the reduction of sulfate to inorganic sulfide, is run in the back direction in the absence of AMP, an N^5-sulfite

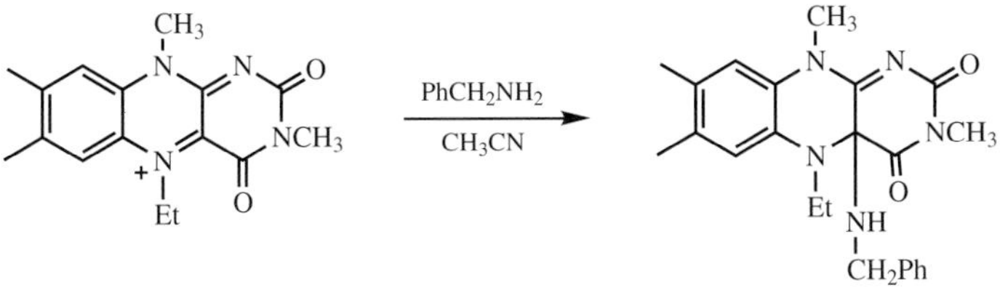

SCHEME 3.40 Nonenzymatic reaction of benzylamine with N^5-ethylflavin.

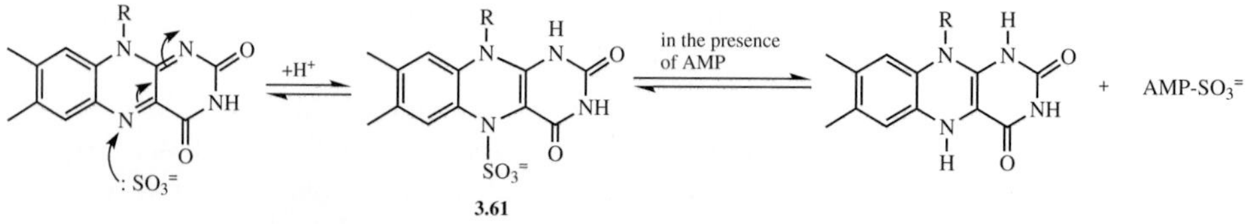

SCHEME 3.41   Reverse reaction catalyzed by AMP–sulfate reductase.

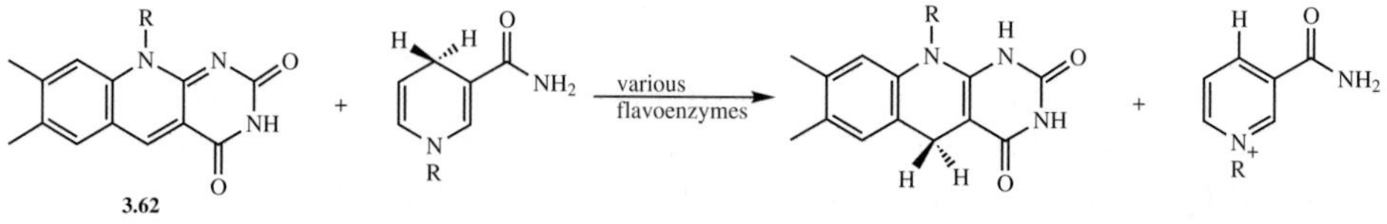

SCHEME 3.42   NADH-dependent reduction of 5-deazaflavin by various flavoenzymes.

adduct with the flavin (**3.61**) can be deduced on the basis of changes in the absorption spectra (Scheme 3.41).[44]

Other early evidence for addition to N^5 in flavoenzymes comes from experiments with an analogue of flavin called *5-deazaflavin* (**3.62**, Scheme 3.42). If 5-deazaflavin is reconstituted into a variety of apoenzymes that normally require FAD or FMN, and these enzymes are incubated with NADH, reduced 5-deazaflavin is generated.[45] However, the consensus of enzymologists now seems to be that 5-deazaflavin is really an analogue of the pyridine nucleotide coenzymes rather than of flavins (Figure 3.10), because it strongly favors two–electron reactions rather than one–electron reactions and therefore is an inappropriate substitute for flavin coenzymes.[46]

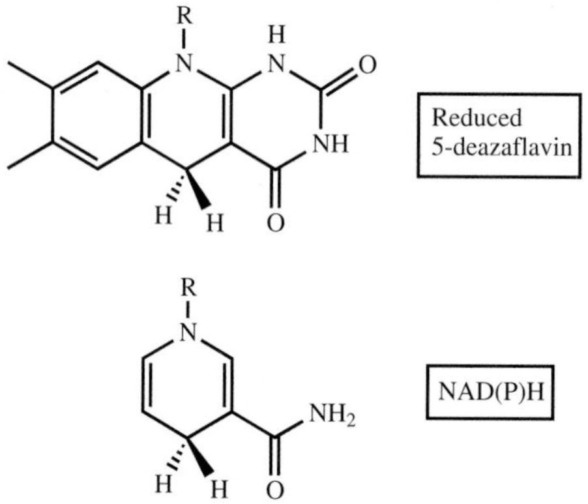

FIGURE 3.10   Comparison of reduced 5-deazaflavin with reduced nicotinamide.

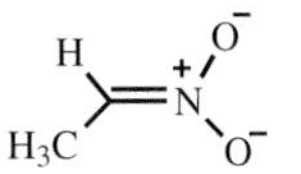

3.63

An alternative to direct nucleophilic attack by a substrate anion is an electron transfer mechanism from the substrate anion to the flavin. However, kinetic isotope effect studies by Kurtz and Fitzpatrick with D-amino acid oxidase suggest that direct nucleophilic attack on flavin is favored.[47] The average of the secondary deuterium isotope effects for the reaction of nitroalkane anions with DAAO, determined over the pH range 6–11, was found to be 0.84 ± 0.02, an inverse isotope effect, indicating significant sp^2 to sp^3 rehybridization in the transition state. The nitroethane anion in solution has an sp^2-hybridized α-carbon (**3.63**). In contrast, formation of a radical from the carbanion (sp^3 to sp^2 hybridization) would generate a small and normal secondary kinetic isotope effect. This supports a covalent carbanionic mechanism (pathway a, Scheme 3.43), not a single-electron transfer mechanism from the carbanion (pathway b), followed by either radical combination (pathway c) or second electron transfer (pathway d).

The evidence for a carbanion mechanism for DAAO appears fairly substantial. However, the X-ray crystal structure of DAAO complexed with the inhibitor benzoate has revealed that there is no base at the active site juxtaposed for deprotonation of the substrate to give the carbanion, but the substrate $C\alpha$-hydrogen does appear to be in line for a direct hydride transfer to the N^5 of the flavin.[48]

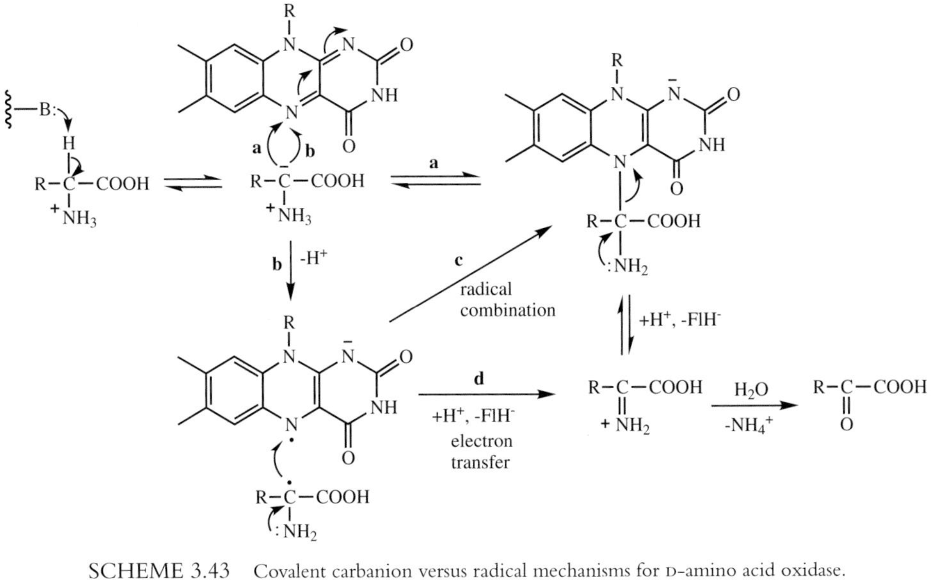

SCHEME 3.43   Covalent carbanion versus radical mechanisms for D-amino acid oxidase.

X-ray analysis of reduced DAAO complexed to iminotryptophan (the oxidized product of Trp) confirms the preceding finding.[49] Therefore, if the carbanion mechanism is correct, there must be a conformational change in the enzyme that exposes an active-site base or there is some other mechanism for deprotonation of the substrate.

b. Carbanion Followed by Two One-Electron Transfers

Some enzymes may initiate a carbanion mechanism but then proceed by an electron transfer (radical) pathway. This is shown in Scheme 3.43 (pathway b) as a possibility for D-amino acid oxidase and in Scheme 3.44 for an alcohol dehydrogenase (X in **3.64** is a carbanion-stabilizing group). Evidence for this kind of mechanism initially comes from chemical model studies in the presence of base and an oxidized flavin (Scheme 3.45).[50] Compounds **3.66** and **3.67** are radical coupling products that would be derived from the initial carbanion (**3.65**).

Enzymatic evidence for this kind of mechanism is now available from studies with general acyl-CoA dehydrogenase, a family of enzymes that catalyze the oxidation of acyl-CoA derivatives (**3.68,** Scheme 3.46) to the corresponding α,β-unsaturated acyl-CoA compound (**3.69**). Compound **3.70,** a mechanism-based inactivator (see Chapter 2, II.C.2.a) of general acyl-CoA dehydrogenase, was originally believed to inactivate this enzyme by the mechanism in Scheme 3.47, leading to flavin adduct **3.71**.[51] The first experiment that provided evidence for a radical intermediate was the incubation of general acyl-CoA dehydrogenase with each enantiomer of **3.70** (the innermost carbon of the cyclopropyl group is a stereogenic center).[52] Both isomers inactivate the enzyme. This suggests a radical fragmentation because only the *pro-R* proton is removed, but ring opening is not stereospecific. Furthermore, when tritium is incorporated into each isomer, identical partition ratios are obtained for both, suggesting that the geometry of the intermediate is not important to cyclopropyl ring cleavage.[53] This is, again, consistent with cyclopropylcarbinyl radical chemistry, giving the stabilized radical **3.72** (Scheme 3.48). Because there is now some question, at least with DAAO, about whether an initial deprotonation occurs, a hydrogen atom abstraction mechanism that goes directly to the carbon radical should be considered.

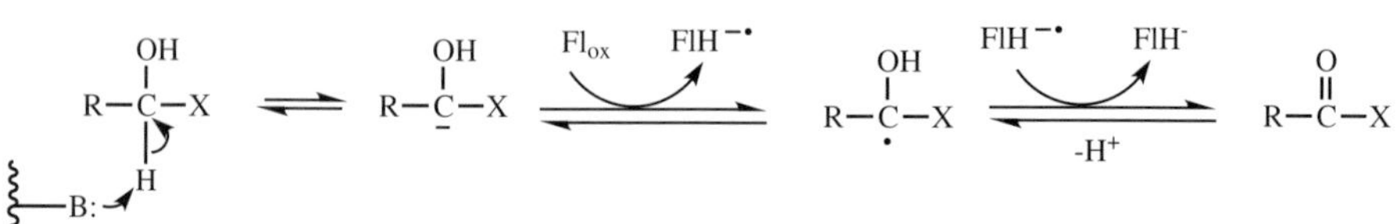

SCHEME 3.44   Mechanism for alcohol dehydrogenase involving a carbanion followed by two one-electron transfers.

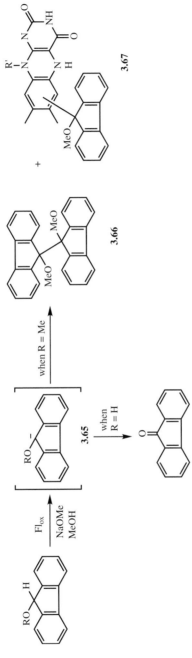

SCHEME 3.45 Chemical model studies for a mechanism involving a carbanion followed by two one–electron transfers.

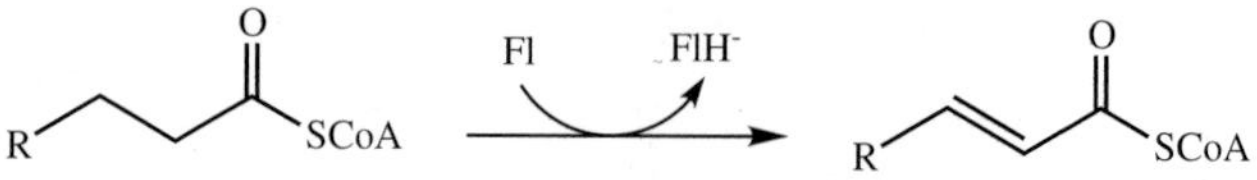

3.68   **3.69**

SCHEME 3.46   Reaction catalyzed by general acyl-CoA dehydrogenase.

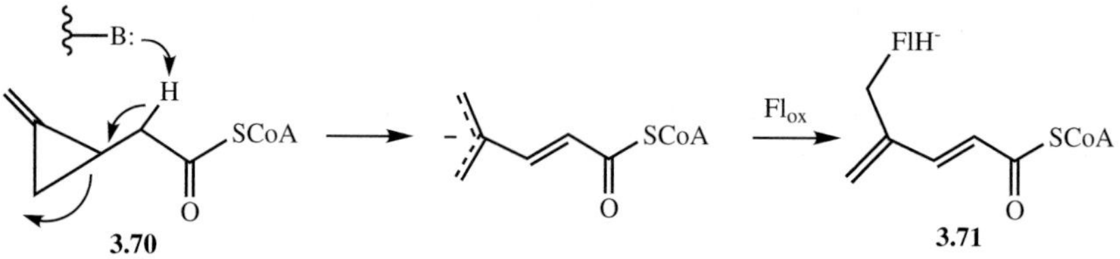

3.70   **3.71**

SCHEME 3.47   Initial mechanism proposed for mechanism-based inactivation of general acyl-CoA dehydrogenase by (methylenecyclopropyl)acetyl-CoA.

very fast—no
stereospecificity
(* is either R- or S)

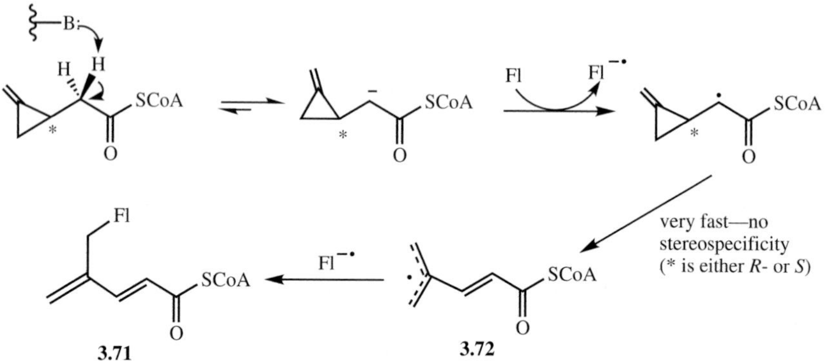

3.71   **3.72**

SCHEME 3.48   Electron transfer mechanism for inactivation of general acyl-CoA dehydrogenase by (methylenecyclopropyl)acetyl-CoA.

Further evidence for a radical intermediate is the isolation of **3.73** after inactivation of general acyl-CoA dehydrogenase by **3.70**; a mechanism that accounts for the formation of **3.73** from the stabilized radical **3.72** is given in Scheme 3.49.[54]

c. One-Electron Mechanism

Evidence for a one-electron (radical) flavin mechanism comes from a variety of experiments with monoamine oxidase.[55] Some possible mechanisms for monoamine oxidase (MAO), which exists in two isozymic forms called MAO A and

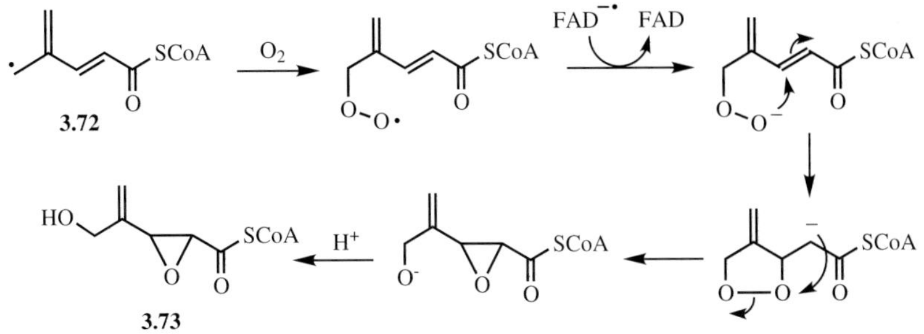

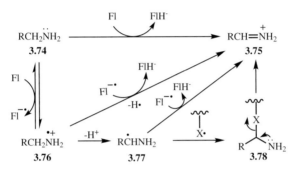

SCHEME 3.49   Mechanism proposed for formation of **3.73** during oxidation of (methylenecyclo-propyl)acetyl-CoA by general acyl-CoA dehydrogenase.

SCHEME 3.50   Possible mechanisms for monoamine oxidase-catalyzed oxidation of amines.

MAO B and which catalyzes the degradation of biogenic amines (**3.74**) to their cor-responding imines (**3.75**), are summarized in Scheme 3.50. All these mechanisms are initiated by a single electron transfer from the amine nonbonded electron pair to the flavin to give amine radical cation **3.76**. Loss of a proton would give carbon radical **3.77**, which could either transfer a second electron to give the imine (**3.75**) or undergo radical combination with an active-site radical to give a covalent inter-mediate **3.78**, which would break down, by elimination of the active-site group, to give **3.75**. An alternative to the initial single electron transfer followed by proton abstraction to give **3.77** is a hydrogen atom abstraction from **3.74** to go directly to **3.77**, which some favor.[56] The active-site radical (X·) could be either the flavin semi-quinone generated in the first step or an amino acid radical, produced by a hydro-gen atom transfer from the amino acid to the flavin semiquinone (Scheme 3.51).

Evidence for the first electron transfer step to give an amine radical cation comes from the finding that a variety of cyclopropyl amines are mechanism-based inacti-vators of MAO B that undergo ring cleavage reactions, leading to covalent attach-ment to the enzyme. The basis for these experiments is the high propensity of

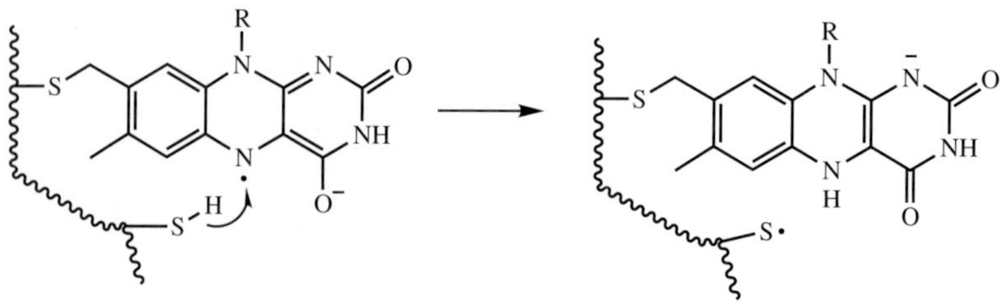

SCHEME 3.51   Mechanism proposed for generation of an active-site amino acid radical during monoamine oxidase-catalyzed oxidation of amines.

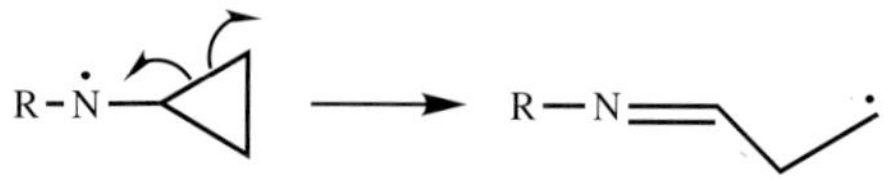

SCHEME 3.52   Cyclopropylaminyl radical rearrangement.

cyclopropylaminyl radicals to undergo homolytic cleavage (Scheme 3.52), as is the case for cyclopropylcarbinyl radicals (see III.A.2).[57] For example, 1-phenylcyclopropyl amine (**3.79**) was shown by radiolabeling experiments to form two covalent adducts to MAO, one a flavin adduct (**3.82**) and one a cysteine adduct (**3.84**; it is not known if this is the iminium or ketone **3.86**) (Scheme 3.53).[58] The flavin adduct could be derived from a one-electron transfer of an amine nonbonded electron to the flavin to give **3.80,** followed by cyclopropyl ring cleavage (**3.81**) and attachment of the flavin semiquinone to the primary radical (pathway a). Evidence for the structure of **3.82** or **3.83** was obtained from the results of various organic reactions carried out on the radiolabeled adduct (Scheme 3.54). Treatment with sodium boro[^{3}H]hydride resulted in the incorporation of about one equivalent of tritium relative to the control using native enzyme. A Baeyer–Villiger oxidation with trifluoroperoxyacetic acid (to give the phenyl ester), followed by treatment with hydroxide, released the radioactivity as ^{14}PhOH. Incubation of the radiolabeled enzyme with 0.5 N hydroxide produced [^{14}C]acrylophenone (**3.85**). Attachment to the flavin was demonstrated by proteolytic digestion and chromatography. The radioactive fractions also had the absorption spectrum of a reduced flavin; UV–visible spectral shifts at different pH values suggested that the attachment was at the N^5-position of the flavin.

In addition to the flavin adduct a much less stable cysteine adduct (**3.84**) was isolated (pathway b). This could arise from a competition with the reaction that gives flavin adduct **3.82**. Prior to attachment of **3.81** to the flavin semiquinone, an active-site cysteine residue could transfer its thiol hydrogen atom to the flavin semiquinone, and the resulting thiyl radical could combine with **3.81** to give **3.84**.

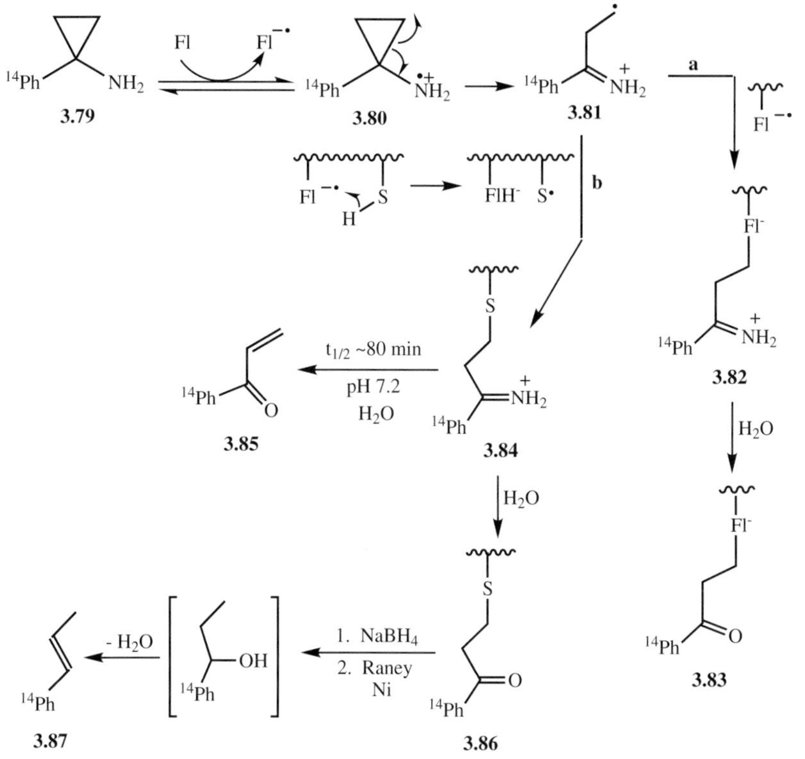

SCHEME 3.53   Mechanisms proposed for inactivation of monoamine oxidase by 1-phenylcyclo-propylamine.

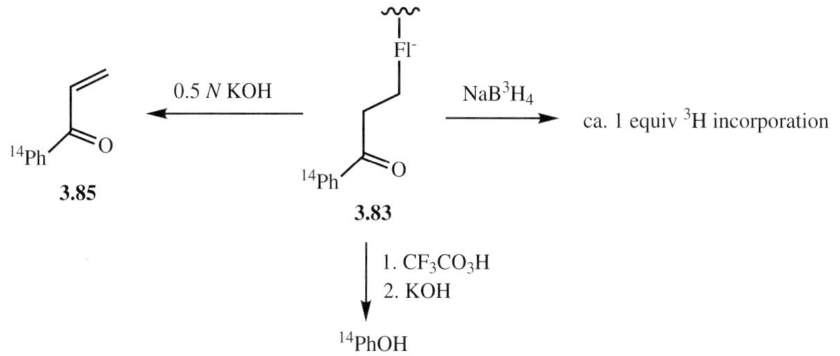

SCHEME 3.54   Chemical reactions to characterize the structure of the flavin adduct formed on inactivation of monoamine oxidase by 1-phenylcyclopropylamine.

Unlike the flavin adduct, this adduct is unstable and undergoes *spontaneous* elimination to give acrylophenone (**3.85**). The identity of **3.84** was established by treatment with sodium borohydride followed by Raney nickel.[59] The product isolated, β-methylstyrene (**3.87**), is consistent with the formation of a structure such as **3.86**; the fact that Raney nickel only reduces C−S bonds supports the formation of a cysteine adduct. Also, cysteine titration showed that there is one less free cysteine in the enzyme after sodium borohydride reduction of **3.86** than there is in the native enzyme (after sodium borohydride reduction).

Because the primary structures of MAO A and MAO B are known,[60] *peptide mapping,* the separation of peptides (usually by HPLC) formed on proteolytic digestion of the enzyme, is a reasonable approach to determining the site of attachment of an inactivator. This approach can be carried out with or without a radiolabeled inactivator. The enzyme is inactivated and proteolytically digested, and the peptides are separated by HPLC. If a radiolabeled inactivator is used, the radiolabeled peptides are isolated (ideally, only one peptide is radiolabeled), repurified, and submitted for Edman degradation. If an unlabeled inactivator is used, a control is run in which an aliquot of the enzyme goes through all of the same procedures as those for the inactivated enzyme, except no inactivator is added. HPLC peptide maps from the inactivated and uninactivated enzymes (after proteolytic digestion) are compared. Ideally, a new peptide will appear in the inactivated enzyme that is not present in the uninactivated enzyme (and a peptide in the uninactivated map will be present that is not in the inactivated one). The amino acid sequence of the unique peptide determined by Edman degradation should match a peptide in the known primary sequence except for one amino acid, which was modified by the inactivator. This modified amino acid will not be identifiable because it is not any of the natural amino acids, but it can be deduced from the missing amino acid by comparison with the known primary sequence of the enzyme. Peptide mapping of MAO B with unlabeled *N*-cyclopropyl-α-methylbenzylamine (**3.88**), followed by sodium borohydride reduction to stabilize the adduct, using the procedure described, gave a unique peptide with the sequence **3.89**.[61] The X is the modified amino acid, and it corresponds to Cys-365. To confirm the peptide structure, matrix-assisted laser desorption/ionization−time of flight (MALDI-TOF) mass spectrometry was carried out; the mass (1010.7 amu) exactly corresponds to that of the peptide minus a proton from the cysteine SH group plus 59 for a $CH_2CH_2CH_2OH$ group (total 1010.55 amu), which could arise by the mechanism shown in Scheme 3.55.

3.88   Lys-Leu-X-Asp-Leu-Tyr-Ala-Lys

                                              3.89

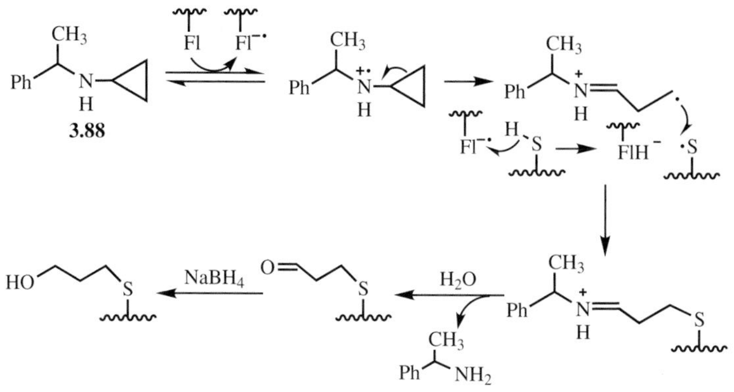

SCHEME 3.55   Mechanism proposed for inactivation of monoamine oxidase by N-cyclopropyl-α-methylbenzylamine.

MALDI-TOF mass spectrometry[62] and *electrospray ionization mass spectrometry*[63] are becoming routine tools for determining where an inactivator becomes attached to an enzyme. The total mass of the inactivated enzyme can be compared to that of the native enzyme to determine the mass difference after inactivation. This gives you an idea of what modification may have occurred. Instead of using the procedure described, the new peaks in the peptide map can be run directly from the HPLC into an electrospray ionization mass spectrometer for mass analysis. With *tandem mass spectrometry* (two mass spectrometers connected in tandem), the modified peptide can be fragmented further for sequence analysis. If the primary structure of the enzyme is known, the sequence of this peptide can be matched to the corresponding sequence in the enzyme. This pinpoints the exact residue that has been modified and also determines the mass of the modification.

Further evidence for an initial single electron transfer from the amine to the flavin in MAO was provided by the observation that 1-phenylcyclobutylamine (**3.90**) is converted to 2-phenyl-1-pyrroline (**3.93**), a known one-electron rearrangement, presumably proceeding via intermediates **3.91** and **3.92** (Scheme 3.56, pathway b).[64] When the reaction was carried out in the presence of a radical spin trap (such as the nitrone shown in Scheme 3.56), the electron paramagnetic spectrum (EPR) showed a triplet of doublets centered at $g = 2.00$, indicative of a spin-trapped organic radical.[65] The enzyme also was inactivated by formation of a flavin adduct (**3.94**, pathway a), also consistent with the formation of **3.91.**

Support for a proton transfer mechanism (**3.76** to **3.77** in Scheme 3.50) rather than the hydrogen atom transfer mechanism (**3.76** to **3.75**) was provided with a cubane radical probe molecule (**3.95**, Scheme 3.57).[66] Compound **3.95** is an inactivator of MAO and produces two major metabolites, **3.97** and another product that was isolated and shown by NMR and mass spectrometry *not* to have a cubane structure and *not* to be a homocubane. This can be rationalized by an intermediate

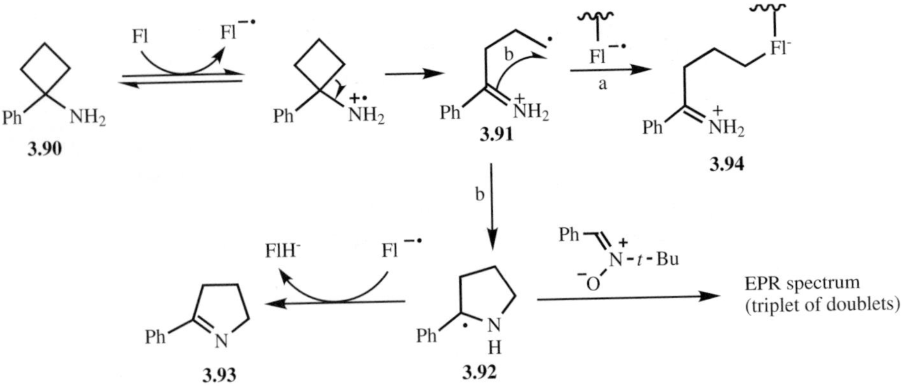

SCHEME 3.56   Mechanism proposed for monoamine oxidase-catalyzed oxidation of 1-phenylcy-clobutylamine and inactivation of the enzyme.

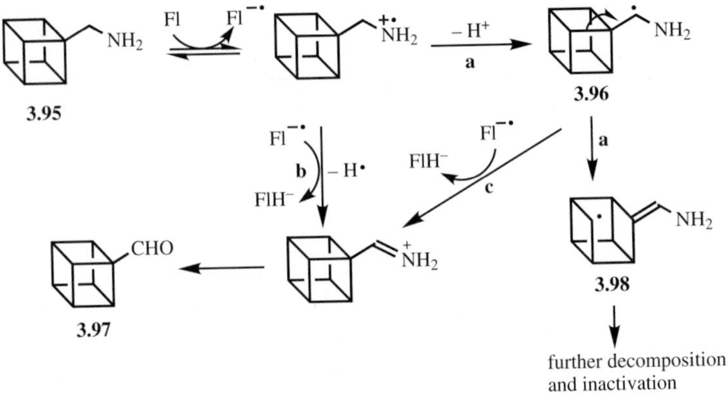

SCHEME 3.57   Oxidation of (aminomethyl)cubane by monoamine oxidase.

such as **3.96** or the corresponding carbanion. Inactivation and the formation of the noncubane-containing product could occur from **3.98.**

To differentiate a radical from a carbanion intermediate, an approach was taken based on the report that radicals adjacent to an epoxide lead to C—C bond cleavage of the epoxide (Scheme 3.58A), but the corresponding carbanions lead to C—O bond cleavage (Scheme 3.58B).[67] MAO was incubated with cinnamylamine-2,3-epoxide (**3.99**) as a test for the intermediacy of a radical or carbanion; the only products formed, glycolaldehyde and benzaldehyde, are C—C bond cleavage products (Scheme 3.59), supporting a radical mechanism.[68]

Additional evidence to support an α-radical intermediate in MAO-catalyzed reactions comes from the reactions of MAO with *cis*- and *trans*-5-(aminomethyl)-

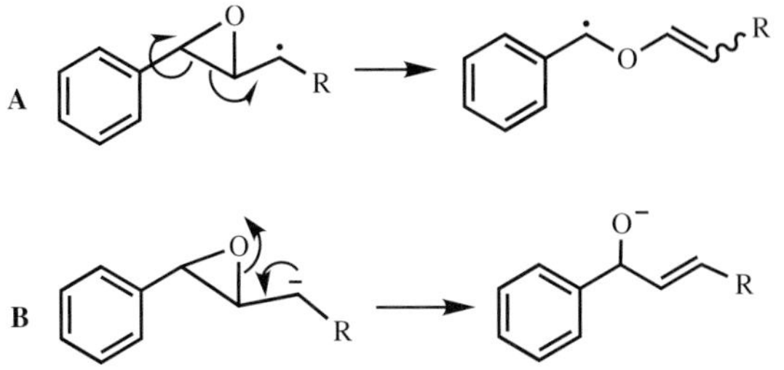

SCHEME 3.58   Reactions to differentiate a radical from a carbanion intermediate.

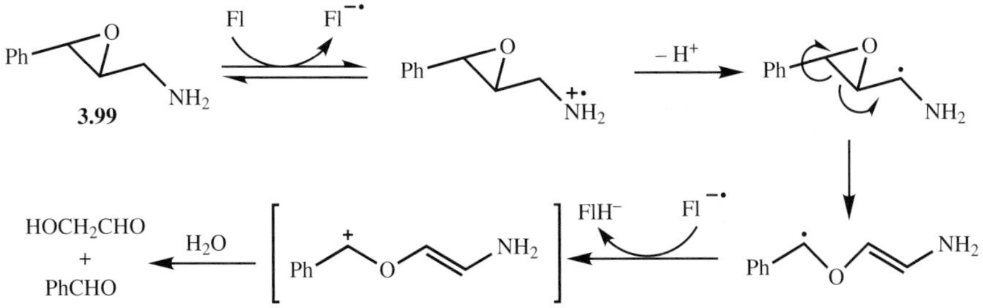

SCHEME 3.59   Mechanism proposed for monoamine oxidase-catalyzed oxidation of cinnamyl-amine-2,3-epoxide.

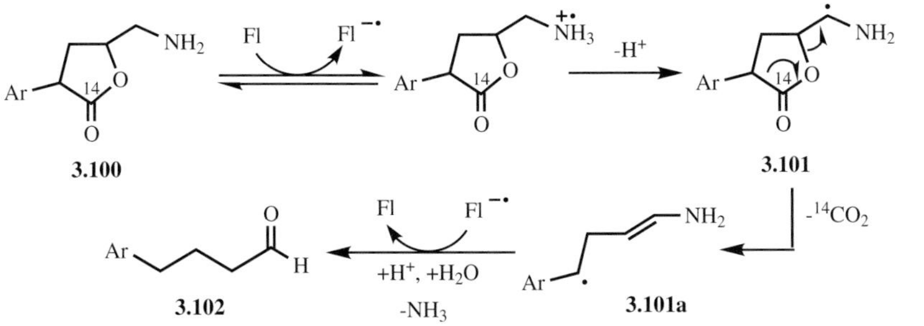

SCHEME 3.60   Mechanism proposed for monoamine oxidase-catalyzed decarboxylation of *cis-* and *trans*-5-(aminomethyl)-3-(4-methoxyphenyl)-2-[^{14}C]dihydrofuran-2(3H)-one.

3-(4-methoxyphenyl)-2-[^{14}C]dihydrofuran-2(3H)-one (**3.100**), both of which re-sulted in the production of $^{14}CO_2$ (Scheme 3.60).[69] This result, seemingly, can only be rationalized by the formation of **3.101**; a carbanionic intermediate would lead to elimination of the carboxylate without decarboxylation (carboxyl radicals rap-idly undergo decarboxylation to the corresponding radical, but decarboxylation of a carboxylate anion requires the formation of a stabilized carbanion after decarboxylation, which is not the case here). Another product isolated in this reaction (starting from **3.100**, except with ^{14}C in the aryl group instead of at the carbonyl) is **3.102**, which is in the same oxidation state as **3.100**.[70] This supports a reversible electron transfer mechanism in which the flavin can accept an electron from the amine, and then, following decarboxylation, return an electron to the resultant radical (**3.101a**).

The final question in the mechanisms shown in Scheme 3.50 is whether **3.77** produces **3.75** directly by second electron transfer or if it first forms a covalent intermediate (**3.78**). Evidence for a possible covalent intermediate with at least some substrates was obtained from studies of the inactivation of MAO B by oxazolidinones (Scheme 3.61).[71] Inactivation of MAO B by either (R)- or (S)-3-[^{3}H]aryl-5-(methylaminomethyl)-2-oxazolidinone (**3.103**; x = 3, y = 12) led to the incorporation of one equivalent of tritium per enzyme after denaturation, suggesting a covalent adduct formed. (R)- and (S)-3-[^{14}C-*carbonyl*]Aryl-5-(methylaminomethyl)-2-oxazolidinone (**3.103**; x = 1, y = 14) also inactivated MAO B with the incorporation of one equivalent of radioactivity after denaturation. The fact that both the tritium in the side chain and the ^{14}C in the oxazolidinone ring become attached stoichiometrically to the enzyme suggests that the entire molecule is part of the enzyme adduct (**3.104**). The stability of **3.104** may be the result of the electron-withdrawing character of the oxazolidone ring; the corresponding lactone[72] and lactam[73] also inactivate MAO, presumably by forming corresponding stable adducts.[74] This suggests that the covalent intermediate could be stabilized

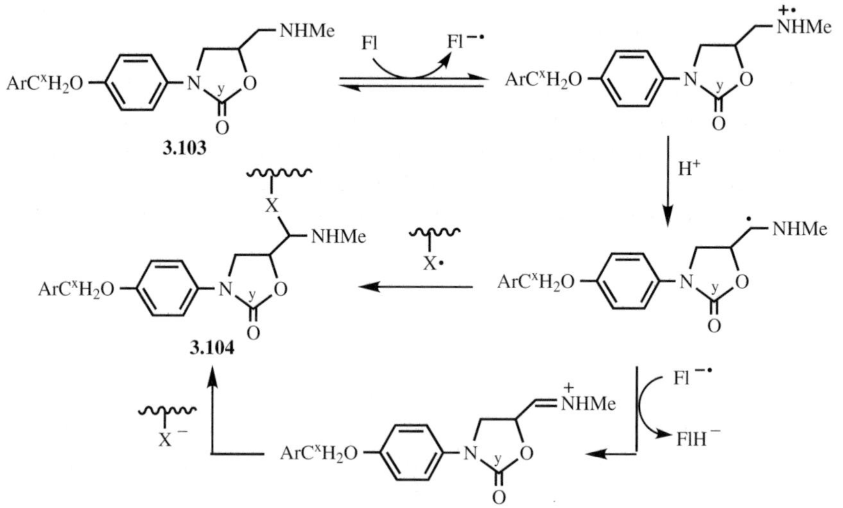

SCHEME 3.61   Mechanism proposed for inactivation of monoamine oxidase by (R)- or (S)-3-[^{3}H]aryl-5-(methylaminomethyl)-2-oxazolidinone.

by appropriate design of substrates with electron-withdrawing character, which is the case.[75] It is not clear if covalent adduct formation (Scheme 3.50, **3.78**) occurs with all substrates or if this result is an artifact of the special nature of substrate analogues that have electron-withdrawing groups, but it seems to occur.

Because no direct spectroscopic observation of the initial electron transfer from the amine to the flavin to give the flavin semiquinone and the amine radical cation has been made,[76] and deuterium isotope effect studies do not support their formation,[77] it has been proposed that MAO proceeds by a direct hydrogen atom abstraction of **3.74** to give **3.77** (Scheme 3.50). However, it is difficult to rationalize the experiments depicted in Schemes 3.53, 3.55, and 3.56 without the formation of the initial amine radical cation formation. Furthermore, there does not appear to be a radical present at the active site that is capable of a direct hydrogen atom abstraction.[78]

Another mechanism that has been suggested for MAO, based on chemical model studies, is a nucleophilic mechanism (Scheme 3.62).[79] However, the model reactions only proceeded under very harsh conditions and with primary amines (as is expected for a nucleophilic mechanism because of steric hindrance to attack) or with primary and secondary amines under mild conditions with an N^5-ethyl-substituted flavin analogue. Given that MAO catalyzes the oxidation of primary, secondary, and tertiary amines—even highly sterically hindered amines[80]—a nucleophilic mechanism is unlikely. Furthermore, the facts that **3.95** (Scheme 3.57) resulted in cubane destruction, that **3.99** (Scheme 3.59) gave no C–O bond cleavage products, and that **3.100** (Scheme 3.60) led to loss of $^{14}CO_2$, all argue against a nucleophilic mechanism, even with primary amines.

d. Hydride Mechanism

The first step in bacterial cell wall peptidoglycan biosynthesis, catalyzed by uridine diphosphate-N-acetylglucosamine (UDP-GlcNAc) enolpyruvyl transferase (called MurA), is discussed in Chapter 6 (Section VI). The product of that enzyme-catalyzed reaction, enolpyruvyl-UDP-GlcNAc (**3.105**), is reduced by an NADPH and flavin-dependent enzyme, UDP-N-acetylenolpyruvylglucosamine reductase

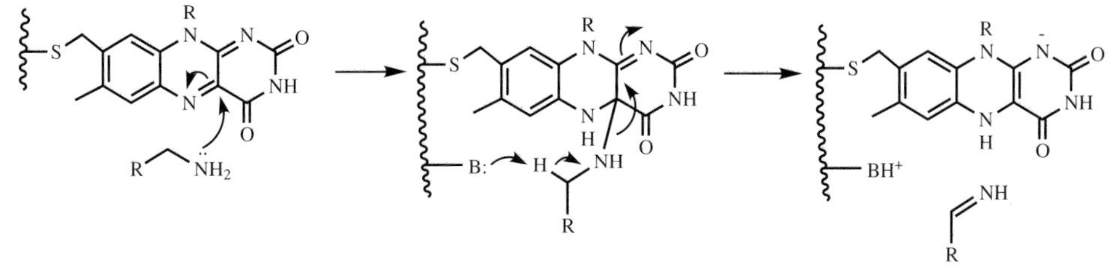

SCHEME 3.62   Nucleophilic mechanism for monoamine oxidase.

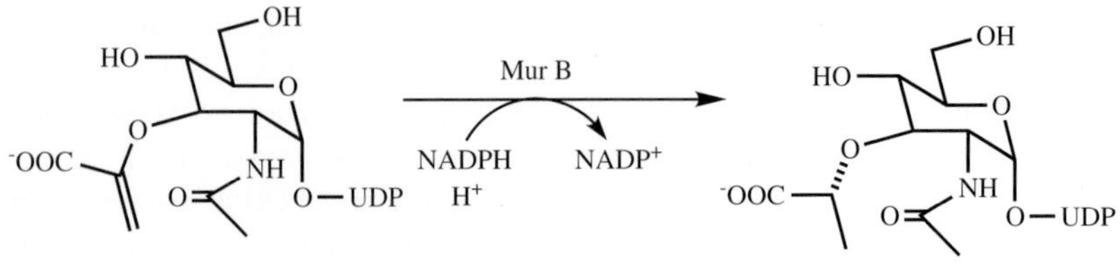

SCHEME 3.63   Reaction catalyzed by UDP-*N*-acetylenolpyruvylglucosamine reductase (MurB).

(or MurB), to give UDP-*N*-acetylmuramic acid (**3.106**) (Scheme 3.63). The proposed reduction mechanism[81] involves initial reduction of the FAD cofactor by
NADPH (because a reduced flavin is highly prone to oxidation, it is typically formed
at the active site of the enzyme by *in situ* reduction with a nicotinamide cofactor, at
the time it is needed), followed by hydride transfer from reduced FAD via a Michael addition to the double bond of the α,β-unsaturated carboxylate (Scheme 3.64).
Evidence for a hydride mechanism, rather than a radical mechanism, was obtained
in a stereochemical study.[82] To monitor the stereochemistry of hydride addition, an
altered substrate had to be used, namely, (*E*)-enolbutyryl-UDP-GlcNAc (**3.107**)
with NADP^{2}H in ^{2}H$_2$O. The product isolated was the (2*R*,3*R*)-dideuterio product **3.108,** the product of stereospecific *anti*-addition of deuterium from NADP^{2}H
and ^{2}H$_2$O (Scheme 3.65); a radical mechanism would not be expected to be stereospecific. The stereochemistry of the product was determined by a *chemoenzymatic*
(partly chemical, partly enzymatic) route. First, the enzymatic product (**3.108**) was
degraded to 2-hydroxybutyrate (**3.109**) by treatment with sodium hydroxide, then
with alkaline phosphatase (to hydrolyze the phosphate group), then with sodium
hydoxide again (Scheme 3.66). The absolute stereochemistry at C-2 of **3.109**
was determined to be *R* because it was a substrate for the enzyme D-lactate dehydrogenase, but not for L-lactate dehydrogenase. To determine if the product
was 2*R*,3*R* or 2*R*,3*S*, both diastereomers were synthesized from 2-ketobutyrate
(Scheme 3.67). A comparison of the proton NMR spectra of the enzymatic product and each of the diastereomers allowed for an unambiguous assignment of **3.109**
(and therefore of **3.108**) as the 2*R*,3*R*-isomer. Given that stereochemistry, the
enzyme-catalyzed reaction appears to proceed as shown in Scheme 3.68, in which
the hydride is delivered to the equivalent of the *re*-face of enolpyruvyl–UDP–
GlcNAc (**3.105**).

A hydride mechanism also has been proposed for the flavoenzyme dihydroorotate dehydrogenase.[83] Dehydrogenation of dihydroorotate (**3.110,** Scheme 3.69)
exhibited deuterium isotope effects on both expelled hydrogens. This indicates that
an E1cB mechanism is not appropriate and that a concerted process is reasonable.
The reverse reaction would deliver a hydride from reduced flavin to orotate.

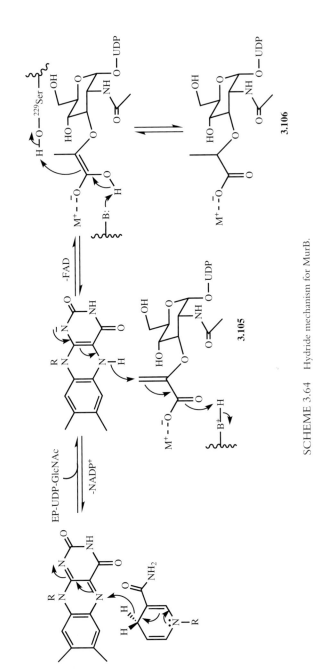

SCHEME 3.64 Hydride mechanism for MurB.

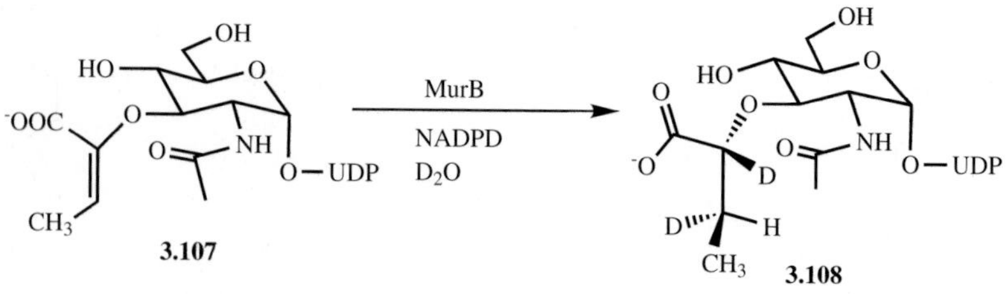

SCHEME 3.65   MurB-catalyzed reduction of (E)-enolbutyryl-UDP-GlcNAc with NADP^{2}H in ^{2}H$_2$O.

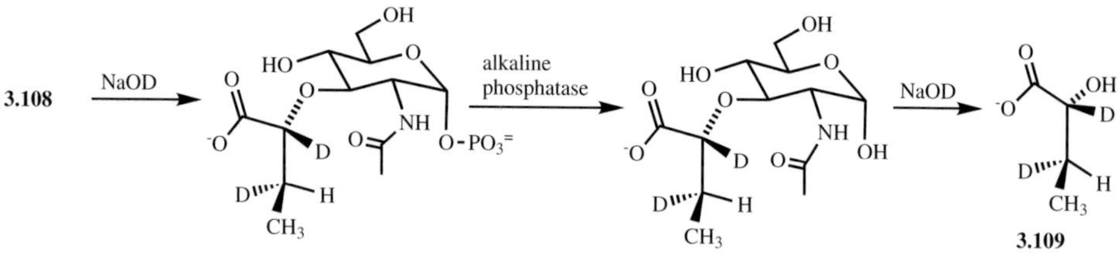

SCHEME 3.66   Conversion to 2-hydroxybutyrate of the product formed from MurB-catalyzed reduction of (E)-enolbutyryl-UDP-GlcNAc with NADP^{2}H in ^{2}H$_2$O.

(2R, 3R)-2,3-[^{2}H$_2$]-2-hydroxybutyrate

(2R, 3S)-2,3-[^{2}H$_2$]-2-hydroxybutyrate

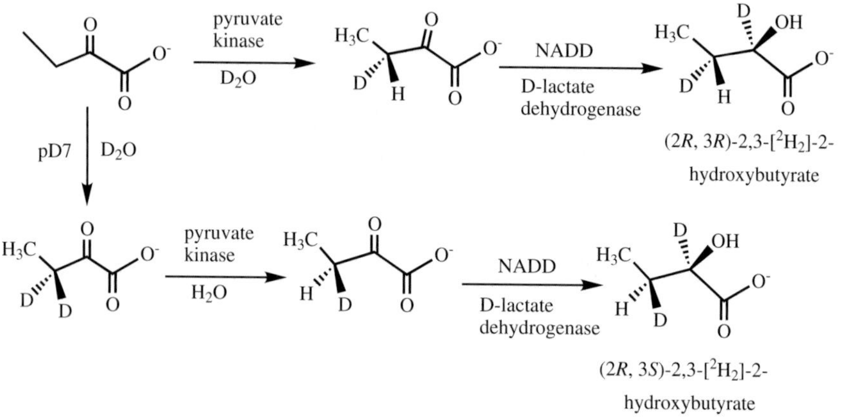

SCHEME 3.67   Syntheses of two diastereomers of 2,3-[^{2}H$_2$]hydroxybutyrate.

4. Artificial Flavoenzymes

Artificial enzymes (also called *semisynthetic enzymes* or *synzymes*) can be made by modification of the active site of known enzymes, thereby changing the catalytic

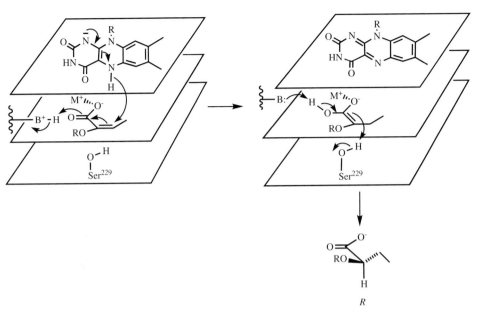

SCHEME 3.68   Stereochemistry of the MurB-catalyzed reduction of (*E*)-enolbutyryl-UDP-GlcNAc.

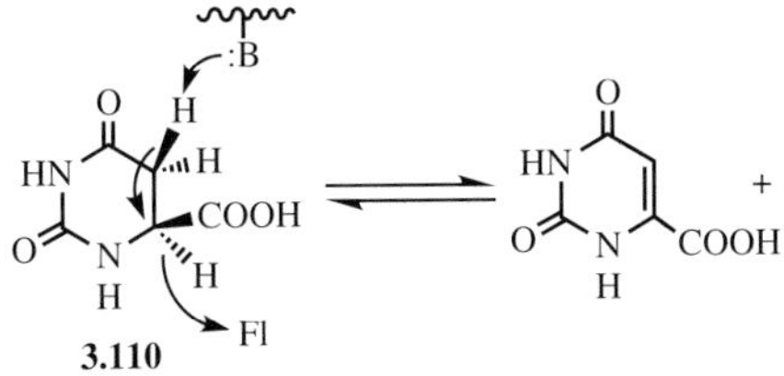

SCHEME 3.69   Reaction catalyzed by dihydroorotate dehydrogenase.

activities of those enzymes. Typically, a coenzyme is covalently attached to the active-site cysteine or serine of a protease. In this way the polypeptide of the enzyme is used as the binding site for the new catalytic activity produced by the attached coenzyme.[84] For example, a flavin cofactor was attached to the active site of the enzyme papain by alkylation of active-site Cys-25 (**3.111**, Scheme 3.70; flavopapain).[85] This synzyme catalyzes the oxidation of dihydronicotinamide to nicotinamide with concomitant reduction of the flavin, a reaction illustrated in Scheme 3.64 and discussed again in Chapter 4 (Section II.B.1). This reaction is important for the *in vivo* conversion of oxidized to reduced flavin.

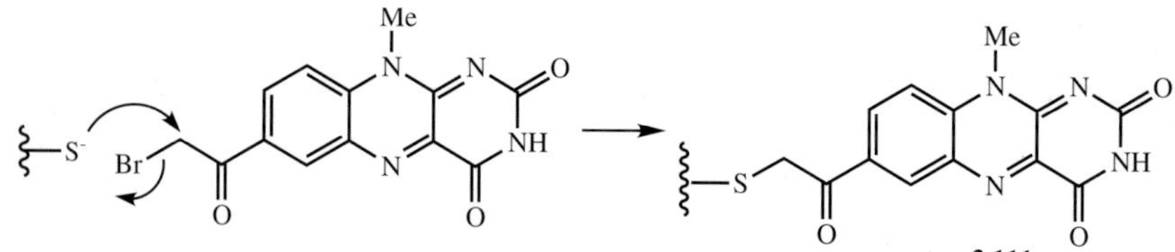

SCHEME 3.70   Synthesis of flavopapain.

5. A Lesson Learned from Flavin Oxidases

Urate oxidase catalyzes the oxidation of urate (**3.112**; this compound is drawn up-side down and backward from the way it normally is drawn so that the analogy to the structure of a flavin is more apparent—*vide infra*) to allantoin (**3.114**) with con-comitant conversion of molecular oxygen to hydrogen peroxide (Scheme 3.71). There are no cofactors, neither metal ions nor organic cofactors, at least in the soybean enzyme,[86] so this appears, at first glance, to be an exception to the general statement that enzymes that catalyze redox chemistry need the assistance of cofac-tors to function. Each of the seven carbons of urate was labeled by Tipton and co-workers with ^{13}C in seven different analogues, and ^{13}C NMR studies with urate oxidase were carried out to determine the formation of any intermediates during turnover.[87] 5-Hydroxyisourate (**3.113**) was observed as the intermediate; experi-ments with $H_2^{18}O$ showed that the 5-hydroxyl group was derived from water, not from molecular oxygen. To rationalize the formation of **3.113**, Tipton and co-workers recognized that urate looks somewhat like reduced flavin (compare **3.112** with **3.52** in Scheme 3.32) and may catalyze its own oxidation, as reduced flavin does when it is oxidized (see Scheme 3.33). Using the mechanism in Scheme 3.33 as a model, the mechanism in Scheme 3.72 was proposed for the oxidation of urate. Stopped-flow absorbance and fluorescence spectrophotometry were used to study soybean urate oxidase under single-turnover conditions; two intermediates detected are the dianion of **3.112** and the proposed hydroperoxy intermediate.[87a]

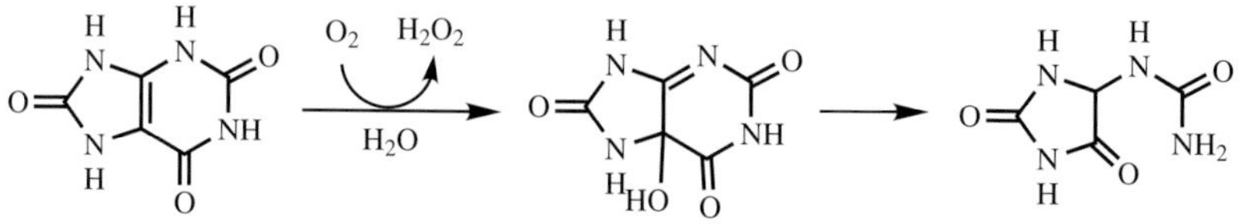

SCHEME 3.71   Reaction catalyzed by urate oxidase.

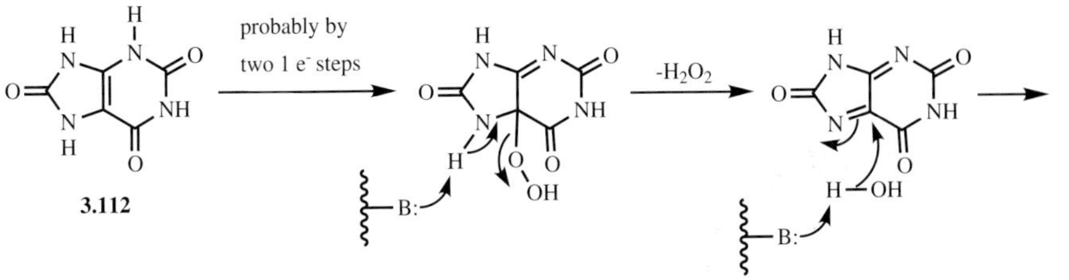

SCHEME 3.72   Possible mechanism for the urate oxidase-catalyzed oxidation of urate.

C. Quinone-Containing Coenzymes

1. Pyrroloquinoline Quinone Coenzymes

In the late 1970s a new coenzyme was isolated from bacterial primary alcohol dehydrogenase and was originally called *methoxatin* (**3.115**).[88] Other names that have evolved for this coenzyme are pyrroloquinoline quinone, PQQ, and coenzyme PQQ.[89] PQQ-dependent enzymes are the first members of a class of enzymes referred to as *quinoproteins,* enzymes that have quinone-containing coenzymes.[90] This coenzyme also was reported to be associated with a human enzyme, placental lysyl oxidase,[91] and another mammalian enzyme, bovine plasma amine oxidase,[92] but, as you will see later (III.C.2 and III.C.3), different quinocofactors are involved with these enzymes. It is now known that eukaryotic proteins do not contain covalently bound PQQ.

Two reasonable mechanisms for the oxidation of alcohols by PQQ-dependent dehydrogenases, such as methanol dehydrogenase[93a] and glucose dehydrogenase[93b] are shown in Scheme 3.73; mechanism A depicts an addition-elimination mechanism, and mechanism B is a direct hydride transfer mechanism. A crystal structure to 1.5 Å resolution of soluble glucose dehydrogenase from *Acinetobacter calcoaceticus* with PQQ and the competitive inhibitor methyl hydrazine bound shows addition of the inhibitor to the C-5 carbonyl, indicating the preferred electrophilicity of that

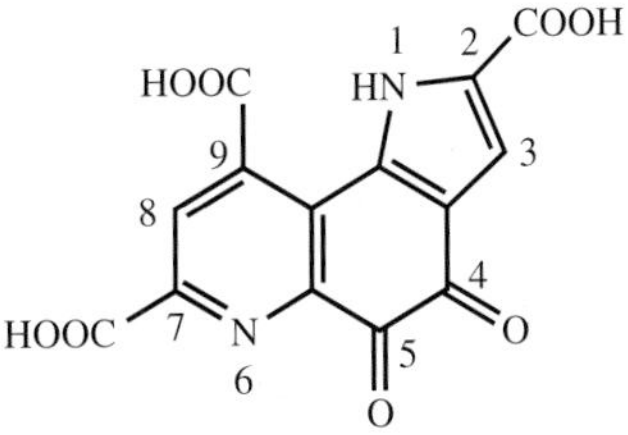

3.115

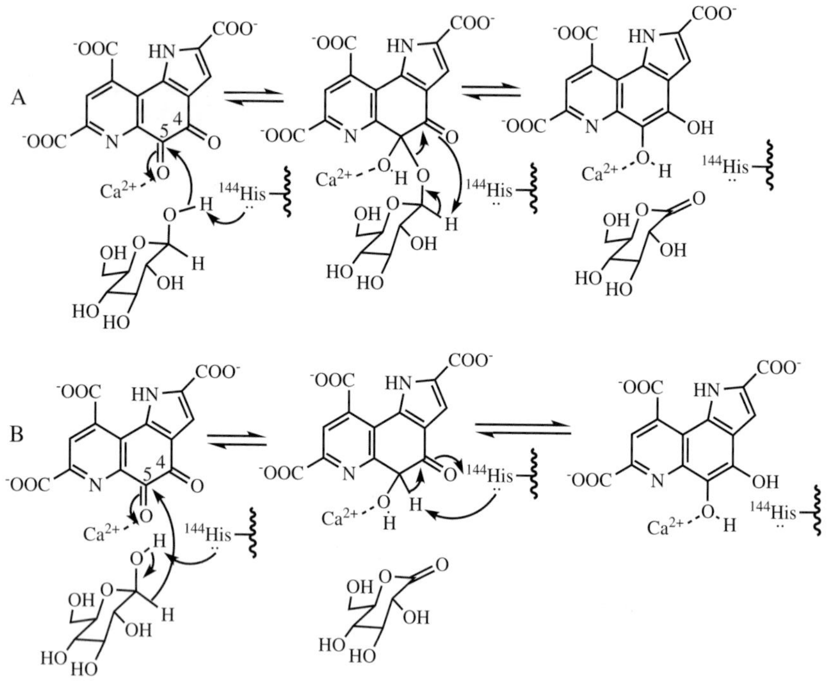

SCHEME 3.73   Possible mechanisms for the glucose dehydrogenase-catalyzed oxidation of glucose.

site relative to C-4.[93b] The crystal structure at 1.9 Å resolution of the same enzyme with reduced PQQ and glucose bound revealed the substrate poised directly above the reduced cofactor C-5 atom.[93c] For the addition-elimination mechanism to occur, three favorable interactions would have to be broken and various rearrangements in structure would have to occur. The orientation of the glucose, however, is well suited for direct hydride transfer from the glucose C-1 atom to the PQQ C-5 atom. This supports the hydride mechanism (Scheme 3.73B). The active-site Ca^{2+} probably polarizes the C-5 carbonyl of PQQ.[93d]

2. 6-Hydroxydopa (Topa Quinone) Coenzymes

One of the earliest, and most commonly, studied quinoprotein is bovine plasma amine oxidase, an enzyme that catalyzes the oxidation of primary amines to aldehydes.[94] It was shown to contain Cu^{2+} and therefore is a member of the *copper amine oxidase* family of enzymes. Much that is known about the mechanism of this enzyme was revealed before researchers recognized that it is *not* a PQQ enzyme.

Evidence to support a Schiff base (imine) mechanism with this enzyme is depicted in Scheme 3.74 (note that when this was proposed, researchers thought the cofactor was PQQ, as written, but it is not).[95] Because 1 equivalent of $[^{14}C]$ is bound

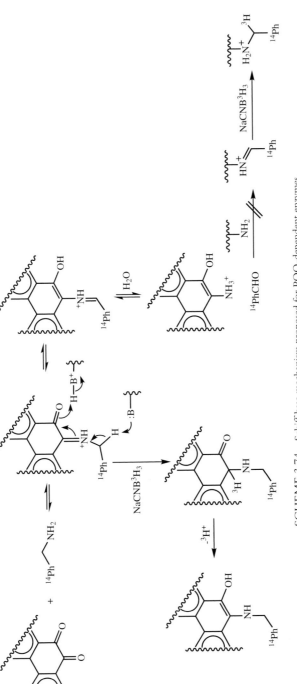

SCHEME 3.74   Schiff base mechanism proposed for PQQ-dependent enzymes.

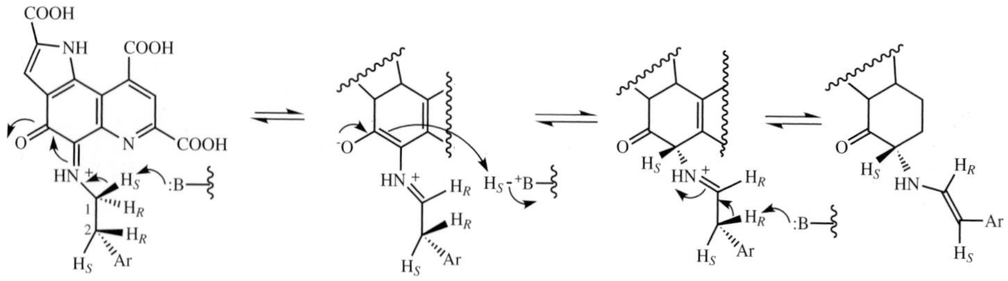

SCHEME 3.75   Stereochemistry of the reaction catalyzed by plasma amine oxidase (with PQQ incorrectly shown as the cofactor).

to the enzyme, but no tritium is incorporated on $NaCNB^3H_3$ reduction, it excludes a mechanism that involves oxidation to $^{14}PhCHO$ followed by Schiff base formation with a lysine residue and then $NaCNB^3H_3$ reduction. A mechanism such as that would result in the incorporation of tritium and ^{14}C into the enzyme.

When $PhCD_2NH_2$ was used as the substrate, a kinetic isotope effect indicated that C–H bond cleavage was fully rate determining. When dopamine was used as the substrate (Scheme 3.75; again note that PQQ is not the real cofactor), the same pK_a values for oxidation of the substrate and exchange of the C-2 proton were obtained, suggesting that the same base may be involved in both processes.[96] When plasma amine oxidase is incubated with $[1(R)-^2H, 2(R)-^3H]$- and $[1(S)-^2H, 2(R)-^3H]$dopamines, the *syn* hydrogens are removed.[97] This supports a one-base mechanism in which a 1,3-prototrophic shift of H_S occurs from C-1 of the substrate to the cofactor, followed by exchange of H_R from C-2, as shown in Scheme 3.75.

The most exciting surprise was the discovery by Klinman and co-workers[98] that bovine plasma amine oxidase does *not* contain PQQ as a cofactor. Instead, oxidized 6-hydroxydopa (referred to as *topa quinone* or TPQ), covalently bound to a pentapeptide (**3.116**), was isolated from a proteolytic digestion of the enzyme. The pentapeptide was characterized by Edman degradation, by mass spectrometry, and by UV–visible, resonance Raman, and NMR spectroscopy. The *para*-quinone moiety of **3.116** can catalyze the same chemistry as depicted for PQQ in Scheme 3.74, which is why it was originally confused for PQQ. Resonance Raman spectroscopy of the hydrazones of the intact cofactor and the cofactor isolated by proteolysis of

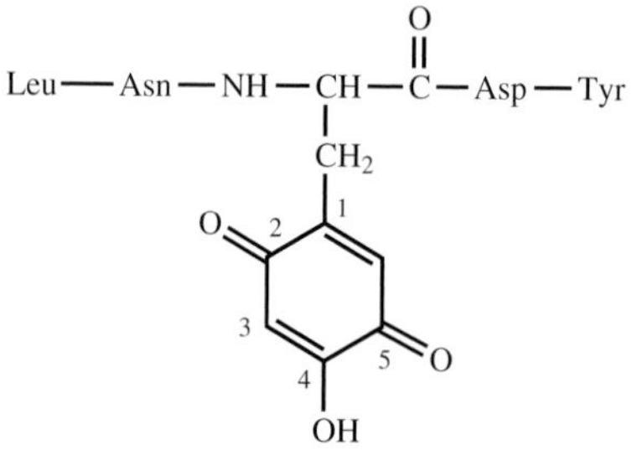

3.116

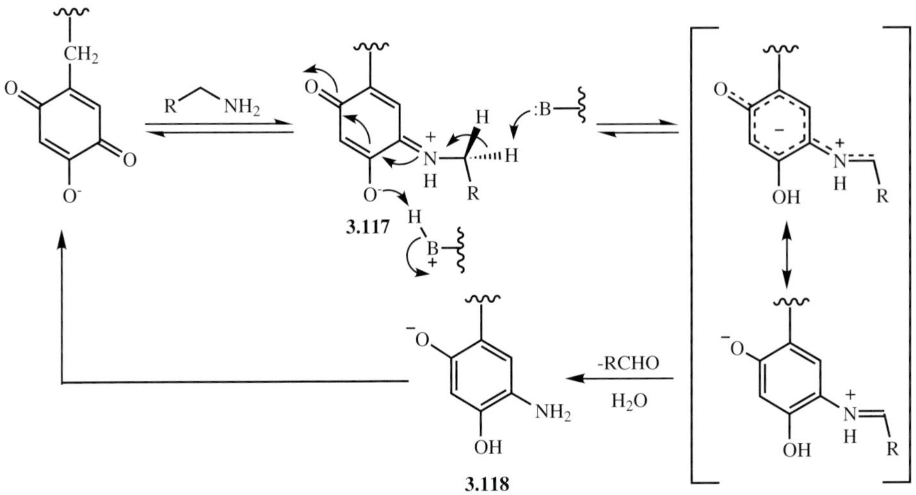

SCHEME 3.76 Plasma amine oxidase-catalyzed amine oxidation with topa quinone shown as the cofactor.

copper amine oxidases from several sources was consistent with that of TPQ.[99] In fact, Klinman and co-workers have suggested that possibly all copper-dependent oxidases are topa quinone enzymes.[100]

A structure−activity relationship (Hammett) study using p-substituted benzyl-amines was carried out, and it was found from the Hammett plot for the bond cleavage step that $\rho = 1.47 \pm 0.27$.[101] Remember (see Section III.B.3.a of this chapter) that a positive Hammett slope (ρ) means that the intermediate has carbanionic character, consistent with the mechanism in Scheme 3.76. Rapid-scanning stopped-flow spectroscopy showed relaxations at 310, 340, and 480 nm, attributed to an enzyme−substrate Schiff base (**3.117**, 340 nm) and to the reduced cofactor (**3.118**, 310 and 480 nm).[102]

Mure and Klinman synthesized analogues of topa quinone (**3.119−3.129**) as chemical models to study the mechanism of this class of enzymes in more detail.[103] First, it was found that nucleophiles preferentially attack the model compounds at C-5 (see **3.116** for numbering). Benzylamine is oxidized by **3.119−3.123** in acetonitrile at room temperature to give N-benzylbenzaldehyde imine (**3.130**). The p-quinones with a bulky substituent at C-1 (**3.119−3.121**) are most efficient. Compounds **3.124−3.127** are inactive. Compounds **3.128** and **3.129** have very low catalytic activity. These results demonstrate a substantial role for the 2-hydroxyl group in preventing the formation of Michael adducts with the substrate amino group and in facilitating reoxidation of the amino intermediates. The bulky substituent prevents dimerization of the Schiff base intermediate.

An explanation for the preferable nucleophilic attack of substrates at C-5 of topa quinone became apparent as a result of a resonance Raman spectroscopic investi-

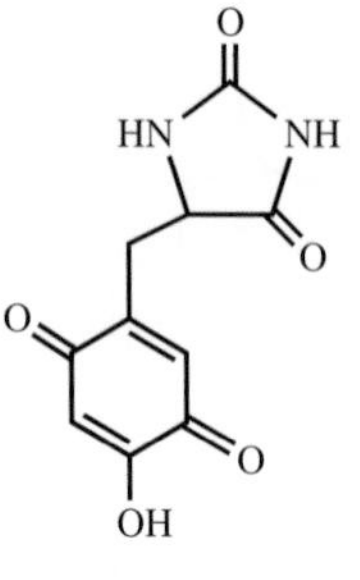

3.119

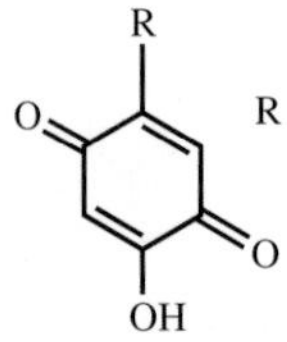

3.120 R = t-Bu
3.121 i-Pr
3.122 Et
3.123 Me

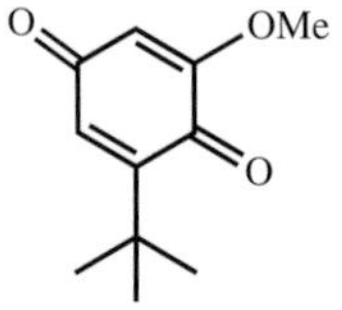

3.124 R = OMe
3.125 Me
3.126 H

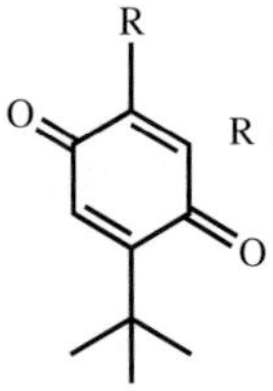

3.127

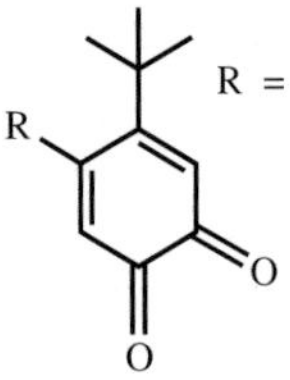

3.128 R = H
3.129 OMe

3.130

gation of the carbonyl character at C-2, C-4, and C-5 of topa quinone in the enzyme phenylethylamine oxidase from *Arthrobacter globiformis*.[104] Carbonyl stretching frequencies with ^{18}O-labeled topa quinone indicate that the C—O bond at C-5 has far greater double-bond character than those at C-2 or C-4, and therefore is more reactive.

Support for an aminotransferase-type mechanism (see Chapter 9, Section IV.C.1) from a Schiff base structure of topa quinone (**3.117**) comes from additional model studies.[105] Various amines were shown to deprotonate the C-4 hydroxyl group of the topa quinone analogues (Scheme 3.77). This deprotonation directs the addition of the amine to the C-5 carbonyl. Furthermore, the ionization state of the C-4 hydroxyl group may control the differential stabilities of the substrate and product Schiff base complexes. The C-4 hydroxyl group also increases the redox potential of the product in the conversion of the reduced cofactor back to topa quinone. On the basis of reactions of the topa analogues with different amines, by

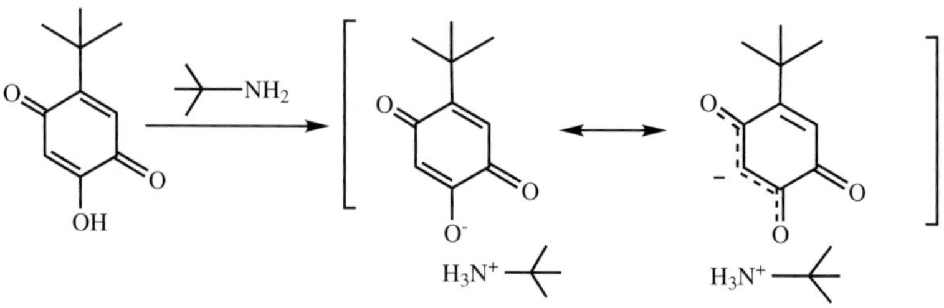

SCHEME 3.77   Chemical model study for the mechanism of topa quinone-dependent enzymes.

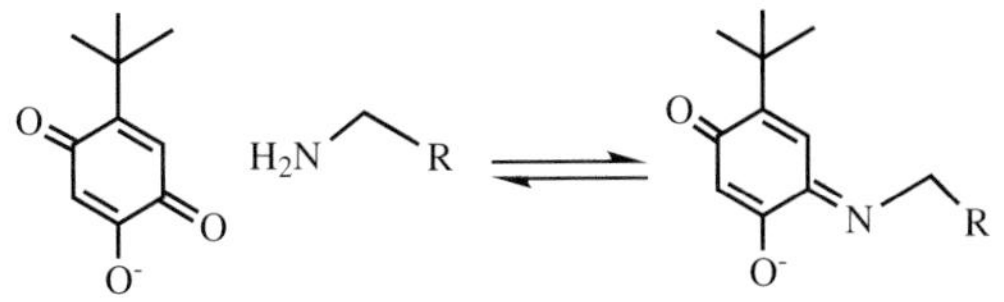

SCHEME 3.78 Chemical model study to show the reaction of amines at the C-3 carbonyl of topa quinone.

two-dimensional NMR studies, and optical spectral studies, the preferred form of the cofactor is the *para*-quinone (Scheme 3.78). Based on all these model studies, Klinman proposed the more detailed mechanism for TPQ-dependent enzymes given in Scheme 3.79.[106] Under anaerobic conditions the amino group of the substrate remains bound to the protein after substrate oxidation, in support of **3.131** as the cofactor product from substrate oxidation.[107]

Oxidation of **3.131** back to topa quinone has been shown to occur in two one-electron steps with the intermediacy of topa semiquinone (**3.132,** Scheme 3.80).[108] The radical, believed to be the Cu(I)–semiquinone in equilibrium with the Cu(II)–hydroquinone, has been observed by EPR spectroscopy with several copper amine oxidases. As a model for this radical, the semiquinone of 6-hydroxydopamine was generated, but the EPR spectrum appeared to be somewhat different from that of the radical produced by the oxidases.[109] After spontaneous reaction with primary amines, however, the EPR spectrum was converted to the same spectrum as that observed in the enzyme reaction. Therefore, the radical is a substrate–cofactor radical. With ^{15}N-labeled substrate the expected changes in the EPR spectrum for this kind of complex was observed.[110] With the isotopically labeled substrate affected, the EPR spectrum is further evidence that the amino group of the substrate remains bound until after cofactor oxidation.

Resonance Raman spectroscopic studies of the TPQ-dependent copper amine oxidase from *Escherichia coli* support the EPR results and indicate that substrate reduction of the enzyme under anaerobic conditions leads to a stable amine-substituted semiquinone (**3.132,** Scheme 3.80).[111] An X-ray crystal structure of a copper amine oxidase from *E. coli* fully supports the proposed mechanism.[112] The crystal structure of the copper-dependent amine oxidase from gram-positive *Arthrobacter globiformis* clearly shows a channel running from the molecular surface of the enzyme to the active site, presumably the entrance for substrates to the active site.[113]

Topa quinone is not a natural amino acid, so where does it come from? The precursor is a specific tyrosine residue in the highly conserved sequence Thr–X–X–Asn–Tyr–(TPQ)–Asp/Glu–Tyr of TPQ enzymes. The generation of TPQ occurs in a posttranslational, self-processing reaction that requires both Cu(II) and O_2.[114] On the basis of EPR and circular dichroism spectra and kinetic studies, a mechanism for the biosynthesis of topa quinone from tyrosine was proposed by Dooley and co-workers;[115] however, I prefer to draw the mechanism as shown in Scheme 3.81.

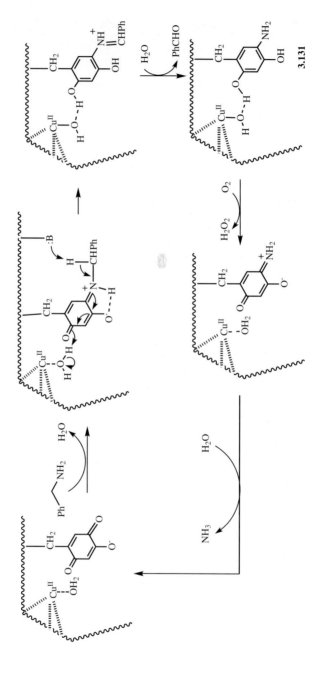

3.131

SCHEME 3.79 Detailed mechanism proposed for topa quinone–dependent enzymes.

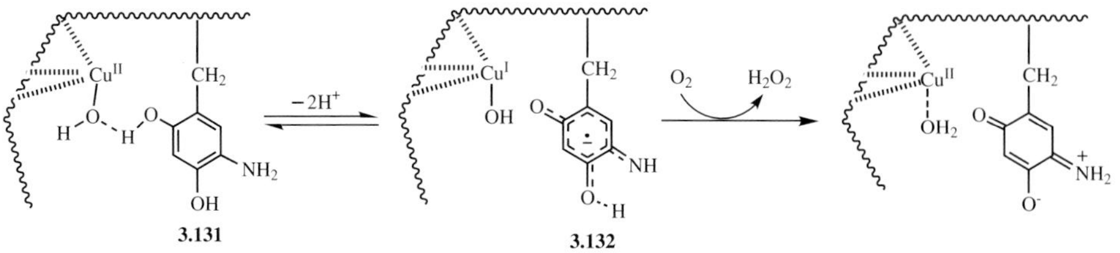

SCHEME 3.80 Mechanism proposed for reoxidation of reduced topa quinone.

To date, topa quinone is the only quinocofactor that has been found to be ubiquitous, occurring in bacteria, yeast, plants, and mammals.

3. Tryptophan Tryptophylquinone Coenzyme

Methylamine dehydrogenase is another quinoprotein, but the structure of its cofactor was shown—by resonance Raman,[116] by chemical and NMR analyses,[117] and by X-ray analysis[118]—to be tryptophan tryptophylquinone (TTQ) (**3.133**). A series of *para*-substituted benzylamines was studied as substrates for methylamine dehydrogenase, and a Hammett plot for log k_3 (conversion of substrate to product) versus σ_{para} exhibited a positive slope, suggesting a carbanionic intermediate.[119] Deuterium isotope effects and stopped-flow kinetic studies indicate that a mechanism similar to that for the TPQ-dependent enzyme, plasma amine oxidase (see Scheme 3.79), was most reasonable.[120] The major difference between the TTQ amine dehydrogenases and the TPQ amine oxidases is the rate of product release; product release is much slower with the TTQ enzymes.[121]

4. Lysine Tyrosylquinone Coenzyme

Another new quinone cofactor was identified for lysyl oxidase. Proteolytic digestion of lysyl oxidase led to the isolation of two hexapeptides that were connected

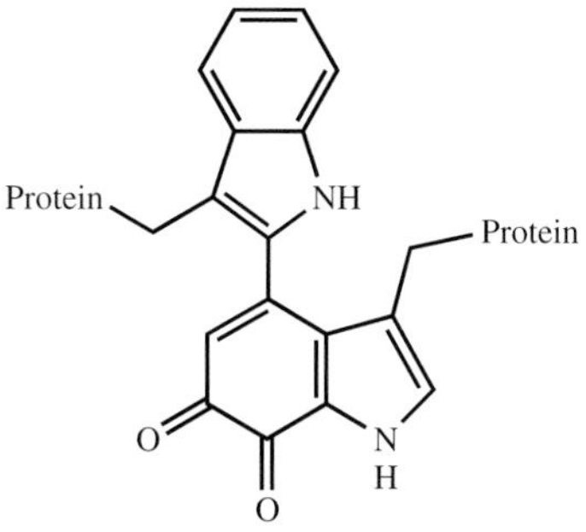

3.133

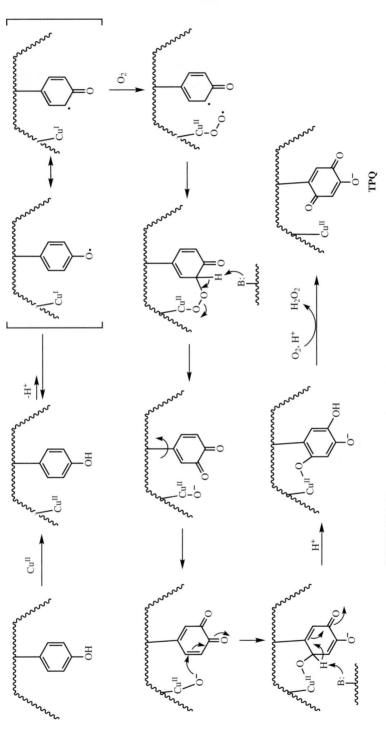

SCHEME 3.81 Mechanism proposed for the biosynthesis of topa quinone from tyrosine.

Asp-Thr-(modified Tyr)-Asn-Ala-Asp

Val-Ala-Glu-Gly-His-(modified Lys)

3.134

Val-Ala-Glu-Gly-His∿NH

Asp-Thr∿N–CHCONH∿Asn-Ala-Asp

H–C—CH$_2$CH$_2$CH$_2$CH$_2$—NH

C—OH

3.135

(**3.134**).[122] Its structure is derived from the crosslinking of the ϵ-amino group of a peptidyl lysine with the modified side chain of a tyrosine residue. Based on rational chemical reasoning, coupled with an exact mass from mass spectrometry, Edman sequencing, UV–visible spectroscopy and resonance Raman spectroscopy, and chemical model studies, the structure of the modified peptide was concluded to be **3.135,** designated as lysine tyrosylquinone (LTQ). α,α-Dideuterio-n-butylamine is oxidized by lysyl oxidase with a kinetic isotope effect on V_{max}, supporting a carbanionic mechanism that is the same as that for PPQ- and TPQ-dependent enzymes.[123]

D. Other Redox Enzymes

1. Enzymes Containing Amino Acid Radicals

A growing new class of enzymes uses active-site amino acid radicals to catalyze reactions; some contain the amino acid radicals in the native state.[124] From the X-ray crystal structure, galactose oxidase—which catalyzes the oxidation of primary alcohols to aldehydes with molecular oxygen—was shown to have as its cofactors a Cu(II) ion and a covalently bound cysteine crosslinked to a tyrosine radical (Tyr-272) at the active site (**3.136,** Scheme 3.82).[125] Molecular modeling studies of the active-site binding of possible transition-state structures have been used to rationalize the specificity of the enzyme (D-galactose is an excellent substrate with *pro-S* stereospecificity, but the epimer, D-glucose, does not even bind) and to support a radical C—H bond cleavage step.[126] Both concerted and stepwise mechanisms for alcohol oxidation appear reasonable (Scheme 3.82). No EPR signal is observed because of the strong antiferromagnetic exchange between these two paramagnetic centers. Spectral studies by Whittaker and co-workers of small-model compounds for the tyrosine–cysteine adducts and the corresponding copper complexes supported the proposed active-site structures.[127]

Mechanism-based inactivation of galactose oxidase occurs on incubation with

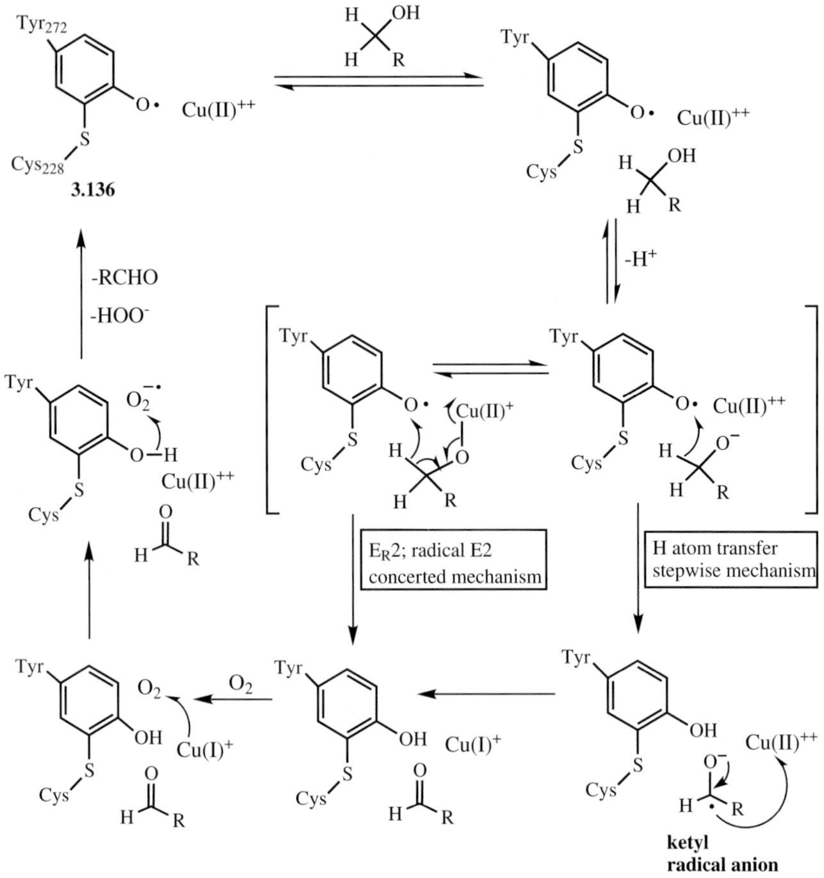

SCHEME 3.82   Mechanism proposed for galactose oxidase using a covalently bound cysteine cross-linked tyrosine radical.

the quadricyclane **3.137** or the norbornadiene **3.138** (Scheme 3.83); inactivated enzyme is completely reactivated by one-electron oxidants known to reactivate the one-electron reduced inactive form.[128] When $[\alpha,\alpha\text{-}^2H_2]$**3.137** was used, a kinetic isotope effect (k_H/k_D) of 6 on the inactivation rate constant was observed. These results support a mechanism involving rate-determining hydrogen atom abstraction to a ketyl radical intermediate (**3.139**), which leads to inactivation by producing the one-electron-reduced inactive form of the enzyme. Further evidence for a mechanism involving a ketyl radical anion comes from an investigation with a series of β-haloethanol analogues (XCH_2CH_2OH, X = F, Cl, Br, I), which also produced the one-electron-reduced inactive enzyme form.[129] The mechanism, based on prior work by Tanner and co-workers on the rearrangement of α-halo ketyl radical anions,[130] is similar to that shown in Scheme 3.83.

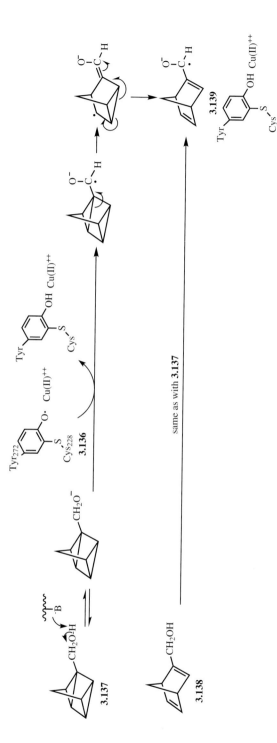

SCHEME 3.83 Mechanism–based inactivation of galactose oxidase by hydroxymethylquadricyclane and hydroxymethylnorbornadiene.

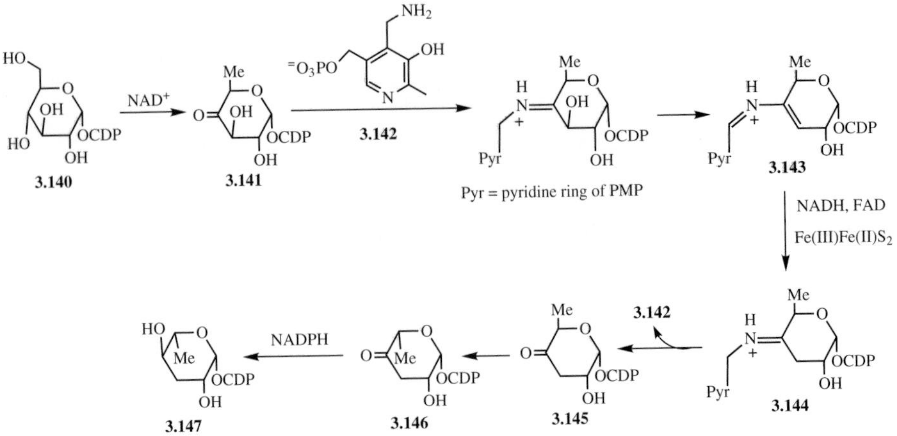

SCHEME 3.84   Reaction catalyzed by CDP-6-deoxy-L-*threo*-D-*glycero*-4-hexulose-3-dehydratase (also called E_1) and CDP-6-deoxy-$\Delta^{3,4}$-glucoseen reductase (also called E_3).

2. Iron–Sulfur Clusters and Pyridoxamine 5′-Phosphate (PMP) in Redox Reactions

3,6-Dideoxyhexoses are found in the O-antigen component of the lipopolysaccharide of a number of gram-negative bacteria and are responsible for bestowing immunogenic properties to the organisms.[131] Of the eight possible 3,6-dideoxyhexose isomers, to date only five have been discovered in nature; the biosynthesis of ascarylose (3,6-dideoxy-L-*arabino*-hexopyranose; **3.147**, Scheme 3.84) by *Yersinia pseudotuberculosis* V is the most well studied.[132] The biosynthetic pathway to ascarylose starts with cytidine 5′-phosphate (CDP)-D-glucose (**3.140**), which is converted with an NAD$^+$-dependent enzyme (see Chapter 10, Section I.A.2) to CDP-4-keto-6-deoxy-D-glucose (or CDP-6-deoxy-L-*threo*-D-*glycero*-4-hexulose; **3.141**). A pyridoxamine 5′-phosphate (PMP, **3.142**)-dependent and iron–sulfur cluster [2Fe–2S] containing enzyme (see later for structure), CDP-6-deoxy-L-*threo*-D-*glycero*-4-hexulose 3-dehydratase, also referred to as E_1, catalyzes the deoxygenation of **3.141** to PMP-bound CDP-4-keto-3,6-dideoxy-D-glucose (**3.144**), which undergoes hydrolysis to CDP-4-keto-3,6-dideoxy-D-glucose (**3.145**). E_1 is coupled to a second enzyme, CDP-6-deoxy-L-*threo*-D-*glycero*-4-hexulose 3-dehydratase reductase (formerly called CDP-6-deoxy-$\Delta^{3,4}$-glucoseen reductase), or E_3 (the en-

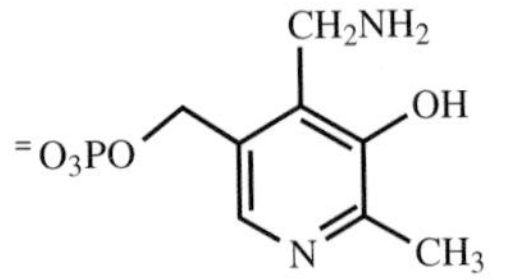

3.142

zyme with which this section is concerned), which is required for the conversion of **3.143** to **3.145**. E_3 uses several cofactors including NADH, FAD, and an iron–sulfur cluster, Fe(III)Fe(II)S_2, in an electron transfer process. The biosynthesis is completed by epimerization of **3.145** to **3.146,** which is reduced with a NADPH-dependent enzyme to asarylose.

PMP is a cofactor that we will discuss in Chapter 9 (Section IV.C.1) when we consider aminotransferase (isomerization) reactions. Typically, it is involved in carbanionic reactions; however, Thorson and Liu have EPR evidence with the E_1/E_3 system that PMP may be involved in an unprecedented two one-electron reduction of CDP-6-deoxy-$\Delta^{3,4}$-glucoseen (**3.143**, Scheme 3.84).[133]

Iron–sulfur clusters have been established by X-ray crystallography to exist as [2Fe–2S] (**3.148**), [3Fe–4S] (**3.149**), and [4Fe–4S] (**3.150**) forms (**3.149** and **3.150** are distorted cubanelike structures); they have redox potentials (E_m) in the range of -240 to -460 mV for **3.148**, -50 to -420 mV for **3.149**, and 0 to -645 mV for **3.150**.[134] Originally, they were believed to be involved only in single electron transfer reactions, but now it is known that they also catalyze dehydration reactions (see Chapter 10, Section I.A.3), stabilize protein structure, regulate metabolic pathways, act as a biological sensor of iron, O_2, and superoxide, and are involved in the formation of protein-bound radicals.[135] The reaction catalyzed by E_1 is discussed in more detail in Chapter 10 (Section I.A.3), so we will consider E_3 here. Some early experiments from the Strominger lab[136] showed that when (4*R*)- and (4*S*)-[4-^{3}H]NADH are incubated with the enzyme complex, the tritium is not transferred to either the sugar or to the PMP, but rather is released into the medium. Interestingly, there is no stereospecificity in the transfer of the tritium from the NADH because tritium is released at both the *pro-R* and *pro-S* positions of the NADH. When the reaction is run in ^{3}H$_2$O, one tritium is incorporated into the product. Liu and co-workers found that C-3 deoxygenation of **3.141** is not complete in the absence of E_3, but other electron transfer proteins, such as diaphorase and methane monooxygenase reductase, can substitute for E_3 and give the 3,6-dideoxyhexose (**3.145**).[137] This suggests that reduction by E_3 occurs by two one-electron transfers rather than by a typical NADH hydride transfer, which is consistent with the fact that no tritium from the NADH is incorporated into either the product or the PMP.

The iron–sulfur cluster of E_3 receives electrons, one at a time, from the reduced flavin (the FAD is first reduced by NADH in a two-electron (hydride) process) and then passes these electrons, via the iron–sulfur cluster of E_1, to **3.143**.[138] The order

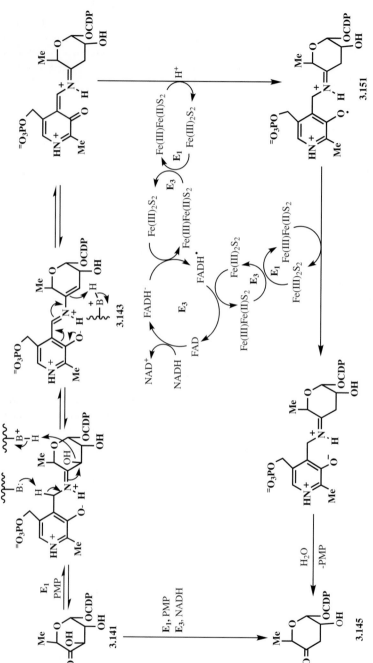

SCHEME 3.85   Mechanism proposed for the reduction of CDP-6-deoxy-$\Delta^{3,4}$-glucoseen by E_1 and E_3.

of electron transfer follows from a UV titration of E_1 with NADH, which confirms that reduction of E_1 only occurs in the presence of E_3.[139] Three possible related mechanisms were contemplated, but EPR analysis of the radical generated during catalysis is most consistent with an oxygen radical at the 3-position of PMP. That being the case, then the mechanism in Scheme 3.85 is most reasonable because it accounts for an O-3 radical (**3.151**, Scheme 3.85).

3. Molybdoenzymes and Related Tungstoenzymes

Enzyme-catalyzed hydroxylation reactions are ubiquitous in nature, but generally these enzymes contain flavin, pterin, heme, or nonheme iron cofactors, all of which catalyze hydroxylation from molecular oxygen as the source of the hydroxyl oxygen atom and are known as monooxygenases (see Chapter 4, Section II.B). In contrast to these enzymes, a family of enzymes that uses oxomolybdenum[140] or tungsten[141] cofactors catalyzes hydroxylation with incorporation of the hydroxyl group of water.[142] For example, liver sulfite oxidase contains a molybdopterin (**3.152**) cofactor held at the active site by coordination of Cys-207 (and possibly another amino acid) to the molybdenum atom,[143] *Rhodobacter sphaeroides* dimethyl sulfoxide reductase contains a bis(molybdopterin guanine dinucleotide) (**3.153**) cofactor,[144] and *Pyrococcus furiosus* aldehyde ferredoxin oxidoreductase has the related tungsten cofactor (**3.154**).[145] A possible mechanism for sulfite oxidation is shown in Scheme 3.86.

4. Pyruvoyl- and Selenocysteine-Dependent Reduction

There are two important amino acid reductases that use an active-site pyruvoyl group as a cofactor.[146] Pyruvoyl-dependent enzymes are discussed further in Chapter 8 (Section V.B). Inhibition of an enzyme with cyanide, hydroxylamine, phenylhydrazine, or sodium borohydride typically indicates the presence of a catalytically active carbonyl group, such as in PLP (see Chapter 8, Section V.A), PQQ (see Section III.C.1), or pyruvate. The identity of the pyruvoyl-containing active site can be revealed from several experiments. The pyruvoyl group is typically bound to the enzyme in an amide linkage; mild acid hydrolysis (1.5 N HCl, 100 °C, 1.5 h) produces pyruvate, which can be assayed with NADH-dependent lactate dehydrogenase.[147] Because dehydroalanyl-containing enzymes (see Chapter 10, Section I.C.1) also yield pyruvate on hydrolysis, these two classes of enzymes must be differentiated by another process. This can be accomplished by reduction with [³H]NaBH₄ followed by strong acid hydrolysis (6 N HCl, 110 °C, 24 h); pyruvoyl enzymes produce [³H]lactate, whereas dehydroalanyl enzymes give [³H]alanine.

Among PLP-, PQQ-, and pyruvoyl-dependent enzymes, the former two can be excluded as relevant coenzymes on the basis of the absorption spectrum of the enzyme; there is no absorbance above 300 nm for a pyruvoyl enzyme. Treatment of the pyruvoyl enzyme with ammonia and sodium cyanoborohydride followed

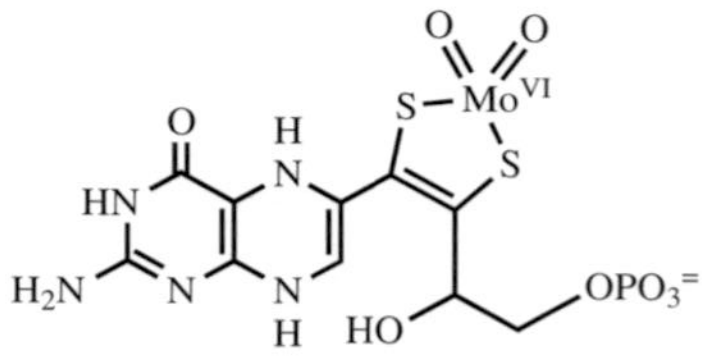

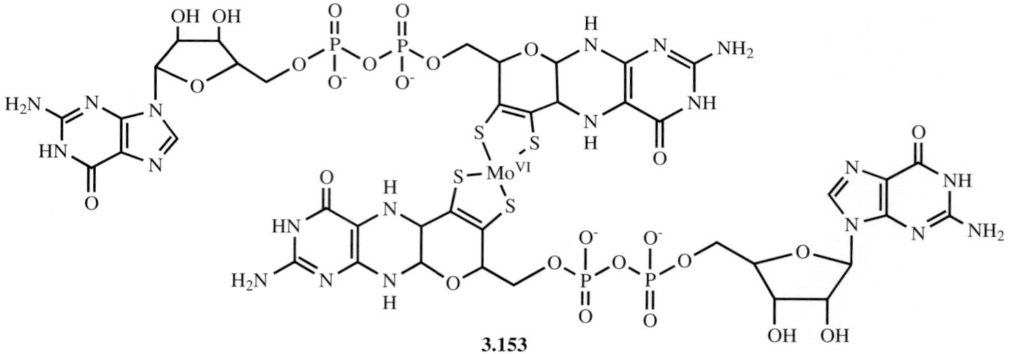

3.153

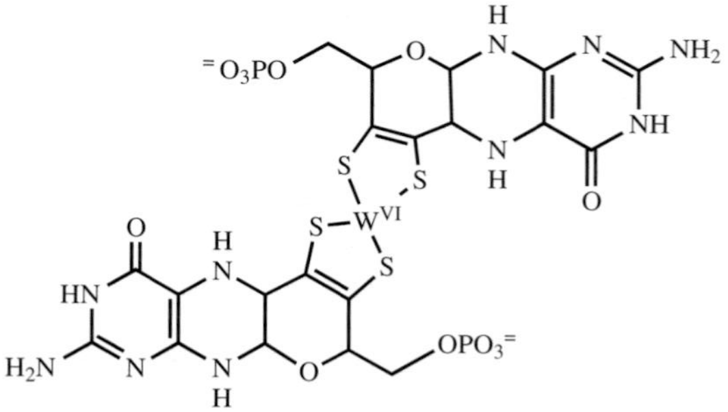

3.154

3.152

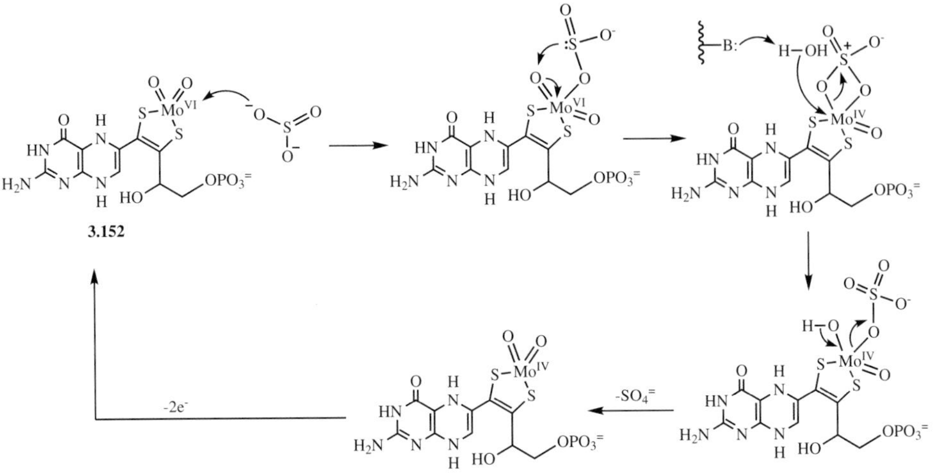

SCHEME 3.86   Mechanism proposed for sulfite oxidase.

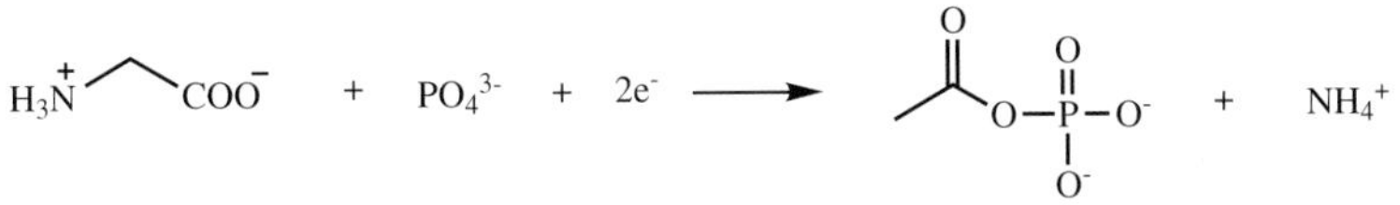

SCHEME 3.87   Reaction catalyzed by glycine reductase.

by acid hydrolysis produces alanine in the amino acid sequence that was not present prior to this treatment;[148] this would not occur with PLP- or PQQ-dependent enzymes.

Glycine reductase from *Clostridium sticklandii* is a multienzyme complex consisting of three proteins, referred to as protein A, B, and C, and catalyzes the two-electron reduction of glycine with inorganic phosphate to give acetyl phosphate and ammonium ion (Scheme 3.87).[149] Protein A contains a selenocysteine residue and two cysteine residues.[150] Protein B contains a covalently bound pyruvate cofactor. In the presence of protein B, protein A reacts with glycine to form a covalent protein A–*Se*-(carboxymethyl)selenocysteine intermediate (**3.155,** Scheme 3.88).[151] This covalent intermediate produces acetyl phosphate on addition of protein C and inorganic phosphate. Several experiments by Arkowitz and Abeles support the formation of acetyl phosphate as a product and suggest the intermediacy of an acetyl enzyme (**3.157**).[152] There is no exchange of [^{14}C]glycine with acetate, indicating that free acetate is not a precursor of acetyl phosphate. [^{18}O]Glycine loses the ^{18}O in the product, but the ^{18}O content of unreacted glycine remains unchanged after about half is converted to product. This suggests that an acyl enzyme (**3.157**) is an intermediate, and this acyl enzyme reacts with inorganic phosphate to give acetyl

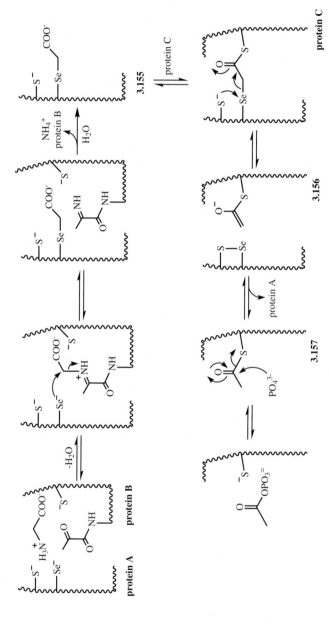

SCHEME 3.88 Mechanism proposed for the glycine reductase complex.

phosphate. The acetyl enzyme was isolated by the reaction of protein C and protein A–*Se*-(carboxymethyl)selenocysteine (generated by the reaction of glycine with proteins A and B) or by incubation of protein C with [³H]acetyl phosphate; the acetyl enzyme by either route is kinetically competent in the formation of acetyl phosphate with inorganic phosphate. When [¹⁴C]acetyl enzyme is denatured with trichloroacetic acid and redissolved in urea, radioactivity remains associated with the protein. At pH 11.5, the radioactivity is released at a rate comparable to the rate of hydrolysis of thioesters. Incubation of the labeled enzyme with hydroxylamine (which reacts rapidly with thioesters) releases all of the radioactivity, and treatment with potassium borohydride releases the radioactivity as [¹⁴C]ethanol. These results suggest that an enzyme thiol (cysteine residue) is acetylated. Finally, proteins A and C together catalyze the exchange of tritium from ³H₂O into acetyl phosphate, supporting the formation of an enol of the acetyl enzyme (**3.156**) as an intermediate. These results are consistent with the mechanism in Scheme 3.88. The alkylation of selenocysteine in the second step may seem unusual, but selenide is one of the most powerful nucleophiles known. The mechanism for the reaction of **3.155** with the thiol of protein C is unknown, but there must be some type of activation of the carboxylate for it to occur. A mechanism similar to that shown in Scheme 3.88 was proposed for the D-proline reductase-catalyzed conversion of D-proline to 5-aminovaleric acid, except without the formation of an acyl enzyme intermediate.[153]

5. Reduction with No Cofactors

Hydrogenases catalyze reactions with molecular hydrogen either as substrate or product. All but one hydrogenase studied so far are metalloenzymes. They all contain iron–sulfur clusters, and most also contain nickel.[154] The hydrogenase from a methanogenic archaebacterium, which catalyzes the reversible reduction of N^5,N^{10}-methenyl tetrahydromethanopterin (**3.158**) with molecular hydrogen to N^5,N^{10}-methylene tetrahydromethanopterin (**3.159**) and a proton contains neither nickel nor an iron–sulfur cluster and is metal free (Scheme 3.89).[155] This

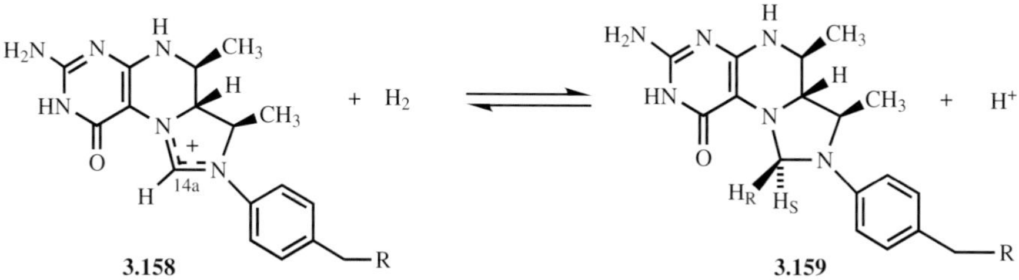

3.158   **3.159**

SCHEME 3.89 Reduction of N^5,N^{10}-methenyl tetrahydromethanopterin to N^5,N^{10}-methylene tetrahydromethanopterin catalyzed by the hydrogenase from a methanogenic archaebacterium.

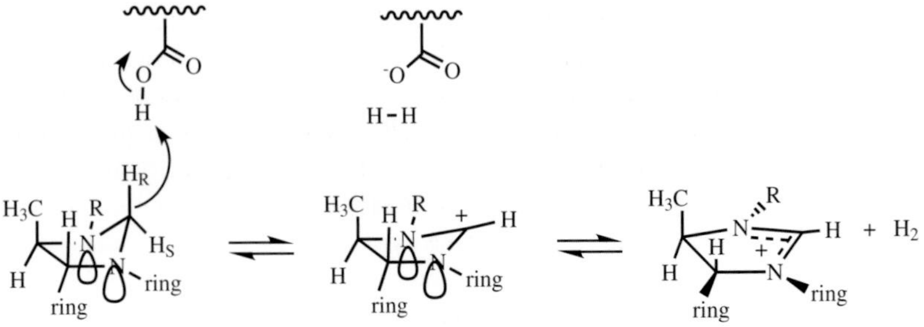

SCHEME 3.90   Mechanism proposed for oxidation of N^5,N^{10}-methylene tetrahydromethanopterin to N^5,N^{10}-methenyl tetrahydromethanopterin (reverse of the reaction in Scheme 3.89).

reaction is similar to the reaction catalyzed by the NADP$^+$-dependent N^5,N^{10}-methylene tetrahydrofolate dehydrogenase that we will discuss in Chapter 12 (Section I.B). The $\Delta G° = -1.3$ kcal/mol. The *pro-R* hydrogen is stereospecifically transferred to and from the product with exchange by solvent; H_2/H^+ exchange does not occur in the absence of substrate. A mechanism consistent with these observations is shown in Scheme 3.90. This may appear to be an unusual mechanism, but similar reactions are known to occur in superacid solutions. For example, per-hydro-3a,6a,9a-triazaphenalene (**3.161**) reacts with tetrafluoroboric acid (HBF$_4$) at 110 °C to give the *ortho* amide (**3.162**) and H_2 quantitatively (Scheme 3.91).[156] This reaction is assisted by the nitrogen lone pair orbitals *antiperiplanar* to the C—H bond that is ejected, thereby providing the necessary stereoelectronic effect. In the non-enzymatic reaction, however, there is no exchange with DBF$_4$, and the reaction is irreversible. To account for the reversibility of the enzyme reaction, intermediate **3.160** (Scheme 3.90) is proposed, in which the carbocation formed is not initially resonance stabilized (high energy). In the back direction (**3.158** → **3.159**), a conformational change in the enzyme could drive **3.158** to **3.160**, a superelectrophile, which could facilitate cleavage of the H—H bond. H_2 reacts with carbocations to give the corresponding saturated hydrocarbon,[157] and alkanes can undergo H/D exchange in superacid solution with retention of configuration;[158] both reactions support the mechanisms in Scheme 3.90.

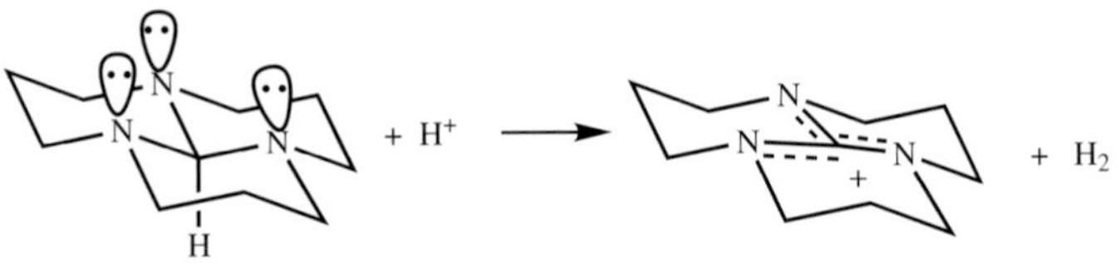

SCHEME 3.91   Reaction of perhydro-3a,6a,9a-triazaphenalene with tetrafluoroboric acid.

Ab initio calculations carried out on the mechanism of the metal-free hydrogenase from *M. thermoautotrophicum* indicate that the reaction has a strong dependence on base strength; the optimal base is predicted to be a primary amine, such as the ϵ-amino group of a lysine residue.[159] These calculations also predict that no initial distortion of the formamidinium cation from planarity is required.

Next we will consider enzymes that incorporate one oxygen atom from molecular oxygen into molecules, known as *monooxygenases*.

NOTE THAT PROBLEMS AND SOLUTIONS RELEVANT TO EACH CHAPTER CAN BE FOUND IN APPENDIX II.

REFERENCES

1. (a) Fan, C.; Teixeira, M.; Moura, J.; Moura, I.; Huynh, B.-H.; Le Gall, J.; Peck, H. D., Jr.; Hoffman, B. M. *J. Am. Chem. Soc.* **1991,** *113,* 20. (b) Silverman, R. B.; Cesarone, J. M.; Lu, X. *J. Am. Chem. Soc.* **1993,** *115,* 4955. (c) Rose, I. A.; Kuo, D. J. *Biochemistry* **1989,** *28,* 9579. (d) Rose, I. A.; Fung, W.-J.; Warms, J. V. B. *Biochemistry* **1990,** *29,* 4312.

2. (a) Rose, I. A. *Biochim. Biophys. Acta* **1957,** *25,* 214. (b) Franzen, V. *Chem. Ber.* **1956,** *89,* 1020. (c) Franzen, V. *Chem Ber.* **1957,** *90,* 623.

3. (a) Hall, S. S.; Doweyko, A. M.; Jordan, F. *J. Am. Chem. Soc.* **1976,** *98,* 7460. (b) Hall, S. S.; Doweyko, A. M.; Jordan, F. *J. Am. Chem. Soc.* **1978,** *100,* 5934.

4. (a) Chari, R. V. J.; Kozarich, J. W. *J. Biol. Chem.* **1981,** *256,* 9785. (b) Kozarich, J. W.; Chari, R. V. J.; Wu, J. C.; Lawrence, T. L. *J. Am. Chem. Soc.* **1981,** *103,* 4593.

5. (a) Dolphin, D.; Poulson, R.; Avramovic, O. "Pyridine Nucleotide Coenzymes," Parts A and B, Wiley: New York, 1987. (b) Oppenheimer, N. J.; Handlon, A. L. In "The Enzymes," 3rd ed., Sigman, D. S., Ed. vol. 20, Academic Press: San Diego, 1992, pp. 453–505.

6. Preiss, J.; Handler, P. *J. Biol. Chem.* **1958,** *233,* 493.

7. Newcomb, M. *Tetrahedron* **1993,** *49,* 1151.

8. (a) Suckling, C. J. *J. Chem. Soc. Chem. Commun.* **1982,** 1146. (b) Tatsumisago, M.; Matsubayashi, G-E.; Tanaka, T.; Nishigaki, S.; Nakatsu, K. *J. Chem. Soc. Chem. Commun.* **1982,** 121. (c) MacInnes, I.; Nonhebel, D. C.; Orszulik, S. T.; Suckling, C. J. *J. Chem. Soc. Perkin Trans. 1* **1983,** 2777.

9. Laurie, D.; Lucas, E.; Nonhebel, D. C.; Suckling, C. J.; Walton, J. C. *Tetrahedron* **1986,** *42,* 1035.

10. Tanner, D. D.; Stein, A. R. *J. Org. Chem.* **1988,** *53,* 1642.

11. Aleixo, L. M.; deCarvelho, M.; Moran, P. J. S.; Rodrigues, J. A. R. *Bioorg. Med. Chem. Lett.* **1993,** *3,* 1637.

12. See any basic organic chemistry text for a more complete set of rules for *R,S* nomenclature.

13. (a) Vennesland, B. *Fed. Proc.* **1958,** *17,* 1150. (b) Westheimer, F. H. In *Pyridine Nucleotide Coenzymes,* Part A, Dolphin, D.; Avramovic, O.; Poulson, R., Eds., Wiley: New York, 1987, p. 253.

14. (a) Benner, S. A. *Experientia* **1982,** *38,* 633. (b) Oppenheimer, N. J.; Arnold, L. J.; Kaplan, N. O. *Biochemistry* **1978,** *17,* 2613.

15. Young, L.; Post, C. B. *Biochemistry* **1996,** *35,* 15129.

16. Garavito, R. M.; Rossmann, M. G.; Argos, P.; Eventoff, W. *Biochemistry,* **1977,** *16,* 5065. (b) Sem, D. S.; Kasper, C. B. *Biochemistry,* **1992,** *31,* 3391.

17. (a) Nambiar, K. P.; Stauffer, D. M.; Kolodziej, P. A.; Benner, S. A. *J. Am. Chem. Soc.* **1983,** *105,* 5886. (b) Benner, S. A.; Nambiar, K. P.; Chambers, G. K. *J. Am. Chem. Soc.* **1985,** *107,* 5513.

18. Fall, R.; Murphy. S. E. *J. Am. Chem. Soc.* **1984,** *106,* 3033.

19. You, K-S. *CRC. Crit. Rev. Biochem.* **1985,** *17,* 313.

20. Benner, S. A.; Nambiar, K. P.; Chambers, G. K. *J. Am. Chem. Soc.* **1985,** *107,* 5513.

21. (a) Almarsson, O.; Karaman, R.; Bruice, T. C. *J. Am. Chem. Soc.* **1992**, *114*, 8702. (b) Almarsson, O.; Bruice, T. C. *J. Am. Chem. Soc.* **1993**, *115*, 2125.

22. Schneider-Bernlöhr, H.; Adolph, H-W.; Zeppezauer, M. *J. Am. Chem. Soc.* **1986**, *108*, 5573.

23. Liu, H.; Wallace, K. K.; Reynolds, K. A. *J. Am. Chem. Soc.* **1997**, *119*, 2973.

24. Grunwald, J.; Wirz, B.; Scollar, M. P.; Klibanov, A. M. *J. Am. Chem. Soc.* **1986**, *108*, 6732.

25. Brunhuber, N. M. W.; Blanchard, J. S. *CRC Crit. Rev. Biochem. Mol. Biol.* **1994**, *29*, 415.

26. Stillman, T. J.; Baker, P. J.; Britton, K. L.; Rice, D. W. *J. Mol. Biol.*, **1993**, *234*, 1131.

27. Teng, H.; Segura, E.; Grubmeyer, C. *J. Biol. Chem.* **1993**, *268*, 14182.

28. Blatter, E. E.; Abriola, D. P.; Pietruszko, R. *Biochem. J.* **1992**, *282*, 353.

29. Corbier, C.; Della Seta, F.; Branlant, G. *Biochemistry* **1992**, *31*, 12532.

30. Wang, W.; Papov, V. V.; Minakawa, N.; Matsuda, A.; Biemann, K.; Hedstrom, L. *Biochemistry* **1996**, *35*, 95.

31. (a) Klepp, J.; Fallert-Muller, A.; Grim, K.; Hull, W. E.; Rétey, J. *Eur. J. Biochem.* **1990**, *192*, 669. (b) Klepp, J.; Rétey, J. *Eur. J. Biochem.* **1989**, *185*, 615.

32. Schubert, C.; Röttele, H.; Spraul, M.; Rétey, J. *Angew. Chem. Int. Ed. Engl.* **1995**, *34*, 652.

33. Schubert, C.; Rétey, J. *Bioorg. Chem.*, **1994**, *22*, 368.

34. (a) Walsh, C. *Acc. Chem. Res.* **1980**, *13*, 148. (b) Bruice, T. C. *Acc. Chem. Res.* **1980**, *13*, 256. (c) Ghisla, S.; Massey, V. *Eur. J. Biochem.* **1989**, *181*, 1.

35. (a) Chlumsky, L. J.; Sturgess, A. W.; Nieves, E.; Jorns, M. S. *Biochemistry* **1998**, *37*, 2089. (b) Singer, T. P.; Edmondson, D. E. *Methods Enzymol.* **1980**, *66*, 253. (c) Edmondson, D. E.; Kenney, W. C.; Singer, T. P. *Methods Enzymol.* **1978**, *53*, 449.

36. (a) Kearney, E. B.; Salach, J. I.; Walker, W. H.; Seng, R. L.; Kenney, W.; Zeszotek, E.; Singer, T. P. *Eur. J. Biochem.* **1971**, *24*, 321. (b) Steenkamp, D. J.; Denney, W. C.; Singer, T. P. *J. Biol. Chem.* **1978**, *253*, 2812.

37. Bruice, T. C. *Israel J. Chem.* **1984**, *24*, 54.

38. Hammett, L. P. *Physical Organic Chemistry*, McGraw-Hill: New York, 1940.

39. Neims, A. H.; DeLuca, D. C.; Hellerman, L. *Biochemistry* **1966**, *5*, 203.

40. Walsh, C. T.; Schonbrunn, A.; Abeles, R. H. *J. Biol. Chem.* **1971**, *246*, 6855.

41. Dang, T-Y.; Cheung, Y-F.; Walsh, C. *Biochem. Biophys. Res. Commun.* **1976**, *72*, 960.

42. (a) Yokoe, Y.; Bruice, T. C. *J. Am. Chem. Soc.* **1975**, *97*, 450. (b) Loechler, L. E.; Hollocher, T. C. *J. Am. Chem. Soc.* **1975**, *97*, 3235.

43. (a) Ball, S.; Bruice, T. C. *J. Am. Chem. Soc.* **1980**, *102*, 6498. (b) Kim, J. M.; Bogdan, M. A.; Mariano, P. S. *J. Am. Chem. Soc.* **1993**, *115*, 10591.

44. Michaels, G. B.; Davidson, J. T.; Peck, H. D., Jr. *Biochem. Biophys. Res. Commun.* **1970**, *39*, 321.

45. Fisher, J.; Spencer, R.; Walsh, C. *Biochemistry* **1976**, *15*, 1054.

46. (a) Hemmerich, P.; Massey, V.; Fenner, H. *FEBS Lett.* **1977**, *84*, 5. (b) Walsh, C. *Acc. Chem. Res.* **1986**, *19*, 216.

47. Kurtz, K. A.; Fitzpatrick, P. F. *J. Am. Chem. Soc.* **1997**, *119*, 1155.

48. Mattevi, A.; Vanoni, M. A.; Todone, F.; Rizzi, M; Teplyakov, A.; Coda, A.; Bolognesi, M.; Curti, B. *Proc. Natl. Acad. Sci. USA* **1996**, *93*, 7496.

49. Todone, F.; Vanoni, M. A.; Mozzarelli, A.; Bolognesi, M.; Coda, A.; Curti, B.; Mattevi, A. *Biochemistry* **1997**, *36*, 5853.

50. (a) Novak, M.; Bruice, T. C. *J. Am. Chem. Soc.* **1977**, *99*, 8079. (b) Williams, R. F.; Bruice, T. C. *J. Am. Chem. Soc.* **1976**, *98*, 7752. (c) Bruice, T. C.; Taulane, J. P. *J. Am. Chem. Soc.* **1976**, *98*, 7769. (d) Bruice, T. C.; Yano, Y. *J. Am. Chem. Soc.* **1975**, *97*, 5263. (e) Bruice *Proc. Natl. Acad. Sci. USA* **1975**, *72*, 1763.

51. Wenz, A.; Thorpe, C. Ghisla, S. *J. Biol. Chem.* **1981**, *256*, 9809.

52. Lenn, N. D.; Shih, Y.; Stankovich, M. T.; Liu, H-w. *J. Am. Chem. Soc.* **1989**, *111*, 3065.

53. Lai, M-t.; Liu, L-d.; Liu, H-w. *J. Am. Chem. Soc.* **1991**, *113*, 7388.

54. (a) Lai , M-t; Liu, H-w. *J. Am. Chem. Soc.* **1992**, *114*, 3160. (b) Lai, M-t.; Oh, E.; Liu, H-w. *J. Am. Chem. Soc.* **1993**, *115*, 1619.

55. (a) Silverman, R. B. In *Advances in Electron Transfer Chemistry,* vol. 2, Mariano, P. S., Ed., JAI Press: Greenwich, CT, 1992, pp. 177–213. (b) Silverman, R. B. *Acc. Chem. Res.* **1995,** *28,* 335.

56. (a) Walker, M. C.; Edmondson, D. E. *Biochemistry* **1994,** *33,* 7088. (b) Miller, J. R.; Edmondson, D. E.; Grissom, C. B. *J. Am. Chem. Soc.* **1995,** *117,* 7830. (c) Jonsson, T.; Edmondson, D. E.; Klinman, J. P. *Biochemistry* **1994,** *33,* 14871.

57. (a) Maeda, Y.; Ingold, K. U. *J. Am. Chem. Soc.* **1980,** *102,* 328. (b) Musa, O. M.; Horner, J. H.; Shahin, H.; Newcomb, M. *J. Am. Chem. Soc.* **1996,** *118,* 3862. (c) Newcomb, M. *Tetrahedron* **1993,** *49,* 1151.

58. Silverman, R. B.; Zieske, P. A. *Biochemistry* **1985,** *24,* 2128.

59. Silverman, R. B.; Zieske, P. A. *Biochem. Biophys. Res. Commun.* **1986,** *135,* 154.

60. Bach, A. W. J.; Lan, N. C.; Johnson, D. L.; Abell, C. W.; Bembenek, M. E.; Kwan, S-W.; Seeburg, P. H.; Shin, J. C. *Proc. Natl. Acad. Sci. USA* **1988,** *85,* 4934.

61. Zhong, B.; Silverman, R. B. *J. Am. Chem. Soc.* **1997,** *119,* 6690.

62. Burlingame, A. L.; Boyd, R. K.; Gaskell, S. J. *Anal. Chem.* **1996,** *68,* 599R. (b) Fitzgerald, M. C.; Siuzdak, G. *Chem. Biol.* **1996,** *3,* 707. (c) Roepstorff, P. *Curr. Opin. Biotechnol.* **1997,** *8,* 6.

63. (a) Fitzgerald, M. C.; Siuzdak, G. *Chem. Biol.* **1996,** *3,* 707. (b) Siuzdak, G. *Proc. Natl. Acad. Sci. USA* **1994,** *91,* 11290.

64. Silverman, R. B.; Zieske, P. A. *Biochemistry* **1986,** *25,* 341.

65. Yelekçi, K.; Lu, X.; Silverman, R. B. *J. Am. Chem. Soc.* **1989,** *111,* 1138.

66. Silverman, R. B.; Zhou, J. P.; Eaton, P. E. *J. Am. Chem. Soc.* **1993,** *115,* 8841.

67. (a). Dickinson, J. M.; Murphy, J. A.; Patterson, C. W.; Wooster, N. F. *J. Chem Soc., Perkin Trans.* **1990,** *1,* 1179. (b) Johns, A.; Murphy, J. A.; Patterson, C. W.; Wooster, N. F. *J. Chem. Soc. Chem. Commun.* **1987,** 1238.

68. Silverman, R. B.; Lu, X.; Zhou, J. J. P.; Swihart, A. *J. Am. Chem. Soc.* **1994,** *116,* 11590.

69. Silverman, R. B.; Zhou, J. J. P.; Ding, C. Z.; Lu, X. *J. Am. Chem. Soc.* **1995,** *117,* 12895.

70. Lu, X.; Silverman, R. B. *J. Am. Chem. Soc.* **1998,** *120,* 10583.

71. Gates, K. S.; Silverman, R. B. *J. Am. Chem. Soc.* **1990,** *112,* 9364.

72. Ding, Z.; Silverman, R. B. *J. Med. Chem.* **1992,** *35,* 885.

73. Ding, C. Z.; Silverman, R. B. *J. Enz. Inhib.* **1992,** *6,* 223.

74. Silverman, R. B.; Ding, C. Z. *J. Am. Chem. Soc.* **1993,** *115,* 4571.

75. (a) Annan, N.; Silverman, R. B. *J. Med. Chem.,* **1993,** *36,* 3968–3970. (b) Silverman, R. B.; Hawe, W. P. *J. Enz. Inhib.,* **1995,** *9,* 203–215.

76. Miller, J. R.; Edmondson, D. E.; Grissom, C. B. *J. Am. Chem. Soc.* **1995,** *117,* 7830.

77. Walker, M. C.; Edmondson, D. E. *Biochemistry* **1994,** *33,* 7088.

78. DeRose, V. J.; Woo, J. C. G.; Hawe, W. P.; Hoffman, B. M.; Silverman, R. B.; Yelekçi, K. *Biochemistry* **1996,** *35,* 11085.

79. (a) Hamilton, G. A. In *Progress in Bioorganic Chemistry,* vol. 1, Kaiser, E. T.; Kezdy, F. J., Eds., Wiley: New York, 1971, p. 83. (b) Kim, J.-M.; Bogdan, M. A.; Mariano, P. S. *J. Am. Chem. Soc.* **1993,** *115,* 10591. (c) Kim, J.-M.; Hoegy, S. E.; Mariano, P. S. *J. Am. Chem. Soc.* **1995,** *117,* 100.

80. Silverman, R. B.; Lu, X. *J. Am. Chem. Soc.* **1994,** *116,* 4129.

81. Benson, T. E.; Marquardt, J. L.; Marquardt, A. C.; Etzkorn, F. A.; Walsh, C. T. *Biochemistry* **1993,** *32,* 2024.

82. Lees, W. J.; Benson, T. E.; Hogle, J. M.; Walsh, C. T. *Biochemistry* **1996,** *35,* 1342.

83. Hines, V.; Johnston, M. *Biochemistry* **1989,** *28,* 1227.

84. Kaiser, E. T.; Lawrence, D. S. *Science* **1984,** *226,* 505.

85. Slama, J. T.; Radziejewski, C.; Oruganti, S. R.; Kaiser, E. T. *J. Am. Chem. Soc.* **1984,** *106,* 6778.

86. Kahn, K.; Tipton, P. A. *Biochemistry* **1997,** *36,* 4731.

87. Kahn, K.; Serfozo, P.; Tipton, P. A. *J. Am. Chem. Soc.* **1997,** *119,* 5435.

87a. Kahn, K.; Tipton, P. A. *Biochemistry* **1998,** *37,* 11651.

88. Salisbury, S. A.; Forrest, H. S.; Cruse, W. B. T.; Kennond, O. *Nature* **1979,** *280,* 843.

89. (a) Duine, J. A.; Frank, J.; Jongejan, J. A. *Adv. Enzymol.* **1987,** *59,* 169. (b) Duine, J. A. *Eur. J. Biochem.* **1991,** *200,* 271.

90. Duine, J. A.; Frank, J.; Verwiel, J.; Verwiel, P. E. J. *Eur. J. Biochem.* **1980,** *108,* 187.

91. (a) Williamson, P. R.; Moog, R. S.; Dooley, D. M.; Kagan, H. M. *J. Biol. Chem.* **1986,** *261,* 16302. (b) van der Meer, R. A.; Duine, J. A. *Biochem. J.* **1986,** *239,* 789.

92. Duine, J. A.; Jongejan, J. A. *Annu. Rev. Biochem.* **1989,** *58,* 403.

93a. (a) Blake, C. C. F.; Ghosh, M.; Harlos, K.; Avezoux, A.; Anthony, C. *Nature Struct. Biol.* **1994,** *1,* 102. (b) Xia, Z.-x.; Dai, W.-w.; Zhang, Y.-f; White, S. A.; Boyd, G. D.; Mathews, F. S. *J. Mol. Biol.* **1996,** *259,* 480.

93b. Oubrie, A.; Rozeboom, H. J.; Dijkstra, B. W. *Proc. Natl. Acad. Sci. USA* **1999,** *96,* 11787.

93c. Oubrie, A.; Rozeboom, H. J.; Kalk, K. H.; Olsthoorn, A. J. J.; Duine, J. A.; Dijkstra, B. W. *EMBO J.* **1999,** *18,* 5187.

93d. Zheng, Y.-J.; Bruice, T. C. *Proc. Natl. Acad. Sci. USA* **1997,** *94,* 11881.

94. Klinman, J. P.; Mu, D. *Annu. Rev. Biochem.* **1994,** *63,* 299.

95. Hartmann, C.; Klinman, J. P. *J. Biol. Chem.* **1987,** *262,* 962.

96. Farnum, M.; Palcic, M.; Klinman, J. P. *Biochemistry* **1986,** *25,* 1898.

97. Farnum, M. F.; Klinman, J. P. *Biochemistry* **1986,** *25,* 6028.

98. Janes, S. M.; Mu, D.; Wemmer, D.; Smith, A. J.; Kaur, S.; Maltby, D.; Burlingame, A. L.; Klinman, J. P. *Science* **1990,** *248,* 981.

99. Brown, D. E.; McGuirl, M. A.; Dooley, D. M.; Janes, S. M.; Mu, D.; Klinman, J. P. *J. Biol. Chem.* **1991,** *266,* 4049.

100. Janes, S. M.; Palcic, M. M.; Scaman, C. H.; Smith, A. S.; Brown, D. E.; Dooley, D. M.; Mure, M.; Klinman, J. P. *Biochemistry* **1992,** *31,* 12147.

101. Hartmann, C.; Klinman, J. P. *Biochemistry* **1991,** *30,* 4605.

102. Hartmann, C.; Brzovic, P.; Klinman, J. P. *Biochemistry* **1993,** *32,* 2234.

103. (a) Mure, M.; Klinman, J. P. *J. Am. Chem. Soc.* **1993,** *115,* 7117. (b) Mure, M.; Klinman, J. P. *J. Am. Chem. Soc.* **1995,** *117,* 8698.

104. Nakamura, N.; Moënne-Loccoz, P.; Tanizawa, K.; Mure, M.; Suzuki, S.; Klinman, J. P.; Sanders-Loehr, J. *Biochemistry* **1997,** *36,* 11479.

105. Mure, M.; Klinman, J. P. *J. Am. Chem. Soc.* **1995,** *117,* 8707.

106. Klinman, J. P. *J. Biol. Chem.* **1996,** *271,* 27189.

107. Janes, S. M.; Klinman, J. P. *Biochemistry* **1991,** *30,* 4599.

108. (a) Finazzi-Agro, A.; Rinaldi, A.; Floris, G.; Rotolio, G. *FEBS Lett.* **1984,** *176,* 378. (b) Dooley, D. M.; McGuirl, M. A.; Peisaeh, J.; McCracken, J. *FEBS Lett.* **1987,** *214,* 274. (c) Dooley, D. M.; McGuirl, M. A.; Brown, D. E.; Turowski, P. N.; McIntire, W. S.; Knowles, P. F. *Nature* **1991,** *349,* 262. (d) Warncke, K.; Babcock, G. T.; Dooley, D. M.; McGuirl, M. A.; McCracken, J. *J. Am. Chem. Soc.* **1994,** *116,* 4028.

109. Pedersen, J. Z.; El-Sherbini, S.; Finazzi-Agrò, A.; Rotilio, G. *Biochemistry* **1992,** *31,* 8.

110. McCracken, J.; Peisach, J.; Cote, C. E.; McGuirl, M. A.; Dooley, D. M. *J. Am. Chem. Soc.* **1992,** *114,* 3715.

111. Moënne-Loccoz, P.; Nakamura, N.; Steinebach, V.; Duine, J. A.; Mure, M.; Klinman, J. P.; Sanders-Loehr, J. *Biochemistry,* **1995,** *34,* 7020.

112. Parsons, M. R.; Convery, M. A.; Wilmot, C. M.; Yadav, K. D. S.; Blakley, V.; Corner, A. S.; Phillips, S. E. V.; McPherson, J. J.; Knowles, P. F. *Structure* **1995,** *15,* 1171.

113. Wilce, M. C. J.; Dooley, D. M.; Freeman, H. C.; Guss, J. M.; Matsunami, H; McIntyre, W. S.; Ruggiero, C. E.; Tanizawa, K.; Yamaguchi, H. *Biochemistry* **1997,** *36,* 16116.

114. (a) Cai, D.; Klinman, J. P. *Biochemistry* **1994,** *33,* 7647. (b) Matsuzaki, R.; Fukui, T.; Sato, H.; Ozaki, Y.; Tanizawa, K. *FEBS Lett.* **1994,** *351,* 360.

115. Ruggiero, C. E.; Smith, J. A.; Tanizawa, K.; Dooley, D. M. *Biochemistry* **1997,** *36,* 1953.

116. (a) Backes, G.; Davidson, V. L.; Huitema, F.; Duine, J. A.; Sanders-Loehr, J. *Biochemistry* **1991,** *30,* 9201. (b) Davidson, V. L. In *Principles and Applications of Quinoproteins,* Davidson, V. L., Ed., Dekker: New York, 1993, pp. 73–95.

117. (a) McIntyre, W. S.; Wemmer, D. E.; Chistoserdov, A.; Lidstrom, M. E. *Science* **1991,** *252,* 817. (b) McIntire, W. S.; Bates, J. L., Brown, D. E.; Dooley, D. M. *Biochemistry* **1991,** *30,* 125–133.

118. Chen, L.; Mathews, F. S.; Davidson, V. L.; Huizinga, E. G.; Vellieux, F. M. D.; Duine, J. A.; Hol, W. G. L. *FEBS Lett.* **1991,** *287,* 163.

119. Davidson, V. L.; Jones, L. H., Graichen, M. E. *Biochemistry* **1992,** *31,* 3385.

120. Brooks, H. B.; Jones, L. H.; Davidson, V. L. *Biochemistry,* **1993,** *32,* 2725.

121. Hyun, Y.; Davidson, V. L. *Biochemistry,* **1995,** *34,* 816.

122. Wang, S. X.; Mure, M.; Medzihradszky, K. F.; Burlingame, A. L.; Brown, D. E.; Dooley, D. M.; Smith, A. J.; Kagan, H. M.; Klinman, J. P. *Science* **1996,** *273,* 1078.

123. Williamson, P. R.; Kagan, H. M. *J. Biol. Chem.* **1987,** *262,* 8196.

124. (a) Stubbe, J. *Annu. Rev. Biochem.* **1989,** *58,* 257. (b) Pedersin, J. Z.; Finazzi-Agro, A. *FEBS Lett.* **1993,** *325,* 53.

125. (a) Ito, N.; Phillips, S. E. V.; Stevens, C.; Ogel, Z. B.; McPherson, M. J.; Keen, J. N.; Yadav, K. D. S.; Knowles, P. F. *Nature (London)* **1991,** *350,* 87. (b) Ito, N.; Phillips, S. E. V.; Yadav, K. D. S.; Knowles, P. F. *J. Mol. Biol.* **1994,** *238,* 794.

126. Wachter, R. M.; Branchaud, B. P. *J. Am. Chem. Soc.* **1996,** *118,* 2782.

127. Whittaker, M. M.; Chuang, Y-Y.; Whittaker, J.W. *J. Am. Chem. Soc.* **1993,** *115,* 10029.

128. Branchaud, B. P.; Montague-Smith, M. P.; Kosman, D. J.; McLaren, F. R. *J. Am. Chem. Soc.* **1993,** *115,* 798.

129. Wachter, R. M.; Montague-Smith, M. P.; Branchaud, B. P. *J. Am. Chem. Soc.* **1997,** *119,* 7743.

130. (a) Tanner, D. D.; Chen, J. J.; Chen, L.; Luelo, C. *J. Am. Chem. Soc.* **1991,** *113,* 8074. (b) Tanner, D. D.; Xie, G.-J.; Hooz, J.; Yang, C.-M. *J. Org. Chem.* **1993,** *58,* 7138.

131. (a) Lindberg, B. *Adv. Carbohydr. Chem. Biochem.* **1990,** *48,* 279. (b) Williams, N. R.; Wander, J. D. In *The Carbohydrates: Chemistry and Biochemistry,* vol. 1B, Pigman, W.; Horton, D., Eds., Academic Press: Orlando, 1970, p. 761.

132. Liu, H.-w.; Thorson, J. S. *Annu. Rev. Microbiol.* **1994,** *48,* 223.

133. Thorson, J. S.; Liu, H.-w. *J. Am. Chem. Soc.* **1993,** *115,* 12177.

134. Commack, R. In *Advances in Inorganic Chemistry: Iron–Sulfur Proteins,* vol. 38, Commack, R., Ed., Academic Press: San Diego, 1992, pp. 281–322.

135. (a) Flint, D. H.; Allen, R. M. *Chem. Rev.* **1996,** *96,* 2315. (b) Frey, P. A.; Reed, G. A. *Adv. Enzymol.* **1993,** *66,* 1. (c) Beinert, H.; Holm, R. H.; Münck, E. *Science* **1997,** *277,* 653.

136. Rubenstein, P. A.; Strominger, J. L. *J. Biol. Chem.* **1974,** *249,* 3782.

137. Weigel, T. M.; Miller, V. P.; Liu, H.-W. *Biochemistry* **1992,** *31,* 2140.

138. Miller, V. P.; Thorson, J. S.; Ploux; Lo, S. F.; Liu, H. W. *Biochemistry* **1993,** *32,* 11934.

139. Thorson, J. S.; Liu, H.-w. *J. Am. Chem. Soc.* **1993,** *115,* 7539.

140. Hille, R. *Chem. Rev.* **1996,** *96,* 2757.

141. Johnson, M. K.; Rees, D. C.; Adams, M. W. W. *Chem. Rev.* **1996,** *96,* 2817.

142. Hille, R. *Biochim. Biophys. Acta* **1994,** *1184,* 143.

143. Garrett, R. M.; Rajagopalan, K. V. *J. Biol. Chem.* **1996,** *271,* 7387.

144. Hilton, J. C.; Rajagopalan, K. V. *Arch. Biochem. Biophys.* **1996,** *325,* 139.

145. Johnson, J. L.; Rajagopalan, K. V.; Mukund, S.; Adams, M. W. *J. Biol. Chem.* **1993,** *268,* 4848.

146. van Poeljie, P. D.; Snell, E. E. *Annu. Rev. Biochem.* **1990,** *59,* 29.

147. (a) Riley, W. D.; Snell, E. E. *Biochemistry* **1968,** *7,* 3520. (b) Gutmann, I.; Wahlefeld, A. W. In *Methods of Enzymatic Analysis;* Bergmeyer, H. U., Ed., Academic Press: New York, 1974, pp. 1464–1468.

148. Huynh, Q. K.; Vaaler, G. L.; Recsei, P. A.; Snell, E. E. *J. Biol. Chem.* **1984,** *259,* 2826.

149. Arkowitz, R. A.; Abeles, R. H. *Biochemistry* **1991,** *30,* 4090.

150. Cone, J. E.; Del Rio, R. M.; Davis, J. N.; Stadtman, T. C. *Proc. Natl. Acad. Sci. USA* **1976,** *73,* 2659.

151. Arkowitz, R. A.; Abeles, R. H. *J. Am. Chem. Soc.* **1990,** *112,* 870.

152. Arkowitz, R. A.; Abeles, R. H. *Biochemistry* **1989,** *28,* 4639.

153. Arkowitz, R. A.; Dhe-Paganon, S.; Abeles, R. H. *Arch. Biochem. Biophys.* **1994,** *311,* 457.

154. Albracht, S. P. J. *Biochim. Biophys. Acta,* **1994,** *1188,* 167.

155. Berkessel, A.; Thauer, R. K. *Angew. Chem. Int. Ed. Engl.,* **1995,** *34,* 2247.

156. Erhardt, J. M.; Wuest, J. D. *J. Am. Chem. Soc.,* **1980,** *102,* 6363.

157. Bickel, A. F.; Gaasbeck, C. J.; Hogeveen, H.; Oelderck, J. M.; Platteeuw, J. C. *J. Chem. Soc. Chem. Commun.,* **1967,** 634.

158. (a) Hogeveen, H.; Gaasbeck, C. J. *Recl. Trav. Chim. Pays-Bas,* **1968,** *87,* 319. (b) Olah, G. A.; Hartz, N.; Rasul, G.; Surya Prakash, G. K. *J. Am. Chem. Soc.,* **1995,** *117,* 1336.

159. Teles, J. H.; Brode, S.; Berkessel, A. *J. Am. Chem. Soc.* **1998,** *120,* 1345.

CHAPTER 4

Monooxygenation

I. GENERAL

In the last chapter we discussed enzymes that catalyze oxidations of molecules. In those cases where molecular oxygen was involved, neither oxygen atom of O_2 was incorporated into the substrate molecule; the O_2 acted only as an oxidizing agent. In this chapter, one oxygen atom of molecular oxygen will end up in a product. Enzymes that catalyze the incorporation of one oxygen atom from molecular O_2 into the substrate are called *monooxygenases;* the other oxygen atom is incorporated into H_2O. Typical reactions catalyzed by monooxygenases are shown in Table 4.1. All of these reactions can be explained as reactions related to the action of a hydroperoxide, but as we will see, some of these reactions will require a more reactive oxidant.

Internal monooxygenases catalyze the oxygenation of a substrate without requiring an external reducing agent—the substrate itself is also the reducing agent. *External monooxygenases* require an external reducing agent to activate the enzyme for reaction.

II. FLAVIN-DEPENDENT HYDROXYLASES

A. No Reducing Agent Required

1. Lactate Oxidase, an Oxidative Decarboxylation Reaction

Lactate oxidase from *Mycobacteria* is a flavoenzyme that catalyzes the oxidative decarboxylation of lactate (**4.1**) to acetate, carbon dioxide, and water (Scheme 4.1). No external reducing agent is required, because the substrate acts as the reducing agent. The following experiments were carried out to elucidate the mechanism. In

TABLE 4.1   Typical Reactions Catalyzed by Monooxygenases

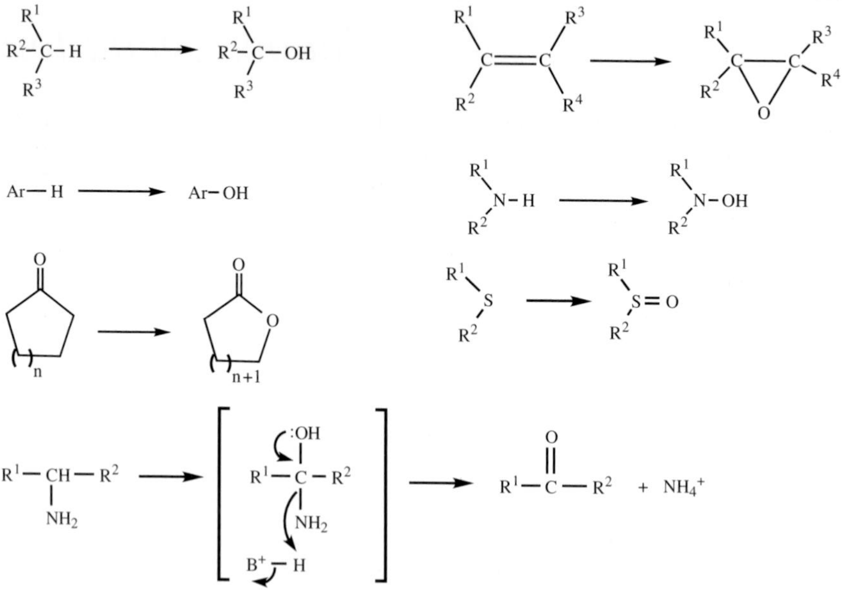

a one-turnover experiment (enzyme concentration equivalent to or in excess over substrate), under anaerobic conditions the product obtained is pyruvate (**4.2**) and the flavin is reduced (Scheme 4.2); therefore, the enzyme is acting like an oxidase (see Chapter 3, Section III.B), not like a monooxygenase.[1] If the reduced enzyme is incubated with [^{14}C]pyruvate, and then O_2 is added, the normal reaction products are obtained (Scheme 4.3); therefore, pyruvate is an intermediate.[2] However, if

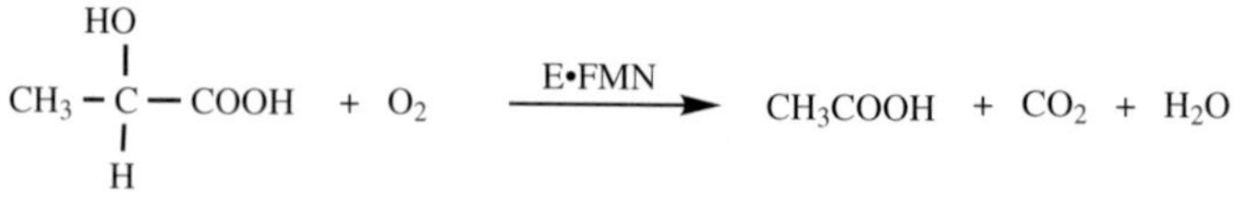

$$CH_3-\overset{\overset{\displaystyle HO}{|}}{\underset{\underset{\displaystyle H}{|}}{C}}-COOH \;+\; O_2 \xrightarrow{\text{E•FMN}} CH_3COOH \;+\; CO_2 \;+\; H_2O$$

4.1

SCHEME 4.1   Reaction catalyzed by lactate oxidase from *Mycobacteria*.

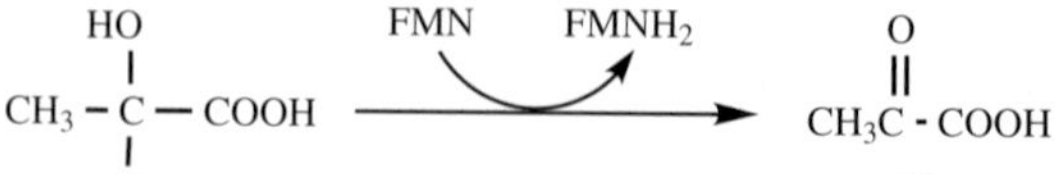

$$CH_3-\overset{\overset{\displaystyle HO}{|}}{\underset{\underset{\displaystyle H}{|}}{C}}-COOH \xrightarrow[\text{FMN} \quad \text{FMNH}_2]{} CH_3\overset{\overset{\displaystyle O}{||}}{C}\text{-}COOH$$

4.2

SCHEME 4.2   The lactate oxidase reaction under anaerobic conditions.

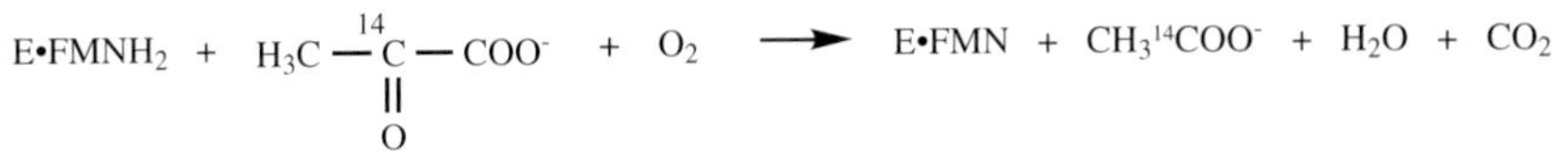

$$E \cdot FMNH_2 \ + \ H_3C - \overset{14}{\underset{\underset{O}{\|}}{C}} - COO^- \ + \ O_2 \ \longrightarrow \ E \cdot FMN \ + \ CH_3{}^{14}COO^- \ + \ H_2O \ + \ CO_2$$

SCHEME 4.3   Reaction of reduced lactate oxidase with pyruvate and oxygen.

the reduced enzyme is first treated with O_2, then with [^{14}C]pyruvate, the pyruvate remains unchanged, but hydrogen peroxide is produced. This result supports a mechanism such as flavin-dependent oxidases (see Scheme 3.33) in which H_2O_2 is generated by the enzyme but is released in the absence of substrate. When $^{18}O_2$ is utilized, one ^{18}O ends up in the acetate carboxyl group, and the other as $H_2{}^{18}O$, as expected for a monooxygenase.[3] A model study was carried out to test the ability of hydrogen peroxide to convert pyruvate to acetate, carbon dioxide, and water; it does.[4]

The preceding experiments support the mechanisms in Scheme 4.4. Oxidation of lactate by FMN can be accomplished by one of the mechanisms discussed in Chapter 3 (Section III.B.3.a, Scheme 3.43 and Section III.B.3.b, Scheme 3.44) to give pyruvate and reduced flavin (FMNH$_2$). Reaction of FMNH$_2$ with O_2 would give the flavin C^{4a}-hydroperoxide (**4.3**) via one of the mechanisms discussed in Chapter 3 (Section III.B.2, Scheme 3.33). Intermediate **4.3** could convert pyruvate to acetate, carbon dioxide, and water by two related pathways. Pathway a is the direct reaction of **4.3** with pyruvate to give the flavin-pyruvate peroxide (**4.4**); decarboxylative elimination gives acetate, carbon dioxide, and the flavin hydroxide (**4.5**). Elimination of water gives FMN. Pathway b starts with elimination of hydrogen peroxide from the flavin to give an enzyme-bound hydrogen peroxide that can react with pyruvate, yielding the pyruvate peroxide (**4.6**); decarboxylative elimination produces acetate, carbon dioxide, and water.

As evidence for the C^{4a}-hydroperoxide intermediate (**4.3**), compound **4.7** was synthesized and was shown to bind tightly to lactate oxidase, D–amino–acid oxidase, and *p*-hydroxybenzoate hydroxylase; spectral properties were similar to those of the transient oxygen adduct intermediate, presumably the flavin hydroperoxide.[5]

In this case the flavin hydroperoxide acts as a nucleophile. This is reasonable when the substrate is electrophilic, such as a ketone. Next, however, we will see that when the substrate is a nucleophile, the flavin hydroperoxide acts as an electrophile!

B. Hydroxylases Requiring an External Reducing Agent

1. O$_2$ Activation

Molecular oxygen is activated by the flavin, and the flavin is activated by a reducing source, namely, either NADH or NADPH (see Chapter 3, Section III.A, Scheme 4.5), which converts the flavin to its reduced form. If a hydroxylase is incubated with the reduced pyridine nucleotide cofactor, and then the reduced

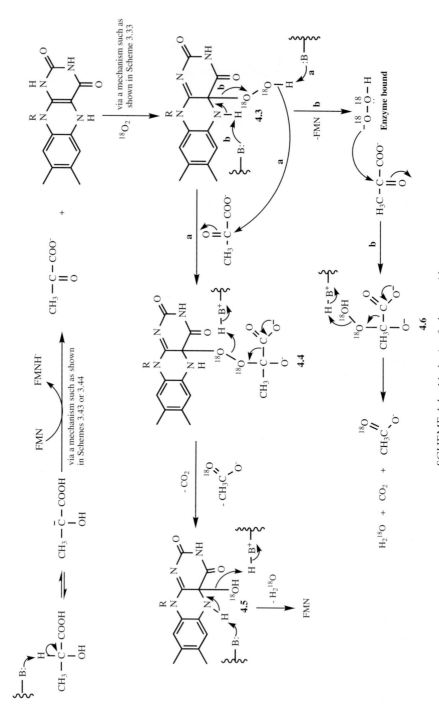

SCHEME 4.4 Mechanism for lactate oxidase.

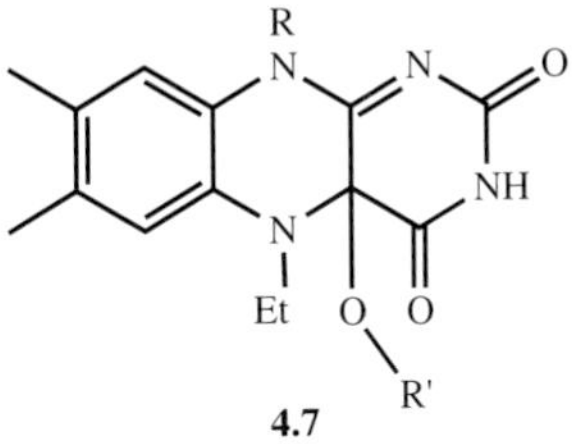

4.7

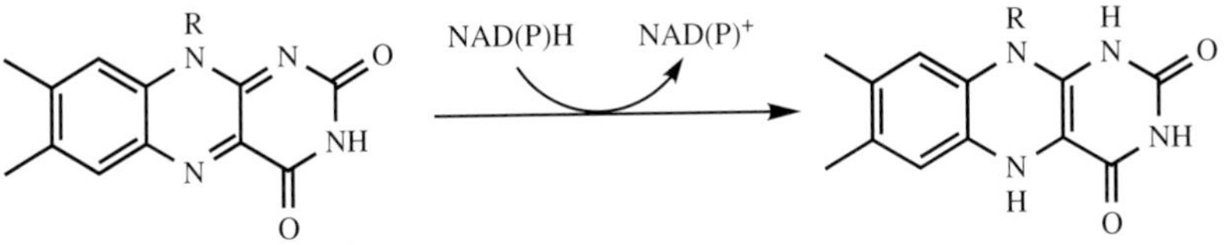

SCHEME 4.5   NAD(P)H reduction of flavin.

enzyme is given a compound that binds but does not undergo hydroxylation, molecular oxygen is reduced to hydrogen peroxide. We have seen this reaction before, namely, in the conversion of reduced flavin to oxidized flavin via the flavin C^{4a}-hydroperoxide during the second half of an oxidase-catalyzed reaction (see Chapter 3, Section III.B.2, Scheme 3.33 or **4.3** in Scheme 4.4). Activated oxygen, then, is most likely in the form of the flavin hydroperoxide.

2. Oxygen Transfer Mechanism

Two classes of reactions utilize a flavin hydroperoxide intermediate (**4.3**, Scheme 4.4): those in which the substrate is nucleophilic and those in which the substrate is electrophilic. An example where the substrate is electrophilic was mentioned in the last section (pyruvate). Let's now consider nucleophilic substrates, in which case the flavin hydroperoxide acts as an electrophile. A variety of mechanisms were proposed initially, on the basis of chemical model studies,[6] but none eventually proved relevant. On the basis of stopped-flow spectroscopic evidence for the formation and decay of C^{4a}-flavin hydroperoxide anion, C^{4a}-flavin hydroperoxide, and C^{4a}-flavin hydroxide intermediates (boxed intermediates in Scheme 4.6) with p-hydroxybenzoate hydroxylase,[7] phenol hydroxylase,[8] and p-hydroxyphenylacetate 3-hydroxylase,[9] Massey and co-workers proposed the mechanism shown in Scheme 4.6 (this is the reaction catalyzed by p-hydroxyphenylacetate 3-hydroxylase, but it is applicable to all the monooxygenases that utilize nucleophilic substrates).[10] Note that in this mechanism the electrophilic hydroxyl group of the flavin peroxide accepts the nucleophile (in this case, the aromatic ring of p-hydroxyphenylacetate is the nucleophile) with concomitant heterolytic cleavage of the weak O−O bond of

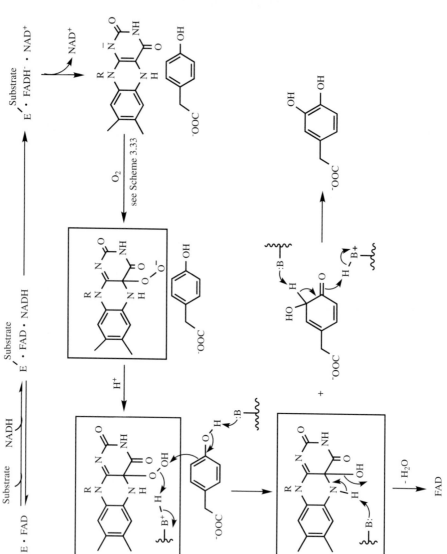

SCHEME 4.6 Mechanism proposed for flavin-dependent hydroxylases.

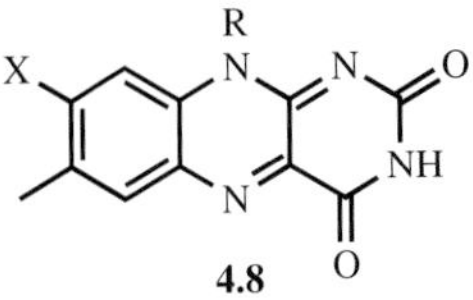

4.8

the peroxide. With p-hydroxybenzoate hydroxylase, a series of 8-substituted FAD analogues (**4.8**) was reconstituted with the apoenzyme, and the rate of hydroxylation was determined as a function of the inductive effects of the substituents of the flavin.[11] A Hammett plot of the ratio of the log of the rate constant for hydroxylation by the 8-substituted flavins relative to the rate constant with FAD versus the calculated pK_a for the C^{4a}-flavin hydroxide intermediates of the 8-substituted flavin species gave a straight line with a slope of -0.5. This linear free-energy relationship with a negative ρ value (cationic) is consistent with an electrophilic aromatic substitution mechanism, as is shown in Scheme 4.6.

As noted in Section II.A.1, the mechanism is different when electrophilic substrates are involved, although there is still one atom of oxygen from O_2 incorporated into the product and one atom into water. In this case, the flavin hydroperoxide acts as the nucleophile and attacks the electrophilic substrate. Bacterial luciferase, which catalyzes the oxidation of long-chain aldehydes to the corresponding carboxylic acids with concomitant emission of light provides an example of a monooxygenase that requires an external reducing agent and utilizes an electrophilic substrate (Scheme 4.7). Low-temperature (-30 °C) enzymatic experiments in mixed aqueous–organic solvents (known as *cryoenzymology*) were utilized to isolate an enzyme-bound C^{4a}-peroxyFMN, which, on warming in the presence of substrate, resulted in the formation of product with emission of light at the usual 10% quantum efficiency.[12] A flavin C^{4a}-hydroxyl intermediate was identified spectrophotometrically and by fluorescence emission spectroscopy right after light emission, and the mechanism shown in Scheme 4.8 was proposed.[13] However, when the bioluminescence of the enzyme reaction is measured with FMN analogues having various substituents at the 8-position, the rate of decay of light emission increases with the decrease in the one-electron oxidation potentials of the FMN analogues.[14] On the basis of these results and a comparison with model compounds using UV–visible spectroscopy, cyclic voltammetry, and redox potentials, it was determined that the reaction is compatible with an intramolecular *chemically*

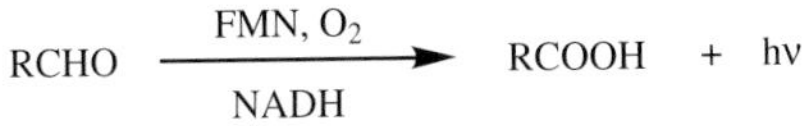

$$\text{RCHO} \xrightarrow[\text{NADH}]{\text{FMN, O}_2} \text{RCOOH} + h\nu$$

SCHEME 4.7   Reaction catalyzed by bacterial luciferase.

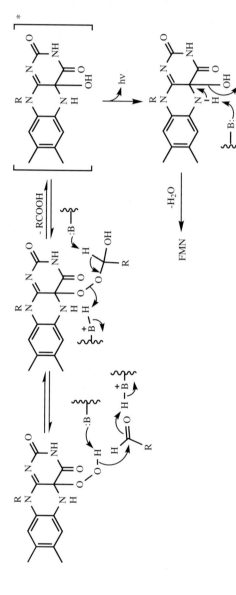

SCHEME 4.8   Nucleophilic mechanism for bacterial luciferase.

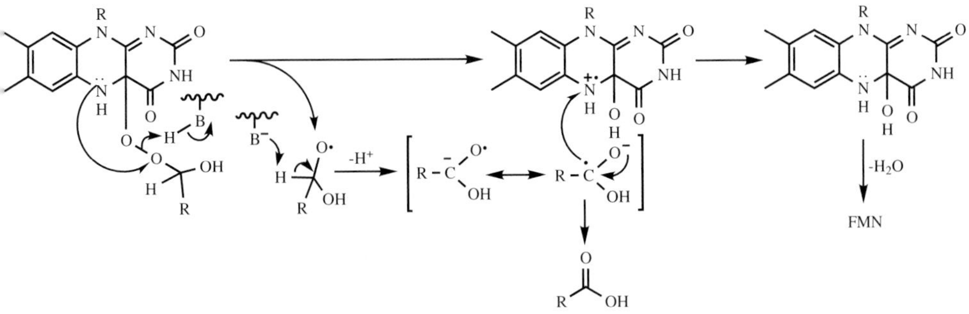

SCHEME 4.9   Chemically initiated electron exchange luminescence (CIEEL) mechanism for bacterial luciferase.

initiated electron exchange luminescence (CIEEL) mechanism (Scheme 4.9).[15] These results are incompatible with a Baeyer–Villiger oxidation mechanism[16] (see Section II.B.3), because in that case the rate should increase with redox potential.[17]

Raushel, Baldwin, and co-workers proposed another electron transfer-type mechanism for this enzyme reaction, namely, a dioxirane mechanism.[18,19] A Hammett plot of k_X/k_H versus σ_p for bacterial luciferase reconstituted with a series of 8-substituted flavins (**4.8**) has a ρ value of -4, suggesting that the reaction is facilitated by increased electron density at the reaction center. This result also is incompatible with a Baeyer–Villiger mechanism, because reactions of benzaldehydes with substituted perbenzoic acids, where H is the migrating group of the benzaldehyde, show small positive ρ values (0.2–0.6). The data are consistent with the formation of a C^{4a}-hydroxyflavin radical cation intermediate, such as that shown in the CIEEL mechanism (Scheme 4.9) and the dioxirane mechanism (Scheme 4.10). Oxidation potential experiments similar to those described earlier gave the same results, namely, that the rate is fastest with FMN analogues of the lowest oxidation potential (easiest to transfer an electron). Therefore, in the rate-determining step there is a decrease in the electron density at the reaction center of the flavin, favoring electron-donating groups, and supporting Scheme 4.10 (as well as Scheme 4.9).

3. Ketone Monooxygenases, an Example of a Baeyer–Villiger Oxidation

A Baeyer–Villiger oxidation[20] is the reaction of a ketone or aldehyde with a peracid to give an ester or carboxylic acid, respectively (Scheme 4.11).[21] The choice of the R versus R′ group for the [1,2]-shift is determined by the *migratory aptitude* of the two groups; in general, the more electron-donating group migrates (except in the case of hydrogen).

A family of flavin-dependent monooxygenases appears to catalyze Baeyer–Villiger oxidations.[22] A C^{4a}-FAD hydroperoxide intermediate was detected during the cyclohexanone oxygenase-catalyzed reaction.[23] Free hydrogen peroxide from

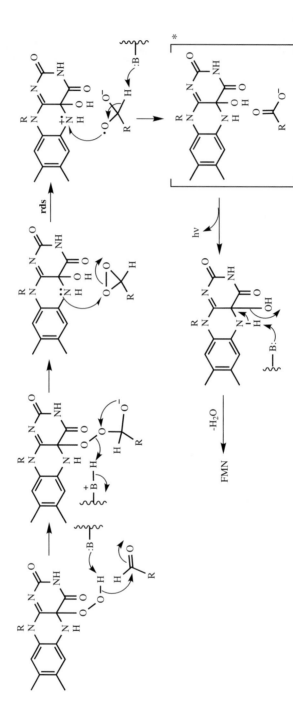

SCHEME 4.10 Dioxirane mechanism for bacterial luciferase.

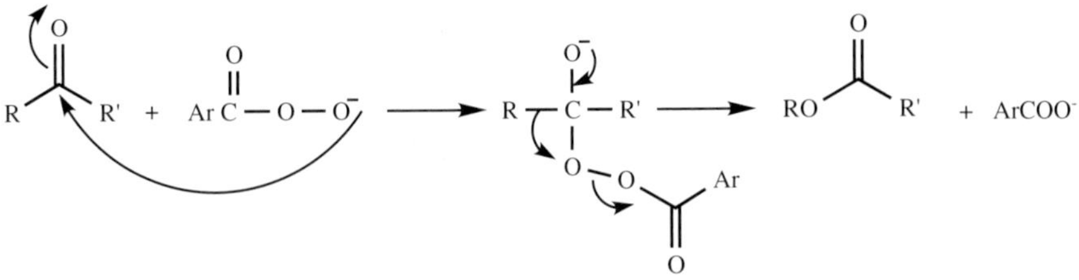

SCHEME 4.11   Baeyer–Villiger oxidation of ketones.

the breakdown of the flavin hydroperoxide is not responsible for product formation, because substitution of the flavin by hydrogen peroxide does not lead to product. One ^{18}O atom of $^{18}O_2$ is incorporated into the product lactone, and one ^{18}O goes into water (Scheme 4.12). Other reactions catalyzed by cyclohexanone oxygenase (Scheme 4.13) show the same migratory aptitudes as do the nonenzymatic reactions.[24]

The Baeyer–Villiger reaction goes with retention of configuration of the migrating group; so does the cyclohexanone oxygenase-catalyzed reaction (Scheme 4.14).[25] The migratory aptitude generally follows the order $3° > 2° > 1° >$ Me;

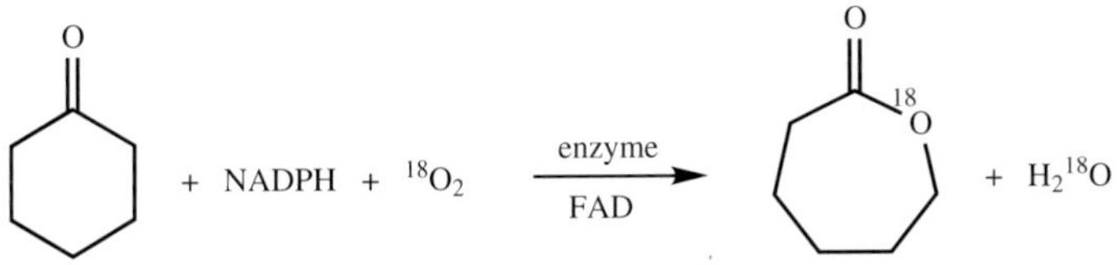

SCHEME 4.12   Reaction catalyzed by cyclohexanone oxygenase.

RCHO ⟶ RCOOH
R = alkyl

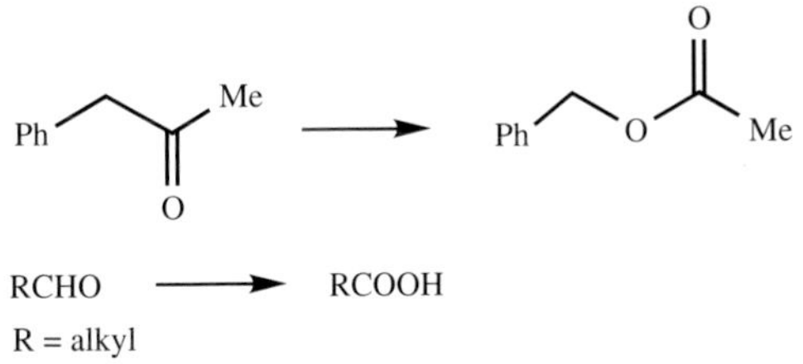

SCHEME 4.13   Other reactions catalyzed by cyclohexanone oxygenase.

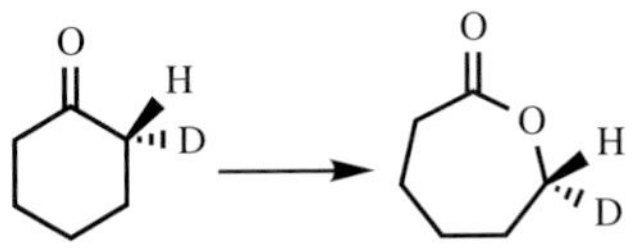

SCHEME 4.14   Cyclohexanone oxygenase proceeds with retention of configuration.

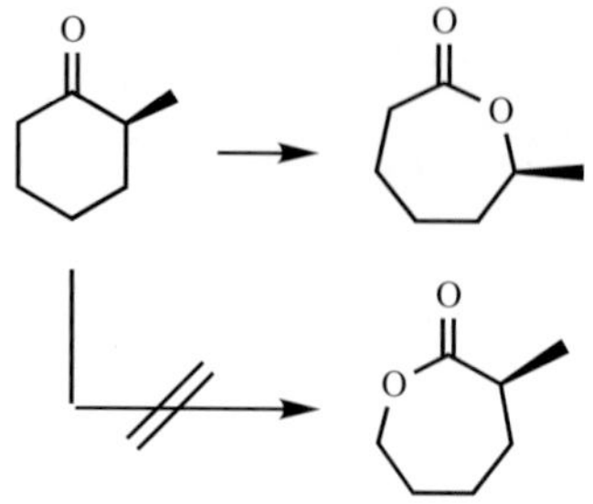

SCHEME 4.15   Migratory aptitude of cyclohexanone oxygenase-catalyzed reaction.

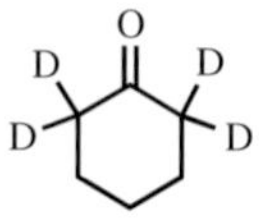

4.9

this is the same order observed with cyclohexanone oxygenase (Scheme 4.15). Also, when 2,2,6,6-tetradeuteriocyclohexanone (**4.9**) is incubated with cyclohexanone oxygenase, the lactone is obtained with no loss of deuterium. All of these results are consistent with a Baeyer–Villiger reaction mechanism in which the flavin hydroperoxide has the same role as does the peracid utilized nonenzymatically (Scheme 4.16).

Whitesides, Walsh, and co-workers utilized immobilized cyclohexanone oxygenase to convert a variety of cyclic ketones on 30–80 mmol scales in high yields to the corresponding lactones; the migratory aptitudes parallel those in the corresponding nonenzymatic Baeyer–Villiger oxidations.[26]

Boranes also undergo nonenzymatic Baeyer–Villiger reactions with peracids to give the corresponding alcohols (after hydrolysis) with retention of configuration. Cyclohexanone oxygenase catalyzes the same reaction (Scheme 4.17).[27]

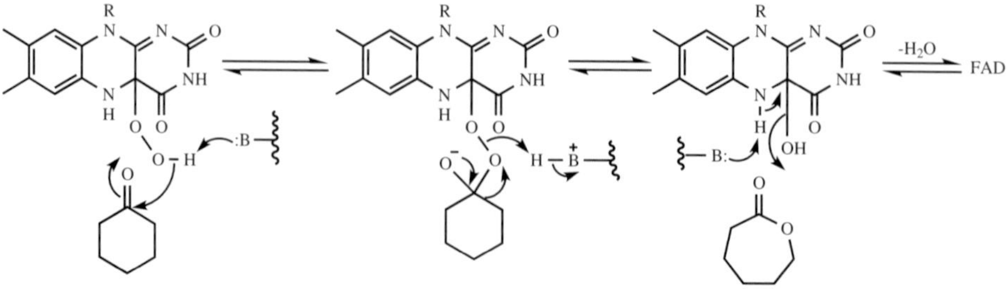

SCHEME 4.16   Baeyer–Villiger-type mechanism proposed for cyclohexanone oxygenase.

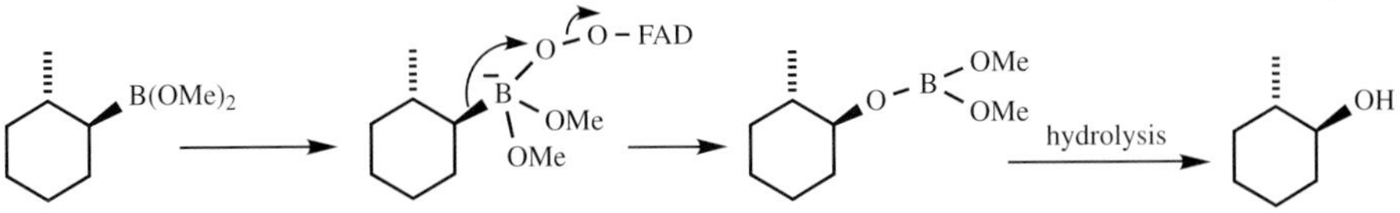

SCHEME 4.17   Reaction of cyclohexanone oxygenase with boranes.

Other enzymatic reactions mimic the Baeyer–Villiger reaction. The ketone monooxygenases from *Pseudomonas* and *Acinetobacter* organisms were utilized to catalyze the oxidation of **4.10** to **4.11** and **4.12** (Scheme 4.18). When $R_1 = R_2 = CH_3$, the ratio of **4.11** to **4.12** is 1:20, and when $R_1 = H$ and $R_2 = CH_3$, the ratio is 1:1, respectively.[28] Both of these ratios are exactly the same as those obtained in the nonenzymatic reaction with a peracid. Note also that the products reflect the expected migratory aptitudes. The enantioselectivities of the Baeyer–Villiger reaction catalyzed by the *Acinetobacter calcoaceticus* and *Pseudomonas putida* enzymes were found to be quite high, and, unexpectedly, complementary! When $R^1 = R^2 = H$ (**4.10**), the *Acinetobacter* enzyme gave **4.13** (1*S*,5*R*) and **4.14** (1*R*,5*S*) (Scheme 4.19) in >95% ee,[29] whereas the *Pseudomonas* enzyme gave **4.15** (1*R*,5*S*) and **4.16** (1*S*,5*R*)

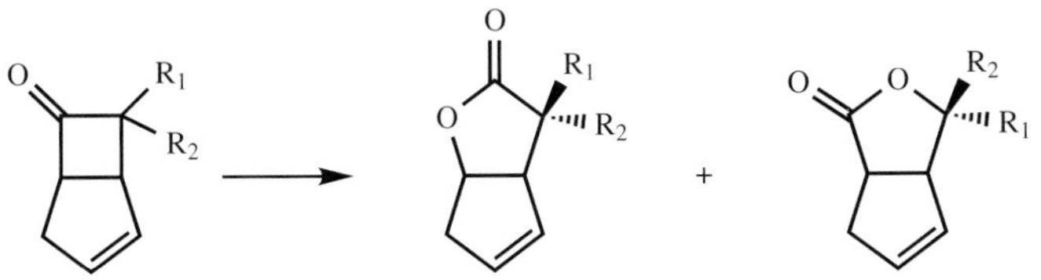

SCHEME 4.18   Reactions catalyzed by ketone monooxygenase.

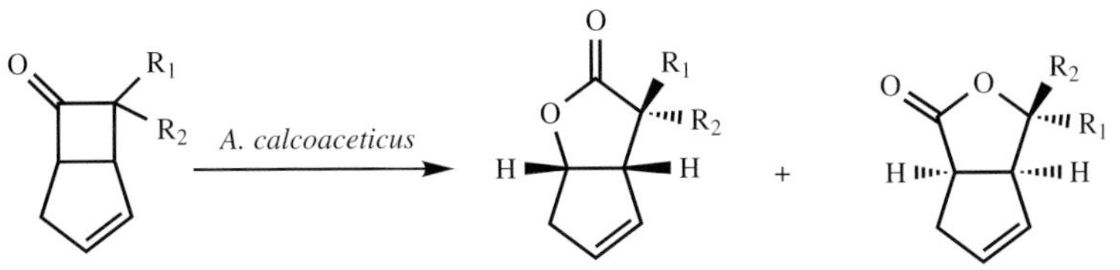

$R_1 = R_2 = H$ (1*S*,5*R*)   (1*R*,5*S*)

$R_1 = H, R_2 = CH_3$ (1*S*,5*S*)   (1*R*,5*S*)

SCHEME 4.19   Reactions catalyzed by the ketone monooxygenase from *A. calcoaceticus*.

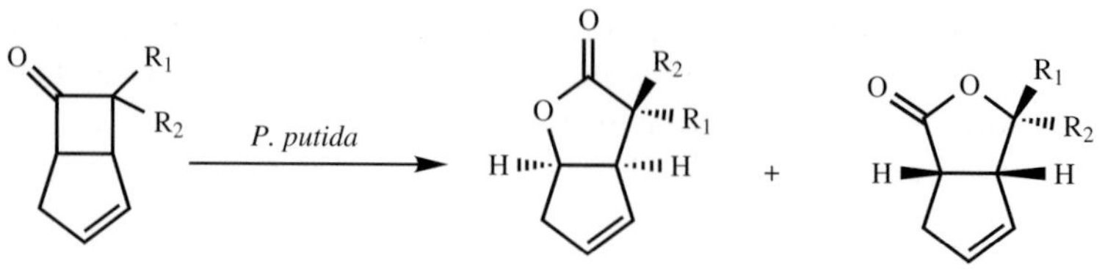

4.10 4.15 4.16

	4.15	4.16
$R_1 = R_2 = H$	(1*R*,5*S*)	(1*S*,5*R*)
$R_1 = H$, $R_2 = CH_3$	(1*R*,5*R*)	(1*S*,5*R*)

SCHEME 4.20 Reactions catalyzed by the ketone monooxygenase from *P. putida*.

(Scheme 4.20) in 50% ee and >95% ee, respectively.[30] Likewise, when $R^1 = H$ and $R^2 = CH_3$, the *Acinetobacter* enzyme gave **4.13** (1*S*,5*S*) and **4.14** (1*R*,5*S*) in >95% ee,[31] but the *Pseudomonas* enzyme gave **4.15** (1*R*,5*R*) and **4.16** (1*S*,5*R*) in >95% ee. All of these results point strongly to a Baeyer–Villiger-type mechanism for ketone monooxygenases.

III. PTERIN-DEPENDENT HYDROXYLASES

A. General

Pterin-dependent hydroxylases[31a] get their name from the fact that their cofactor has a pteridine ring system (**4.17**). Although relatively few enzymes utilize a pterin cofactor, three very important enzymes, phenylalanine hydroxylase, tyrosine hydroxylase, and tryptophan hydroxylase, all require tetrahydrobiopterin (**4.18**) and Fe^{2+} for activity. These aromatic amino acid hydroxylases are involved in the bio-

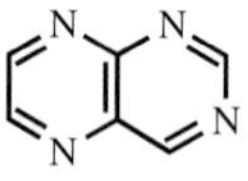

4.17

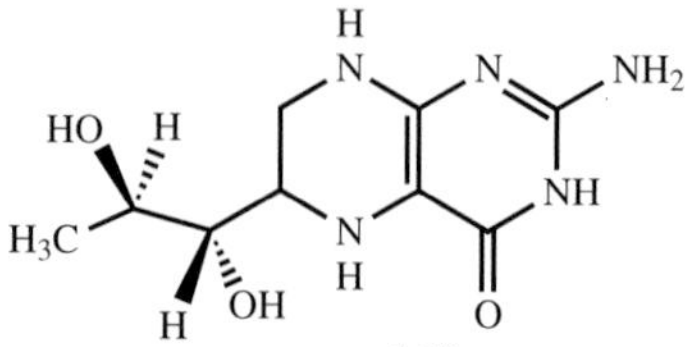

4.18

synthesis and metabolism of the neurotransmitters dopa, norepinephrine, epineph-rine, and serotonin.[32,33] The crystal structures of the C-terminal domains of phenyl-alanine hydroxylase[34] and tyrosine hydroxylase[35] show that the catalytic domains of these enzymes are very similar. Dysfunction of phenylalanine hydroxylase causes the most common *inborn error of metabolism,* namely, *phenylketonuria.* This autosomal recessive trait leads to the inability of the individual to metabolize phenylalanine, which causes mental retardation. You may have noticed that products containing Nutrasweet® have the warning "PHENYLKETONURICS: Contains Phenylala-nine." Nutrasweet is the methyl ester of L-aspartyl-L-phenylalanine.

In addition to the three aromatic amino acid hydroxylases, there are a few other enzymes that require tetrahydrobiopterin for activity, namely, glyceryl-ether monooxygenase,[36] the only enzyme known to cleave unsaturated glyceryl ethers, mandelate 4-monooxygenase,[37] benzoate 4-hydroxylase,[38] anthranilate 3-mono-oxygenase,[39] and nitric oxide synthase, which catalyzes the conversion of L-arginine to the important second messenger, nitric oxide.[40]

B. Mechanism

Pterin-dependent enzymes (except for nitric oxide synthase) catalyze aromatic hydroxylation reactions, similar to a reaction catalyzed by the flavin-dependent hy-droxylases. This is not surprising, if you compare the structures of reduced flavin and tetrahydrobiopterin (Scheme 4.21).

Biopterin (**4.19**) is enzymatically reduced by the NADPH-dependent pteridine reductase to quinonoid dihydrobiopterin (**4.20**). A comparison of the structure of **4.20** with that of oxidized flavin confirms the similarity. Likewise, reduction of **4.20** by another NADPH-dependent enzyme, dihydropteridine reductase, gives tetrahydrobiopterin (**4.21**), which is related in structure to reduced flavin. I have drawn tetrahydrobiopterin in a less important tautomeric form (**4.21a**) to empha-size the structural similarity to reduced flavin.

The overall reaction catalyzed by phenylalanine hydroxylase is shown in Scheme 4.22; after each catalytic cycle, the oxidized pterin (dihydrobiopterin) is released, so it can be reduced back to tetrahydrobiopterin. If it were not for the X group, this would appear to be the same reaction as that catalyzed by the flavin hydroxylases. When a *para*-labeled substrate is utilized with flavin hydroxylase, the label gets washed out. However, if you carry out other experiments such as those utilized to characterize the mechanism of flavin hydroxylases (II.B.2), similar results are obtained. Therefore, it is reasonable to suggest that a pterin hydroperoxide in-termediate forms (**4.22**). The fact that the X group ends up in the adjacent posi-tion indicates that a [1,2]-shift takes place. The research group that discovered this was working at the National Institutes of Health (NIH),[41] so the phenomenon soon became known as the *NIH shift.* Although this shift was originally discovered from work with pterin-dependent enzymes, little mechanistic work on the NIH shift has

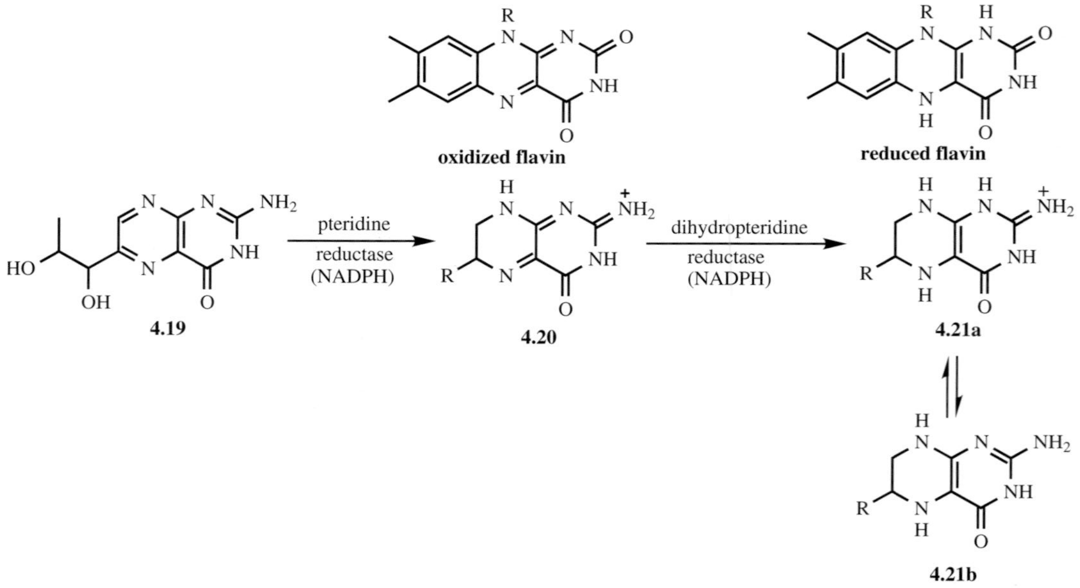

SCHEME 4.21 Comparison of the structures of dihydrobiopterin and tetrahydrobiopterin with oxidized flavin and reduced flavin.

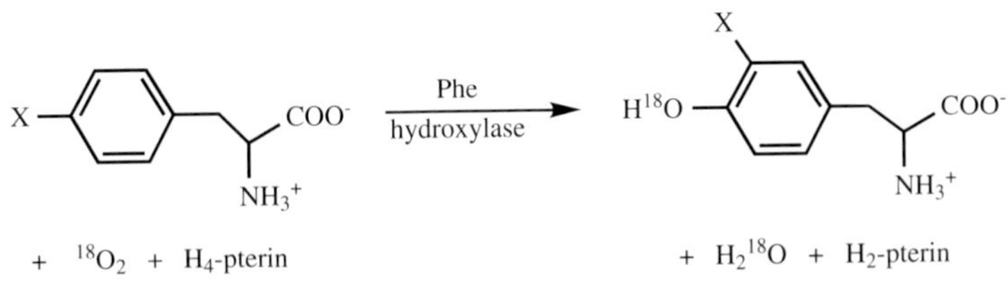

$$X = {}^2H, {}^3H, Cl, Br, alkyl$$

SCHEME 4.22   Reaction catalyzed by phenylalanine hydroxylase.

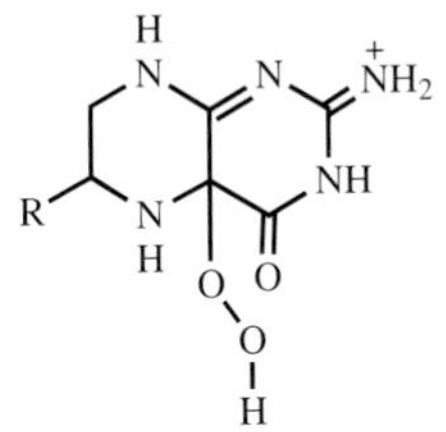

4.22

been done with these enzymes, partly because of their high substrate specificity. However, it was found that cytochrome P450, a heme–dependent enzyme (see Section IV), also catalyzes a NIH shift, and this enzyme has very broad specificity; therefore, most of the studies on the NIH shift will be discussed in the next section. Nonetheless, Scheme 4.23 gives a possible mechanism.

There is some question as to whether the arene oxide (**4.23**) is a true intermediate or is off the normal pathway. Evidence for its existence is that dihydrophenylalanine (**4.24**) is converted to the corresponding epoxide (**4.25**) by phenylalanine hydroxylase (Scheme 4.24).[42] It also is known that incubation of tyrosine hydroxylase with phenylalanine gives, in addition to tyrosine (i.e., 4-hydroxyphenylalanine), 3-hydroxyphenylalanine, which could be rationalized as further evidence for an arene oxide intermediate (Scheme 4.25). However, opening of the epoxide and [1,2]-hydrogen shift will be subject to a kinetic isotope effect. If both tyrosine and 3-hydroxyphenylalanine arise from the common arene oxide intermediate, then incubation with [4-^{2}H]phenylalanine should result in a decrease in the amount of tyrosine formed and an increase in the amount of 3-phenylalanine (the deuterium must migrate to make tyrosine, but not to make 3-phenylalanine). However, Fitzpatrick found that the isotope effects observed with both [4-^{2}H]- and [3,5-^{2}H$_2$]phenylalanine are independent of the ratio of tyrosine and 3-phenylalanine produced.[43] These results are not what is predicted for partitioning of an arene oxide intermediate and suggest a direct attack of an oxygenating species at either the 4- or 3-position of the substrate. Further support for this conclusion comes from a study by Hillas and Fitzpatrick of a series of *para*-substituted (X = F, Br, Cl,

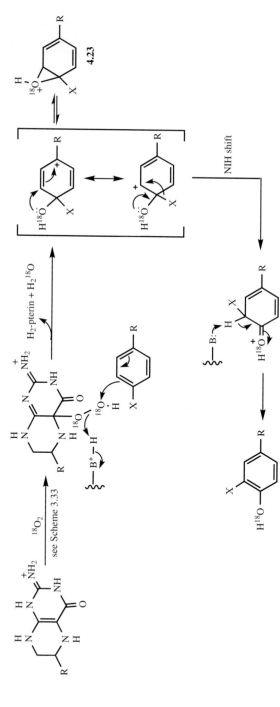

SCHEME 4.23 Mechanism of the reaction catalyzed by tetrahydrobiopterin–dependent monooxygenases.

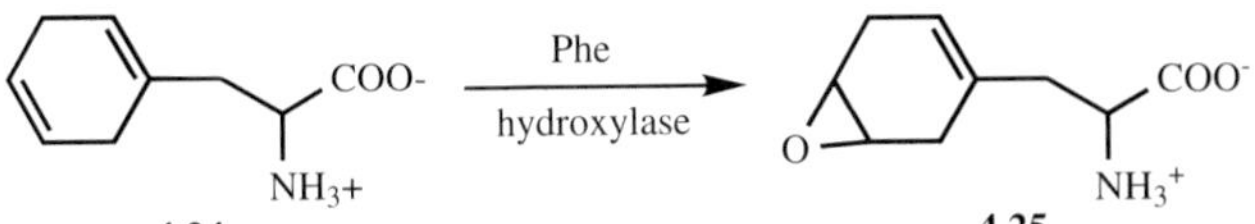

SCHEME 4.24   Reaction of dihydrophenylalanine with phenylalanine hydroxylase.

CH_3, OCH_3) phenylalanines.[44] As the size of the substituent increases, the site of hydroxylation switches from the 4- to the 3-position. The amount of product formed correlates well with the σ parameter of the substituent. The ρ value of about -5 is consistent with a highly electron-deficient transition state during attack of a reactive oxygenating intermediate on the amino acid substrate. The simplest rationalization for the intermediate would be a cation-like species; the large negative ρ value is not consistent with the formation of a neutral radical. In addition to aromatic hydroxylation Hillas and Fitzpatrick found products of benzylic C-hydroxylation; Benkovic and co-workers[44a] also observed C-hydroxylation, but of a cycloalkane ring. These aliphatic hydroxylations are reactions more typical of heme-dependent enzymes (see the next section) than of flavoenzymes and also suggest a role for the Fe^{2+} in pterin-dependent enzymes. Either **4.25a** or **4.25b** could account for these reactions; **4.22** is not capable of aliphatic hydroxylation, and Fe^{2+} and H_2O_2 (from decomposition of **4.22**) gives hydroxyl radicals, which do not undergo NIH shifts. Intermediates **4.25a** and **4.25b** also may be involved in aromatic hydroxylation, given that the NIH shifts occur with migratory aptitudes consistent with a cation-like intermediate (Scheme 4.26, where Y is either H (from **4.22**) or Fe^{2+} (from **4.25a** or **4.25b**).

IV. HEME-DEPENDENT MONOOXYGENASES

A. General

In 1958 Klingenberg[45] and Garfinkel[46] independently reported a pigment in liver microsomes, which, when carbon monoxide was added, exhibited a carbon monoxide difference spectrum with a very strong absorption band maximum at 450 nm. Six years later this pigment was identified as a protein containing *heme,* ferric protoporphyrin IX (**4.26**), and this protein was termed "P450."[47] Liver microsomal P450, called *cytochrome P450,* actually exists in a variety of forms, that is, different gene products known as *isozymes.* About 500 different P450 isozymes, now referred to as the *cytochrome P450 superfamily,* have been reported. Cytochromes P450 are important enzymes in the hydroxylation of endogenous substrates as well as the monooxygenation of a wide range of *xenobiotics,* that is, drugs and other chemicals foreign to living organisms. It is the P450s that protect us from the toxins in the air we breathe and the foods/chemicals that we eat.

Most P450s are membrane bound, and therefore difficult to purify. A soluble,

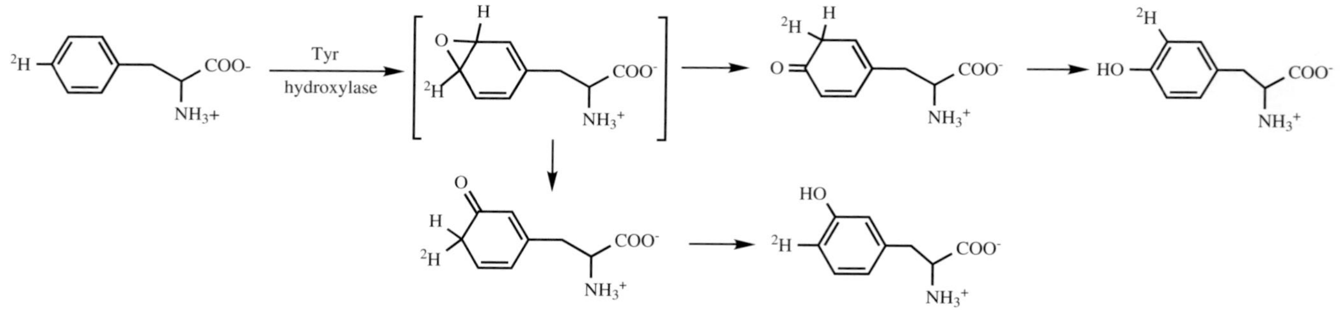

SCHEME 4.25 Arene oxide mechanism proposed for tetrahydrobiopterin-dependent monooxygenases.

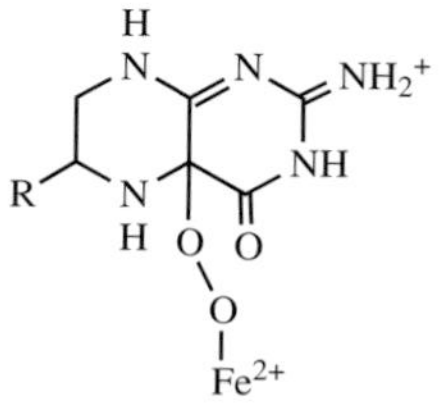

4.25a

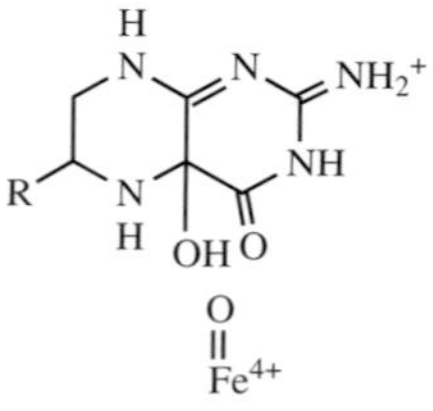

4.25b

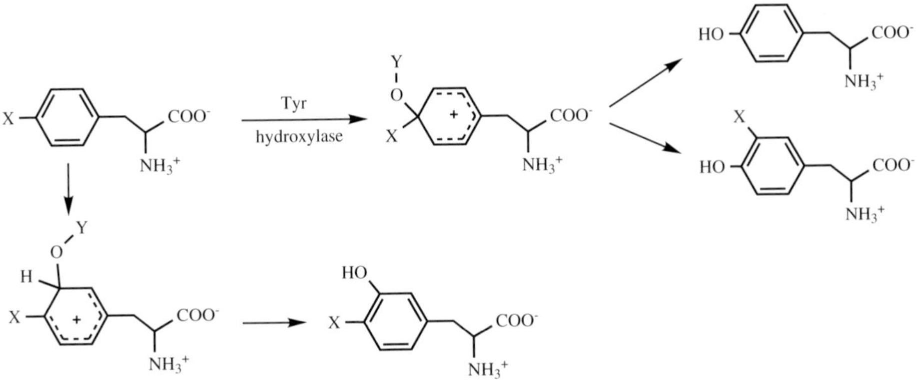

SCHEME 4.26   Cationic mechanism proposed for tetrahydrobiopterin-dependent monooxygenases.

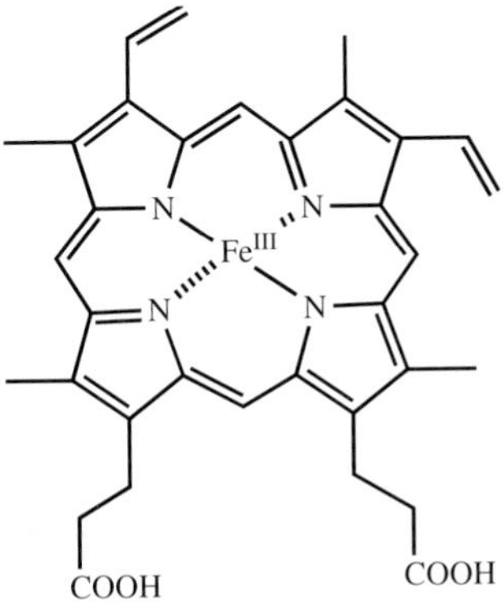

4.26

camphor-inducible (various chemicals can induce genes to produce new P450 iso-zymes that degrade those chemicals), bacterial P450 (P450$_{cam}$, or, by the new no-menclature that is based on sequence identity of the isozymes,[48] CYP1O1 or P450 1O1) was purified,[49] and a high–resolution crystal structure was obtained.[50] The structure shows that the heme is bound to the protein via coordination of Cys-357 to the bottom axial position of the heme iron. Typically, the heme of heme enzymes is coordinated either to a cysteine or a histidine residue of the protein.

The reactions catalyzed by heme-dependent monooxygenases include alkane (alkyl) hydroxylation, alkene epoxidation, arene hydroxylation and epoxidation, dealkylation of amines, sulfides, and ethers (N-, S-, and O–dealkylation, respec-tively), oxygenation of amines (N-oxidation) and sulfides (S-oxidation), conversion of primary amines to aldehydes or ketones and ammonia (oxidative deamination), conversion of halides to aldehydes or ketones and HX (oxidative dehalogenation), oxidation of alcohols to aldehydes and aldehydes to carboxylic acids, dehydrogena-tion (unsaturation), and (anaerobic) reductions of a variety of functional groups. All of these reactions require a reducing agent, either NADH or NADPH, as well as the oxidizing agent O_2 (except for the anaerobic reactions). Because of this dichotomy of requiring both a reducing agent and an oxidizing agent, these enzymes are called *mixed-function oxidases*.

B. Molecular Oxygen Activation

The function of the heme is to activate molecular oxygen to a species that can insert into various molecules. The proposed catalytic cycle for cytochrome P450, including some hypothesized intermediates, is shown in Scheme 4.27.[51] When a ligand (such as water) is bound to the top axial position, the ferric iron is in an almost planar low-spin state (**4.27**; the parallelogram is the abbreviated notation for the heme's porphyrin ring with the pyrrole nitrogens at each corner). On substrate (R–H) binding to the ferric heme enzyme (**4.27**), the top ligand is released, which results in the conversion of the iron to a 5-coordinate high-spin state in which the iron is out of the plane of the porphyrin ring (**4.28**). Conversion of the iron from a low-spin to a high-spin state results in a large increase in the redox potential of the heme from -330 mV to -173 mV (versus NHE). Therefore, substrate binding induces electron transfer. One electron is needed to reduce the ferric heme to ferrous heme (**4.29**). That electron originates from NADH or, in more cases, NADPH. However, as noted in Chapter 3 (III.A), the nicotinamide coenzymes strongly favor two-electron transfers. Therefore, two electrons are transferred by the NAD(P)H to an electron transport protein, which contains either flavin or flavin and iron–sulfur cluster (Fe_2S_2), coenzymes, which can transfer single electrons (Chapter 3 III.B. and III.D.2, respectively). Microsomal P450 systems utilize a single flavoenzyme, NADPH-cytochrome P450 reductase, which contains both FAD and FMN. In Scheme 4.27 two electrons are shown passing from NADPH to FAD to give $FADH^-$, which then passes the two electrons to FMN to give $FMNH^-$; one of these electrons, then, is transferred to the ferric heme, converting it to ferrous heme (**4.29**) with oxidation of $FMNH^-$ to FMN semiquinone. Molecular oxygen binds to ferrous heme to give the low-spin heme peroxy radical (**4.30**). Another electron is transferred to **4.30** to give heme peroxy anion (**4.31**), which can be protonated to heme peroxide (**4.32**). A second proton is transferred to the distal oxygen, which activates the O–O bond for cleavage with loss of water, and generates a very high energy oxo-ferryl heme (**4.33**). Four different resonance structures are shown: **4.33a** is an Fe^{III}-oxene, **4.33b** is an Fe^{IV}-oxo radical, **4.33c** is an Fe^{V}-oxo, and **4.33d,** in which one electron from the porphyrin ring is transferred to the iron because of the exceedingly high (nonexistent?) energy state of Fe^{V}, is an Fe^{IV}-oxo-porphyrin π-radical cation. Calculations by Loew et al.[52] support **4.33b** as the active oxygen species.

The proton source for conversion of **4.31** to **4.32** or **4.32** to **4.33** has been thought to be Thr-252 (either directly or via a water molecule) in the case of CYP1O1 (P450$_{cam}$); mutation of this residue to valine or alanine produces a mutant

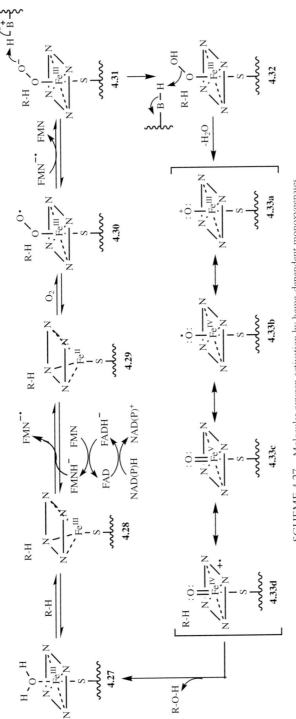

SCHEME 4.27   Molecular oxygen activation by heme–dependent monooxygenases.

that cannot hydroxylate camphor, but neither substrate binding nor the rate of the first electron transfer to the heme is affected.[53] The proton source for the oxygenated heme in P450$_{BM-3}$ was initially thought to be Thr-268 (either directly or via a water molecule).[54] However, mutation of Thr-252 with the unnatural amino acid O-methylthreonine gave a mutant that retained about a third of the monooxygenase activity.[55] This suggests that the oxygen atom of threonine, not the hydroxyl proton, is involved in the proton transfer somehow, possibly by interacting with a water molecule that is bound to Asp-251; mutation of Asp-251 dramatically decreases the rate of catalysis.[56] Intermediates **4.27−4.30** have been isolated in stable form and are well characterized.[57]

The electrophilic nature of the activated heme species (**4.33**) is evident when substrates are utilized in which carbon atoms have equal access to the active-site heme. In these cases reaction occurs preferentially at the more substituted (tertiary > secondary > primary) carbon.[58] Also, monosubstituted aromatic compounds are hydroxylated mainly at the *ortho-* and *para-*positions, as expected for electrophilic aromatic substitution reactions.

C. Mechanistic Considerations

In this section we will consider mechanisms for six different reactions catalyzed by the P450 superfamily: alkane hydroxylation, alkene epoxidation, arene hydroxylation, nitrogen and sulfur oxygenation, dealkylation of amines, sulfides, and ethers, and carbon−carbon bond cleavage. There are many similarities to these mechanisms, but the outcomes differ because of the structures of the substrates.

1. Alkane Hydroxylation

Hydroxylation of unactivated alkanes (or alkyl groups) was initially proposed to be a concerted insertion process because of the commonly observed retention of configuration and the very small kinetic isotope effects obtained ($k_H/k_D < 2$).[59] However, Hjelmeland[60] and Groves[61] later showed that the isotope effect on benzylic hydroxylation was much larger ($k_H/k_D > 11$). The difference in these results stems from the fact that the earlier studies utilized *intermolecular isotope effects,* obtained by comparing the reaction rates with unlabeled substrates versus the same reaction with deuterium-labeled substrates. However, if C−H bond cleavage is not the rate-determining step, no isotope effect will be observed. The later studies were done by measuring *intramolecular isotope effects.* In this case a molecule is used that has two identical and symmetrical sites that can undergo hydroxylation. On one site deuteriums are incorporated, and the other has hydrogens, then the amount of C−D versus C−H bond cleavage is measured. Using this approach only the isotope effect on C−H bond cleavage will be determined. The large isotope effect suggests complete breakage of the C−H bond during catalysis. It also is inconsistent with a

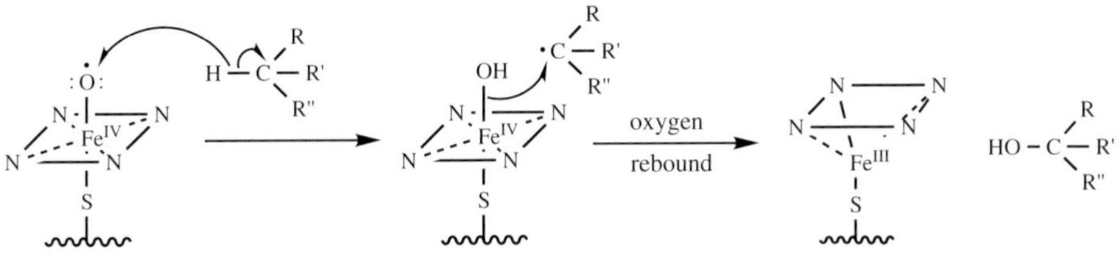

SCHEME 4.28   Two-step radical mechanism with oxygen rebound for alkane oxygenation by heme-dependent monooxygenases.

direct carbocation mechanism. The very low acidity of the hydrogens that are removed negates a carbanion mechanism. What is left is a two-step radical process, in which the activated heme species (**4.33b**, for example) abstracts a hydrogen atom from the substrate to give a carbon radical and a heme-bound hydroxyl radical (**4.34**), followed by combination of those two radicals; the second step, hydroxylation of the carbon radical, is known as *oxygen rebound*[62] (Scheme 4.28).

Further evidence for a two-step mechanism is the stereochemical scrambling observed when deuterated substrates are incubated with P450. When all-*exo*-2,3,5,6-tetradeuterionorbornane (**4.35**) is incubated with CPY2B4, four norbornanols are produced (**4.36–4.39**) (Scheme 4.29).[63] It is apparent that the reaction does not proceed with retention of configuration, and this supports a radical intermediate.

A common approach to the detection of radical intermediates in enzyme-catalyzed reactions is one that utilizes modified substrates containing groups that can undergo rapid radical rearrangements, such as cyclopropylcarbinyl radical rearrangements (see Chapter 3, Section III.A.2). Because the rates of radical rearrangements are well known,[64] the existence and lifetime of the radical intermediate in the enzyme-catalyzed reaction can be estimated from the ratio of unrearranged to rearranged products (Scheme 4.30). These *radical clocks*, substrates containing a

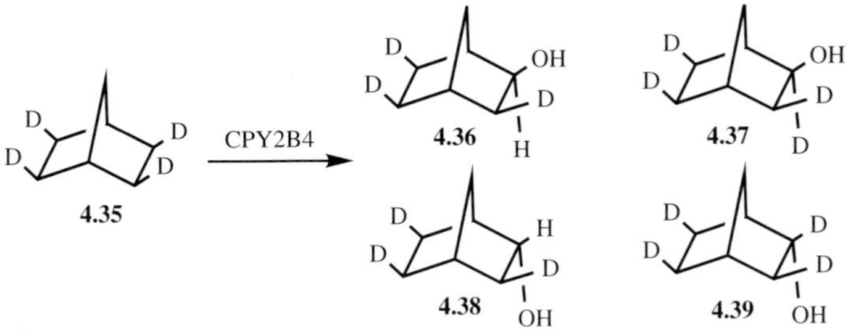

SCHEME 4.29   Products from the reaction of all *exo*-2,3,5,6-tetradeuterionorbornane with the CPY2B4 isozyme of cytochrome P450.

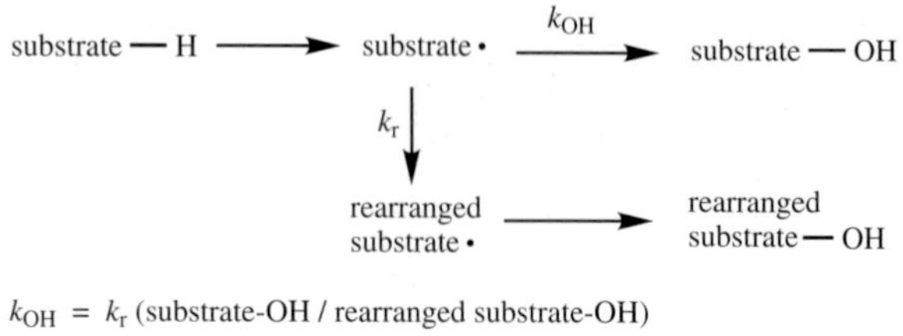

$$k_{OH} = k_r \text{ (substrate-OH / rearranged substrate-OH)}$$

SCHEME 4.30   Radical clock approach for determination of reaction rates in radical rearrangement reactions.

group whose rate of rearrangement is known, are typically highly strained carbo-cyclic structures. For example, a series of cyclopropane-containing compounds was tested as substrates for CPY2B4. Rearranged products are only observed with the substrates that, on hydrogen atom abstraction, produce a cyclopropylcarbinyl radical known to undergo nonenzymatic rearrangement with a rate constant $\geq 2.0 \times 10^9 \text{ s}^{-1}$ (Scheme 4.31 gives one example in the series).[65] The oxygen rebound step (k_{OH}) can be estimated from the ratio of unrearranged to rearranged alcohols to be $2.4 \times 10^{11} \text{ s}^{-1}$ for a primary carbon radical intermediate and an order of magnitude slower for rebound to a secondary carbon radical.

Further support for a radical intermediate was obtained with the ultrafast radical clock probe, *trans*-1-methyl-2-phenylcyclopropane (**4.40**, Scheme 4.32),[66] whose methylene radical rearranges at the remarkable rate of $3 \times 10^{11} \text{ s}^{-1}$. Cytochrome P450-catalyzed oxidation gives three products (**4.42**−**4.44**); pathway a results in the formation of the cyclopropylcarbinyl radical intermediate (**4.41**) that either undergoes oxygen rebound to give **4.42** or rearrangement followed by oxygen rebound to **4.43**. Formation of **4.43** is strong evidence for a radical intermediate. Pathway b accounts for the formation of the third product, the aromatic hydroxylation product **4.44** (see IV.C.3 for the mechanism of this reaction). The perdeu-

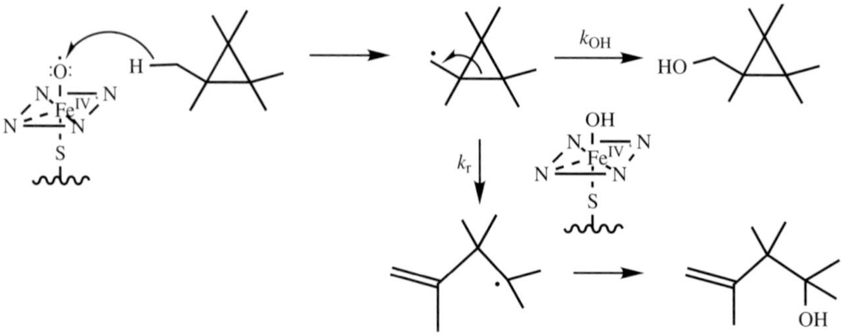

SCHEME 4.31   Cytochrome P450-catalyzed monooxygenation of a cyclopropane analogue.

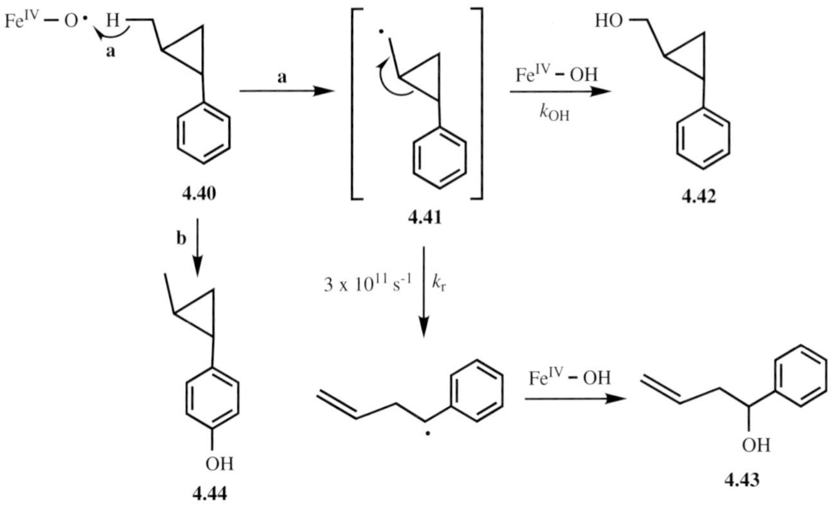

SCHEME 4.32   Cytochrome P450-catalyzed oxidation of *trans*-1-methyl-2-phenylcyclopropane.

teriomethyl analogue leads to an increase in the oxidation at the phenyl ring (pathway b), a process termed *metabolic switching,* that is, deuteration at one site changes the partition between two metabolic pathways because of the isotope effect on C–H bond cleavage at the first site. A combined primary and secondary deuterium isotope effect of 12.5 is observed for the perdeuteriomethyl analogue. With the use of a kinetic analysis reported by Hanzlik and co-workers,[67] the primary and secondary deuterium isotope effects were calculated to be 7.87 and 1.26, respectively. This large primary isotope effect indicates extensive C–H bond stretching in the transition state, and the large secondary isotope effect suggests that the conversion of the sp^3 carbon (CH_3) to the sp^2 carbon ($\cdot CH_2$) is well advanced in the transition state.

A correlation of the nonenzymatic isotope effects on the hydrogen atom abstraction of various substrates by *tert*-butoxy radical (known to react by a hydrogen atom abstraction mechanism[68]) with the enzymatic isotope effects on these same substrates by several cytochrome P450s gave a straight line with unit slope.[69] These results support a radical mechanism for P450.

Experiments with other ultrafast radical clocks, however, have brought into question the existence of a true radical intermediate. Compound **4.45** (Scheme 4.33) gives a very low amount of rearranged products, even though the ring-opening rate is 3×10^{11} s^{-1} (the same as **4.40**); the calculated k_{OH} would be 1.4×10^{13} s^{-1}, which is faster than decomposition of a transition state (6×10^{12} s^{-1}).[70] This, of course, assumes that in the constrained space of the active site the rearrangement can occur at the same rate as in solution. To rationalize these data, a carbocation mechanism, where the carbocation is formed *after* the oxidation step, was suggested.

A hypersensitive radical probe substrate, (*trans, trans*-2-*tert*-butoxy-3-phenyl-cyclopropyl)methane (**4.46,** Scheme 4.34), based on a similar probe designed by

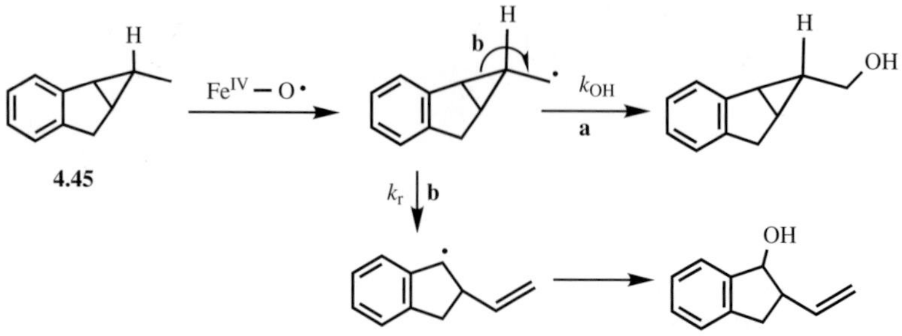

SCHEME 4.33   Another ultrafast radical clock reaction catalyzed by cytochrome P450.

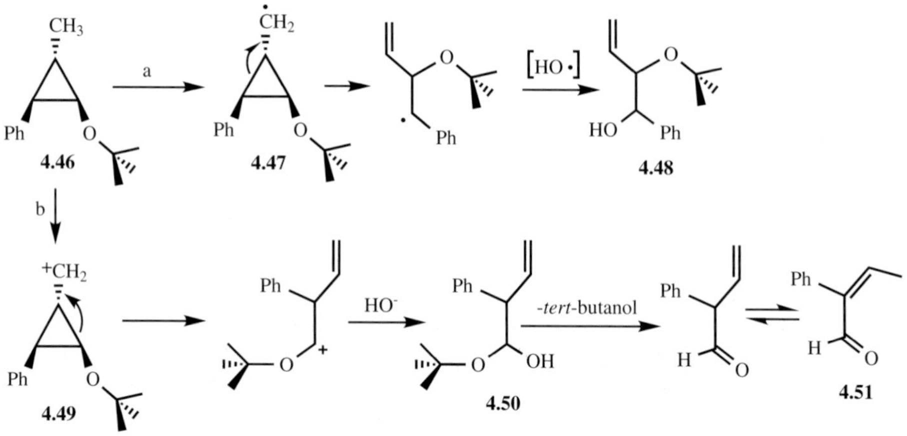

SCHEME 4.34   A hypersensitive radical probe substrate to differentiate a radical from a cation intermediate generated by cytochrome P450.

Newcomb and Chestney for nonenzymatic reactions,[71] was designed for P450 and permits discrimination between radical and cationic intermediates.[72] Abstraction of a hydrogen atom from **4.46** to give radical **4.47** was shown to result in about 98% cleavage of the cyclopropyl C−C bond that connects the methyl and phenyl groups (pathway a); oxygen rebound would give **4.48**. On the basis of the model study, cation formation (**4.49**) is expected to give exclusive cleavage of the cyclopropyl C−C bond that connects the methyl to the *tert*-butoxyl group (pathway b); hydrolysis of that cation gives **4.50,** which would decompose to **4.51**. CPY2B1-catalyzed oxidation of the methyl group of **4.46** gave mostly unrearranged hydroxylation product, but it also gave two minor rearranged hydroxylation products, one from a radical intermediate (**4.48**) and one from a cation intermediate (**4.51**). Most of the rearranged products were thought to be derived from a cationic intermediate, apparently produced during the course of the hydroxylation reaction after the ini-

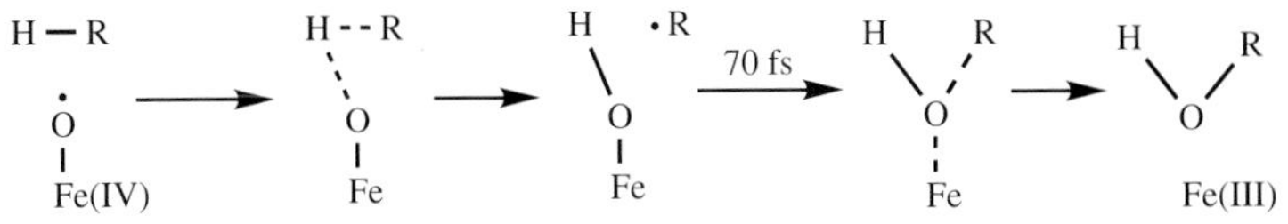

SCHEME 4.35   A concerted, but nonsynchronous, mechanism proposed for cytochrome P450.

tial radical "intermediate" forms. The radical species in the hydroxylation reaction, however, was calculated to have a lifetime of only 70 fs (yes, femtoseconds). That being the case, the species would not be a true intermediate, but rather a part of a reacting ensemble, which defines this as a concerted reaction. It is believed that the small amount of radical rearrangement occurs because the insertion reaction is nonsynchronous with C—H bond cleavage leading to C—O bond formation, and there is a competition between rearrangement and hydroxyl radical capture. The short radical lifetime would require the oxygen atom to be within bonding distance of the carbon at the instant of hydrogen abstraction, which would be a "side on" approach (Scheme 4.35).

Other experiments with ultrafast radical probes[73] do *not* support a mechanism in which a transient radical is converted to a cation, or formation of an "agostic complex,"[74] or a cationic rearrangement via solvolysis of a protonated alcohol that is produced by insertion of an OH^+ species from iron-complexed hydrogen peroxide. The conclusion reached using mutants lacking the active-site threonine[75] (the residue believed to donate a proton in going from **4.31** to **4.33** in Scheme 4.27) and other studies with ultrafast radical probes,[76] was that more than one electrophilic oxidizing species appears to be produced during P450-catalyzed oxidations. A hypothesis of Shaik and others that is consistent with these contradictory results involves two pathways having multiple states for the high-energy heme complex.[77] One reaction pathway involves two reactivity states, a high-spin ground state and a low-spin excited state. Another competing pathway has the same spin state throughout the reaction. Generally, the two-state pathway involves concerted reactions that conserve stereochemistry, because the high-spin state, which does not favor concerted reactions, is energetically unfavorable compared with the low-spin state, which prefers concerted reactions via insertion intermediates (R—[Fe]—OH, followed by reductive elimination to [Fe] + R—OH). Hydrogen atom abstraction and oxygen rebound via loosely bound intermediates such as [Fe]—OH/R·, are also quite facile; these reactions also lead to a concerted, but nonsynchronous, pathway having very short-lived radical character. The single-state pathway gives stepwise mechanisms. Therefore, all the reaction mechanisms just discussed are accommodated by this multistate hypothesis; spin inversion probability, then, determines the distribution of mechanistic outcomes. The competition between these pathways depends on the strength of the C—H bond to be broken and the size of the substituent. The two-state pathway is sensitive to steric hindrance. This hypothesis satisfies the numerous inconsistencies in the results from various research groups with regard to the mechanism of alkane hydroxylation. The multitude of mechanisms may all be possible, but the substrate structure may determine which one(s) become predominant in that particular reaction.

2. Alkene Epoxidation

Cytochrome P450 oxygenation of alkenes leads to epoxides, most often with retention of configuration.[78] A mechanism, similar to the mechanism described for alkane hydroxylation (Scheme 4.28), can be drawn (Scheme 4.36); instead of abstraction of a hydrogen atom from the alkane, this is a radical addition to the π-bond of the alkene to give **4.52**, which collapses to the epoxide and high-spin ferric heme. Spectroscopic evidence in a model compound indicated that an intermediate is generated on the way to epoxide formation; Groves suggested either a π-complex (**4.53**) or an oxametallocycle (**4.54**).[79] Other model studies by Bruice and co-workers[80] suggest that alkene epoxidation may involve a charge transfer complex between the iron-oxo species and the alkene, followed by a concerted epoxidation. As in the case of alkane hydroxylation, there is some ambiguity from radical clock substrates regarding formation of a true radical intermediate with alkene epoxidation. Incubation of **4.55** (Scheme 4.37) with P450 gave no ring opened products, only the corresponding epoxide (**4.56**), suggesting that a radical intermediate is not formed or is very short lived.[81] Again, however, this may be a substrate-dependent multiple-pathway reaction.

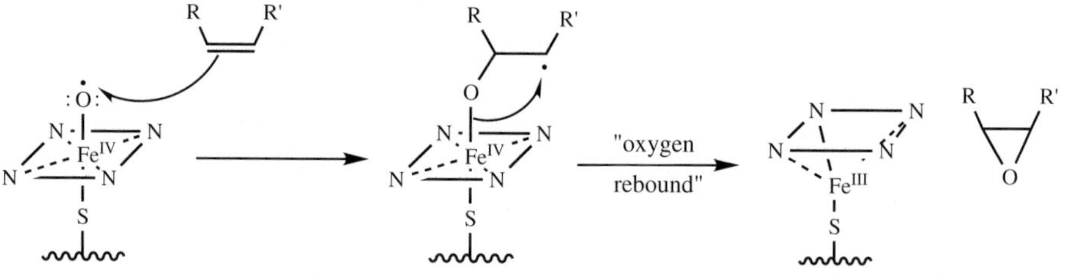

4.52

SCHEME 4.36   Two-step radical mechanism with oxygen rebound for alkene oxygenation by heme-dependent monooxygenases.

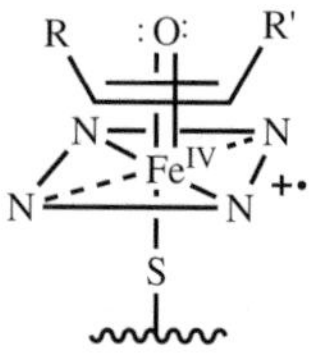

4.53 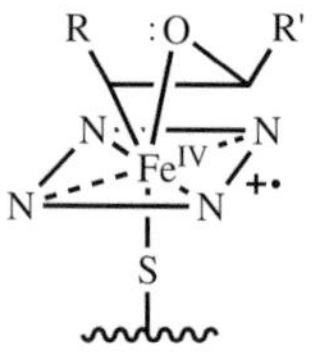   **4.54**

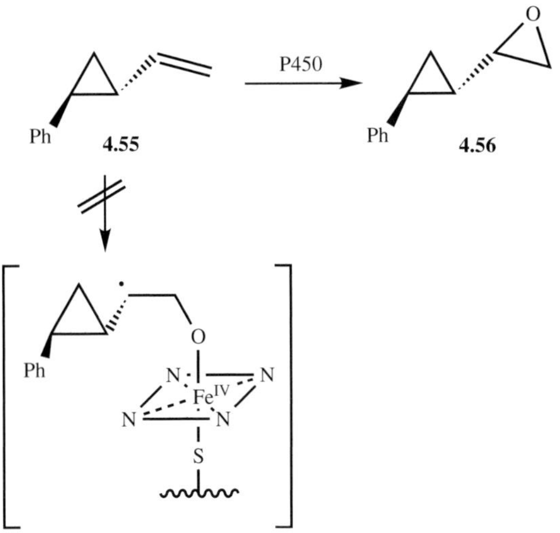

SCHEME 4.37   Cytochrome P450-catalyzed epoxidation of *trans*-1-phenyl-2-vinylcyclopropane.

3. Arene Hydroxylation

In the discussion of pterin-dependent enzymes (III.B) an unusual migration of *para*-substituted groups to the *meta*-position was noted, known as the NIH shift. Because of the specificity of these enzymes, further elucidation of the mechanism of pterin-dependent enzymes and the NIH shift will have to come from analogy to the results of studies with cytochrome P450. However, as in the case of alkane and alkene hydroxylation, there are many contradictory results.

Evidence in support of an arene oxide intermediate comes from the isolation of the first arene oxide, naphthalene 1,2-oxide (**4.57**) during P450 oxidation of naphthalene (Scheme 4.38).[82] The absence of an intermolecular isotope effect on hydroxylation of perdeuterated naphthalene is consistent with rate-determining arene oxide formation. Decomposition of [1-^{2}H]- and [2-^{2}H]naphthalene 1,2-oxides yields the same amount of deuterium in their respective 1-naphthol products, which suggests a common [2-^{2}H, 2-^{1}H]cyclohexadienone intermediate (**4.58**) from the arene oxide (Scheme 4.39).[83] Formation of a cyclohexadienone intermediate could come

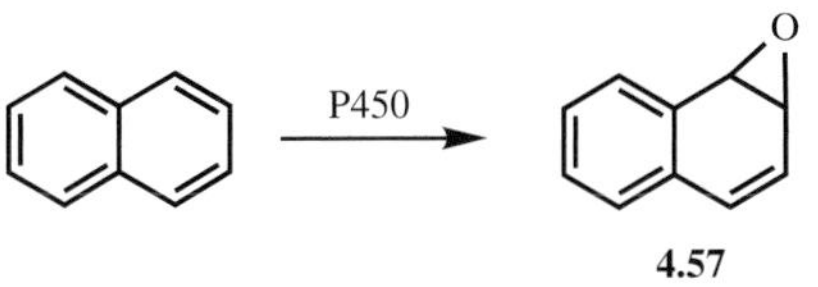

4.57

SCHEME 4.38   Cytochrome P450-catalyzed formation of an arene oxide.

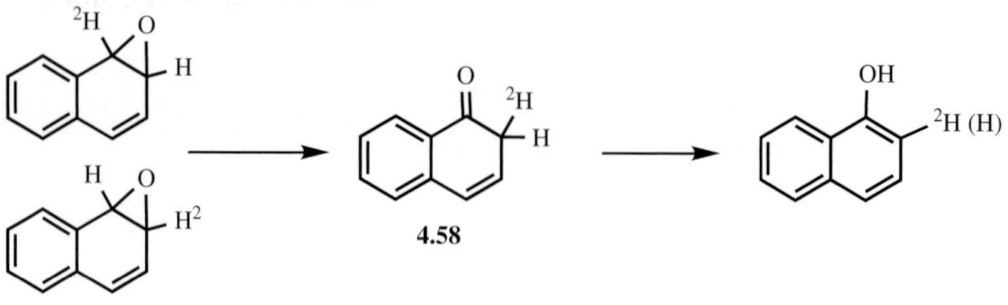

SCHEME 4.39   Evidence for a common intermediate in the oxygenation of naphthalene.

from a concerted (pathway a) or stepwise (pathway b) mechanism (Scheme 4.40). Two pieces of data rule out the concerted pathway: There is no isotope effect on naphthalene oxide consumption when deuterium is incorporated (C–H bond cleavage is not in the isomerization step), and a Hammett plot shows a large negative ρ value with various substituted arene oxides, indicating that positive charge develops in the transition state.[84]

However, other evidence suggests that an arene oxide is not necessarily along the pathway at all (although it may be a product from an alternative pathway). One product of P450-catalyzed oxidation of chlorobenzene is *meta*-hydroxylated chlorobenzene (*m*-chlorophenol). This product is not observed by any nonenzymatic reactions with either the 3,4- or 2,3- arene oxide of chlorobenzene.[85]

The involvement of a single electron transfer step in arene hydroxylation is supported by a Hammett relationship between the rates of *meta*-hydroxylation of monohalogen-substituted benzene by CPY2B1 and the σ^+ values (which correlate with the $E_{1/2}$ values for benzene) of the substituents.[86]

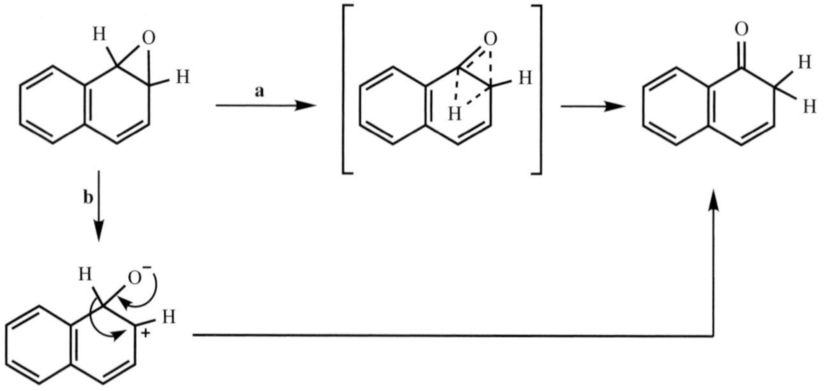

SCHEME 4.40   Concerted (pathway a) and stepwise (pathway b) mechanisms for the potential conversion of an arene oxide to a cyclohexadienone.

The compositions of hydroxylated products of 20 different deuterated mono-substituted benzenes were determined, and the isotope effects were measured.[87] These results support both direct hydroxylation (in which no arene oxide or NIH shift occurs) and indirect hydroxylation (NIH shift) mechanisms. The k_H/k_D for hydroxylation at each ring position is 4.0 ± 0.2, which supports a cyclohexadienone intermediate when a NIH shift is involved. Further isotope effect studies on the cytochrome P450-catalyzed hydroxylation of deuterated chlorobenzenes indicate that initial arene oxide formation or a charge transfer step are *not* viable mechanisms.[88]

High-energy intermediates, instead of arene oxides, are supported by studies with ammonia monooxygenase of *Netrosomonas europaea*.[89] Aromatic hydroxylations of a series of benzene derivatives with *ortho/para*-directing substituents give the predicted phenolic products, primarily *para*-substituted. In contrast, aromatic compounds with *meta*-directing substituents give a mixture of *meta*- and *para*-substituted phenols. This is what is expected of a mechanism involving electrophilic addition of the heme oxo species to the aromatic ring to give either a radical or carbocation intermediate. With halobenzenes, the values for migration and retention of deuterium during hydroxylation (the NIH shift) are nearly identical when the deuterium is either at the site of hydroxylation or the adjacent site, indicating a possible cyclohexadienone common intermediate without formation of an arene oxide.

Taking all of these results into consideration, the most likely mechanism is electrophilic addition of an active triplet-like oxygen atom to the π-system of the substrate to give either a tetrahedral intermediate radical (**4.59a**) or cation (**4.59b**) (Scheme 4.41). However, 1,2-shifts of hydrogen and alkyl radicals are not energetically favorable (and are generally rare) because this is a three-electron transition state (the two electrons in the C–H bond and the one electron of the adjacent radical) with only one bonding level, and therefore the third electron must be placed in an antibonding orbital. To avoid that problem, an electron could be transferred from the cyclohexadienyl radical to the Fe^{IV}, resulting in a carbocation species bound to Fe^{III} (**4.59b**). A 1,2-shift of hydrogen to a carbocation is a highly favorable process, leading to the cyclohexadienone (pathway c), which would tautomerize (with preferential proton abstraction) to the phenol. Formation of the arene oxide (pathway b) may not be directly relevant to normal catalysis, but may be an alternative pathway for the high-energy species **4.59a**.

4. Nitrogen and Sulfur Oxygenation

Compounds containing nitrogen and sulfur atoms undergo a cytochrome P450-catalyzed oxygenation to the corresponding *N*- or *S*-oxide, respectively, particularly when there are no α-hydrogens (when there are α-hydrogens, generally another product is formed; see IV.C.5). The mechanism is believed to involve electron transfer followed by oxygen rebound (Scheme 4.42 shows the reaction with a sulfide). Support for the electron transfer mechanism for sulfides is that the rates

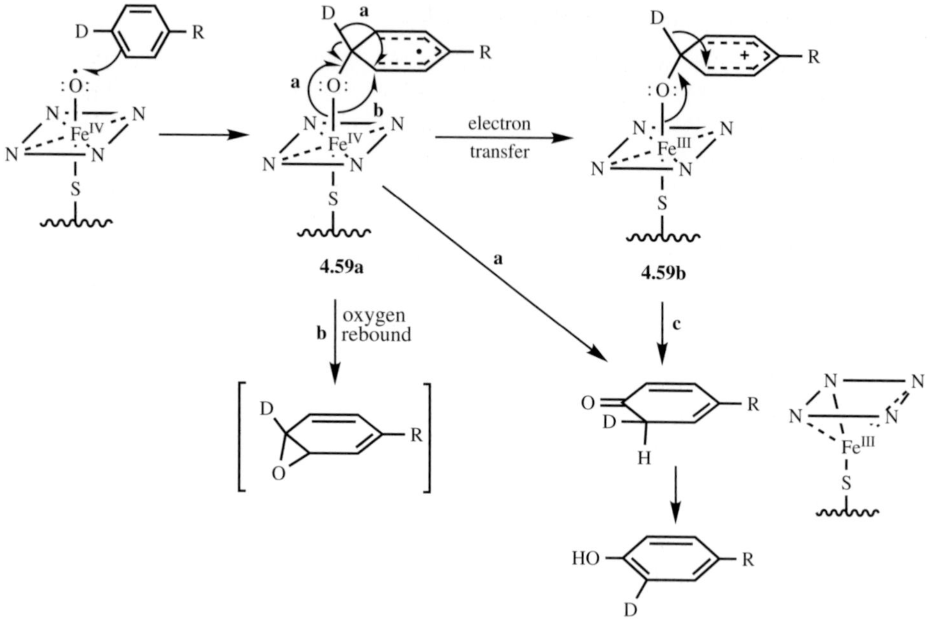

SCHEME 4.41   Mechanism proposed for heme-dependent oxygenation of aromatic compounds.

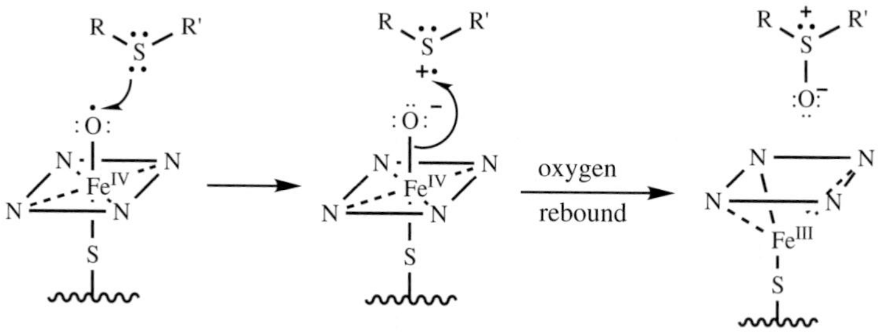

SCHEME 4.42   Mechanism proposed for heme-dependent oxygenation of sulfides.

of sulfoxide formation (log k_{cat}) for a series of *para*-substituted thioanisoles exhibit linear free-energy relationships with the one-electron oxidation potentials of these compounds as well as with the σ^+ values of the substitutents.[90] Ethers, in contrast, do not undergo oxygenation, because the more electronegative oxygen atom has a much higher oxidation potential. Amines and amides, however, can be N-oxygenated, but this reaction is generally significant only when there are no α-hydrogens. If α-hydrogens are present, the predominant reaction is N-dealkylation.

5. Dealkylation of Amines, Sulfides, and Ethers

Substituted amines having α-hydrogens readily undergo cytochrome P450-catalyzed N-dealkylation by a mechanism that is initiated by single-electron transfer, as was shown for N- and S-oxygenation in Scheme 4.42. From the radical cation (**4.60**), however, α-proton abstraction is favored,[91] giving the α-radical (**4.61**); oxygen rebound at that stage gives a carbinolamine (**4.62**), which can readily decompose to the lower amine and the aldehyde (Scheme 4.43). Even for amines that undergo N-dealkylation, however, small amounts of N-oxygenation products can be detected.[92] Because of the stable nature of sulfur radical cations, S-oxygenation predominates, even for sulfides with α-hydrogens, although S-dealkylation does occur.[93]

Ethers do not undergo one-electron oxidation, yet they readily exhibit O-dealkylation reactions. To rationalize these observations a direct hydrogen atom abstraction/oxygen rebound mechanism (or the concerted mechanism) must be important (Scheme 4.44). Dinnocenzo and co-workers have suggested that N-dealkylation may also proceed by a direct hydrogen atom abstraction mechanism.[94] Most likely both mechanisms are relevant; the one that predominates depends on the oxidation potential of the amine and the C−H bond dissociation energy.

6. Carbon−Carbon Bond Cleavage

The last type of cytochrome P450 reaction that we will discuss is one that leads to cleavage of a carbon−carbon bond. Several members of the cytochrome P450 superfamily of enzymes catalyze reactions leading to the cleavage of a methyl group from a steroid ring. One interesting example of this reaction is aromatase, the enzyme that catalyzes the conversion of the androgen androstenedione (**4.63**) to the estrogen estrone (**4.64**) (Scheme 4.45). Aromatase catalyzes the aromatization of the A ring of the steroid (the far left-hand ring in **4.63**). Note that this reaction, as well as the others in this family, requires three molecules of oxygen and three molecules of NADPH. The first two molecules of O_2 and NADPH appear to be consumed in standard "alkane" hydroxylation reactions, converting the C-19 methyl group first to the corresponding alcohol (**4.65**), then to the aldehyde (**4.67**), via the hydrate (**4.66**) (Scheme 4.46). The fate of each atom involved in this reaction is shown in Scheme 4.46; the formic acid produced was converted to the benzyl ester, and this ester was examined for ^{18}O incorporation by mass spectrometry. The enzyme can process the alcohol (**4.65**) or the aldehyde (**4.67**) to estrone and formic acid with the consumption of only two or one equivalent, respectively, of O_2 and NADPH, indicating that these compounds are true intermediates. Many of the early labeling studies provided confusing, if not contradictory results.[95] On the basis of many elegant labeling studies, Akhtar and co-workers suggested three possible mechanisms (Scheme 4.47).[96] Akhtar is the first to suggest that the heme peroxide (mechanism 1) may be directly involved in heme reactions rather than being

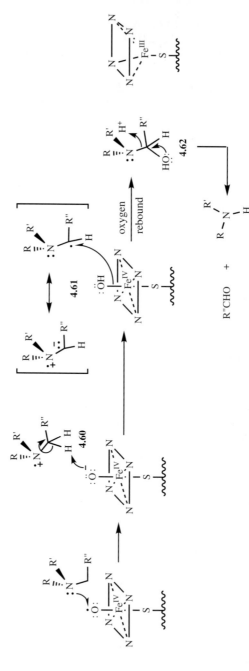

SCHEME 4.43   Mechanism proposed for heme-dependent oxygenation of amines.

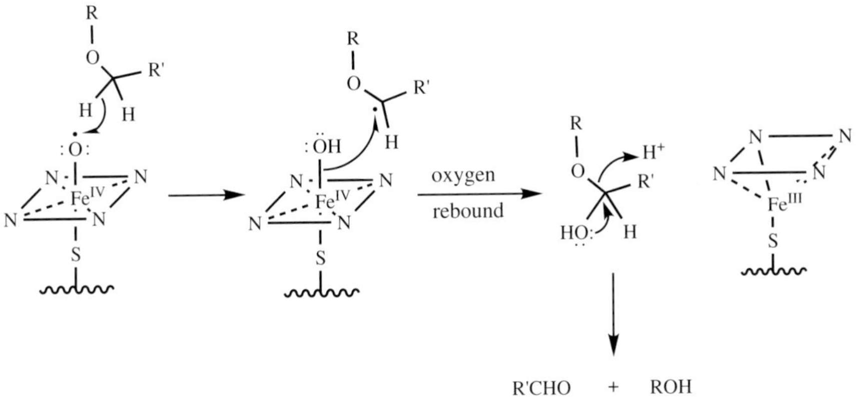

SCHEME 4.44   Mechanism proposed for heme-dependent oxygenation of ethers.

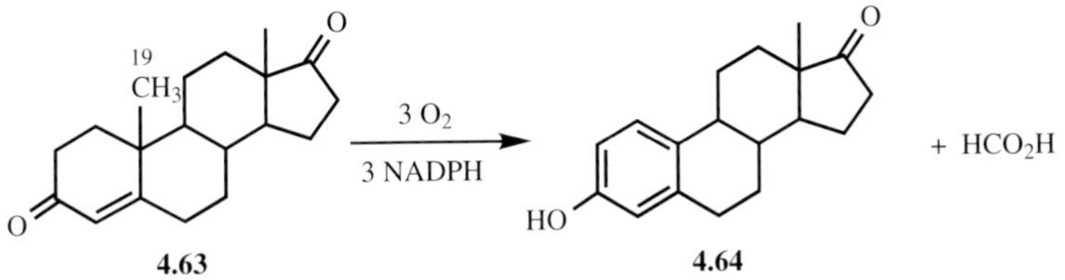

SCHEME 4.45   Reaction catalyzed by aromatase.

converted first to the heme oxo radical (mechanisms 2 and 3). Because the substrate (the aldehyde) is electrophilic, it is reasonable that the heme species should be nucleophilic, such as the heme peroxy anion. This is reminiscent of the change in reactivity of flavin hydroperoxides to nucleophiles when an electrophilic substrate (such as a ketone) is involved (see II.B.3).

To demonstrate that the heme peroxide is involved, not the heme oxo species, Akhtar and co-workers carried out a clever series of experiments.[97] For most enzymes in the P450 superfamily that catalyze a C−C bond cleavage reaction, the intermediacy of either a heme peroxide (Fe^{III}−OOH) or a heme oxo ($Fe^{IV}O \cdot$) species would give the same result, that is, loss of a hydrogen atom from the α-position with respect to the scissile bond and incorporation of an atom of O from O_2 into one of the cleaved fragments. However, by using a specifically deuterated substrate, Akhtar found that in the formation of 17α-hydroxyandrogen (**4.69**) from pregnenolone (**4.68**), catalyzed by P450$_{17\alpha}$, the 16α, 16β, and 17 hydrogens are retained in the product, whereas an atom of oxygen from O_2 is incorporated (Scheme 4.48). Therefore, the mechanism with heme peroxide can be differentiated from that with heme oxo because in order for Fe^{IV}−O$\cdot$ to promote fragmentation without cleaving

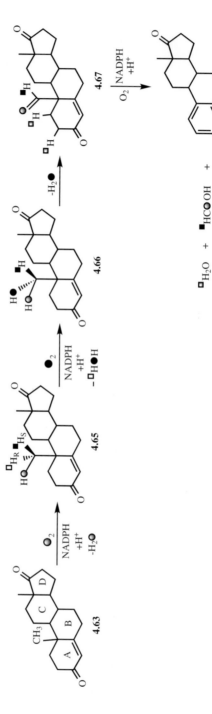

SCHEME 4.46   Fate of the atoms during aromatase-catalyzed conversion of androstenedione to estrone.

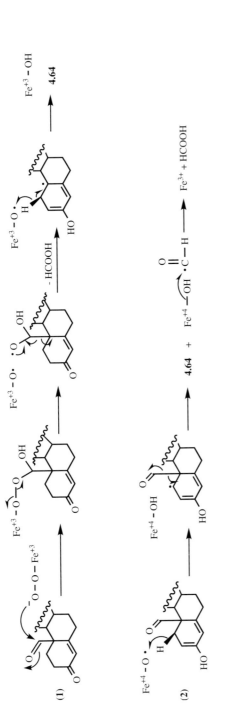

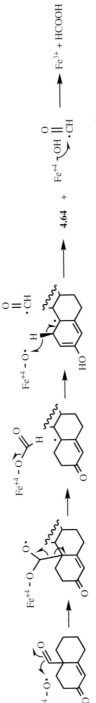

SCHEME 4.47 Three possible mechanisms for the last step in the aromatase-catalyzed oxygenation of androstenedione.

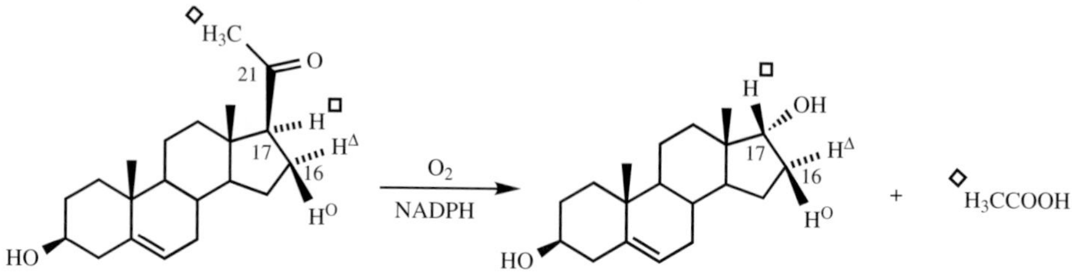

SCHEME 4.48   Oxygenation of pregnenolone, catalyzed by an isozyme of cytochrome P450.

the 16α, 16β, or 17 C–H bond, it must abstract one of the C_{21} methyl hydrogens, generating the C_{21} radical (**4.70**, Scheme 4.49) and ketene (**4.71**). When this re-action was carried out in D_2O, no deuterium was found in the acetate, which ex-cludes the heme oxo species as the intermediate. The observed results can be ra-tionalized by a nucleophilic mechanism (Scheme 4.50).

Site-directed mutagenesis of P450 1O1[98] and P450 2B4[99] further supports heme peroxide as the active species for deformylation and heme oxo species for hydrox-ylation reactions. From a variety of structural studies, there appears to be evidence that Thr-252 in P450 1O1 and Thr-302 in P450 2B4 are the active-site residues that donate a proton to the heme peroxide species in its conversion to the heme oxo species (see Scheme 4.27). If that is the case, then mutation of that residue should decrease reactions that require heme oxo species and increase reactions that require heme peroxide. In fact, mutation of Thr-252 in P450 1O1 (T252A) gives a mutant devoid of camphor hydroxylation activity, even though substrate binding and the rate of the first electron transfer to the heme center is unaffected. Mutation of T302A in P450 2B4 gives a mutant with decreased activity toward hydroxylation (a reaction that would require heme oxo), but increased activity toward deformyla-tion, suggesting the importance of the heme peroxide when a carbonyl-containing substrate (as in deformylation) is involved.

Although a heme peroxide intermediate was supported for the third oxygenation step of the P450 enzyme, lanosterol 14α-methyl demethylase, which demethylates lanosterol (**4.72**) to the corresponding diene (**4.74**, Scheme 4.51), a different mech-anism following heme peroxide addition to the carbonyl was proposed. Conditions were found for the isolation of a suspected intermediate, the 14α-formyloxy ana-logue (**4.73**). Synthesis and incubation of this proposed intermediate with lanos-terol 14α-methyl demethylase gave **4.74**.[100] This provides further evidence for addition of $Fe^{III}OO^-$ to the carbonyl, but it supports a Baeyer–Villiger rearrange-ment (Scheme 4.51), earlier proposed by Akhtar,[101a] rather than a radical dissocia-tion mechanism (Scheme 4.52).[101b] Akhtar, however, prepared the 19-formyloxy derivative of androstenedione (**4.75**), the corresponding Baeyer–Villiger rearrange-ment intermediate for aromatase, and found that on incubation with aromatase,

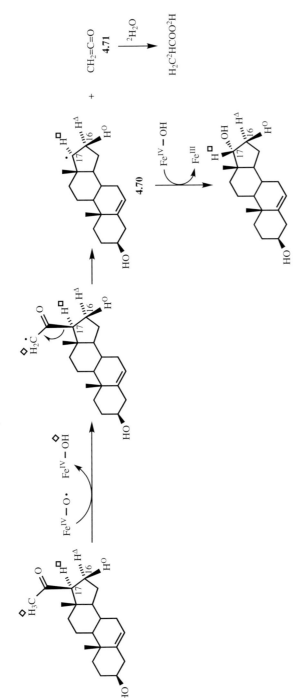

SCHEME 4.49   Hydrogen atom abstraction mechanism, using a heme iron oxo species, for the P450$_{17\alpha}$–catalyzed oxygenation of pregnenolone.

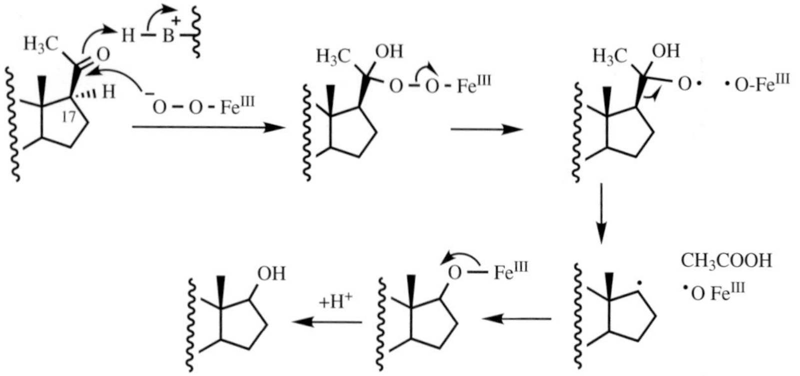

SCHEME 4.50 Nucleophilic mechanism, using heme peroxy anion followed by a radical decomposition of the heme peroxide, for the P450$_{17\alpha}$-catalyzed oxygenation of pregnenolone.

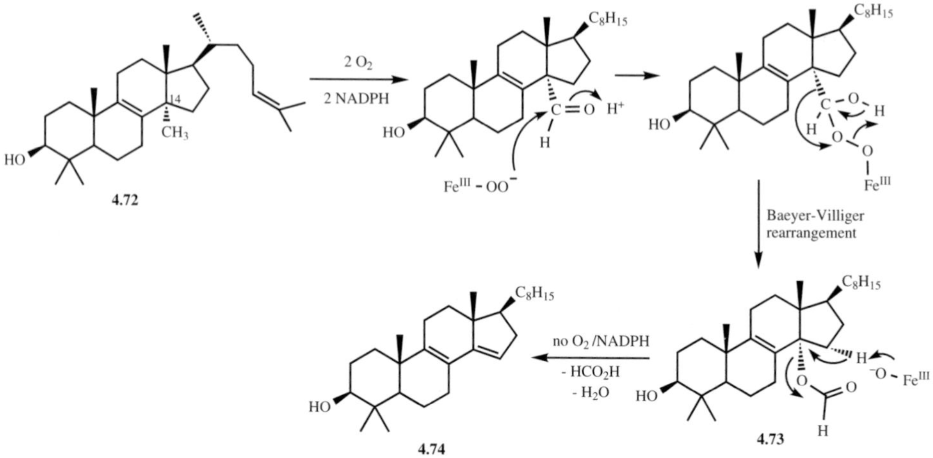

SCHEME 4.51 Nucleophilic mechanism, using heme peroxy anion followed by a Baeyer–Villiger rearrangement, for the lanosterol 14α-methyl demethylase-catalyzed oxygenation of lanosterol.

no estrone product was obtained.[102] Either aromatase and P450$_{17\alpha}$ have different mechanisms than lanosterol 14α-methyl demethylase or the formyloxy intermediate is not released in the case of aromatase and P450$_{17\alpha}$ (the evidence against the corresponding formyloxy intermediate with aromatase is only the fact that it is not a substrate).

Another refinement of the aromatase mechanism comes from model studies.[103] A comparison of the products of chemical oxidation of the C-19 formyl dienol ether (**4.76**) with the C-19 formyl ketone (**4.77**) showed that **4.76** gave the aromatized product (at only about 1/500th the rate of the enzymatic reaction), but **4.77** did not (Scheme 4.53). On this basis, it was proposed that the third oxidation step

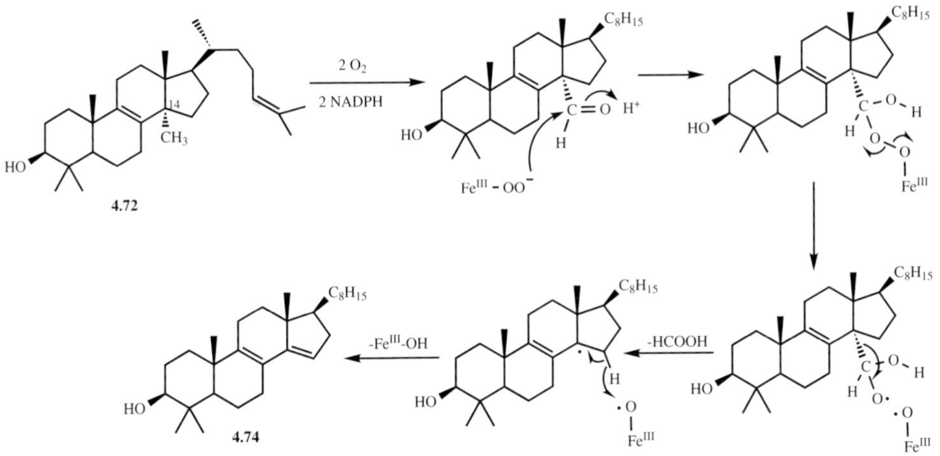

SCHEME 4.52   Nucleophilic mechanism, using heme peroxy anion followed by a radical decomposition of the heme peroxide, for the lanosterol 14α-methyl demethylase-catalyzed oxygenation of lanosterol.

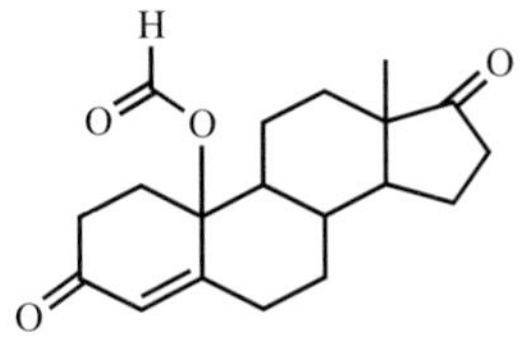

4.75

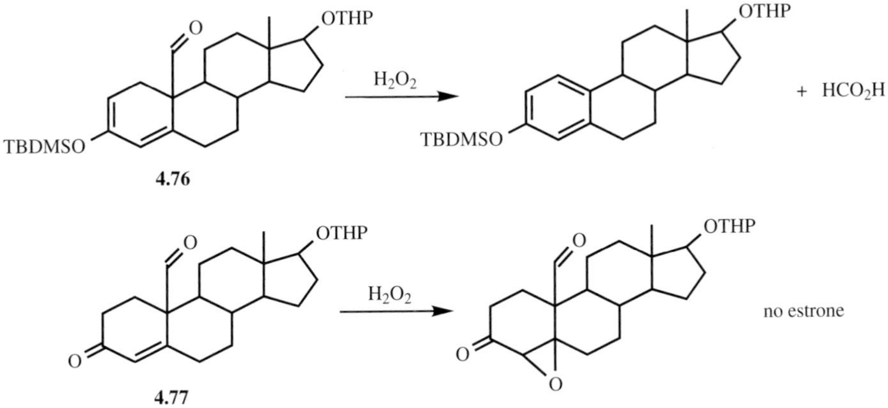

SCHEME 4.53   Model studies on the mechanism of aromatase.

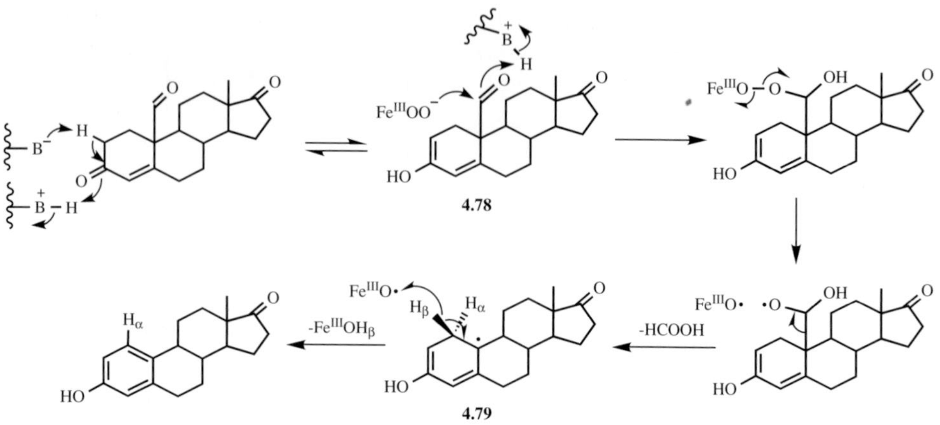

SCHEME 4.54   Mechanism proposed for aromatase initiated by dienol formation.

of aromatase proceeds via the dienol **4.78** (Scheme 4.54; this also could be written as the nucleophilic mechanism with Baeyer–Villiger rearrangement as shown in Scheme 4.51). Furthermore, when the model compound is selectively deuterated at the 1-position, and $H_2{}^{18}O_2$ is the oxidant, the 1β-hydrogen (**4.79**) is exclusively lost (with an isotope effect >2), and one ^{18}O atom is incorporated into the formic acid produced (Scheme 4.54), as in the case of aromatase.

V. NONHEME IRON OXYGENATION

A. Methane Monooxygenase

As we just saw, hydroxylation of alkanes is generally carried out by heme-containing enzymes in the cytochrome P450 superfamily. A similar reaction is catalyzed by methanotrophic bacteria, except these enzymes do not contain heme. Instead, they

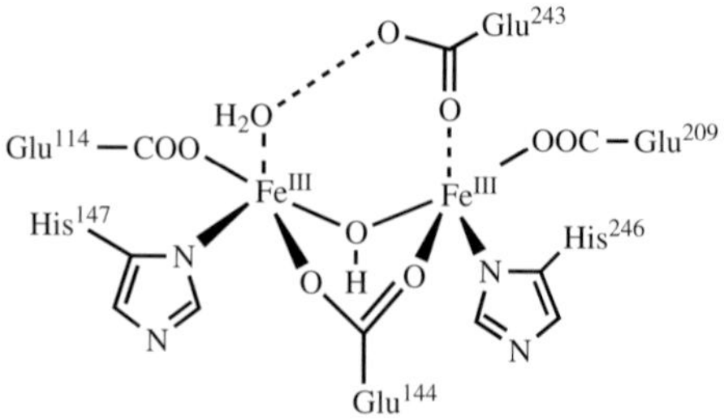

4.80

contain a binuclear iron cluster (**4.80**). Soluble methane monooxygenase from *Methylococcus capsulatus* catalyzes the conversion of methane (and a variety of saturated, unsaturated, linear, branched, and cyclic hydrocarbons) to methanol (or the corresponding hydrocarbon alcohols) in the presence of O_2 and NADPH (the second oxygen atom of O_2 becomes water).[104] Transient kinetic analysis of the reaction has revealed at least five and probably six intermediates.[105] Mössbauer spectroscopy showed that the species with which the substrate reacts is a diamagnetic cluster, suggesting that this intermediate may contain two antiferromagnetically coupled high-spin Fe^{IV} atoms.[106] This could arise from reduction of the native-state binuclear ferric cluster (**4.81**, Scheme 4.55) by NADPH to the ferrous form (**4.82**) to which O_2 binds, forming a diferric oxygen complex (**4.83**). XAS and Mössbauer spectroscopy support **4.83a**, not **4.83b**.[106a] Protonation with loss of water would give a binuclear Fe^{IV} oxo species, which can be depicted in a variety of ways, but **4.84a** resembles the heme oxo species. This is known as Compound Q, which appears to be the active iron species that converts the hydrocarbon to an enzyme-bound product. A variety of mechanisms can be drawn,[107] including hydrogen atom abstraction and oxygen rebound and a nonsynchronous, concerted insertion mechanism, both described for cytochrome P450 (Schemes 4.28 and 4.35, Section IV.C.1). The exceedingly large deuterium kinetic isotope effect (50–100!) is consistent with complete breakage of the C–H bond in the oxygenation step of the reaction;[108] however, a variety of radical clock substrate probes[109] and labeling experiments[110] indicate that, if a radical is formed, its lifetime is less than 150 fs. Studies with soluble methane monooxygenase and either methylcubane or the hypersensitive cyclopropane-based probe to distinguish radical and cation intermediates (see **4.46**, Scheme 4.34) indicate that a cation, not a radical, intermediate is important.[110a]

VI. COPPER-DEPENDENT OXYGENATION

A. Dopamine β-Monooxygenase

Dopamine β-monooxygenase (also known as *dopamine hydroxylase*), which is important in controlling the concentrations of the neurotransmitters/hormones dopamine (**4.85**) and norepinephrine (**4.86**), catalyzes the reaction shown in Scheme 4.56. The source of the electrons is ascorbic acid, and the optimal activity occurs with two Cu^{II} per enzyme subunit.[111] Klinman and co-workers utilized rapid mixing, acid quench techniques to demonstrate that there was a separate binding site on the enzyme for the reductant and for the substrate or product and that each of the copper ions performs a separate function.[112] One Cu^{II} catalyzes electron transfer from ascorbate, and the other Cu^{II} catalyzes the insertion of oxygen into the substrate from O_2.[113] Therefore, this enzyme is a *mononuclear metallomonooxygenase*.

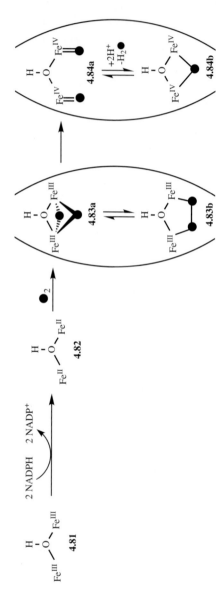

SCHEME 4.55　Binuclear ferric cluster of methane monooxygenase.

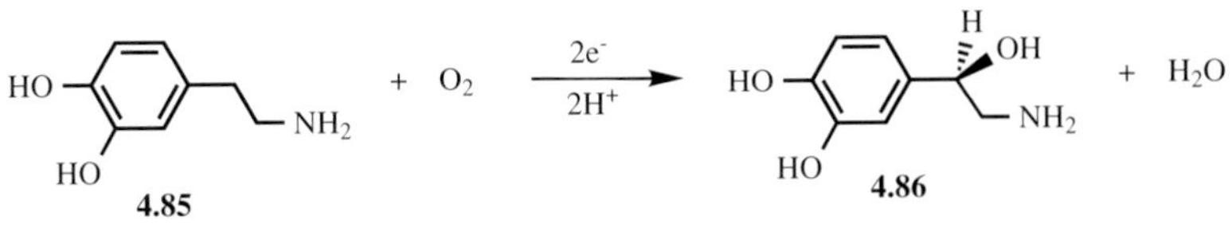

SCHEME 4.56   Reaction catalyzed by dopamine β-monooxygenase.

A Hammett plot of log k for C–H bond cleavage versus σ_p for a series of p-substituted phenylethylamines gave a $\rho = -1.5$. This small negative value is consistent with either a radical or a carbocation intermediate, but the fit is better to σ_p than to σ_p^+ (σ_p for resonance-stabilized carbocations), suggesting a radical mechanism with a polarized transition state.[114]

Isotope studies do not support a simple hydroperoxide mechanism in which cleavage of the hydroperoxyl O–O bond and cleavage of the C–H bond of substrate occur together. In this case, the magnitude of the ^{18}O isotope effect should increase as the substrate becomes less reactive because of the required formation of a transition state with more extensive O–O bond cleavage to give a more reactive oxygen radical. A mechanism in which copper hydroperoxide is reductively cleaved prior to C–H abstraction predicts an increase in bond order to oxygen as the transition state for C–H cleavage becomes more productlike, and this would lead to a decrease in the ^{18}O isotope effect with decreasing substrate reactivity, which is what is observed. A mechanism consistent with these results, proposed by Klinman and co-workers, is shown in Scheme 4.57.[115] This mechanism depicts an intermediate with four unpaired spins, two spins from Cu^{II} (one at each of the two Cu^{II} binding sites), one spin from the Cu^{II}–O·, and one from the tyrosyl radical. If the tyrosyl radical is between the two Cu^{II} sites and is spin coupled, it may be diamagnetic, which is consistent with the inability of Klinman and co-workers to detect Cu^{II} signals in the EPR spectrum. Scheme 4.57 may seem circuitous. Why should the initial O–O bond cleavage (**4.87**) lead to tyrosine hydrogen atom abstraction, then in a second step the tyrosinyl radical abstracts the β-hydrogen atom from dopamine (**4.88**)? It would be more economical to have the O–O cleavage in the first step lead directly to dopamine hydrogen atom abstraction with formation of Cu^{II}–O· and water. However, this additional step was proposed to accommodate the isotope effect results and other kinds of reactions that this enzyme has been observed to catalyze, such as sulfoxidation, oxygenative ketonization, olefin oxidation, alkyne oxidation, selenooxidation, and oxygenative N-dealkylation.[116]

The reactions described in this chapter incorporate one of the oxygen atoms of O_2 into the substrate; the second oxygen atom is converted to water. In the next chapter we will see enzymes that catalyze the incorporation of both atoms of oxygen from O_2 into substrate molecules. These enzymes are called *dioxygenases*.

NOTE THAT PROBLEMS AND SOLUTIONS RELEVANT TO EACH CHAPTER CAN BE FOUND IN APPENDIX II.

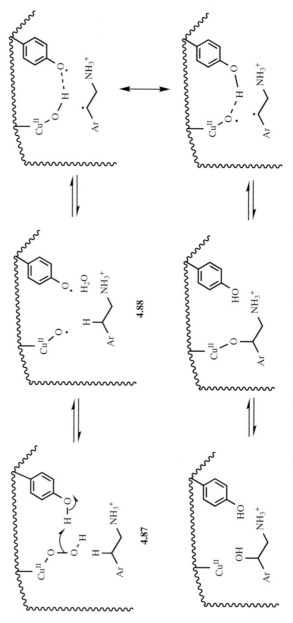

SCHEME 4.57 Mechanism proposed for dopamine β-monoxygenase.

REFERENCES

1. Sutton, W. B. *J. Biol. Chem.* **1957**, *226*, 395.
2. Lockridge, O.; Massey, V.; Sullivan, P. A. *J. Biol. Chem.* **1972**, *247*, 8097.
3. Dole, M.; Milner, D. C.; Williams, T. F. *J. Am. Chem. Soc.* **1957**, *79*, 4809.
4. Hamilton, G. In *Progress in Bioorganic Chemistry*, vol. 1, Kaiser, T.; Kezdy, F., Eds., Wiley: New York, 1971, p. 83.
5. Ghisla, S.; Entsch, B.; Massey, V.; Husein, M. *Eur. J. Biochem* **1977**, *76*, 139.
6. (a) Hamilton, G. In *Progress in Bioorganic Chemistry*, vol. 1, Kaiser, T.; Kezdy, F., Eds., Wiley: New York, 1971, p. 83. (b) Orf, H. W.; Dolphin, D. *Proc. Natl. Acad. Sci. USA* **1974**, *71*, 2646. (c) Rastetter, W. H.; Gadek, T. R.; Tane, J. P.; Frost, J. W. *J. Am. Chem. Soc.* **1979**, *101*, 2228.
7. (a) Entsch, B.; Ballou, D. P.; Massey, V. *J. Biol. Chem.* **1976**, *251*, 2550. (b) Schopfer, L. M.; Wessiak, A.; Massey, V. *J. Biol. Chem.* **1991**, *266*, 13080.
8. (a) Maeda-Yorita, K.; Massey, V. *J. Biol. Chem.* **1993**, *268*, 4134. (b) Taylor, M. G.; Massey, V. *J. Biol. Chem.* **1990**, *265*, 13687.
9. Arunachalam, U.; Massey, V.; Miller, S. M. *J. Biol. Chem.*, **1994**, *269*, 150.
10. Massey, V. *J. Biol. Chem.* **1994**, *269*, 22459.
11. Ortiz-Maldonado, M.; Ballou, D. P.; Massey, V. In *Flavins and Flavoproteins 1996*, Stevenson, K. J.; Massey, V.; Williams, C. H., Jr., Eds.; University of Calgary Press: Calgary, 1997, p. 323.
12. (a) Hastings, J. W.; Balny, C.; LePeuch, C.; Douzou, P. *Proc. Natl. Acad. Sci. USA* **1973**, *70*, 3468. (b) Hastings, J. W.; Blany, C. *Biochemistry* **1976**, *14*, 4719. (c) Kemal, C.; Bruice, T. C. *Proc. Natl. Acad. Sci. USA* **1976**, *73*, 995.
13. Kurfuerst, M.; Macheroux, P.; Ghisla, S.; Hastings, J. W. *Biochim. Biophys. Acta.* **1987**, *924*, 104.
14. Eckstein, J. W.; Hastings, J. W.; Ghisla, S., *Biochemistry* **1993**, *32*, 404.
15. (a) Macheroux, P.; Ghisla, S.; Kurfürst, M.; Hastings, J. W. In *Flavins and Flavoproteins*, Edmondson, D. E.; McCormick, D. B., Eds., Walter deGruyter: New York, 1984, pp. 37–40. (b) Mager, H. I. X.; Sazou, D.; Liu, Y. H.; Tu, S.-C.; Kadish, K. M. *J. Am. Chem. Soc.* **1988**, *110*, 3759.
16. Ahrens, M.; Macheroux, P.; Eberhard, A.; Ghisla, S.; Branchaud, B.; Hastings, J. W. *Photochem. Photobiol.* **1991**, *54*, 295.
17. Bruice, T. C.; Noar, J. B.; Ball, S. S.; Venkataram, U. V. *J. Am. Chem. Soc.* **1983**, *105*, 2452.
18. Raushel, F. M.; Baldwin, T. O. *Biochem. Biophys. Res. Commun.* **1989**, *164*, 1137.
19. Francisco, W. A.; Abu-Soud, H. M.; Topgi, R.; Baldwin, T. O.; Raushel, F. M. *J. Biol. Chem.* **1996**, *271*, 104.
20. Baeyer, A.; Villiger, V. *Ber. Deut. Chem. Ges.* **1899**, *32*, 3625.
21. Hudlicky, M. In *Oxidations in Organic Chemistry*, American Chemical Society: Washington, 1990, pp. 186–195.
22. Walsh, C.; Chen, Y.-C. J. *Angew. Chem. Int. Ed. Engl.* **1988**, *27*, 333.
23. Ryerson, C. C.; Ballou, D. P.; Walsh, C. *Biochemistry* **1982**, *21*, 2644.
24. Branchaud, B. P.; Walsh, C. T. *J. Am. Chem. Soc.* **1985**, *107*, 2153.
25. Schwab, J. M.; Li, W.-b.; Thomas, L. P. *J. Am. Chem. Soc.* **1983**, *105*, 4800.
26. Abril, O.; Ryerson, C. C.; Walsh, C.; Whitesides, G. M. *Bioorg. Chem.* **1989**, *17*, 41.
27. Latham, Jr., J. A.; Walsh, C. *J. Chem. Soc. Chem. Commun.* **1986**, 527.
28. Carnell, A. J.; Roberts, S. M.; Sik, V.; Willetts, A. J. *J. Chem. Soc. Chem. Commun.* **1990**, 1438.
29. Alphand, V.; Archelas, A.; Furstoss, R. *Tetrahedron Lett.* **1989**, *30*, 3663.
30. Grogan, G.; Roberts, S. M.; Willetts, A. J. *J. Chem. Soc. Chem. Commun.* **1993**, 699.
31. Carnell, A. J.; Roberts, S. M.; Sik, V.; Willetts, A. J. *J. Chem. Soc. Perkin Trans.* **1991**, *1*, 2385.
31a. Fitzpatrick, P. F. *Annu. Rev. Biochem.* **1999**, *68*, 355.
32. Dix, T. A.; Benkovic, S. J. *Acc. Chem. Res.* **1988**, *21*, 101.
33. Kappock, T. J.; Caradonna, J. P. *Chem. Rev.* **1996**, *96*, 2659.
34. Erlandsen, H.; Fusetti, F.; Martinez, A.; Hough, E.; Flatmark, T.; Stevens, R. C. *Nature Struct. Biol.* **1997**, *4*, 995.

35. Goodwill, K. E.; Sabatier, C.; Marks, C.; Raag, R.; Fitzpatrick, P. F.; Stevens, R. C. *Nature Struct. Biol.* **1997,** *4,* 578.

36. Tietz, A.; Lindberg, M.; Kennedy, E. P. *J. Biol. Chem.* **1964,** *239,* 4081.

37. Bhat, S. C.; Vaidyanathan, C. S. *Arch. Biochem. Biophys.* **1976,** *176,* 314.

38. Reddy, C. C.; Vaidyanathan, C. S. *Arch. Biochem. Biophys.* **1976,** *177,* 488.

39. Sreeleela, N. S.; SubbaRao, P. V.; Premkumar, R.; Vaidyanathan, C. S. *J. Biol. Chem.* **1969,** *244,* 2293.

40. Kerwin, J. F., Jr.; Lancaster, J. R., Jr.; Feldman, P. L. *J. Med. Chem.* **1995,** *38,* 4343.

41. (a) Guroff, G.; Levitt, M.; Daly, J.; Udenfriend, S. *Biochem. Biophys. Res. Commun.* **1966,** *25,* 253. (b) Jerina, D. M.; Daly, J. W. *Science* **1974,** *185,* 573.

42. Miller, R. J.; Benkovic, S. J. *Biochemistry* **1988,** *27,* 3658.

43. Fitzpatrick, P. F. *J. Am. Chem. Soc.* **1994,** *116,* 1133.

44a. Carr, R. T.; Balasubramanian, S.; Hawkins, P. C. D.; Benkovic, S. J. *Biochemistry* **1995,** *34,* 7525.

44. Hillas, P. J.; Fitzpatrick, P. F. *Biochemistry* **1996,** *35,* 6969.

45. Klingenberg, M. *Arch. Biochem. Biophys.* **1958,** *75,* 376.

46. Garfinkel, D. *Arch. Biochem. Biophys.* **1958,** *77,* 493.

47. Omura, T.; Sato, R. *J. Biol. Chem.* **1964,** *239,* 2370.

48. Nebert, D. W.; Nelson, D. R.; Coon, M. J.; Estabrook, R. W.; Feyereisen, R.; Fujii-Kuriyama, Y.; Gonzalez, F. J.; Guengerich, F. P.; Gunsalus, I. C.; Johnson, E. F.; Loper, J. C.; Sato, R.; Waterman, M. R.; Waxman, D. J. *DNA Cell Biol.* **1991,** *10,* 1.

49. Gunsalus, I. C.; Sligar, S. G. *Adv. Enzymol. Relat. Areas Mol. Biol.* **1978,** *47,* 1.

50. Poulos, T. L.; Finzel. B. C.; Howard, A. J. *J. Mol. Biol.* **1987,** *195,* 687.

51. (a) Sono, M.; Roach, M. P.; Coulter, E. D.; Dawson, J. H. *Chem. Rev.* **1996,** *96,* 2841. (b) Ortiz de Montellano, P. R., Ed., *Cytochrome P-450,* 2nd ed., Plenum: New York, 1995. (c) Guengerich, F. P.; Macdonald, T. L., *FASEB J.* **1991,** *4,* 2453.

52. Loew, G. H.; Kert, C. D.; Hjelmeland, L. M.; Kirchner, R. F. *J. Am. Chem. Soc.* **1977,** *99,* 3534.

53. (a) Raag, R.; Martinis, S. A.; Sligar, S. G.; Poulos, T. L. *Biochemistry* **1991,** *30,* 11420. (b) Aikens, J.; Sligar, S. G. *J. Am. Chem. Soc.* **1994,** *116,* 1143.

54. Yeom, H.; Sligar, S. G.; Li, H.; Poulos, T. L.; Fulco, A. J. *Biochemistry* **1995,** *34,* 14733.

55. Kimata, Y.; Shimada, H.; Hirose, T.; Ishimura, Y. *Biochem. Biophys. Res. Commun.* **1995,** *208,* 96.

56. (a) Shimada, H.; Makino, R.; Imai, M.; Horiuchi, T.; Ishimura, Y. In *International Symposium on Oxygenases and Oxygen Activation;* Nozaki, M; Yamamoto, S.; Ishimura, Y., Eds., Yamada Science Foundation, pp. 133–136. (b) Gerber, N. C.; Sligar, S. G. *J. Am. Chem. Soc.* **1992,** *114,* 8742.

57. (a) Dawson, J. H.; Sono, M. *Chem. Rev.* **1987,** *87,* 1255. (b) Dawson, J. H. *Science* **1988,** *240,* 433.

58. White, R. E. *Pharmacol. Ther.* **1991,** *49,* 21.

59. (a) Shapiro, S.; Piper, J. U.; Caspi, E. *J. Am. Chem. Soc.* **1982,** *104,* 2301. (b) Hamberg, M; Bjorkhem, I. *J. Biol. Chem.* **1971,** *246,* 7411.

60. Hjelmeland, L. M.; Aronow, L.; Trudell, J. R. *Biochem. Biophys. Res. Commun.* **1977,** *76,* 541.

61. Groves, J. T.; McClusky, G. A.; White, R. E.; Coon, M. J. *Biochem. Biophys. Res. Commun.* **1978,** *81,* 154.

62. Groves, J. T.; McClusky, G. A. *J. Am. Chem. Soc.* **1976,** *98,* 859.

63. Groves, J. T. *J. Chem. Educ.* **1985,** *62,* 928.

64. (a) Newcomb, M. *Tetrahedron* **1993,** *49,* 1151. (b) Newcomb, M.; Tanaka, N.; Bouvier, A.; Tronlhe, C.; Horner, J. H.; Musa, O. M.; Martinez, F. N. *J. Am. Chem. Soc.* **1996,** *118,* 8505.

65. Atkinson, J. K.; Ingold, K. U. *Biochemistry* **1993,** *32,* 9209.

66. Atkinson, J. K.; Hollenberg, P. F.; Ingold, K. U.; Johnson, C. C.; Le Tadic, M-H; Newcomb, M.; Putt, D. A. *Biochemistry* **1994,** *33,* 10630.

67. (a) Hanzlik, R. P.; Ling, K-H. *J. Am. Chem. Soc.* **1985,** *107,* 7164. (b) Hanzlik, R. P.; Ling, K-H. *J. Org. Chem.* **1990,** *55,* 3992. (c) Hanzlik, R. P.; Ling, K-H. *J. Am. Chem. Soc.* **1993,** *115,* 9363.

68. Karki, S. B.; Dinnocenzo, J. P.; Jones, J. P.; Korzekwa, K. R. *J. Am. Chem. Soc.* **1995,** *117,* 3657.

69. Manchester, J. I.; Dinnocenzo, J. P.; Higgins, L.; Jones, J. P. *J. Am. Chem. Soc.* **1997,** *119,* 5069.

70. Newcomb, M.; Le Tadic, M.-H.; Putt, D. A.; Hollenberg, P. F. *J. Am. Chem. Soc.* **1995**, *117*, 3312.

71. Newcomb, M.; Chestney, D. L. *J. Am. Chem. Soc.* **1994**, *116*, 9753.

72. Newcomb, M.; Le Tadic-Biadatti, M.-H.; Chestney, D. L.; Roberts, E. S.; Hollenberg, P. F. *J. Am. Chem. Soc.* **1995**, *117*, 12085.

73. Toy, P. H.; Newcomb, M.; Hollenberg, P. F. *J. Am. Chem. Soc.* **1998**, *120*, 7719.

74. Collman, J. P.; Chien, A. S.; Eberspacher, T. A.; Brauman, J. E. *J. Am. Chem. Soc.* **1998**, *120*, 425.

75. Vaz, A. D. N.; McGinnity, D. F.; Coon, M. J. *Proc. Natl. Acad. Sci. USA* **1998**, *95*, 3555.

76. Toy, P. H.; Newcomb, M.; Coon, M. J.; Vaz, A. D. N. *J. Am. Chem. Soc.* **1998**, *120*, 9718.

77. Shaik, S.; Filatov, M.; Schröder, D.; Schwarz, H. *Chem. Eur. J.* **1998**, *4*, 193.

78. (a) Kunze, K. L.; Mangold, B. L. K.; Wheeler, C.; Beilan, H. S.; Ortiz de Montellano, P. R. *J. Biol. Chem.* **1983**, *258*, 4202. (b) Watanabe, T.; Akamatsu, K. *Biochem. Pharmacol.* **1974**, *23*, 1079.

79. Groves, J. T.; Watanabe, Y. *J. Am. Chem. Soc.* **1986**, *108*, 507.

80. (a) Ostovic, D.; Bruice, T. C. *J. Am. Chem. Soc.* **1989**, *111*, 6511. (b) Ostovic, D.; Bruice, T. C. *Acc. Chem. Res.* **1992**, *25*, 314.

81. Miller, V. P.; Fruetel, J. A.; Ortiz de Montellano, P. R. *Arch. Biochem. Biophys.* **1992**, *298*, 697.

82. Jerina, D. M.; Daly, J. W.; Witkop, B. *J Am. Chem. Soc.* **1968**, *90*, 6525.

83. Boyd, D. R.; Daly, J. W.; Jerina, D. M. *Biochemistry* **1972**, *11*, 1961.

84. Kasperek, G. J.; Bruice, T. C. *J. Chem. Soc. Chem. Commun.* **1972**, 784.

85. (a) Selander, H. G.; Jerina, D. M.; Daly, J. W. *Arch. Biochem. Biophys.* **1975**, *168*, 309. (b) Selander, H. G.; Jerina, D. M.; Piccolo, D. E.; Berchtold, G. A. *J. Am. Chem. Soc.* **1975**, *97*, 4428.

86. Burka, L. T.; Plucinski, T. M.; MacDonald, T. L. *Proc. Natl. Acad. Sci. USA* **1983**, *80*, 6680.

87. Hanzlik, R. P.; Hogberg, K.; Judson, C. M. *Biochemistry* **1984**, *23*, 3048.

88. (a) Korzekwa, K. R.; Trager, W.; Gouterman, M.; Spangler, D.; Loew, G. H. *J. Am. Chem. Soc.* **1985**, *107*, 4273. (b) Korzekwa, K. R.; Swinney, D. C.; Trager, W. F. *Biochemistry* **1989**, *28*, 9019.

89. Vannelli, T.; Hooper, A. B. *Biochemistry* **1995**, *34*, 11743.

90. (a) Watanabe, Y.; Iyanaqi, T.; Oae, S. *Tetrahedron Lett.* **1982**, *23*, 533. (b) Watanabe, Y. *Bull. Chem. Soc. Jpn.* **1982**, *55*, 188.

91. Guengerich, F. P.; Okazaki, O.; Seto, Y.; Macdonald, T. L. *Xenobiotica* **1995**, *25*, 689.

92. Seto, Y.; Guengerich, F. P. *J. Biol. Chem.* **1993**, *268*, 9986.

93. Oae, S.; Mikami, A.; Matsuura, T.; Ogawa-Asana, K.; Watanabe, Y.; Fujimora, K.; Iyanagi, T. *Biochem. Biophys. Res. Commun.* **1985**, *131*, 567.

94. (a) Dinnocenzo, J. P.; Karki, S. B.; Jones, J. P. *J. Am. Chem. Soc.* **1993**, *115*, 7111. (b) Karki, S. B.; Dinnocenzo, J. P.; Jones, J. P.; Korzekwa, K. R. *J. Am. Chem. Soc.* **1995**, *117*, 3657.

95. Wright, J. N.; Akhtar, M. *Steroids* **1990**, *55*, 142.

96. Akhtar, M.; Wright, J. N. *Nat. Prod. Rep.* **1991**, *8*, 527.

97. Akhtar, M.; Corina, D.; Miller, S.; Shyadehi, A. Z.; Wright, J. N. *Biochemistry* **1994**, *33*, 4410.

98. Raag, R.; Martinis, S. A.; Sligar, S. G.; Poulos, T. L. *Biochemistry* **1991**, *30*, 11420.

99. Vaz, A. D. N.; Pernecky, S. J.; Raner, G. M.; Coon, M. J. *Proc. Natl. Acad. Sci. USA* **1996**, *93*, 4644.

100. Fischer, R. T.; Trzaskos, J. M.; Magolda, R. L.; Ko, S. S.; Brosz, C. S.; Larsen, B. *J. Biol. Chem.* **1991**, *266*, 6124.

101. (a) Akhtar, M. *Bioorg. Chem.* **1977**, *6*, 529. (b) Shyadehi, A. Z.; Lamb, D. C.; Kelly, S. L.; Kelly, D. E.; Schunck, W.-H.; Wright, J.-N.; Corina, D.; Akhtar, M. *J. Biol. Chem.* **1996**, *271*, 12445.

102. (a) Akhtar, M.; Njar, V. C.; Wright, J. N. *J. Steroid Biochem.* **1993**, *44*, 375. (b) Akhtar, M.; Calder, M. R.; Corina, D. L.; Wright, J. N. *Biochem. J.* **1982**, *201*, 569.

103. (a) Cole, P. A.; Robinson, C. H. *J. Am. Chem. Soc.* **1988**, *110*, 1284. (b) Cole, P. A.; Robinson, C. H. *J. Am. Chem. Soc.* **1991**, *113*, 8130.

104. Lipscomb, J. D. *Annu. Rev. Microbiol.* **1994**, *48*, 371.

105. (a) Liu, Y.; Nesheim, J. C.; Lee, S. K.; Lipscomb, J. D. *J. Biol. Chem.* **1995**, *270*, 24662. (b) Lee, S. K.; Nesheim, J. C.; Lipscomb, J. D. *J. Biol. Chem.* **1993**, *268*, 21569.

106. Lee, S. K.; Fox, B. G.; Froland, W. A.; Lipscomb, J. D.; Munck, E. *J. Am. Chem. Soc.* **1993**, *115*, 6450.

106a. Hwang, J.; Krebs, C.; Huynh, B. H.; Edmondson, D. E.; Theil, E. C.; Penner-Hahn, J. E. *Science* **2000**, *287*, 122.

107. Wallar, B. J.; Lipscomb, J. D. *Chem. Rev.* **1996**, *96*, 2625.

108. Nesheim, J. C.; Lipscomb, J. D. *Biochemistry* **1996**, *35*, 10240.

109. (a) Liu, K. E.; Johnson, C. C.; Newcomb, M.; Lippard, S. J. *J. Am. Chem. Soc.* **1993**, *115*, 939. (b) Choi, S. Y.; Eaton, P. E.; Hollenberg, P. F.; Liu, K. E.; Lippard, S. J.; Newcomb, M.; Putt, D. A.; Upadhyaya, S. P.; Xiong, Y. *J. Am. Chem. Soc.* **1996**, *118*, 6547. (c) Valentine, A. M.; LeTadic-Biadatti, M.-H.; Toy, P. H.; Newcomb, M.; Lippard, S. J. *J. Biol. Chem.* **1999**, *274*, 10771.

110. Valentine, A. M.; Wilkinson, B.; Liu, K. E.; Komar-Panicucci, Priestley, N. D.; Williams, P. G.; Morimoto, H.; Floss, H. G.; Lippard, S. J. *J. Am. Chem. Soc.* **1997**, *119*, 1818.

110a. Choi, S.-Y.; Eaton, P. E.; Kopp, D. A.; Lippard, S. J.; Newcomb, M.; Shen, R. *J. Am. Chem. Soc.* **1999**, *121*, 12198.

111. Klinman, J. P.; Krueger, M.; Brenner, M.; Edmondson, D. E. *J. Biol. Chem.* **1984**, *259*, 3399.

112. Brenner, M. C.; Murray, C. J.; Klinman, J. P. *Biochemistry* **1989**, *28*, 4656.

113. Brenner, M. C.; Klinman, J. P. *Biochemistry* **1989**, *28*, 4664.

114. Miller, S. M.; Klinman, J. P. *Biochemistry* **1985**, *24*, 2114.

115. Tian, G.; Berry, J. A.; Klinman, J. P. *Biochemistry* **1994**, *33*, 226.

116. Stewart, L. C.; Klinman, J. P. *Annu. Rev. Biochem.* **1988**, *57*, 551.

CHAPTER 5

Dioxygenation

I. GENERAL

Dioxygenases catalyze the incorporation of both atoms of O_2 into substrates. When both oxygen atoms are incorporated into the same molecule, it is referred to as an *intramolecular dioxygenase;* when the two atoms of oxygen are incorporated into two products, it is an *intermolecular dioxygenase.* The intermolecular dioxygenases generally require α-ketoglutarate, which is converted into succinate by oxidative decarboxylation with the incorporation of one atom of oxygen from O_2 into the succinate. Most of these enzymes require a nonheme form of iron. We will first look into examples of intramolecular dioxygenases, then discuss the intermolecular type.

II. INTRAMOLECULAR DIOXYGENASES

A. Catechol Dioxygenases

Catechol dioxygenases are environmentalists' friends, because they are important in the degradation of a wide range of naturally occurring and human-made aromatic compounds found in soil, including insecticides, weed management agents, and other potential carcinogens.[1] This family of enzymes, which cleaves the aromatic ring of catechols, is divided into two general classes, the *intradiol-* and *extradiol-cleaving enzymes,* depending on whether the aromatic ring is broken between the catechol hydroxyl groups (intradiol) or next to one of the catechol hydroxyls (extradiol) (Figure 5.1). The extradiol-cleaving enzymes are further divided into *proximal* and *distal* extradiol dioxygenases based on whether they cut the aromatic ring at a point closer to (proximal) a ring substituent or farther from (distal) the substituent. High-resolution structural information is now available for the intradiol dioxygenases[2] and the extradiol dioxygenases.[3] The intradiol catechol dioxygenases

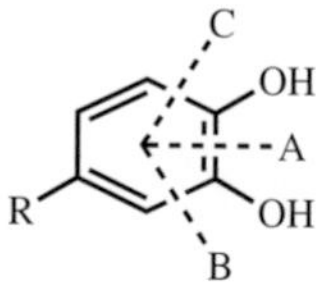

A — intradiol cleavage
B — proximal extradiol cleavage
C — distal extradiol cleavage

FIGURE 5.1   Different types of aromatic cleavages for catechol dioxygenases.

are a bright burgundy-red color because they contain nonheme Fe^{III} bound to two tyrosine ligands, and they also exhibit an EPR signal. The extradiol-cleaving enzymes utilize nonheme Fe^{II}, and therefore are colorless and do not show EPR signals.[4]

1. Intradiol Catechol Dioxygenases

Much more is known about the mechanism of the intradiol enzymes than the extradiol ones because there is a multitude of spectroscopic properties of Fe^{III} enzymes and because crystal structures of several intradiol complexes are available. The lack of spectroscopic evidence for the intermediacy of Fe^{II} species[5] suggests that the substrate activation mechanism involves coordination of the catechol to the Fe^{III} center, which activates the catechol for direct attack by O_2.[6] Kinetic studies are consistent with this suggestion, and indicate that substrate binds first, then O_2.[7] Experiments carried out in the presence of $^{18}O_2$ and in $H_2^{18}O$ demonstrate that the two atoms of oxygen that are incorporated into product come from $^{18}O_2$, not from $H_2^{18}O$.[8] A mechanism for protocatechuate 3,4-dioxygenase, shown in Scheme 5.1, is based on the model studies by Que and co-workers,[9] on the crystal structures of the anaerobic protocatechuate 3,4-dioxygenase·3,4-dihydroxybenzoate complex, the aerobic complexes with two inhibitor analogues of the substrate, ternary complexes of enzyme·inhibitor·cyanide,[10] and crystal structures of the enzyme complexed with seven competitive inhibitors.[11] Dissociation of Tyr-447 (5.1) is thought to provide the active-site base for initiation of substrate binding and generation of the substrate dianion (5.2),[12] which leads to the formation of chelate 5.3. Binding of oxygen and electron transfer from the substrate to the ferric ion gives 5.4. Radical combination generates the hypothetical peroxy radical adduct (5.5); transfer of an electron back from the ferrous ion to the peroxy radical, followed by coordination of the peroxide to the ferric ion, gives 5.6. A Criegee rearrangement,[13] as demonstrated in biomimetic reactions,[14] produces the cyclic anhydride bound to the ferric complex (5.7). Hydrolysis of the anhydride gives the enzyme-bound tricarboxylate (5.8). Attack of water and the active-site Tyr-447 hydroxyl releases β-carboxy-cis,cis-muconate (5.9) as the product and regenerates the active site.

The alternative dioxetane mechanism[15] (Scheme 5.2, pathway b; drawn for catechol 1,2-dioxygenase with 1,2,3-trihydroxybenzene (5.10) as substrate) is not supported because incubation of the enzyme with $^{18}O_2$ and substrate gives α-hydroxymuconic acid (5.12) with loss of some ^{18}O (the gray O in 5.12 represents

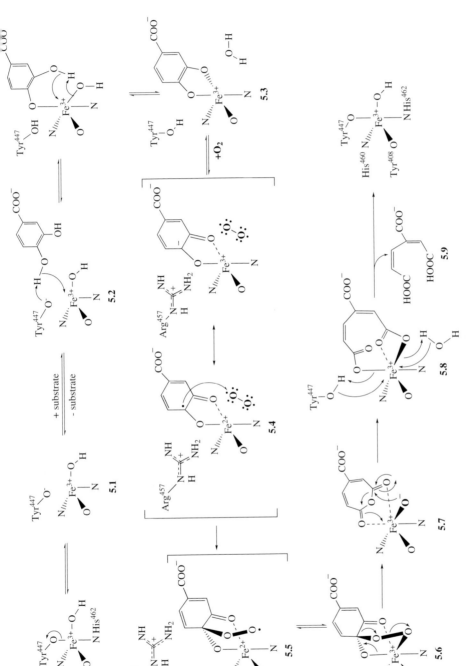

SCHEME 5.1 A mechanism for intradiol dioxygenases using protocatechuate 3,4-dioxygenase as an example.

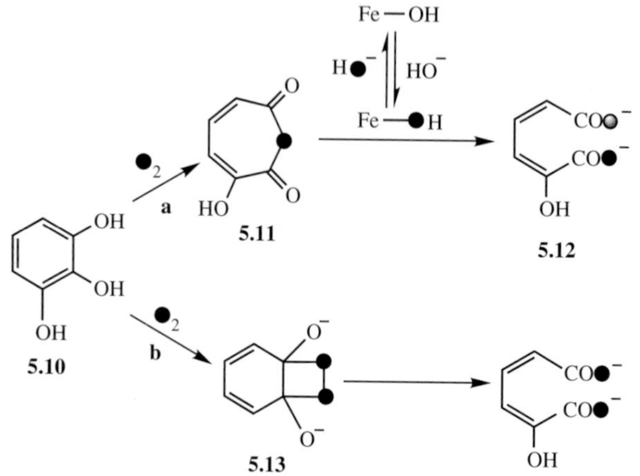

SCHEME 5.2   Criegee versus dioxetane mechanisms for intradiol dioxygenases.

less than one ^{18}O) from one of the carboxylates.[16] A dioxetane (**5.13**) mechanism (pathway b) requires both ^{18}O atoms to be retained in the carboxylates, but in a Criegee mechanism (pathway a) incorporation of the two oxygen atoms of O_2 occurs in two discrete steps via the anhydride (**5.11**) followed by subsequent hydrolysis by iron-bound hydroxide, which could exchange with unlabeled water.

2. Extradiol Catechol Dioxygenases

Extradiol-cleaving catechol dioxygenases are much more common than the intradiol counterparts, but mechanistic evidence has lagged for the former class of enzymes because spectroscopic handles are not available to probe Fe^{II} species. Experimental evidence for a Criegee rearrangement of a peroxy intermediate in the enzyme-catalyzed reaction was obtained from ^{18}O-labeling studies similar to those described for the intradiol catecholases (see Scheme 5.2).[17] On the basis of those results Bugg and co-workers proposed the mechanism shown in Scheme 5.3 for 2,3-dihydroxyphenylpropionate 1,2-dioxygenase. Following coordination of the substrates to the nonheme iron (**5.14**), single-electron transfer from the catechol dianion to molecular oxygen results in the semiquinone (**5.15**). The superoxide so generated can recombine at one of two different sites. Nucleophilic attack (Michael addition) by **5.15c** at the carbon adjacent to the oxygen-bearing carbon of the ring gives **5.16**; electron transfer of the ring radical to the peroxy radical gives the peroxide (**5.17**). Alternatively, **5.15b** could undergo radical combination at the oxygen-bearing carbon of the ring, generating **5.18**. Either intermediate, **5.17** or **5.18**, could undergo a Criegee rearrangement to give the lactone **5.19**. Hydrolysis of **5.19** by enzyme-bound Fe^{II} hydroxide gives the product (**5.20**). Evidence for

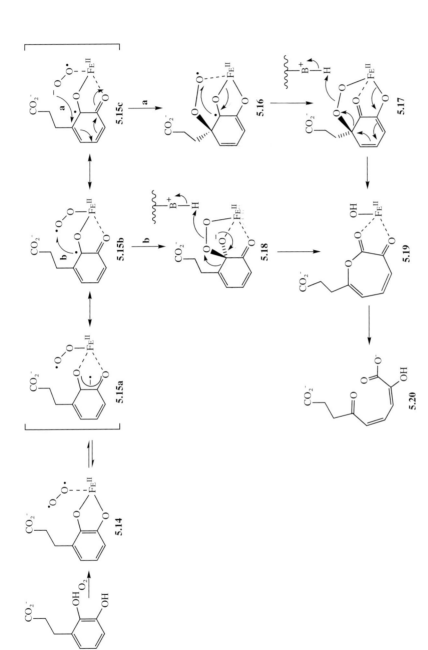

SCHEME 5.3   A mechanism for extradiol dioxygenases using 2,3–dihydroxyphenylpropionate 1,2–dioxygenase as an example.

the lactone hydrolysis step comes from the observation that a related saturated lactone undergoes enzyme-catalyzed hydrolysis.

Lipscomb, Que, and co-workers suggest that there may be a difference between the site of oxygen addition to the aromatic ring depending on whether the enzyme is an intradiol or an extradiol catechol dioxygenase.[18] In the case of the intradiol catechol dioxygenase, an electrophilic oxygen (triplet oxygen) reacts with the aromatic ring at the oxygen-bearing carbon (see Scheme 5.1), whereas extradiol catechol dioxygenases proceed by an enzyme-bound superoxide, which has a nucleophilic oxygen and can undergo Michael addition at the carbon adjacent to the oxygen-bearing carbon of the ring, leading to extradiol cleavage (pathway a, Scheme 5.3).

Bugg and co-workers obtained evidence for the intermediacy of a substrate semiquinone with two diastereomeric cyclopropyl-containing substrates (**5.21** and **5.22**, Scheme 5.4).[19] Analogue **5.21** gave a 94:6 mixture of **5.23** to **5.24** and analogue **5.22** gave a 90:10 ratio of the same products. These results can be rationalized on the basis of a reversible radical cyclopropyl ring cleavage via **5.25**. Because there is no proton exchange, a reversible ring cleavage is reasonable.

B. Prostaglandin H Synthase (Cyclooxygenase)

Prostaglandins are hormones that affect nerve transmission, blood circulation, inflammatory responses, and smooth muscle contraction. Nonsteroidal antiinflammatory drugs (NSAIDs), such as aspirin and ibuprofen, inhibit prostaglandin H syn-

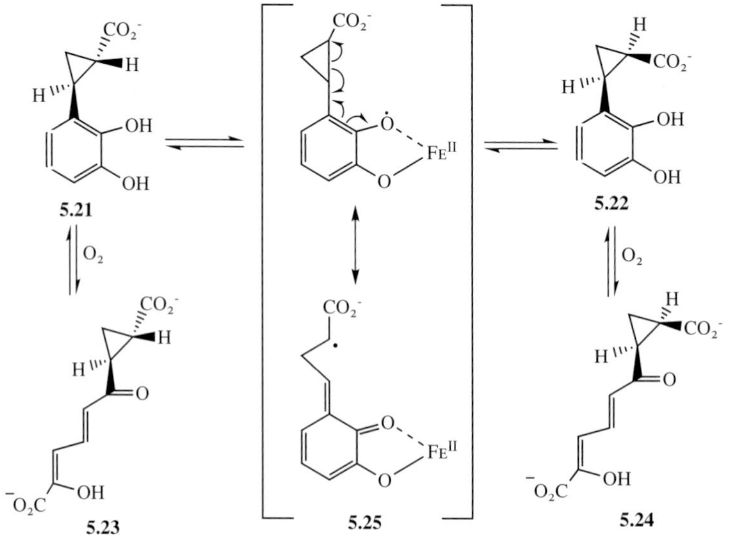

SCHEME 5.4   Evidence for a semiquinone intermediate in extradiol catechol dioxygenases.

thase, also called cyclooxygenase.[20] In 1991 it was shown that there are at least two isozymes of cyclooxygenase, referred to as cyclooxygenase-1 (or COX-1) and cyclooxygenase-2 (or COX-2).[21] COX-1, the constitutive form of the enzyme, is responsible for the physiological production of prostaglandins and is important in maintaining tissues in the stomach lining. COX-2, induced by cytokines in inflammatory cells, is responsible for the elevated production of prostaglandins during inflammation and is associated with inflammation, pain, and fever. Aspirin and other NSAIDs inhibit both COX-1 and COX-2, leading to stomach irritation (and, in some cases, to ulcers).[22]

Selective inhibition of COX-2 should give a NSAID without stomach lining complications.[23] Both Monsanto/Searle and Merck have highly selective COX-2 inhibitors that were recently approved for the drug market. X-ray crystal structures of both isoforms of the enzyme are known. COX-1[24] has an active-site isoleucine (Ile-523) and COX-2[25] has a valine (Val-523) at that position. Because valine is smaller than isoleucine, inhibitors with a substituent bulky enough to provide steric hindrance to Ile-523, but not to Val-523 show selectivity for COX-2. This resulted in compounds with IC_{50} values (inhibitor concentration that gives 50% inhibition in the presence of a given concentration of substrate) in the low nM range and selectivities of 10^3–10^4 for inhibition of COX-2 versus COX-1.[26] Site-directed mutagenesis of Ile-523 in COX-1 and Val-523 in COX-2 demonstrated that they were important to the selective inhibitor binding.[27]

Prostaglandin H synthase is a bifunctional heme-containing enzyme that catalyzes the conversion of arachidonic acid (**5.26**, Scheme 5.5) (or the corresponding molecule without the 5-double bond) to prostaglandin G (**5.27**; PGG_2, with the 5-double bond and PGG_1 without the 5-double bond) then to prostaglandin

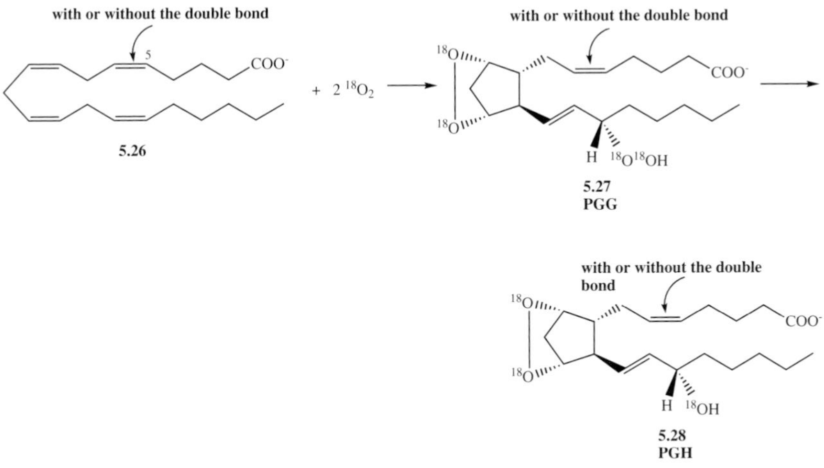

SCHEME 5.5   Reactions catalyzed by prostaglandin H synthase (cyclooxygenase).

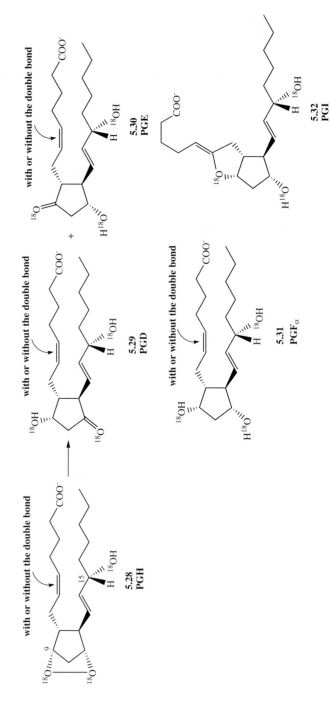

SCHEME 5.6   Conversion of prostaglandin H to other prostaglandins.

H (**5.28**, PGH$_2$ with the 5-double bond and PGH$_1$ without the 5-double bond); conversion of **5.26** to **5.27** represents the cyclooxygenase activity of the enzyme, and conversion of **5.27** to **5.28** is the peroxidase activity, both catalyzed by the same enzyme.[28] Various other prostaglandin synthases catalyze the conversion of PGH$_2$ to PGD$_2$ (**5.29**), PGE$_2$ (**5.30**), PGF$_{2\alpha}$ (**5.31**), and the prostacyclin PGI$_2$ (**5.32**) (Scheme 5.6); the corresponding PGX$_1$ series is made from PGH$_1$. Thromboxane synthases convert PGH$_2$ into thromboxanes.

Early mechanistic studies showed that 8,11,14-eicostrienoic acid (**5.33**; arachidonic acid without the 5-double bond) in the presence of crude enzyme and $^{18}O_2$ gives PGE$_1$ with three atoms of ^{18}O incorporated (Scheme 5.7).[29] When a mixture of $^{18}O_2$ and $^{16}O_2$ is used, it can be shown by mass spectrometry that both ring oxygens are either ^{18}O or ^{16}O, indicating that both of these oxygens come from the same molecule of O$_2$.

The hydrogens at C-8, C-11, and C-12 of 8,11,14-eicostrienoic acid (**5.33**) are retained in the prostaglandins.[30] Whereas the hydrogen at C-9 is lost in PGE$_1$, it is retained in PGF$_{1\alpha}$. Therefore, PGF$_{1\alpha}$ cannot be derived from PGE$_1$. Using [13R-^{3}H,3-^{14}C]- and [13S-^{3}H,3-^{14}C]8,11,14-eicostrienoic acid, Hamberg and Samuelsson[31] found that only the *pro-S* hydrogen at C-13 is removed. In 1973 and 1974 Samuelsson and co-workers reported that they isolated the endoperoxide/hydroperoxide (**5.27**, PGG$_2$) and that an endoperoxide isomerase converts it to the PGE series and an endoperoxide reductase converts it to the PGF series.[32]

Hydroxylated product **5.34** was isolated as a by-product of the reaction, suggesting that endoperoxide formation occurs prior to hydroperoxidation at C-15 (Scheme 5.8). Further evidence that endoperoxidation occurs prior to lipid

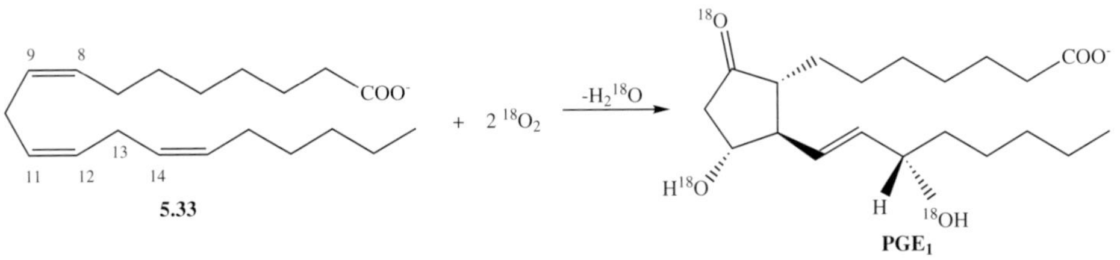

SCHEME 5.7   Prostaglandin H synthase-catalyzed incorporation of $^{18}O_2$ into 8,11,14-eicostrienoic acid.

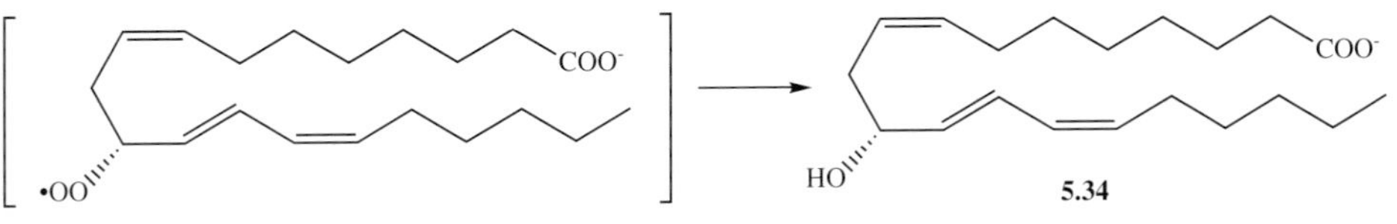

SCHEME 5.8   A hydroxylated by-product isolated from the prostaglandin H synthase-catalyzed reaction.

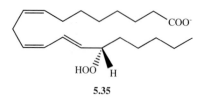

5.35

hydroperoxidation is that when the hydroperoxide **5.35** is incubated with the enzyme, no PGE_1 is generated. On the basis of the preceding results, two mechanisms can be drawn (Scheme 5.9). Pathway a involves initial abstraction of the C-13 *pro-S* hydrogen atom by an enzyme radical (discussed later) to give a highly resonance-stabilized radical (**5.36**) followed by addition of triplet oxygen to give **5.37**. Ring closure and addition of the second molecule of oxygen gives PGG_1. Pathway b starts with endoperoxidation to the cyclopentane ring without abstraction of the ^3H to give **5.38**. It is shown as a radical mechanism because a conrotatory [2 + 2 + 2] cycloaddition with singlet oxygen is thermally forbidden. Evidence to support pathway a is provided by the fact that an intermediate was isolated having only one molecule of O_2 incorporated into it and no tritium. According to pathway b, the molecule containing one O_2 incorporated (**5.38**) still has its tritium.

From the crystal structure, the cyclooxygenase active site consists of a long, narrow channel lined almost entirely with hydrophobic amino acid residues that could bind arachidonic acid appropriately for hydrogen atom abstraction.[33] Arg-120 appears to hold the carboxylate in the proper orientation. Ser-530, the serine residue to which aspirin becomes attached during inactivation, is located in the cyclooxygenase active site.

The peroxidase active site is located adjacent to the cyclooxygenase active site. The peroxidase activity serves two functions: It is involved in the generation of the protein radical that abstracts the C-13 hydrogen atom and initiates the cyclooxygenase activity, and it reduces PGGs to PGHs. A ferric heme coenzyme is located in the peroxidase active site. The heme that is activated by a peroxide (PGG after the first catalytic cycle, but some other peroxide needs to initiate the first catalytic cycle), is oxidized to heme Fe^{IV}-oxo-porphyrin π-radical cation, also known as Compound I, the low-spin ferryl iron with a porphyrin radical cation (see Chapter 4, Section IV.B). Intermediates in prostaglandin H synthase resembling Compounds I and II of horseradish peroxidase have been detected by electronic absorption spectroscopy.[34]

The heme oxo species is thought to be responsible for abstraction of a hydrogen atom from Tyr-385, which would be the enzyme radical that abstracts the C-13 hydrogen atom from arachidonic acid in Scheme 5.9 (X· in pathway a). The peroxy radical in **5.37** (Scheme 5.9) that abstracts an enzyme hydrogen atom (XH) could propagate turnover if XH is Tyr-385 (i.e., if the two X groups are the same

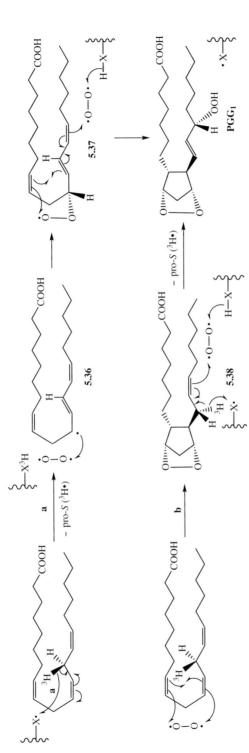

SCHEME 5.9   Two possible mechanisms for prostaglandin H synthase.

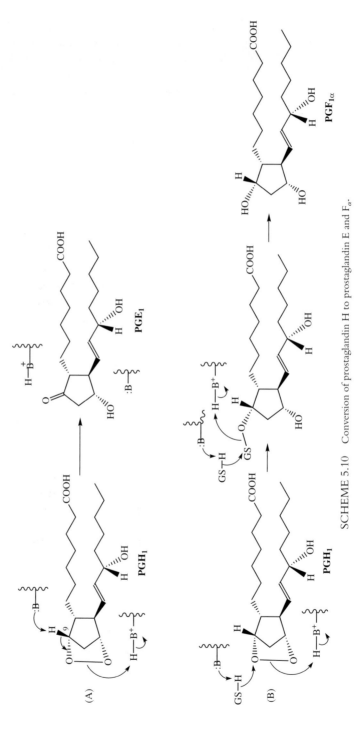

SCHEME 5.10   Conversion of prostaglandin H to prostaglandin E and F$_\alpha$.

tyrosine residue). This is Ruf's *"branched chain" mechanism,* which depicts the interplay between the peroxidase and cyclooxygenase activities of prostaglandin H synthase.[35] Evidence for implication of an active-site tyrosine radical is the observance of an EPR signal (under anaerobic conditions) that looks like known enzyme tyrosine radicals. Mutation of Tyr-385 to Phe gives a mutant that has peroxidase activity, but no cyclooxygenase activity. The crystal structure shows that Tyr-385 is located in close proximity to the heme and in between the heme-binding site and the arachidonic acid-binding site. This allows Tyr-385 to undergo hydrogen atom abstraction by the heme to give the tyrosine radical, which is able to abstract the C-13 hydrogen atom from arachidonate.

The PGH series is then converted to the PGE and PGF_α series by an isomerase and a glutathione-dependent reductase, respectively (Scheme 5.10). Note that the C-9 hydrogen is lost in the conversion of PGH_1 to PGE_1 (A), but it is not lost in the conversion of PGH_1 to $PGF_{1\alpha}$ (B), as noted earlier.

III. INTERMOLECULAR DIOXYGENASES

A. α-Keto Acid-Dependent Dioxygenases

The α-keto acid-dependent family of enzymes is important in the biosynthesis of various biological compounds.[36] In addition to hydroxylation of proline-containing peptides (prolyl hydroxylase) and lysine (lysyl hydroxylase), these enzymes are involved in the biosynthesis of penicillins,[37] cephalosporins,[38] and clavulanic acid and related compounds.[39]

The key feature of this class of enzymes is its dependence on nonheme Fe^{II}, O_2, ascorbic acid, and an α-keto acid, usually α-ketoglutarate. The ascorbic acid appears to be required to keep the iron in the reduced Fe^{II} oxidation state and to protect the enzymes from oxidative self-inactivation.[40] In general, these enzymes catalyze oxidation of an unactivated C−H bond to give either hydroxylated products or oxidative cyclization products and water (Scheme 5.11). One of the two atoms of oxygen from molecular oxygen is incorporated into the oxidatively decarboxylated α-keto acid product (succinic acid, when the α-keto acid is α-ketoglutarate); the other atom of oxygen either goes into the substrate when hydroxylation occurs or into water when a ring cyclization occurs.

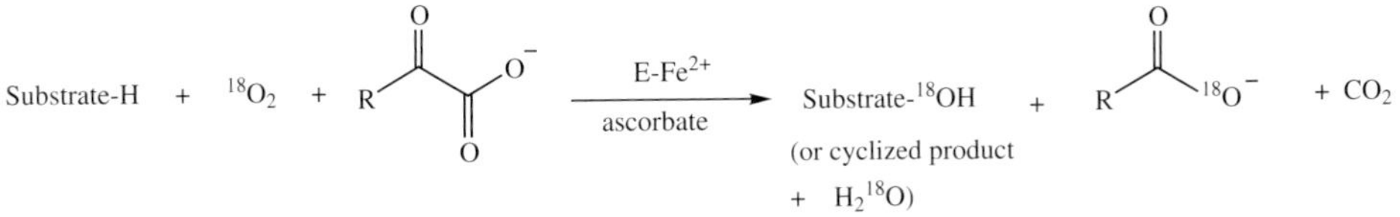

Substrate-H $+$ $^{18}O_2$ $+$ R—C(=O)—C(=O)—O⁻ $\xrightarrow[\text{ascorbate}]{E\text{-}Fe^{2+}}$ Substrate-^{18}OH $+$ R—C(=O)—^{18}O⁻ $+$ CO_2
(or cyclized product + $H_2{}^{18}O$)

SCHEME 5.11   Reaction catalyzed by α-keto acid-dependent dioxygenases.

Persuccinic acid, the peroxy acid of succinic acid, does not appear to be the oxidant in at least the hydroxylation of proline residues, because it cannot substitute for α-ketoglutarate and O_2 in prolyl hydroxylase and does not inhibit the binding of α-ketoglutarate.[41] The hydroxylation of unactivated C–H bonds is strongly reminiscent of the reaction catalyzed by the heme-dependent monooxygenases (see Chapter 4, Section IV.C.1) and peroxidases. It is reasonable to propose, then, that these nonheme Fe^{II}-dependent enzymes proceed by a mechanism related to that of the heme-dependent enzymes. Activation of the Fe^{II} and O_2 could occur by reaction with α-ketoglutarate to give a high energy $Fe^{IV}=O$ species (**5.39**) as shown in Scheme 5.12. Note, however, that this is an Fe^{IV}-oxo species, one electron reduced from the $Fe^{V}=O$ species in heme-dependent monooxygenases, so the iron-oxo species formed with α-ketoglutarate enzymes would not be quite as reactive as the heme iron-oxo species. Based on circular dichroism spectroscopic studies with the iron- and α-ketoglutarate-dependent enzyme, clavaminate synthase (see below), the ferrous iron is shown coordinated to the α-ketoglutarate as a bidentate ligand to the carbonyl and carboxylate groups[42] prior to the formation of the activated iron species. The $Fe^{IV}=O$ species generated could be involved in hemelike oxygenation reactions, such as hydroxylations. The mechanism in Scheme 5.12 is not very different from the mechanism proposed by Siegel in 1979 for prolyl hydroxylase (Scheme 5.13).[43]

Another interesting example of an α-ketoglutarate-dependent enzyme reaction, the clavaminate synthase-catalyzed formation of the oxazole ring in the biosynthesis of clavulanic acid, could proceed as shown in Scheme 5.14.[44] Note that this ring closure reaction involves oxidations (not oxygenations) of the molecule, and the second atom of oxygen from molecular oxygen ends up in water rather than in substrate molecule. Therefore, it is not, strictly speaking, a dioxygenase.

Further support for the mechanism of α-ketoglutarate-dependent dioxygenases resembling that of the heme-dependent enzymes is provided by studies of thymine hydroxylase from *Rhodotorula glutinis,* which catalyzes the oxidation of thymine (**5.40**) to the corresponding alcohol (5-(hydroxymethyl)uracil), aldehyde (5-formyluracil), and carboxylic acid (5-carboxyuracil) in three successive reactions (Scheme 5.15).[45] Each step involves stoichiometric consumption of O_2 and α-ketoglutarate and formation of CO_2 and succinate. There is a deuterium isotope effect of 2.08 on V_{max} from (trideuteriomethyl) thymine. The *pro-S* hydrogen is stereospecifically removed from [5'-^{2}H]5-(hydroxymethyl)uracil with no isotope effect. In addition to these reactions, the enzyme catalyzes epoxidation of 5-vinyluracil (**5.41**), the oxidation of 5-(methylthio)uracil (**5.42**) to the corresponding sulfoxide and sulfone, the hydroxylation of an unactivated C–H bond of 5,6-dihydrothymine (**5.43**), and the *N*-demethylation of 1-methylthymine (**5.44**) to formaldehyde and thymine. With $^{18}O_2$ it was shown that one ^{18}O atom is incorporated into succinate and one ^{18}O atom into the other product. These reactions are very similar to those of heme-dependent cytochrome P-450 and suggest a similar mechanism.

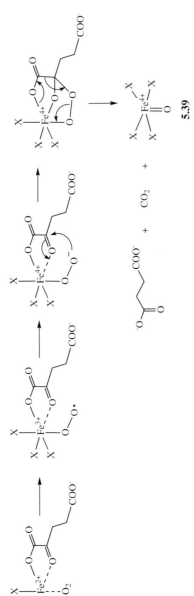

5.39

$$2L + Fe^{2+} + O_2 \rightleftharpoons \underset{L}{\overset{L}{FeO_2}} +$$

L = ascorbate

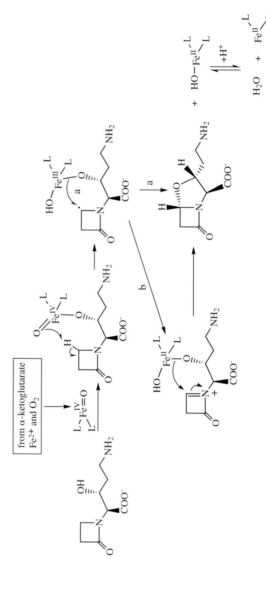

SCHEME 5.14   Mechanisms for clavaminate synthase–catalyzed ring closure.

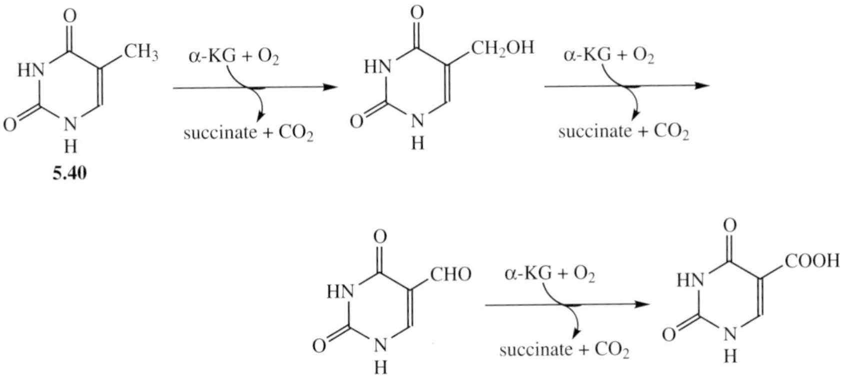

SCHEME 5.15   Reactions catalyzed by thymine hydroxylase.

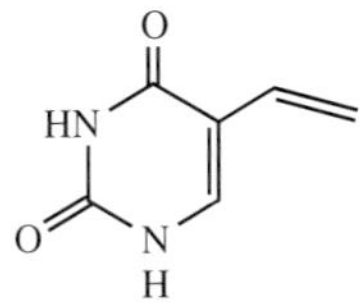

Another similarity of the thymine hydroxylase-catalyzed reactions to those of heme-dependent enzymes is the inactivation of thymine hydroxylase by 5-ethynyluracil (**5.45,** Scheme 5.16).[46] Inactivation leads to the formation of 5-(carboxymethyl)uracil (**5.49**) and uracil-5-acetylglycine (**5.50;** the glycine comes from the buffer utilized) and to attachment of **5.45** to an active site phenylalanine residue. Tryptic digestion of the enzyme inactivated with [2-^{14}C]-5-[1',2'-^{13}C$_2$]**5.45** followed by NMR analysis of the labeled peptide revealed the formation of a 7-carboxylated norcaradiene moiety. To account for all the experimental observations, Stubbe and co-workers suggested the mechanism in Scheme 5.16, modeled after the elegant work of Ortiz de Montellano and co-workers[47] on the mechanism of inactivation of cytochrome P450 (a family of heme enzymes) by acetylenes. Epoxidation of **5.45** would give **5.46,** which should undergo rearrangement to α-keto carbene **5.47.** This intermediate could then rearrange (pathway a) to ketene **5.48;** reaction of this ketene with water would give **5.49** or with glycine (present in the buffer) would give **5.50.** Alternatively, intermediate **5.47** could insert into the phenylalanine residue (pathway b), leading to 7-formyl norcaradiene **5.51,** which is oxidized further by the enzyme to **5.52.** Evidence to support this proposal comes from an experiment in which ^{18}O$_2$ is utilized as the oxidizing agent;

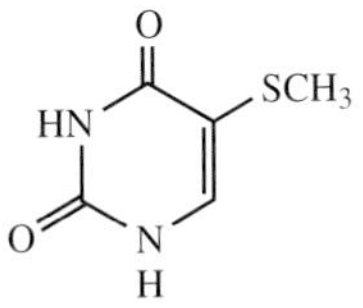

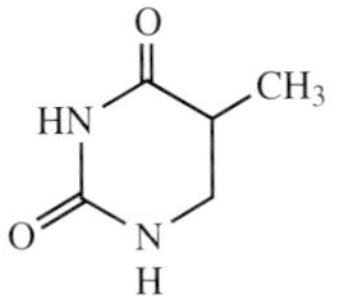

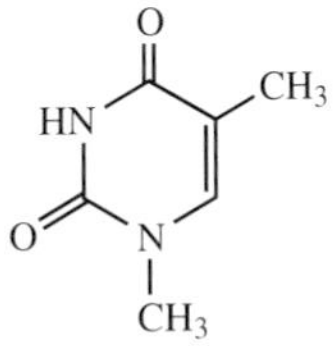

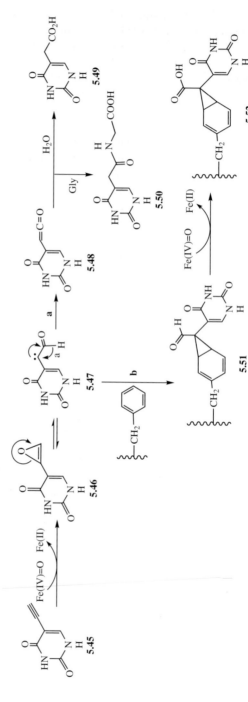

SCHEME 5.16 Inactivation of thymine hydroxylase by 5-ethynyluracil.

FAB mass spectrometry of the tryptic-digested inactivated enzyme corresponds to the addition of four mass units to the labeled peptide compared with the same experiment that utilized $^{16}O_2$. Therefore, both oxygens in the carboxylate come from molecular oxygen. Because α-ketoglutarate-dependent dioxygenases incorporate only one atom of oxygen into the substrate (and a second oxygen atom into succinate), the addition of two oxygen atoms into the product suggests that two oxygenation steps may be involved.

An interesting intramolecular version of an α-keto acid-dependent dioxygenase is the reaction catalyzed by (p-hydroxyphenyl)pyruvate dioxygenase, which converts (p-hydroxyphenyl)pyruvate (**5.53**) into homogentisic acid (**5.54,** Scheme 5.17).[48] If $^{18}O_2$ is utilized, then one atom of ^{18}O becomes incorporated into the carboxyl group, but only 0.3 atom of ^{18}O is incorporated into the new OH group. However, it was found that the new hydroxyl group exchanges with solvent; if the reaction is carried out in $H_2^{18}O$, 0.7 atom of ^{18}O is incorporated into the hydroxyl group.

A model study was carried out to ascertain a reasonable mechanism.[49] Incubation of (p-hydroxyphenyl)pyruvate with singlet oxygen leads to the loss of carbon dioxide and the formation of **5.56;** treatment of **5.56** at pH 12 gives homogentisic acid (Scheme 5.18). It was assumed that **5.56** was derived from **5.55.**

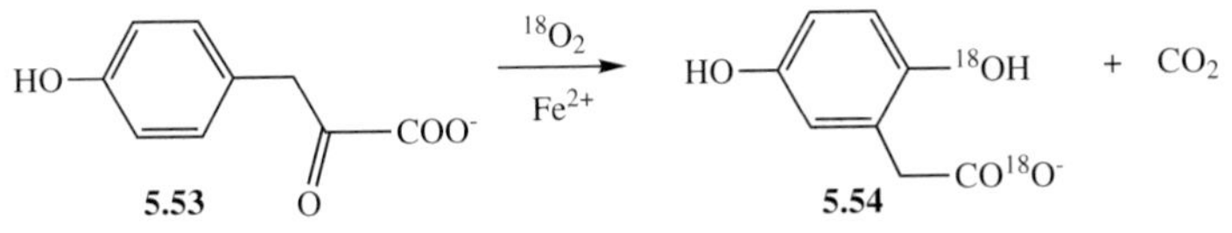

SCHEME 5.17   Reaction catalyzed by (p-hydroxyphenyl)pyruvate dioxygenase.

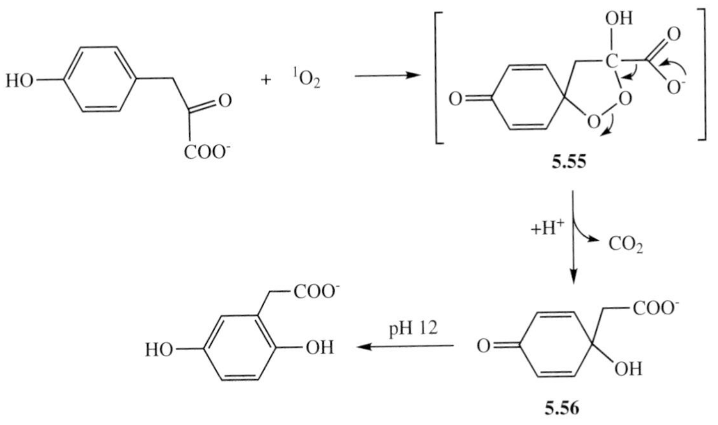

SCHEME 5.18   Model study for the mechanism of (p-hydroxyphenyl)pyruvate dioxygenase.

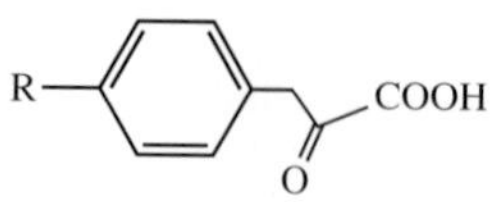

$$R-\bigcirc\!\!\!-CH_2-\underset{\underset{O}{\parallel}}{C}-COOH$$

5.57

This model was disproven, however, on the basis of the following experiments. If (*p*-hydroxyphenyl)pyruvate dioxygenase is incubated with **5.56,** no homogentisic acid is produced.[50] An experiment was carried out to determine if **5.56** is a released intermediate during turnover. This was done by incubating the enzyme with radiolabeled substrate (*p*-hydroxyphenyl)pyruvate) and unlabeled intermediate (**5.56**), quenching the reaction, then looking for radiolabeled intermediate formation. If it is found, then the radiolabeled intermediate—derived from (*p*-hydroxyphenyl)pyruvate—is released during turnover. No ^{14}C was detected in the isolated **5.56;**[51] therefore it can be concluded that, **5.56** is not a free intermediate, or it never builds up to any extent, or it is not an intermediate at all.[52] Furthermore, the enzyme also accepts (*p*-fluorophenyl)pyruvate (**5.57,** R = F) and phenylpyruvate (**5.57,** R = H) as substrates, which is not consistent with the formation of an intermediate such as **5.56,** as required for the rearrangement to homogentisic acid shown in Scheme 5.18.

As with thymine hydroxylase, (*p*-hydroxyphenyl)pyruvate dioxygenase also catalyzes sulfoxidations, similar to heme-dependent cytochrome P450,[53] lending support for an iron-oxo or iron-hydroxy species. A mechanism combining that drawn for α-ketoglutarate-dependent enzymes (Scheme 5.12) and that described for hydroxylation of aromatic compounds by heme-dependent enzymes (see Chapter 4, Scheme 4.41) seems most reasonable (Scheme 5.19). The iron-oxo species (**5.58**) also could account for exchange of [18]O prior to incorporation into the substrate.

These last three chapters have dealt with different oxidation and oxygenation reactions catalyzed by enzymes. Because of the need to transfer electrons during these reactions, unusual coenzymes are required. We next turn our attention to enzyme-catalyzed substitution reactions, which require only nucleophiles and electrophiles and, therefore, do not require organic coenzymes.

NOTE THAT PROBLEMS AND SOLUTIONS RELEVANT TO EACH CHAPTER CAN BE FOUND IN APPENDIX II.

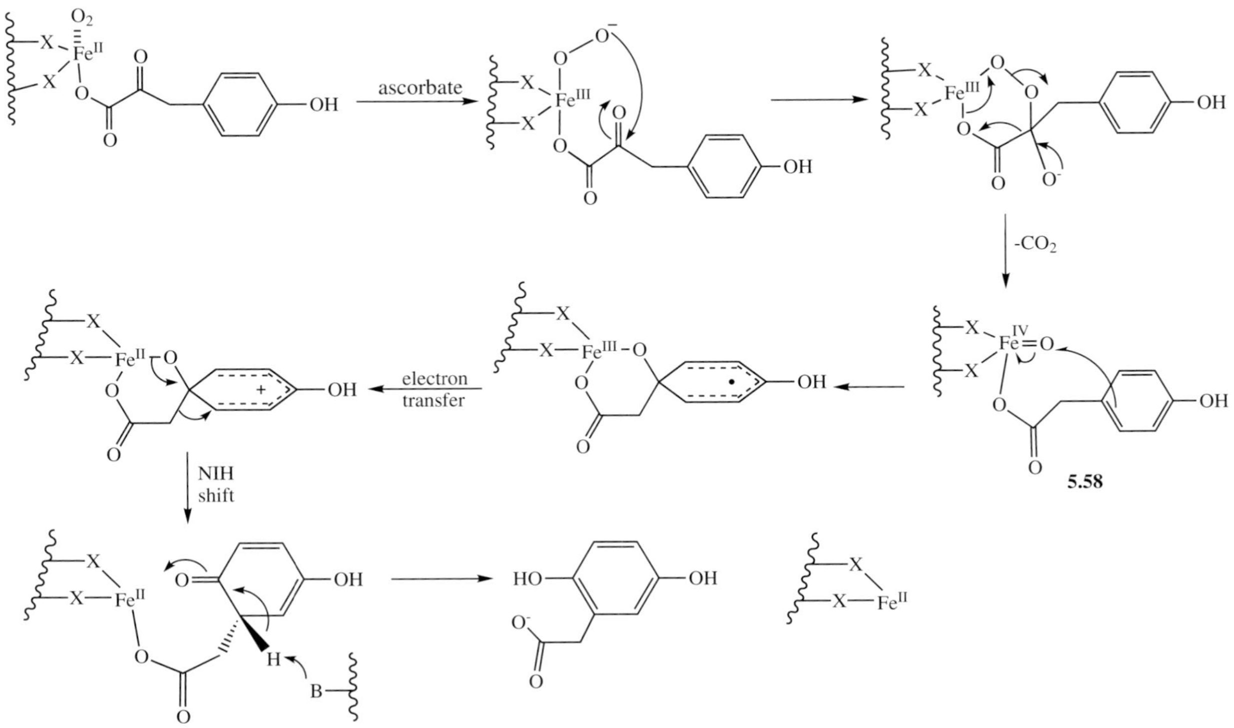

SCHEME 5.19 Proposed mechanism for (p-hydroxyphenyl)pyruvate dioxygenase.

5.58

REFERENCES

1. (a) *Microbial Degradation of Organic Molecules,* Gibson, D. T., Ed., Marcel Dekker: New York, 1984. (b) Feig, A. L.; Lippard, S. J. *Chem. Rev.* **1994,** *94,* 759.
2. Ohlendorf, D. H.; Lipscomb, J. D.; Weber, P. C. *Nature* **1988,** *336,* 403.
3. (a) Han, S.; Eltis, L. D.; Timmis, K. N.; Muchmore, S. W.; Bolin, J. T. *Science* **1995,** *270,* 976. (b) Senda, T.; Sugiyama, K.; Narita, H.; Yamamoto, T.; Kimbara, K.; Fukuda, M.; Sato, M.; Yano, K.; Mitsui, Y. *J. Mol. Biol.* **1996,** *255,* 735.
4. Lipscomb, J. D.; Orville, A. M. *Metal Ions Biol. Syst.* **1992,** *28,* 243.
5. (a) Que, L. Jr.; Lipscomb, S. D.; Zimmermann, R.; Munck, E.; Orme-Johnson, N. R.; Orme-Johnson, W. H. *Biochim. Biophys. Acta* **1976,** *452,* 320. (b) Walsh, T. A.; Ballou, D. P.; Mayer, R.; Que, L., Jr. *J. Biol. Chem.* **1983,** *258,* 14422. (c) Bull, C.; Ballou, D. P.; Otsuka, S. *J. Biol. Chem.* **1981,** *256,* 12681.
6. Que, L., Jr.; Lipscomb, J. D.; Münck, E.; Wood, J. M. *Biochim. Biophys. Acta* **1977,** *485,* 60.
7. Nozaki, M. In *Molecular Mechanisms of Oxygen Activation,* Hayaishi, O., Ed., Academic Press: New York, 1974, pp. 135–165.
8. Hayaishi, O.; Katagiri, M.; Rothberg, S. *J. Am. Chem. Soc.* **1955,** *77,* 5450.
9. (a) Jang, H. G.; Cox, D. D.; Que, L. Jr. *J. Am. Chem. Soc.* **1991,** *113,* 9200. (b) Que, L., Jr.; Kolanczyk, R. C.; White, L.S. *J. Am. Chem. Soc.* **1987,** *109,* 5373. (c) Cox, D. D.; Que, L., Jr. *J. Am. Chem. Soc.* **1988,** *110,* 8085.
10. Orville, A. M.; Lipscomb, J. D.; Ohlendorf, D. H. *Biochemistry* **1997,** *36,* 10052.
11. Orville, A. M.; Elango, N.; Lipscomb, J. D.; Ohlendorf, D. H. *Biochemistry* **1997,** *36,* 10039.
12. Frazee, R. W.; Orville, A. M.; Dolbeare, K. B.; Yu, H.; Ohlendorf, D. H.; Lipscomb, J. D. *Biochemistry* **1998,** *37,* 2131.
13. Criegee, R. *Angew. Chem. Int. Ed. Engl.* **1975,** *14,* 745.
14. (a) White, L. S.; Nilsson, P. V.; Pignolet, L. H.; Que, L. Jr. *J. Am. Chem. Soc.* **1984,** *106,* 8312. (b) Koch, W. O.; Krüger, H.-J. *Angew. Chem. Int. Ed. Engl.* **1995,** *34,* 2671.
15. Hayaishi, O.; Katagiri, M.; Rothberg, S. *J. Am. Chem. Soc.* **1955,** *77,* 5450.
16. Mayer, R. J.; Que, L., Jr. *J. Biol. Chem.* **1984,** *259,* 13056.
17. Sanvoisin, J.; Langley, G. J.; Bugg, T. D. H. *J. Am. Chem. Soc.* **1995,** *117,* 7836.
18. Shu, L.; Chiou, Y.-M.; Orville, A. M.; Miller, M. A.; Lipscomb, J. D.; Que, L., Jr. *Biochemistry* **1995,** *34,* 6649.
19. Spence, E. L.; Langley, G. J.; Bugg, T. D. H. *J. Am. Chem. Soc.* **1996,** *118,* 8336.
20. Higgs, G. A.; Higgs, E. A.; Moncada, S. In *Comprehensive Medicinal Chemistry,* vol. 2; Hansch, C.; Sammes, P. G.; Taylor, J. B., Eds.; Pergamon Press: Oxford, 1990, pp. 147-173.
21. Smith, W. L.; Garavito, R. M.; DeWitt, D. L. *J. Biol. Chem.* **1996,** *271,* 33157.
22. Bjorkman, D. J. *Am. J. Med.* **1996,** *101 (Suppl. A),* 25S.
23. Tally, J. J. *Exp. Opin. Ther. Patents* **1997,** *7,* 55.
24. Picot, D.; Loll, P. J.; Garavito, R. M. *Nature* **1994,** *367,* 243.
25. Kurumbail, R. G.; Stevens, A. M.; Gierse, J. K.; McDonald, J. J.; Stegeman, R. A.; Pak, J. Y.; Gildehaus, D.; Miyashiro, J. M.; Penning, T. D.; Seibert, K.; Isakson, P. C.; Stallings, W. C.; *Nature* **1996,** *384,* 644.
26. (a) Khanna, I. K.; Weier, R. M.; Yu, Y.; Collins, P. W.; Miyashiro, J. M.; Koboldt, C. M.; Veenhuizen, A. W.; Currie, J. L.; Seibert, K.; Isakson, P. C. *J. Med. Chem.* **1997,** *40,* 1619. (b) Khanna, I. K.; Weier, R. M.; Yu, Y.; Xu, X. D.; Koscyk, F. J.; Collins, P. W.; Koboldt, C. M.; Veenhuizen, A. W.; Perkins, W. E.; Casler, J. J.; Masferrer, J. L.; Zhang, Y. Y.; Gregory, S. A.; Seibert, K.; Isakson, P. C. *J. Med. Chem.* **1997,** *40,* 1634.
27. Gierse, J. K.; McDonald, J. J.; Hauser, S. D.; Rangwala, S. H.; Koboldt, C. M.; Seibert, K. *J. Biol. Chem.* **1996,** *271,* 15810.
28. (a) Smith, W. L.; Marnett, L. J.; Dewitt, D. L. *Pharmacol. Ther.* **1991,** *49,* 153. (b) Smith, W. L.; Marnett, L. J. *Biochim. Biophys. Acta* **1991,** *1083,* 1.

29. Samuelsson, B. *J. Am. Chem. Soc.* **1965,** *87,* 3011.

30. Klenberg, D.; Samuelsson, B. *Acta Chem. Scand.* **1965,** *19,* 534.

31. Hamberg, M.; Samuelsson, B. *J. Biol. Chem.* **1967,** *242,* 5336.

32. Hamberg, M.; Samuelsson, B. *Proc. Natl. Acad. Sci. USA* **1973,** *70,* 899; Hamberg, M.; Svensson, J.; Wakabayashi, T.; Samuelsson, B. *Proc. Natl. Acad. Sci. USA* **1974,** *71,* 345.

33. Picot, D.; Loll, P. J.; Garavito, M. *Nature* **1994,** *367,* 243.

34. (a) Lambeir, A.-M.; Markey, C. M.; Dunford, H. B.; Marnett, L. J. *J. Biol. Chem.* **1985,** *260,* 14894. (b) Dietz, R.; Nastainczyk, W. Ruf, H.H. *Eur. J. Biochem.* **1988,** *171,* 321.

35. (a) Karthein, R.; Dietz, R.; Nastainczyk, W.; Ruf, H. H. *Eur. J. Biochem.* **1988,** *171,* 313. (b) Tsai, A.-L.; Kulmacz, R.J.; Palmer, G. *J. Biol. Chem.* **1995,** *270,* 10503.

36. Kivirikko, K. L.; Myllyla, R.; Pihlajaniemi, T. *FASEB J.* **1989,** *3,* 1609.

37. (a) Scott, R. A.; Wang, S.; Eidsness, M. K.; Kriauciunas, A.; Frolik, C. A.; Chen, V. J. *Biochemistry* **1992,** *31,* 4596. (b) Baldwin, J.; Bradley, M. *Chem. Rev.* **1990,** *90,* 1079.

38. (a) Baldwin J. E.; Abraham, E. *Nat. Prod. Rep.* **1988,** *5,* 129. (b) Baldwin, J. E.; Adlington, R. M.; Crouch, N. P.; Keeping, J. W.; Leppard, S. W.; Pitlik, J.; Schofield, C. J.; Sobey, W. J.; Wood, M. E. *J. Chem. Soc. Chem. Commun.* **1991,** 768.

39. (a) Elson, S. W.; Baggaley, K. H.; Davison, M.; Fulston, M.; Nicholson, N. H.; Risbridger, G. D.; Tyler, J. W. *J.Chem. Soc. Chem. Commun.* **1993,** 1212. (b) Townsend, C. A. *Biochem. Soc. Trans.* **1993,** *21,* 208. (c) Baldwin, J. E.; Adlington, R. M.; Crouch, N. P.; Drake, D. J.; Fujishima, Y.; Elson, S. W.; Baggaley, K. H. *J. Chem. Soc. Chem. Commun.* **1994,** 1133.

40. (a) Majamaa, K.; Hanauske-Abel, H. M.; Gunzler, V.; Kivirikko, K. I. *Eur. J. Biochem.* **1984,** *138,* 239. (b) Myllyla, R.; Majamaa, K.; Gunzler, V.; Hanauske-Abel, H. M.; Kivirikko, K. I. *J. Biol. Chem.* **1984,** *259,* 5403.

41. Counts, D. F.; Cardinale, G. J.; Udenfriend, S. *Proc. Natl. Acad. Sci. USA* **1978,** *75,* 2145.

42. Pavel, E. G.; Zhou, J.; Busby, R. W.; Gunsior, M.; Townsend, C. A.; Solomon, E. I. *J. Am. Chem. Soc.* **1998,** *120,* 743.

43. Siegel, B. *Bioorg. Chem.* **1979,** *8,* 219.

44. Townsend, C. A. *Biochem. Soc. Trans.* **1993,** *21,* 208.

45. Thornburg, L. D.; Lai, M.-t.; Wishnok, J. S.; Stubbe, J. *Biochemistry* **1993,** *32,* 14023.

46. (a) Thornburg, L. D.; Stubbe, J. *Biochemistry* **1993,** *32,* 14034. (b) Lai, M.-t.; Wu, W.; Stubbe, J. *J. Am. Chem. Soc.* **1995,** *117,* 5023.

47. (a) Ortiz de Montellano, P. R.; Kunze, K. L. *Arch. Biochem. Biophys.* **1981,** *209,* 710. (b) Ortiz de Montellano, P. R.; Komives, E. A. *J. Biol. Chem.* **1985,** *260,* 3330.

48. Lindblad, B.; Lindstedt, G.; Lindstedt, S. *J. Am. Chem. Soc.* **1970,** *92,* 7446.

49. (a) Saito, I.; Yamane, M.; Shimazu, H.; Matsuura, T.; Cahnmann, H. J. *Tetrahedron Lett.* **1975,** 641. (b) Saito, I.; Chujo, Y.; Shimazu, H.; Yamane, M.; Matsuura, T.; Cahnmann, H. J. *J. Am. Chem. Soc.* **1975,** *97,* 5272.

50. Nakai, C.; Nozaki, M.; Hayaishi, O.; Saito, I.; Matsuura, T. *Biochem. Biophys. Res. Commun.* **1975,** *67,* 590.

51. Schweizer, J. ; Lathell, R.; Hecker, E. *Experientia* **1975,** *31,* 1267.

52. Cleland, W. W. *Biochemistry* **1990,** *29,* 3194.

53. Pascal, R. A., Jr.; Oliver, M. A.; Chen, Y.-C. J. *Biochemistry* **1985,** *24,* 3158.

CHAPTER 6

Substitutions

I. S$_N$1

A. Farnesyl Diphosphate Synthase and Related Enzymes

Farnesyl diphosphate synthase, an enzyme in the sterol biosynthetic pathway, catalyzes the sequential addition of isopentenyl diphosphate (**6.1**, Scheme 6.1; PPO in all figures stands for diphosphate) to dimethylallyl diphosphate (**6.2**) to give geranyl diphosphate (**6.3**), and then catalyzes the addition of isopentenyl diphosphate to geranyl diphosphate (**6.3**) to give farnesyl diphosphate (**6.4**).[1] Both steps are irreversible. These kinds of enzymes are referred to as *prenyltransferases,* enzymes that catalyze alkylation of electron-rich substrates by the hydrocarbon moieties of isoprenoid allylic diphosphates.

A Hammett study (Chapter 3, Section III.B.3) with substituted substrates supported a carbocation intermediate.[2] First, a few background studies.[3] 2-Fluorogeranyl diphosphate (**6.5**) was found to be a substrate for farnesyl diphosphate

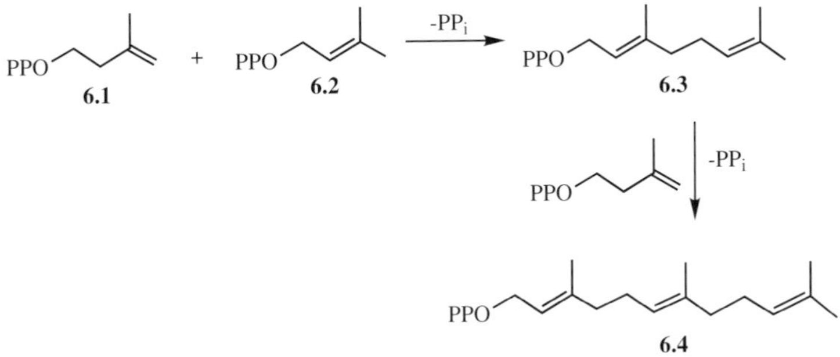

SCHEME 6.1   Reactions catalyzed by farnesyl diphosphate synthase.

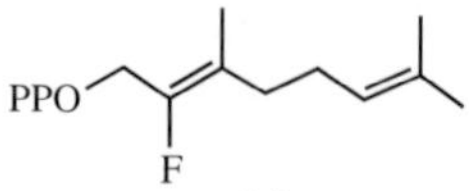

6.5

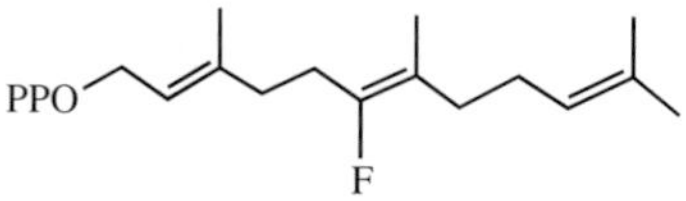

6.6

synthase with a K_m value almost the same as that for geranyl diphosphate (**6.3**, Scheme 6.1); however, the k_{cat} for the formation of 6–fluorofarnesyl diphosphate (**6.6**) was 8.4×10^{-4} times that for the conversion of geranyl diphosphate to farnesyl diphosphate (**6.4**, Scheme 6.1), indicating a strong inductive effect of the fluorine atom on the rate-determining step. As a test of the mechanism involved, two nonenzymatic model reactions were carried out, one known to proceed via a carbocation intermediate and the other a S_N2 reaction. The rate of solvolysis (a carbocation mechanism) of geranyl methanesulfonate (**6.7**, X = H) was compared to that for 2–fluorogeranyl methanesulfonate (**6.7**, X = F); the fluorinated analogue was solvolyzed at a rate 4.4×10^{-3} times that of the nonfluorinated compound. This rate difference is similar to the difference in the enzymatic rate difference for the fluorinated and nonfluorinated substrates. Next a S_N2 reaction of geranyl chloride and 2–fluorogeranyl chloride with cyanide ion was carried out; the rate of the 2-fluoro analogue was two times *faster* than that of the nonfluorinated compound. The similarity of the enzymatic rate difference to that in the carbocation model study, and of the large difference to that in the S_N2 model, support a carbocation intermediate in the enzymatic reaction.

Poulter and co-workers extended this study to three other fluorinated analogues of geranyl diphosphate (**6.8–6.10**).[2] All had K_m values with farnesyl diphosphate synthase similar to that for geranyl diphosphate, but the rates for **6.8**, **6.9**, and **6.10**

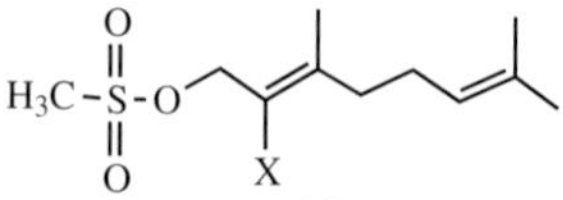

6.7

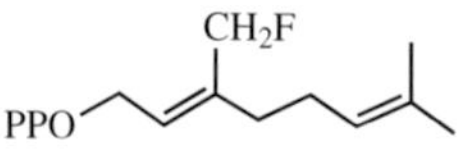

6.8

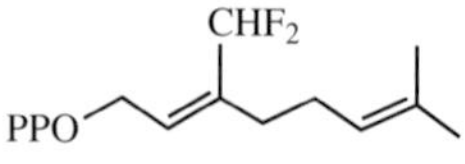

6.9

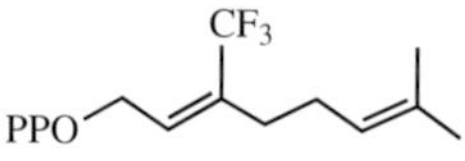

6.10

were 1.75×10^{-2}, 1.90×10^{-6}, and 3.62×10^{-7}, respectively, times that of the rate with geranyl diphosphate. The nonenzymatic solvolysis rates for the corresponding methanesulfonates of the fluorinated analogues were 7.7×10^{-4}, 2.2×10^{-6}, and 4.0×10^{-7}, respectively, times the rate for geranyl methanesulfonate. A Hammett plot for the enzymatic rates versus the nonenzymatic solvolysis rates for the fluorinated methanesulfonates relative to geranyl methanesulfonate, gives a linear correlation, indicating a similarity in mechanisms between these two reactions, namely, a carbocation mechanism (Scheme 6.2). Initial ionization of **6.2** or **6.3** would produce carbocation **6.11**. Alkene addition to the carbocation, followed by deprotonation, gives the product. Poulter and co-workers provided further confirmation of the carbocation mechanism with two bisubstrate analogues (**6.12** and **6.13**, Scheme 6.3).[4] Both compounds were found to be catalytically active. Upon incubation with farnesyl diphosphate synthase, substrate **6.12** gives cyclized products **6.16** and **6.18**, and substrate **6.13** produces **6.16**, **6.18**, and **6.19**. These products are consistent with carbocation intermediates **6.14**, **6.15**, and **6.17**.

With regard to the stereochemistry of the reaction catalyzed by farnesyl diphosphate synthase (Figure 6.1), the allylic substrate (**6.21**) is added to the *si* face of the double bond in **6.20** (*si* is on top as drawn in **6.20**; see Chapter 3.III.A.3 for a discussion of stereochemical nomenclature), and the *pro-R* proton is removed from C-2 of **6.20** with concomitant formation of the double bond at the C-2–C-3 position of **6.20** to give geranyl diphosphate. Therefore, it is a *syn* addition/elimination.

In the presence of PP$_i$ and the absence of the substrate nucleophile **6.1** (Scheme 6.1), a water molecule can serve as an alternate substrate, in which case water becomes the prenyl acceptor of the cation formed from **6.2** (or **6.3**); this acceptance leads to an autocatalytic hydrolysis of the substrate (Scheme 6.4). In H$_2$^{18}O, the oxygen in the product alcohol is ^{18}O-labeled, and C-1 is inverted, as in the normal condensation reaction.[5]

Farnesyl diphosphate (**6.4**) is the common acyclic precursor of the more than 300 cyclic sesquiterpenes that have been characterized to date.[6] Sesquiterpenes are

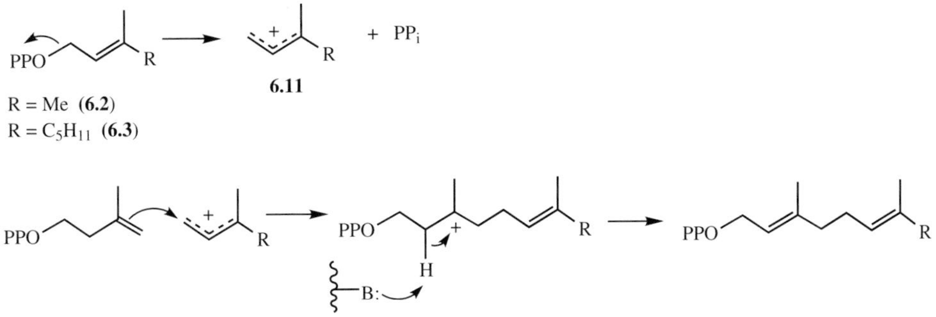

R = Me (**6.2**)
R = C$_5$H$_{11}$ (**6.3**)

SCHEME 6.2   Carbocation mechanism for farnesyl diphosphate synthase.

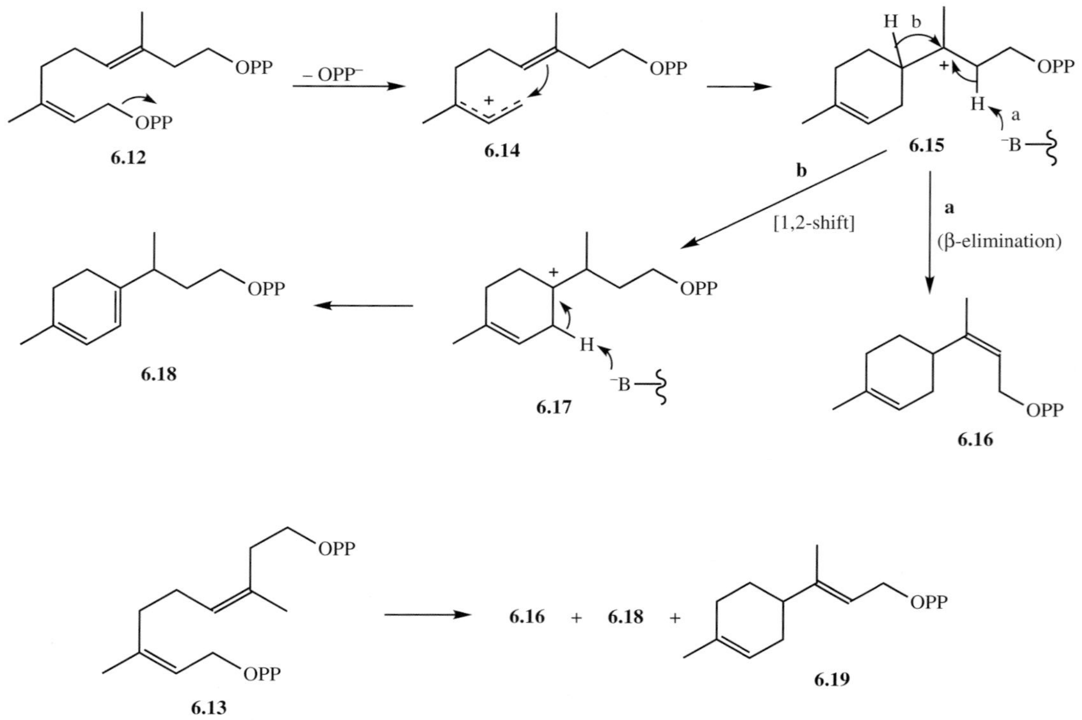

SCHEME 6.3 Possible carbocation mechanisms for the farnesyl diphosphate synthase–catalyzed cyclizations of **6.12** and **6.13**.

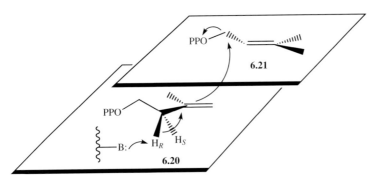

FIGURE 6.1   Stereochemistry of farnesyl diphosphate synthase.

a group of natural products secreted by marine and terrestrial plants, fungi, and microorganisms, some of which exhibit useful medicinal properties. Most sesquiterpene cyclization reactions proceed through variations of a common mechanism (Scheme 6.5 provides an example of cyclization of farnesyl diphosphate by pentalenene synthase to humulene (**6.22**), protoilludyl cation (**6.23**), and pentalenene (**6.24**)) that involves ionization of farnesyl diphosphate followed by nucleophilic attack of one of the remaining π bonds of the substrate on the resultant allylic cation; subsequent cationic rearrangements are terminated by either deprotonation or attack by an exogenous nucleophile such as water. The crystal structure of pentalenene synthase at 2.6 Å resolution reported by Christianson and co-workers with farnesyl diphosphate bound reveals important active-site structures and interactions that may have far-reaching consequences regarding not only this reaction but other enzyme-catalyzed reactions involving carbocation intermediates.[7] Farnesyl diphosphate is centered about an axis defined by Phe-77 and Asn-219, which are optimally located to stabilize highly reactive carbocation intermediates through quadrupole–charge and dipole–charge interactions, respectively (Figure 6.2). Other residues capable of comparable interactions with carbocations at other sites on the substrate molecules also were identified.

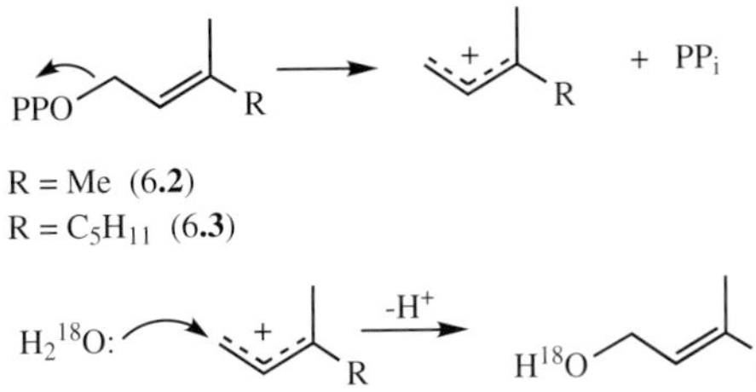

R = Me (**6.2**)
R = C$_5$H$_{11}$ (**6.3**)

SCHEME 6.4   Farnesyl diphosphate synthase-catalyzed reaction of **6.2** or **6.3** in the absence of **6.1**.

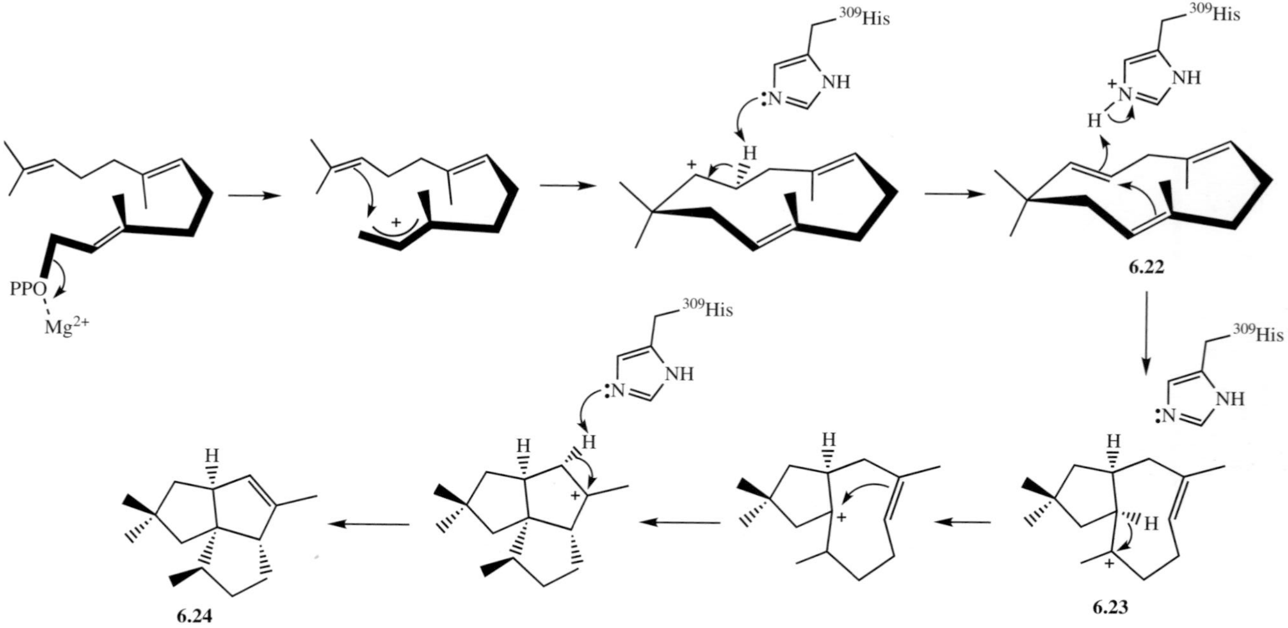

6.22

6.23

6.24

SCHEME 6.5   Reaction catalyzed by pentalenene synthase.

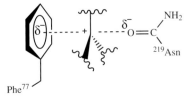

Phe^{77}

FIGURE 6.2   Stabilization of carbocation intermediates by active-site phenylalanine and asparagine residues.

II. S_N1/S_N2

A. Disaccharide Phosphorylases

Phosphorylases catalyze substitution reactions in which inorganic phosphate functions as the nucleophile (Scheme 6.6). The reaction catalyzed by disaccharide phosphorylases is shown in Scheme 6.7; a disaccharide (**6.25**) is converted to the corresponding sugar$_A$-1-phosphate (**6.26**) and sugar$_B$ (**6.27**). If ^{18}O is incorporated into the disaccharide at the anomeric carbon, the product sugar phosphate **6.26** contains no ^{18}O; all of the isotope is found in the other sugar (**6.27**). On the basis of the stereochemical results with various disaccharides (Table 6.1), it appears that the reaction proceeds by two different mechanisms depending on which disaccharide is the substrate. It seems reasonable to conclude that for cellobiose and maltose a S_N2 displacement by HPO_4^{2-} gives the inverted configuration and that for sucrose either S_N1 with exclusive front side attack (the enzyme could block back side attack) or two S_N2 reactions (double inversion) could be important. However,

$$R-O-R' \quad + \quad {}^-O-\overset{\overset{O}{\|}}{\underset{\underset{OH}{|}}{P}}-O^- \quad \longrightarrow \quad RO-\overset{\overset{O}{\|}}{\underset{\underset{O^-}{|}}{P}}-O^- \quad + \quad R'OH$$

SCHEME 6.6   Reaction catalyzed by phosphorylases.

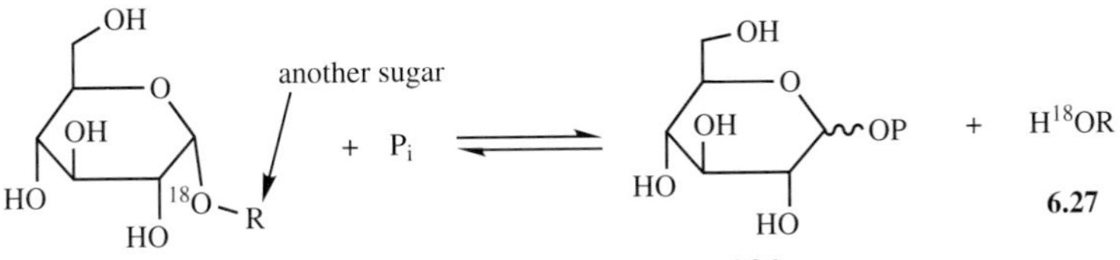

6.25   another sugar   + P$_i$ ⇌   **6.26**   + H^{18}OR

6.27

SCHEME 6.7   Reaction catalyzed by disaccharide phosphorylases.

TABLE 6.1   Stereochemistry of the Reactions Catalyzed by Various
Disaccharide Phosphorylases

	C-1 Configuration of Disaccharide	C-1 Configuration of Phosphorylated Product
Cellobiose phosphorylase	β	α
Maltose phosphorylase	α	β
Sucrose phosphorylase	α	α

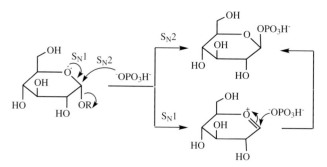

SCHEME 6.8 S_N2 versus stereospecific S_N1 reaction.

inversion of stereochemistry could be rationalized either as a S_N2 reaction or as a S_N1 reaction followed by stereospecific attack from the opposite side (Scheme 6.8).

No partial exchange reactions occur with cellobiose and maltose phosphorylases, consistent with a direct S_N2 reaction; with sucrose phosphorylase, however, ^{14}C in fructose (the alcohol product, **6.27**) is incorporated into the sucrose in the presence of unlabeled sucrose and in the absence of inorganic phosphate.[8] This suggests a double S_N2 displacement mechanism. If sucrose (**6.28**) is labeled with ^{14}C in the glucosyl group, a ^{14}C-labeled protein can be isolated under certain conditions (pH 3, denaturation). If the ^{14}C label is in the fructosyl group, no ^{14}C-protein can be isolated. The identity of the residue on the enzyme to which the

glucosyl   fructosyl

6.28

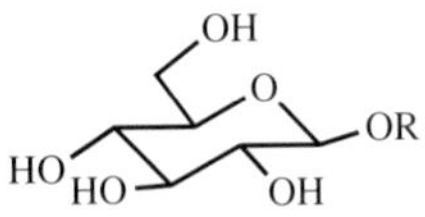

6.29

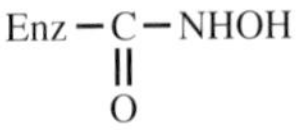

$$\text{Enz} - \overset{\overset{\displaystyle}{\|}}{\underset{\displaystyle O}{C}} - \text{NHOH}$$

6.30

radiolabel becomes bound was determined by various experiments: (1) the glucosyl enzyme was treated with methanol, and only glucose (**6.29**, R = H) was produced, not methyl glucoside (**6.29**, R = CH_3); (2) the glucosyl enzyme bond was found to be very sensitive to base; (3) glucose was released from the enzyme by treatment with alkaline NH_2OH, producing **6.30**; at pH 6 the reaction is too slow for the enzyme adduct to be a thioester. These experiments can be rationalized in terms of an ester linkage to the active-site residue, that is, an enzyme aspartate or glutamate is involved (**6.31**, Scheme 6.9). However, the fact that the covalent adduct (**6.31**) can be isolated does not necessarily require it to be a true intermediate in the enzyme-catalyzed reaction. It could be the product of a side reaction as a result of the conditions utilized to isolate the covalent intermediate. For example, the true intermediate could be **6.32**, in which the carboxylate group provides electrostatic stabilization, and then this intermediate collapses to **6.31** (pathway a) during the isolation process.

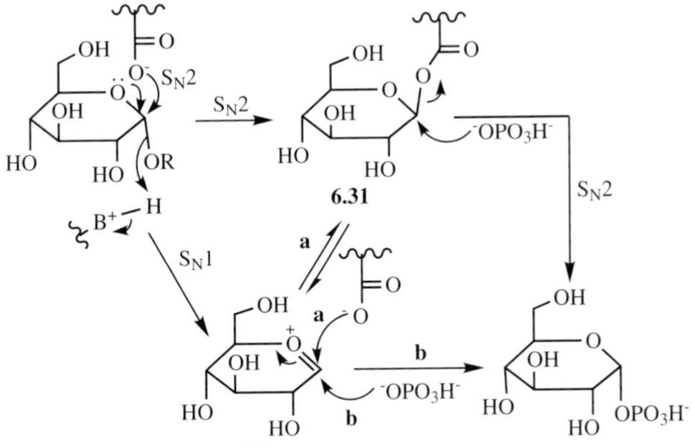

SCHEME 6.9   Disaccharide phosphorylase reactions involving an active-site carboxylate.

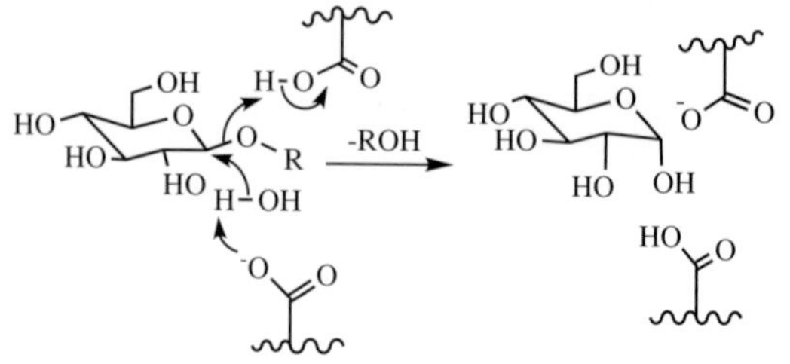

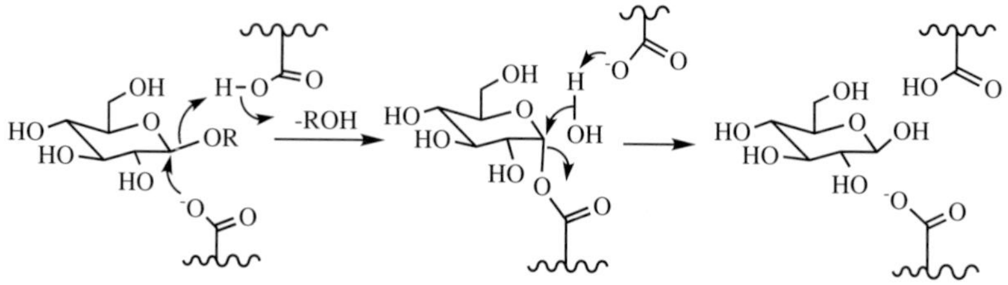

SCHEME 6.10   Two mechanisms for reactions catalyzed by β-glycosidases.

B. β-Glycosidases

β-Glycosidases catalyze the hydrolysis of oligosaccharides by two distinct mechanisms (Scheme 6.10).[9] One family of enzymes catalyzes a direct displacement (S_N2-like), leading to inversion of stereochemistry at the anomeric carbon (**A**), and the other utilizes a double displacement mechanism, involving a covalent intermediate, which gives retention of configuration (**B**). Both mechanisms may involve oxocarbenium ion–like transition states, and both involve two carboxylic acid residues. In mechanism A, however, one carboxylic acid acts as an acid and the other as a base, but in mechanism B one functions as an acid and a base and the other as a nucleophile. Also, to allow for binding of water these two residues need to be farther apart for mechanism A than for mechanism B. Mutation of Glu-358 to Ala (E358A) in a β-glucosidase from *Agrobacterium faecalis,* which normally catalyzes the reaction by mechanism B, generated a mutant whose k_{cat} was 10^7-fold lower than that of the wild-type enzyme.[10] However, addition of azide as an alternative nucleophile in lieu of the second carboxylate residue (the bottom one) increased the k_{cat} 10^5-fold and produced the α-glucosyl azide. Therefore, by site-specific mutation and use of a good nucleophile, the mechanism can be converted from B into mechanism A.

To differentiate a S_N2 from a S_N1 mechanism for β-glycosidases, and to determine if the covalent intermediate is kinetically competent, Withers and co-workers[11] prepared 2-deoxy-2-fluoro analogues of substrates (**6.33**). Because fluorine is more electronegative than hydroxyl, an oxocarbenium ion intermediate would be destabilized relative to direct S_N2 displacement, and therefore the fluorine substitution would slow down a S_N1 reaction but accelerate S_N2 (as in the model de-

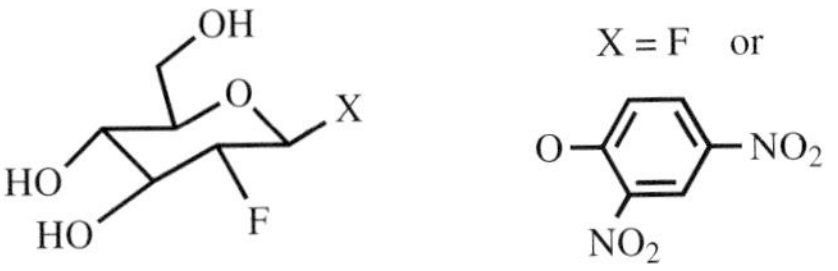

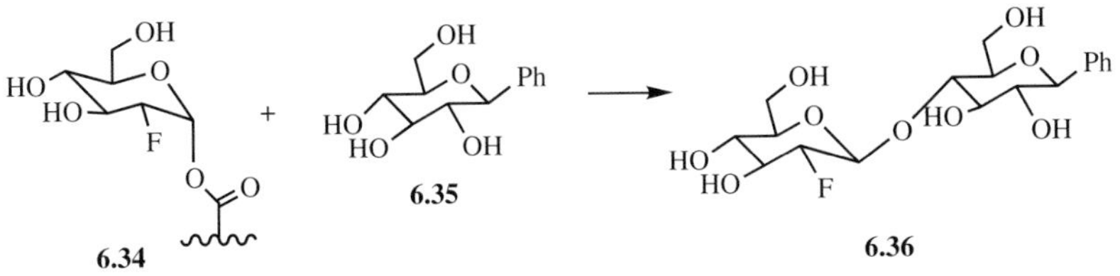

SCHEME 6.11   Reaction of **6.33**-inactivated β-glucosidase with **6.35**.

scribed on p. 252). It would slow down a S$_N$1 reaction but accelerate S$_N$2 (as in the model described on p. 252). It also would stabilize a covalent adduct. By using an excellent leaving group, such as fluoride or 2,4-dinitrophenolate, the first step is accelerated, and hydrolysis of the enzyme adduct becomes the rate-determining step. This allows the intermediate to be isolated, and its stereochemistry elucidated by ^{19}F NMR spectrometry. Proteolysis of the inactivated enzyme showed that Glu-358 of β-glucosidase is labeled. Incubation of the labeled enzyme (**6.34**, Scheme 6.11) with **6.35** gave **6.36**, the substitution product of **6.34** and **6.35**, at the same rate as from **6.33**. Therefore, the intermediate (**6.34**) is kinetically competent. When 2-deoxy substrates (no substituent at C-2) are utilized, the V_{max}/K_m is increased over the natural sugar; this suggests an intermediate with carbenium (oxocarbenium) ion character.

Replacement of Glu-358 in β-glucosidase by Asn and Gln results in inactive enzyme; replacement by Asp gives a mutant with a rate constant for the formation of the glucosyl enzyme 2500 times lower than that of the native enzyme. Withers and co-workers have shown that the binding of ground-state inhibitors is little affected by the mutation, but transition-state inhibitor binding is greatly affected, consistent with the role of Glu-358 being involved in transition–state stabilization and not in substrate binding.[12]

A kinetic analysis of β-glucosidase is consistent with a two-step mechanism involving a glucosyl enzyme intermediate. Secondary deuterium isotope effect studies indicate a large degree of bond cleavage at the transition state. With poorer substrates, a mechanism more like S$_N$2 is suggested in the formation of the glucosyl enzyme, whereas deglucosylation is more like S$_N$1 (Scheme 6.12).[13] This reaction thus appears to have both S$_N$2 and S$_N$1 character. A similar study of the α-glucosidase from *Saccharomyces cerevisiae* with 5-fluoroglycosyl fluorides revealed Asp-214 as the catalytic nucleophile.[14]

As a model for the reaction of active-site nucleophiles with disaccharides, Banait

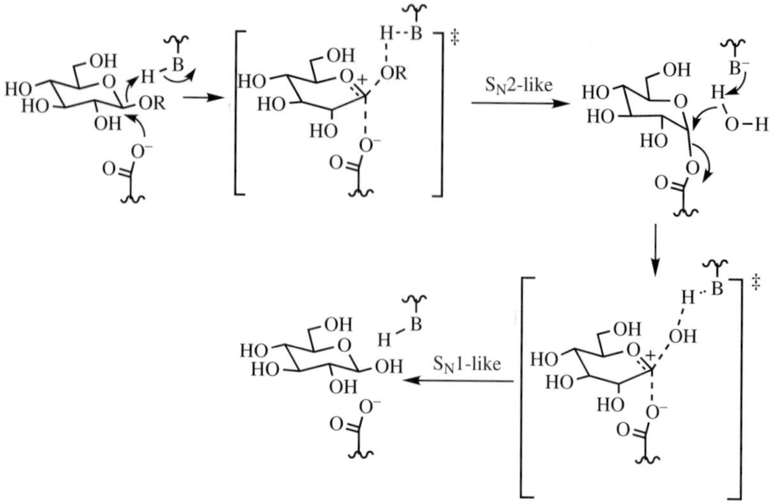

SCHEME 6.12   Both S_N2- and S_N1-like character of β-glucosidase.

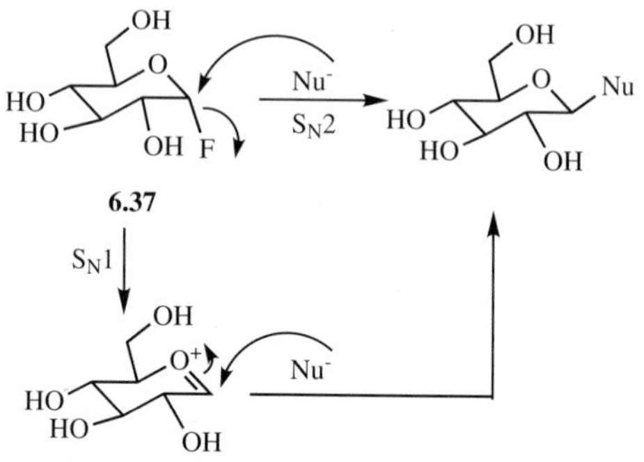

SCHEME 6.13   Chemical model for the reaction catalyzed by β-glucosidases.

and Jencks studied the mechanism of the reaction of nucleophiles with α-D-glucopyranosyl fluoride (**6.37**, Scheme 6.13).[15] The rate $= k\,[\alpha\text{-gluF}][\text{Nu}^-]$ with complete inversion of stereochemistry; therefore, a S_N2 mechanism is favored. The oxocarbenium ion is generated by solvolysis in the absence of strong nucleophiles, but its lifetime is short, even in H_2O.

III. S_N2

Microsomal epoxide hydrolase converts epoxides into glycols; two S_N2 mechanisms were considered (Scheme 6.14; general base (A) or nucleophilic (B) mecha-

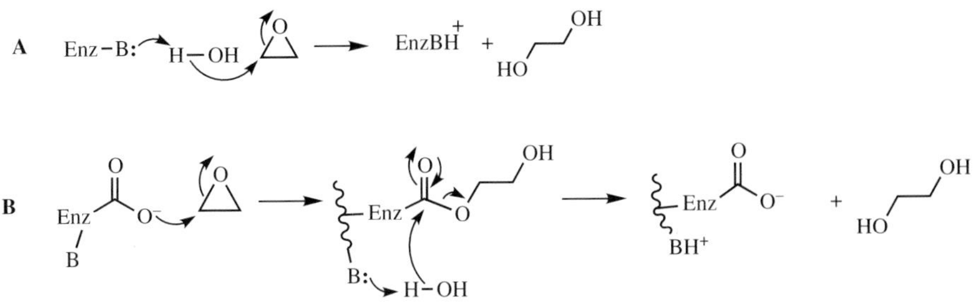

SCHEME 6.14   Two mechanisms for epoxide hydrolase.

nisms). Lacourciere and Armstrong found that a single-turnover experiment (excess enzyme) in H$_2$^{18}O gave *no* ^{18}O into the product glycol, but enzyme labeled with ^{18}O in an aspartate carboxylate incorporates ^{18}O into product, even in H$_2$^{16}O.[16] Conversely, tryptic digestion of soluble epoxide hydrolase incubated with substrate in H$_2$^{18}O gave a peptide fragment containing Asp-333; electrospray ionization mass spectrometry showed the incorporation of ^{18}O into this aspartic acid residue.[17] The unique feature of these results is the transfer of an oxygen atom from the enzyme to the product, which is consistent with mechanism B, but not mechanism A.

Further evidence for the formation of a covalent intermediate in the hydrolysis of epoxides, catalyzed by epoxide hydrolase, is the isolation of the covalent intermediate (**6.39**, Scheme 6.15).[18] Murine-soluble epoxide hydrolase was incubated with [C$_{10}$–^{3}H]juvenile hormone III (**6.38**) for 3 seconds followed by addition of 10% acetic acid for 3 seconds and then an acetone quench at −20 °C (to precipitate the enzyme). After exhaustive washing, tritium was still bound to the protein. Treatment of the tritium-labeled protein with LiAlH$_4$ gave the triol **6.40**, and treatment with sodium hydroxide released the diol acid **6.41**. These results are consistent with an aspartate adduct (**6.39**).

A related epoxide-opening reaction was catalyzed by a catalytic antibody. A 6-*endo-tet* ring closure of an epoxy alcohol (**6.42**, Scheme 6.16) to a tetrahydropyran (**6.43**)—formally a violation of Baldwin's rules of ring closure, which predict that a 5-*exo-tet* ring closure is favored[19]—was catalyzed by a catalytic antibody, indicating that catalytic antibodies may have utility in catalyzing unfavorable reactions.[20] *Ab initio* calculations were carried out to determine the transition-state structures for the 6-*endo-tet* versus 5-*exo-tet* processes.[21] The 5-*exo-tet* pathway is favored by 1.8 kcal/mol (96:4 product ratio at 25 °C). The catalytic antibody must therefore lower the 6-*endo-tet* activation energy 3.6 kcal/mol more than it lowers the 5-*exo-tet* activation energy to get a >96% yield of the 6-*endo-tet* product, as was observed.

Another enzyme that appears to catalyze a substrate alkylation of an active-site nucleophile is haloalkane dehalogenase from *Xanthobacter autotrophicus,* which catalyzes the hydrolytic cleavage of carbon−halogen bonds in a broad range of haloalkanes. On the basis of its X-ray crystal structure it was suspected that Asp-124 was involved. Site-directed mutagenesis (D124A, D124G, and D124E) gave mutants

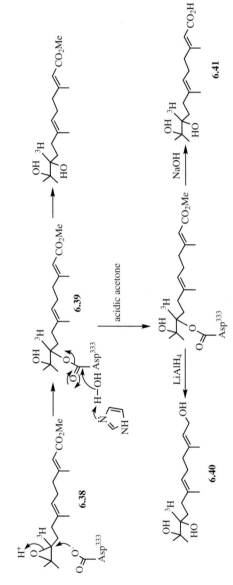

SCHEME 6.15   Covalent intermediate isolated during reaction catalyzed by epoxide hydrolase.

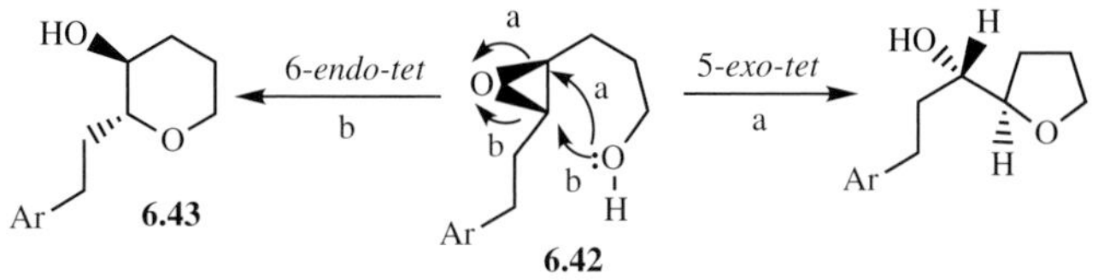

SCHEME 6.16   A catalytic antibody-catalyzed 6-*endo-tet* ring closure.

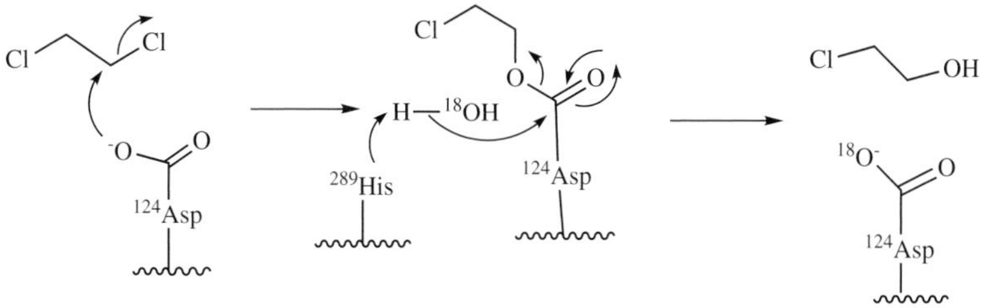

SCHEME 6.17   Reaction catalyzed by haloalkane dehalogenase.

with no enzyme activity.[22] Incubation of 1,2-dichloroethane with haloalkane de-halogenase in $H_2{}^{18}O$ gave 2-chloroethanol containing ^{18}O *and* ^{18}O in the Asp-124. The mechanism in Scheme 6.17 depicts how ^{18}O from $H_2{}^{18}O$ is incorporated into the enzyme. Every succeeding turnover would lead to incorporation of ^{18}O into the 2-chloroethanol product.

Different stages in the reaction pathway of the haloalkane dehalogenase-catalyzed reaction were trapped and visualized by soaking crystals of the enzyme with the sub-strate, 1,2-dichloroethane, under various pH and temperature conditions.[23] Incu-bation of the enzyme and substrate at pH 5 and 4 °C facilitated the determination of the crystal structure with 1,2-dichloroethane bound; warming to room temper-ature led to determining the crystal structure with Asp-124 alkylated (with release of Cl⁻); at pH 6 and room temperature, the crystal structure of the product bound to the enzyme was determined. These crystal structures firmly establish the two-step catalytic mechanism shown in Scheme 6.17.

IV. S_N2'

Isochorismate synthase, which catalyzes the isomerization of chorismate (**6.44**, Scheme 6.18) to isochorismate (**6.45**), appears to utilize a S_N2' mechanism, that is, a vinylogous S_N2 mechanism (Scheme 6.19).[24] In $H_2{}^{18}O$ the enzyme catalyzes the incorporation of ^{18}OH into isochorismate.

Two other enzymes that catalyze related reactions are anthranilate synthase, which

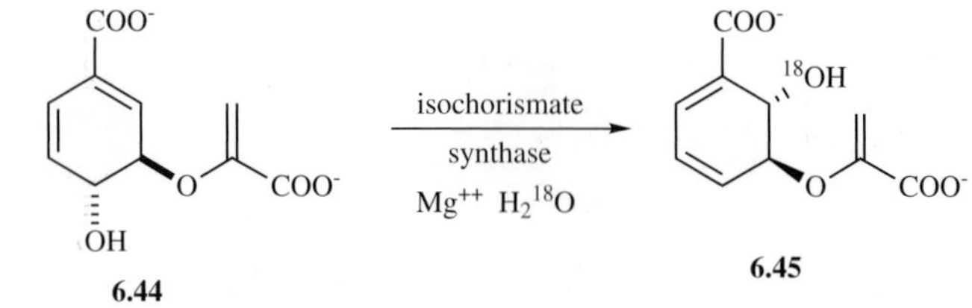

SCHEME 6.18   Reaction catalyzed by isochorismate synthase.

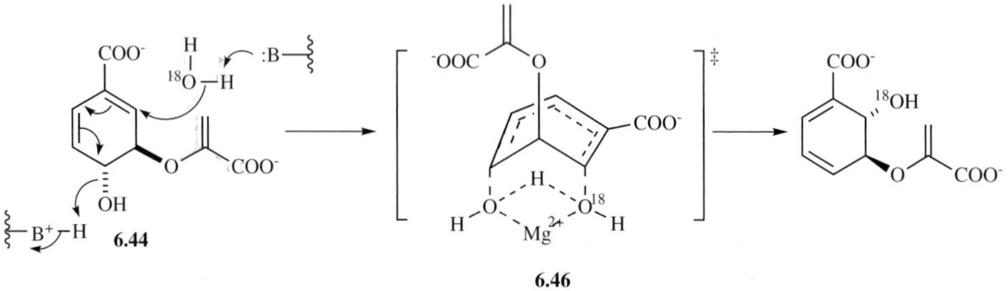

SCHEME 6.19 S_N2' mechanism for isochorismate synthase.

converts chorismate into anthranilate (**6.48**; Scheme 6.20) and *p*-aminobenzoic acid (PABA) synthase, which catalyzes the conversion of chorismate into PABA (**6.50**; Scheme 6.21);[25] **6.47** and **6.49** are believed to be intermediates in the respective reactions. Compounds **6.47**[26] and **6.49**[27] were synthesized and shown to be kineti-

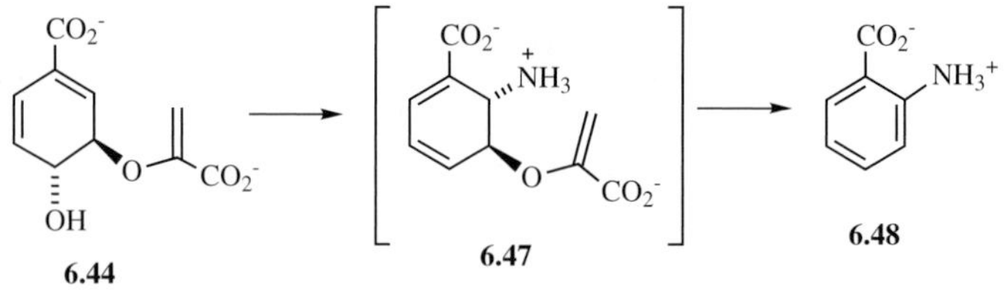

SCHEME 6.20   Reaction catalyzed by anthranilate synthase.

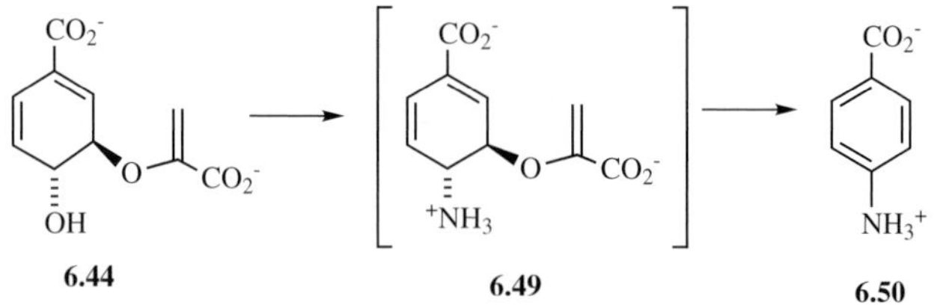

SCHEME 6.21   Reaction catalyzed by *p*-aminobenzoic acid synthase.

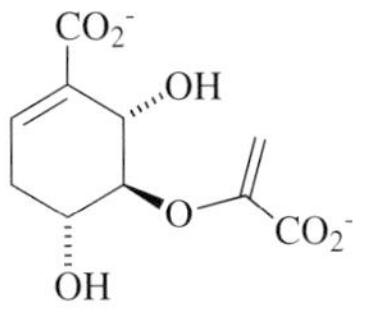

6.51 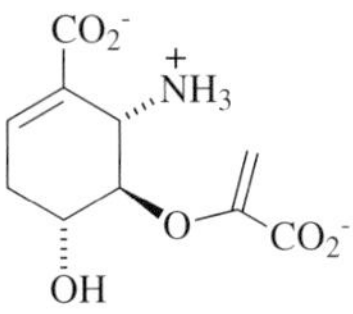   **6.52** 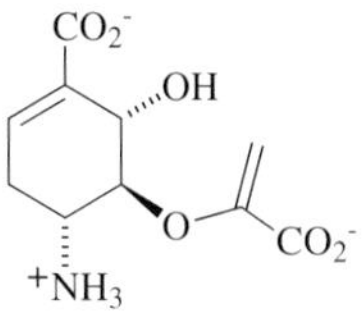   **6.53**

cally competent in the formation of anthranilate and PABA, respectively. Compound **6.47** also was isolated in an *Escherichia coli* auxotrophic mutant,[28] and **6.49** was observed by NMR spectroscopy during enzyme turnover.[29] Bartlett and coworkers designed a series of compounds (**6.51–6.53**) to mimic the putative transition state for isochorismate synthase, anthranilate synthase, and *p*-aminobenzoate synthase, respectively.[30] Compounds **6.51–6.53**, in the all-axial conformation (**6.54**), mimic the proposed transition-state structure for substrates bound to isochorismate synthase and anthranilate synthase (compare **6.55** with **6.46** in Scheme 6.19). All three compounds are competitive inhibitors of the three enzymes, but they bind strongly to isochorismate synthase and anthranilate synthase and only weakly to *p*-aminobenzoate synthase. This supports the proposed transition state for isochorismate synthase and anthranilate synthase, but suggests that *p*-aminobenzoate synthase functions by a different mechanism.

Mattia and Ganem proposed and synthesized a common intermediate for anthranilate, PABA, and isochorismate synthases (**6.56**, Scheme 6.22).[31] This intermediate would arise from a reversible *syn*-allylic rearrangement of the hydroxyl

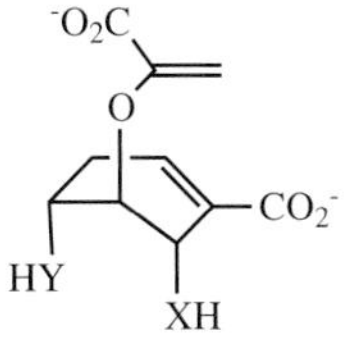

6.54

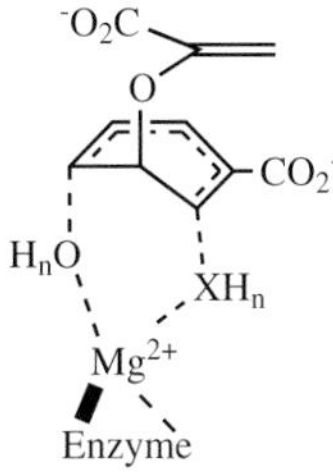

6.55

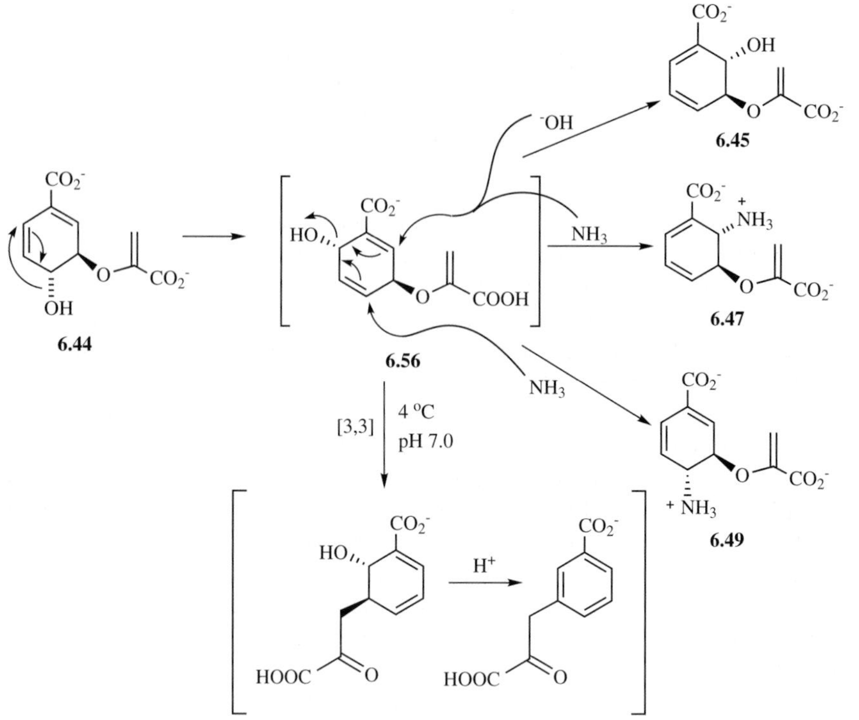

SCHEME 6.22   Possible intermediate in the biosynthesis of isochorismate, anthranilate, PABA and *meta*-substituted aromatic amino acids.

group of chorismate (**6.44**). Hydroxide or ammonia could convert this intermediate to **6.45**, **6.47**, or **6.49** by S_N2' reactions. Compound **6.56** also was shown to undergo a rapid [3,3]-sigmatropic rearrangement, much faster than chorismate or isochorismate; acid-catalyzed dehydration of the rearranged product led to the *meta*-substituted aromatic compound, which could be an alternative mechanistic rationale in the biosynthesis of *meta*-substituted aromatic amino acids in higher plants.

V. S_NAr: NUCLEOPHILIC AROMATIC SUBSTITUTION

Glutathione *S*-transferase catalyzes the nucleophilic addition of the tripeptide glutathione (**6.57**, GSH) to electrophiles, usually by S_N2, Michael addition, or S_NAr mechanisms. Glutathione, a potent nucleophile because of the thiol group in its structure, is a molecule ubiquitous in mammalian tissues for the purpose of protecting the cells from potentially toxic electrophiles that are ingested, inhaled, or generated by xenobiotic metabolism. The pKa of GSH bound to glutathione *S*-transferase was found to be only 6.2,[31a] suggesting that it is anionic. Armstrong and

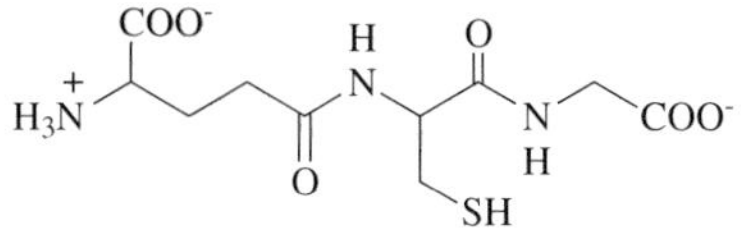

6.57

coworkers proposed that an active site tyrosine residue stabilized the thiolate by hydrogen bonding.[31b] By mutation of the active site tyrosine by replacement with fluorinated tyrosine residues, Schultz and coworkers provided support for Armstrong's hypothesis.[31c] The same conclusion was reached from *ab initio* MO theory calculations.[31d] A generalized reaction for nucleophilic aromatic substitution by GSH is shown in Scheme 6.23.

Hammett plots (see Chapter 3.III.B.3.a) of log k_c (the rate constant for unassisted ionization) versus σ^- (a σ constant utilized when there is a resonance interaction between the reaction site that becomes electron rich and a substituent that can stabilize a negative charge by resonance, such as a nitro group) for a series of 4-substituted 1-chloro-2-nitrobenzenes give ρ values of $+1.2$, $+1.9$, and $+2.5$ for GSH, *N*-acetyl GSH, and γ-Glu-Cys, respectively, as compared to $+3.4$ for the specific base-catalyzed reaction with GSH.[32] This result indicates the importance of the peptide structure for correct orientation of the thiol, and suggests that one important contribution the enzyme may make in catalysis is to lower the pK_a of the bound GSH. It also has been found that the 1-fluoro analogue is a better substrate than the 1-chloro analogue. Unlike S$_N$2 reactions, which proceed at a faster rate when chloride is the leaving group rather than fluoride, S$_N$Ar reactions are faster when fluoride is the leaving group. This is because the rate-determining step is the initial nucleophilic addition, and fluoride is more electron withdrawing than chloride, which activates the benzene ring for addition by nucleophiles better than chloride does.

The orange-red σ-complex formed on addition of glutathione to 1,3,5-trinitrobenzene (**6.58**, Scheme 6.24; in general, this initial σ-complex is called a *Meisenheimer complex*) bound to glutathione *S*-transferase was observed by Armstrong

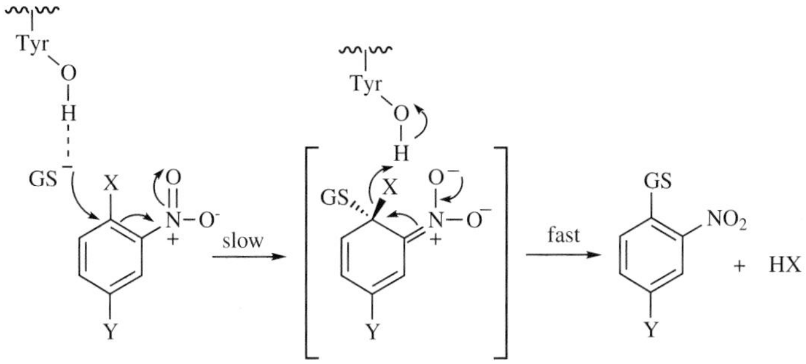

SCHEME 6.23   Reaction catalyzed by glutathione *S*-transferase.

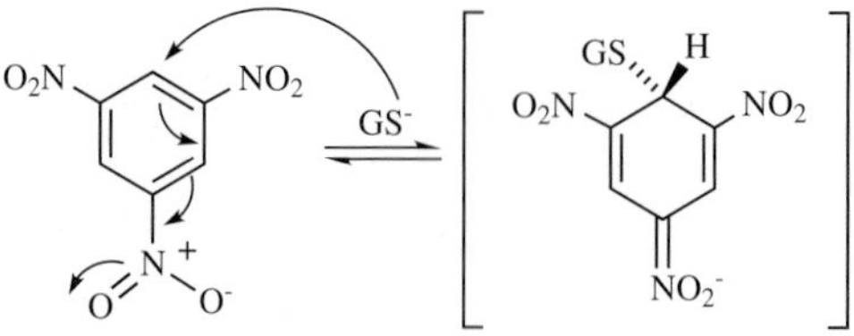

6.58

SCHEME 6.24   Glutathione S-transferase-catalyzed reaction of glutathione with 1,3,5-trinitro-benzene.

and co-workers spectroscopically.[33] Formation of the σ-complex also was observed in single crystals. When crystals of GSH S-transferase were soaked with GSH and trinitrobenzene, they turned bright orange, indicating formation of the σ-complex. Crystals were grown of enzyme, intermediate-enzyme (using a transition-state analogue), and product–enzyme complexes with GSH transferase. X-ray crystal structures (1.9–2.0 Å resolution) were obtained for each complex.[34] These structures represent "snapshots" of the catalytic process and reveal binding interactions that are important to catalysis. From the geometries of the intermediate and product complexes, a reaction coordinate motion during catalysis can be proposed.

Another enzyme that appears to be catalyzing a S_NAr reaction is 4-chlorobenzoyl-CoA dehalogenase, which catalyzes the hydrolysis of 4-chlorobenzoyl-CoA (**6.59**, X = Cl; Scheme 6.25). However, when X is varied, the k_{cat} follows the order Br > Cl > F, which is opposite the rate found for S_NAr reactions in solution and which is consistent with a S_N2 reaction.[35] Either the enzyme accelerates the first step (addition of water to the benzene ring), thereby making the second step (elimination of X$^-$) the rate-determining step, or it is a different mechanism, such as radical mechanisms $S_{RN}1$ or $S_{ON}2$.

Alternative mechanisms were considered. An aryne mechanism (Scheme 6.26) was ruled out by the absence of a deuterium isotope effect on the reaction (and the absurdly high pK_a of the proton being removed).

$S_{RN}1$ and $S_{ON}2$ mechanisms (Scheme 6.27) were excluded because of lack of evidence for a metal ion or organic cofactor. Therefore, the S_NAr mechanism appears to be the most reasonable.

Attempts were made to detect the Meisenheimer complex using a potent com-

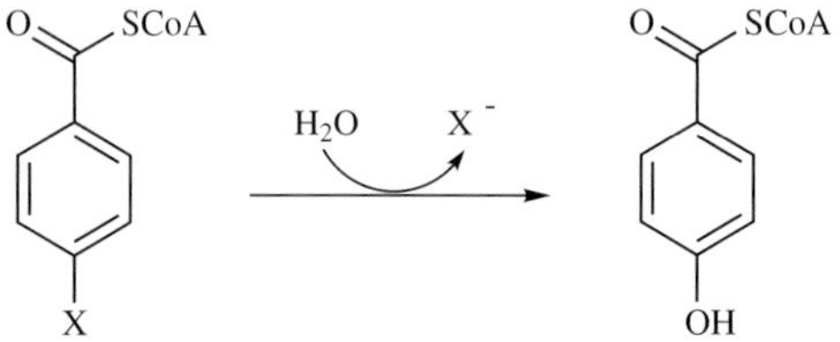

6.59

SCHEME 6.25   Reaction catalyzed by 4-chlorobenzoyl-CoA dehalogenase.

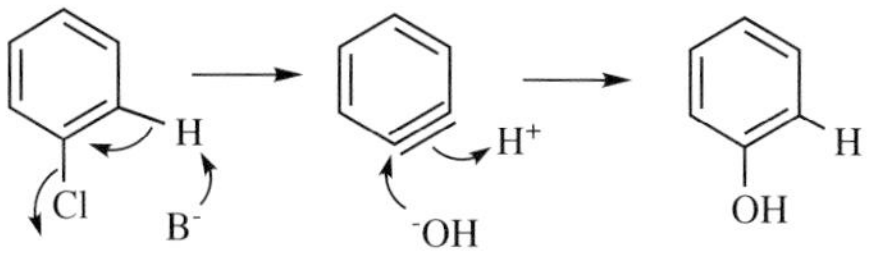

SCHEME 6.26   Aryne mechanism for 4-chlorobenzoyl-CoA dehalogenase.

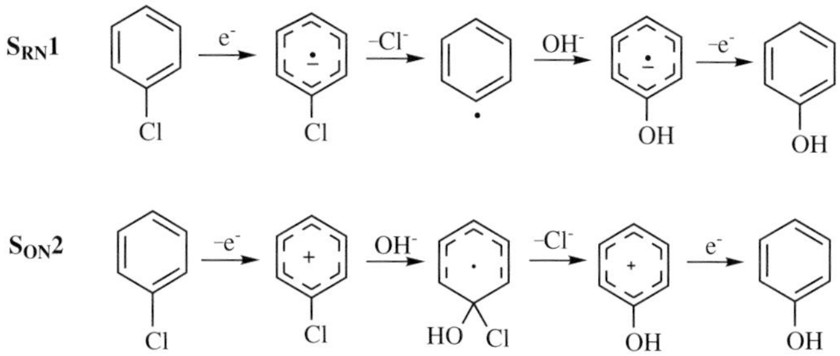

SCHEME 6.27   Radical mechanisms for 4-chlorobenzoyl-CoA dehalogenase.

petitive inhibitor, 4-trifluoromethylbenzoyl-CoA (**6.59**, X = CF$_3$, Scheme 6.25; K_i = 50 μM), having a *para* electron-withdrawing group that is not a leaving group, namely, a trifluoromethyl group.[36] The idea is to determine if the hydroxide addition product (the Meisenheimer complex) could be detected, given that the trifluoromethyl group would activate the compound for addition, but the intermediate could not be eliminated to product. Compound **6.59** (X = CF$_3$) was incubated with the enzyme; however, no increase in absorbance in the visible region was observed, suggesting that either only a small concentration of the Meisenheimer complex forms or a complex that has an absorbance masked by the protein absorbance forms.

An alternative possibility, that the enzyme might catalyze displacement of Cl$^-$ with an enzyme nucleophile, was considered. This type of covalent mechanism was discussed earlier for haloalkane dehalogenase[37] and epoxide hydrolase.[38] Hydroxylamine does not substitute for H$_2$O in the reaction catalyzed by 4-chlorobenzoyl-CoA dehalogenase. At high concentrations of NH$_2$OH (100 mM), enzyme inhibition occurs; removal of the excess NH$_2$OH leads to reactivation of the enzyme; apparently the enzyme can slowly hydrolyze the hydroxamic acid produced and restore the active-site carboxylate.

Support for a covalent catalytic mechanism comes from rapid-quench kinetics experiments by Dunaway-Mariano and co-workers, which show a buildup of a covalent intermediate during catalysis; when the reaction is carried out with [^{14}C] 4-chlorobenzoyl-CoA, an intermediate containing radioactively labeled enzyme is formed that coincides with the formation of the covalent intermediate determined by the rapid-quench experiments.[39] To further investigate the possibility of a mechanism involving a covalent aryl enzyme intermediate, the enzyme reaction was

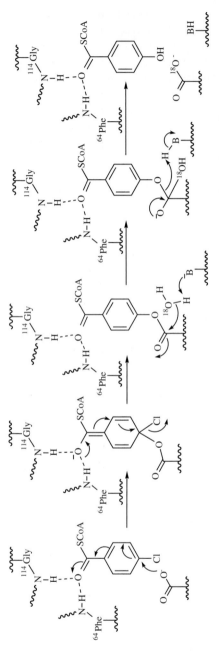

SCHEME 6.28   Reaction catalyzed by 4-chlorobenzoyl-CoA dehalogenase in the presence of $H_2^{18}O$.

carried out in $H_2^{18}O$ under single-turnover conditions. As shown in Scheme 6.28, and discussed in Section III for epoxide hydrolase, if the nucleophile is an active-site carboxylate residue (Asp or Glu), the oxygen atom of the hydroxyl group in the product after the first turnover should be derived from the enzyme nucleophile (^{16}O) rather than from solvent (^{18}O). If the reaction occurs via general base catalysis, the hydroxyl will be derived from solvent. Incubation of excess enzyme in $H_2^{18}O$ led to incorporation of only a small amount of ^{18}O into the product, consistent with covalent catalysis by an active-site carboxylate group.

Weak nucleophiles, such as carboxylates, typically do not participate readily in nucleophilic aromatic substitution reactions, particularly when the aromatic ring is only activated by weak activating groups such as a carbonyl, so how does the enzyme activate the aromatic ring for nucleophilic attack by a carboxylate residue? UV-visible, Raman, and ^{13}C NMR spectroscopy by Carey, Dunaway-Mariano, and co-workers of enzyme-bound substrate or product analogues containing ^{13}C or ^{18}O in the carbonyl group indicate that the environment of the enzyme active site induces a significant reorganization of the benzoyl ring π-electrons and polarizes the thioester carbonyl.[40] This polarization, which would strongly activate the aromatic ring for nucleophilic attack, is believed to arise from two sources: a dipolar electrostatic effect of active-site residues 114–121 (depicted by + + in Scheme 6.28) and by hydrogen bonding of an active-site residue to the substrate carbonyl oxygen. Both of these effects are enhanced by active-site hydrophobic residues. Active-site mutants were prepared, and it appears that hydrogen bonding and hydrophobicity are, indeed, important for catalysis and binding.[41]

VI. ELECTROPHILIC SUBSTITUTION (ADDITION/ELIMINATION)

5-Enolpyruvylshikimate-3-phosphate (EPSP) synthase catalyzes the synthesis of EPSP (**6.62**) from shikimate 3-phosphate (**6.60**) and phosphoenolpyruvate (**6.61**, PEP) (Scheme 6.29).[42] This is the enzyme that is the target of the herbicide

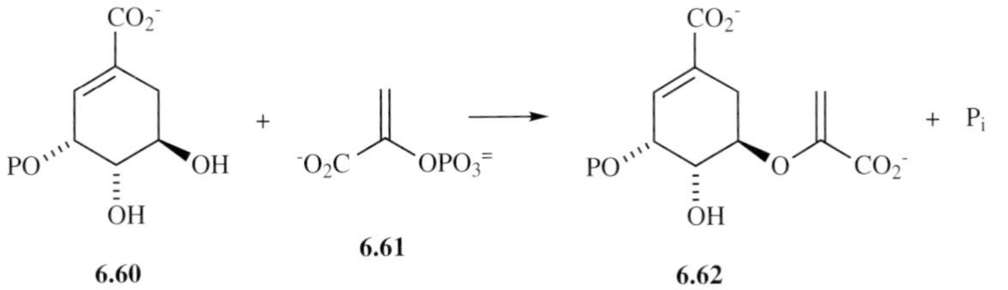

SCHEME 6.29   Reaction catalyzed by 5-enolpyruvylshikimate-3-phosphate (EPSP) synthase.

$$^{-}OOCCH_2\overset{+}{N}H_2CH_2PO_3^{=}$$

6.63

glyphosate (**6.63**; Roundup™). Let's consider four different mechanisms for EPSP synthase (Scheme 6.30; ROH is **6.60**). Mechanism 1 is a concerted addition/elimination mechanism; acid-catalyzed addition of the alcohol to the alkene (Mar-

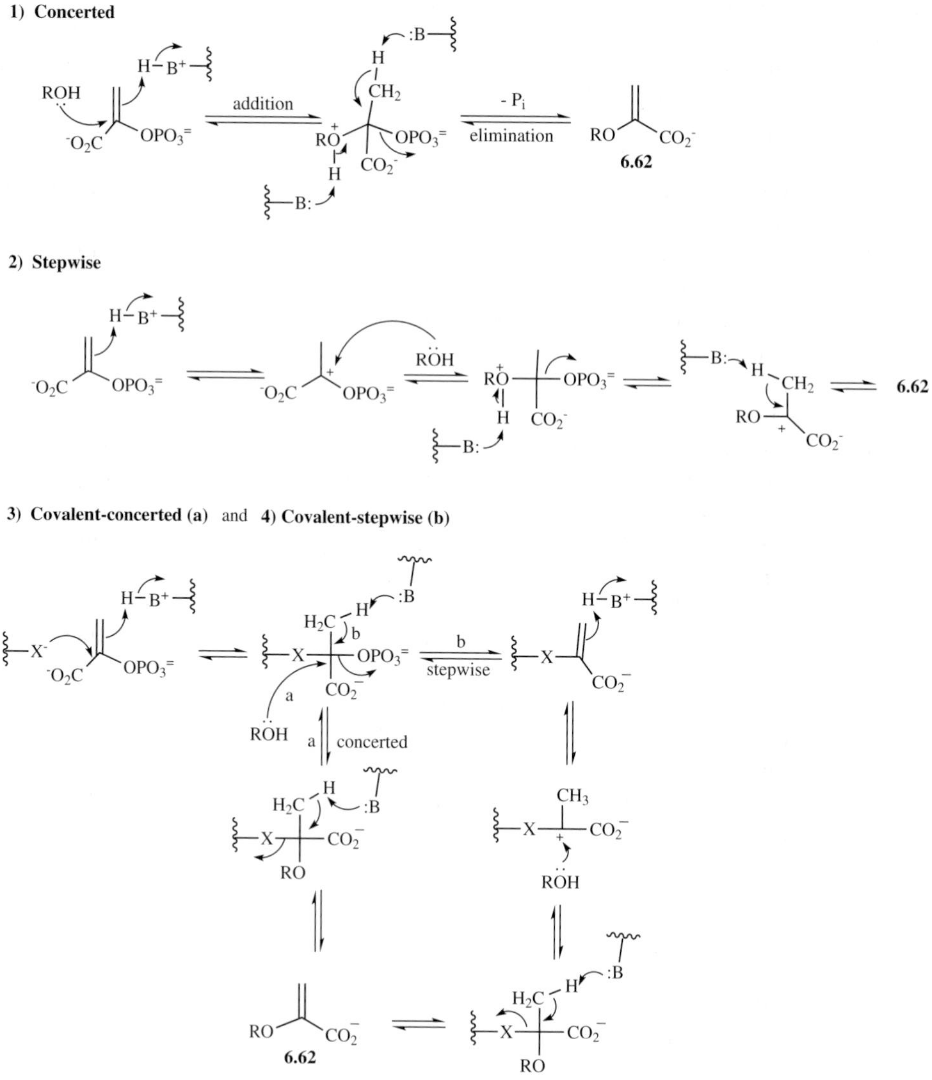
1) Concerted
addition
- Pi
elimination
6.62
2) Stepwise
6.62
3) Covalent-concerted (a) and 4) Covalent-stepwise (b)
b
stepwise
a
concerted
6.62

SCHEME 6.30 Four possible mechanisms for EPSP synthase.

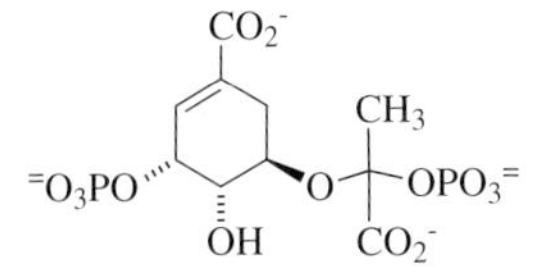

6.64

kovnikov addition) is followed by elimination of inorganic phosphate. The same mechanism, except written as a stepwise process, is shown in mechanism 2. The third mechanism involves addition of an active-site nucleophile followed by a concerted displacement of the phosphate by the alcohol and elimination of the active-site nucleophile (mechanism 3, pathway a). Mechanism 4 is the corresponding stepwise addition and elimination mechanism (pathway b).

With labeled substrates and using a rapid quench with neat Et_3N to keep the solution slightly basic, Anderson and Sikorski were able to isolate and characterize the tetrahedral intermediate (**6.64**).[43] This intermediate was incubated with EPSP synthase and found to be kinetically competent in the formation of EPSP.[44] This excludes mechanisms 3 and 4. A rapid-quench steady-state kinetic analysis using enzyme in excess indicated only one intermediate along pathway, consistent with mechanism 1.[45]

(Z)-3-Fluorophosphoenolpyruvate (**6.65**, Scheme 6.31), but not the (E)-isomer, was shown to be a pseudosubstrate for EPSP synthase.[46] It is a pseudosubstrate because the addition step occurs to give the fluorinated tetrahedral intermediate (**6.66**), which is a stable product; it does not break down to give (Z)-9-fluoro-EPSP (**6.67**). To see if EPSP synthase can convert **6.67** to **6.66** in the presence of inorganic phosphate (i.e., the reverse reaction), **6.67** was synthesized.[47] It is not a substrate for EPSP synthase in the reverse direction. This suggests that in the normal mechanism there is probably high carbocation character (**6.69**, Scheme 6.32) in the breakdown of the tetrahedral intermediate (**6.68**) and that the fluorine destabilizes the formation of this carbocation. The fact that **6.65** is a substrate for the first step (the addition step) suggests that probably not much carbocation character is generated in the PEP during the addition step; carbocation **6.70** (Scheme 6.33) would not be stabilized much by the phosphate oxygen, so addition probably occurs before much carbocation character develops. This result also is consistent

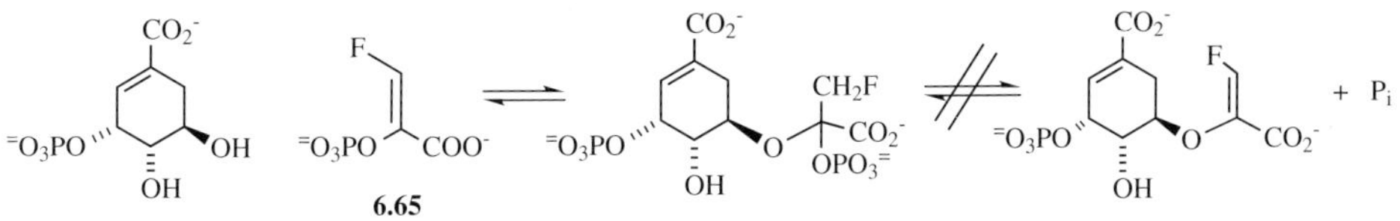

SCHEME 6.31   EPSP synthase-catalyzed reaction of shikimate-3-phosphate and (Z)-3-fluoroPEP.

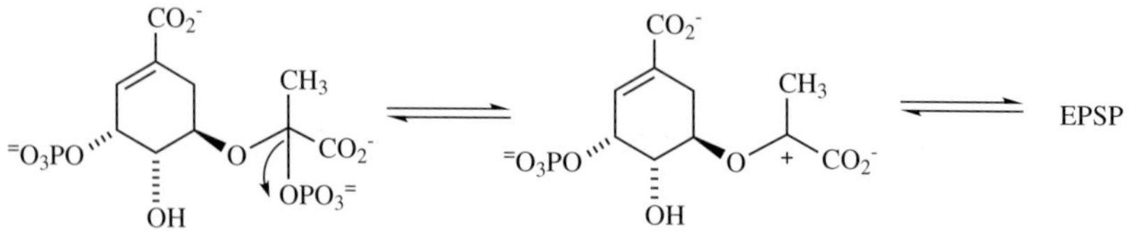

SCHEME 6.32 Carbocation character in the reaction catalyzed by EPSP synthase.

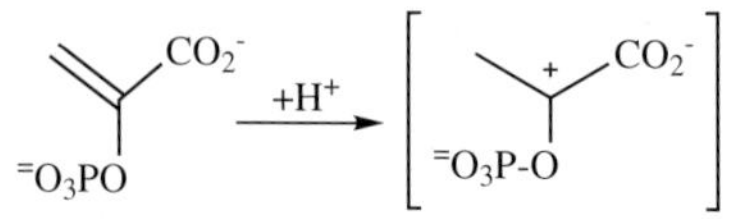

SCHEME 6.33 Comparison of the structure of the herbicide glyphosate with a proposed intermediate in the reaction catalyzed by EPSP synthase.

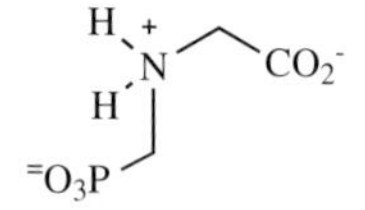

with the kinetic analysis of the herbicide glyphosate (**6.63**), which had long been thought to be a transition-state (or intermediate) analogue of EPSP synthase because it is potentially similar in structure to the purported PEP carbocation intermediate (**6.70**). The kinetic analysis indicates that glyphosate does not obey the kinetic criteria for a transition-state analogue, again suggesting that the carbocation is not important in the addition step.[48]

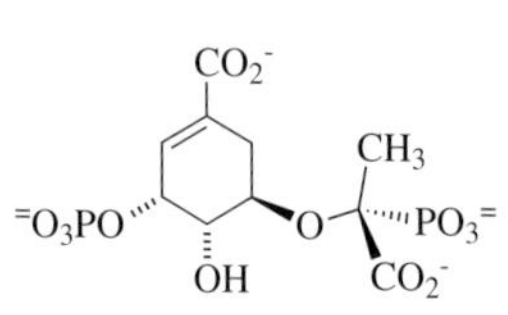

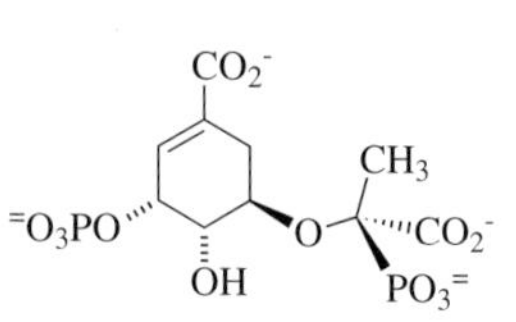

Alberg and Bartlett synthesized epimeric transition-state analogues **6.71** and **6.72** on the basis of the structure of the isolated intermediate **6.64**.[49] For stability purposes, a phosphonate group is generally substituted for a phosphate group when designing transition-state analogues. The K_i values for **6.71** and **6.72** are 15 nM and 1130 nM, respectively; the K_i for glyphosate is 400 nM. On the basis of the K_i values for **6.71** and **6.72**, Alberg and Bartlett concluded that the intermediate binds with the stereochemistry shown by isomer **6.71**.

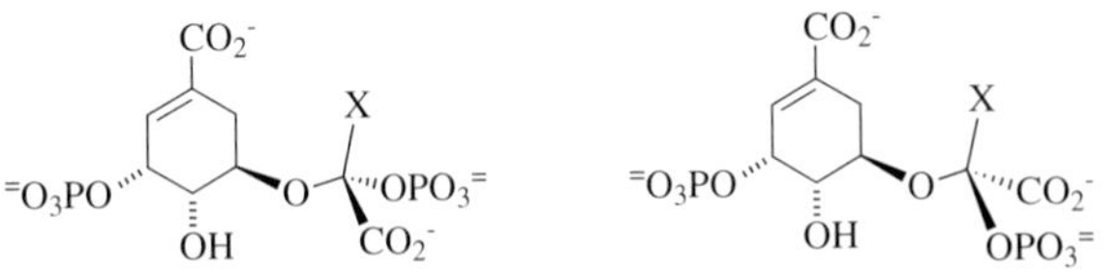

When an electron-withdrawing group is appended to a molecule near a leaving group, it stabilizes that molecule. Therefore, to synthesize a phosphate analogue instead of a phosphonate, electron-withdrawing groups (X = CH_2F, CHF_2, and CF_3) were substituted for the methyl group in **6.64**, namely, **6.73** and **6.74**.[50] Surprisingly, the analogues with the stereochemistry opposite that of the most potent of the phosphonate compounds (**6.71**) were more potent than the ones with the same stereochemistry as **6.71**. Compound **6.74**, where X = CHF_2, is the most potent inhibitor of EPSP synthase to date, having a K_i = 4 nM. Apparently the phosphate and carboxylate groups have interchangeable binding sites, so for some structures the phosphate group binds in the carboxylate binding site and vice versa.

An enzyme related to EPSP synthase in the chemical reaction that it catalyzes is uridine diphosphate-N-acetylglucosamine enolpyruvyl transferase (also called MurA or, sometimes, MurZ), which transfers enolpyruvate from PEP to the 3-hydroxyl group of UDP-N-acetylglucosamine (**6.75**, Scheme 6.34). This is the first committed step in bacterial cell wall peptidoglycan biosynthesis, the reaction of UDP-GlcNAc (**6.75**) with PEP to give enolpyruvyl UDP-GlcNAc (**6.76**) and inorganic phosphate. The kinetic mechanism suggests a tetrahedral phospholactyl-UDP-GlcNAc intermediate not covalently bound to the enzyme.[51] Incubation of the enzyme with either [^{14}C]PEP or [^{32}P]PEP gives stoichiometric labeling of the enzyme. The NMR spectrum of the enzyme labeled with [2-^{13}C]PEP is consistent with a tetrahedral phospholactyl enzyme adduct attached to C-2 of PEP. Rapid chemical quench analysis under single-turnover conditions using [^{32}P]PEP in the

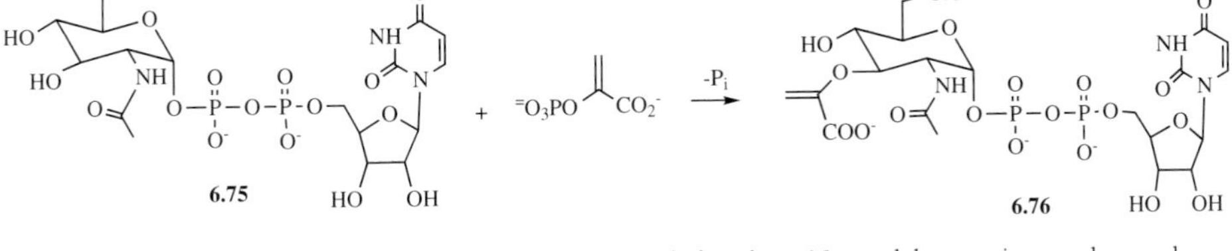

SCHEME 6.34 Reaction catalyzed by uridine diphosphate-N-acetylglucosamine enolpyruvyl transferase (MurA).

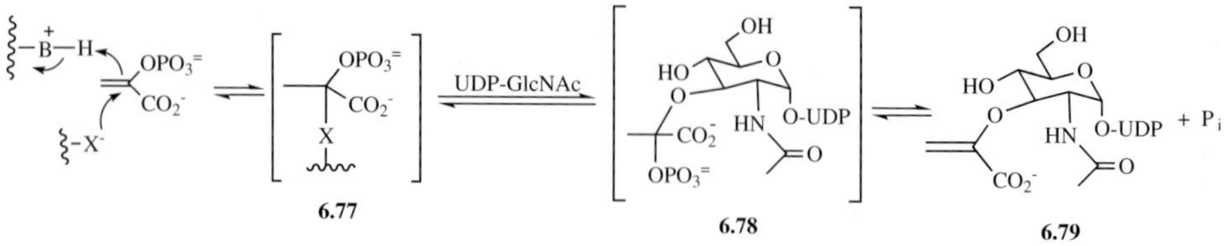

SCHEME 6.35   One possible mechanism for the reaction catalyzed by MurA.

presence of UDP-GlcNAc demonstrated that the covalent pholactyl enzyme adduct appears and decays at a rate competent with enzyme catalysis. Substrate-trapping experiments demonstrate that the covalent pholactyl enzyme adduct (**6.77**), when incubated with UDP-GlcNAc, is competent to form the noncovalent tetrahedral pholactyl-UDP-GlcNAc intermediate (**6.78**) and product (**6.79**), consistent with the pathway shown in Scheme 6.35.[52]

Evans and co-workers used time-resolved solid-state NMR spectroscopy[53] to detect the two intermediates in this reaction, one covalent (**6.77**) and the other noncovalent (**6.78**).[54] Wanke and Amrhein had earlier proposed a covalent adduct to Cys-115 (**6.77**, X = Cys), from studies with [^{32}P]PEP and [^{14}C]PEP.[55]

(E)-(**6.80**)- and (Z)-3-Fluorophosphoenolpyruvate (**6.65**) inactivate UDP-GlcNAc enolpyruvyl transferase by forming two stable adducts, one covalent (**6.81**) and one noncovalent (**6.82**) (Scheme 6.36).[56] The kinetics of formation indicate that **6.81** and **6.82** are formed by parallel pathways from the ternary complex enzyme·UDP-GlcNAc·(Z)-FPEP, although there is a fivefold kinetic preference for formation of **6.82**. Experiments with ^{32}P-labeled FPEP and D_2O indicate that **6.81** and **6.82** interconvert directly without C−H bond cleavage or FPEP formation. The decreased rate of formation of products suggests that the fluorine destabilizes an oxocarbenium intermediate. Also, a comparison of the kinetic data with FPEP and the rapid-quench data for the normal reaction with PEP[57] supports a mechanism in which formation and decomposition of the tetrahedral intermediate, as well as the interconversion between the covalent and noncovalent complexes, involve an oxocarbenium ion intermediate. The kinetic data on formation of **6.82** demonstrate that its formation does not require prior formation of **6.81**. Therefore a branching catalytic mechanism was proposed (Scheme 6.37) in which the cova-

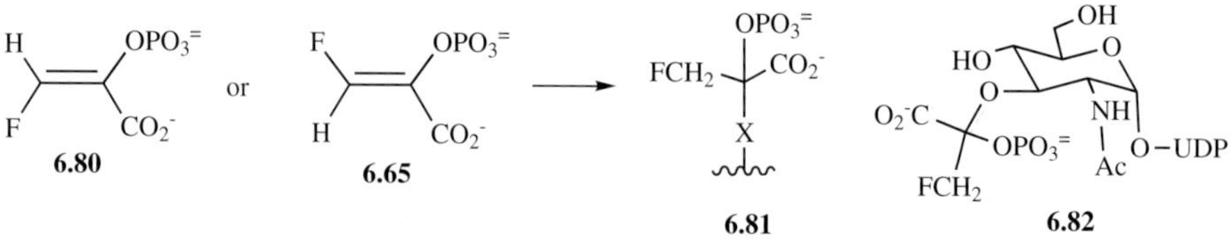

SCHEME 6.36   Inactivation of MurA by (E)- and (Z)-3-fluoroPEP.

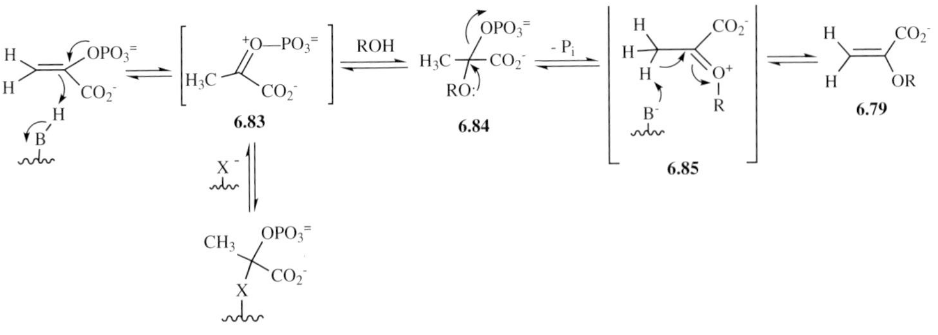

SCHEME 6.37   More consistent mechanism for the reaction catalyzed by MurA.

lent adduct (**6.86**) is off the main pathway and derives from addition of an active-site nucleophile to the first oxocarbenium intermediate (**6.83**). The primary pathway is direct formation and breakdown of the noncovalent intermediate (**6.84**), leading to another oxocarbenium intermediate (**6.85**), which tautomerizes to the product (**6.79**).

As shown in Scheme 6.31, EPSP synthase catalyzes the addition of (Z)-3-fluoroPEP (**6.65**) to shikimate-3-phosphate (**6.60**) to give a stable fluorotetrahedral intermediate (**6.66**). The same intermediate stability was observed in the MurA-catalyzed reaction of **6.65** with UDP-GlcNAc (**6.75**, Scheme 6.34); unlike EPSP synthase, however, both (E)- and (Z)-3-fluoroPEP react to give stable fluorinated tetrahedral intermediates. The stability of these intermediates was utilized to determine the stereochemistry of their formation, and, therefore, the stereochemistry of the addition step. The MurA-catalyzed condensation of (E)- and (Z)-3-fluoroPEP with UDP-GlcNAc was carried out by Walsh and co-workers in D_2O in the presence of alkaline phosphatase, pyruvate carboxylase, and malate dehydrogenase.[58] The alkaline phosphatase was included to convert the stable fluorinated tetrahedral intermediates (**6.87**, Scheme 6.38) into [3-^{2}H]3-fluoropyruvate (**6.88**) and UDP-GlcNAc. A coupled reaction of pyruvate carboxylase and malate dehydrogenase was utilized for stereospecific conversion of **6.88** to (2R,3R)-3-fluoromalates (**6.89**), analyzed by ^{19}F NMR for either ^{1}H or ^{2}H attached to C-3. Starting with **6.80**, [3-^{2}H]3-fluoromalate (**6.89E**) is the product and from **6.65**, [3-^{1}H]3-fluoromalate (**6.89Z**) forms. These results establish the R-configuration for **6.87E** and the S-configuration for **6.87Z**. Therefore, addition of D^+ is to the 2-re face of 3-fluoroPEP (the bottom face as drawn for both **6.80** and **6.65**). Extrapolating these results to the MurA reaction with PEP, proton addition to C-3 of PEP would proceed at the same face of the double bond, namely, the 2-si face (remember, without the fluorine the group priorities change, even though they are in the same position).

The crystal structure of the C115A mutant of *Escherichia coli* MurA complexed

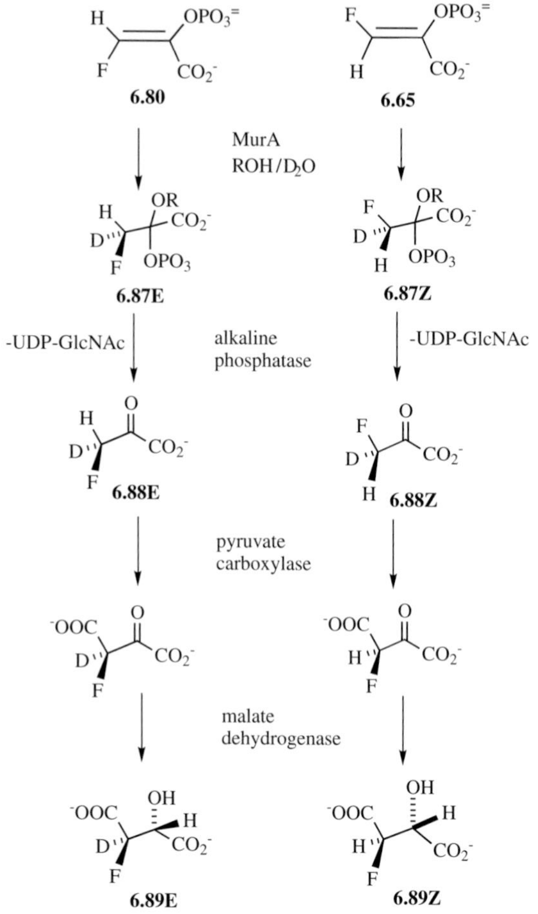

SCHEME 6.38   Determination of the stereochemistry of the reaction catalyzed by MurA.

with **6.82** (from the reaction of UDP–GlcNAc and (Z)–3–fluoroPEP; see Scheme 6.36) was determined at 2.8 Å resolution.[59] The mutant was utilized to avoid attack of Cys–115 on C-2 of the (Z)–3–fluoroPEP in competition with formation of **6.82**. The absolute configuration of the adduct at C-2 is 2R. Therefore, addition of the 3′-hydroxyl group of UDP–GlcNAc is to the 2-si face of (Z)–3–fluoroPEP (which corresponds to the 2-re face of PEP). Given the earlier observation that in D_2O the addition of D^+ is to the 2-si face of PEP, the addition across the double bond of PEP is anti. Because the overall stereochemical course is either anti/syn or syn/ anti,[60] it is now apparent that the stereochemistry of elimination of the proton from C-3 and P_i from C-2 of the tetrahedral intermediate is syn (Scheme 6.39).

The same stereochemical study using (E)–(**6.80**) and (Z)–3–fluoroPEP (**6.65**) was carried out for EPSP synthase to compare with the results with MurA.[61] EPSP

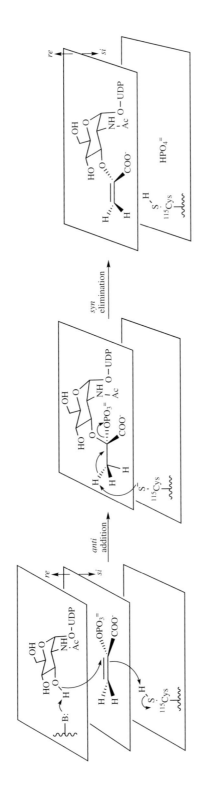

SCHEME 6.39   Stereochemistry of the reaction catalyzed by MurA.

synthase utilizes the same stereochemistry as MurA, namely, addition of solvent-derived proton to the 2-*re* face of 3-fluoroPEP (2-*si* of PEP).

VII. ELECTROPHILIC AROMATIC SUBSTITUTION

Alkylation of aromatic rings is called Friedel–Crafts alkylation.[62] Scheme 6.40 depicts the general Friedel–Crafts chemical reaction. Although the reaction of alkyl halides and aluminum chloride with substituted benzenes is not found in nature, Friedel–Crafts alkylation reactions of alkyl diphosphates are common. In the biosynthesis of vitamin K, coenzyme Q, and other quinones, the hydrocarbon side chain is attached to the benzene ring with a Friedel–Crafts-type reaction[63] (Scheme 6.41). Dimethylallyltryptophan synthase[64] catalyzes the isoprenyl alkylation of tryptophan via an apparent Friedel–Crafts reaction (Scheme 6.42).

An electrophilic heteroaromatic substitution is catalyzed by porphobilinogen deaminase,[65] an enzyme in tetrapyrrole biosynthesis, leading to porphyrins (needed for heme-dependent enzymes—see Chapter 4.IV) and to corrins (needed for co-

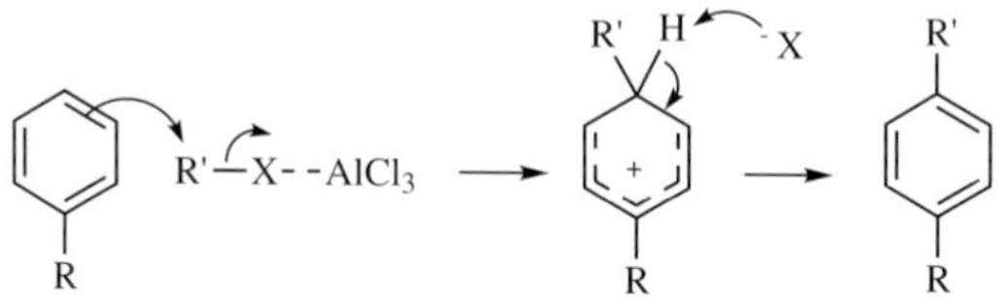

SCHEME 6.40   Friedel–Crafts reaction.

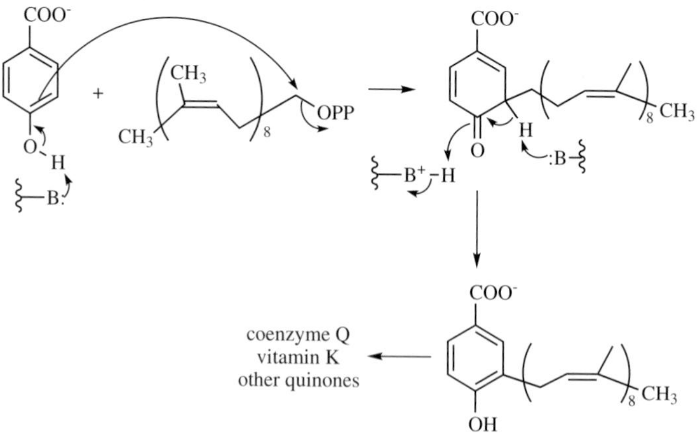

SCHEME 6.41   Enzymatic Friedel–Crafts reactions.

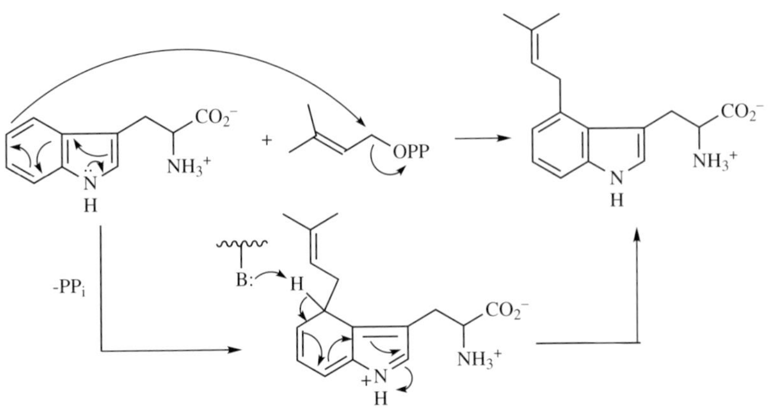

SCHEME 6.42   Friedel–Crafts reaction catalyzed by dimethylallyltryptophan synthase.

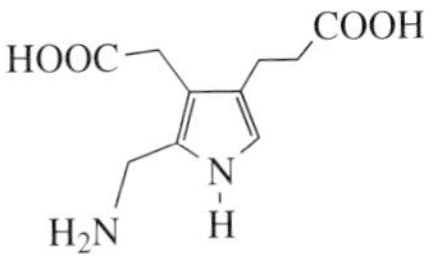

6.90 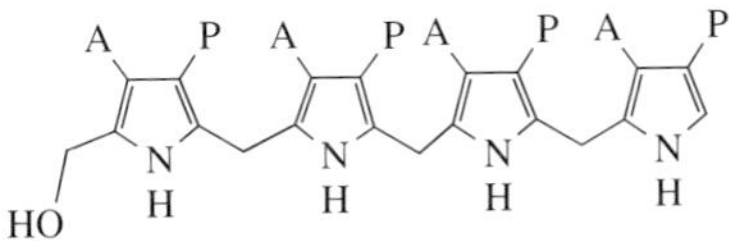   **6.91**

enzyme B_{12}-dependent enzymes—see Chapter 13.III.B). Porphobilinogen deaminase catalyzes the condensation of four units of porphobilinogen (**6.90**), with the loss of 4 equivalents of ammonia, to give hydroxymethylbilane (**6.91**), the substrate for uroporphyrinogen-III synthase, which catalyzes the cyclization of the tetrapyrrole. Three mechanisms for the initial elimination of ammonia from **6.90** are shown in Scheme 6.43. To test these mechanisms, substrate analogues **6.92–6.96** were synthesized. Compound **6.92** is a substrate, although it does not produce the tetrapyrrole. Nonetheless, the fact that it undergoes condensation suggests that E2′ and E1cB mechanisms are unlikely. Compounds **6.93** and **6.95** are excellent substrates (**6.95** is a substrate for the reverse reaction), whereas **6.94** and **6.96** are not substrates. These results strongly suggest that a carbocation (E1) intermediate is important because the methyl group (**6.93** and **6.95**) stabilizes it, and the trifluoromethyl group (**6.94** and **6.96**) destabilizes it. However, the strong electron-withdrawing effect of the trifluoromethyl group decreases the pK_a of the amino group (the pK_a was determined to be 5); therefore, at the pH of the reaction, the amino group is not protonated, and thus is not a leaving group, so that result is not as clearcut as the others. Nonetheless, these results support a mechanism involving a carbocation intermediate (Scheme 6.44).

Another possible Friedel–Crafts alkylation reaction is catalyzed by histidine

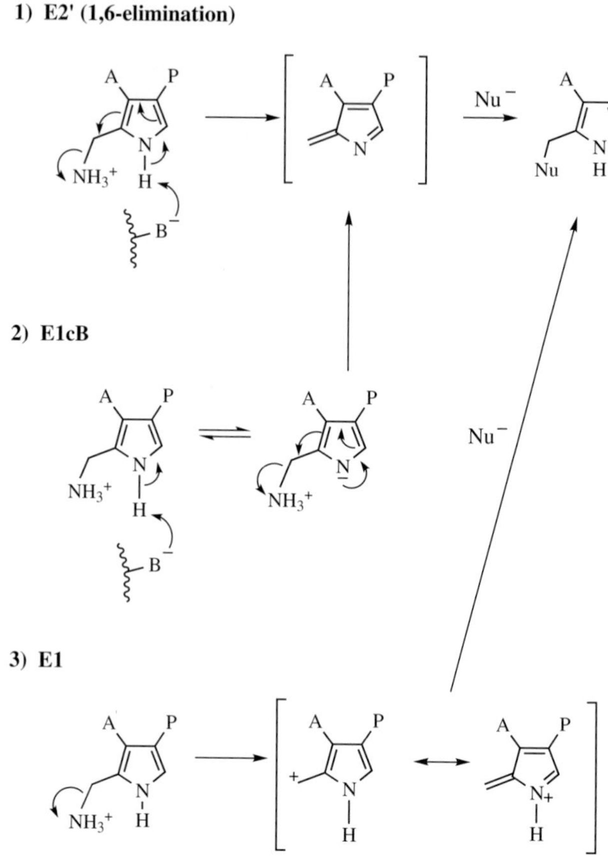

1) E2' (1,6-elimination)

2) E1cB

3) E1

SCHEME 6.43   Possible mechanism for the reaction catalyzed by porphobilinogen deaminase.

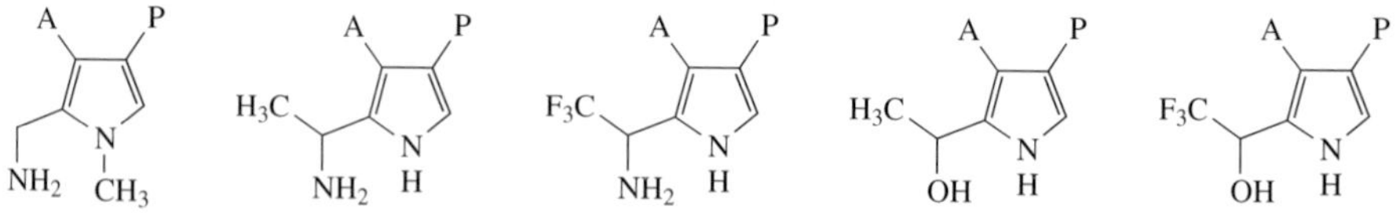

| 6.92 | 6.93 | 6.94 | 6.95 | 6.96 |

ammonia–lyase and phenylalanine ammonia–lyase, discussed in Chapter 10 (Section I.C.1). Next we see how enzymes catalyze carboxylation reactions.

NOTE THAT PROBLEMS AND SOLUTIONS RELEVANT TO EACH CHAPTER CAN BE FOUND IN APPENDIX II.

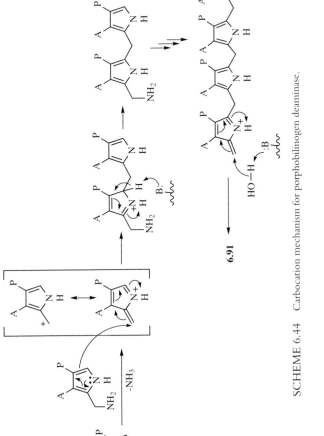

SCHEME 6.44   Carbocation mechanism for porphobilinogen deaminase.

REFERENCES

1. Poulter, C. D. In *Biochemistry of Cell Walls and Membranes in Fungi,* Kuhn, P. J.; Trinci, A. P. J.; Jung, M. J.; Goosey, M. W.; Copping, L. G., Ed.; Springer-Verlag: Berlin, 1990, pp. 169–188.
2. Poulter, C. D.; Wiggins, P. L.; Le, A. T. *J. Am. Chem. Soc.* **1981,** *103,* 3926.
3. Poulter, C. D.; Argyle, J. C.; Mash, E. A. *J. Biol. Chem.* **1978,** *253,* 7227.
4. (a) Davisson, V. J.; Neal, T. R.; Poulter, D. C. *J. Am. Chem. Soc.* **1993,** *115,* 1235. (b) Davisson, V. J.; Poulter, C. D. *J. Am. Chem. Soc.* **1993,** *115,* 1245.
5. Poulter, C. D.; Rilling, H. A. *Acc. Chem. Res.* **1978,** *11,* 307.
6. (a) Cane, D. E. *Chem. Rev.* **1990,** *90,* 1089. (b) Cane, D. E. *Acc. Chem. Res.* **1985,** *18,* 220.
7. Lesburg, C. A.; Zhai, G.; Cane, D. E.; Christianson, D. W. *Science* **1997,** *277,* 1820.
8. Mieyal, J.; Abeles, R. In *The Enzymes,* vol. 7, 3rd ed., Boyer, P., Ed., Academic Press: New York, 1972, p. 515.
9. Sinnott, M. L. *Chem. Rev.* **1990,** *90,* 1171.
10. Wang, Q.; Graham, R. W.; Trimbur, D.; Warren, R. A. J.; Withers, S. G. *J. Am. Chem. Soc.* **1994,** *116,* 11594.
11. (a) Street, I. P.; Kempton, J. B.; Withers, S. G. *Biochemistry* **1992,** *31,* 9970. (b) Withers, S. G.; Warren, R. A. J.; Street, I. P.; Rupitz, K.; Kempton, J. B.; Aebersold, R. *J. Am. Chem. Soc.* **1990,** *112,* 5887. (c) Withers, S. G.; Street I. P. *J. Am. Chem. Soc.* **1988,** *110,* 8551. (d) Withers, S. G.; Street, I. P.; Bird, P.; Dolphin, D. H. *J. Am. Chem. Soc.* **1987,** *109,* 7530.
12. Withers, S. G.; Rupitz, K.; Trimbur, D.; Warren, R. A. J. *Biochemistry* **1992,** *31,* 9979.
13. Kempton, J. B.; Withers, S. G. *Biochemistry* **1992,** *31,* 9961.
14. McCarter, J. D.; Withers, S. G. *J. Biol. Chem.* **1996,** *271,* 6889.
15. Banait, N. S.; Jencks, W. P. *J. Am. Chem. Soc.* **1991,** *13,* 7951.
16. Lacourciere, G. M.; Armstrong, R. N. *J. Am. Chem. Soc.* **1993,** *115,* 10466.
17. Borham, B.; Jones, A. D.; Pinot, F.; Grant, D. F.; Kurth, M. J.; Hammock, B. D. *J. Biol. Chem.* **1995,** *270,* 26923.
18. Hammock, B. D.; Pinot, F.; Beetham, J. K.; Grant, D. F.; Arand, M. E.; Oesch, F. *Biochem. Biophys. Res. Commun.* **1994,** *198,* 850.
19. (a) Baldwin, J. E. *J. Chem. Soc. Chem. Commun.* **1976,** 734. (b) Baldwin, J. E.; Kruse, L. I. *J. Chem. Soc. Chem. Commun.* **1977,** 233; (c) Baldwin, J. E.; Lusch, M. J. *Tetrahedron* **1982,** *38,* 2939.
20. Janda, K. D.; Shevlin, C. G.; Lerner, R. A. *Science* **1993,** *259,* 490.
21. Na, J.; Houk, K. N.; Shevlin, C. G.; Janda, K. D; Lerner, R. A. *J. Am. Chem. Soc.* **1993,** *115,* 8453.
22. Pries, F.; Kingma, J.; Pentenga, M.; van Pouderoyen, G.; Jeronimus-Strathingh, C. M.; Bruins, A. P.; Janssen, D. B. *Biochemistry* **1994,** *33,* 1242.
23. Verschueren, K. H. G.; Seljee, F.; Rozeboom, H. J.; Kalk, K. H.; Dijkstra, B. W. *Nature* **1993,** *363,* 693.
24. (a) Walsh, C. T.; Liu, J.; Rusnak, F.; Sakaitani, M. *Chem. Rev.* **1990,** *90,* 1105. (b) Liu, J.; Quinn, N.; Berchtold, G. A.; Walsh, C. T. *Biochemistry* **1990,** *29,* 1417.
25. Walsh, C. T.; Erion, M. D.; Walts, A. E.; Delany, Jr. J. J.; Berchtold, G. A. *Biochemistry* **1987,** *26,* 4734.
26. (a) Teng, C. Y. P.; Ganem, B. *J. Am. Chem. Soc.* **1984,** *106,* 2463. (b) Policastro, P. P.; Au, K. G.; Walsh, C. .; Berchtold, G. A. *J. Am. Chem. Soc.* **1984,** *106,* 2443–4.
27. Teng, C. Y. P.; Ganem, B.; Doktor, S. Z.; Nichols, B. P.; Bhatnagar, R. K.; Vining, L. C. *J. Am. Chem. Soc.* **1985,** *107,* 5008.
28. Morollo, A. A.; Bauerle, R. *Proc. Natl. Acad. Sci. USA* **1993,** *90,* 9983.
29. Anderson, K. S.; Kati, W. M.; Ye, Q. Z.; Liu, J.; Walsh, C. T.; Benesi, A. J.; Johnson, K. A. *J. Am. Chem. Soc.* **1991,** *113,* 3198.
30. Kozlowski, M. C.; Tom, N. J.; Seto, C. T.; Sefler, A. M.; Bartlett, P. A. *J. Am. Chem. Soc.* **1995,** *117,* 2128.
31. Mattia, K. M.; Ganem, B. *J. Org. Chem.* **1994,** *59,* 720.

31a. Bjornestedt, R.; Stenberg, G.; Widersten, M.; Board, P. G.; Sinning, I.; Jones, T. A.; Mannervik, B. *J. Mol. Biol.* **1995**, *247*, 765.

31b. Liu, S.; Zhang, P.; Ji, X.; Johnson, W. W.; Gilliland, G. L.; Armstrong, R. N. *J. Biol. Chem.* **1992**, *267*, 4296.

31c. Thorson, J. S.; Shin, I.; Chapman, E.; Stenberg, G.; Mannervik, B.; Schultz, P. G. *J. Am. Chem. Soc.* **1998**, *120*, 451.

31d. Zheng, Y.-J.; Ornstein, R. L. *J. Am. Chem. Soc.* **1997**, *119*, 1523.

32. Chen, W.-J.; Graminski, G. F.; Armstrong, R. N. *Biochemistry* **1988**, *27*, 647.

33. Graminski, G. F.; Zhang, P.; Sesay, M. A.; Ammon, H. L.; Armstrong, R. N. *Biochemistry* **1989**, *28*, 6252.

34. Ji, X.; Armstrong, R. N.; Gilliland, G. L. *Biochemistry* **1993**, *32*, 12949.

35. Crooks, G. P.; Copley, S. D. *J. Am. Chem. Soc.* **1993**, *115*, 6422.

36. Crooks, G. P.; Xu, L.; Barkley, R. M.; Copley, S. D. *J. Am. Chem. Soc.* **1995**, *117*, 10791.

37. (a) Verschueren, K. H. G.; Seljee, F.; Rozeboom, H. J.; Kalk, K. H.; Dijkstra, B. W. *Nature* **1993**, *363*, 693. (b) Pries, F.; Kingma, J.; Pentenga, M.; van Pouderoyen, G.; Jeronimus-Stratingh, C. M.; Bruins, A. P.; Janssen, D. B. *Biochemistry* **1994**, *33*, 1242.

38. Lacourciere, G. M.; Armstrong, R. N. *J. Am. Chem. Soc.* **1993**, *115*, 10466.

39. Yang, G.; Liang, P.-H.; Dunaway-Mariano, D. *Biochemistry* **1994**, *33*, 8527.

39a. Benning, M.; Taylor, K. L.; Liu, R. Q.; Yang, G.; Xiang, H.; Wesenberg, G.; Dunaway-Mariano, D.; Holden, H. M. *Biochemistry* **1996**, *35*, 8103.

40. (a) Taylor, K. L.; Liu, R.-A; Liang, P-H; Price, J.; Dunaway-Mariano, D.; Tonge, P. J.; Clarkson, J.; Carey, P. R. *Biochemistry* **1995**, *34*, 13881. (b) Clarkson, J.; Tonge, P. J.; Taylor, K. L.; Dunaway-Mariano, D.; Carey, P. R. *Biochemistry* **1997**, *36*, 10192.

41. Taylor, K. L.; Xiang, H.; Liu, R.-Q.; Yang, G.; Dunaway-Mariano, D. *Biochemistry* **1997**, *36*, 1349.

42. Anderson, K. S.; Johnson, K. A. *Chem Rev.* **1990**, *90*, 1131.

43. Anderson, K. S.; Sikorski, J. A., *J. Am. Chem. Soc.* **1988**, *110*, 6577.

44. Anderson, K. S.; Johnson, K. A. *J. Biol. Chem.* **1989**, *265*, 5567.

45. Anderson, K. S.; Sikorski, J. A.; Johnson, K. A. *Biochemistry* **1988**, *27*, 7395.

46. Walker, M. C.; Jones, C. R.; Somerville, R. L.; Sikorski, J. A. *J. Am. Chem. Soc.* **1992**, *114*, 7601.

47. Seto, C. T.; Bartlett, P. A. *J. Org. Chem.* **1994**, *59*, 7130.

48. Sammons, R. D.; Gruys, K. J.; Anderson, K. S.; Johnson, K. A.; Sikorski, J. A. *Biochemistry* **1995**, *34*, 6433.

49. Alberg, D. G.; Bartlett, P. A., *J. Am. Chem. Soc.* **1989**, *111*, 2337.

50. Alberg, D. G.; Lauhon, C. T.; Nyfeler, R.; Fassler, A.; Bartlett, P. A. *J. Am. Chem. Soc.* **1992**, *114*, 3535.

51. Marquardt, J. L.; Brown, E. D.; Walsh, C. T.; Anderson, K. S. *J. Am. Chem. Soc.* **1993**, *115*, 10398.

52. Brown, E. D.; Marquardt, J. L.; Lee, J. P.; Walsh, C. T.; Anderson, K. S. *Biochemistry* **1994**, *33*, 10638.

53. Evans, J. N. S. *Time-Resolved Solid-State NMR of Enzyme−Substrate Interactions.* In *Encyclopedia of NMR,* Wiley: New York, 1995.

54. Ramilo, C.; Appleyard, R. J.; Wanke, C.; Krekel, F.; Amrhein, N.; Evans, J. N. S. *Biochemistry* **1994**, *33*, 15071.

55. Wanke, C.; Amrhein, N. *Eur. J. Biochem.* **1993**, *218*, 861.

56. Kim, D. H.; Lees, W. J.; Haley, T. M.; Walsh, C. T. *J. Am. Chem. Soc.* **1995**, *117*, 1494.

57. Brown, E. D.; Marquardt, J. L.; Lee, J. P.; Walsh, C. T.; Anderson, K. S. *Biochemistry* **1994**, *33*, 10638.

58. Kim, D. H.; Lees, W. J.; Walsh, C. T. *J. Am. Chem. Soc.* **1995**, *117*, 6380.

59. Skarzynski, T.; Kim, D. H.; Lees, W. J.; Walsh, C. T.; Duncan, K. *Biochemistry* **1998**, *37*, 2572.

60. Lees, W. J.; Walsh, C. T. *J. Am. Chem. Soc.* **1995**, *117*, 7329.

61. Kim, D. H.; Tucker-Kellogg, G. W.; Lees, W. J.; Walsh, C. T. *Biochemistry* **1996**, *35*, 5435.

62. March, J. *Advanced Organic Chemistry,* 4th ed., Wiley: New York, 1992, pp. 534−539.

63. Roberts, R. M.; Khalaf, A. A. *Friedel−Crafts Alkylation Chemistry,* Marcel Dekker: New York, 1984.

64. Gebler, J. C.; Woodside, A. B.; Poulter, C. D. *J. Am. Chem. Soc.* **1992**, *114*, 7354.

65. Pichon, C.; Clemens, K. R.; Jaccobson, A. R.; Scott, A. I. *Tetrahedron* **1992**, *48*, 4687.

Carboxylations

I. GENERAL CONCEPTS

In general, a carbanion (or carbanionic character) must be generated at the carbon where carboxylation is to occur. Therefore, either a relatively acidic proton must be removed or a protected enolate must be deprotected so that the carbanion formed is somewhat stabilized. Keep in mind, however, that enzymes are capable of removing protons that are not highly acidic (see Chapter 1, Section II.C). Furthermore, metal ion complexation of the oxygen of the keto and enol forms can increase the acidity of an adjacent C–H bond by 4–6 orders of magnitude; for example, complexation with Mg^{2+} lowers the pK_a at C-3 of oxaloacetate from 13 to 9.[1] Enolates of α-keto acids are effectively stabilized by metal complexation.

Carbon dioxide (CO_2) is an excellent electrophile for carboxylation, but at physiological pH it is in very low concentration; its predominant form is bicarbonate (HCO_3^-), which actually is a nucleophile. To convert bicarbonate into an electrophile, it must be activated either by phosphorylation or by dehydration back to CO_2.

Some of the carboxylation reactions that are discussed in this chapter will utilize CO_2 as the carboxylating agent, and others will utilize bicarbonate. In general, all enzymes utilize CO_2 except for phosphoenolpyruvate carboxylase (see Section III.A) and the biotin-dependent enzymes, which utilize bicarbonate. It is possible to determine which of these carboxylating agents is the substrate for the enzyme-catalyzed reaction. The reaction of carbon dioxide with water to give bicarbonate ($CO_2 + H_2O \rightleftharpoons H_2CO_3$) is a "slow" reaction; it takes more than a minute at 15 °C! If you put CO_2 into the enzyme reaction at a concentration approximating its K_m value, and utilize sufficient enzyme so that a significant amount of product is produced in the first few seconds, there are two possible outcomes (Figure 7.1). If the rate of product formation starts rapidly, then levels to a slower steady-state rate (Figure 7.1A), the substrate is CO_2. The rate diminishes as the CO_2 is converted to

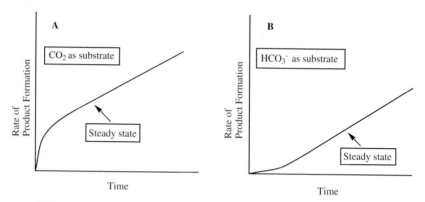

FIGURE 7.1   Test for whether CO_2 or HCO_3^- is the substrate for a carboxylase.

HCO_3^-, then the equilibrium of bicarbonate to carbon dioxide determines the steady-state rate. If the rate gradually increases, then levels at a faster steady-state rate (Figure 7.1B), the substrate is bicarbonate. The rate rises while CO_2 is being converted into HCO_3^-, then the equilibrium between carbon dioxide and bicarbonate determines the steady-state rate. The same experiment can be done starting with bicarbonate, with the opposite result.

Another way to test whether CO_2 or bicarbonate is the substrate for the enzyme-catalyzed reaction is to repeat the preceding experiment, except in the presence of the enzyme carbonic anhydrase (carbonate dehydratase), which catalyzes the rapid hydrolysis of CO_2 to bicarbonate, and determine the effect of added carbonic anhydrase on the reaction rate.[2] If CO_2 is the substrate, the rate should be greatly diminished (no initial burst), but if HCO_3^- is the substrate, the rate should be greatly enhanced (no initial lag).

Although the deprotonation step of the substrate may be a somewhat poor equilibrium, the carboxylation of the enolate formed is generally favorable. It is not unusual for enzymatic reactions to circumvent thermodynamics by coupling unfavorable equilibria to favorable equilibria.

II. CARBON DIOXIDE AS THE CARBOXYLATING AGENT

A. Phosphoenolpyruvate (PEP) Carboxykinase

The reaction catalyzed by phosphoenolpyruvate (PEP) carboxykinase—the reaction of phosphoenolpyruvate (**7.1**), CO_2,[3] and a nucleotide diphosphate to give oxaloacetate (**7.2**) and the corresponding nucleotide triphosphate—is shown in Scheme 7.1. Note that the name indicates that both a carboxylation and a kinase

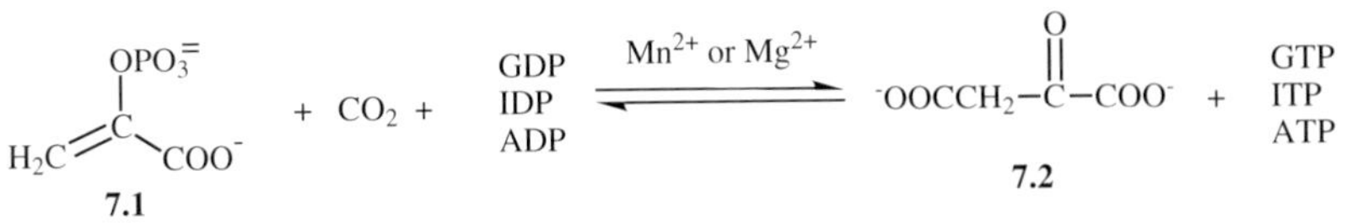

$$\underset{\textbf{7.1}}{\overset{\displaystyle OPO_3^=}{\underset{H_2C}{\overset{\shortmid}{\diagup}}\overset{\displaystyle C}{\diagdown COO^-}}} + CO_2 + \underset{ADP}{\overset{GDP}{IDP}} \underset{}{\overset{Mn^{2+}\ or\ Mg^{2+}}{\rightleftharpoons}} \underset{\textbf{7.2}}{\overset{\displaystyle O}{\ ^-OOCCH_2-\overset{\parallel}{C}-COO^-}} + \underset{ATP}{\overset{GTP}{ITP}}$$

SCHEME 7.1   Reaction catalyzed by PEP carboxykinase.

reaction (see Chapter 2, Section II.D.3) are occurring. Because CO_2 is electrophilic, it does not need to be activated.

1. Evidence for Mechanism

If the enzyme reaction is run with CO_2 in $H_2{}^{18}O$ instead of in H_2O, no ^{18}O is incorporated into products. This observation supports a reaction that utilizes CO_2, not HCO_3^-; if bicarbonate were involved, the CO_2 would have to react first with the $H_2{}^{18}O$, producing $[^{18}O]$bicarbonate, which would lead to ^{18}O incorporation into the product. To do such an experiment, the enzyme has to be in a concentration high enough that the reaction is over before isotopic equilibrium is reached; thus the reaction time must be short.

In the absence of CO_2, PEP reacts with adenosine diphosphate (ADP) to give pyruvate (**7.3**) and ATP (Scheme 7.2). In this case, a proton is the electrophile in place of CO_2, and the enzyme is acting simply as a kinase.

No partial exchange reactions (see Chapter 2, Section II.C.2.b) are observed. For example, if [^{14}C]pyruvate is added with substrate, no [^{14}C]oxaloacetate is observed, implying that either a concerted mechanism occurs or no intermediates (such as enolpyruvate) leave the active site until all of the products are formed.

If the enzyme reaction is carried out in D_2O in the presence of malate dehydrogenase, which catalyzes the reduced nicotinamide adenine dinucleotide (NADH)-dependent reduction (see Chapter 3, Section III) of oxaloacetate (**7.2**) to malate (**7.4,** Scheme 7.3), no deuterium is found in the malate. The malate dehydroge-

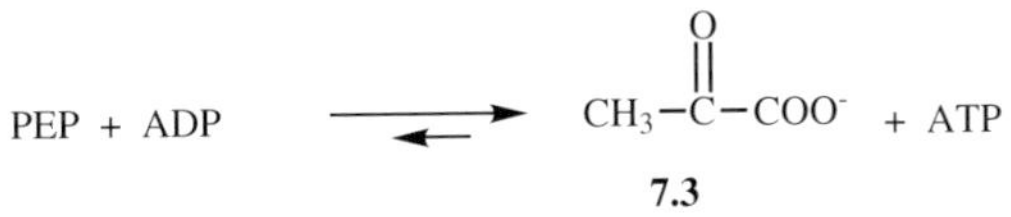

$$PEP + ADP \ \underset{\longleftarrow}{\overset{\longrightarrow}{}} \ \underset{\textbf{7.3}}{\overset{\displaystyle O}{CH_3-\overset{\parallel}{C}-COO^-}} + ATP$$

SCHEME 7.2   PEP carboxykinase-catalyzed reaction of PEP with ADP.

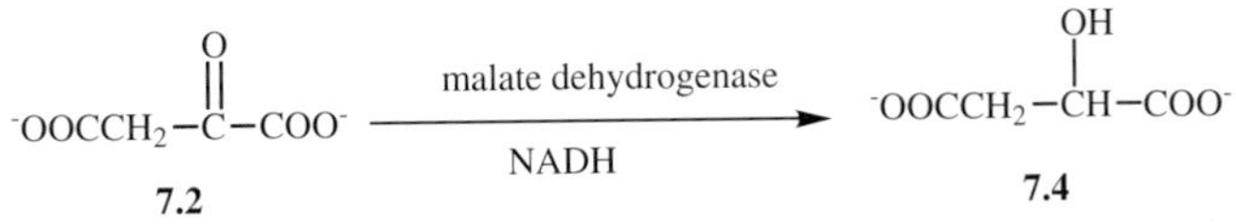

$$\underset{\textbf{7.2}}{\overset{\displaystyle O}{^-OOCCH_2-\overset{\parallel}{C}-COO^-}} \xrightarrow[\text{NADH}]{\text{malate dehydrogenase}} \underset{\textbf{7.4}}{\overset{\displaystyle OH}{^-OOCCH_2-\overset{\shortmid}{C}H-COO^-}}$$

SCHEME 7.3   Reduction of oxaloacetate by malate dehydrogenase.

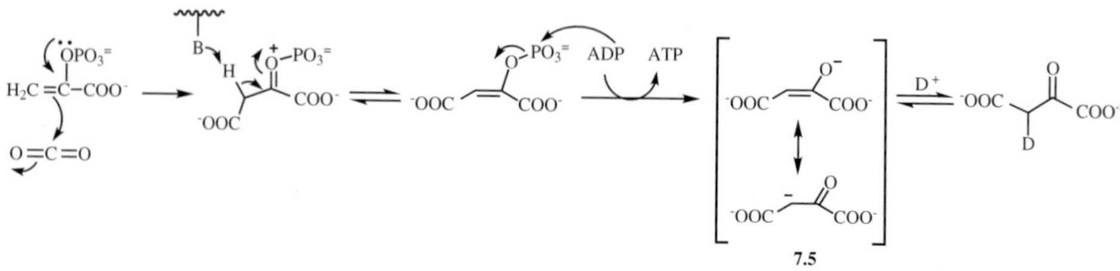

SCHEME 7.4   Hypothetical mechanism for PEP carboxykinase that involves the enolate of oxaloacetate.

nase is added to reduce the oxaloacetate as soon as it is generated and to prevent both nonenzymatic decarboxylation and exchange of the oxaloacetate methylene protons, which are somewhat acidic because they are adjacent to a carbonyl group. This experiment excludes a mechanism that involves the formation of the enolate of oxaloacetate (**7.5**, Scheme 7.4), unless the enolate is protonated in the active site by an acid residue that cannot exchange with the solvent. Actually, the possibility of a proton not exchanging with the medium is not such a crazy notion. Many enzymes meet this criterion.[4] As mentioned in Chapter 3 (Section II), the proton donor group may be buried in a hydrophobic pocket of the active site to which the solvent has no access.

The following experiment by Frey and co-workers excluded a covalent catalytic mechanism for PEP carboxykinase to give a phosphoenzyme.[5] (R_P)-Thio-GTP containing a chiral γ-thiophosphate group (**7.6**, Scheme 7.5; the subscript P refers to the stereochemistry around the phosphorus atom) was incubated with PEP carboxykinase in the presence of oxaloacetate; the thiophosphoenolpyruvate (**7.7**) that was isolated had S_P stereochemistry. Inversion of stereochemistry means that there cannot be phosphorylation of the enzyme followed by GDP displacement of the phosphate group from the enzyme (Scheme 7.6), because that would be two inversions of stereochemistry, and the product would retain its stereochemistry (i.e., R_P). Direct phosphorylation of GDP would give inversion of stereochemistry.

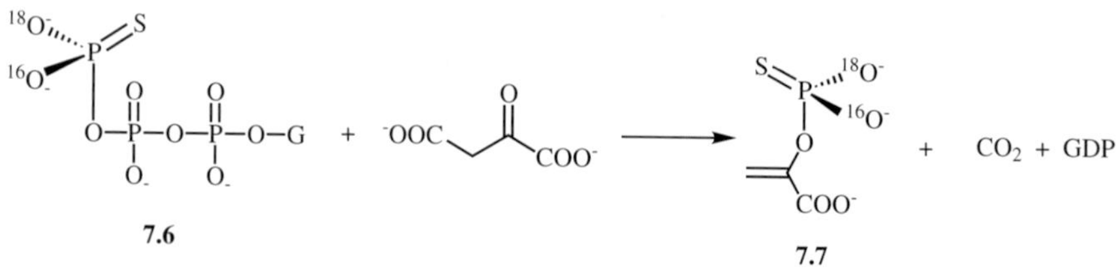

SCHEME 7.5   Stereochemistry of the reaction catalyzed by PEP carboxykinase.

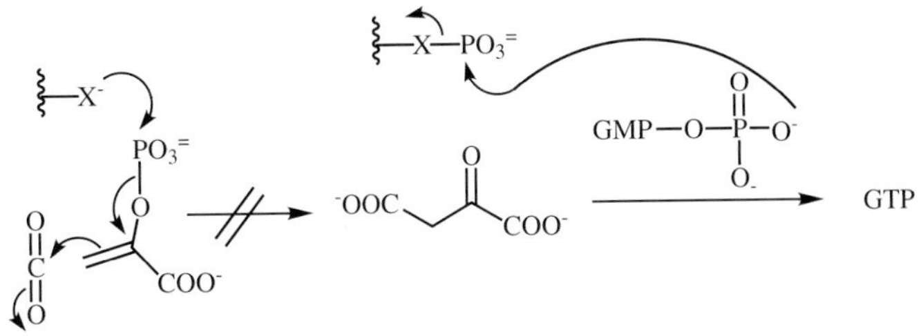

SCHEME 7.6   Inconsistent double-inversion mechanism for PEP carboxykinase.

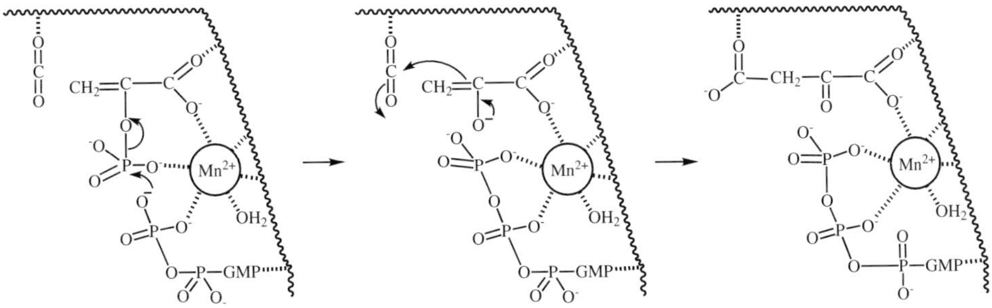

SCHEME 7.7   Concerted mechanism for PEP carboxykinase.

2. Mechanism

On the basis of the mechanistic evidence just presented, a concerted mechanism, or more likely a stepwise mechanism (Scheme 7.7) without release of intermediates, is most reasonable. The D268N mutant of the *Escherichia coli* enzyme, run in the reverse direction, retains the decarboxylation of oxaloacetate activity, but phosphoryl transfer is abolished,[6] supporting a stepwise process. Evidence for a binding site for the CO_2 comes from a carbon isotope effect study, by Arnelle and O'Leary, on carboxylation; the isotope effect increases with decreasing PEP concentration.[7] This indicates that CO_2 must bind prior to PEP binding; otherwise, varying the PEP concentration would not affect C–C bond formation. Various residues at the active site have been identified from crystal structures solved at 1.8 Å resolution as being important to catalysis.[8]

B. Phosphoenolpyruvate Carboxytransphosphorylase

1. Evidence for Mechanism

This enzyme catalyzes the same reaction as PEP carboxykinase, except that inorganic phosphate is the donor instead of a nucleotide diphosphate, and diphosphate (PP_i) is the product instead of nucleotide triphosphate (Scheme 7.8).

$$PEP + CO_2 + P_i \xrightarrow{\text{Mn}^{2+} \text{ or } Mg^{2+}} OAA + PP_i$$

SCHEME 7.8   Reaction catalyzed by PEP carboxytransphosphorylase.

Essentially the same results have been obtained with this enzyme as with PEP carboxykinase. No ^{18}O from $H_2^{18}O$ is incorporated into product (therefore CO_2 is the carboxylating agent). In the absence of CO_2, a proton can accept the proposed enol pyruvate intermediate to give pyruvate and diphosphate. Protonation, however, is not stereospecific, which indicates that in the absence of CO_2 the enolpyruvate is released from the enzyme prior to protonation; stereospecific protonation would have indicated an enzyme-catalyzed protonation. No partial isotope exchange reactions are observed, suggesting either a concerted reaction or retention of the bound initial product until after the remaining substrates bind to the active site and react to give products. A nonconcerted version of the carboxylation mechanism is shown in Scheme 7.9. So, why should two similar enzyme-catalyzed reactions, PEP carboxykinase and PEP carboxytransphosphorylase, be discussed? There are two reasons: to show concerted and nonconcerted versions of the same mechanism, and to introduce another typical stereochemical problem with its solution. The problem is determining the stereochemistry of carboxylation. To understand the stereochemical outcome, we need to discuss one more stereochemical principle, this time dealing with substituted alkenes. In Chapter 3 (Section III.A.3) we discussed rules for naming the stereochemistry of simple alkene addition reactions in which one face is the *re* face and the other is the *si* face. What if you have a di-, tri-, or tetra-substituted alkene? Consider (Z)-1-bromo-1-propene (**7.8**, Figure 7.2), and apply the following nomenclature rules.[9]

Assign priorities to the groups attached to each carbon of the double bond as in the R/S system. If the priority sequence from highest to lowest is clockwise from

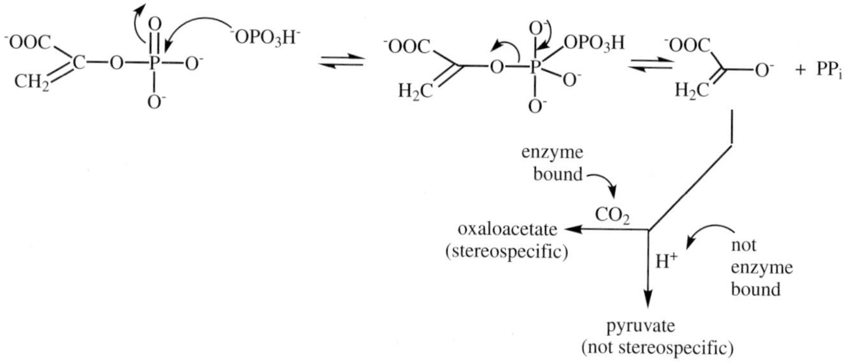

SCHEME 7.9   Nonconcerted mechanism for PEP carboxytransphosphorylase.

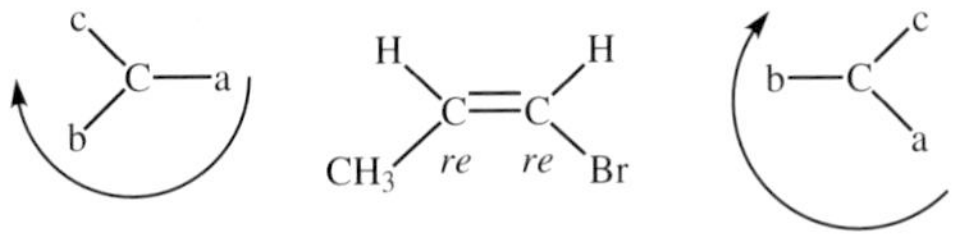

7.8

FIGURE 7.2   Alkene stereochemistry nomenclature rules for (Z)-1-bromo-1-propene (**7.8**).

the top, then the top is the *re* face for that atom; if the sequence is counterclockwise, the top is the *si* face for that atom. If the face is *re* at both of the trigonal atoms of the double bond, then this side is the *re-re* face, the other is the *si-si* face. This makes the top face for (Z)-1-bromo-1-propene, as drawn in **7.8**, the *re-re* face and the bottom face *si-si* (the priorities are shown as a > b > c). What if one atom is *re* and the other is *si*? Then, according to the rules, one face is the *re-si* face and the other is the *si-re* face. But which is which? Consider (E)-1-bromo-1-propene (**7.9**, Figure 7.3). To establish which side is *re-si* and which is *si-re,* disregard the double bond and assign priorities to the four substituents attached to the two trigonal carbon atoms. Cite first the side of the double bond that has the trigonal atom bearing the highest-priority ligand. In this case, viewed from the top, the left side is *si* and the right side is *re,* but that does not mean this is the *si-re* face (if I had drawn the same molecule with the methyl and bromine groups switched, it would still be (E)-1-bromo-1-propene, but the left side would be *re* and the right side *si*). The highest-priority group is the bromine, so no matter how we draw the compound, the carbon with the bromine is named first. Therefore, drawn as **7.9**, the top face is the *re-si* face.

Getting back to the question of stereochemistry of carboxylation of PEP carboxytransphosphorylase, let's consider the two possible outcomes, namely, top face and bottom face addition of CO_2. To differentiate these faces, we need to produce a stereogenic center at the carboxylation site. This will not occur unless the two methylene hydrogens of PEP are different. Labeling one of the hydrogens with an isotope does not change the chemistry but does provide a means of producing a stereogenic product. Scheme 7.10 shows both carboxylation outcomes starting with (Z)-[3-^{3}H]PEP (**7.10**) as substrate.[10] The general approach is to run the carboxylation reaction in the presence of other enzymes, whose stereochemical outcomes are already known, which can convert into a stable known compound the

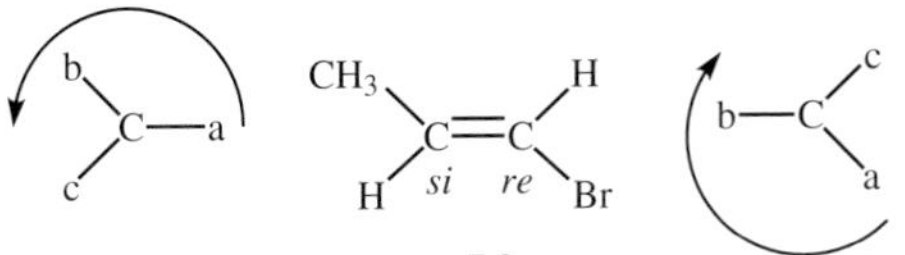

7.9

FIGURE 7.3   Alkene nomenclature rules for (E)-1-bromo-1-propene (**7.9**).

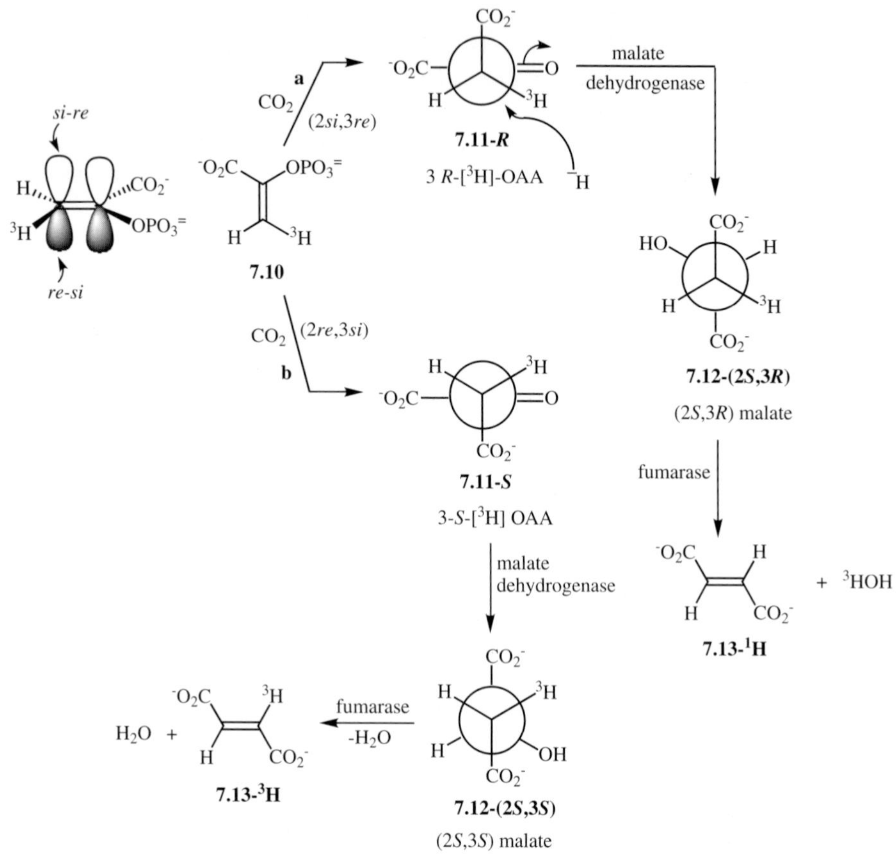

SCHEME 7.10   Two possible stereochemical outcomes for carboxylation of PEP catalyzed by PEP carboxytransphosphorylase.

enzymatic product of the enzyme reaction whose stereochemistry is unknown. The stereochemistry of the stable product can then be determined. In this case, malate dehydrogenase and fumarase were added to the PEP carboxytransphosphorylase enzyme reaction. Malate dehydrogenase, a NADH–dependent enzyme (see Chapter 3, Section III), reduces the ketone carbonyl of oxaloacetate (**7.11**) stereospecifically to 2S–malate (**7.12**), and fumarase is known to catalyze an *anti*–elimination of H_2O from malate to fumarate (**7.13**). The malate dehydrogenase is added to prevent nonenzymatic washout of the tritium from oxaloacetate (the tritium is adjacent to a carbonyl) and epimerization at the 3-position. If carboxylation occurs at the 2-S,3-R face (pathway a; top face), then the oxaloacetate produced will be 3R–[3-^{3}H]OAA (**7.11**-R). Malate dehydrogenase reduction would give the (2S, 3R)–[3-^{3}H]malate (**7.12**-(2S, 3R)). To get *anti*–elimination and still produce fumarate, the tritium has to be *anti* to the hydroxyl group. Therefore, elimination releases the tritium as ^3H_2O and produces fumarate without a tritium (**7.13**-^1H). Following *re-si* face addition (pathway b, bottom face) through the same sequence gives (2*S*,3*S*)-[3-^3H]malate, or **7.12**-(2*S*, 3*S*); elimination by fumarase puts the protium *anti* to the hydroxyl group and produces H_2O and tritiated fumarate (**7.13**-^3H). A stereochemical experiment should never be carried out on just one isomer; for consistency of results, both isomers must be used. Starting with (*E*)-[3-^3H]PEP, the opposite result should be obtained if the two isomers bind the same. With (*Z*)-[3-^3H]PEP (**7.10**), 98% of the tritium is released as ^3H_2O; with (*E*)-[3-^3H]PEP, 98% of the tritium remains in the fumarate. Working back from the product, that translates into carboxylation from the *si* face of PEP (pathway a). Therefore, the PEP binds to the enzyme so that the *re* face is inaccessible to the CO_2 (presumably blocked by the enzyme). When this type of experiment is carried out with other PEP carboxylases, the same result—that is, addition to the *si* face of PEP—occurs, which may be an evolutionary preference.

Note that in PEP carboxylations, the P–O bond of PEP is broken, but in the EPSP synthase- and MurA-catalyzed reactions of PEP with shikimate-3-phosphate and UDP-*N*-acetylglucosamine, respectively (see Chapter 6, Section VI), it was the C–O bond of PEP that broke. This indicates the versatility of PEP in enzyme-catalyzed reactions.

C. Vitamin K-Dependent Carboxylation of Proteins

Here is another excellent example of how organic chemical model studies can shed light on an unusual enzymological mechanistic problem and lead to a series of enzymological experiments that support the mechanistic hypothesis.[11] Vitamin K_1 is an obligatory substrate for a carboxylase that activates proteins in the blood-clotting cascade. Carboxylation converts selected glutamate residues in the blood-clotting cascade proteins to γ-carboxyglutamate (Gla) residues (Scheme 7.11), which enable the proteins to bind calcium. The bound calcium ions interact with the phosphate groups of phospholipids on the membrane surfaces of blood platelets, endothelial cells, and vascular cells, thereby holding the blood-clotting proteins, which are proteases, to the cell surfaces (Figure 7.4). Binding of the blood-clotting proteins to the membrane surfaces organizes the proteins in the clotting cascade to promote cleavage of the blood-clotting zymogens to their enzyme active forms. For example, carboxylation of prothrombin leads to binding of the Ca^{2+}–prothrombin complex to factors Va and Xa at the membrane surface. Cleavage of prothrombin by factor Xa produces thrombin, which cleaves fibrinogen to fibrin, thereby clotting the blood.

The unusual aspect of this carboxylation reaction is that carboxylation requires a second substrate, namely, reduced vitamin K_1 (**7.15,** 2-methyl-3-phytyl-1,4-naphthohydroquinone; Scheme 7.11), which is generated from vitamin K_1

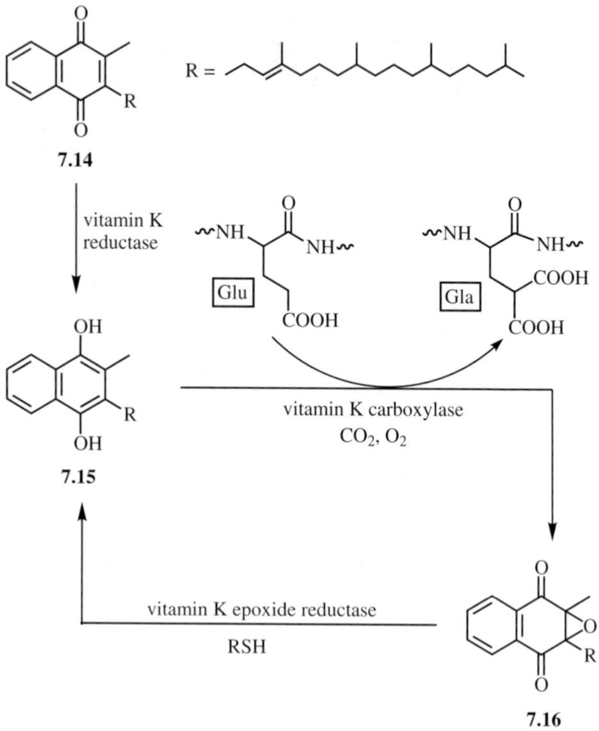

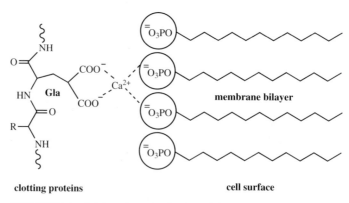

SCHEME 7.11   Vitamin K cycle for carboxylation of proteins.

(**7.14**) by the enzyme vitamin K reductase. Reduced vitamin K_1 is converted to vitamin K_1 2,3-epoxide (**7.16**), in the presence of CO_2 and O_2, concomitant with carboxylation of the clotting factors' Glu residues. Vitamin K epoxide reductase, which catalyzes the reduction of **7.16** back to **7.15,** appears to require only a thiol.

FIGURE 7.4   Calcium-dependent binding of clotting proteins to cell surfaces.

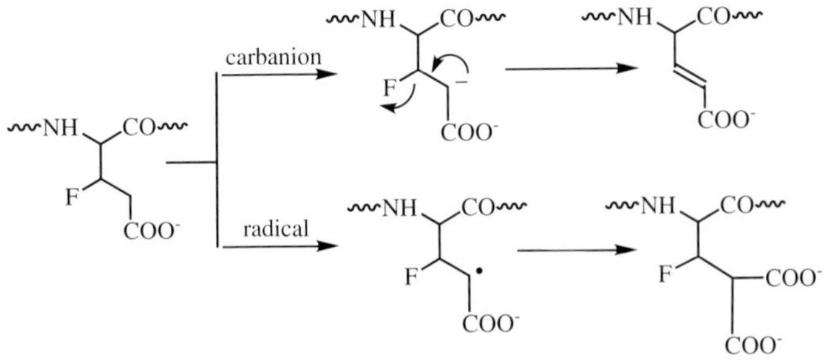

SCHEME 7.12   Carbanion versus radical mechanisms for vitamin K carboxylase.

Consider a carbanion and a radical mechanism for carboxylation of the glutamate residues. To differentiate these mechanisms, Marquet and co-workers[12] synthesized (erythro)- and (threo)-3-fluoroglutamate, incorporated them into a peptide known to be a substrate for vitamin K carboxylase, then looked for enzyme-catalyzed release of fluoride ion (Scheme 7.12). If a carbanion is generated at C-4, then the fluoride ion should be eliminated; if a radical mechanism is involved, no fluoride elimination should occur. In fact, the erythro isomer resulted in fluoride ion elimination, but no fluoride ion elimination was observed with the threo isomer. An enzyme-catalyzed deprotonation is expected to be stereospecific, so the fact that only one diastereomer leads to elimination is consistent with this stereospecific (enzyme-catalyzed) reaction.

Because a carbanion retains its stereochemical integrity (unless it is resonance stabilized), a carbanionic intermediate should lead to a product with either inversion or retention of stereochemistry, whereas a radical, which loses its stereochemical integrity when formed, would give a racemic product. (2S,4R)-4-Fluoroglutamate was incorporated into a pentapeptide (**7.17**, Scheme 7.13), which was incubated with vitamin K carboxylase, and the product stereochemistry was determined by chemical modification and mass spectrometry.[13] Carboxylation proceeded stereospecifically with inversion of configuration, which suggests the mechanism shown in Scheme 7.14. This is further evidence for a carbanionic intermediate.

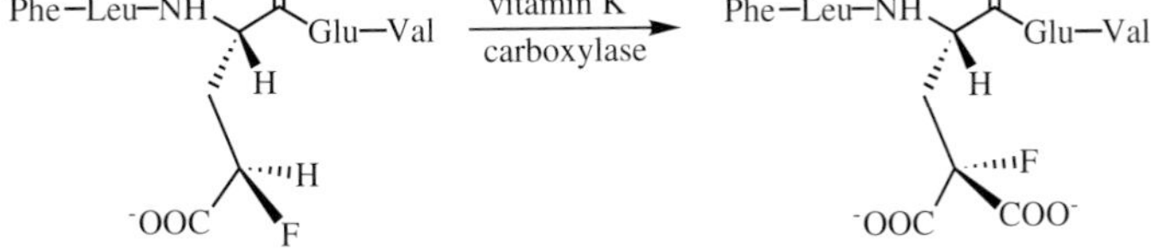

7.17

SCHEME 7.13   Stereochemical outcome of vitamin K carboxylase-catalyzed carboxylation of (2S,4R)-4-fluoroglutamate.

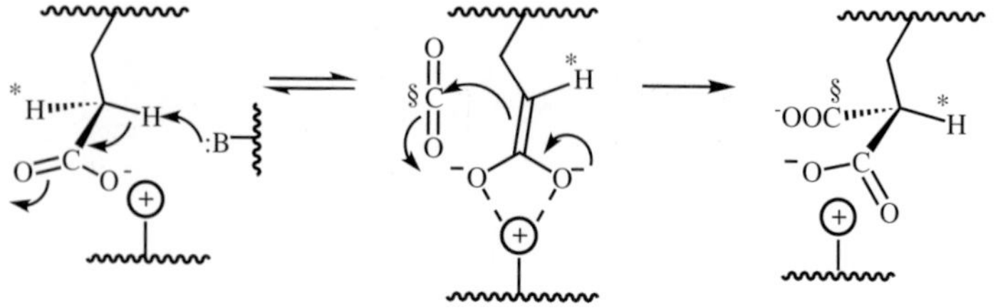

SCHEME 7.14   Proposed vitamin K carboxylase-catalyzed carboxylation of glutamate residues via a carbanionic intermediate.

The only problem with the mechanism in Scheme 7.14 is that reduced vitamin K_1 is required for activity, and this cosubstrate does not appear in the mechanism. Here is where the organic model studies become important. Removal of the C-4 proton of glutamate residues is not easy because of the relatively high pK_a of that proton—estimated to be about 22.[14] Dowd and co-workers[15] proposed that the function of the reduced vitamin K_1 is to act as a strong base, which removes the C-4 proton from the glutamate residues. However, it is not evident from the structure of reduced vitamin K_1 how it would act as a strong base. As a chemical model, phenolate **7.18** (a model for reduced vitamin K_1) was allowed to react, in the presence of oxygen, with diethylhexane-1,6-dioate (**7.20**, Scheme 7.15). The Dieckmann condensation product (**7.21**) was produced. The mechanism shown in Scheme 7.15 is consistent with this reaction, given that no Dieckmann condensation occurs in the absence of oxygen. It was proposed that the relatively weak base **7.18** is transformed into a strong base (**7.19**), which is responsible for deprotona-

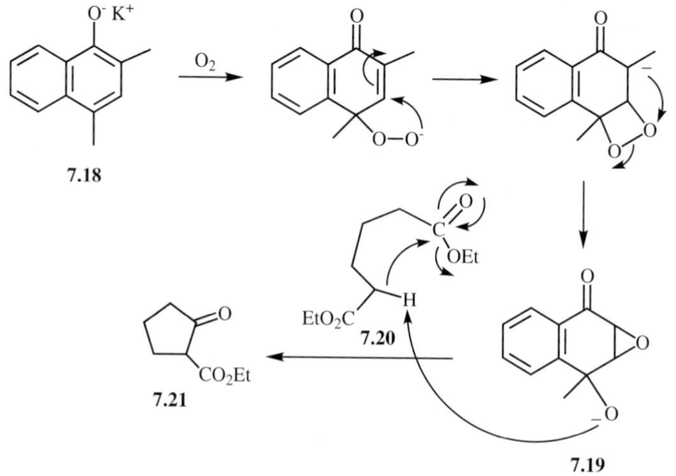

SCHEME 7.15   Chemical model study for the activation of vitamin K_1 as a base.

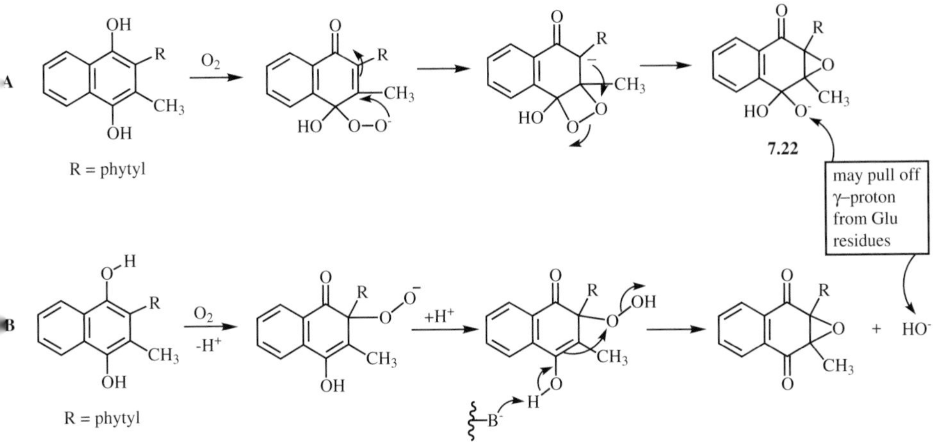

SCHEME 7.16   Proposed mechanisms for activation of vitamin K_1 as a base.

tion of **7.20.** This model suggests that the function of the reduced vitamin K_1 in the vitamin K carboxylase reaction may be to generate a base strong enough to deprotonate the C-4 proton of the glutamate residues, and supports **7.22** (Scheme 7.16A) as that base. This process of conversion of a weak base into a strong base has been termed the *base strength amplification mechanism*. An alternative mechanism that generates a potentially strong base (hydroxide ion), particularly in a hydrophobic environment, is shown in Scheme 7.16B.[16]

Once plausible mechanisms are hypothesized and model reactions are run to support the chemistry, the next step is to test the hypothesis in the enzyme system. The two mechanisms in Scheme 7.16 could be differentiated by an appropriate labeling experiment. Walsh and co-workers ran the vitamin K carboxylase reaction under an atmosphere of $^{18}O_2$, and 0.95 mol atom of ^{18}O was detected in the epoxide of the resultant vitamin K epoxide, as predicted from both of the mechanisms in Scheme 7.16, but 0.05 mol atom of ^{18}O also was detected in the quinone oxygen.[17] No ^{18}O is incorporated into vitamin K epoxide when $H_2^{18}O$ replaces H_2O and $^{16}O_2$ is utilized in place of $^{18}O_2$. The observation of even 0.05 atom ^{18}O in the quinone supports the Dowd mechanism (Scheme 7.16A). Dowd and co-workers[18] also carried out this same experiment and found 0.17 mol atom of ^{18}O incorporated into the quinone carbonyl in addition to the 1 atom of ^{18}O in the epoxide. This result suggests that most of the ^{18}O is washed out in the process, which is reasonable, because the ketal in **7.22** would eliminate hydroxide in conversion to the ketone carbonyl. A model reaction involving treatment of vitamin K hydroquinone monoanion with $^{18}O_2$ in THF gave vitamin K epoxide with one atom of ^{18}O at the epoxide oxygen and 0.34 atom of ^{18}O at the carbonyl oxygen, again supporting the dioxetane mechanism (Scheme 7.16A).[19]

The converse experiment also was carried out to see if washout occurs from the

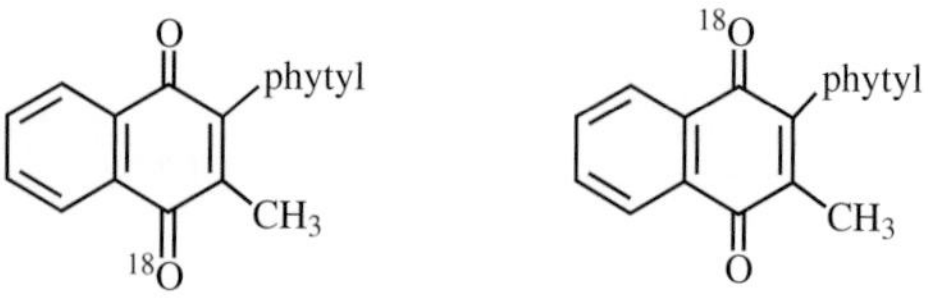

7.23　　　　　　　　**7.24**

vitamin K quinone and to determine which ketone, the one next to the methyl group or the one next to the phytyl group, is involved in the enzyme-catalyzed reaction. Incubation of rat liver microsomes under an atmosphere of $^{16}O_2$ with vitamin K regiospecifically labeled with ^{18}O in each carbonyl (**7.23** and **7.24**) resulted in the loss of 0.17 mol atom of ^{18}O from the oxygen adjacent to the methyl group and no loss of ^{18}O from the oxygen next to the phytyl group.[20] Therefore, initial electrophilic attack of the vitamin K hydroquinone by O_2 occurs at the carbon next to the methyl group.

The $\Delta H°$ for conversion of reduced vitamin K_1 to vitamin K_1 epoxide was estimated to be -57.5 kcal/mol by a calorimetric measurement of the heat of oxygenation of potassium-2,4-dimethylnaphthoxide (**7.25**, Scheme 7.17).[21] This result confirms the earlier prediction by Dowd and co-workers of -62 kcal/mol for this reaction,[22] and is consistent with the proposal that the reaction with O_2 can convert a weakly basic anion of reduced vitamin K_1 into a strong base.

Because most of the ^{18}O is released during the enzyme-catalyzed reaction, Dowd and co-workers modified the base strength amplification mechanism.[23] In the modified version (Scheme 7.18), a weak base (an active-site cysteine residue) removes the hydroquinone proton from reduced vitamin K, which reacts with O_2, leading to the strong base, the ketal anion (**7.26a** and/or **7.26b**). But this anion is not the base that removes the glutamate proton; rather, the elimination product, hydroxide ion, is proposed to be the strong base involved (as in Scheme 7.16B also). Therefore, the function of vitamin K appears to be to convert O_2 into hydroxide ion in a hydrophobic environment (vitamin K is very hydrophobic) where it can deprotonate the glutamate residues. This would be even more effective than aque-

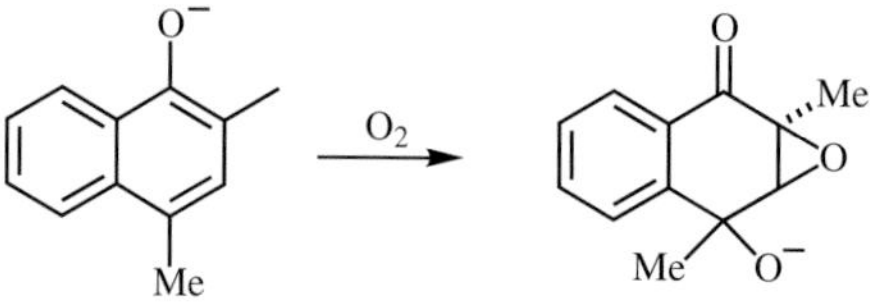

7.25

SCHEME 7.17　　Chemical model to estimate the $\Delta H°$ for conversion of reduced vitamin K_1 to vitamin K_1 epoxide.

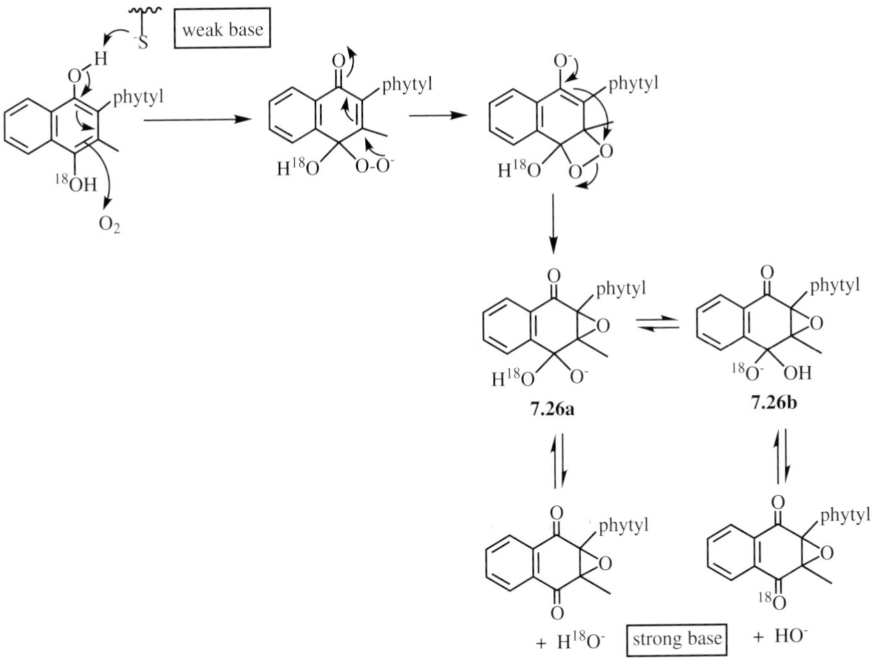

SCHEME 7.18   Modified base strength amplification mechanism for vitamin K carboxylase.

ous hydroxide, because bases are known to be stronger in hydrophobic solvents than in aqueous media.[24]

III. BICARBONATE AS THE CARBOXYLATING AGENT

A. Phosphoenolpyruvate Carboxylase

Phosphoenolpyruvate (PEP) carboxylase requires Mg^{2+}, and the carboxylating agent is HCO_3^-. The carboxylating agent was determined by incubation of the enzyme with PEP and $HC^{18}O_3^-$ and showing that no $H_2^{18}O$ is formed. If CO_2 were the carboxylating agent, the $HC^{18}O_3^-$ would have had to lose $H^{18}O^-$, and this would have been detected as $H_2^{18}O$. Of course, this experiment cannot be carried out under normal conditions. It has to be run with a high enzyme concentration for a short incubation time and at alkaline pH so that the bicarbonate is not spontaneously converted to CO_2 and HO^-. Furthermore, when $HC^{18}O_3^-$ is used, the oxaloacetate product (**7.27**) contains two ^{18}O atoms in the carboxylate group, and the inorganic phosphate contains one ^{18}O atom (Scheme 7.19).[25]

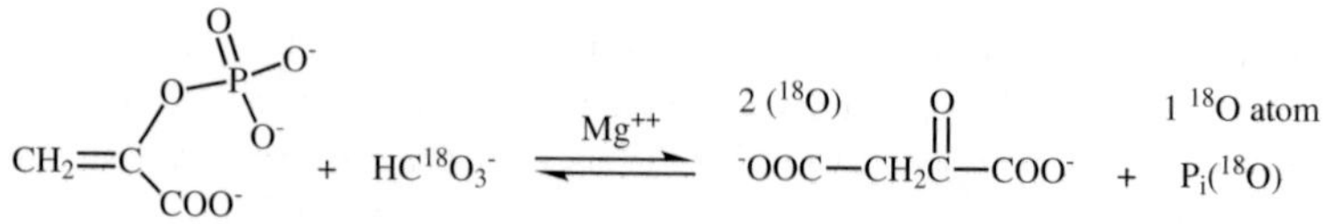

SCHEME 7.19   Reaction catalyzed by PEP carboxylase.

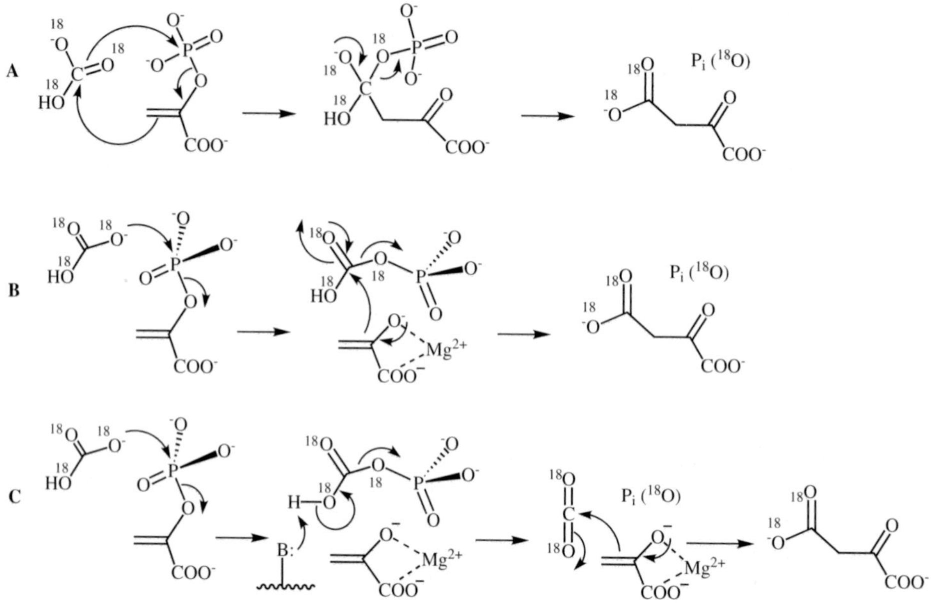

SCHEME 7.20   Concerted (**A**), stepwise associative (**B**), and stepwise dissociative (**C**) mechanisms for PEP carboxylase.

Consider three mechanisms, one concerted (**A**, Scheme 7.20), one stepwise associative (**B**), and one stepwise dissociative (**C**). No partial exchange reactions have been detected; for example, incubation in the presence of [^{14}C]pyruvate does not lead to incorporation of ^{14}C into the oxaloacetate (or the PEP). The conclusion was that this is a concerted reaction, because the stepwise reactions have enolpyruvate as the intermediate. However, a tightly bound intermediate would not exchange with added compounds, so no conclusion can be made on the basis of this experiment alone.

There appears to be much more evidence in favor of a stepwise mechanism. O'Leary and co-workers carried out an isotope effect study at various pH values and found that the carbon isotope effect for HCO_3^-, k^{12}/k^{13}, is 1.0029 ± 0.0005.[26] The precision of this result may surprise you. Can this experiment really be accurate to the fourth decimal place? Actually, it can. The numbers are determined by an isotope ratio mass spectrometer, which measures the relative abundance of $^{13}C^{16}O_2$ and $^{12}C^{17}O^{16}O$ to $^{12}C^{16}O_2$, and this instrument is capable of that kind of accuracy. The rate of carboxylation was found to be independent of pH (7.5–10), but the isotope effect decreased with increasing pH. This is not possible with a concerted process in which everything happens in one step, and the pH affects both the rate and the isotope effect. Furthermore, for other concerted processes k^{12}/k^{13} has been found to be >1.03, which is quite different from this experimental value of 1.0029.

In addition to the carboxylation of PEP, PEP carboxylase catalyzes the hydrolysis of PEP (to pyruvate and P_i).[27] The rate of hydrolysis is metal dependent. The active-site metal was hypothesized to bind to the incipient enolate oxygen, thus stabilizing the proposed enolate intermediate. The more stablized the enolate, the more hydrolysis was observed, because the intermediate was less reactive toward carboxylation. Stabilization of an enolate intermediate suggests a stepwise mechanism.

A stereochemical experiment by Hanson and Knowles also suggests a stepwise mechanism.[28] $[S-^{16}O,^{17}O]$Thiophosphoenolpyruvate (**7.28**) was substituted for PEP as the substrate, the reaction was performed in $H_2^{18}O$, and the stereochemistry of the inorganic $[^{16}O,^{17}O,^{18}O]$thiophosphate (**7.29**) product was determined. The concerted mechanism consists of a six-electron, *suprafacial* sigmatropic rearrangement, which is known to proceed with retention of configuration. The stepwise mechanisms involve a direct S_N2 displacement of P_i (inversion of configuration). The product was found to have complete inversion of stereochemistry (as is shown in **7.29**), supporting a stepwise mechanism.

Because the stereochemistry requires that the transition state is linear at phosphorus, the carbonyl carbon in the intermediate carboxyphosphate following transfer is quite far from C-3 of the enolate; a conformational change would be needed to place the two carbon atoms near each other. The most reasonable way to accomplish this is for the carboxyphosphate to decompose by base-catalyzed deprotonation to carbon dioxide and inorganic phosphate (mechanism C, Scheme 7.20).[29] This brings the CO_2 above the plane of the metal ion–stabilized enolate and within bonding distance of C-3. Further evidence for the dissociative mechanism (mechanism C) is that when the enzyme reaction is carried out with methylPEP and $HC^{18}O_3^-$, more than one atom of ^{18}O is incorporated into the P_i produced,[30] and substrate recovered after partial reaction has ^{18}O in the nonbridging positions of the phosphate group;[31] this isotope scrambling suggests reversible formation of CO_2 and P_i at the active site.

7.28   **7.29**

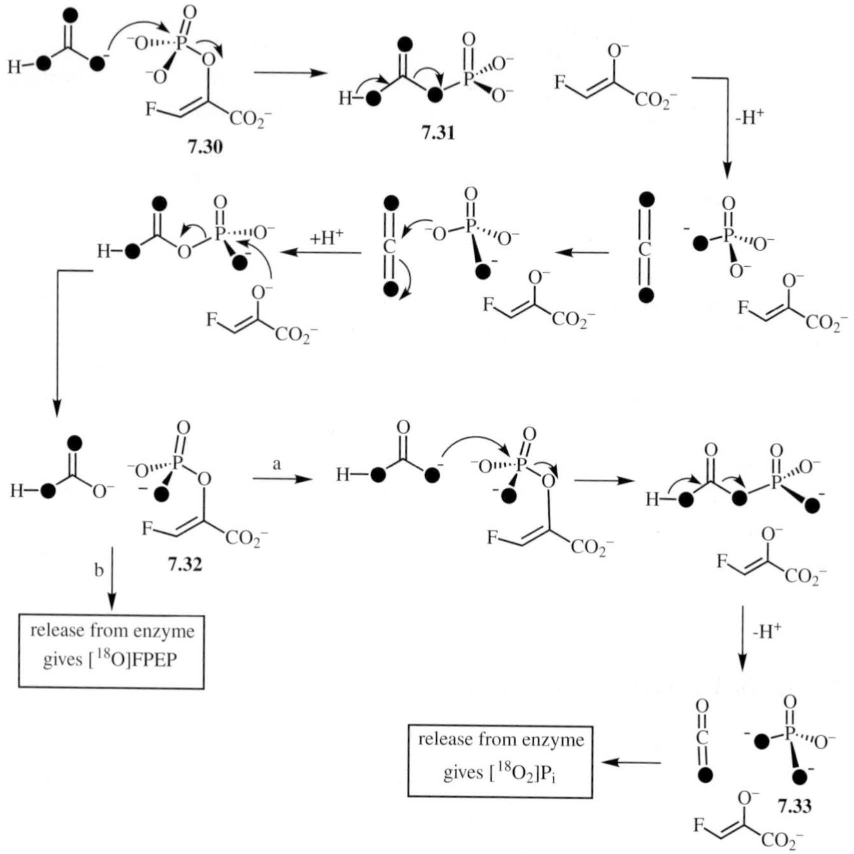

SCHEME 7.21   Further evidence for a stepwise mechanism for PEP carboxylase.

One last (rather sophisticated) experiment, which demonstrates the stepwise dissociative nature of the mechanism for PEP carboxylase, stems from the observation that (E)- and (Z)-3-fluorophosphoenolpyruvate (**7.30**, Scheme 7.21) are substrates for the enzyme.[32] Both isomers partition between carboxylation and hydrolysis; carboxylation is 3% with the Z-isomer and 86% with the E-isomer, and the percentage of carboxylation is metal dependent, as in the case of PEP. The different percentages of carboxylation for the isomers may result from hydrogen bonding to the F atom at the active site, which gives the enolate a different orientation toward carboxylation. Carboxylation occurs at the 2-*re* face, which corresponds to the 2-*si* face of PEP (remember, the fluorine atom switches the nomenclature, not the stereochemistry). When the enzyme is incubated with $HC^{18}O_3^-$, the ^{18}O is incorporated into residual (E)- and (Z)-3-fluoroPEP (**7.32**), and the inorganic phosphate product is multilabeled with ^{18}O (**7.33**). Because ^{18}O is incorporated into

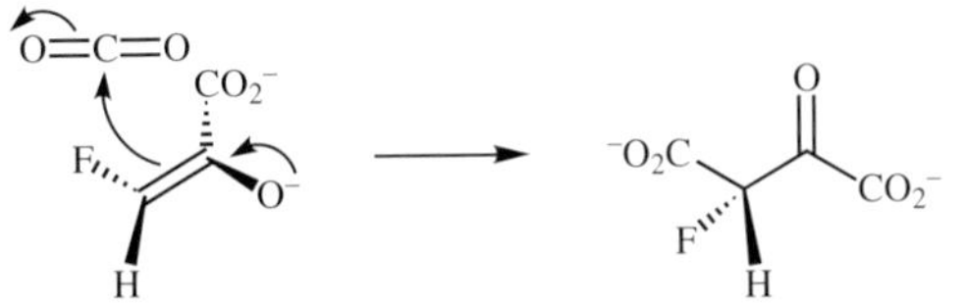

SCHEME 7.22   Reaction of the enolate of (E)-3-fluoroPEP with CO_2.

inorganic phosphate, carboxyphosphate (**7.31**) was likely to have been formed. The $^{13}(V/K)$ isotope effect on carboxylation is 1.009 ± 0.006 for the (E)-isomer and 1.049 ± 0.003 for the (Z)-isomer. The isotope exchange data and the isotope effect results are most consistent with attack of the enolate of 3-fluoropyruvate on CO_2 rather than on carboxyphosphate, as shown for the (E)-isomer in Scheme 7.22. If carboxylation occurred by attack on carboxyphosphate, the partitioning between carboxylation and hydrolysis would not follow an irreversible step, as was observed, and the isotope effect for the (E)-isomer would not be suppressed to the extent observed. The corresponding dissociative mechanism (Scheme 7.23), which bypasses the carboxyphosphate intermediate, also could be relevant.

B. Biotin-Dependent Carboxylases

For a relatively small number of multisubunit enzymes, d-biotin (**7.34**, Scheme 7.24) is the coenzyme that is required for catalytic carboxylation.[33] In all biotin-dependent carboxylases examined, the coenzyme is attached covalently to the active site by an amide linkage to the ϵ-amino group of a lysine residue (**7.35**). The long arm connecting the heterocyclic ring to the enzyme may be for mobility, to allow the biotin to reach into different subunits. The linkage reaction connecting the coenzyme to the protein requires ATP and gives AMP and PP_i. Synthetic biotinyl-AMP will replace biotin and ATP in the linkage reaction. Running the enzyme reaction with $HC^{18}O_3^-$ gives P_i with one ^{18}O atom and product with 2 ^{18}O atoms, supporting bicarbonate, not CO_2, as the substrate.[34]

1. Reactions Catalyzed by Biotin-Dependent Carboxylases

All of the reactions catalyzed by biotin-dependent enzymes (Figure 7.5) require bicarbonate and ATP for activation of the substrate. One diagnostic method for determining whether an enzyme contains biotin is to treat it with avidin, a protein from egg white that has a very high affinity for biotin ($K_D = 1.3 \times 10^{-15}$ M; the strongest interaction known between a ligand and a protein[35]), even when it is bound to the enzyme. If the enzyme activity is inhibited by avidin, biotin is likely required.

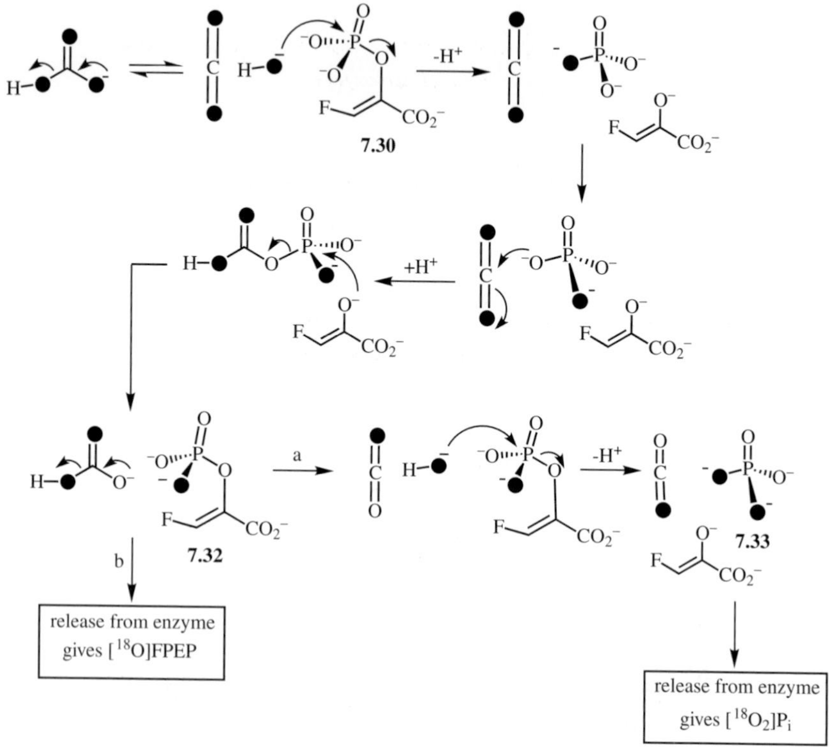

SCHEME 7.23   Alternative dissociative mechanism that accounts for the ^{18}O-containing products from the PEP carboxylase-catalyzed reaction of (Z)-3-fluoroPEP with $HC^{18}O_3^-$.

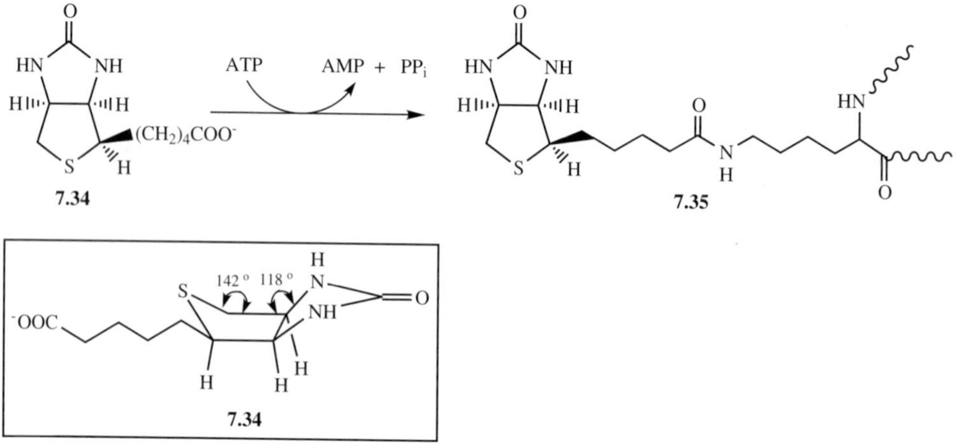

SCHEME 7.24   Covalent attachment of d-biotin to an active-site lysine residue.

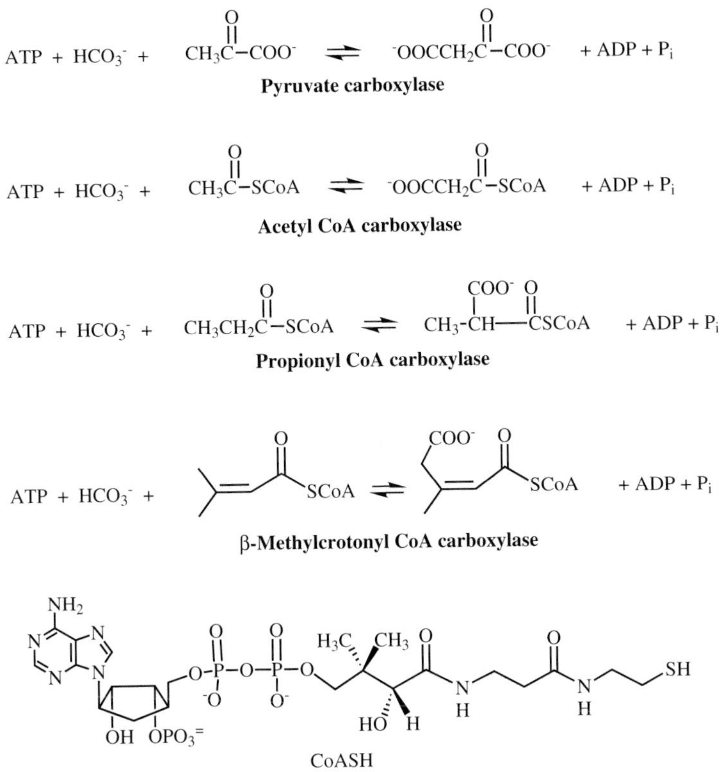

FIGURE 7.5   Reactions catalyzed by biotin-dependent carboxylases.

2. Mechanism of Biotin-Dependent Carboxylases

a. Partial Exchange Reactions

Exchange of $^{32}P_i$ with ATP requires the addition of ADP, HCO_3^-, and a divalent metal ion (M^{2+}), but not substrate or product (Scheme 7.25). This suggests that the ATP activates the bicarbonate, but it is not clear if biotin is involved in the activation. The simplest way to depict this activation is direct phosphorylation of bicarbonate (Scheme 7.26), but it may require biotin.

$$ATP + {}^{32}P_i \xrightleftharpoons[\quad\quad]{\substack{M^{+2} \\ HCO_3^- \\ ADP}} AT^{32}P + P_i$$

SCHEME 7.25   Partial exchange reaction of $^{32}P_i$ into ATP with biotin-dependent carboxylases.

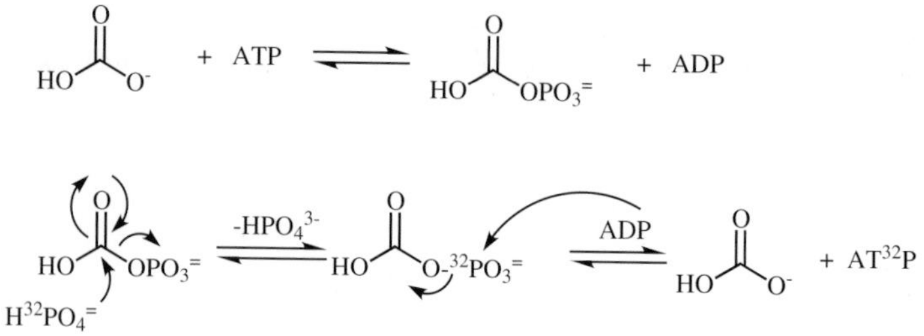

SCHEME 7.26   Mechanism for partial exchange of $^{32}P_i$ into ATP with biotin-dependent carboxylases.

Exchange of $[^{14}C]ADP$ with ATP requires HCO_3^- and M^{2+} (Scheme 7.27). Again, carboxyphosphate and ADP may be discrete intermediates, as shown in Scheme 7.28, but biotin may be involved instead, in which case these may not be discrete intermediates. So far, the results are ambiguous, and the picture may not get much clearer over the next several pages.

Incubation of the enzyme with $[^{14}C]$product in the presence of substrate, bicarbonate, ATP, and M^{2+} gives $[^{14}C]$substrate, indicating that the reaction is reversible.

b. Enzyme-Bound Intermediate

Incubation of pyruvate carboxylase with $H^{14}CO_3^-$, M^{2+}, and ATP in the absence of pyruvate gives a carboxylated enzyme; treatment of the carboxylated enzyme with pyruvate gives $[^{14}C]$oxaloacetate (Scheme 7.29). Up to this point, it is not clear if the X in this scheme is biotin or an enzyme nucleophile.[36] When $[^{14}C]$bicarbonate is utilized, the covalently bound ^{14}C is unstable to acid (pH 4.5), even at 0 °C; 97% of the bound ^{14}C is released as $^{14}CO_2$ in 25 minutes. However, Lynen and co-workers found that in 0.033 N KOH at 0 °C, only 7% of the ^{14}C is released.[37] The ^{14}C-labeled carboxylated enzyme was purified by gel filtration (see

$$[^{14}C]\text{-ADP} + \text{ATP} \xrightleftharpoons{HCO_3^-/M^{2+}} [^{14}C]\text{-ATP} + \text{ADP}$$

SCHEME 7.27   Partial exchange reaction of $[^{14}C]ADP$ into ATP with biotin-dependent carboxylases.

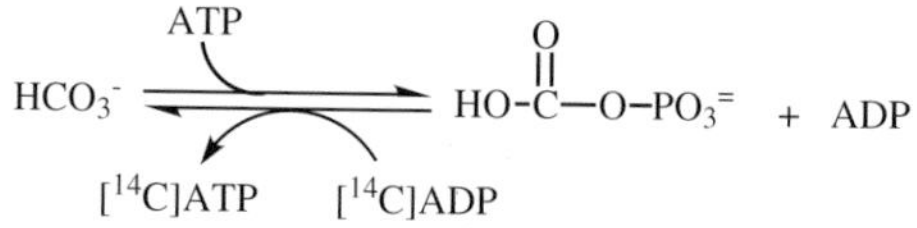

SCHEME 7.28   Mechanism for partial exchange reaction of $[^{14}C]ADP$ into ATP with biotin-dependent carboxylases.

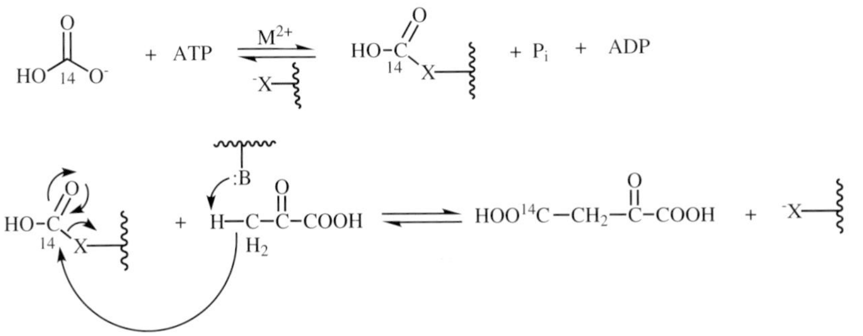

SCHEME 7.29   Pyruvate carboxylase-catalyzed incorporation of ^{14}C from $H^{14}CO_3^-$ into the enzyme.

Chapter 2, Section II.A.1.c) and was found to be stabilized by treatment with di-azomethane, which is utilized to convert carboxylic acids into methyl esters. The ^{14}C-labeled enzyme was then digested with two proteases, trypsin and papain, and the radioactivity was found to migrate with the biotin; treatment of the labeled bi-otin analogue with biotinidase, which catalyzes the hydrolysis of the biotin from the lysine residue, to which it is attached, gave N-methoxycarbonylbiotin (**7.36,** Scheme 7.30).[38] An X-ray crystal structure of **7.36** confirmed its identity.[39]

c. Possible Mechanisms for Formation of Carboxybiotin

Consider six mechanisms for the formation of carboxybiotin (Figures 7.6 and 7.7). The rapid exchange of the N—H protons of biotin indicates that initial forma-tion of the conjugate base of biotin is a reasonable intermediate in any mechanism that involves biotin carboxylation.[40]

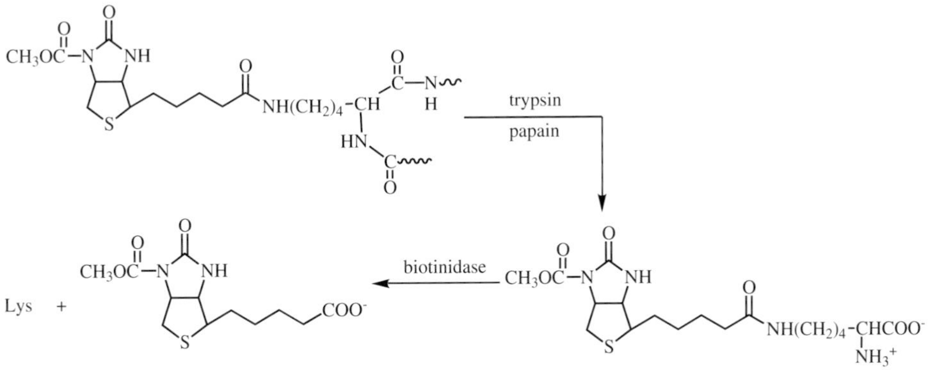

SCHEME 7.30   Isolation of N-methoxycarbonylbiotin from the reaction catalyzed by pyruvate car-boxylase followed by diazomethane trapping of the N-carboxybiotin.

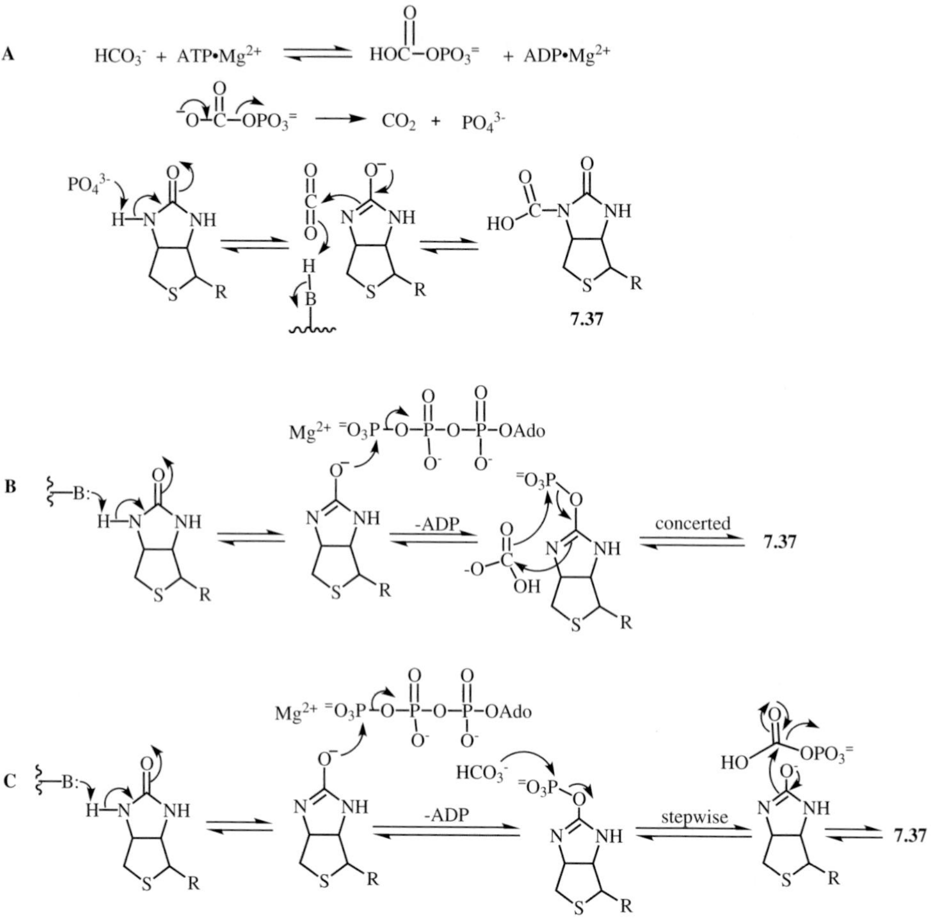

FIGURE 7.6   Three possible mechanisms for formation of N^1-carboxybiotin.

Several attempts to gain evidence for the formation of carboxyphosphate as an intermediate, which would support mechanisms **A** and **C** in Figure 7.6, were unsuccessful. Two experiments, however, with the biotin carboxylase subunit of acetyl-CoA carboxylase from *Escherichia coli*, the subunit responsible for the carboxylation of biotin, may provide support for the formation of carboxyphosphate. Climent and Rubio[41] observed that biotin carboxylase also catalyzes the hydrolysis of ATP (an ATPase activity) in the presence of bicarbonate but in the absence of biotin. This may be evidence for a partial reaction between ATP and bicarbonate, leading to carboxyphosphate, which then undergoes hydrolysis to bicarbonate and inorganic phosphate (Scheme 7.31). However, the rate of this reaction is only 0.5% that of the reaction in the presence of biotin, and, although this ATPase reaction possesses

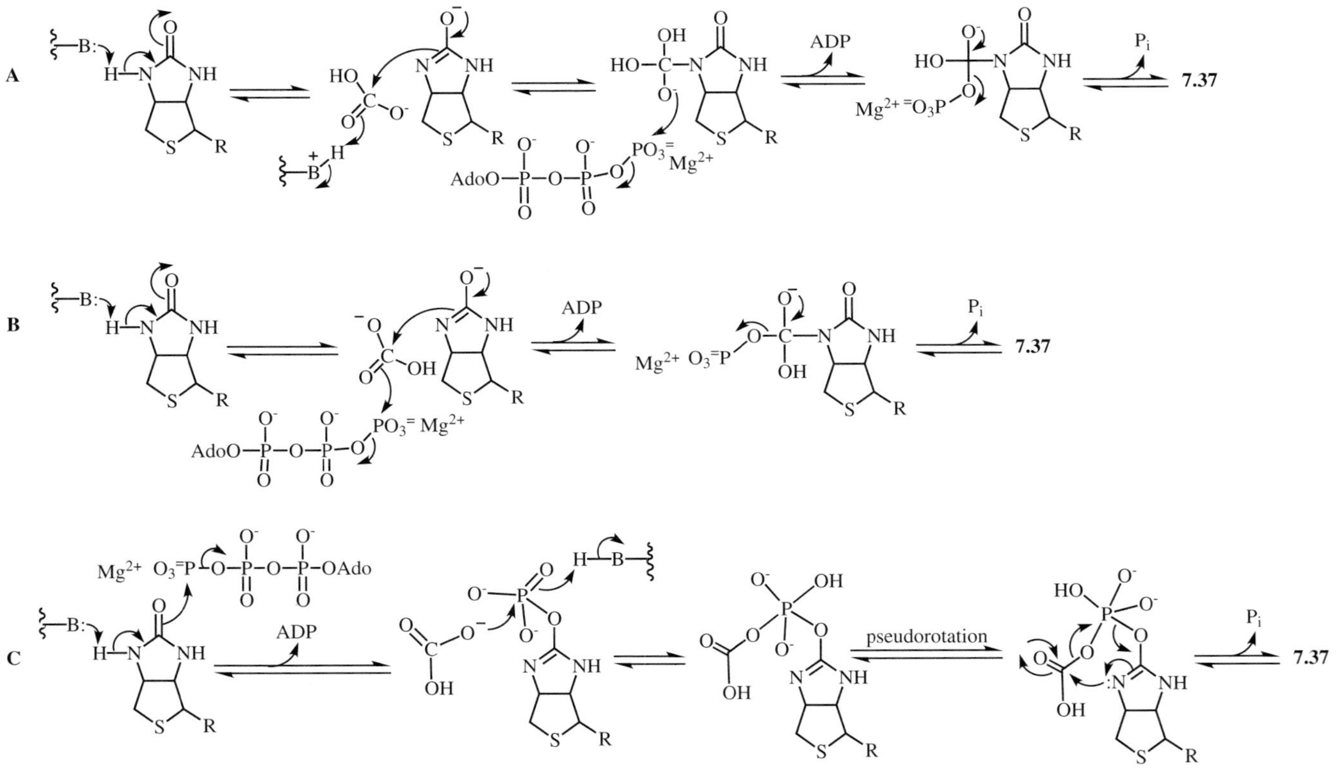

FIGURE 7.7   Three more mechanisms for the formation of N^1-carboxybiotin.

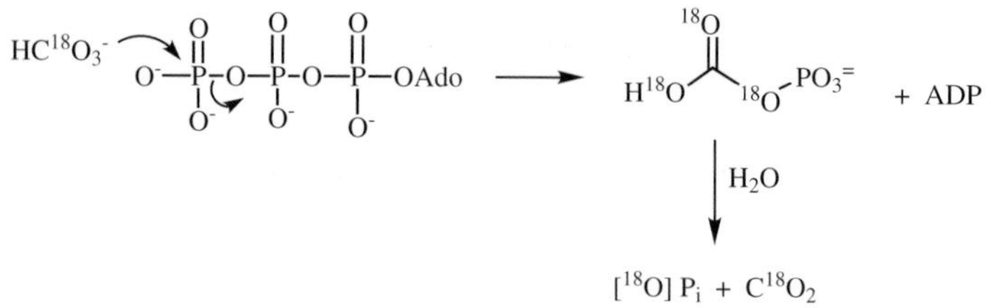

SCHEME 7.31 Mechanism for the formation of carboxyphosphate in the reaction catalyzed by acetyl-CoA carboxylase.

similar properties to those of the normal reaction, it may not be relevant to the reaction in the presence of biotin. The ATPase reaction also was carried out with [^{18}O]bicarbonate, and one ^{18}O atom was incorporated into the P_i produced.[42] Again, if this reaction is relevant to the biotin-dependent reaction, then it provides evidence for the intermediacy of carboxyphosphate.

Only two of the six mechanisms in Figures 7.6 and 7.7 have carboxyphosphate as an intermediate, mechanisms A and C in Figure 7.6. However, mechanism C does not generate the carboxyphosphate until after the biotin has reacted with the ATP. The experiments with biotin carboxylase just noted were carried out in the absence of biotin, so if they are relevant to the reaction in the presence of biotin, then mechanism C cannot be correct and only mechanism A (Figure 7.6) is applicable to the carboxylation of biotin.

But what if the reactions with the biotin carboxylase subunit are not relevant to the reaction in the presence of biotin? Elegant chemical model studies were carried out in support of mechanism A of Figure 7.7.[43,44] Kluger has provided a variety of cogent arguments to support all of the mechanisms in Figure 7.7 as being reasonable—that is, if carboxyphosphate is not relevant.[45]

d. Structure of Carboxybiotin

All six of the mechanisms in Figures 7.6 and 7.7 represent the carboxylated biotin as N^1-carboxybiotin (**7.37**, Figure 7.6); however, early chemical model studies did not support this structure. For example, there was no reaction of the cyclic urea **7.38** (Scheme 7.32) with reactive esters such as p-nitrophenyl acetate (**7.39**) or the acyl imidazolium **7.40**;[46] the reactive esters are intended to mimic carboxyphosphate. Therefore, a urea is apparently not a good nucleophile.

However, another model study, shown in Scheme 7.33,[47] utilized an intramolecular reaction of a urea with a carbonyl to increase reactivity. This reaction led to cyclization, but the more reactive carbonyl electrophiles (the phenyl esters) produced a urea oxygen attack (route b), whereas the less reactive carbonyls, methyl ester and amide (route a), gave the expected N-urea attack. Route a, nitrogen attack, represents the thermodynamic product, and route b leads to the kinetic prod-

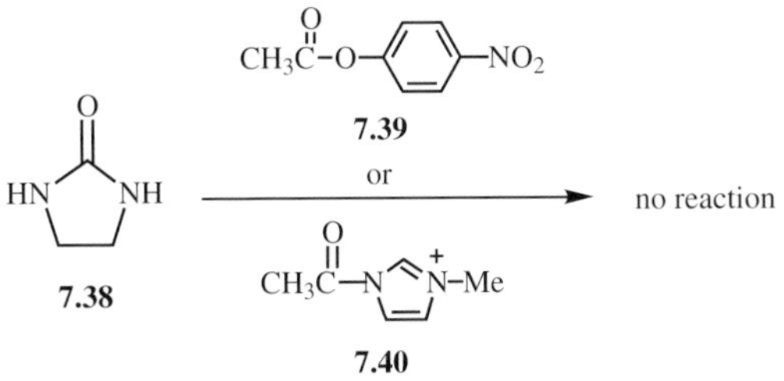

SCHEME 7.32   Failed chemical model study for the carboxylation of biotin.

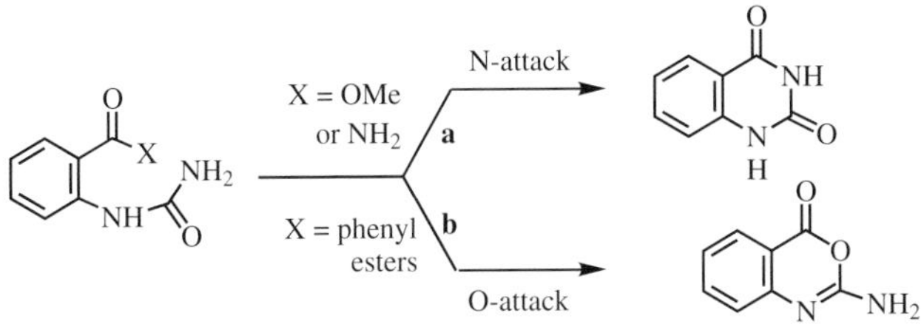

SCHEME 7.33   Chemical model study that suggests that *O*-carboxybiotin may be a viable intermediate.

uct. The fact that the more reactive carbonyls give acylation of the urea oxygen, rather than of the urea nitrogen, suggested that the same may be occurring with biotin-dependent enzymes, namely, that the biotin urea oxygen may be acylated by carboxyphosphate (that is, if carboxyphosphate is an intermediate) (Scheme 7.34)— an idea originally proposed by Lynen et al.[48] This notion would be intriguing, except that the biotin after bicarbonate carboxylation and diazomethane treatment had already been shown by X-ray crystallography to be N^1-methoxycarbonylbiotin (**7.36**, Scheme 7.30). So how can it be proposed, seven years after the structure was established, that the mechanism proceeds via *O*-acylation instead of *N*-acylation of the biotin? Bruice and co-workers proposed that N^1-methoxycarbonylbiotin was

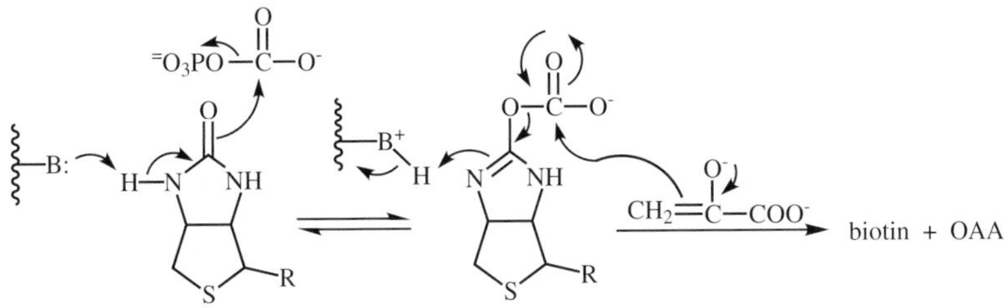

SCHEME 7.34   Possible mechanism for the biotin-catalyzed carboxylation of pyruvate via *O*-carboxybiotin.

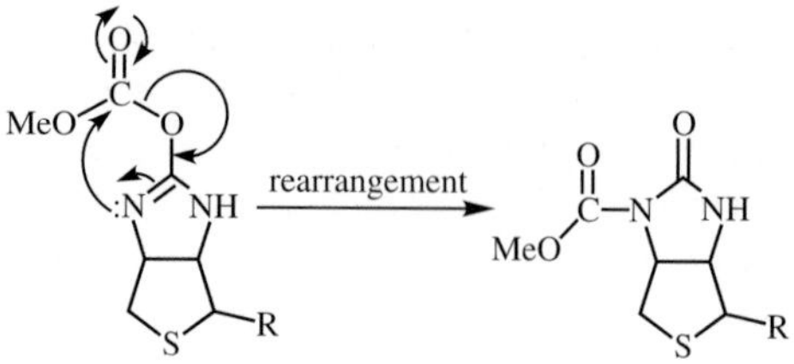

SCHEME 7.35   Alternative rationale for the formation of N-methoxycarbonylbiotin from O-carboxybiotin on diazomethane trapping of the carboxybiotin.

an artifact of the isolation reaction; upon diazomethane treatment, O-methoxycar-bonylbiotin (**7.41**) may have rearranged to the thermodynamically more stable N^1-methoxycarbonylbiotin (**7.36**, Scheme 7.35).

In response to this suggestion, Lane and co-workers carried out several clever experiments with acetyl-CoA carboxylase to discredit this hypothesis.[49] First, [^{14}C]carboxybiotin was isolated from the incubation of the enzyme with [^{14}C]bi-carbonate (remember, carboxybiotin is relatively stable in basic solution). A syn-thetic standard of N^1-carboxybiotin (**7.42**) was prepared by saponification of N^1-methoxycarbonylbiotin (**7.36**). The site of carboxylation of this standard must be at the nitrogen because the precursor was shown to be N-acylated by X-ray crys-tallography, and it cannot rearrange back to the oxygen during saponification, be-cause the thermodynamic downhill direction is going from O- to N-acylation. The compound isolated from the enzyme reaction was then shown to undergo nonenzymatic first-order decarboxylation over a range of pH values from 5 to 9 at rates identical with the synthetic standard, suggesting that the enzymatic product is the same as the standard.

A more convincing experiment that was carried out was the synthesis of N^1-carboxybiotin and its use as a substrate for acetyl-CoA carboxylase in two different reactions. First, it was able to substitute for biotin, bicarbonate, and ATP (in the presence of Mg^{2+}) in the conversion of acetyl-CoA to oxaloacetate CoA. It also is a substrate for the reverse reaction with ADP and P_i (Scheme 7.36). These exper-

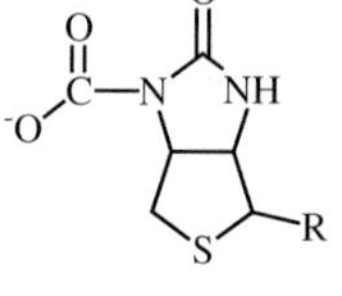

7.42

$$CO_2^--biotin + ADP + P_i \; \overset{Mg^{2+}}{\rightleftharpoons} \; ATP + biotin + HCO_3^-$$

SCHEME 7.36   Acetyl-CoA carboxylase-catalyzed reaction of N-carboxybiotin with ADP and P_i.

iments therefore rule out the possibility of *O*-carboxybiotin as the intermediate. In the case of the reaction run in reverse, according to the *principle of microscopic reversibility* (the mechanism in the reverse direction of a reversible reaction must be identical to that in the forward direction, only reversed), if *O*-carboxybiotin were the active intermediate, then the more stable N^1-carboxybiotin would have to rearrange first to *O*-carboxybiotin in the reverse reaction before it carboxylated the P_i, and this process would be thermodynamically uphill.

e. Mechanisms for Transfer of CO_2 from N^1-Carboxybiotin to Substrates

Again, consider several mechanistic possibilities, both concerted and stepwise (Figure 7.8). Evidence for the concerted pathway is that carboxylation of propionyl

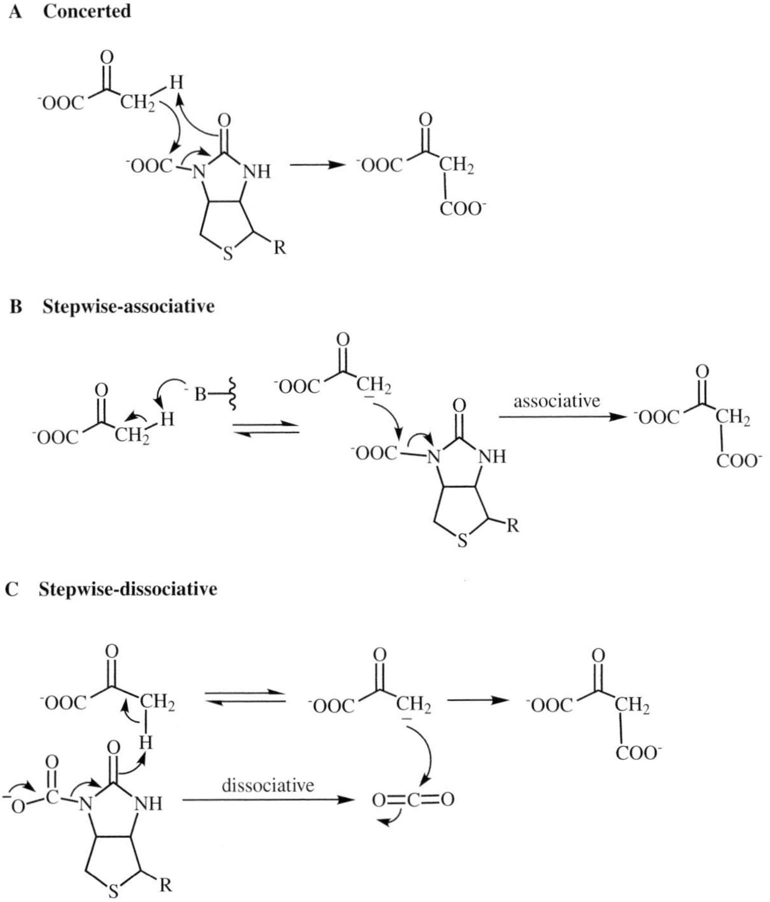
A Concerted
B Stepwise-associative
associative
C Stepwise-dissociative
dissociative

FIGURE 7.8 Possible mechanisms for transfer of CO_2 from N'-carboxybiotin to substrates.

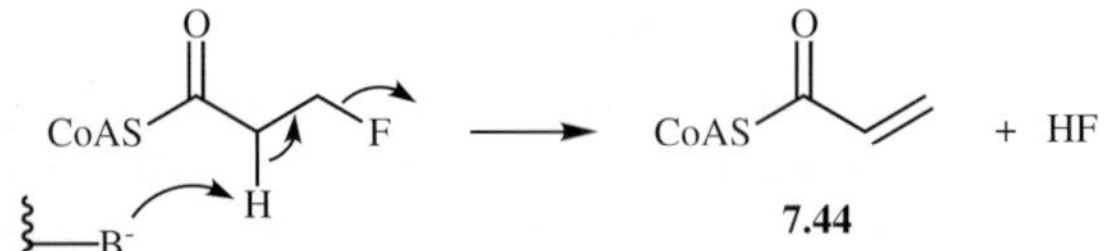

SCHEME 7.37   Transcarboxylase and propionyl–CoA carboxylase-catalyzed elimination of HF from β-fluoropropionyl–CoA.

CoA by propionyl-CoA carboxylase proceeds with retention of configuration at carbon, consistent with a concerted pathway.[50] Interestingly, all biotin-dependent carboxylases have been shown to proceed with retention of configuration at the site of carboxylation. However, this does not rule out a stepwise pathway, if there is stereospecific carboxylation in the second step.

Two experiments support a stepwise mechanism. Transcarboxylase and propionyl-CoA carboxylase catalyze the elimination of HF from β-fluoropropionyl-CoA (**7.43**, Scheme 7.37).[51] Although no carboxylation products are observed, ATP is required and is hydrolyzed to ADP + P_i. Furthermore, the rate of ADP formation is equal to the rate of acrylyl CoA (**7.44**) formation. This indicates that the enzyme is performing all of its catalytic functions with **7.43** except carboxylation and supports a carbanionic intermediate.

Another experiment supporting a stepwise mechanism utilizes a sensitive mass spectrometric analysis termed a *double-isotope fractionation test*.[52] In the carboxylation reaction, a C–H bond is broken and a C–C bond is formed. If the reaction is concerted, then it should show both ^2H and ^{13}C primary kinetic isotope effects (provided both are rate-determining steps). If the reaction is stepwise, it should show either one primary kinetic isotope effect (either ^2H- or ^{13}C-isotope effect), if the carbanion intermediate partitions unevenly, or both isotope effects, if the carbanion partitions evenly.[53] Furthermore, if the reaction is concerted, there will be no change in the ^{13}C isotope effect between pyruvate and [CD_3]pyruvate, but for a stepwise mechanism the deuterium will change the relative free energies of the two transition states and make the ^{13}C-sensitive step less rate limiting than for pyruvate (i.e., lower the ^{13}C isotope effect). O'Keefe and Knowles determined the natural abundance ^{13}C kinetic isotope effect [$^{13}(V/K)$] on carboxylation of pyruvate, catalyzed by transcarboxylase, to be 1.0227 ± 0.0008, and for [CD_3]pyruvate they determined the effect to be 1.0141 ± 0.001,[54] which is in excellent agreement with the predicted value of 1.0136 for a stepwise mechanism.

As in the case of PEP carboxylase (Section III.A), the more effective mechanism is the dissociative mechanism (Figure 7.8C), which generates the reactive carboxylating agent CO_2. Model studies by Kluger and co-workers indicate that N^1-carboxybiotin, especially in a hydrophobic environment, is highly unstable, and should readily produce biotin and CO_2.[55] If that is the case in the enzyme-catalyzed reaction, then all of the bicarbonate-dependent carboxylation reactions proceed via an activated carboxylate form that decomposes to CO_2 at the site of the acceptor.

NOTE THAT PROBLEMS AND SOLUTIONS RELEVANT TO EACH CHAPTER CAN BE FOUND IN APPENDIX II.

REFERENCES

1. Leussing, D. L. *Adv. Inorg. Biochem.* **1982**, *4,* 171.
2. Cooper, T. G.; Tchen, T. T.; Wood, H. G.; Benedict, C. R. *J. Biol Chem.* **1968**, *243,* 3857.
3. Jomain-Baum, M.; Schramm, V. L. *J. Biol. Chem.* **1978**, *253,* 3648.
4. (a) Fan, C.; Teixeira, M.; Moura, J.; Moura, I.; Huynh, B.-H.; Le Gall, J.; Peck, H. D., Jr.; Hoffman, B. M. *J. Am. Chem. Soc.* **1991**, *113,* 20. (b) Silverman, R. B.; Cesarone, J. M.; Lu, X. *J. Am. Chem. Soc.* **1993**, *115,* 4955. (c) Rose, I. A.; Kuo, D. J. *Biochemistry* **1989**, *28,* 9579. (d) Rose, I. A.; Fung, W.-J.; Warms, J. V. B. *Biochemistry* **1990**, *29,* 4312.
5. Sheu, K-F.; Ho, H.-T.; Nolan, L. D.; Markovitz, P.; Richard, J. P.; Utter, M. F.; Frey, P. A. *Biochemistry* **1984**, *23,* 1779.
6. Hou, S.-Y.; Chao, Y.-P.; Liao, J. C. *J. Bacteriol.* **1995**, *177,* 1620.
7. Arnelle, D. R.; O'Leary, M. H. *Biochemistry* **1992**, *31,* 4363.
8. Matte, A.; Tari, L. W.; Goldie, H.; Delbaere, L. T. J. *J. Biol. Chem.* **1997**, *272,* 8105.
9. Hanson, K. R. *J. Am. Chem. Soc.* **1966**, *88,* 2731.
10. Rose, I. A.; O'Connell, E. L.; Noce, P.; Utter, M. F.; Wood, H. G.; Willard, J. M.; Cooper, T. G.; Benziman, M. *J. Biol. Chem.* **1969**, *244,* 6130.
11. Dowd, P.; Hershline, R.; Ham, S. W.; Naganathan, S. *Nat. Prod. Rep.* **1994**, *11,* 251.
12. Vidal-Cros, A.; Gaudry, M.; Marquet, A. *Biochem. J.* **1990**, *266,* 749.
13. Dubois, J.; Dugave, C.; Foures, C.; Kaminsky, M.; Tabet, J.-C.; Bory, S.; Gaudry, M.; Marquet, A. *Biochemistry* **1991**, *30,* 10506.
14. Zheng, Y.-J.; Bruice, T. C. *J. Am. Chem. Soc.* **1998**, *120,* 1623.
15. Dowd, P.; Ham, S. W.; Geib, S. J. *J. Am. Chem. Soc.* **1991**, *113,* 7734.
16. (a) Suttie, J. W.; Larson, A. E.; Canfield, L. M.; Carlisle, T. L. *Fed. Proc. Am. Soc. Exp. Biol.* **1978**, *37,* 2605. (b) Zheng, Y.-J.; Bruice, T. C. *J. Am. Chem. Soc.* **1998**, *120,* 1623.
17. Kuliopulos, A.; Hubbard, B. R.; Lam, Z.; Koski, I. J.; Furie, B.; Furie, B. C.; Walsh, C. T. *Biochemistry* **1992**, *31,* 7722.
18. Dowd, P.; Ham, S. W.; Hershline, R. *J. Am. Chem. Soc.* **1992**, *114,* 7613.
19. Ham, S. W.; Yoo, J. S. *J. Chem. Soc. Chem. Commun.* **1997**, 929.
20. Naganathan, S.; Hershline, R.; Ham, S. W.; Dowd, P. *J. Am. Chem. Soc.* **1993**, *115,* 5839.
21. (a) Flowers, R. A., II; Naganathan, S.; Dowd, P.; Arnett, E. M.; Ham, S. W. *J. Am. Chem. Soc.* **1993**, *115,* 9409.; (b) Arnett, E. M.; Dowd, P.; Flowers, R. A., II; Ham, S. W.; Naganathan, S. *J. Am. Chem. Soc.* **1992**, *114,* 9209.
22. Dowd, P.; Ham, S. W.; Geib, S. J. *J. Am. Chem. Soc.* **1991**, *113,* 7734.
23. (a) Dowd, P.; Hershline, R.; Ham, S. W.; Naganathan, S. *Science* **1995**, *269,* 1684. (b) Naganathan, S.; Hershline, R.; Ham, S. W.; Dowd, P. *J. Am. Chem. Soc.,* **1994**, *116,* 9831.
24. Parker, A. J. *Adv. Org. Chem.* **1965**, *5,* 1.
25. (a) Maruyama, H.; Easterday, R. L.; Chang, H. C.; Lane, M. D. *J. Biol. Chem.* **1966**, *241,* 2405. (b) O'Leary, M. H.; Hermes, J. D. *Anal. Biochem.* **1987**, *162,* 358.
26. O'Leary, M. H.; Rife, J. E.; Slater, J. D. *Biochemistry* **1981**, *20,* 7308.
27. Ausenhus, S. L.; O'Leary, M. H. *Biochemistry* **1992**, *31,* 6427.
28. Hanson, D. E.; Knowles, J. R. *J. Biol. Chem.* **1982**, *257,* 14795.
29. Chollet, R.; Vidal, J.; O'Leary, M. H. *Annu. Rev. Plant Physiol. Plant Mol. Biol.* **1996**, *47,* 273.

30. Fujita, N.; Izui, K.; Nishino, T.; Katsui, H. *Biochemistry* **1984**, *23*, 1774.
31. O'Leary, M. H. *The Enzymes,* vol. 20, Sigman, D. S., Ed., Academic Press: San Diego, p. 235.
32. Jane, J. W., Urbauer, J. L.; O'Leary, M. H.; Cleland, W. W. *Biochemistry* **1992**, *31*, 6432.
33. Knowles, J. R. *Annu. Rev. Biochem.* **1989**, *58*, 195.
34. Kaziro, Y.; Hase, L. F.; Boyer, P. D.; Ochoa, S. *J. Biol. Chem.* **1962**, *237*, 1460.
35. Green, N. M. *Adv. Protein Chem.* **1975**, *29*, 85.
36. (a) Scrutton, M. C.; Keech, D. B.; Utter, M. F. *J. Biol. Chem.* **1965**, *240*, 574. (b) Lynen, F.; Knappe, J.; Lorch, E.; Jütting, G.; Ringelmann, E.; Lachance, J.-P. *Biochem Z.* **1961**, *335*, 123.
37. Lynen, F.; Knappe, J.; Lorch, E.; Jütting, G.; Ringelmann, E. *Angew. Chem.* **1959**, *71*, 481.
38. Knappe, J.; Wenger, B.; Weigand, U.; Lynen, F. *Biochem Z.,* **1963**, *337*, 232.
39. Bonnemere, C.; Hamilton, J. A.; Steinrauf, L. K.; Knappe, J. *Biochemistry* **1965**, *4*, 240.
40. Perrin, C. L.; Dwyer, T. J. *J. Am. Chem. Soc.* **1987**, *109*, 5163.
41. Climent, I.; Rubio, V. *Arch. Biochem. Biophys.* **1986**, *251*, 465.
42. Ogita, T.; Knowles, J. R. *Biochemistry* **1988**, *27*, 8028.
43. (a) Taylor, S. D.; Kluger, R. *J. Am. Chem. Soc.* **1993**, *115*, 867. (b) Kluger; R.; Taylor, S. D. *J. Am. Chem. Soc.* **1991**, *113*, 996.
44. Blagoeva, I. B.; Pojarlie, F. F.; Tashev, D. T. *J. Chem. Soc. Perkin Trans.* 2 **1989**, 347.
45. Kluger, R. *Chem. Rev.* **1990**, *90*, 1151.
46. Caplow, M. *J. Am. Chem. Soc.* **1965**, *87*, 5774.
47. Bruice, T. C.; Hegarty, A. F. *Proc. Natl. Acad. Sci. USA* **1970**, *65*, 805.
48. Lynen, F.; Knappe, J.; Lorch, E.; Jütting, G.; Ringelmann, E. *Angew. Chem.* **1959**, *71*, 481.
49. Guchhait, R. B.; Polakis, S. E.; Hollis, D.; Fenselau, C.; Lane, M. D. *J. Biol. Chem.* **1974**, *249*, 6646.
50. Rétey, J.; Lynen, F. *Biochem Z.* **1965**, *342*, 256.
51. Stubbe, J.; Fish, S.; Abeles, R. H. *J. Biol. Chem.* **1980**, *255*, 236.
52. Hermes, J. D.; Roeske, C. A.; O'Leary, M. H.; Cleland, W. W. *Biochemistry* **1982**, *21*, 5106.
53. O'Keefe, S. J.; Knowles, J. R. *J. Am. Chem. Soc.* **1986**, *108*, 328.
54. O'Keefe, S. J.; Knowles, J. R. *Biochemistry* **1986**, *25*, 6077.
55. Rahil, J.; You, S.; Kluger, R. *J. Am. Chem. Soc.* **1996**, *118*, 12495.

Decarboxylation

I. GENERAL

The primary driving force for decarboxylation is either the formation of carbon dioxide (an entropic effect) and a stabilized carbanion, generally stabilized by resonance, or the involvement in an elimination reaction (Scheme 8.1). The first two of these characteristics are satisfied by β-keto acids.

II. β-KETO ACIDS

The ketone carbonyl in β-keto acids (**8.1**) can stabilize the incipient carbanion by enolate formation (Scheme 8.2). Decarboxylation is facilitated by initial protonation of the ketone carbonyl, as evidenced by the fact that decarboxylation of β-keto acids occurs much more rapidly in acidic medium than in base. We have already seen this phenomenon: Carboxybiotin was much more stable at high pH than at low pH, and raising the pH was the way that carboxybiotin was first isolated (see Chapter 7, Section III.B.2.b). Westheimer and Jones proposed that the reason for the ease of decarboxylation in acidic medium was the formation of a six-membered

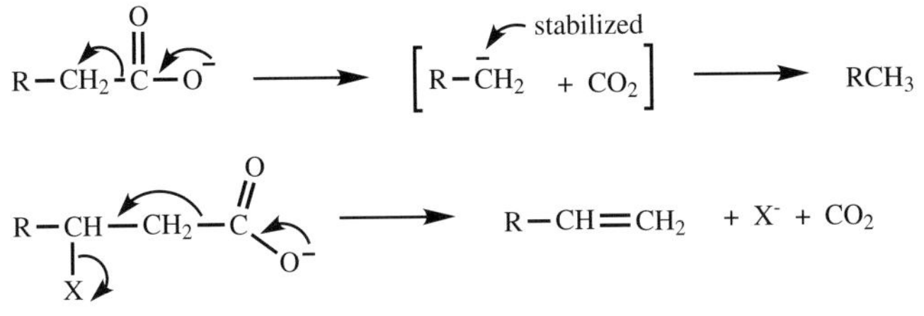

$$R-CH_2-\overset{\overset{\displaystyle O}{\|}}{C}-O^- \longrightarrow \left[R-\overset{-}{C}H_2 + CO_2 \right] \longrightarrow RCH_3$$

stabilized

$$R-CH-CH_2-\overset{\overset{\displaystyle O}{\|}}{C}\diagdown_{O^-} \longrightarrow R-CH=CH_2 + X^- + CO_2$$
$$\underset{X}{|}$$

SCHEME 8.1   Decarboxylation reactions.

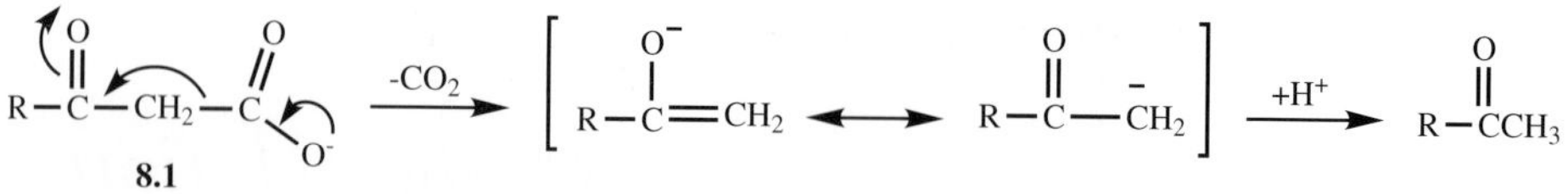

SCHEME 8.2   Decarboxylation of β-keto acids.

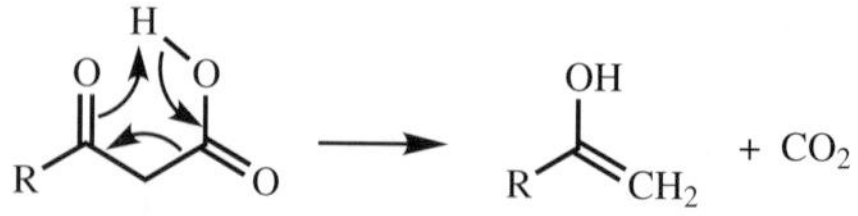

SCHEME 8.3   Cyclic transition state for decarboxylation of β-keto acids.

transition state (**8.2,** Scheme 8.3).[1] Reaction rates of β-keto acid decarboxylation do not vary much with a change from a nonpolar to a polar solvent and are not subject to acid catalysis.[2] Protonation of the β-carbonyl group directly would facilitate decarboxylation; however, the pK_a of a protonated carbonyl is approximately -7, which means that only very strong acids could protonate a carbonyl oxygen to any extent. Unlike the very weakly basic carbonyl oxygen, the nitrogen of a Schiff base (an imine; **8.3**) can be protonated easily; it is difficult to measure the pK_a of aliphatic ketone imines because of their hydrolytic instability, but the pK_a values of benzylidene imines are near neutrality[3] (Scheme 8.4). Schiff base formation would permit facile protonation of the imine and lower the energy of decarboxylation. It was Pedersen[4] who suggested that amine catalysis of β-keto acid decarboxylation was the result of Schiff base formation. Note that in the mechanism in Scheme 8.4 the starting amine is regenerated in the product; it therefore acts as a catalyst. This is the ideal situation for an enzyme-catalyzed reaction.

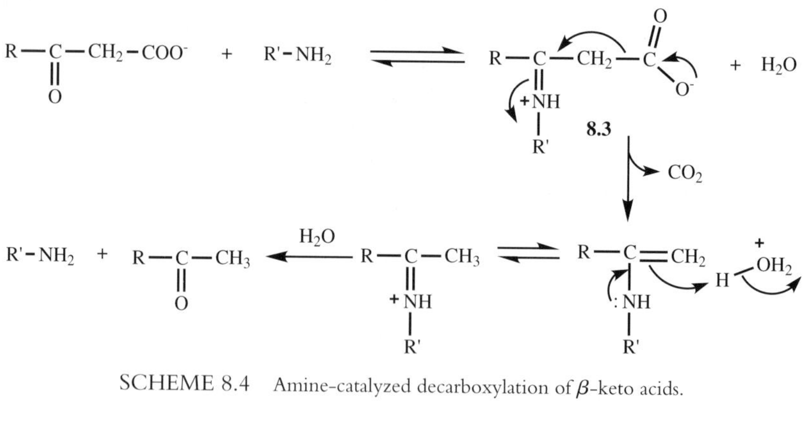

SCHEME 8.4   Amine-catalyzed decarboxylation of β-keto acids.

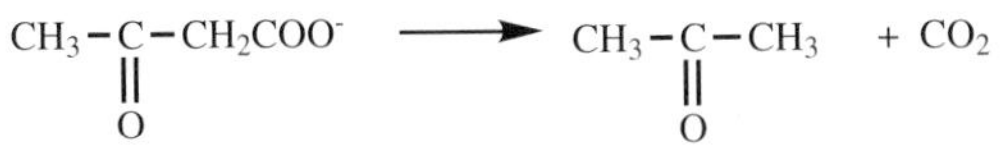

$$CH_3-C-CH_2COO^- \longrightarrow CH_3-C-CH_3 + CO_2$$
$$\underset{O}{\overset{\|}{}} \qquad\qquad \underset{O}{\overset{\|}{}}$$

8.4

SCHEME 8.5   Reaction catalyzed by acetoacetate decarboxylase.

A. Schiff Base Mechanism

Acetoacetate decarboxylase catalyzes the decarboxylation of acetoacetic acid (**8.4**) to acetone and carbon dioxide (Scheme 8.5). This enzyme was very much in demand during World War I, because at the time it was involved in the primary commercial process to make acetone, which was needed for airplane wing glue. Westheimer suggested that this enzyme could utilize Schiff base catalysis, much like amine catalysis of β-keto acid decarboxylation in solution.[5]

1. Evidence for Mechanism

When [β-^{18}O]acetoacetate was incubated with acetoacetate decarboxylase in $H_2^{16}O$, all of the ^{18}O was released as $H_2^{18}O$ (in the absence of enzyme, none was released) (Scheme 8.6). Conversely, when unlabeled acetoacetate was incubated with acetoacetate decarboxylase in $H_2^{18}O$, the acetone generated contained ^{18}O in the carbonyl. If the reaction was carried out in ^2H_2O, deuterium was incorporated into the acetone. However, because the control experiment initially showed significant incorporation of deuterium into the acetone as well, the reaction had to be carried out with a large amount of enzyme and with a short incubation period so that sufficient enzyme-generated acetone could be produced faster than nonenzymatic deuterium exchange.[6]

Westheimer and co-workers made another observation that led to the elucidation of the mechanism: The pK_a of the active-site lysine residue (now known to be Lys-115 in *Clostridium acetobutylicum*) was 5.9, which is 4.5 pK_a units lower than the pK_a of lysine in solution![7] Westheimer suggested that the pK_a of Lys-115 was low because of its proximity to Lys-116. As discussed in Chapter 1 (Section II.C), the pK_a values of active-site amino acids can be quite different from amino acids in solution; the pK_a of a base can be lowered by the presence of another protonated base adjacent to it. Twenty-five years after this hypothesis was made, it was brought to an elegant test—ironically, by two former Westheimer graduate students (John

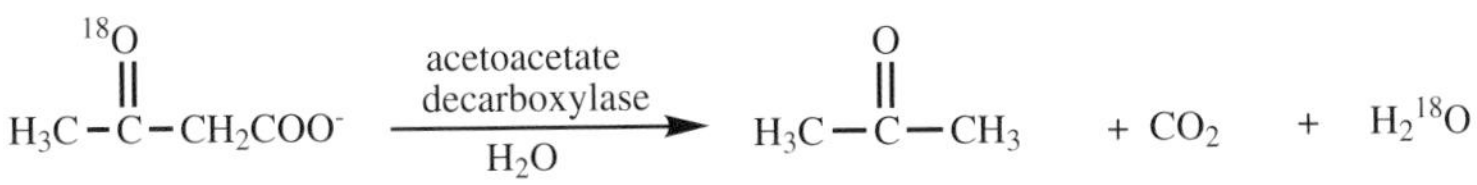

$$\underset{\overset{\|}{O}}{\overset{^{18}O}{}}$$
$$H_3C-C-CH_2COO^- \xrightarrow[\text{H}_2\text{O}]{\substack{\text{acetoacetate} \\ \text{decarboxylase}}} H_3C-\underset{\overset{\|}{O}}{C}-CH_3 + CO_2 + H_2^{18}O$$

SCHEME 8.6   Fate of the ketone oxygen in the reaction catalyzed by acetoacetate decarboxylase.

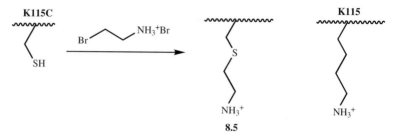

SCHEME 8.7   Reaction of the K115C mutant of acetoacetate decarboxylase with 2-bromoethyl-amine.

Gerlt and George Kenyon), one of which had been in the Westheimer group when the original work was just being completed (John Gerlt).[8] Lys–115 was mutated to a cysteine and to a glutamine (K115C and K115Q). Both mutants were catalytically inactive at pH 5.95, the pH optimum of the wild-type enzyme, presumably because there no longer was a protonated base adjacent to K116. The activity of K115C, however, was restored by aminoethylation of the cysteine residue to give K115aminoethylcysteine (**8.5**, Scheme 8.7), which is about the same size as lysine and mimics a lysine residue (see the comparison in Scheme 8.7). Substitutions for Lys–116 (K116C, K116N, K116R) gave mutants with lower activities than the wild type. The pK_a of Lys-115 in the K116R mutant is similar to that of wild type (arginine is a protonated base, just like lysine), but the pK_a values of Lys-115 in the K116C and K116N mutants were >9.2, presumably because these mutants do not have adjacent protonated bases like the wild-type lysine. Alkylation of the cysteine of K116C with 2-bromoethylammonium bromide (just like in Scheme 8.7) resulted in both significant enzyme activation and restoration of the pK_a of Lys-115 to 5.9. These results beautifully support Westheimer's hypothesis that placing two lysine residues adjacent to each other can lower their pK_a values.

Although the ^{18}O washout (and wash-in) experiment appears to be unrelated to the deuterium incorporation experiment and the lysine pK_a experiments, in fact, they all support a mechanism that involves Schiff base formation (Scheme 8.8; [^{18}O] acetoacetate and D_2O are utilized to show the results of those experiments). Formation of **8.6** occurs on loss of the acetoacetate carbonyl oxygen as water. Decarboxylation of **8.6** produces an enamine (**8.7**), which can undergo protonation to **8.8** (deuterium is incorporated); hydrolysis of **8.8** gives back acetone and the active-site lysine.

To further establish a Schiff base mechanism, Fridovich and Westheimer incubated [3-^{14}C]acetoacetate with acetoacetate decarboxylase; during incubation the enzyme was quenched with sodium borohydride (this reagent has no effect on the enzyme in the absence of substrate because there are no electrophilic centers that can be reduced).[9] The result was a ^{14}C-labeled enzyme. Acid hydrolysis of the radioactive protein produced [^{14}C]isopropyllysine (**8.9**, Scheme 8.9). When NaB^{3}H$_4$

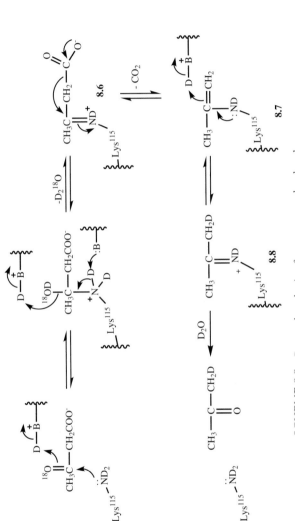

SCHEME 8.8 Proposed mechanism for acetoacetate decarboxylase.

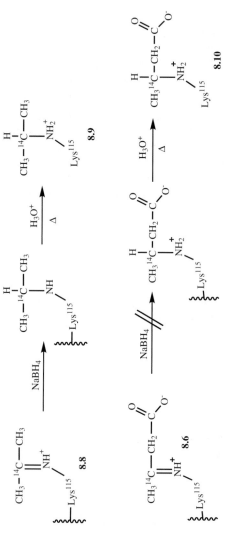

SCHEME 8.9 Fate of NaBH₄ reduction during the reaction catalyzed by acetoacetate decarboxylase.

was utilized, ³H was incorporated into the isopropyl group.[10] These results indicate that decarboxylation occurs very fast and prior to borohydride reduction; if reduction occurred first, decarboxylation would be prohibited and **8.10** would have resulted. As a confirmation of this result, it was found that the same adduct (**8.9**) is obtained if the enzyme is treated with acetone and NaB^3H_4.

B. Metal Ion–Catalyzed Mechanism

Some decarboxylases have metal ion cofactors and do not utilize Schiff base covalent catalysis. These enzymes, therefore, do not catalyze the washout of ^{18}O from the carbonyl of the substrate and are not inactivated by coincubation of the substrate with sodium borohydride. However, deuterium from ^2H_2O is incorporated into the product. In these cases, the metal ion acts as a "superproton" to catalyze decarboxylation without covalent catalysis (Scheme 8.10).

An example of this kind of catalysis is the Mg^{2+}-catalyzed decarboxylation of (S)-α-acetolactate (**8.11**) by α-acetolactate decarboxylase from *Klebsiella aerogenes*, which proceeds with inversion of stereochemistry to give (R)-acetoin (**8.12**, Scheme 8.11).[11]

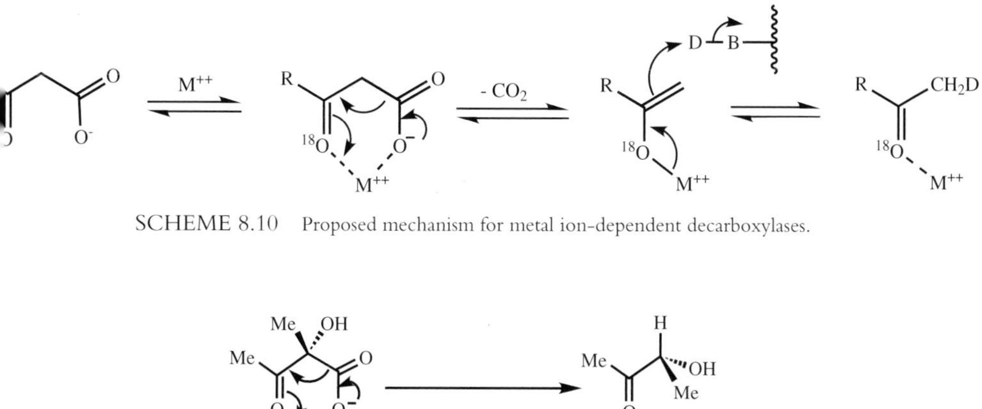

SCHEME 8.10   Proposed mechanism for metal ion-dependent decarboxylases.

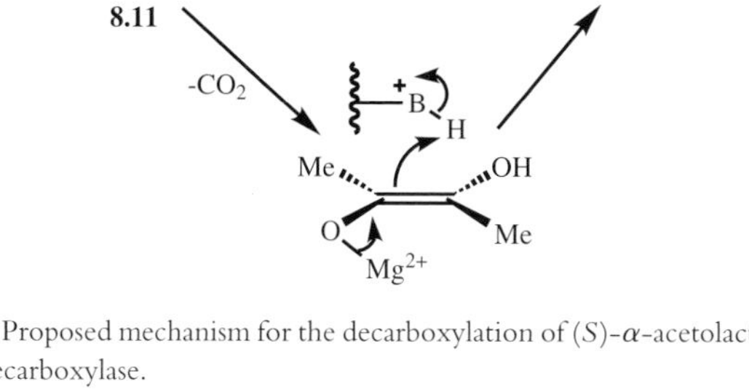

SCHEME 8.11   Proposed mechanism for the decarboxylation of (S)-α-acetolactate (**8.11**) catalyzed by α-acetolactate decarboxylase.

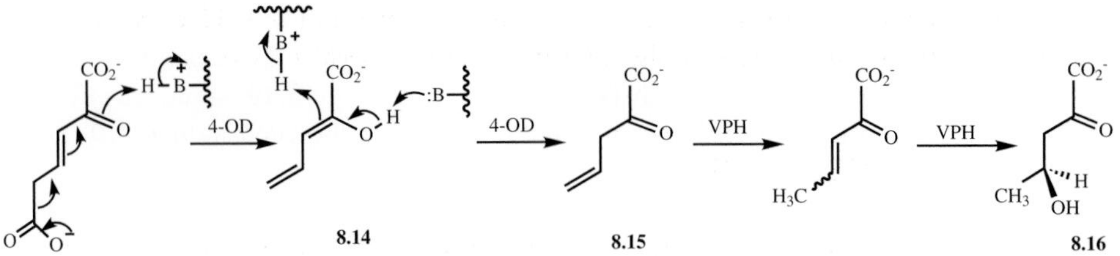

8.13 8.14 8.15 8.16

SCHEME 8.12 Reactions catalyzed by 4-oxalocrotonate decarboxylase.

4-Oxalocrotonate decarboxylase utilizes a metal ion cofactor (either Mn^{2+} or Mg^{2+}) to catalyze a vinylogous *β*-keto acid decarboxylation.[12] With the aid of specifically labeled 2-oxo-3-hexenedioic acid (**8.13**), intermediates, and water, Lian and Whitman showed that the 4-oxalocrotonate decarboxylase (4-OD)/vinylpyruvate hydratase (VPH) complex proceeds by the vinylogous decarboxylation mechanism shown in Scheme 8.12. Intermediates **8.14** and **8.15** were detected by UV and NMR spectroscopy. (5*S*)-[5-^{2}H]**8.13** is converted by 4-oxalocrotonate decarboxylase to (4*E*)-[5-^{2}H]**8.14** (Scheme 8.13), and incubation of **8.14** in D_2O with 4-oxalocrotonate decarboxylase gives (3*S*)-[3-^{2}H]**8.15** (Scheme 8.14). Therefore, the loss of CO_2 and the incorporation of the deuteron occur on the same side (the bottom side in these schemes) of the dienol intermediate. An enzyme that

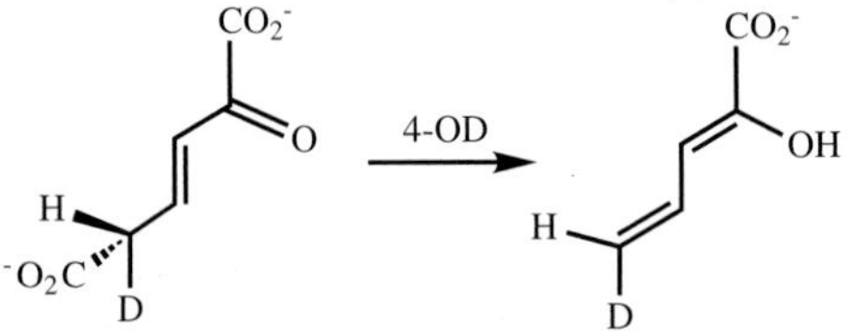

(5*S*)-[5-^{2}H]-8.13 **(4*E*)-[5-^{2}H]-8.14**

SCHEME 8.13 4-Oxalocrotonate decarboxylase-catalyzed decarboxylation of (5*S*)-[5-^{2}H]2-oxo-3-hexenedioic acid.

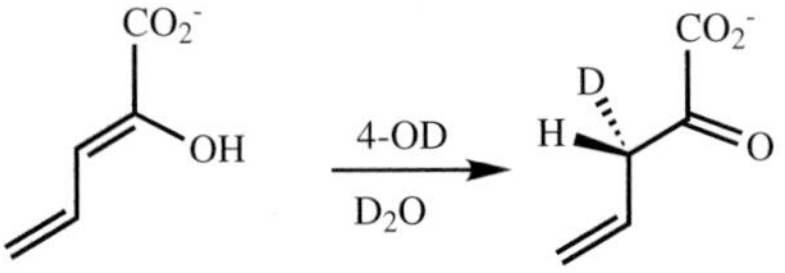

8.14 **(3*S*)-[3-^{2}H]-8.15**

SCHEME 8.14 4-Oxalocrotonate decarboxylase-catalyzed incorporation of deuterium from solvent into proposed intermediate **8.14**.

catalyzes a reaction similar to the first part of the conversion of **8.15** to **8.16** (4-oxalocrotonate tautomerase) is discussed in Chapter 9 (Section IV.B.1).

III. β-HYDROXY ACIDS

As discussed earlier, decarboxylation requires an adjacent electron sink (such as the ketone carbonyl of a β-keto acid) to stabilize the incipient carbanion. A *β-hydroxy acid* does not have this characteristic. Therefore the enzyme must generate a stabilizing group prior to decarboxylation.

A. Isocitrate Dehydrogenase

One of the enzymes in the Krebs cycle, isocitrate dehydrogenase, is a typical example of this transformation. The name of the enzyme gives away the solution to the conundrum. If it is a dehydrogenase, it must be catalyzing a redox reaction. In fact, it requires $NADP^+$ for activity and converts isocitrate (**8.17**) to α-ketoglutarate (**8.19**); it also requires a divalent metal ion, such as magnesium ion, to catalyze the decarboxylation (Scheme 8.15).[13] Although this mechanism is reasonable, there is no evidence for the oxalosuccinate intermediate (**8.18**); oxalosuccinate is converted to α-ketoglutarate by isocitrate dehydrogenase in the absence of the coenzyme, but the reaction is not kinetically competent. Also, oxalosuccinate cannot be trapped with sodium borohydride during the enzyme-catalyzed reaction. If the enzyme is incubated with [^{14}C]isocitrate, then oxalosuccinate is added, the radioactivity in

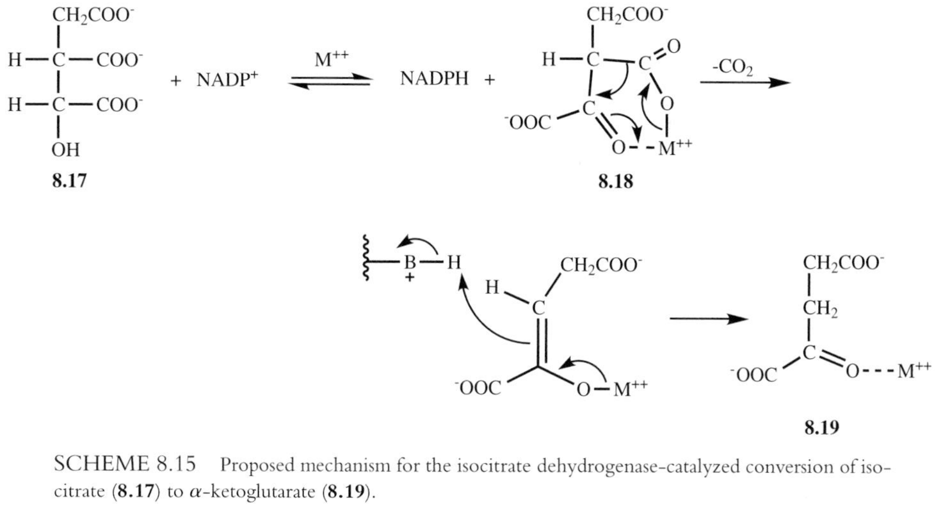

SCHEME 8.15   Proposed mechanism for the isocitrate dehydrogenase-catalyzed conversion of isocitrate (**8.17**) to α-ketoglutarate (**8.19**).

the α-ketoglutarate is not diluted. Therefore there is no partial exchange. The conclusion to date, then, must be that if oxalosuccinate is a finite intermediate, it either must remain tightly bound before decarboxylation occurs or else the oxidation and decarboxylation steps are concerted.

B. 6-Phosphogluconate Dehydrogenase

Not all decarboxylations must be either Schiff base- or metal ion-catalyzed. 6-Phosphogluconate dehydrogenase catalyzes the conversion of 6-phosphogluconate (**8.20**) to ribulose-5-phosphate (**8.21**); as a β-hydroxy acid, an oxidation must first occur (Scheme 8.16). No metal ions have been detected in the enzyme, so an experiment was carried out to test for a Schiff base mechanism. Using [2-^{18}O]ribulose-5-phosphate and running the 6-phosphogluconate dehydrogenase reaction in the back direction, it was shown that the 6-phosphogluconate isolated completely retained the ^{18}O. Therefore, it can only be a Schiff base mechanism if the H_2^{18}O released by Schiff base formation remains bound to the active site and is utilized by the enzyme to hydrolyze the imine after decarboxylation. This is highly unlikely, especially because there are no metal ions to bind to the water molecule and keep it in place. In lieu of a metal ion, Topham and Dalziel have proposed that there is an active-site proton that activates the β-keto acid (**8.22**) generated by NADP$^+$ oxidation of the β-hydroxy acid (Scheme 8.17).[14]

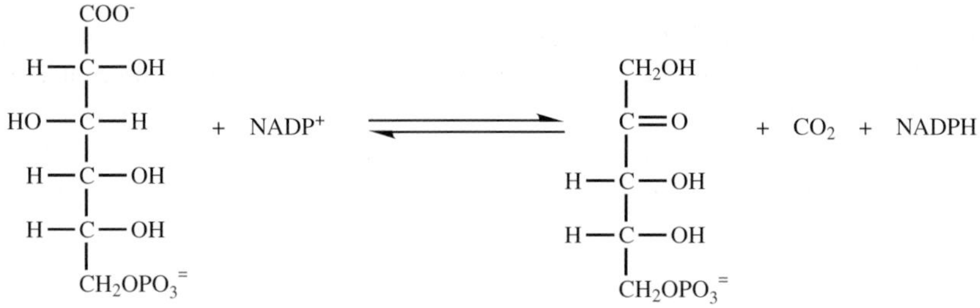

8.20   **8.21**

SCHEME 8.16   Reaction catalyzed by phosphogluconate dehydrogenase.

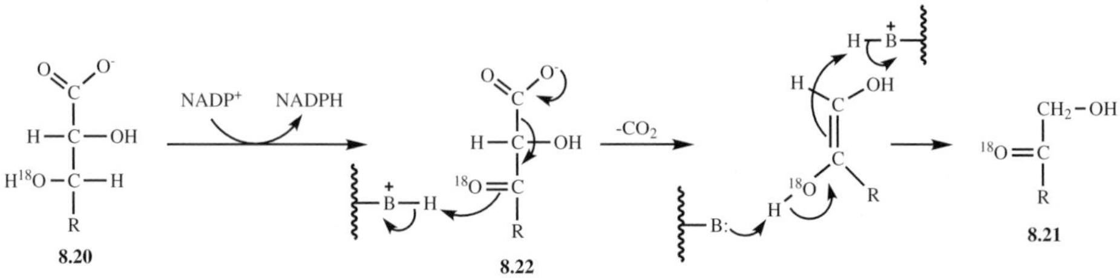

8.20   **8.22**   **8.21**

SCHEME 8.17   Proposed mechanism for the reaction catalyzed by phosphogluconate dehydrogenase.

IV. *α*-KETO ACIDS

A. General

Although *β*-hydroxy acids have no electron sink to stabilize an incipient carbanion generated by decarboxylation, the enzyme can generate the electron sink (the ketone carbonyl) by oxidation of the alcohol. With *α-keto acids* (**8.23**) an electron sink is present, but it is not located in a position that is useful for resonance stabilization of a carbanion generated by decarboxylation (**8.24**, Scheme 8.18). Once again, you could think that enzymes have special powers to defy chemical principles, or you could ask what the enzyme could do to stabilize the carbanion. To accomplish this feat, enzymes that catalyze the decarboxylation of *α*-keto acids require a coenzyme. In this case the coenzyme is *thiamin diphosphate* (TDP; also abbreviated in the literature as TPP, which stands for thiamin pyrophosphate, the older name for thiamin diphosphate; **8.26**); the numbering scheme for TDP is shown in Scheme 8.19. TDP is the coenzyme derived from vitamin B$_1$ or *thiamin* (**8.25**) by the action of a kinase (Scheme 8.19); a diet deficient in vitamin B$_1$ results in the disease beriberi.[15] As you will see, the function of the TDP is to extend the *α*-keto group to an electron sink that is beta to the carboxylate and to thereby stabilize the incipient carbanion.

B. Chemistry and Properties of Thiamin Diphosphate

All of the chemistry of TDP occurs at the thiazolium ring, so it will be abbreviated as **8.27**. In 1958 Breslow found by NMR spectrometry that the C$_2$ proton of TDP exchanges in neutral D$_2$O with a half-life of 20 minutes at room temperature.[16] The acidity of this proton has been attributed to an electrostatic interaction (ylide formation, **8.28**), delocalization (carbene formation, **8.29**), and d-p orbital overlap

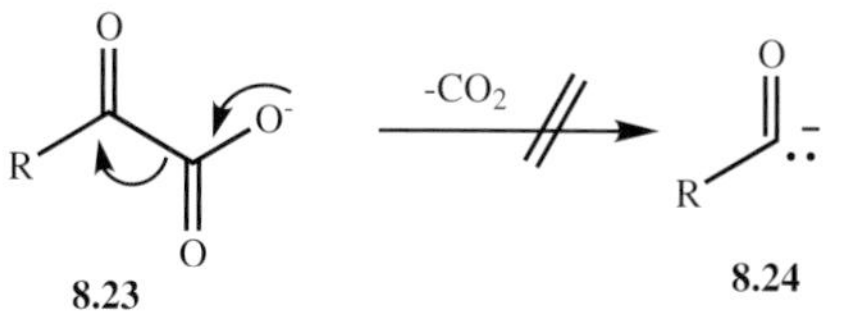

8.23   −CO$_2$ **8.24**

SCHEME 8.18   Improbable decarboxylation of *α*-keto acids.

8.25   ATP   AMP   **8.26**

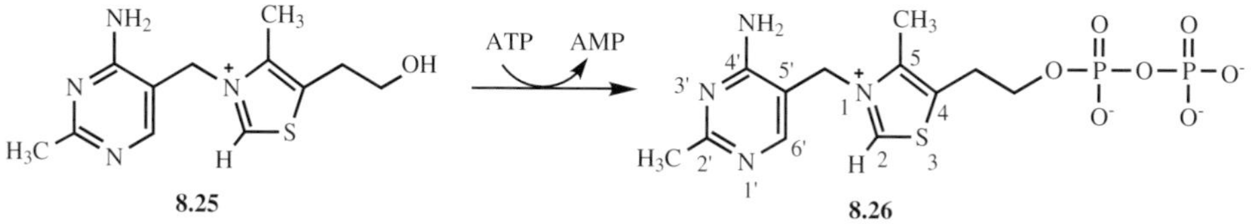

SCHEME 8.19   Diphosphorylation of thiamin.

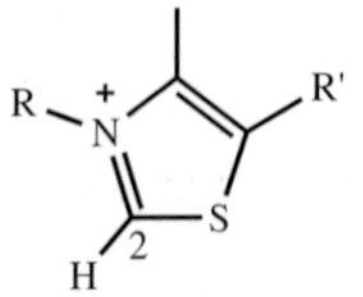

8.27

with the sulfur atom (**8.30**, Scheme 8.20). Evidence for the potential carbene character of the deprotonated thiazolium ring comes from the crystal structure of an
imidazolium stabilized carbene (**8.31a**), which Arduengo et al. showed to be a
true carbene with negligible ylide character (**8.31b**).[17] Hopmann and Brugnoni determined the pK_a of the thiazolium ring of thiamin to be 12.7.[18] Jencks and co-
workers[19] measured the pK_a of a series of thiamin analogues and found the pK_a of
thiamin to be 18; they also concluded that the aminopyrimidinyl group has no effect, other than an inductive effect, on the C_2 proton exchange rate. Bordwell and
Satish measured the equilibrium acidity of the N-methylthiazolium ring in DMSO
solvent to be 16.5.[20]

The mechanism for deprotonation of thiamin diphosphate was studied by a
variety of techniques. X-ray crystallography (at 2.5 Å resolution) of a thiamin
diphosphate–dependent dimeric enzyme (transketolase; see Chapter 11, Section
I.B.4) shows that the cofactor is bound at the interface between the two subunits.[21]
The thiazolium ring interacts with residues from both subunits, and the pyrimidine
ring is in a hydrophobic pocket made up mostly of aromatic side chains. Based
on this crystal structure, and the thermodynamics and kinetics of deprotonation at
C_2, which were measured by NMR spectroscopy, by H/D exchange, with modi-

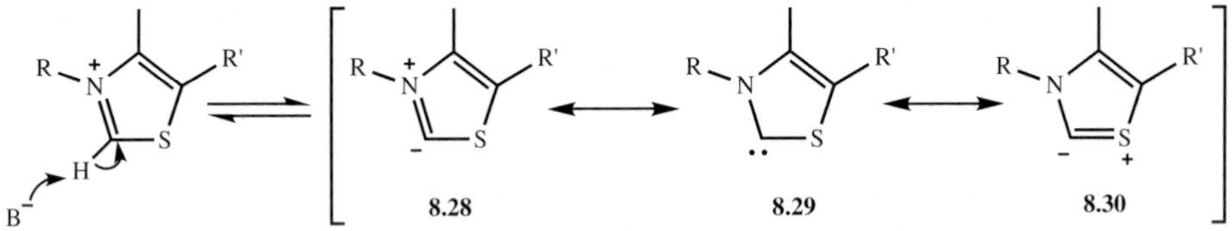

8.28   **8.29**   **8.30**

SCHEME 8.20   Resonance stabilization of thiazolium ylide.

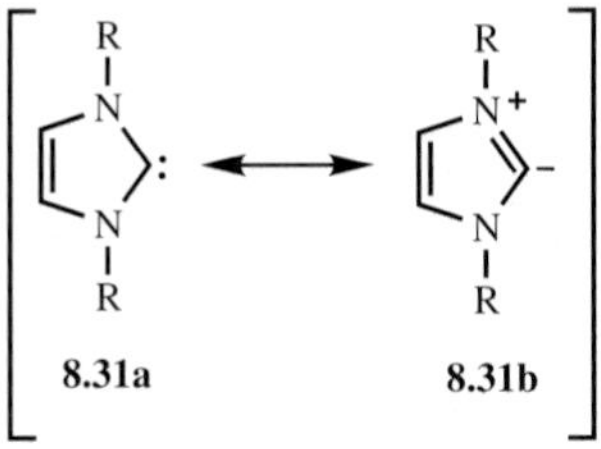

8.31a   **8.31b**

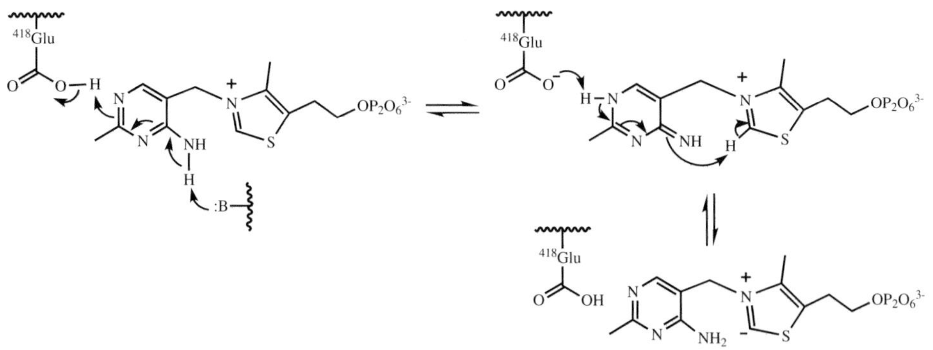

SCHEME 8.21   Proposed mechanism for autodeprotonation of thiamin diphosphate.

fied coenzyme analogues, and by site-specific mutants of pyruvate decarboxylase and transketolase,[22] it can be concluded that the interaction of an active-site glutamate side chain (Glu-418; this glutamate is highly conserved in thiamin diphosphate-dependent enzymes) with N-1′ of the pyrimidine ring is essential for rate acceleration. Furthermore, this hydrogen-bonding interaction at N-1′ appears to activate the 4′-amino group as a base for removal of the C_2 proton of the thiazolium ring (Scheme 8.21). The 2.5 Å crystal structure shows His-481 close to the 4′-amino group of the pyrimidine ring. However, mutation of His-481 does not alter the rate of deprotonation. Coenzyme analogues having no N-1′ nitrogen[23] or N-4′ amino group are inactive, even though the 2.0 Å resolution crystal structures of the modified coenzymes bound to the enzyme show that the binding orientations are the same as that for bound thiamin diphosphate. The corresponding modified co-enzyme lacking the N-3′ nitrogen, however, is active. Because the enzyme can catalyze the deprotonation of C_2 by several orders of magnitude beyond the rate of the overall reaction, a concerted mechanism is excluded.

Chemical model studies were carried out to provide evidence for the carbene nature of thiazolium salts.[24] Compound **8.32** was treated with the base DBN, and a carbene (**8.33**) was proposed as the initial product, because the researchers assumed that the symmetrical dimer **8.34** subsequently formed, even though it was not isolated (Scheme 8.22). What was isolated was the unsymmetrical rearranged dimer

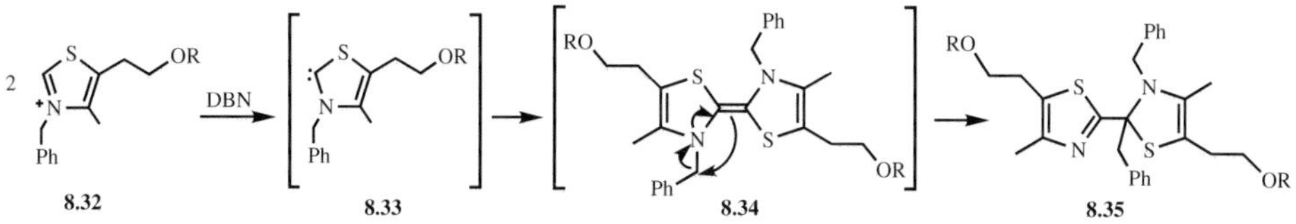

SCHEME 8.22   Model studies to determine the carbene nature of thiazolium salts.

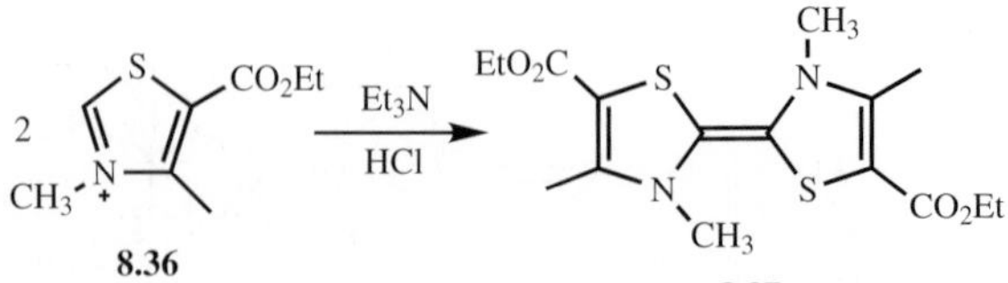

8.36 8.37

SCHEME 8.23 Dimerization of a thiazolium salt in base.

8.35. Likewise, thiazolium **8.36** dimerized in base to give the symmetrical dimer **8.37,** which was isolated (Scheme 8.23). However, Jordan and co-workers[25] isolated a different unsymmetrical dimer (**8.38**) as an intermediate on the way to the symmetrical dimer (**8.39;** *syn* and *anti*), which is then converted to the rearranged unsymmetrical dimer **8.40** (Scheme 8.24). The isolation of **8.38,** then, supports an addition mechanism from the ylide rather than a carbene dimerization mechanism (Scheme 8.25). No evidence could be obtained for carbene character in the reaction intermediate with the use of carbene-trapping reagents.

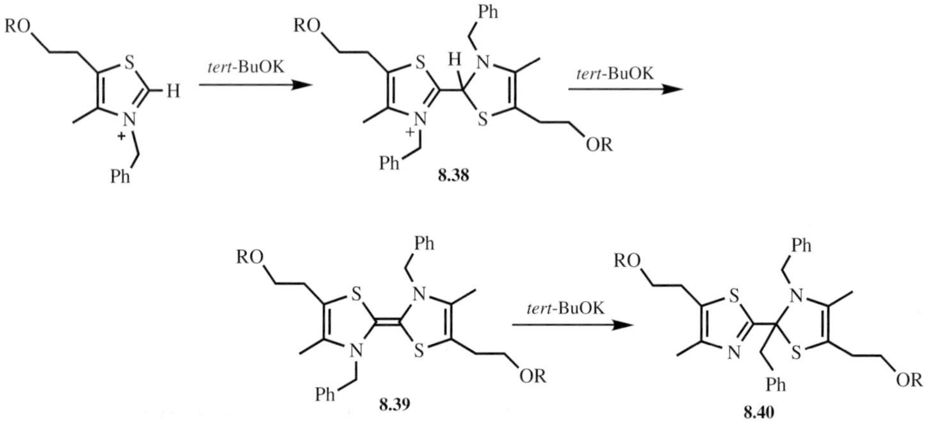

8.38 8.39 8.40

SCHEME 8.24 Intermediates in the base-catalyzed dimerization of a thiazolium salt.

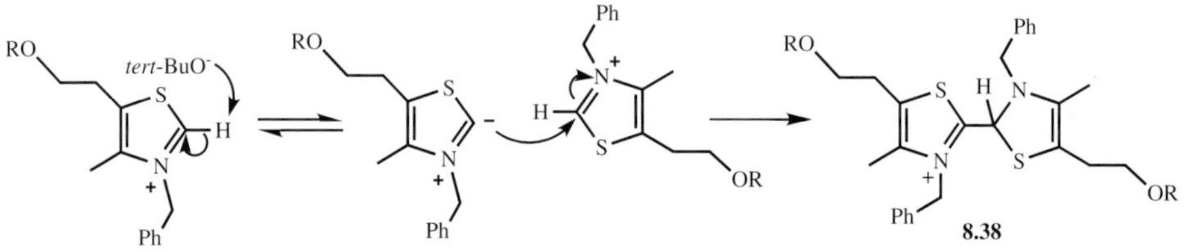

8.38

SCHEME 8.25 Addition mechanism for dimerization of thiazolium salts.

8.41   **8.42**   **8.43**

Given the ubiquity of imidazole-containing molecules in nature, why is the thiazolium ring utilized in thiamin? A possible answer comes from a model study that Duclos and Haake carried out to compare the stabilities and acidities of N-methylthiazolium (**8.41**), N-methyloxazolium (**8.42**), and the N-methylimidazolium ion (**8.43**).[26] Compound **8.42** is 100 times more acidic than **8.41**, so you might think that **8.42** would have been a better choice for the coenzyme heterocycle than **8.41**. On the one hand, however, **8.42** does not chemically catalyze α-keto acid decarboxylations, and it undergoes hydrolysis easily in water at pH 7. On the other hand, **8.41** and **8.43** are stable in water at pH 7, but deprotonation of **8.43** is very slow. It therefore appears that the thiazolium ring really is the best heterocycle for the coenzyme.

C. Mechanism of Thiamin Diphosphate-Dependent Enzymes

Two general families of reactions are catalyzed by TDP: nonoxidative decarboxylation and oxidative decarboxylation of α-keto acids.

1. Nonoxidative Decarboxylation of α-Keto Acids

The two enzymes we will consider both utilize pyruvate (**8.44**) as the substrate; one (pyruvate decarboxylase) catalyzes the conversion of pyruvate to acetaldehyde and carbon dioxide (Scheme 8.26A), and the other (acetolactate synthase) catalyzes the conversion of two molecules of pyruvate to acetolactate (**8.45**), a β-keto acid, which undergoes decarboxylation to acetoin (**8.46**) (Scheme 8.26B).

Before discussing the mechanism of thiamin diphosphate-dependent reactions, first examine the benzoin condensation reaction (Scheme 8.27), a chemical model reaction that mimics the TDP-catalyzed formation of acetoin. The benzoin condensation utilizes cyanide ion as a catalyst for the conversion of two molecules of benzaldehyde to benzoin (**8.47**). If pK_a values were ignored, the simplest way to write a mechanism for this reaction would be the incorrect one shown in Scheme 8.28, formation of benzaldehyde anion followed by addition to the second molecule of benzaldehyde. But the aldehyde proton is not acidic (because the resultant carbanion is not stabilized) and cyanide is a weak base, so this mechanism must be

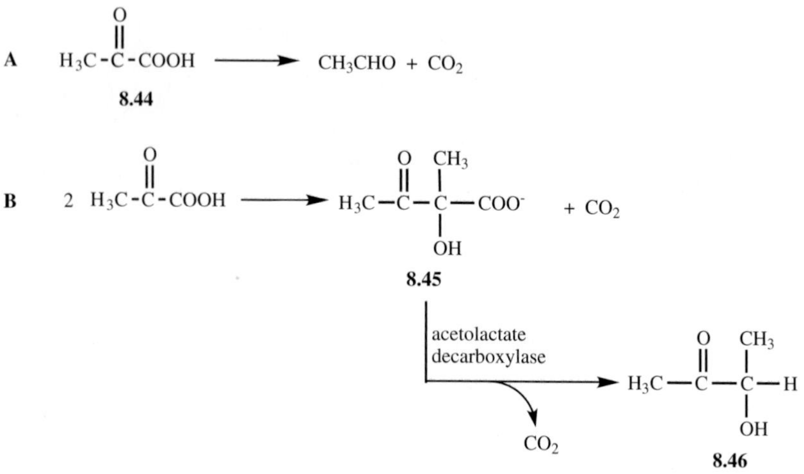

A   H₃C-C-COOH ⟶ CH₃CHO + CO₂

8.44

B   2 H₃C-C-COOH ⟶ H₃C-C-C—COO⁻ + CO₂

8.45

acetolactate decarboxylase

H₃C—C—C—H

CO₂

8.46

SCHEME 8.26   Nonoxidative decarboxylation of α-keto acids: (A) the reaction catalyzed by pyruvate decarboxylase, (B) the reaction catalyzed by acetolactate synthase.

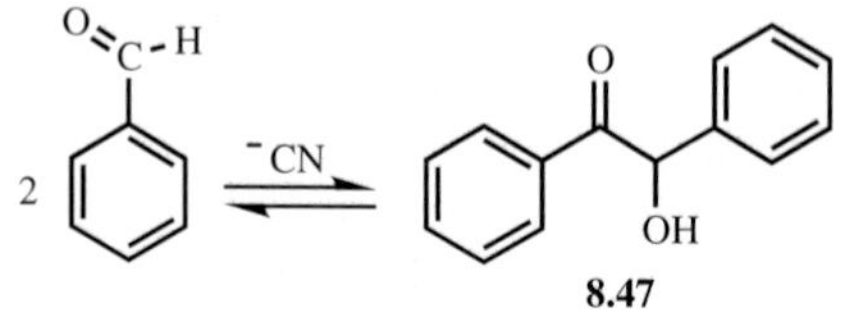

8.47

SCHEME 8.27   Benzoin condensation.

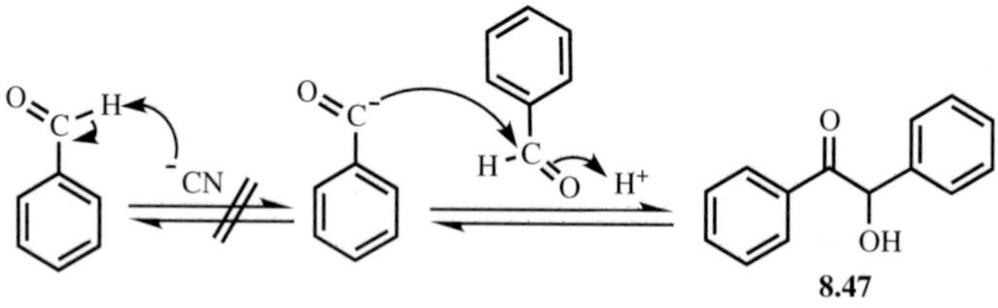

8.47

SCHEME 8.28   Incorrect mechanism for the benzoin condensation.

wrong. However, we would like to be able to make the carbonyl carbon nucleophilic somehow. The mechanism in Scheme 8.29, in which cyanide ion acts as a nucleophile at first (cyanide is an excellent nucleophile), is much more reasonable. In the first step the cyanide ion adds to the carbonyl to give the cyanohydrin (**8.48**). As a result of this addition, what originally was the nonacidic benzaldehyde aldehyde proton becomes an acidic proton in **8.48** because of the resonance stabilization

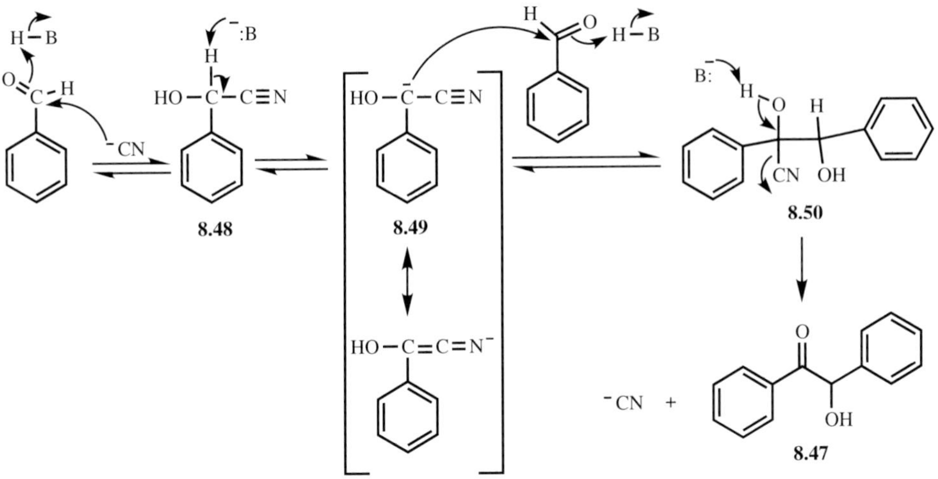

SCHEME 8.29   Mechanism for the benzoin condensation.

of the incipient carbanion (**8.49**) by the cyano group. Addition of this carbanion to the carbonyl of another molecule of benzaldehyde gives the new cyanohydrin (**8.50**); elimination of cyanide ion regenerates the catalyst and produces benzoin (**8.47**). Note how the electrophilic carbon of benzaldehyde (the carbonyl carbon) is converted into the nucleophilic carbon of **8.49** by virtue of the addition of the cyanide ion catalyst. Intermediate **8.49** is chemically equivalent to the benzaldehyde carbanion in Scheme 8.28. Because cyanide is so toxic, it would not be wise for this to be the natural catalyst; instead, TDP is used. Think of TDP as the biological cyanide ion.

a. Acetolactate Synthase

Acetolactate synthase catalyzes the conversion of two molecules of pyruvate to acetolactate (Scheme 8.26B). On the basis of radiolabeling experiments and experiments similar to those described later for pyruvate decarboxylase (III.C.1.b), the mechanism shown in Scheme 8.30, which strongly resembles the benzoin condensation mechanism (Scheme 8.29), is reasonable. Deprotonation of TDP gives the ylide (equivalent to cyanide ion in Scheme 8.29), which adds to the first pyruvate molecule to give **8.51** (equivalent to **8.48** in Scheme 8.29). Decarboxylation (which is equivalent to deprotonation) gives **8.52** (**8.49** in Scheme 8.29). Addition of **8.52** to the second pyruvate molecule gives **8.53** (see **8.50** in Scheme 8.29 for comparison); elimination of TDP ylide from **8.53** (equivalent to loss of cyanide ion from **8.50** in Scheme 8.29) produces acetolactate (**8.45**). Another enzyme, acetolactate decarboxylase catalyzes the decarboxylation of the β-keto acid acetolactate to acetoin (**8.46**).

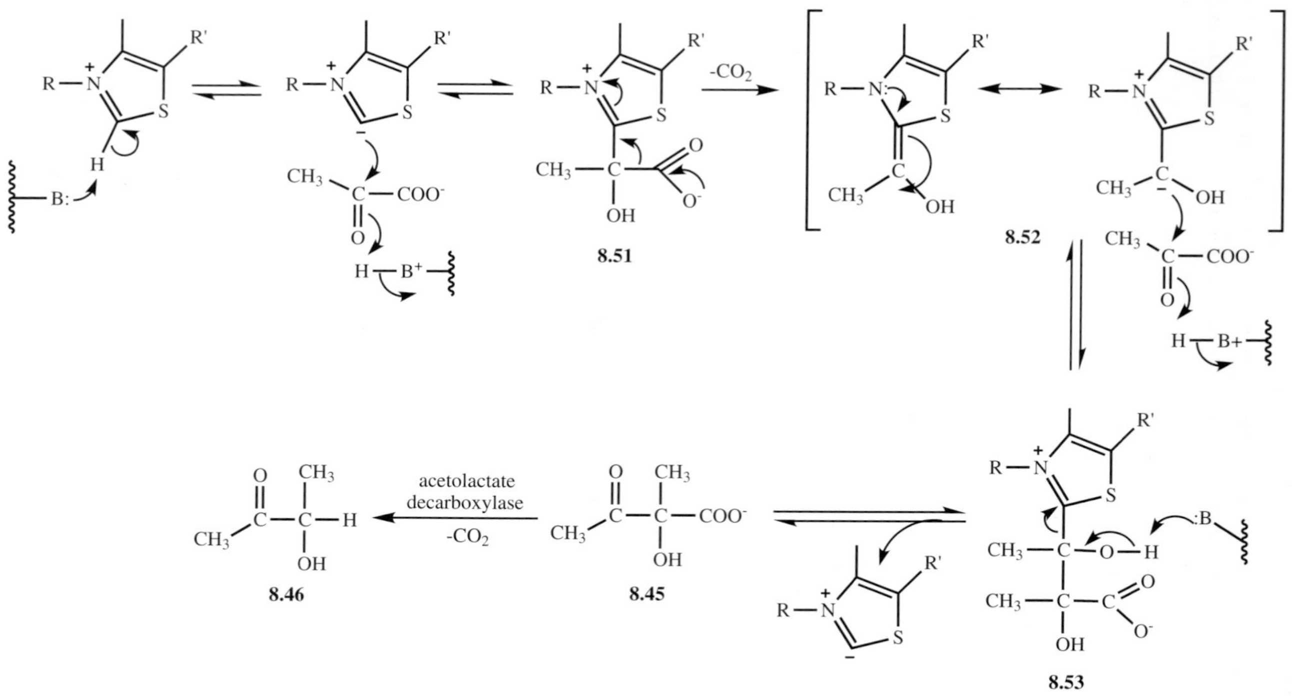

SCHEME 8.30   Proposed mechanism for the reaction catalyzed by acetolactate synthase.

b. Yeast Pyruvate Decarboxylase

The pyruvate decarboxylase reaction is less complicated than that of acetolactate synthase, but it does not mimic the benzoin condensation mechanism as closely because it does not involve a dimerization (Scheme 8.31). Compound **8.51** was synthesized and was found to undergo nonenzymatic decarboxylation to hydroxyethyl TDP (**8.54**). The rate of decarboxylation is considerably accelerated in nonhydroxylic solvents (even in ethanol the rate is much greater than in water), which suggests that the enzyme may have a hydrophobic active site. Hydroxyethyl TDP (**8.54**) also has been synthesized, and it undergoes rapid (kinetically competent) enzyme-catalyzed conversion to acetaldehyde and TDP. Because pyruvate decarboxylase does not catalyze deuterium exchange (in D_2O) of the proton on the carbon to which the TDP is bound, and the hydroxyethyl TDP produced from pyruvate is optically active, it was proposed that decarboxylation and protonation may be concerted.[27] Evidence for a nonconcerted mechanism comes from the isolation of a stable zwitterion (**8.56**) from the pyruvate decarboxylase-catalyzed decarboxylation of **8.55** (Scheme 8.32), but **8.56** is highly stabilized, so **8.55** may not be a representative substrate.[28]

Pyruvate decarboxylase, as well as other enzymes that catalyze reactions proceeding via carbanion intermediates, such as acetolactate synthase, glutamate decarboxylase, and class II aldolase, also catalyze O_2-consuming reactions.[29]

2. Oxidative Decarboxylation of α-Keto Acids

Examples of these types of reactions, shown in Scheme 8.33, include the conversion of pyruvate to acetyl CoA (**A**) and α-ketoglutarate to succinyl CoA (**B**). These enzyme systems are very complex and require five different coenzymes.

a. α-Ketoacid Dehydrogenase Complexes

These α-ketoacid dehydrogenase complexes are multienzyme complexes that function as a catalytic packet. Pyruvate dehydrogenase, for example, consists of three enzymes: pyruvate decarboxylase, dihydrolipoyl transacetylase, and dihydrolipoyl dehydrogenase.

(1) Pyruvate Decarboxylase   This enzyme converts pyruvate with TDP to 2-hydroxyethyl TDP anion (see Scheme 8.31 up to **8.52**); TDP is the first of the five cofactors.

(2) Dihydrolipoyl Transacetylase   Product **8.52** from the pyruvate decarboxylase-catalyzed reaction then reacts with enzyme-bound lipoic acid (**8.57**; cofactor 2; Scheme 8.34) to give intermediate **8.58**, which undergoes elimination to TDP and acetyl lipoamide (**8.59**). Reaction of **8.59** with the third cofactor, CoASH

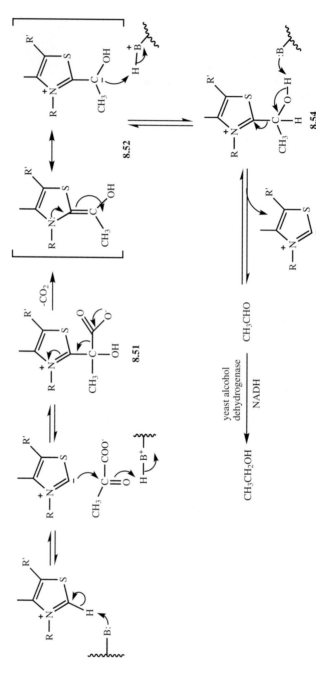

SCHEME 8.31   Proposed mechanism for the reaction catalyzed by pyruvate decarboxylase.

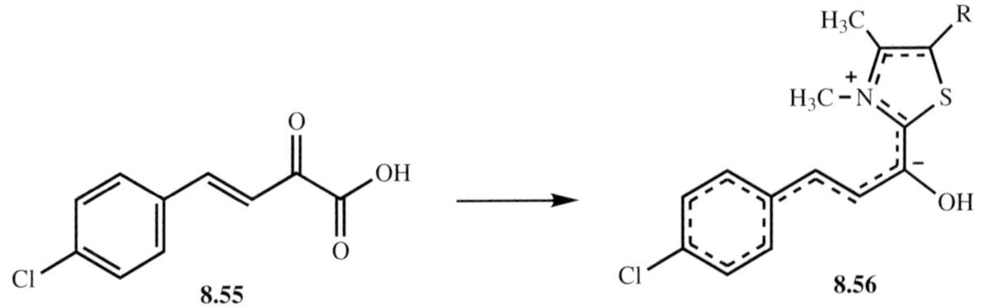

8.55 8.56

SCHEME 8.32 Zwitterion isolated from the reaction of **8.55** with pyruvate decarboxylase.

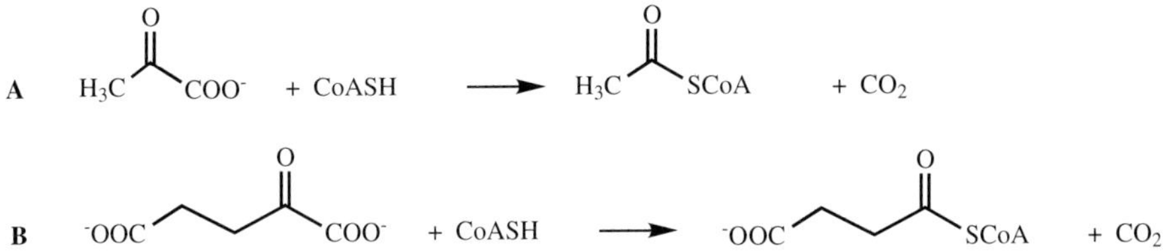

SCHEME 8.33 Examples of oxidative decarboxylation of α-keto acids.

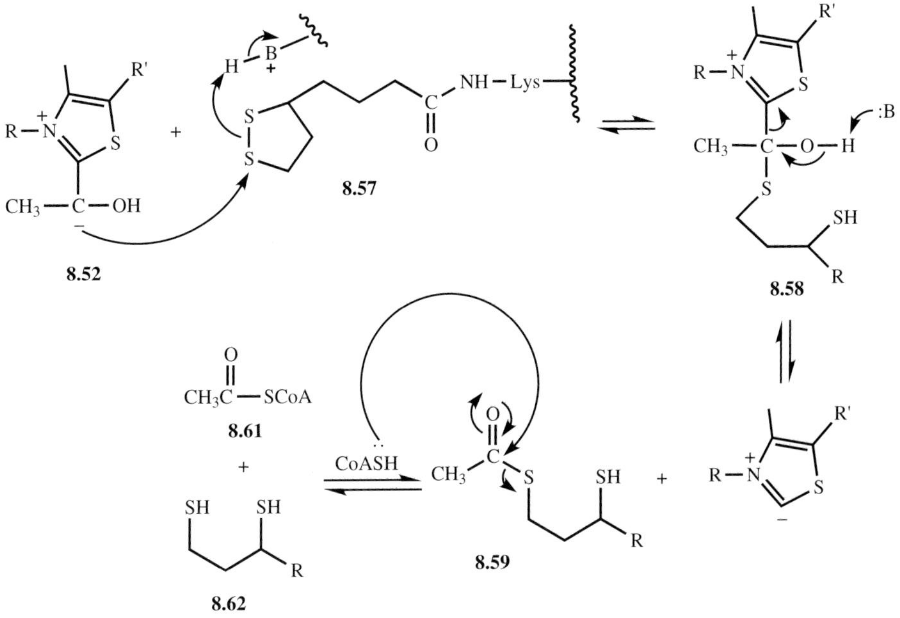

SCHEME 8.34 Proposed mechanism for the reaction catalyzed by dihydrolipoyl transacetylase.

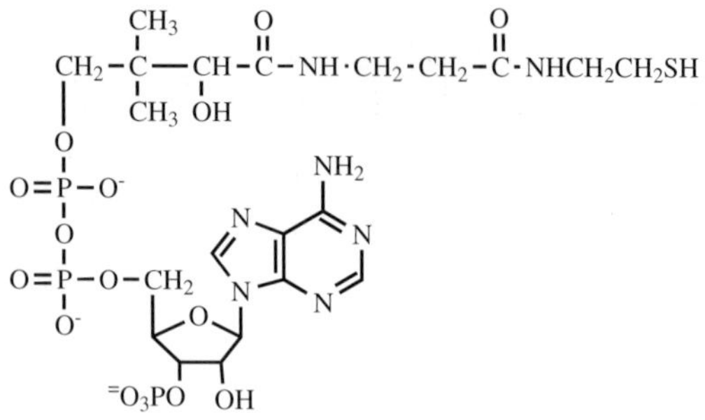

8.60

(see **8.60** for its structure), gives acetyl CoA (**8.61**) and reduced lipoic acid (**8.62**). An equally reasonable alternative mechanism is shown in Scheme 8.35.

Jordan and co-workers proposed an alternative mechanism to the carbanion chemistry described in Schemes 8.34 and 8.35 on the basis of electrochemical model studies of TDP analogues.[30] These results suggest that oxidation of hydroxyethyl TDP could occur by two 1-electron transfers, depending on what oxidizing agent is involved (Scheme 8.36). Some of the TDP oxidative enzymes require flavin or

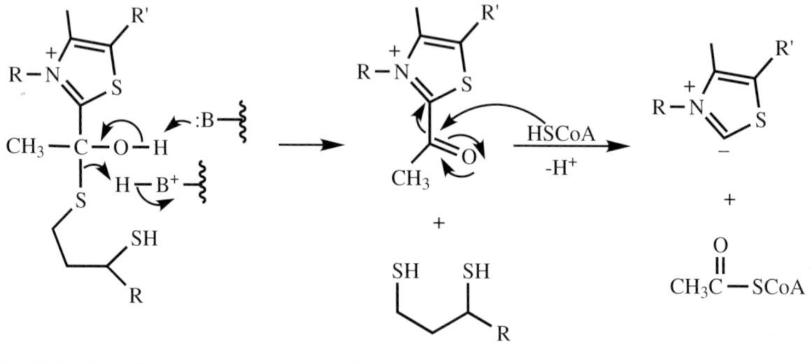

SCHEME 8.35   Alternative proposed mechanism for the reaction catalyzed by dihydrolipoyl transacetylase.

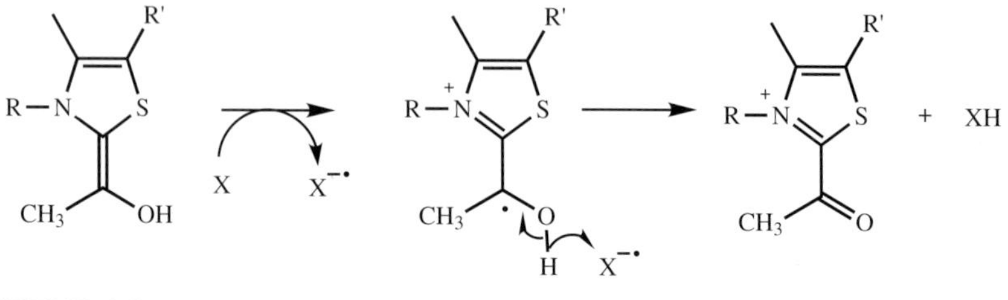

SCHEME 8.36   Proposed one-electron mechanism for the reaction catalyzed by dihydrolipoyl transacetylase.

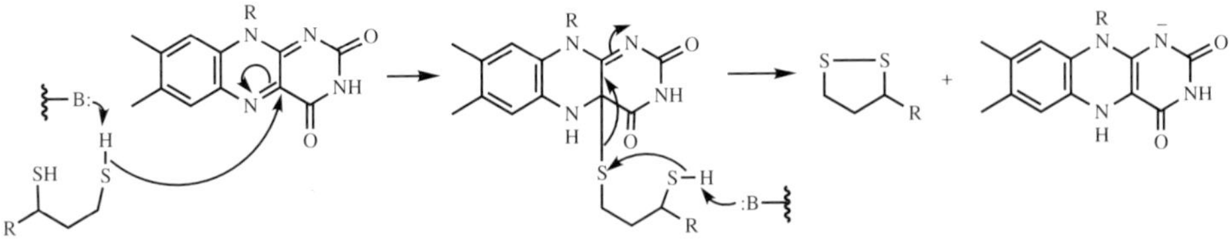

8.62

SCHEME 8.37   Proposed mechanism for the reaction catalyzed by dihydrolipoyl dehydrogenase.

an iron sulfur cluster (Fe_4S_4) (see Chapter 3, Section III.D.2) instead of lipoic acid. The X in Scheme 8.36 could be the lipoamide, flavin, or iron sulfur cluster. Spin-trapping of an intermediate radical and EPR detection during the reaction of pyruvate: ferrodoxin 2-oxidoreductase (contains an Fe_4S_4 cluster) has been reported.[31]

(3) Dihydrolipoyl Dehydrogenase   Cofactor 4 in the pyruvate dehydrogenase complex, FAD, is required to reoxidize the reduced lipoic acid (**8.62**, Scheme 8.37). Model and enzyme studies by Loechler and Hollocher suggest that attack of the sulfhydryl group of reduced lipoic acid occurs at the C-4a position of the flavin ring.[32]

The fifth cofactor is NAD^+, which is needed to reoxidize the reduced flavin generated in Scheme 8.37; because this is a flavin dehydrogenase, molecular oxygen is not involved.

V. AMINO ACIDS

A. Pyridoxal Phosphate-Dependent Decarboxylation

We have seen that decarboxylation requires an electron sink beta to the carboxyl group, such as a β-keto group. Enzymes that catalyze decarboxylation reactions of β-keto acids often form Schiff bases between the β-keto group and the ϵ-amino group of a lysine residue. Amino acids can undergo a similar type of reaction using the NH_2 group of the amino acid and the aldehyde carbonyl of a coenzyme named *pyridoxal 5'-phosphate* (**8.64**, PLP),[33] a cofactor derived from vitamin B_6 (**8.63**, *pyridoxine*, Scheme 8.38). A dietary deficiency of vitamin B_6 leads to dermatitis, infections, and convulsions. The PLP in PLP-dependent enzymes is always found covalently bound to the active site of these enzymes via a Schiff base (imine) linkage to a lysine residue (**8.65**, Scheme 8.39).[34] In addition to securing the PLP in the optimal position at the active site, Schiff base formation activates the carbonyl for nucleophilic attack. Cordes and Jencks[35] found that the reaction between an amine and an imine (to give a new imine; see Scheme 8.40A) is 30 times faster than the reaction of an amine with an aldehyde (Schiff base formation; Scheme 8.40B). This is important to catalysis because the first step in all PLP-dependent catalyzed reactions

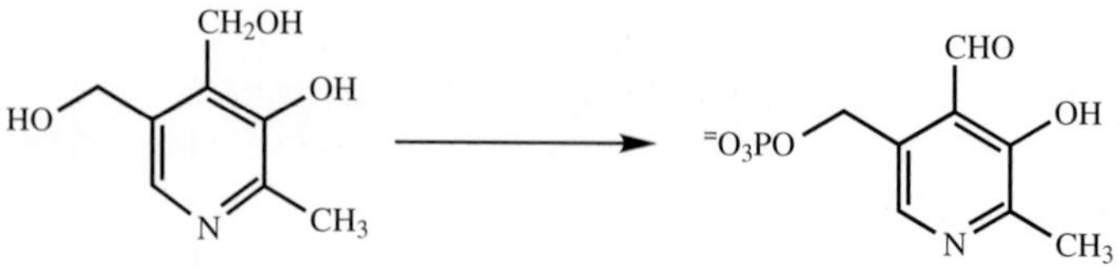

8.63   **8.64**

SCHEME 8.38   Conversion of pyridoxine to pyridoxal-5′-phosphate (PLP).

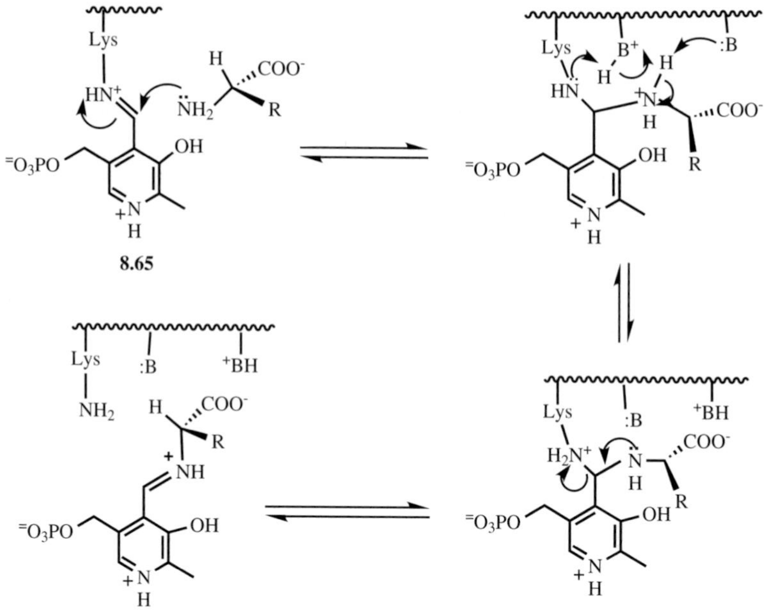

SCHEME 8.39   The first step catalyzed by all PLP-dependent enzymes, the formation of the Schiff base between the amino acid substrate and PLP.

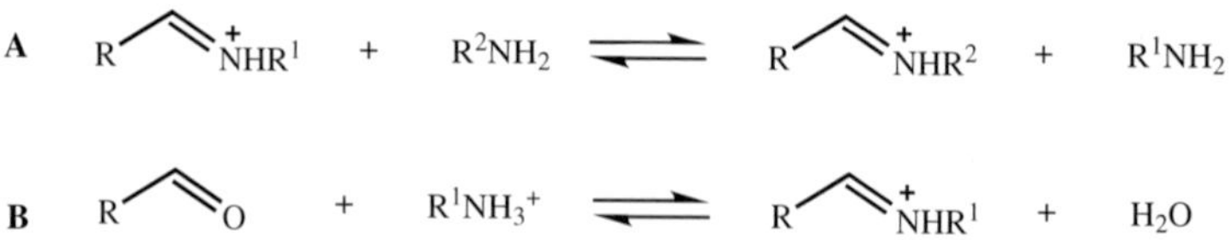

SCHEME 8.40   (A) Reaction of an amine with an imine. (B) Reaction of an amine with an aldehyde.

is a transimination reaction, namely, the conversion of the lysine-PLP imine (**8.65**) to the substrate-PLP imine (**8.66**). In Scheme 8.39 two different bases are shown to be involved in acid/base catalysis; a similar mechanism could be drawn with a single base. Although PLP-dependent enzymes are similar in mechanisms, there are generally only two conserved residues: a lysine bound to the PLP (Lys-355 in *Lactobacillus 30a* ornithine decarboxylase)[36] and an aspartate or glutamate residue, which is hydrogen-bonded to the pyridine ring nitrogen (Asp-316 in ornithine decarboxylase). Generally, there also is an arginine residue that binds to the α-carboxylate of PLP-bound substrate (not identified in ornithine decarboxylase).

Specific decarboxylases are known for more than 10 of the common amino acids. Each catalyzes the conversion of the amino acid to the corresponding amine and carbon dioxide (Scheme 8.41). Aromatic L-amino acid decarboxylase (sometimes referred to as dopa decarboxylase) plays a vital role in converting the antiparkinsonism drug L-dopa to dopamine in the brain; low dopamine levels are characteristic of parkinsonism.

Many bacterial amino acid decarboxylases have low pH optima. It is possible that at least one function of these decarboxylases, which generate amine bases from neutral amino acids, is to neutralize the acidic conditions in the cell. Another function may be to control the bacterial intracellular CO_2 pressure.

Treatment of a PLP-dependent enzyme with $NaBH_4$ followed by hydrolysis of the protein generally gives **8.67** (Scheme 8.42). This demonstrates that the PLP is

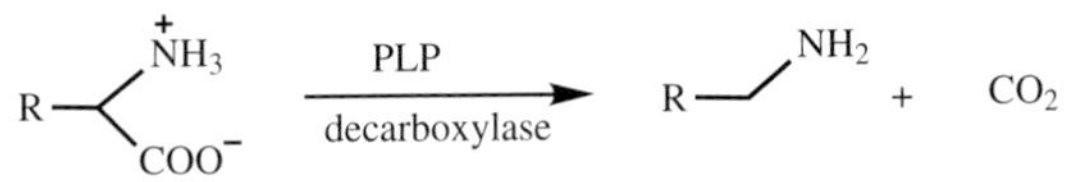

SCHEME 8.41   Reaction catalyzed by PLP-dependent decarboxylases.

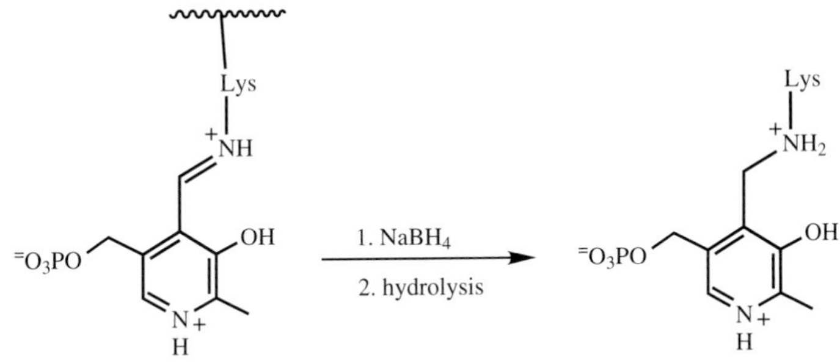

SCHEME 8.42   Reduction and hydrolysis of PLP enzymes. Evidence for the formation of a Schiff base between PLP and an active-site lysine residue.

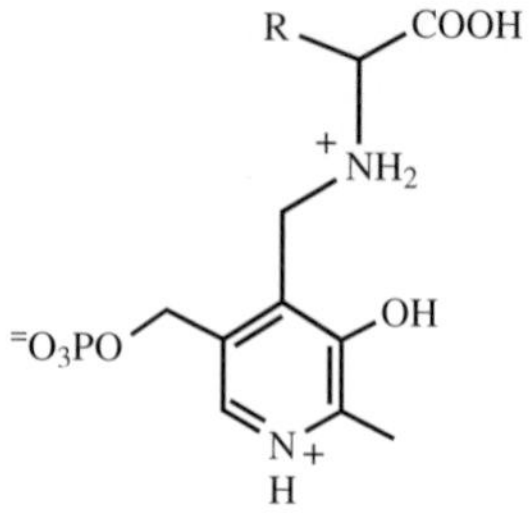

8.68

held at the active site of these enzymes via a Schiff base to the ε-amino group of a lysine residue. If the enzyme is incubated with substrate before NaBH$_4$ reduction, the amino acid substrate becomes reduced onto the PLP (**8.68**).

On decarboxylation, one proton is incorporated stereospecifically from the solvent into the α-position of the product amine. This stereospecific incorporation indicates that protonation is enzyme catalyzed. The α-proton of the amino acid is not exchanged with solvent during enzyme catalysis. If the enzyme reaction is carried out in H$_2$^{18}O, no ^{18}O is found in the CO$_2$ generated. This result excludes a hydrolytic mechanism (Scheme 8.43).

The mechanism shown in Scheme 8.44 for PLP-dependent decarboxylases is consistent with all of the results just described. All the steps are reversible except the decarboxylation step. Much more will be said about PLP-dependent enzymes later in the book (Chapter 9, Sections II.B and IV.C; Chapter 10, Section II.B; Chapter 11, Section II.C; Chapter 12, Section I.A; and Chapter 13, Section III.B.2.d).

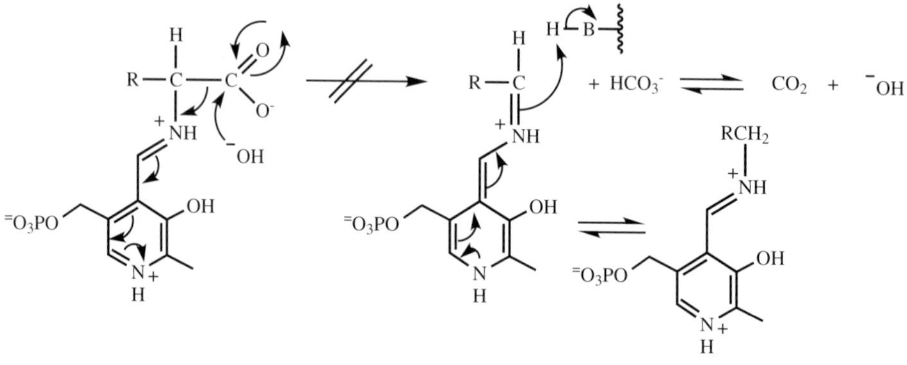

SCHEME 8.43   Incorrect hydrolytic mechanism for PLP-dependent enzymes.

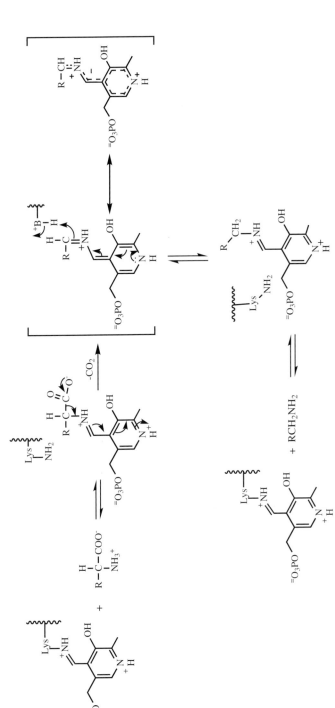

SCHEME 8.44 Proposed mechanism for PLP-dependent decarboxylases.

B. Pyruvoyl-Dependent Decarboxylation

Some amino acid decarboxylases utilize an active-site pyruvoyl group instead of PLP.[37] These enzymes catalyze decarboxylation and reductive cleavage (see Chapter 3, Section III.D.4) reactions of amino acids. Inhibition of an enzyme with cyanide, hydroxylamine, phenylhydrazine, or sodium borohydride typically indicates the presence of a catalytically active carbonyl group, such as in PLP (see Section V.A), PQQ (see Chapter 3, Section III.C.1), or pyruvate. The identity of the pyruvoyl-containing active site can be derived from several experiments. The pyruvoyl group is typically bound to the enzyme in an amide linkage; mild acid hydrolysis (1.5 N HCl, 100 °C, 1.5 h) produces pyruvate, which can be assayed with NADH-dependent lactate dehydrogenase.[38] 4′-Phosphopantothenoylcysteine decarboxylase, however, is bound to the protein in an ester linkage rather than an amide linkage; treatment of this enzyme with 0.1 N NaOH at room temperature for an hour liberates the pyruvate, which can be assayed with lactate dehydrogenase.[39] As mentioned in Chapter 3 (Section III.D.4), because dehydroalanyl-containing enzymes (see Chapter 10, Section I.C.1) also yield pyruvate on hydrolysis, these and pyruvoyl-containing enzymes must be differentiated by another process. This differentiation can be accomplished by reduction with [³H]NaBH₄ followed by strong acid hydrolysis (6 N HCl, 110 °C, 24 h); pyruvoyl enzymes produce [³H]lactate, whereas dehydroalanyl enzymes give [³H]alanine.

PLP and PQQ cofactors can be excluded on the basis of the absorption spectrum of the enzyme, because there is no absorbance above 300 nm for a pyruvoyl enzyme. Treatment of the pyruvoyl enzyme with ammonia and sodium cyanoborohydride followed by acid hydrolysis produces alanine (pathway a, Scheme 8.45) which was not present prior to this treatment.[40] Incubation of the enzyme with the amino acid substrate and sodium borohydride gives two products on acid hydrolysis, N-carboxyethylated amino acid (**8.69,** pathway b) and N-carboxyethylated amine (**8.70,** pathway c). The mechanism for these enzymes is the same as that for the PLP-dependent enzymes (Scheme 8.45).

In some cases, the substrate leads to inactivation of the enzyme; for example, Diaz and Anton[41] have utilized the general mechanism shown in Scheme 8.45 to rationalize the mechanism of inactivation of S-adenosylmethionine decarboxylase by its substrate, leading to substrate-alkylation of Cys-140 (Scheme 8.46). The difference between the turnover and inactivation mechanisms depends on protonation of **8.71** adjacent to the amide carbonyl instead of adjacent to the substrate amino group. This protonation produces **8.72,** which is susceptible to β-elimination because of the excellent leaving group ability of the sulfonium ion to give the α,β-unsaturated iminium ion (**8.73**), which can undergo Michael addition by Cys-140 (**8.74**). Tautomerization and hydrolysis gives alkylated Cys-140 (**8.75**). The structure of the modified cysteine was characterized using [¹⁴C]S-adenosylmethionine, reducing the **8.75** with sodium borohydride, tryptic digesting (to a 15-mer), and Edman sequencing the peptide (**8.76**). The only cycle that released radioactivity

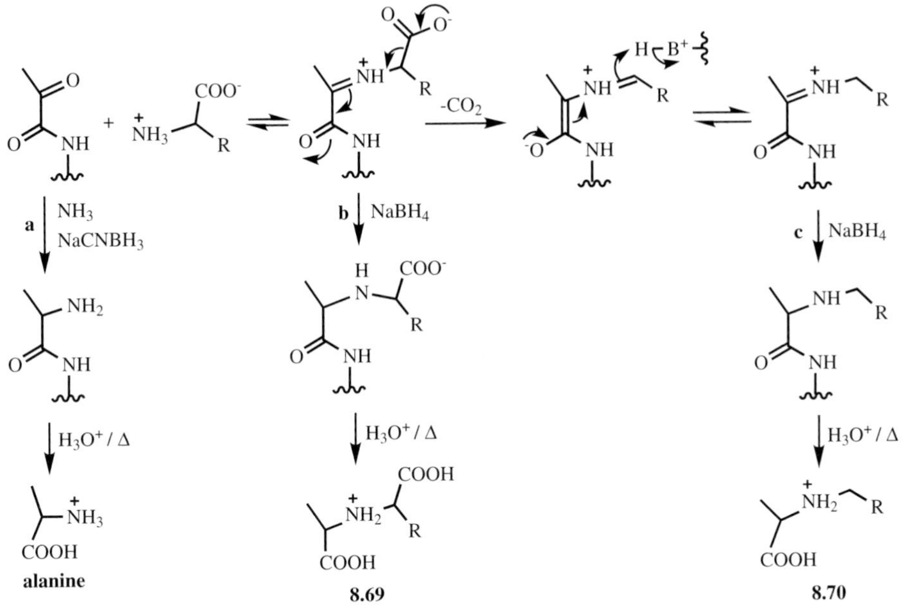

SCHEME 8.45   Proposed mechanism for pyruvoyl-dependent decarboxylases.

corresponded to a modified Cys-140, which had identical chromatographic properties as synthetic S-(3-hydroxypropyl)cysteine. Furthermore, the FAB mass spectrum of the labeled peptide had a m/z consistent with **8.76.**

For histidine decarboxylase from *Lactobacillus 30a,* the pyruvoyl group is biosynthesized from an active-site serine residue.[42] A -Ser-Ser- dipeptide in the prohistidine decarboxylase is cleaved to yield the active-site pyruvoyl group of histidine decarboxylase (Scheme 8.47 gives two possible mechanisms). If ^{18}O is incorporated by site-specific mutation into each of these serine residues, both ^{18}O atoms end up in the carboxylate group of the C-terminal serine of the β-subunit of histidine decarboxylase.

VI. OTHER SUBSTRATES

A. Orotidine 5′-Monophosphate

Two mechanisms for orotidine 5′-monophosphate (OMP; orotidylate) decarboxylase have been suggested (Schemes 8.48[43] and 8.49[44]). The mechanism in Scheme 8.48 was proposed on the basis of two chemical model studies, one that demonstrated that a sulfur nucleophile could undergo Michael addition to N,N-dimethylorotaldehyde and N,N-dimethyl-6-acetyluracil (Scheme 8.50A), and another that

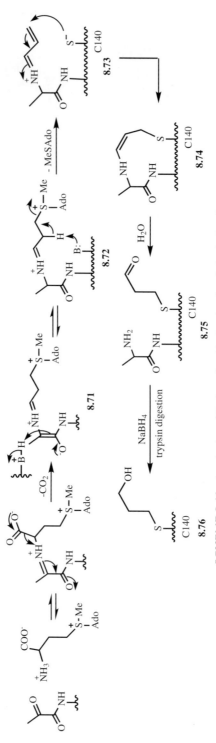

SCHEME 8.46 Inactivation of S-adenosylmethionine decarboxylation by its substrate.

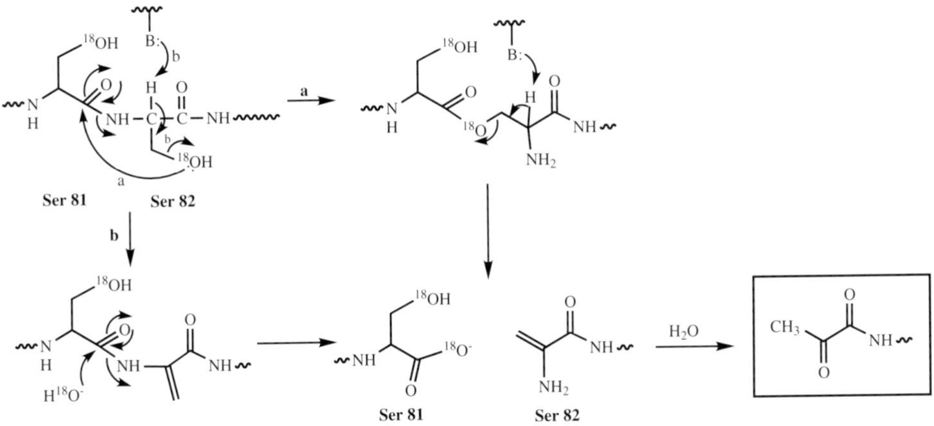

SCHEME 8.47   Biosynthesis of the active-site pyruvoyl group of histidine decarboxylase.

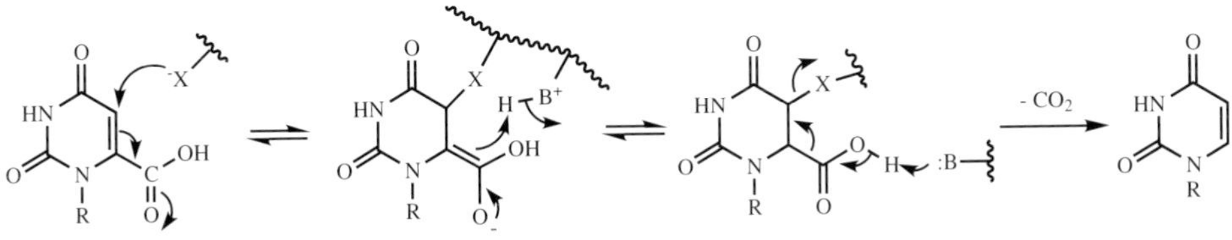

SCHEME 8.48   Proposed addition/elimination mechanism for orotidine 5′-monophosphate decarboxylase.

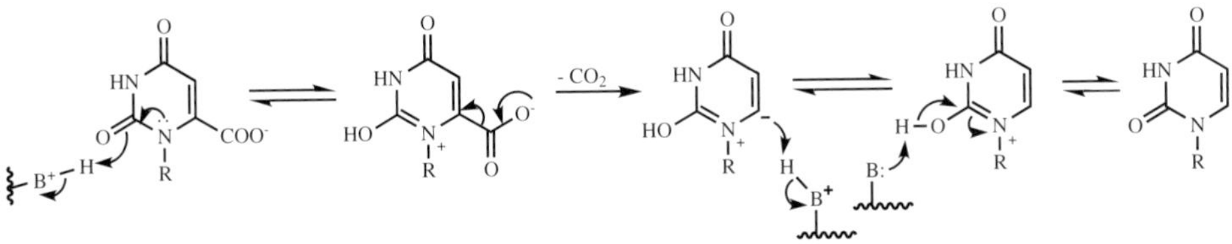

SCHEME 8.49   Proposed zwitterion mechanism for orotidine 5′-monophosphate decarboxylase.

showed that decarboxylative elimination of a model covalent adduct was facile (Scheme 8.50B).

In support of the mechanism in Scheme 8.49, Wolfenden and co-workers[45] incubated the enzyme with a potent inhibitor, [5-^{13}C]1-(phosphoribosyl)barbituric acid (8.77), and saw no rehybridization of the 5,6-double bond in the ^{13}C NMR spectrum. With [5-^{2}H]orotidine monophosphate (8.78), there was no secondary deuterium isotope effect, again suggesting no rehybridization of the 5,6-double bond. The mechanism in Scheme 8.48 involves a change in the hybridization at

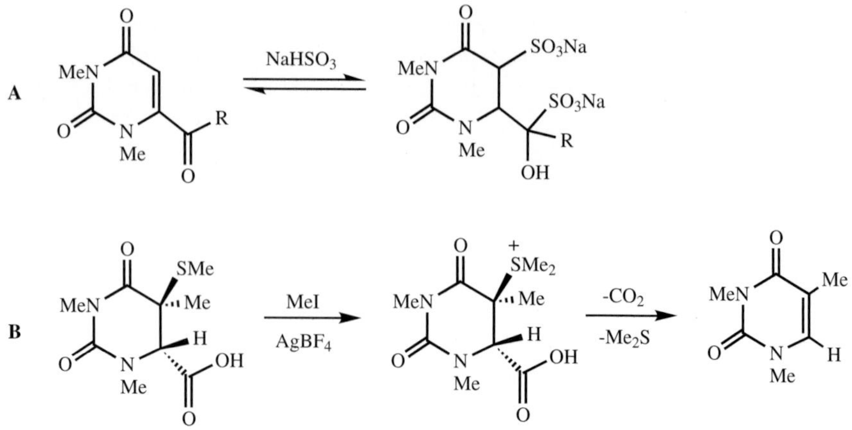

SCHEME 8.50   Model reactions for the addition/elimination mechanism for orotidine 5′-mono-phosphate decarboxylase.

C-5 from sp^2 to sp^3 (and back again) during turnover, whereas the mechanism in Scheme 8.49 does not involve a hybridization change. Other substrates/inhibitors of OMP decarboxylase provide further support for the mechanism in Scheme 8.49. 5-Bromo- and 5-chloroorotidylate (**8.79**, X = Br and Cl) are excellent inhibitors of orotidine 5′-phosphate decarboxylase, and 5-fluoroorotidylate (**8.79**, X = F)

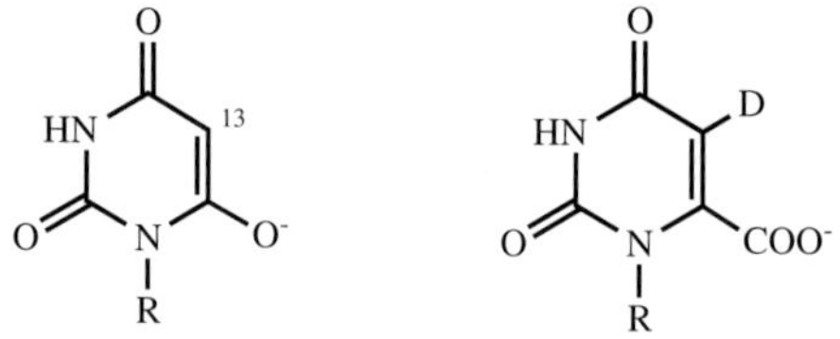

8.77   **8.78**

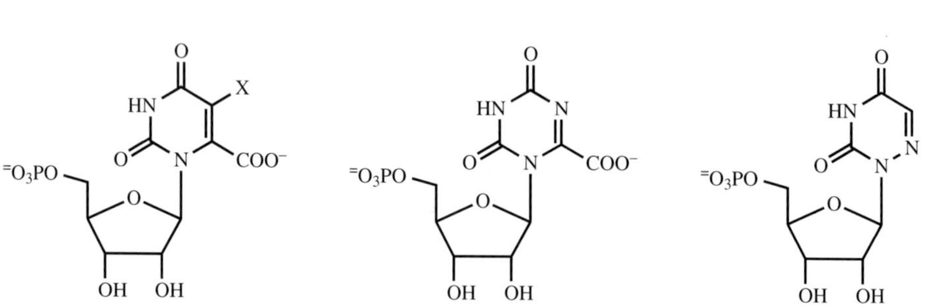

8.79   **8.80**   **8.81**

and 5-azaorotidylate (**8.80**) are excellent substrates. These compounds would disfavor attack at C-5, as in Scheme 8.48. Spectral studies with 6-azauridylate (**8.81**), which would strongly favor C-5 attack, indicate no loss of double-bond character between C-5 and N-6 on binding. These results are not consistent with a mechanism that involves nucleophilic addition at C-5. Additional support for the mechanism in Scheme 8.49 comes from Smiley and Benkovic's[46] ability to raise catalytic antibodies that catalyze the decarboxylation of orotate from a hapten that mimics the transition state for the reaction shown in Scheme 8.49.

As a model for the OMP decarboxylase mechanism in Scheme 8.49, Radzicka and Wolfenden[47] investigated the thermal conversion of 1-methylorotic acid (**8.82**, R = COOH) to 1-methyluracil (**8.82**, R = H) between 140 °C and 200 °C. Extrapolation of the Arrhenius plot to 25 °C gave a nonenzymatic rate constant, $k_{non} = 2.8 \pm 2 \times 10^{-16}$ s^{-1} ($t_{1/2} = 7.8 \times 10^7$ years). Given the turnover number for yeast OMP decarboxylase at pH 6 and 25 °C (39 s^{-1}), there is a 1.4×10^{17}-fold rate enhancement (k_{cat}/k_{non}). The *catalytic proficiency,* a measure of the enzyme's ability to lower the activation barrier for the reaction of a substrate in solution, is the ratio of the second-order rate constant for the same reaction in neutral aqueous solution in the absence of enzyme (k_{non}). This ratio represents the lower limit of the enzyme's affinity for the altered substrate in the transition state. The catalytic proficiency for OMP decarboxylase, then, is $5.6 \times 10^7/2.8 \times 10^{-16} = 2.0 \times 10^{23}$ M^{-1}, which is the largest catalytic proficiency known to date. Because the catalytic proficiency is a measure of the binding at the transition state, proficient enzymes should be especially sensitive to inhibition by transition-state analogue inhibitors.

Lee and Houk[48] carried out quantum mechanical calculations of the OMP decarboxylase-catalyzed reaction; based on these calculations, they predicted that a stabilized carbene intermediate is involved. Decarboxylation by protonation at O-2, as suggested by Beak and Siegel (Scheme 8.49), would give a carbene with separated charges (**8.83**). A more stable species, predicted by these calculations, is the neutral carbene (**8.84**), which could arise from enzyme-catalyzed protonation at O-4 of orotate. This is reminiscent of the proposed acidity of thiamin diphosphate (see Section IV.B); a carbene resonance structure of the thiamin diphosphate zwitterion is important to its stability. Decarboxylation of orotate that has been protonated at O-4 has a very low energy barrier, only +15.5 kcal/mol, whereas decarboxylation of orotate that has been protonated at O-2 (as in Scheme 8.49) is endothermic by +21.6 kcal/mol. Furthermore, gas-phase acidities of protonated

8.82

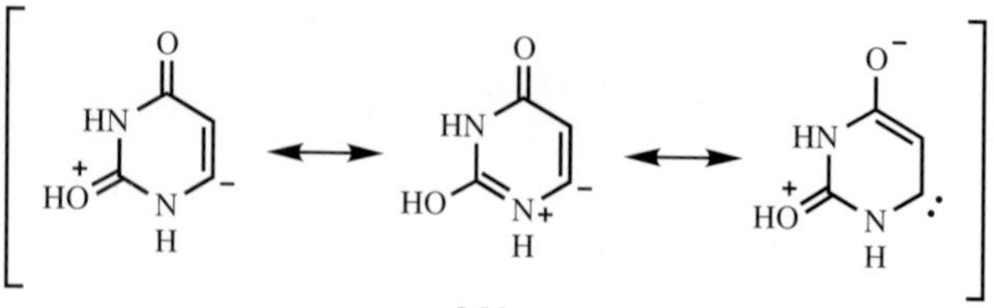

8.83

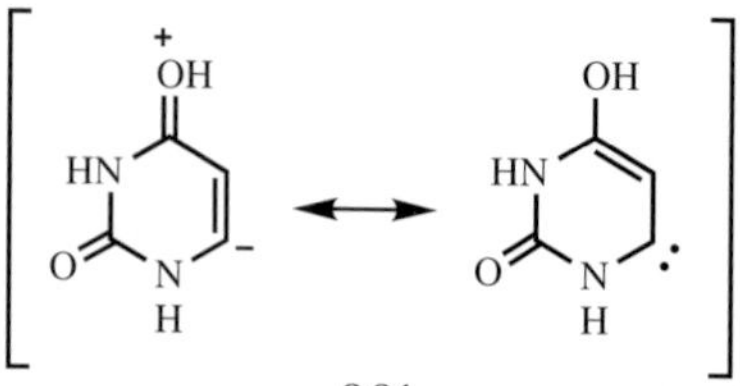

8.84

orotates indicate that the 4-position is, in fact, more basic than the 2-position. For protonation to such a weakly basic site as O-4 of orotidylate to occur, there must be a very nonpolar environment. The concomitant proton transfer and decarboxylation in a nonpolar environment, however, could account for the proton transfer and extraordinary rate acceleration of 10^{17}.

Wolfenden and co-workers utilized flame atomic absorption/emission spectrophotometry to reveal the presence of two atoms of zinc in each monomer of active yeast OMP decarboxylase.[49] Dialysis against EDTA removed the zinc as well as the enzyme activity (dialysis without EDTA had no effect on enzyme activity). However, the same group later found that the recombinant enzyme was devoid of metal ions, but had the same specific activity and mass as the enzyme from yeast.[49a]

B. Mevalonate Diphosphate

Mevalonate diphosphate decarboxylase, an important enzyme in the biosynthesis of cholesterol and other terpenes, catalyzes the decarboxylation of mevalonate diphosphate (**8.85**) to isopentenyl diphosphate (**8.86**, Scheme 8.51). 6-Fluoromevalonate

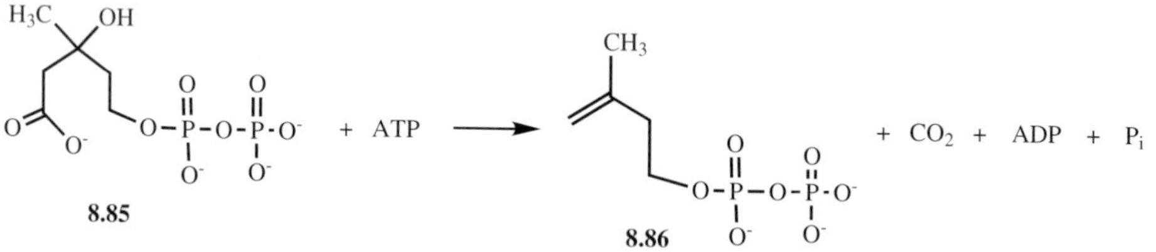

SCHEME 8.51   Reaction catalyzed by mevalonate diphosphate decarboxylase.

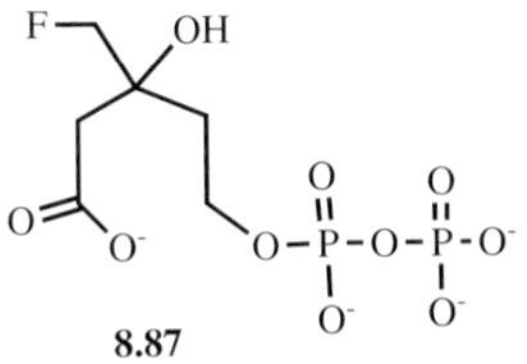

8.87

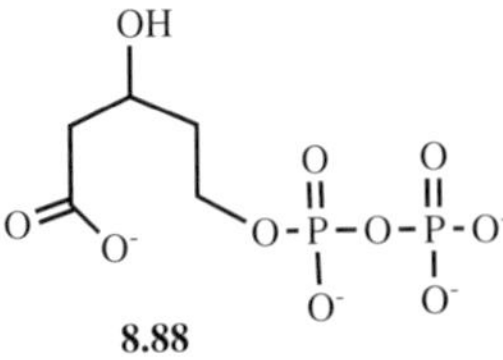

8.88

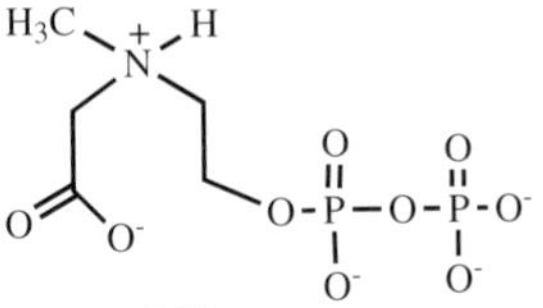

8.89

diphosphate (**8.87**) and 3-hydroxy-5-diphosphopentanoic acid (**8.88**), which are more electron deficient than **8.85,** are much poorer substrates than **8.85,** suggesting a carbocation intermediate.[50] Further support for a carbocationic transition state is the finding that N-methyl-N-carboxymethyl-2-diphosphoethanolamine (**8.89**) is a very potent inhibitor (K_i = 0.75 μM), presumably a transition-state analogue. Consistent with a carbocation intermediate is the mechanism in Scheme 8.52.

In general, decarboxylation requires an "electron sink," that is, a place for the electrons from the carbon-carboxylate bond to be stabilized. β-Keto acids have a β-keto carbonyl group; β-hydroxy acids require initial oxidation of the hydroxyl group to give a β-electron sink (namely, a β-keto group); α-keto acids do not have an electron sink, so they need thiamin diphosphate to extend the α-keto group to a β-electron sink; amino acids also do not have a β-electron sink, so they require pyridoxal 5'-phosphate (or an active-site pyruvoyl group) to form an imine to the amino group and extend the imine π-electrons into the pyridinium ring of PLP; mevalonate diphosphate is another substrate that does not have a β-electron sink, but it is converted into a β-carboxy carbenium ion, and the carbenium ion is the electron sink. The only example in this chapter of a decarboxylation that does not have an obvious electron sink is orotidine 5'-phosphate decarboxylase, but theoretical calculations indicate that for orotidine 5'-phosphate there is a reasonable α-electron sink.

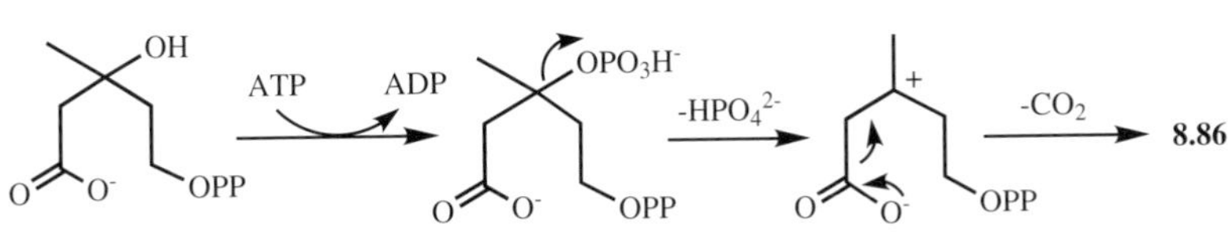

SCHEME 8.52   Proposed carbocation mechanism for mevalonate diphosphate decarboxylase.

Note that problems and solutions relevant to each chapter can be found
in Appendix II.

REFERENCES

1. Westheimer, F. H.; Jones, W. A. *J. Am. Chem. Soc.* **1941**, *63*, 3283.
2. Noyce, D. S.; Metesich, M. A. *J. Org. Chem.* **1967**, *32*, 3243.
3. Bruyneel, W.; Charette, J. J.; De Hoffmann, E. *J. Am. Chem. Soc.* **1966**, *88*, 3808.
4. (a) Pedersen, K. J. *J. Phys. Chem.* **1934**, *38*, 559. (b) Pedersen, K. J. *J. Am. Chem. Soc.* **1938**, *60*, 595.
5. Westheimer, F. H. *Proc. Chem. Soc.* **1963**, 253.
6. (a) Hamilton, G. A.; Westheimer, F. H., *J. Am. Chem. Soc.* **1959**, *81*, 6332. (b) Fridovich, I.; Westheimer, F. H. *J. Am. Chem. Soc.* **1962**, *84*, 3208.
7. (a) Kokesh, F. C.; Westheimer, F. H. *J. Am. Chem. Soc.* **1971**, *93*, 7270. (b) Frey, P. A.; Kokesh, F. C.; Westheimer, F. H. *J. Am. Chem. Soc.* **1971**, *93*, 7266. (c) Schmidt, D. E., Jr.; Westheimer, F. H. *Biochemistry* **1971**, *10*, 1249.
8. Highbarger, L. A.; Gerlt, J. A.; Kenyon, G. L. *Biochemistry* **1996**, *35*, 41.
9. Fridovich, I.; Westheimer, F. H. *J. Am. Chem. Soc.* **1962**, *84*, 3208.
10. Warner, S.; Zerner, B.; Westheimer, F. H. *Biochemistry* **1966**, *5*, 817.
11. Crout, D. H. G.; Littlechild, J.; Mitchell, M. B.; Morrey, S. M. *J. Chem. Soc. Perkin Trans. 1* **1984**, 2271.
12. Lian, H.; Whitman, C. P. *J. Am. Chem. Soc.*, **1994**, *116*, 10403.
13. Hurley, J. H.; Dean, A. M.; Koshland, D. E., Jr.; Stroud, R. M. *Biochemistry* **1991**, *30*, 8671.
14. Topham, C. M.; Dalziel K. *Biochem. J.* **1986**, *234*, 671.
15. (a) Kluger, R. *The Enzymes*, Vol. 20, 3rd ed., Academic Press: San Diego, 1992, p. 271. (b) Kluger, R. *Chem. Rev.* **1987**, *87*, 863.
16. Breslow, R. *J. Am. Chem. Soc.* **1958**, *80*, 3719.
17. Arduengo, A. J., III; Dias, H. V. R.; Dixon, D. A.; Harlow, R. L.; Klooster, W. T.; Koetzle, T. F. *J. Am. Chem. Soc.* **1994**, *116*, 6812.
18. Hopmann, R. F. W.; Brugnoni, G. P. *Nature (New Biol.)* **1973**, *246*, 157.
19. (a) Washabaugh, M. W.; Jencks, W. P. *Biochemistry* **1988**, *27*, 5044. (b) Washabaugh, M. W.; Jencks, W. P. *J. Am. Chem. Soc.* **1989**, *111*, 674. (c) Washabaugh, M. W.; Jencks, W. P. *J. Am. Chem. Soc.* **1989**, *111*, 683.
20. Bordwell, F. G.; Satish, A. V. *J. Am. Chem. Soc.* **1991**, *113*, 985.
21. Lindqvist, Y.; Schneider, G.; Ermler, U.; Sundstrom, M. *EMBO J.* **1992**, *11*, 2373.
22. Kern, D.; Kern, G.; Neef, H.; Tittmann, K.; Killenberg-Jabs, M.; Wikner, C.; Schneider, G.; Hübner, G. *Science* **1997**, *275*, 67.
23. Konig, S.; Schellenberger, A.; Neef, H.; Schneider, G. *J. Biol. Chem.* **1994**, *269*, 10879.
24. Doughty, M. B.; Risinger, G. E. *Bioorg. Chem.* **1987**, *15*, 1.
25. (a) Chen, J.-T.; Jordan, F. *J. Org. Chem.* **1991**, *56*, 5029. (b) Bordwell, F.G.; Satish, A.V.; Jordan, F.; Rios, C. B.; Chung, A. C. *J. Am. Chem. Soc.* **1990**, *112*, 792.
26. Duclos, J. M.; Haake, P. *Biochemistry* **1974**, *13*, 5358.
27. Kluger, R.; Pike, D. C. *J. Am. Chem. Soc.* **1977**, *99*, 4504.
28. Kuo, D. J.; Jordan, F. *J. Biol. Chem.* **1983**, *258*, 13415.
29. Abell, L. M.; Schloss, J. V. *Biochemistry* **1991**, *30*, 7883.
30. Barletta, G.; Chung, A. C.; Rios, C. B.; Jordan, F.; Schlegel, J. M. *J. Am. Chem. Soc.* **1990**, *112*, 8144.
31. Docampo, R.; Moreno, S. N. J.; Mason, R. P. *J. Biol. Chem.* **1987**, *262*, 12417.
32. Loechler, E. L.; Hollocher, T. C. *J. Am. Chem. Soc.* **1975**, *97*, 3235.

33. Dolphin, D.; Poulson, R.; Avramovic, O. *Vitamin B₆: Pyridoxal Phosphate: Chemical, Biochemical, and Medical Aspects*, Parts A and B, Wiley: New York; 1986.

34. John, R. A. *Biochim. Biophys. Acta* **1995**, *1248*, 81.

35. Cordes, E. H.; Jencks, W. P. *Biochemistry* **1962**, *1*, 773.

36. Momany, C.; Ernst, S.; Ghosh, R.; Chang, N. L.; Hackert, M. L. *J. Mol. Biol.* **1995**, *252*, 643.

37. van Poelje, P. D.; Snell, E. E. *Annu. Rev. Biochem.* **1990**, *59*, 29.

38. (a) Riley, W. D.; Snell, E. E. *Biochemistry* **1968**, *7*, 3520. (b) Gutmann, I.; Wahlefeld, A. W. In *Methods of Enzymatic Analysis*, Bergmeyer, H. U., Ed., Academic Press: New York; 1974, pp. 1464–8.

39. Yang, H.; Abeles, R. H. *Biochemistry* **1987**, *26*, 4076.

40. Huynh, Q. K.; Vaaler, G. L.; Recsei, P. A.; Snell, E. E. *J. Biol. Chem.* **1984**, *259*, 2826.

41. Diaz, E.; Anton, D. L. *Biochemistry* **1991**, *30*, 4078.

42. Recsei, P. A.; Huynh, Q. K.; Snell, E. E. *Proc. Natl. Acad. Sci. USA* **1983**, *80*, 973.

43. Silverman, R. B.; Groziak, M. P. *J. Am. Chem. Soc.* **1982**, *104*, 6434.

44. Beak, P; Siegel, B. *J. Am. Chem. Soc.* **1976**, *98*, 3601.

45. Acheson, S. A.; Bell, J. B.; Jones, M. E.; Wolfenden, R. *Biochemistry* **1990**, *29*, 3198.

46. (a) Smiley, J. A.; Benkovic, S. J. *Proc. Natl. Acad. Sci. USA* **1994**, *91*, 8319. (b) Smiley, J. A.; Benkovic, S. J. *J. Am. Chem. Soc.* **1995**, *117*, 3877.

47. Radzicka, A.; Wolfenden, R. *Science* **1995**, *267*, 90.

48. Lee, J. K.; Houk, K. N. *Science* **1997**, *276*, 942.

49. Miller, B. G.; Traut, T. W.; Wolfenden, R. *J. Am. Chem. Soc.* **1998**, *120*, 2666.

49a. Miller, B. G.; Smiley, J. A.; Short, S. A.; Wolfenden, R. *J. Biol. Chem.* **1999**, *274*, 13841.

50. Dhe-Paganon, S.; Magrath, J.; Abeles, R. H. *Biochemistry,* **1994**, *33*, 13355.

CHAPTER 9

Isomerizations

I. GENERAL

An isomerization involves the conversion of one molecule into another molecule with the same formula. In this chapter we will look at enzymes that catalyze the movement of electrons into a different position with concomitant movement of a hydrogen. For convenience, these reactions will be divided into hydrogen shifts to the same carbon (a [1,1]-hydrogen shift), to the adjacent carbon (a [1,2]-hydrogen shift), or to two carbon atoms away (a [1,3]-hydrogen shift).

II. [1,1]-HYDROGEN SHIFT

A [1,1]-hydrogen shift is a more confusing way of saying *racemization* (if the molecule has only one stereogenic center) or *epimerization* (if the molecule has more than one stereogenic center).

A. Amino Acid Racemases That Require No Cofactors

1. Glutamate Racemase and Proline Racemase

Glutamate racemase from *Lactobacillus fermenti,* which catalyzes the interconversion of the enantiomers of glutamate and provides the bacterium with the (*R*)-glutamate it needs for the construction of the peptidoglycan in its cell wall, was shown by Gallo and Knowles to be cofactor independent.[1] Pyridoxal 5′-phosphate (Chapter 8, Section V.A) was excluded as a cofactor by the fact that there is no optical absorption between 330 and 420 nm. Treatment of the enzyme with acid or

base followed (after neutralization) by assay for pyruvate with lactate dehydrogenase (NADH-dependent) showed that no pyruvate was produced. This result excludes a pyruvoyl moiety at the active site (Chapter 8, Section V.B). Finally, EDTA
had no effect on enzyme activity, supporting the lack of a metal ion cofactor. So it
appears that the active-site residues catalyze the racemization.

No ^{18}O is washed out of [1-[^{18}O]carboxyl]glutamate during racemization; this
experiment excludes a possible acyl enzyme intermediate.[2] In D_2O the enzyme catalyzes the incorporation of deuterium into the C-2 position (the stereogenic carbon) of glutamate, thereby supporting a deprotonation/protonation mechanism
instead of an oxidation/reduction mechanism (in which, for example, the α-proton
is removed by NAD$^+$ (see Chapter 3, Section III.A) as a hydride and replaced on
the other side by the same hydride from the incipient NADH). Because two cysteine residues are in the protein (Cys-73 and Cys-184), and reduced thiols stabilize
the enzyme, these residues may be the active-site bases that deprotonate C-2.

There are typically two possible mechanisms for enzyme-catalyzed racemization
(epimerization) that involve proton abstraction, a *one-base mechanism* (Scheme 9.1A)
and a *two-base mechanism* (Scheme 9.1B).[3] In a one-base mechanism, a single active-
site base serves as both the proton abstractor and proton donor. In a two-base
mechanism, one active-site base removes the proton from the substrate, and the
conjugate acid of a second active-site base delivers the proton to the opposite face.
The simplest way to differentiate these mechanisms is on the basis of solvent deuterium incorporation into the substrate and the product. In a one-base mechanism,

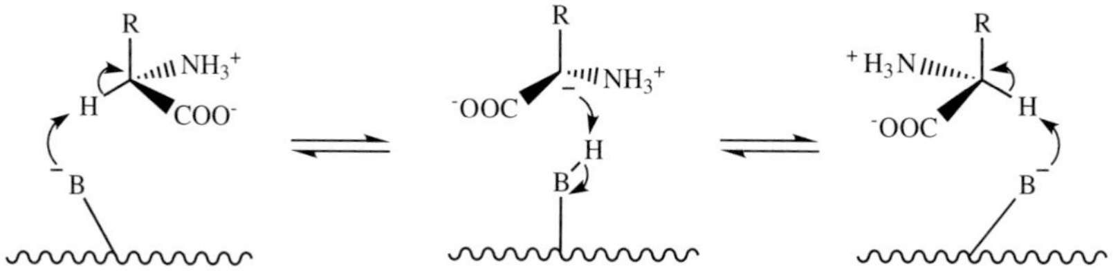

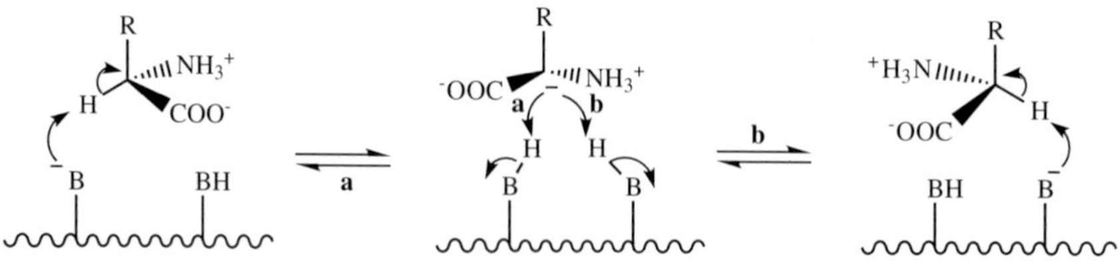

SCHEME 9.1   (**A**) One-base mechanism for racemization (epimerization), (**B**) Two-base mechanism for racemization (epimerization).

the substrate proton is transferred to the product. In a two-base mechanism, the incorporated proton must come solely from the solvent. In the case of glutamate racemase, solvent deuterium is incorporated into the product enantiomer but not into recovered substrate enantiomer, regardless of the direction of the reaction. This experiment supports a two-base mechanism. The two-base mechanism requires that there be two forms of the enzyme that differ by the protonation states of the active-site bases; the enzyme must be in the proper protonation state to convert the appropriate enantiomer to product.

Further support for a two-base mechanism comes from kinetic analyses.[4] Primary kinetic isotope effects on V_{max} were observed in both directions with [2-^{2}H]glutamate. *Competitive deuterium washout* and *double competitive deuterium washout experiments*[5] (see this reference if you are interested in learning about these experiments) gave kinetic isotope effects on V_{max}/K_m of 2.5 for (S)-glutamate and 3.4 for (R)-glutamate. Another common experiment to test for a two-base mechanism is known as an *"overshoot" experiment*[6] in which the racemization is monitored in D$_2$O by circular dichroism (Figure 9.1). When the racemization of one of the enantiomers of glutamate (say, R) is monitored in D$_2$O, the optical rotation, which is initially negative for (R)-glutamate, approaches zero (as racemization is approached), goes beyond zero ("overshoots") and becomes transiently positive, and then returns to the equilibrium value of zero. This curious phenomenon has a ready explanation. Initially, only (R)-glutamate is present (negative optical rotation). As it racemizes in D$_2$O, the (S)-isomer becomes deuterated at C-2. When there is an equal mixture of (R)- and (S)-isomers, the optical rotation is zero (the deuterium has a negligible effect on the optical rotation), but the (R)-isomer is a mixture of (R)-glutamate (unreacted starting material) and (R)-[2-^{2}H]glutamate (which comes from the back

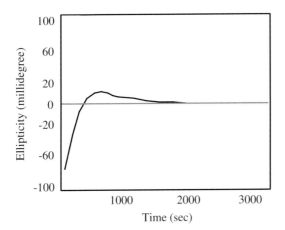

FIGURE 9.1   An "overshoot" experiment with (R)-glutamate to test for a two-base mechanism for glutamate racemase.

reaction of the (S)-[2-²H]-glutamate product), whereas the (S)-isomer is only (S)-[2-²H]glutamate. Because there is a kinetic deuterium isotope effect on C-2 deprotonation, the nondeuterated (R)-glutamate will preferentially be converted to (S)-[2-²H]glutamate, giving a net flux toward an excess of the (S)-isomer and a positive optical rotation (an "overshoot" past an optical rotation of zero). Both enantiomers gradually reach isotopic equilibrium, and the optical rotation returns to zero.

Two mutant enzymes were generated by site-directed mutagenesis, one with C73A and one with C184A. Both mutants were inactive as racemases. However, they both were capable of catalyzing the elimination of HCl from opposite enantiomers of *threo*-3-chloroglutamic acid (**9.1**) to give α-ketoglutarate (**9.2**) (Scheme 9.2), because this reaction only requires a single base. The wild-type enzyme catalyzes the elimination of HCl from both enantiomers of **9.1,** converting it completely to **9.2**. Note that the two catalytic cysteine residues are not near each other in the primary sequence, but must fold around to the active site in the intact en-

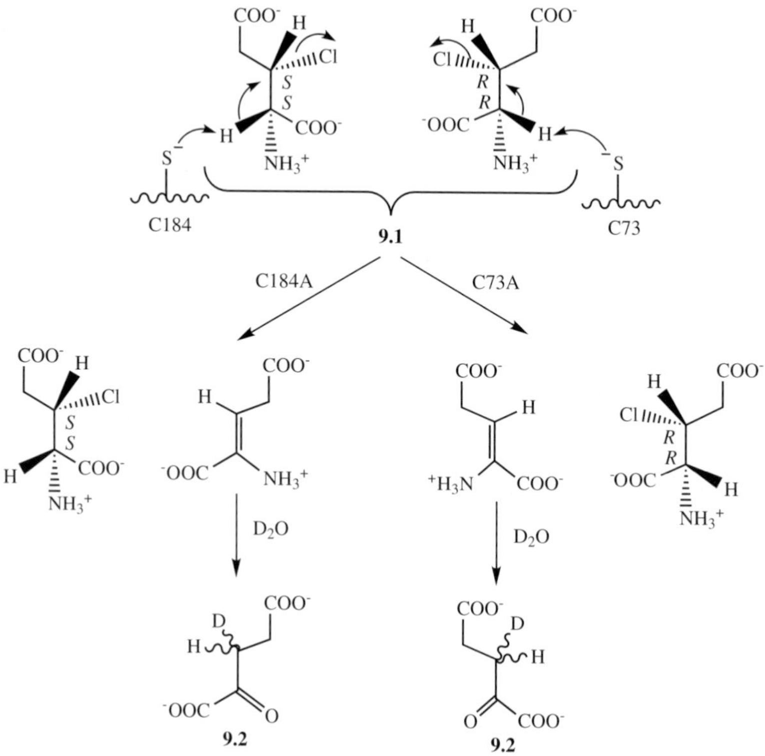

SCHEME 9.2   Elimination of HCl from *threo*-3-chloroglutamic acid by the C73A and C184A mutants of glutamate racemase.

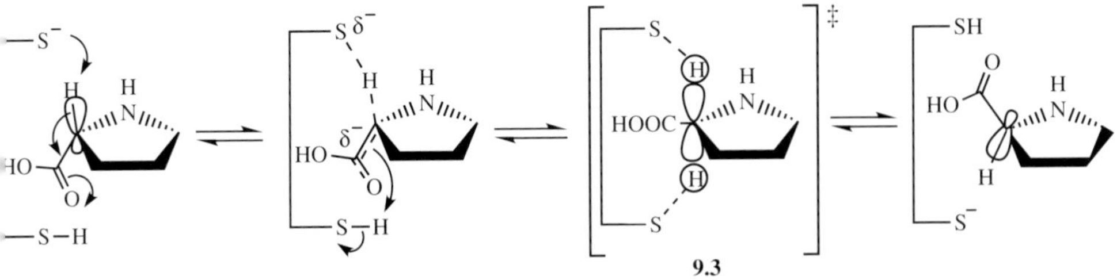

SCHEME 9.3   Proposed mechanism for proline racemase.

zyme. Cys-73 abstracts the C-2 proton from (R)-glutamate, and Cys-184 abstracts the C-2 proton from (S)-glutamate. Another cofactor-independent amino acid racemase that uses a two-base mechanism is aspartate racemase;[7] this enzyme also appears to use two cysteine residues as the bases, as does proline racemase.[8] Cysteine residues are typically involved in nucleophilic mechanisms. These cofactor-independent racemases exemplify the rare utilization of cysteine residues in acid/base catalysis.

In the case of proline racemase, a Zn^{++}-dependent enzyme, the affinity labeling agent iodoacetate, ICH_2COO^-, was shown to inactivate the enzyme, but only after a reducing agent, such as sodium borohydride or a thiol, was added. This suggests the presence of a disulfide linkage (a cystine) that is reduced to two cysteine residues; in fact, two moles of ICH_2COO^- are incorporated after reduction. A mechanism to account for these observations is shown in Scheme 9.3.

As discussed in Chapter 1 (Section I.A), a substrate is most tightly bound to the enzyme at the transition state of the reaction. On that basis, Bernhard and Orgel[9] hypothesized that an inhibitor molecule resembling the transition-state species would be much more tightly bound to the enzyme than would be the substrate or an inhibitor resembling the substrate. Therefore, a potent enzyme inhibitor would be a stable compound whose structure resembles that of the substrate *at a postulated transition state* (or transient intermediate) of the reaction rather than that at the ground state. A compound of this type, which would bind much more tightly to the enzyme, is called a *transition-state analogue inhibitor*. Jencks[10] was the first to suggest the existence of transition-state analogue inhibitors, and cited several possible literature examples; Wolfenden[11] and Lienhard[12] developed the concept further. Based on the hypothesized planar transition-state structure in Scheme 9.3 (**9.3**), 2-carboxypyrrole (**9.4**), designed as a transition-state analogue, was found to be an exceedingly potent *reversible* inhibitor of proline racemase.

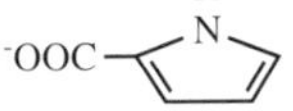

9.4

B. Amino Acid Racemases That Require Pyridoxal-5′-Phosphate

In Chapter 8 (Section V.A), we saw that pyridoxal-5′-phosphate (PLP) is a coenzyme that enzymes use to catalyze the decarboxylation of amino acids to amines and CO_2. That is not the only function of PLP. α–Amino acids also undergo enzyme-catalyzed racemization with PLP-dependent enzymes; in fact, PLP-dependent enzymes are the most common ones that catalyze racemizations of amino acids. These enzymes are highly specific for the amino acid used as its substrate, so alanine racemase does not catalyze the racemization of glutamate. Also, (and at least as fascinating a fact as the substrate specificity), a decarboxylase does not also catalyze racemization of its substrate, so histidine decarboxylase converts L-histidine to histamine without formation of D-histidine, although there are some PLP-dependent enzymes that do catalyze a slow side reaction different from its normal function.

Consider alanine racemase as an example of this type of reaction. The reaction catalyzed by this enzyme is the interconversion of L- and D-alanine (Scheme 9.4) with an equilibrium constant of about 1, typical for racemases, so no matter which enantiomer is used as the substrate, the racemate is produced. Generally, however, the K_m values for D- and L-amino acids differ. For various alanine racemases the K_m^L is $\geq K_m^D$, but the V_{max}^L is $\geq V_{max}^D$, so the ratio $(V_{max}^L/V_{max}^D)(K_m^D/K_m^L)$ is

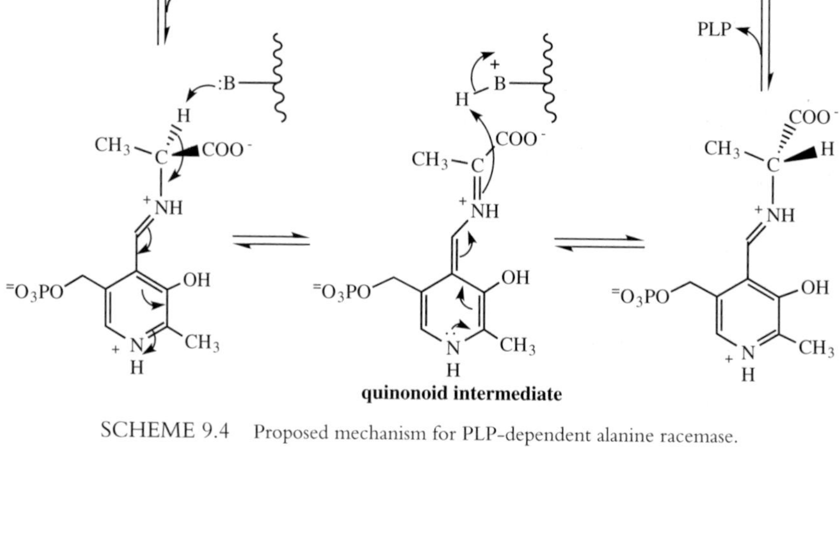

SCHEME 9.4   Proposed mechanism for PLP-dependent alanine racemase.

approximately 1.[13] Alanine racemase is the major source of the D-alanine used in the peptidoglycan of bacterial cell walls, although a eucaryotic alanine racemase (from the fungus *Tolypocladium niveum*) has been isolated as well.[14]

Racemization could occur with one base or with two bases. In general, race-mases (and other isomerases) that proceed via a resonance-stabilized carbanion (as with the PLP-dependent enzymes) operate via a one-base mechanism and those that proceed via a carbanion that is not resonance stabilized proceed via a two-base mechanism.[15] This notion was mentioned in Chapter 1, Section II.C. An impor-tant experiment to test a one-base mechanism is to look for *internal return,* that is, the transfer of the substrate proton to the same carbon atom in the product. Inter-nal return supports a one-base mechanism. An apparent exception to this general-ity is alanine racemase (from several different sources), for which no internal return could be detected.[16] The lack of internal return could be interpreted as a two-base mechanism or a one-base mechanism in which the solvent rapidly exchanges with the protons on the active-site base. The crystal structure of alanine racemase from *Bacillus stearothermophilus* at 1.9 Å resolution shows that Lys-39 is bound to the PLP.[17] Based on this crystal structure, Shaw, Petsko, and Ringe propose that Lys-39 and Tyr-265 act as the acid/base couple in a two-base racemization mechanism. Further support for this two-base mechanism was obtained by Sun and Toney.[17a] They found a difference in the primary- and solvent isotope effects on racemization in each direction with various Arg-219 mutants and also observed a difference in the absor-bance of the quinonoid intermediate in each direction. It is interesting that Arg-219 is the residue that is hydrogen bonded to the pyridine nitrogen of PLP. Typi-cally an aspartate or glutamate residue hydrogen bonds to the pyridine nitrogen.

The PLP enzyme, α-amino-ϵ-caprolactam racemase, does show internal re-turn.[18] Conversion of [α-^{2}H]D-α-amino-ϵ-caprolactam in H_2O and unlabeled D-α-amino-ϵ-caprolactam in ^2H_2O into the L-isomer under single-turnover conditions showed 11–20% incorporation of deuterium from the deuterated substrate (or pro-tium from the unlabeled substrate) into the product. This finding is strong evidence for a one-base mechanism. The low incorporation of substrate proton into prod-uct could be explained by either partial exchange with solvent or the presence of a polyprotic base, such as a lysine residue, in which case once the α-proton is ab-stracted from the substrate, that proton is scrambled with the protons already on the polyprotic base; with lysine, there would be no more than a one-in-three prob-ability that the same proton removed from substrate would be returned to product (there would be two hydrogens and one deuterium on the amino group starting from [α-^{2}H]D-α-amino-ϵ-caprolactam, but a deuterium isotope effect would favor proton transfer).

It is apparent that PLP-dependent enzymes can catalyze decarboxylation reac-tions (Chapter 8, Section V.A) and also racemization reactions, and the two activ-ities are separate. However, for decarboxylation, the C^{α}–COO$^-$ bond is broken, but for racemization the C^{α}–H bond is broken. If an amino acid is heated with PLP in the absence of an enzyme, a mixture of products is obtained, including decarboxylation and racemization products.[19] So how can a PLP-dependent enzyme catalyze a regiospecific cleavage of only one of these bonds?

First, let's compare the mechanism for PLP-dependent decarboxylation (Scheme 8.44) with that for racemization (Scheme 9.4). Note that both are initiated from the same PLP-substrate Schiff base complex. The electrons in the C^{α}–COO^- bond are delocalized into the PLP ring on decarboxylation (Scheme 8.44) just as the electrons in the C^{α}–H bond are delocalized into the PLP ring on deprotonation (Scheme 9.4). That is the function of the PLP, to provide an "electron sink" for delocalization of the electrons in a bond alpha to the double bond of the imine. But there are three bonds with electrons in that position, the C^{α}–COO^- bond, the C^{α}–H bond, and the C^{α}–R bond. The bond that breaks must lie in a plane perpendicular to the plane of the PLP-imine π-electron system. In Figure 9.2 the C–H bond is the one perpendicular to the plane of the π system, that is, parallel to the p-orbitals. This configuration results in maximum σ–π electron overlap (the sp^3 σ orbital of the C–H bond and the p-orbital of the aromatic system), and therefore minimizes the transition-state energy for cleavage of the C^{α}–H bond relative to cleavage of the C^{α}–COO^- or C^{α}–R bonds. The problem for the enzyme to solve, then, is how to control the conformation about the C^{α}–N bond so that only the bond that is to be cleaved is perpendicular to the plane of the π system at the active site of the enzyme. In other words, the enzyme must freeze free rotation about the C^{α}–N–bond! The Dunathan hypothesis[20] gives a rational explanation for how an enzyme could control the C^{α}–N bond rotation (Figure 9.3). A positively charged residue at the active site could form a salt bridge with the carboxylate group of

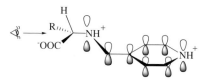

FIGURE 9.2   Stereochemical relationship between the σ-bonds attached to C^{α} and the p-orbitals of the π-system in a PLP-amino acid Schiff base.

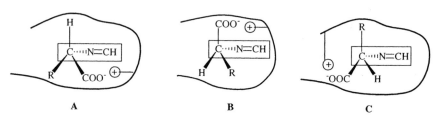

FIGURE 9.3   Dunathan hypothesis for PLP activation of the bonds attached to C^{α} in a PLP-amino acid Schiff base. The rectangles represent the plane of the pyridine ring of the PLP. The angle of viewing is that shown by the eye in Figure 9.2. [Dunathan, H. C. In Meister, A., Ed., *Advances in Enzymology*, vol. 35, p. 79, Copyright © 1971, by John Wiley & Sons, Inc. Reprinted by permission of John Wiley & Sons, Inc.]

the amino acid bound to the PLP. This would make it possible for an enzyme to restrict rotation about the C^{α}–N bond and hold the H (**A**), the COO⁻ (**B**), or the R (**C**) group perpendicular to the plane of the aromatic system (the rectangles in Figure 9.3 are the pyridine ring systems). If the Dunathan hypothesis is accepted, then all of the PLP-dependent enzyme reactions can be readily understood. In fact, crystal structures of PLP-dependent enzymes generally show a group at the active site, often an arginine residue, positioned to bind to the substrate-bound carboxylate group.[21]

Citrobacter freundii tyrosine phenol-lyase is one of the PLP-dependent enzymes that catalyzes a side reaction, namely, the racemization of L- or D-alanine.[22] On the basis of rapid-scan and single-wavelength stopped-flow UV–visible spectrophotometric studies with L- and D-alanine, it was found that there is a common quinonoid intermediate (Scheme 9.4) in the side racemization reaction.

C. Other Racemases

1. Mandelate Racemase

Mandelate racemase, as you can surmise, catalyzes the racemization of mandelic acid (Scheme 9.5). Kenyon, Gerlt, Kozarich and co-workers carried out several experiments to distinguish between a one-base and a two-base mechanism.[23] With mandelate racemase, no substrate-derived α-proton is found in the product in either direction (i.e., there is no internal return). Also, with (R)-mandelate, no α–H exchange with the medium is observed, but with (S)-mandelate there *is* exchange with the medium. These results support a two-base mechanism. On the basis of the crystal structure,[24] it appears that Lys-166 abstracts the α-H from (S)-mandelate, and His-297 abstracts the α-H from (R)-mandelate. Scheme 9.6 rationalizes the difference in the solvent exchange results. The solvent-exchanged deuteriums on Lys-166 become incorporated into the α-position (**A**); the nitrogen of His-297 that removes the proton from (R)-mandelate has no deuterium on it to donate, so the proton removed from substrate is returned without solvent exchange (**B**). When His-297 is changed to asparagine (H297N) by site-directed mutagenesis, the mutant enzyme is capable of exchanging the α-H of (S)-isomer, but not the (R)-isomer (Asn replaces His-297, but Lys-166 is still present)—again, supporting a two-base mechanism.[25] Also, the H297N mutant is capable of catalyzing the elimination of HBr from (S)-p-(bromomethyl)mandelate (**9.5**), but not from the corresponding

R-mandelate   S-mandelate

SCHEME 9.5   Reaction catalyzed by mandelate racemase.

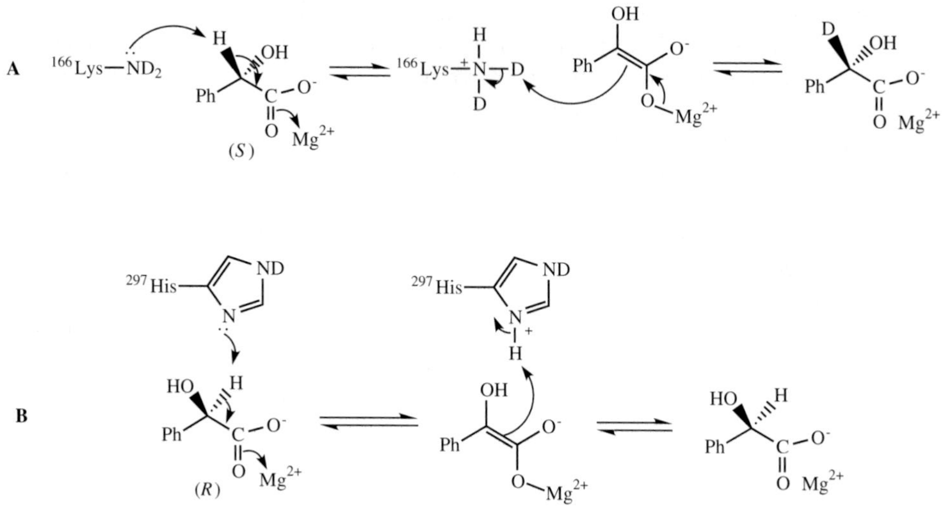

SCHEME 9.6   A two-base mechanism for mandelate racemase that accounts for the deuterium solvent exchange results. Lys-166 acts on the (S)-isomer and His-297 acts on the (R)-isomer.

(R)-isomer, to (p-methyl)benzoylformate (**9.6**) (Scheme 9.7). Likewise, the K166R mutant (Arg in place of Lys-166, but His-297 is still present) catalyzes the elimination of HBr from the (R)-isomer of **9.5**, but not from the (S)-isomer (**9.5**).[26]

The pH dependence of k_{cat} on racemization of both (R)- and (S)-mandelates indicates that the pK_a values of the conjugate acids of Lys-166 and His-297 are both about 6.4, which is normal for a histidine residue, but very low for a lysine residue (the pK_a of Lys in solution is about 10.5). The electrostatic interactions of the nearby cationic groups of Lys-164 and the Mg^{2+} ion in the active site are expected to increase the acidities (lower the pK_a) of the cationic conjugate acids of the acid/base catalysts (see Chapter 1, Section II.C). This rationalizes the observed low pK_a value of Lys-166 but not the "normal" pK_a value of His-297. However, Asp-270 is hydrogen-bonded to the N^δ of His-297, which may compensate for the

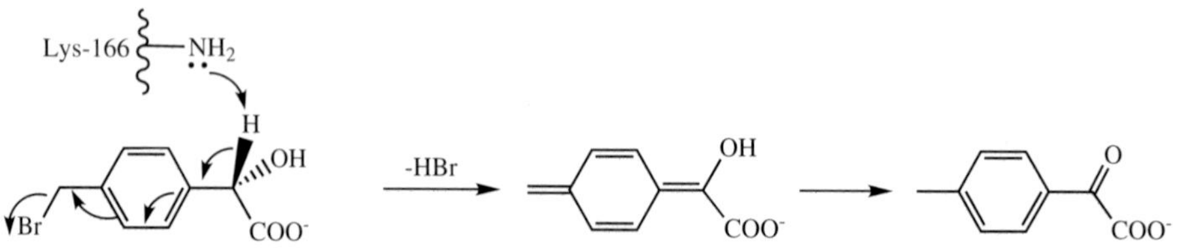

SCHEME 9.7   Elimination of HBr from (S)-p-(bromomethyl)mandelate, catalyzed by the H297N mutant of mandelate racemase.

expected lower pK_a of His-297 and allow it to be normal. To test this, the D270N mutant was made.[27] The crystal structure (with an inhibitor bound in the active site) reveals no significant geometric alterations to the active site. The k_{cat} values for both (R)- and (S)-mandelates, however, are reduced about 10^4-fold. The D270N mutant catalyzes the facile exchange with solvent of the α-proton of (S)- but not (R)-mandelate (the enantiomer from which His-297 removes a proton) and the stereospecific elimination of HBr from **9.5**, but not the (R)-isomer of **9.5**. In accord with the proposal that Asp-270 may influence the pK_a of His-297, no ascending limb is detected in the dependence of k_{cat} on pH in the (R)- to (S)-direction; instead, the k_{cat} decreases from a low pH plateau, as described by a pK_a of 10. In the (S)- to (R)- direction, the dependence of k_{cat} on pH is a bell-shaped curve described by pK_a values of 6.4 and 10. These observations suggest that His-297 and Asp-270 function as a catalytic diad with Asp-270 being at least partially responsible for the normal pK_a of His-297.

D. Epimerases

1. Peptide Epimerization

As we saw in Sections II.A and B, certain enzymes can catalyze the racemization of various amino acids, thereby producing the corresponding D–amino acids. Some peptides, such as the opioid peptides dermorphins[28] and deltorphins[29] and neuroactive tetrapeptides from frog[30] and mollusk,[31] are known to contain D–amino acids. These D–amino acids could be incorporated into the growing peptide chain as D–amino acids, or there may be a posttranslational amino acid epimerization of the L-amino acid-containing peptide. The venom of the *Agelenopsis aperta* spider has been found to contain an enzyme activity that catalyzes the epimerization of specific amino acids within peptide chains.[32] The enzyme isomerizes serine, cysteine, O-methylserine, and alanine residues in the middle of peptide chains with recognition sites for Leu-X-Phe-Ala. Incubation of the enzyme with a peptide substrate having an isomerizable serine residue in $H_2^{18}O$ results in no incorporation of ^{18}O into the product, suggesting that a mechanism involving β-elimination to dehydroalanine (**9.7**) followed by Michael addition of $H_2^{18}O$ is not operative (Scheme 9.8) (unless no exchange of the eliminated water with solvent occurs).

A second mechanism that could be considered is one involving the release of formaldehyde to give an enol (**9.8**), which then reacts with the enzyme-bound formaldehyde on the opposite side (Scheme 9.9). The experiment carried out to eliminate this mechanism showed that the reaction is viable even in the presence of 10 mM hydroxylamine, which rapidly reacts with formaldehyde. However, this experiment assumes that either the formaldehyde can be released from the active site or that hydroxylamine can enter the active site during the reaction; if neither occurs, then hydroxylamine would have no effect.

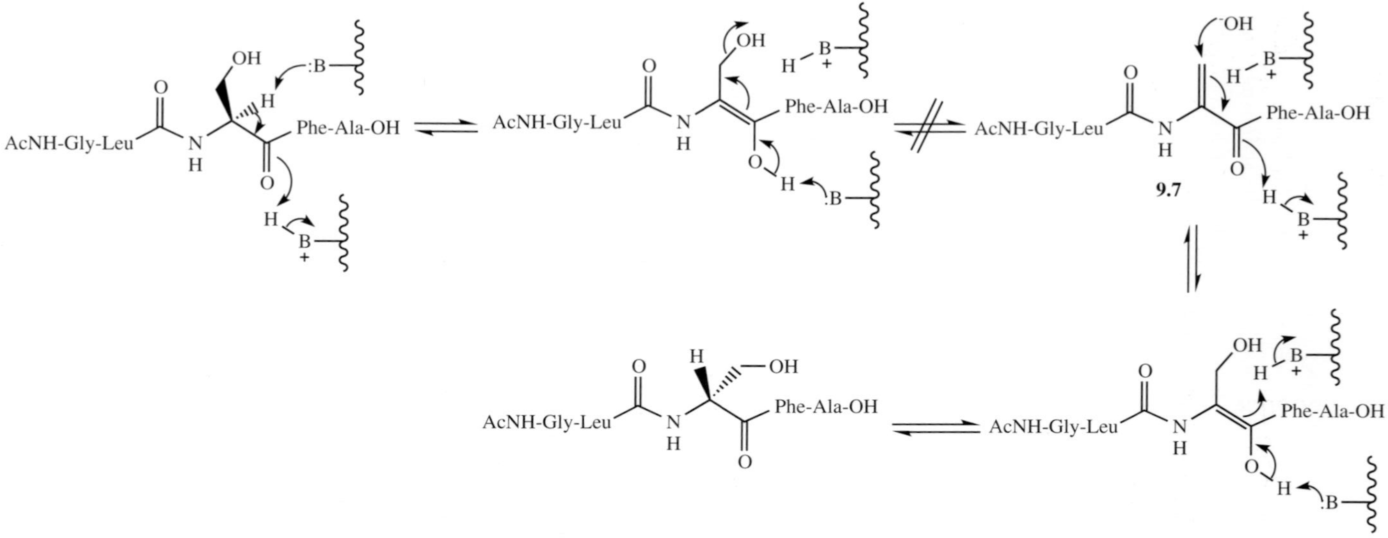

SCHEME 9.8   Incorrect elimination/addition (dehydration–hydration) mechanism for peptide epimerization.

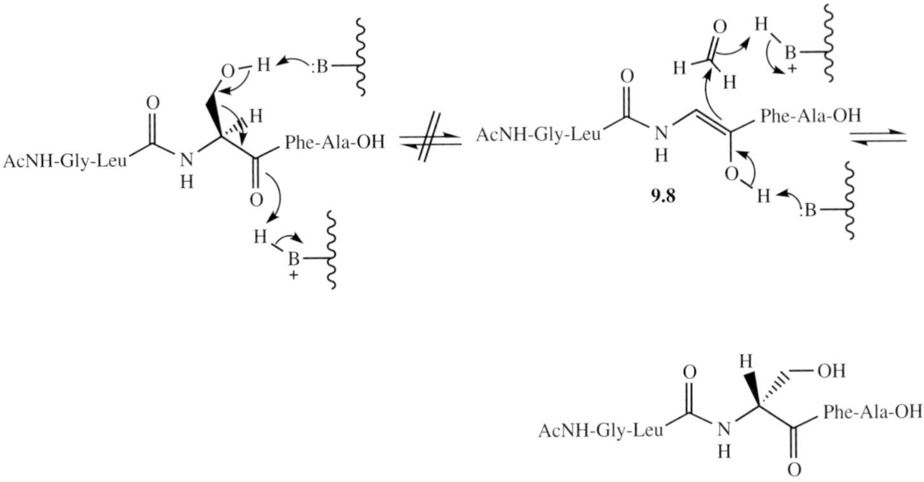

SCHEME 9.9   Incorrect deformylation/reformylation mechanism for peptide epimerization.

The finding that amino acids other than serine, such as alanine, are epimerized definitely eliminates both of the first two mechanisms. A third possibility, which has much support, is a deprotonation/reprotonation mechanism (Scheme 9.10). Electrospray mass spectral analysis during the early part of the reaction shows that

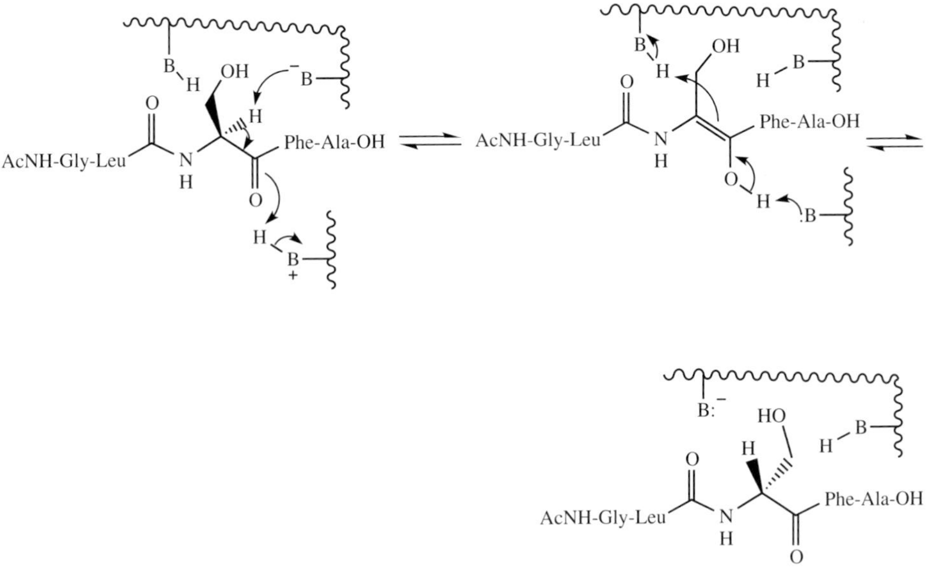

SCHEME 9.10   Deprotonation/reprotonation mechanism for peptide epimerization.

deuterium from ^2H_2O is incorporated into the product but not into the substrate, regardless of which epimer is the substrate. The k_{cat} for the L- to D- direction is 13 times that for the D- to L- direction. The K_m for the L- to D- direction is 7 times larger than for the D- to L- direction. The large difference in the K_m and k_{cat} values between the L- to D- and the D- to L- directions suggests an asymmetry in the active site. The deuterium isotope effect on V_{max} for [α-^2H]-containing substrates in the L- to D- direction is 31, but in the D- to L- direction is only 5. This huge difference in isotope effects on the V_{max} suggests that abstraction of the L-α-proton has a larger effect on the rate in the L- to D- direction than the abstraction of the D-α-proton has in the D- to L- direction. This is consistent with a two-base mechanism in which one base is more efficient at deprotonation and has a larger influence on catalysis than the other base.

2. Epimerization with Redox Catalysis

a. dTDP-L-Rhamnose Synthase

Although some enzymes are multifunctional and catalyze more than one reaction (e.g., isocitrate dehydrogenase and 6-phosphogluconate dehydrogenase; see Chapter 8, Sections III.A and III.B, respectively), an enzyme that is initially thought to be multifunctional often turns out to be part of a cluster of enzymes responsible for multiple steps in a biosynthetic pathway. An example of the latter case is dTDP-L-rhamnose synthase, which was thought to catalyze two epimerizations and a reduction in the conversion of 2′-deoxythymidine diphosphate (dTDP)-4-keto-6-deoxy-D-glucose (9.9) to dTDP-L-rhamnose (9.10); it requires NADPH (see Chapter 3, Section III.A) for the reduction (Scheme 9.11).[33] However, several research groups subsequently found that epimerization and reduction are catalyzed by two different gene products (enzymes), dTDP-4-keto-6-deoxy-D-glucose 3,5-epimerase and dTDP-4-keto-L-rhamnose reductase, respectively.[34] Cleavages of the C–H bond at C-3 and at C-5 show kinetic deuterium isotope effects (k_H/k_D) of 3.4 and 2.0, respectively, indicating that both are partially rate determining.[35] In ^2H_2O as solvent, deuteriums are incorporated into the C-3 and C-5 positions. Partial exchange in ^2H_2O gives molecules with deuterium incorporated at only C-3 or at C-3 and C-5, but none with the deuterium incorporated only at C-5. Therefore, this is an ordered sequential epimerization mechanism (see Appendix I, Section VI.A.2), which suggests that epimerization at C-3 induces a conformational change in the enzyme that allows epimerization at C-5. The hydrogen of NADPH ends up at C-4 of the product,[36] as predicted from the mechanism shown.

b. UDP-Glucose 4-Epimerase

At first sight, the conversion of uridine-5′-diphospho(UDP)-glucose (9.11) to UDP-galactose (9.12) may appear to be a simple epimerization of the C-4 hydroxyl

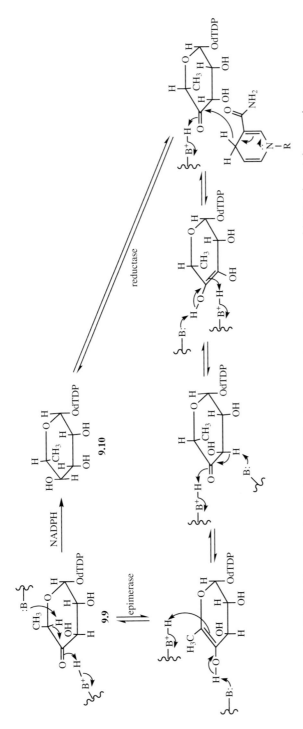

SCHEME 9.11 Proposed mechanism for dTDP-L-rhamnose synthase–catalyzed conversion of dTDP-4-keto-6-deoxy-D-glucose (9.9) to dTDP-L-rhamnose (9.10).

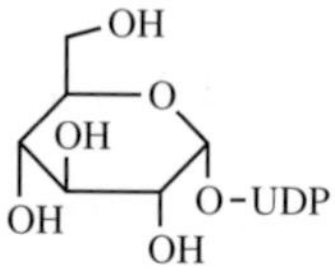

9.11 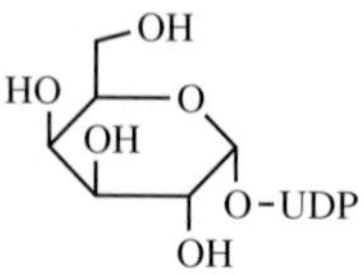   **9.12**

group. However, a deprotonation/reprotonation mechanism would be highly unlikely, because the C-4 proton is not acidic enough, even for an enzyme, to remove.
In fact, it was shown that the enzyme requires NAD^+ as a cofactor, and there is no
exchange with the solvent protons.[37] When NAD^3H is used instead of NAD^+ as the
coenzyme, no ^3H is incorporated into either the substrate or the product; therefore
the enzyme is not functional when the coenzyme is reduced. In $H_2{}^{18}O$ there is no
incorporation of ^{18}O into the sugar. These results suggest that simple epimerization
or dehydration/hydration mechanisms are unlikely. The NAD^+, however, could
oxidize the C-4 hydroxyl group (see Chapter 3, Section III.A). When the epimer-
ase reaction is run backward using an analogue of the suspected intermediate (**9.13**)
with NAD^3H as the coenzyme instead of NAD^+, the tritium is incorporated into
the C-4 position (Scheme 9.12).[38] According to the principle of microscropic reversibility,[39] for a reversible reaction to occur, the mechanism in the reverse direction must be identical to that in the forward direction, only in reverse. This principle points to a ketone as an intermediate in the forward direction. The NAD^+
ends up as NAD^+, and the product is in the same oxidation state as the substrate,
suggesting that the NAD^+ catalyzes the oxidation of the C-4 hydroxyl to a C-4
ketone (**9.14**) with concomitant reduction of the NAD^+ to NADH; the C-4 ketone
is then reduced by the enzyme-bound NADH to the epimer (Scheme 9.13). Di-

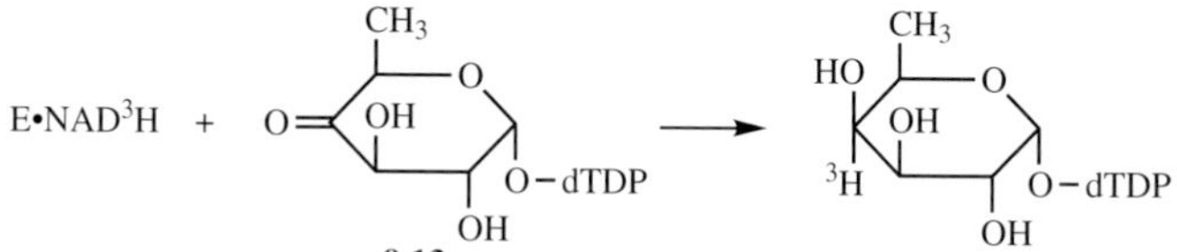

9.13

SCHEME 9.12   Incorporation of tritium from NAD^3H into a derivative of the suspected intermediate of the UDP-glucose 4-epimerase-catalyzed reaction.

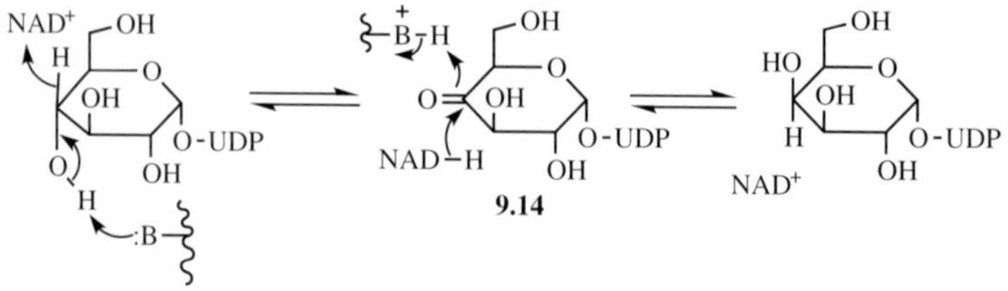

9.14

SCHEME 9.13   Proposed mechanism for reaction catalyzed by UDP-glucose 4-epimerase.

rect evidence for **9.14** comes from an experiment by Maitre and Ankel in which the enzyme was incubated with UDP-galactose and then treated with NaB^3H_4.[40] UDP-glucose (**9.11**) and UDP-galactose (**9.12**) were both isolated containing tritium at the C-4 position. This supports the formation of UDP-4-ketoglucose (**9.14**) as an enzyme-bound intermediate.

For this mechanism to be correct, it appears that the NAD^+ must be above the enzyme-bound substrate, but below **9.14** to deliver the hydride underneath. That sounds like a very large amount of entropy and quite a mobile cofactor! But there is a more energetically feasible way for the cofactor to accomplish the same sequence of events: The cofactor could remain bound, and the sugar ring could simply change conformation (Scheme 9.14).

When protons are not acidic, enzymes often use an oxidation/reduction mechanism to generate a carbonyl next to the proton, thereby acidifying that proton and permitting epimerization. This is reminiscent of the enzymes that catalyze the decarboxylation of β-hydroxy acids (Chapter 8, Section III.A). In those cases the β-hydroxyl group is first oxidized before decarboxylation occurs. Decarboxylation is equivalent to deprotonation; the electrons in the $C-COO^-$ bond are equivalent to the electrons in the $C-H$ bond, as we saw in Section II.B in a comparison of PLP-dependent decarboxylases to PLP-dependent racemases. Therefore, oxidation of an adjacent hydroxyl group to a carbonyl activates both α-decarboxylation as well as α-deprotonation.

An important reaction in the biosynthesis of vitamin C (L-ascorbic acid) in plants[41] is the GDP-D-mannose 3,5-epimerase-catalyzed epimerization of GDP-D-mannose (**9.15**) to GDP-L-galactose (**9.18**) (Scheme 9.15).[42] The mechanism of this enzyme-catalyzed reaction is similar to that for the conversion of dTDP-4-keto-6-deoxy-D-glucose (**9.9**) to dTDP-L-rhamnose (**9.10**) shown in Scheme 9.11, except that **9.15** must first become oxidized prior to epimerization of the adjacent 3,5-carbons. The substrate (**9.15**) is oxidized by NAD^+ to the 4-keto analogue (**9.16**), which acidifies the protons at C-3 and C-5 for enolization and epimerization (**9.17**); once epimerization has occurred, the NADH can add a hydride back to the 4-carbonyl, in this case on the same face from which it was initially removed, to give the product (**9.18**). The overall reaction does not involve a change in oxidation state; an oxidation is carried out followed by a reduction. It is not clear if these reactions are all catalyzed by one enzyme.

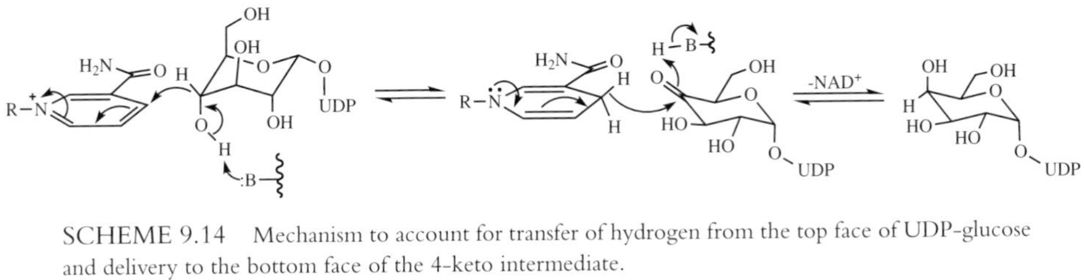

SCHEME 9.14   Mechanism to account for transfer of hydrogen from the top face of UDP-glucose and delivery to the bottom face of the 4-keto intermediate.

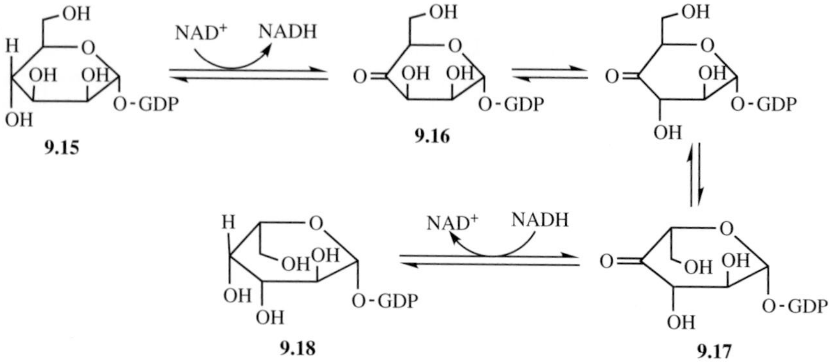

SCHEME 9.15   Mechanistic pathway for the GDP-D-mannose-3,5-epimerase-catalyzed conversion of GDP-D-mannose (**9.15**) to GDP-L-galactose (**9.18**).

III. [1,2]-HYDROGEN SHIFT

A. Aldose–Ketose Isomerases

The reversible isomerization of an α-hydroxy aldehyde (**9.19**) to a β-keto alcohol (**9.20**) (Scheme 9.16), a Lobry de Brun–Alberda von Ekenstein reaction, is an internal redox reaction (the product is in the same oxidation state as the substrate, although the carbonyl and hydroxyl groups have switched places) and therefore does not require a redox cofactor (although, as seen in Scheme 9.13, having the same oxidation states in the substrate and product does not preclude a redox cofactor).

Two mechanisms can be considered, a *cis*-enediol mechanism (Scheme 9.17; written as three-dimensional structures and in Newman projections) and a [1,2]-hydride shift mechanism (Scheme 9.18). These possibilities were considered for the enzyme glyoxalase (see Chapter 3, Section II), in which case a *cis*-enediol mechanism appears to be most consistent with the data. The configuration of enediol intermediates for all of the aldose-ketose isomerases is always *cis*. Isomerases that act on (2*R*)-aldoses transfer a hydrogen at the *re-re* face of the enediol (note that in **9.21** *R* is the remainder of the sugar, so it is second in nomenclature priority) to the *pro-R* position at C-1 of the product ketose; isomerases that act on (2*S*)-aldoses

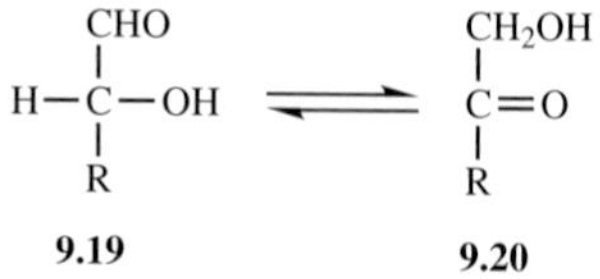

SCHEME 9.16   Reaction catalyzed by aldose–ketose isomerases.

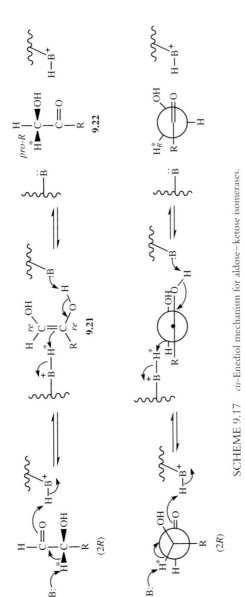

SCHEME 9.17 *cis*-Enediol mechanism for aldose−ketose isomerases.

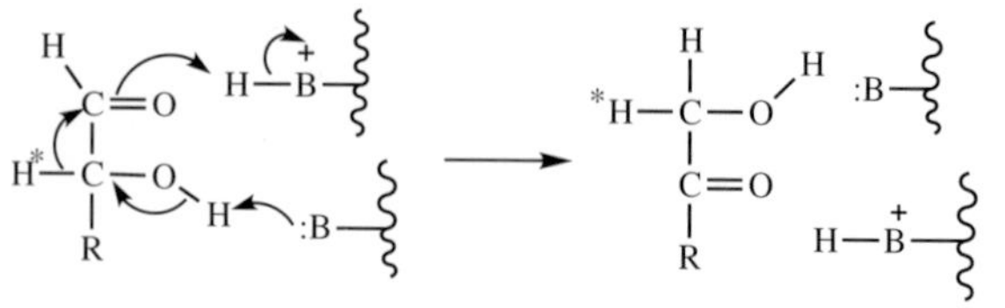

SCHEME 9.18   Hydride transfer mechanism for aldose–ketose isomerases.

transfer a hydrogen at the *si-si* face to the *pro-S* position of the ketose.[43] A *cis*-ene-diol intermediate requires only a single group in the active site to transfer the proton in a *suprafacial* process. A *trans*-enediol intermediate is probably unimportant because that would require a minimum of two electrophilic groups and *antarafacial* hydrogen transfer involving a catalytic base at each of the two diastereotopic faces of the enediol intermediate.

In Scheme 9.17 deprotonation leads to a *cis*-enediol intermediate (**9.21**); a *suprafacial* transfer of the proton gives **9.22,** which could have some of the isotopically labeled hydrogen incorporated if complete exchange with the solvent does not occur. In the hydride mechanism, all of the labeled hydrogen should be incorporated into the product, provided there is no exchange of this proton with solvent (it is an acidic proton). For the metal-independent aldose-ketose isomerases, partial incorporation of the label was observed, which supports an enediol mechanism.[44] The fact that the tritium was observed in a *suprafacial* transfer, indicates a one-base mechanism; in a two-base mechanism a different hydrogen is delivered to the product than is removed from the substrate (see section II.C.1). Crystallographic studies of metal-dependent xylose isomerase support a hydride mechanism.[44a] *Ab initio* and semiempirical MO methods suggest that the *cis*-enediol pathway is favored in the absence of a metal ion, but hydride transfer becomes favored in the presence of a metal ion.[44b]

IV. [1,3]-HYDROGEN SHIFT

A. Enolization

Phenylpyruvate tautomerase catalyzes the enolization of phenylpyruvate and related molecules with stereospecific removal of the *pro-R* hydrogen[45] (Scheme 9.19). Two major conformations of the substrate are possible, *anti* and *syn,* which would lead to two diastereomeric products, *Z-* and *E-*, respectively (Scheme 9.20). As product mimics, the *Z-* and *E-* vinyl fluorides (**9.23** and **9.24,** respectively) and other analogues such as **9.25** and **9.26,** were synthesized by Pirrung et al.[46] Compounds **9.24** and **9.26** (R = H and OH) were much better inhibitors of the enzyme than compounds **9.23** and **9.25,** suggesting that the favored product geometry is *E* and, therefore, *syn* tautomerization to the thermodynamically less stable

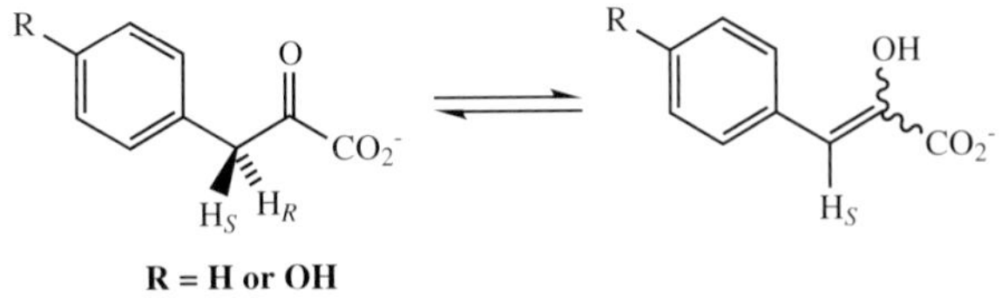

R = H or OH

SCHEME 9.19   Reaction catalyzed by phenylpyruvate tautomerase.

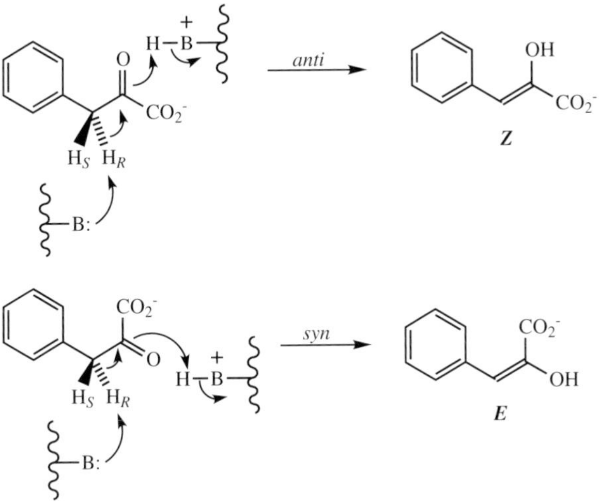

SCHEME 9.20   Conformations of phenylpyruvate that would form Z- and E-enols by phenylpyruvate tautomerase.

E product is the preferred route. Because amide and dicarboxylate enolate mimics are poor inhibitors, a discrete enolate intermediate is probably not involved.

B. Allylic Isomerizations

Allylic isomerizations can occur by a carbanion mechanism (Scheme 9.21), a carbocation mechanism (Scheme 9.22), or a hydride mechanism (Scheme 9.23). A concerted [1,3]-sigmatropic hydride shift (Scheme 9.23), however, is only allowed *antarafacial*.

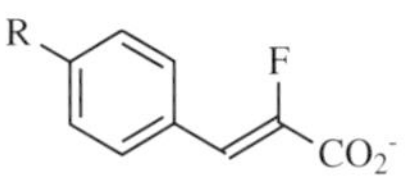

9.23 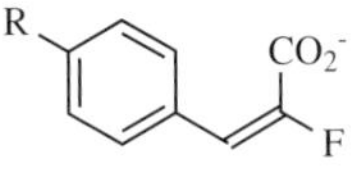   **9.24** 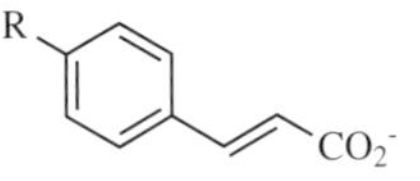   **9.25** 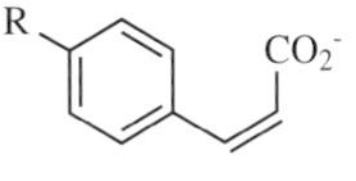   **9.26**

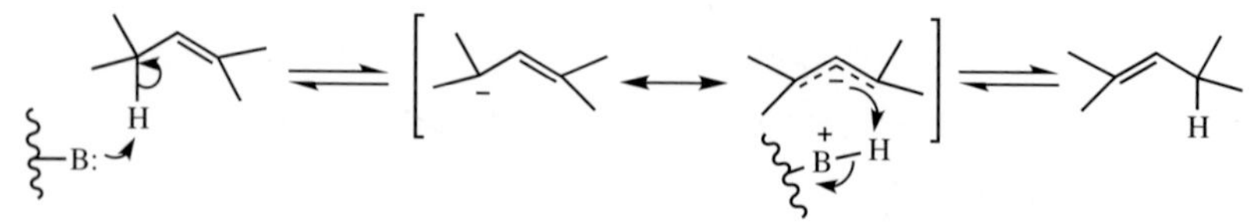

SCHEME 9.21   Carbanion mechanism for allylic isomerases.

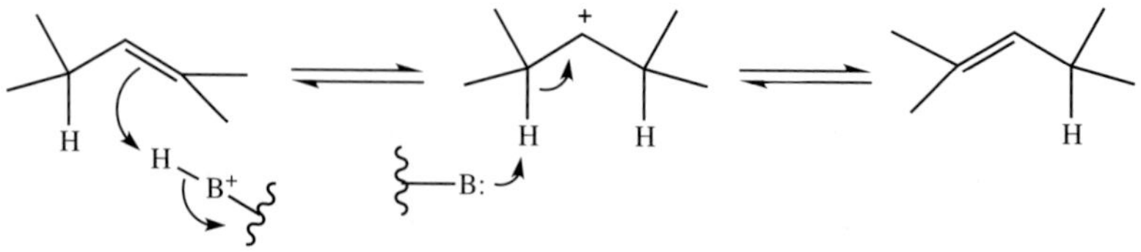

SCHEME 9.22   Carbocation mechanism for allylic isomerases.

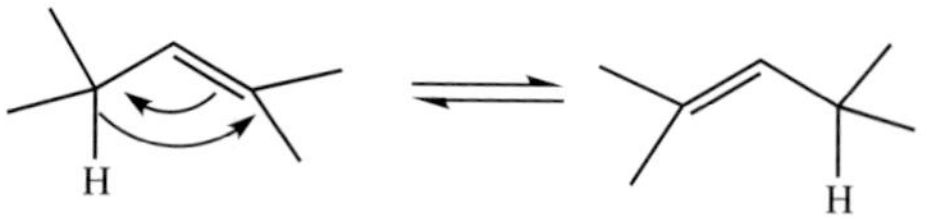

SCHEME 9.23   [1,3]-Sigmatropic hydride shift mechanism for allylic isomerases.

1. Carbanion Mechanism

An example of an enzyme that catalyzes a carbanionic allylic isomerization is 3-oxo-Δ^5-steroid isomerase, which catalyzes the conversion of Δ^5-androstene-3,17-dione (**9.27**) to Δ^4-androstene-3,17-dione (**9.28**) (Scheme 9.24). The k_{cat} (4.4 × 10^6 min^{-1} at pH 7.0 and 25 °C) and specificity constant (k_{cat}/K_m; 2.3 × 10^8 M^{-1}s^{-1}) for **9.27** are one of the largest and approach the diffusion limit.[47] Starting from [4β-^{2}H]Δ^5-androstene-3,17-dione (Scheme 9.24), it was originally thought that only [6β-^{2}H]Δ^4-androstene-3,17-dione is obtained with complete retention of the label,

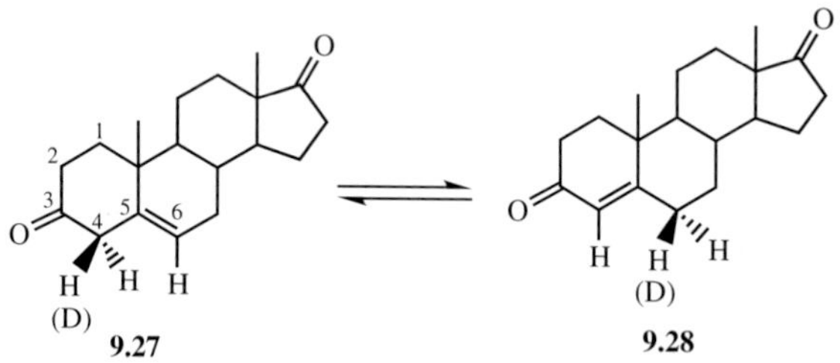

SCHEME 9.24   Reaction catalyzed by 3-oxo-Δ^5-steroid isomerase.

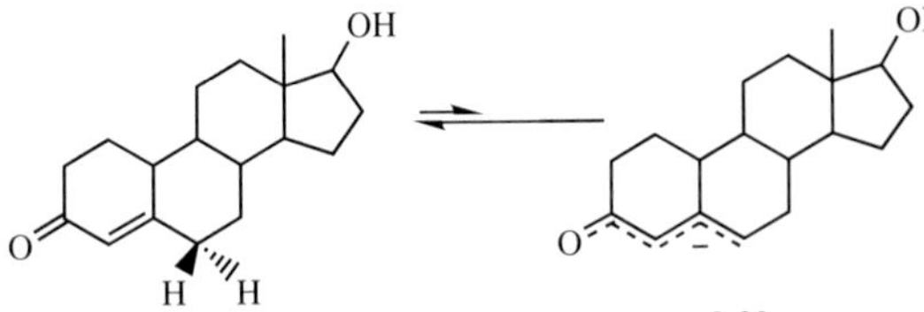

SCHEME 9.25   Enolate formation by 3-oxo-Δ^5-steroid isomerase-catalyzed deprotonation of 19-nortestosterone (**9.29**).

indicating a *suprafacial* transfer of the hydrogen.[48] Later, though, Marquet and co-workers found a temperature-dependent competing abstraction of the 4α-proton, indicating that this reaction is not completely stereospecific.[49] Substrate structure also has a profound effect on the stereospecificity of the isomerization reaction. However, the primary reaction is the *suprafacial* transfer of the hydrogen, so the mechanisms in Schemes 9.22 and 9.23 can be eliminated, because the mechanism in Scheme 9.22 would not give retention of the label in the product, and the mechanism in Scheme 9.23 has to be *antarafacial*.

An intermediate enolate (**9.30**) was originally proposed based on UV–spectrum experiments with 19-nortestosterone (**9.29**), which forms an E·S complex with the enzyme but is not isomerized (Scheme 9.25).[50] However, reinvestigations of these spectral studies indicated that an enolate was not formed; rather, hydrogen bonding was shown to be important.[51] Evidence for an enol intermediate comes from experiments with steroids **9.31** and **9.33** (Scheme 9.26).[52] The product of enzymatic turnover of either **9.31** or the corresponding enol (**9.33**) is **9.32**, whereas the

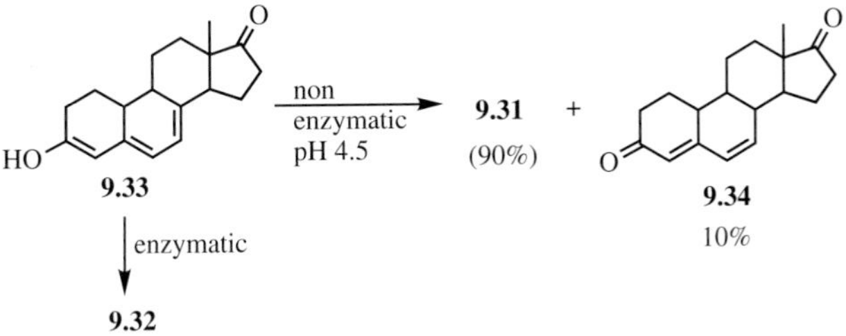

SCHEME 9.26   Evidence for an enol intermediate in the reaction catalyzed by 3-oxo-Δ^5-steroid isomerase.

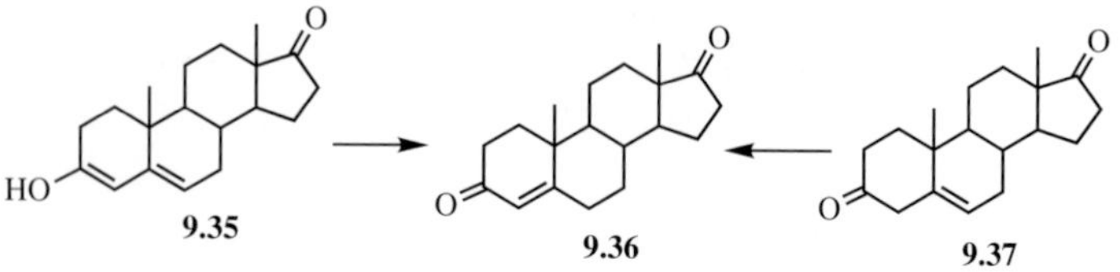

SCHEME 9.27   Further evidence for an enol intermediate in the reaction catalyzed by 3-oxo-Δ⁵-steroid isomerase.

nonenzymatic reaction of **9.33** gives mostly **9.31** and some **9.34**, supporting the enol as the enzymatic intermediate. Stopped-flow spectrophotometry was used to show the formation of the dienolate intermediate. Pollack and co-workers showed that compound **9.35** is converted to **9.36** by the enzyme at the same rate as **9.37** is converted to **9.36** (Scheme 9.27), demonstrating the kinetic competence of the enol intermediate.[53]

On the basis of a series of site-directed mutagenesis experiments, it was proposed that Tyr-14 is the acid that hydrogen bonds to the 3-keto group of the steroid, and Asp-38 is the base that removes the axial C-4β proton concertedly in forming the dienol.[54] Incubation of the substrate with the Y14F mutant reduces the k_{cat} by a factor of $10^{4.7}$, and a D38N mutant has a k_{cat} reduced by a factor of $10^{5.6}$; the double mutant has no measurable catalytic activity.[55] Extensive NOE studies by Mildvan and co-workers indicate an orthogonal arrangement of proton donor and acceptor (**9.38**).[56] To probe the function of Tyr-14 at the active site, Pollack and co-workers investigated the absorption spectra of two model compounds (**9.39** and **9.40**) (Scheme 9.28).[57] The UV spectrum of **9.39** bound to the enzyme is identical to that for **9.39** in neutral solution, suggesting that Tyr-14 does not protonate the amine group at the active site. This is evidence that Tyr-14 does not fully protonate the 3-oxo group of steroid substrates during enzymatic catalysis either. The fluorescence excitation spectrum of **9.40** bound to the enzyme is characteristic of an ionized phenol, even at pH 3.8, suggesting that the pK_a of the phenol is perturbed from 9 to ≤ 3.5 when it is enzyme bound. This is evidence that the pK_a of

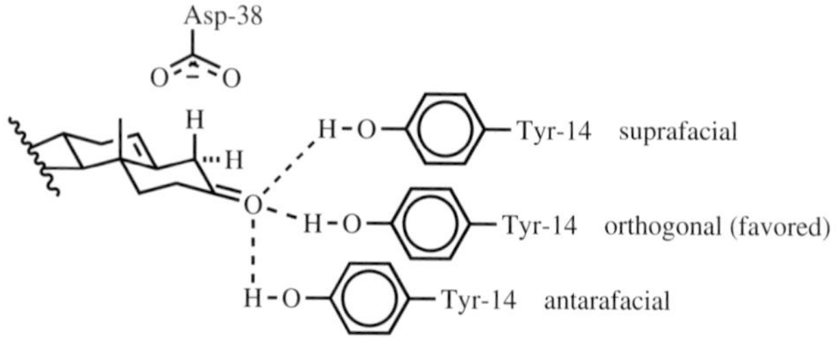

9.38

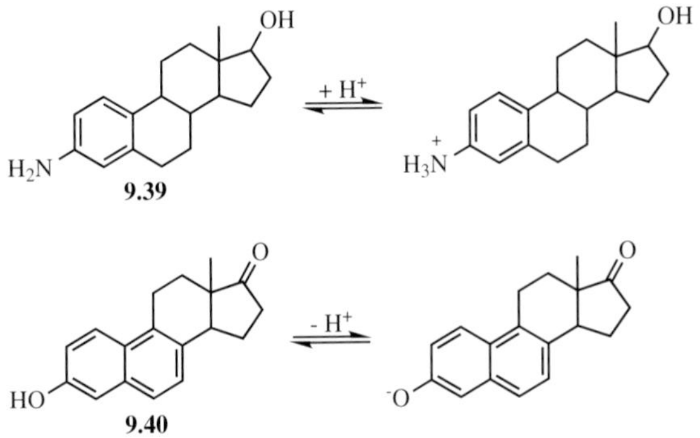

SCHEME 9.28   Reactions designed to investigate the function of Tyr-14 at the active site of 3-oxo-Δ^5-steroid isomerase.

the intermediate dienol is lowered to ≤ 4.5 when the enzyme is bound, and the enzyme acts by an anionic mechanism. The function of Tyr-14, then, would be to stabilize the anion of the dienol by hydrogen bonding rather than by proton transfer. There is no evidence for a cation mechanism involving prior proton donation from Tyr-14 to the 3-oxo carbonyl before abstraction of the 4β-H by Asp-38, again in support of a carbanion mechanism (Scheme 9.29).

The three-dimensional solution structure of the homodimer 3-oxo-Δ^5-steroid isomerase expressed in *E. coli* (28 kDa symmetrical dimer) was solved by Summers, Pollack, and co-workers using heteronuclear multidimensional NMR spectroscopic methods.[58] Each monomer has a hydrophobic cavity. Tyr-14 and Asp-38 are located near the bottom of this cavity and positioned as expected from mechanistic hypotheses. An unexpected acid group, Asp-99, however, is also located in the active site adjacent to Tyr-14, and kinetic and binding studies of the D99A mutant

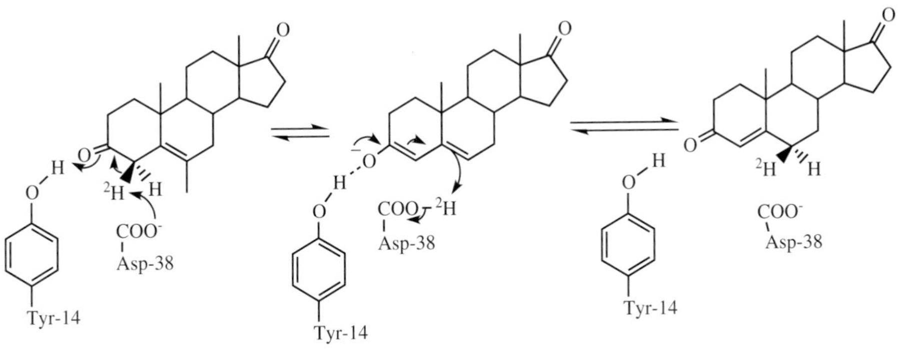

SCHEME 9.29   Mechanism for *suprafacial* transfer of the 4β-proton to the 6β-position of steroids catalyzed by 3-oxo-Δ^5-steroid isomerase.

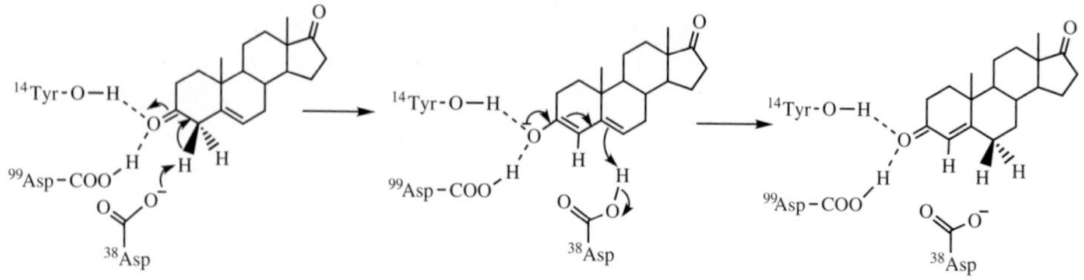

SCHEME 9.30   One mechanism for the function of Asp-99 in the active site of 3-oxo-Δ^5-steroid isomerase.

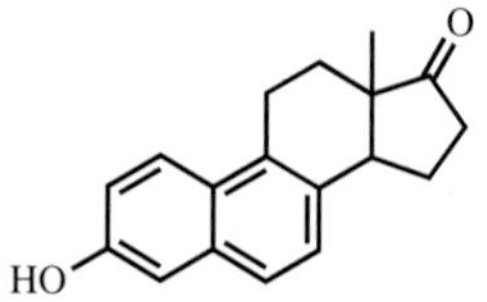

9.41

demonstrate that Asp-99 contributes to catalysis by stabilizing the intermediate. Pollack, Summers, and co-workers propose that the dienolate intermediate is stabilized by two hydrogen bonds, one from the Tyr-14 hydroxyl group and the other from the Asp-99 carboxylate (Scheme 9.30).

The crystal structures of the *Pseudomonas putida* enzyme at 1.9 Å resolution and with equilenin (**9.41**) bound at 2.5 Å resolution were determined by Oh and co-workers,[59] as was the enzyme from *Pseudomonas testosteroni* at 2.3 Å resolution.[60] Tyr-14 and Asp-99 appear to form hydrogen bonds directly with the oxyanion of the bound inhibitor in a completely nonpolar environment of the active site, consistent with the mechanism in Scheme 9.30. Asp-99 is surrounded by six nonpolar residues; consequently its pK_a appears to be elevated to about 9.5.

An alternative mechanism by Mildvan and co-workers (Scheme 9.31)[61] assigns a different role to Asp-99, based on UV resonance Raman spectroscopic studies of Tyr-14[62] and by using heteronuclear multidimensional NMR spectral methods with

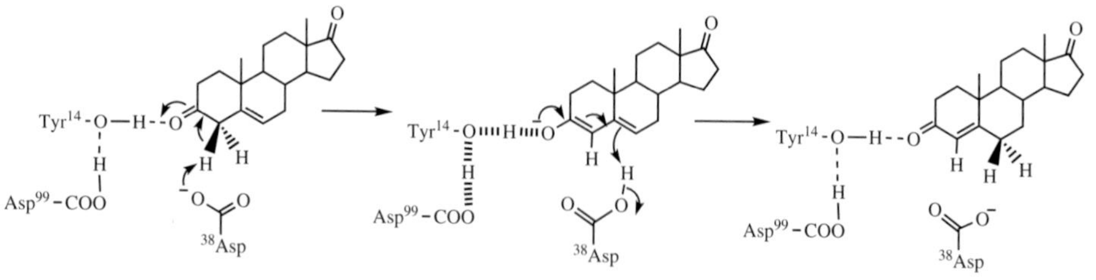

SCHEME 9.31   Another mechanism for the function of Asp-99 in the active site of 3-oxo-Δ^5-steroid isomerase.

the product analogue, 19-nortestosterone hemisuccinate complexed to the homo-dimer.[63] In this mechanism Tyr-14 acts as both a proton donor and a proton acceptor, and a catalytic diad is formed with Asp-99 donating a low-barrier hydrogen bond (see Chapter 1, Section II.C; note that the low-barrier hydrogen bond is denoted as wide vertical dashed lines) to Tyr-14, which strengthens the hydrogen bond from Tyr-14 to the 3-keto oxygen of the steroid. From the NMR spectral results, Tyr-14 is shown to approach Asp-99 and the 3-keto group of the bound steroid. Asp-38 approaches the β-face of the steroid near C4 and C6.

The pK_a of the 4β-proton in solution is 12.7, but the pK_a of Asp-38 is only 4.7.[64] This deprotonation reaction, then, is energetically uphill by about 11 kcal/mol. NMR studies by Mildvan and co-workers showed the generation of a highly deshielded proton (18.15 ppm), consistent with the formation of a low-barrier hydrogen bond to Tyr-14, which could provide at least 7.1 kcal/mol toward the stabilization energy of the dienolate intermediate.[65] Furthermore, if the formation of low-barrier hydrogen bonds among Asp-99, Tyr-14, and the substrate carbonyl occur in concert with the abstraction of the 4β-proton of the substrate by Asp-38, the thermodynamic and kinetic problems of hydrogen bonding from Tyr-14 (pK_a 11.6) to the 3-carbonyl group (pK_a −7) also can be overcome. Mutation of all of the tyrosines other than Tyr-14 to Phe had little or no effect on enzyme activity.[66] The crystal structure of the D99E/D38N mutant complexed with equilenin (**9.41**), however, shows the distance between the Tyr-14 oxygen and the Glu-99 oxygen is too far for a H-bond, but the distance between Glu-99 and the steroid carbonyl oxygen is appropriate for a H-bond, supporting Scheme 9.30 over 9.31.[66a] These results, along with fluorescence and UV spectral data, support a concerted deprotonation by Asp-38 and a strong hydrogen bond by Tyr-14 and Asp-99 to polarize the carbonyl (Scheme 9.30).

The free-energy profile for this reaction was modeled, and four energy barriers were determined—substrate binding, two chemical steps, and dissociation of the product—but no obvious rate-determining step is apparent.[67] The first three steps are comparable in energy, and the product release step is a little lower in energy; therefore, both of the chemical steps seem to be partially rate determining.

Another enzyme that catalyzes an allylic isomerization via a carbanion intermediate is 4-oxalocrotonate tautomerase, which catalyzes the isomerization of (E)-2-oxo-4-hexenedioate (**9.42**) to the α,β-unsaturated ketone (E)-2-oxo-3-hexenedioate (**9.44**), via 2-hydroxymuconate (**9.43**) (Scheme 9.32). With the use

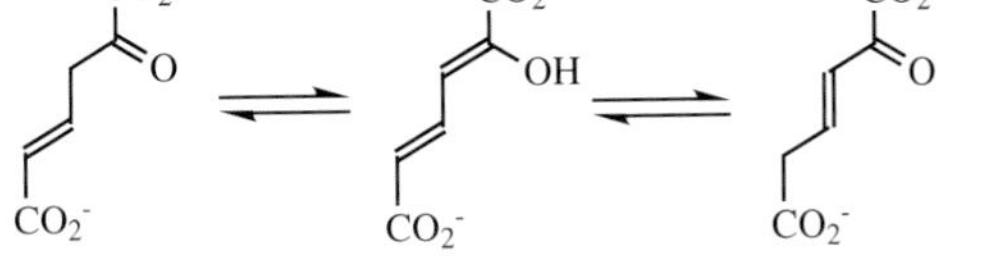

$$CO_2^- \qquad CO_2^- \qquad CO_2^-$$

9.42   **9.43**   **9.44**

SCHEME 9.32   Reaction catalyzed by 4-oxalocrotonate tautomerase.

of specifically deuterated substrates, substrate analogues, and reactions run in D_2O to determine the stereochemistry of deuterium incorporation, Whitman and co-workers conclude that the isomerization of **9.42** to **9.44** is predominantly a *supra-facial* process, suggesting that 4-oxalocrotonate tautomerase proceeds by a one-base mechanism.[68] pH–rate profiles exhibit two important pK_a values, one at 6.2 and one at 9.0.[69] The ^{15}N NMR spectrum of uniformly ^{15}N-labeled enzyme shows that Pro-1 is the pK_a 6.2 group, which is three pK_a units lower than a model compound proline amide in solution (pK_a 9.4). This fact suggests that the active site is very nonpolar or that there is at least one other cationic group nearby. A mechanism is proposed involving general base catalysis by the low pK_a Pro-1 residue concomitant with electrophilic catalysis by a (yet unknown) general acid of pK_a 9.0.

2. Carbocation Mechanism

Isopentenyl diphosphate isomerase, a Mg^{2+}-dependent enzyme in the biosynthetic pathway of terpenes (isoprenoids), catalyzes the interconversion of isopentenyl diphosphate (**9.45**) and dimethylallyl diphosphate (**9.46**) (Scheme 9.33). In the next enzyme-catalyzed reaction in this biosynthetic pathway (farnesyl diphosphate synthase; Chapter 6, Section I.A), **9.45** and **9.46** are joined in a head-to-tail linkage to give geranyl diphosphate (**9.47**), which goes on to give terpenes and steroids (Scheme 9.34). In the isopentenyl diphosphate isomerase reaction, there is no exchange of solvent into the substrate, only into the product. This is consistent with a carbocation mechanism, which is reasonable because a stabilized carbanion cannot be formed. Further evidence for a carbocation mechanism comes from substrate analogue studies.[70] The trifluoromethyl analogue of dimethylallyl diphos-

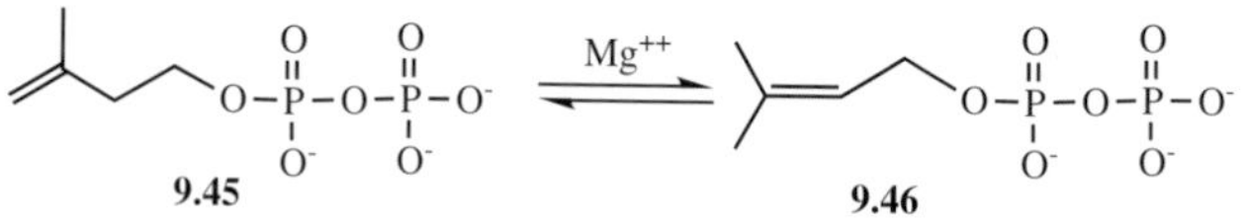

SCHEME 9.33   Reaction catalyzed by isopentenyl diphosphate isomerase.

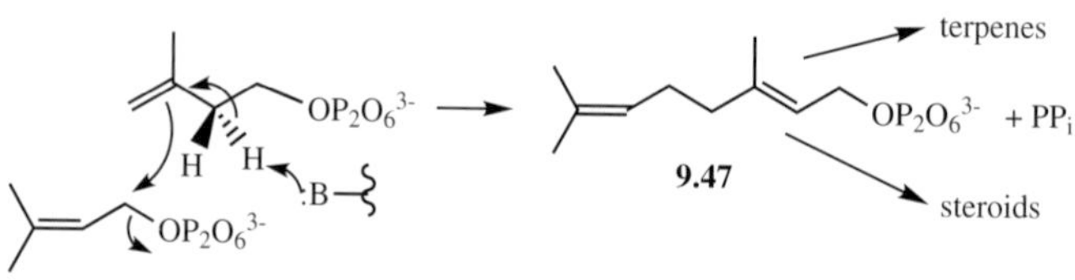

SCHEME 9.34   Reaction catalyzed by farnesyl diphosphate synthase and biosynthesis of terpenes and steroids.

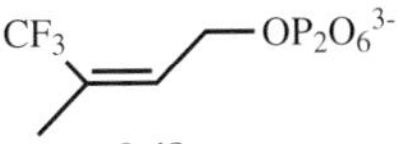

phate (**9.48**) was found to react (in the back reaction) at 1.8×10^{-6} times the rate of **9.46**; the electron-withdrawing fluorines support a carbocation mechanism. Likewise, dimethylammonium ethyl diphosphate (**9.49**) is a potent inhibitor of the enzyme, having a k_{on} of $2.1 \times 10^{6}\ M^{-1}min^{-1}$ and a k_{off} of $3 \times 10^{-5}\ min^{-1}$ ($K_i = k_{off}/k_{on} = 14$ pM!), suggesting that it is a transition-state analogue (see Chapter 2, Section I.B.3) and tight-binding inhibitor (see Appendix I, Section II.A.5). These results point to the formation of a carbocation intermediate (Scheme 9.35), which accounts for the incorporation of solvent hydrogen into the product but not into the substrate.

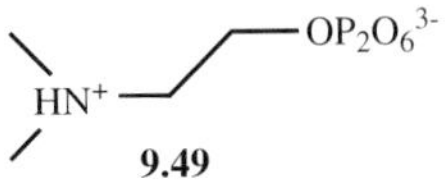

C. Aza-Allylic Isomerizations

Pyridoxal-5′-phosphate (PLP) was discussed earlier with regard to enzymes that catalyze decarboxylations (Chapter 8, Section V.A) and racemizations (II.B). Another reaction for which it acts as a coenzyme is an *aza-allylic isomerization* (Scheme 9.36) of amino acids in which the amino group is transferred to a second substrate molecule; this reaction is known as a *transamination reaction* and the enzymes are called *aminotransferases* (formerly *transaminases*).

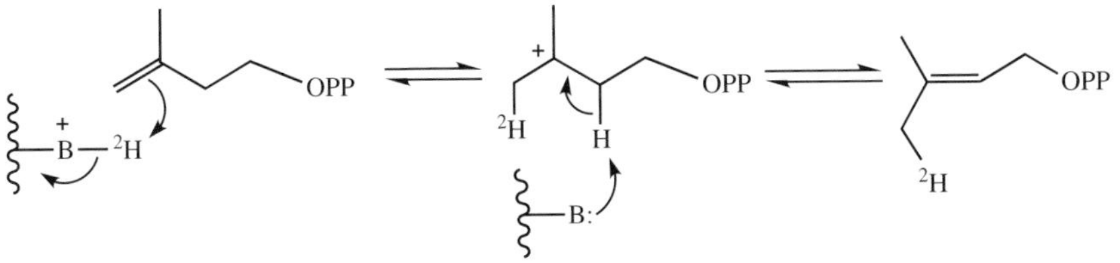

SCHEME 9.35   Proposed mechanism of isopentenyl diphosphate isomerase.

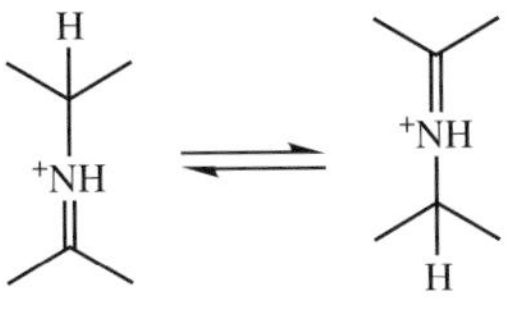

SCHEME 9.36   Aza-allylic isomerization.

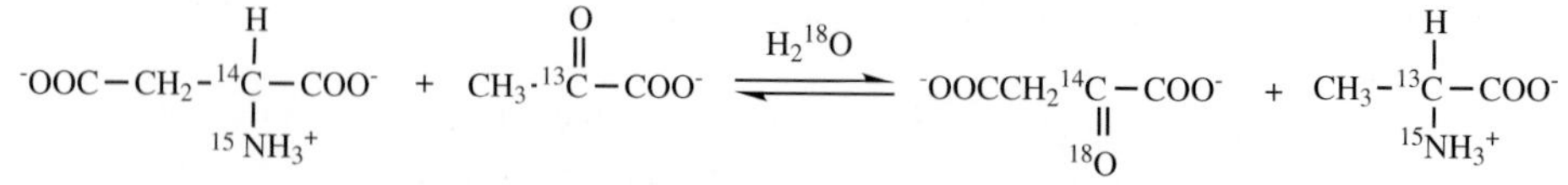

$$^-OOC-CH_2-{}^{14}\overset{\underset{\displaystyle {}^{15}NH_3{}^+}{|}}{\underset{|}{\overset{|}{C}}}-COO^- \;+\; CH_3-{}^{13}\overset{\underset{\displaystyle}{\overset{\displaystyle O}{||}}}{C}-COO^- \;\underset{\xrightarrow{\;\;H_2{}^{18}O\;\;}}{\rightleftharpoons}\; {}^-OOCCH_2{}^{14}\overset{\underset{\displaystyle {}^{18}O}{||}}{C}-COO^- \;+\; CH_3-{}^{13}\overset{\underset{\displaystyle {}^{15}NH_3{}^+}{|}}{\underset{|}{\overset{H}{C}}}-COO^-$$

SCHEME 9.37   Reaction catalyzed by aspartate aminotransferase.

1. L-*Aspartate Aminotransferase (or Aspartate–Pyruvate Aminotransferase)*

Aminotransferases are the most complicated of the PLP-dependent reactions; they involve two substrates going to two products in two half-reactions. Kinetic analyses of these reactions show that they typically follow a Ping Pong Bi–Bi mechanism (Appendix I, VI.A.4). Various PLP-dependent aminotransferases are known for α-, β-, and γ-amino acids. Scheme 9.37 shows the reaction catalyzed by aspartate aminotransferase; various isotopes have been included in this reaction so you can follow where the atoms are going. As we saw earlier, the first step in every PLP-dependent reaction is the same, namely, the formation of the Schiff base between PLP and the amino acid (see Scheme 8.39 in Chapter 8, Section V.A). The first half reaction involves the conversion of the amino acid (aspartate, **9.50**) into the corresponding keto acid (oxalacetate, **9.51**) with concomitant conversion of the PLP into pyridoxamine 5′-phosphate (PMP, **9.52**) (Scheme 9.38). This is an internal redox reaction; the amino acid is oxidized to the keto acid and the oxidized form of the

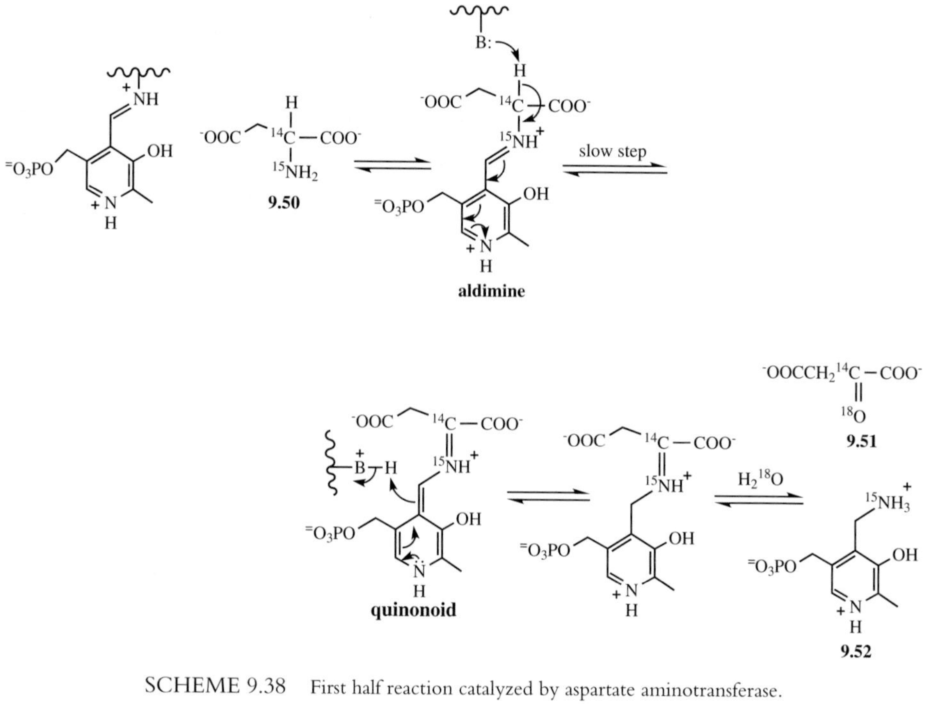

SCHEME 9.38   First half reaction catalyzed by aspartate aminotransferase.

cofactor (PLP) is reduced to PMP. Consequently there is no overall change in oxidation state.

Using X-ray crystallography, Ringe and co-workers elegantly confirmed the general mechanisms for an aminotransferase depicted in Schemes 8.39 and 9.38.[71] Three different crystal structures of D-amino acid aminotransferase—one structure of the native enzyme with PLP bound, one structure with D-alanine reduced on to the PLP (i.e., the aldimine form in Scheme 9.38, but with the imine bond reduced to an amine for stability and to prevent enzymatic turnover), and a third structure of the enzyme with PMP bound[72]—like photographs of the enzyme-catalyzed reaction in progress, clearly delineate the mechanistic pathway shown in Schemes 8.39 and 9.38.

At this point in the mechanistic pathway of an aminotransferase, the enzyme has accomplished its goal of degrading the amino acid to the keto acid. However, the enzyme now has a problem; it is no longer in the correct oxidation state to accept another substrate molecule (amines do not react with amines), so the enzyme is nonfunctional (or, in the vernacular, it is dead). In this case the enzyme has not acted as a catalyst, but rather as a reagent. The enzyme gets around this problem the same way that the flavin-dependent enzymes (see Chapter 3, Section III.B) do. A second substrate—in this case, pyruvate (**9.53**)—enters the active site and converts the PMP back to PLP (Scheme 9.39) with concomitant conversion of the

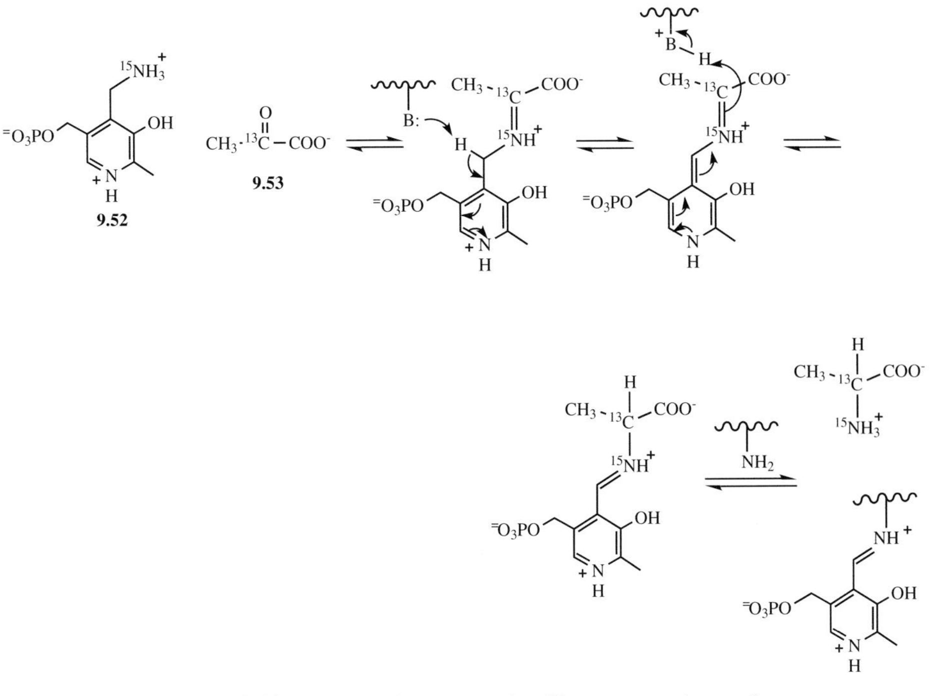

SCHEME 9.39 Second half reaction catalyzed by aspartate aminotransferase.

9.54   9.55

pyruvate to alanine. Notice that the mechanism for the second half reaction is the exact reverse of that for the first half reaction: In Scheme 9.38 the last step (prior to hydrolysis) is protonation of the PMP methylene, and in Scheme 9.39 the first step (after Schiff base formation) is deprotonation of that same methylene. Also note that the amino group of aspartate is transferred first to the cofactor and then from the cofactor to the second substrate; PLP is acting as the amino-transferring agent, hence the name of this class of enzymes. If the keto acid (pyruvate in this case) is omitted from the enzyme assay, the enzyme becomes reversibly inhibited. As soon as the keto acid is added, the enzyme becomes active again. Also, if the keto acid is omitted, one equivalent each of PMP and oxaloacetate can be isolated. With the use of the pseudosubstrate, *erythro-α*-hydroxy-L-aspartate (**9.54**), the highly conjugated quinonoid form (**9.55**) can be observed at 490 nm.[73]

When $[\alpha^{-2}H]$-L-alanine is the substrate with alanine aminotransferase, and the keto acid substrate is omitted, the methylene carbon of the PMP produced has 4% deuterium incorporated (when carried out with nondeuterated substrate in D_2O, 50% of the protium is transferred to the cofactor).[74] This result suggests a single-base mechanism, in which the base first removes the α-proton from the first substrate and then delivers it to the cofactor; this model predicts *suprafacial* stereochemistry for the proton transfer. In fact, Dunathan and Voet observed the same absolute stereochemistry for at least seven PLP-dependent aminotransferases, namely, the proton is removed from the *si* face of the enzyme-bound substrate (**9.56**) and is delivered to the *pro-S* position of the PMP methylene group[75] (**9.57**, Scheme 9.40).

V. *CIS/TRANS* ISOMERIZATIONS

A. Maleylacetoacetate Isomerase

Maleylacetoacetate isomerase catalyzes the *cis/trans* isomerization of maleylaceto-acetate (**9.58**) to fumarylacetoacetate (**9.59**) (Scheme 9.41) in the presence of glutathione (GSH) in mammals and bacteria.[76] The GSH acts as a coenzyme in this enzyme rather than as a reducing agent; other thiols fail to produce enzyme activ-

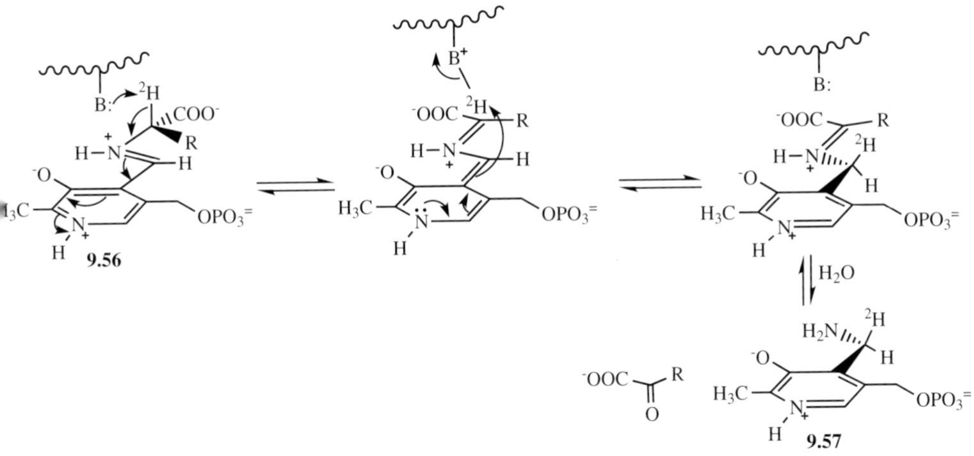

SCHEME 9.40   Stereochemistry of proton transfer in the first step catalyzed by many PLP-dependent aminotransferases.

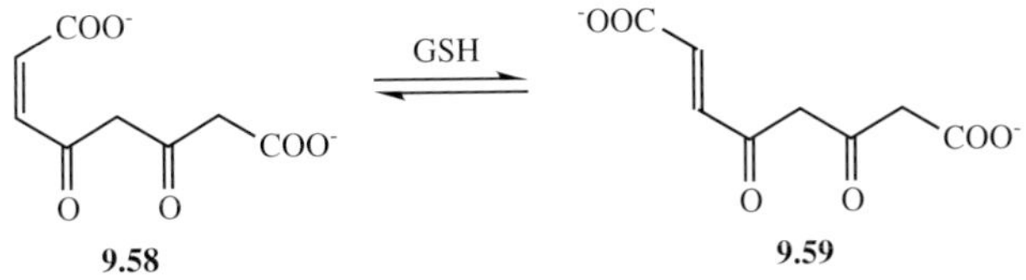

SCHEME 9.41   Reaction catalyzed by maleylacetoacetate isomerase.

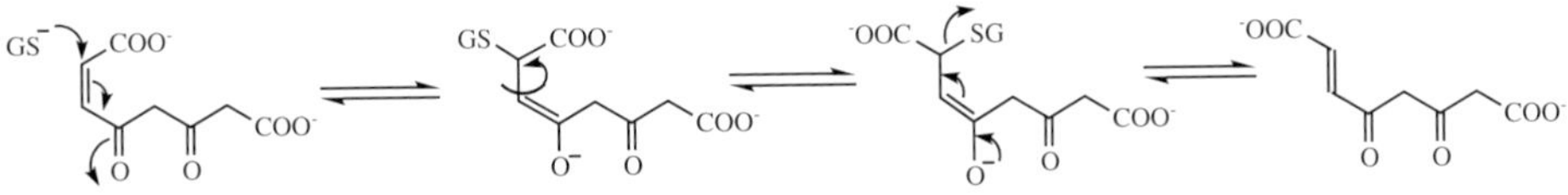

SCHEME 9.42   Proposed mechanism for the reaction catalyzed by maleylacetoacetate isomerase.

ity. It is inhibited by sulfhydryl reagents, and when the enzyme reaction is run in ^2H_2O no deuterium is incorporated into the substrate or product. The mechanism in Scheme 9.42 is consistent with these observations and is consistent with chemical model studies (the enzyme reaction is 10^6 times faster than the nonenzymatic rate).[77] A Schiff base mechanism involving an active-site lysine residue was excluded by incubation of the enzyme with ^{14}C-labeled substrate followed by treatment with sodium borohydride; no radioactivity was covalently bound to the enzyme.

B. Retinol Isomerase

All *trans*-retinol (**9.60**) is converted to 11-*cis*-retinol (**9.64**) by retinol isomerase in a covalent catalytic process; ^{18}O-**9.60** loses the ^{18}O during the conversion (Scheme 9.43).[78] The first step in this sequence is catalyzed by a different enzyme, retinol

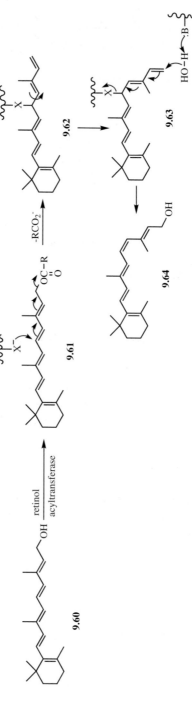

SCHEME 9.43 Proposed mechanism for the reaction catalyzed by retinol isomerase.

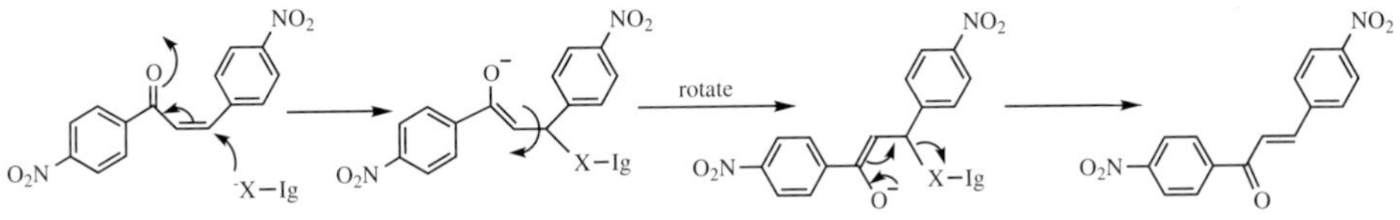

SCHEME 9.44   *Cis/trans* isomerization reaction catalyzed by an antibody (Ig).

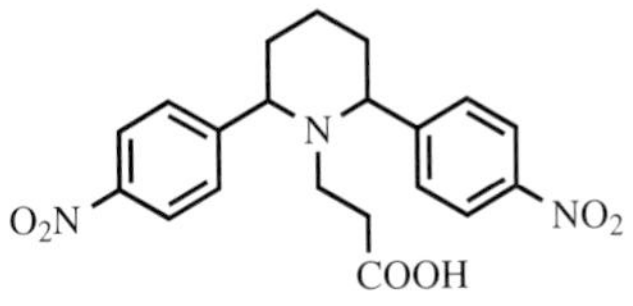

9.65

acyltransferase, which uses dipalmitoylphosphatidylcholine as the acylating agent.[79] Rando and co-workers proposed that an active-site nucleophile undergoes an S_N2' attack on **9.61** to give **9.62**, which then can rotate to **9.63**; Michael addition of water gives **9.64**.

C. Catalytic Antibody-Catalyzed *Cis/Trans* Isomerization

The *cis/trans* isomerization reaction shown in Scheme 9.44 was catalyzed by an antibody generated by Jackson and Schultz from hapten **9.65**, which was coupled to carrier proteins BSA and keyhole limpet hemocyanin.[80] The K_m is 220 μM, and the k_{cat} is 4.8 min^{-1} as compared to the uncatalyzed k_{uncat} of 3.1×10^{-4} min^{-1}.

VI. PHOSPHATE ISOMERIZATION

The *phosphomutases* are a subclass of kinases (see Chapter 2, Section II.D.3) that catalyze an apparent intramolecular phosphoryl transfer (isomerization). Phosphoglucomutase catalyzes the interconversion of α-glucose-6-phosphate (**9.66**) and α-glucose-1-phosphate (**9.67**) (Scheme 9.45). The enzyme only recognizes the α-anomer, not the β-anomer. Actually, the resting state of the enzyme from various

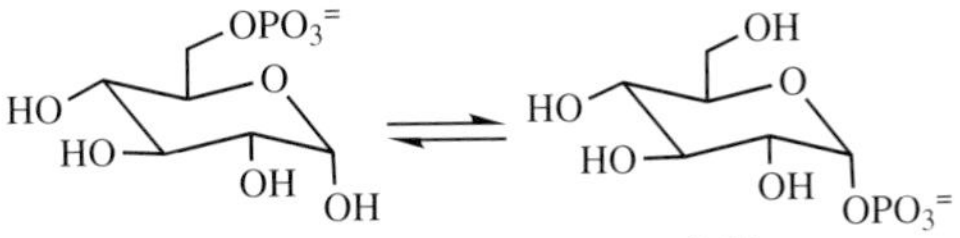

9.66   **9.67**

SCHEME 9.45   Reaction catalyzed by phosphoglucomutases.

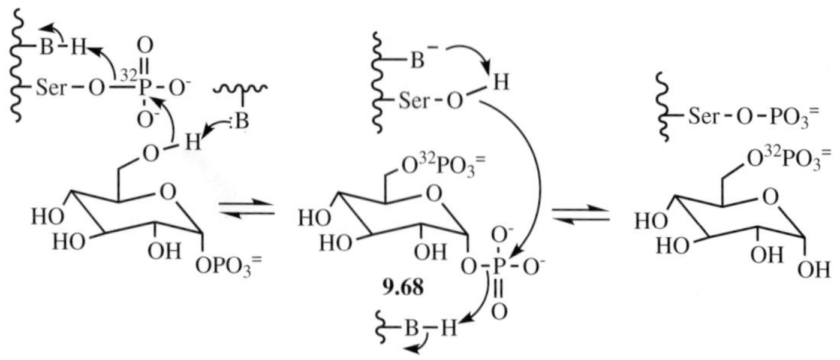

SCHEME 9.46   Proposed mechanism for the reaction catalyzed by phosphoglucomutases.

sources is a phosphorylated form. The dephosphorylated form of the enzyme is inactive toward **9.66** and **9.67,** but can be activated by α-glucose-1,6-diphosphate (**9.68**), which is a tightly bound intermediate in the normal catalytic turnover (Scheme 9.46). To demonstrate that **9.68** is tightly bound, the ^{32}P-phosphorylated enzyme (prepared by incubation of the enzyme with ^{32}P-**9.67**) was incubated with unlabeled **9.67** in the presence of excess unlabeled **9.68**; ^{32}P-**9.66** was produced containing all of the radioactivity.[81] If the intermediate ^{32}P-**9.68** had dissociated from the enzyme, it would have been replaced with the unlabeled **9.68** present; in that case ^{32}P-**9.68** would have been detected, and not all of the ^{32}P would have been found in **9.66.**

Blattler and Knowles have shown phosphoglucomutases to catalyze reactions with overall retention of configuration,[82] which means either no inversion or two inversions occur; the latter has been the mechanism usually drawn for this class of enzymes (Scheme 9.46 is drawn as the back reaction). However, a model reaction was designed by Cullis and Misra to determine if retention of configuration is also produced by dissociative mechanisms, and in fact there is evidence that a dissociative mechanism can proceed with retention, inversion, or racemization (Scheme 9.47).[83] Therefore, it is not clear if the phosphoglucomutases proceed by an associative (Scheme 9.46) or dissociative mechanism.

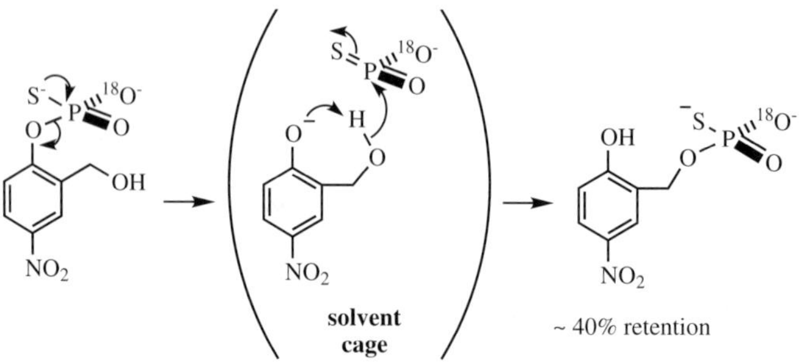

SCHEME 9.47   Model reaction for a dissociative mechanism of phosphomutases.

NOTE THAT PROBLEMS AND SOLUTIONS RELEVANT TO EACH CHAPTER CAN BE FOUND IN APPENDIX II.

REFERENCES

1. Gallo, K. A.; Knowles, J. R. *Biochemistry* **1993**, *32*, 3981.
2. Gallo, K. A.; Tanner, M. E.; Knowles, J. R. *Biochemistry* **1993**, *32*, 3991.
3. (a) Rose, I. A. *Annu. Rev. Biochem.* **1966**, *35*, 23. (b) Cardinale, G. J.; Abeles, R. H. *Biochemistry* **1968**, *7*, 4639.
4. Tanner, M. E.; Gallo, K. A.; Knowles, J. R. *Biochemistry* **1993**, *32*, 3998.
5. Fisher, L. M.; Belasco, J. G.; Bruice, T. W.; Albery, W. J.; Knowles, J. R. *Biochemistry* **1986**, *25*, 2543.
6. Cardinale, G. J.; Abeles, R. H. *Biochemistry* **1968**, *7*, 3970.
7. Yamauchi, T.; Choi, S. Y.; Okada, H.; Yohda, M.; Kumagai, H.; Esaki, N.; Soda, K. *J. Biol. Chem.* **1992**, *267*, 18361.
8. Rudnick, G.; Abeles, R. H. *Biochemistry* **1975**, *14*, 4515.
9. Bernard, S. A.; Orgel, L. E. *Science* **1959**, *130*, 625.
10. Jencks, W. P. In *Current Aspects of Biochemical Energetics*, Kennedy, E. P., Ed., Academic Press: New York, 1966, p. 273.
11. (a) Wolfenden, R. *Annu. Rev. Biophys. Bioengg.* **1976**, *5*, 271. (b) Wolfenden, R. *Nature* **1969**, *223*, 704. (c) Wolfenden, R. *Meth. Enzymol.* **1977**, *46*, 15.
12. (a) Lienhard, G. E. *Science* **1973**, *180*, 149. (b) Lienhard, G. E. *Annu. Repts. Med. Chem.* **1972**, *7*, 249.
13. Adams, E. *Adv. Enzymol. Rel. Areas Mol. Biol.* **1976**, *44*, 69.
14. Hoffmann, K.; Schneider-Scherzer, E.; Kleinkauf, H.; Zocher, R. *J. Biol. Chem.* **1994**, *269*, 12710.
15. Rose, I. A. *Annu. Rev. Biochem.* **1966**, *35*, 23.
16. Faraci, W. S.; Walsh, C. T. *Biochemistry* **1988**, *27*, 3267.
17. Shaw, J. P.; Petsko, G. A.; Ringe, D. *Biochemistry* **1997**, *36*, 1329.
17a. Sun, S.; Toney, M. D. *Biochemistry* **1999**, *38*, 4058.
18. Ahmed, S. A.; Esaki, N.; Tanaka, H.; Soda, K. *Biochemistry* **1986**, *25*, 385.
19. Leussing, D. L. In *Vitamin B₆ Pyridoxal Phosphate*, Part A, Dolphin, D.; Poulson, R.; Avramovic, O., Eds., Wiley: New York, 1986, p. 69.
20. Dunathan, H. C. *Adv. Enzymol.* **1971**, *35*, 79.
21. John, R. A. *Biochim. Biophys. Acta* **1995**, *1248*, 81.
22. Chen, H.; Phillips, R. S. *Biochemistry* **1993**, *32*, 11591.
23. Powers, V. M.; Koo, C. W.; Kenyon, G. L.; Gerlt, J. A.; Kozarich, J. W. *Biochemistry* **1991**, *30*, 9255.
24. Neidhart, D. J.; Howell, P. L.; Petsko, G. A.; Powers, V. M.; Li, R.; Kenyon, G. L.; Gerlt, J. A. *Biochemistry* **1991**, *30*, 9264.
25. Landro, J. A.; Kallarakal, A. T.; Ransom, S. C.; Gerlt, J. A.; Kozarich, J. W.; Neidhart, D. J.; Kenyon, G. L. *Biochemistry* **1991**, *30*, 9274.
26. Kallarakal, A. T.; Mitra, B.; Kozarich, J. W.; Gerlt, J. A.; Clifton, J. G.; Petsko, G. A.; Kenyon, G. L. *Biochemistry* **1995**, *34*, 2788.
27. Schafer, S. L.; Barrett, W. C.; Kallarakal, A. T.; Mitra, B.; Kozarich, J. W.; Gerlt, J. A.; Clifton, J. G.; Petsko, G. A.; Kenyon, G. L. *Biochemistry* **1996**, *35*, 5662.
28. Montecucchi, P. C.; De Castiglione, R.; Piani, S.; Gozzini, L.; Erspamer, V. *Int. J. Peptide Protein Res* **1981**, *17*, 275.
29. Erspamer, V.; Melchiorri, P.; Falconieri-Erspamer, G.; Negri, L.; Corsi, R.; Severini, C.; Barra, D.; Simmaco, M.; Kreil, G. *Proc. Natl. Acad. Sci. USA* **1989**, *86*, 5188.
30. Kamatani, Y.; Minakata, H.; Kenny, P. T.; Iwashita, T.; Watanabe, K.; Funase, K.; Sun, X. P.;

Yongsiri, A.; Kim, K. H.; Novales, E. T.; Kanapi, C. G.; Takeuchi, H.; Nomoto, K. *Biochem. Biophys. Res. Commun.* **1989**, *160*, 1015.

31. Ohta, N.; Kubota, I.; Takao, T.; Shimonishi, Y.; Yasuda-Kamatani, Y.; Minakata, H.; Nomoto, K.; Muneoka, Y.; Kobayashi, M. F. *Biochem. Biophys. Res. Commun.* **1991**, *178*, 486.

32. Heck, S. D.; Faraci, W. S.; Kelbaugh, P. R.; Saccomano, N. A.; Thadeio, P. F.; Volkmann, R. A. *Proc. Natl. Acad. Sci. USA* **1996**, *93*, 4036.

33. Glaser, L. In *The Enzymes,* vol. 6, 3rd ed., Boyer, P., Ed., Academic Press: New York, 1972, p. 355.

34. (a) Marumo, K.; Lindqvist, L.; Verma N.; Weintraub, A.; Reeves, P. R.; Lindberg, A. A. *Eur. J. Biochem.* **1992**, *204*, 539. (b) Köplin, R.; Wang, G.; Hötte, B.; Priefer, U. B.; Pühler, A. *J. Bacteriol.* **1993**, *175*, 7786. (c) Tsukioka, Y.; Yamashita, Y.; Oho, T.; Nakano, Y.; Koga, T. *J. Bacteriol.* **1997**, *179*, 1126.

35. Melo, A.; Glaser, L. *J. Biol. Chem.* **1968**, *243*, 1475.

36. Glaser, L. *Biochim. Biophys. Acta* **1961**, *51*, 169.

37. (a) Anderson, L.; Landel, A. M.; Diedrick, D. F. *Biochim. Biophys. Acta* **1956**, *22*, 573. (b) Kowalsky, A.; Koshland, D. E. *Biochim. Biophys. Acta* **1956**, *22*, 575. (c) Kalckar, H.M.; Maxwell, E.S. *Biochim. Biophys. Acta* **1956**, *22*, 588.

38. (a) Frey, P. A. In *Pyridine Nucleotide Coenzymes;* vol. 2B, Dolphin, D.; Poulson, R.; Avramovic, O., Eds.; Wiley: New York, 1987, p. 462. (b) Glaser, L. In *The Enzymes,* vol. 6, 3rd ed., Boyer, P., Ed., Academic Press: New York, 1972, p. 355.

39. March, J. *Advanced Organic Chemistry,* 4th ed., Wiley: New York; 1992, p. 215.

40. Maitre, U. S.; Ankel, H. *Proc. Natl. Acad. Sci USA* **1971**, *68*, 2660.

41. Wheeler, G. L.; Jones, M. A.; Smirnoff, N. *Nature* **1998**, *393*, 365.

42. Barber, G. A. *J. Biol. Chem.* **1979**, *254*, 7600.

43. Creighton, D. J.; Murthy, N. S. R. K. In *The Enzymes,* vol. 19, 3rd ed., Sigman, D. S.; Boyer, P. D., Eds., Academic Press: San Diego, 1990, p. 323.

44. Hanson, K. R.; Rose, I. A. *Acc. Chem. Res.* **1975**, *8*, 1.

44a. Farber, G. K.; Glasfeld, A.; Tiraby, G.; Ringe, D.; Petsko, G. A. *Biochemistry* **1989**, *28*, 7289.

44b. Zheng, Y.-J.; Merz, K. M., Jr.; Farber, G. K. *Protein Engineering* **1993**, *6*, 479.

45. Rétey, J.; Bartl, K.; Ripp, E.; Hull, W. *Eur. J. Biochem.* **1977**, *72*, 251.

46. Pirrung, M. C.; Chen, J.; Rowley, E. G.; McPhail, A. T. *J. Am. Chem. Soc.* **1993**, *115*, 7103.

47. Batzold, F. H.; Benson, A. M.; Covey, D. F.; Robinson, C. H.; Talalay, P. *Adv. Enzyme Regul.* **1976**, *14*, 243.

48. Malhotra, S. K.; Ringold, H. J. *J. Am. Chem. Soc.* **1965**, *87*, 3228.

49. (a)Viger, A.; Coustal, S.; Marquet, A. *J. Am. Chem. Soc.* **1981**, *103*, 451. (b) Zawrotny, M. E.; Hawkinson, D. C.; Blotny, G.; Pollack, R. M. *Biochemistry* **1996**, *35*, 6438.

50. Wang, S.-F.; Kawahara, F. S.; Talalay, P. *J. Biol. Chem.* **1963**, *238*, 576.

51. (a) Hawkinson, D. C.; Pollack, R. M. *Biochemistry* **1993**, *32*, 694. (b) Austin, J. C.; Zhao, Q.; Jordan, T.; Talalay, P.; Mildvan, A. S.; Spiro, T. G. *Biochemistry* **1995**, *34*, 4441.

52. Bantia, S.; Pollack, R. M. *J. Am. Chem. Soc.* **1986**, *108*, 3145.

53. (a) Hawkinson, D. C.; Eames, C. M.; Pollack, R. M. *Biochemistry* **1991**, *30*, 6956. (b) Zeng, B.; Pollack, R. M. *J. Am. Chem. Soc.* **1991**, *113*, 3838. (c) Pollack, R. M.; Mack, J. P. G.; Eldin, S. *J. Am. Chem. Soc.* **1987**, *109*, 5048.

54. Kuliopulos, A.; Mildvan, A. S.; Shortle, D.; Talalay, P. *Biochemistry* **1989**, *28*, 149.

55. Kuliopulos, A.; Mildvan, A. S.; Shortle, D.; Talalay, P. *Biochemistry* **1990**, *29*, 10271.

56. Kuliopulos, A.; Mullen, G. P.; Xue, L.; Mildvan, A. S. *Biochemistry* **1991**, *30*, 3169.

57. Zeng, B.; Bounds, P. L.; Steiner, R. F.; Pollack, R. M. *Biochemistry* **1992**, *31*, 1521.

58. Wu, Z. R.; Ebrahimian, S.; Zawrotny, M. E.; Thornburg, L. D.; Perez-Alvarado, G. C.; Brothers, P.; Pollack, R. M.; Summers, M. F. *Science* **1997**, *276*, 415.

59. Kim, S. W.; Cha, S.-S.; Cho, H.-S.; Kim, J.-S.; Ha, N.-C.; Cho, M.-J.; Joo, S.; Kim, K. K.; Choi, K. Y.; Oh, B.-H. *Biochemistry* **1997**, *36*, 14030.

60. Cho, H.-S.; Choi, G.; Choi, K. Y.; Oh, B.-H. *Biochemistry* **1998**, *37*, 8325.

61. Zhao, Q.; Abeygunawardana, C.; Gittis, A. G.; Mildvan, A. S. *Biochemistry* **1997,** *36,* 14616.
62. Austin, J. C.; Kuliopulos, A.; Mildvan, A. S.; Spiro, T. G. *Protein Sci.* **1992,** *1,* 259.
63. Massiah, M. A.; Abeygunawardana, C.; Gittis, A. G.; Mildvan, A. S. *Biochemistry* **1998,** *37,* 14701.
64. Hawkinson, D. C.; Pollack, R. M.; Ambulos, Jr., N. P. *Biochemistry* **1994,** *33,* 12172.
65. Zhao, Q.; Abeygunawardana, C.; Talalay, P.; Mildvan, A. S. *Proc. Natl. Acad. Sci. USA* **1996,** *93,* 8220.
66. Li, Y. K.; Kuliopulos, A.; Mildvan, A. S.; Talalay, P. *Biochemistry* **1993,** *32,* 1816.
66a. Choi, G.; Ha, N.-C.; Kim, S. W.; Kim, S.-H.; Park, S.; Oh, B.-H.; Choi, K. Y. *Biochemistry* **2000,** *39,* 903.
67. Hawkinson, D. C.; Eames, T. C. M.; Pollack, R. M. *Biochemistry* **1991,** *30,* 10849.
68. (a) Lian, H; Whitman, C. P., *J. Am. Chem. Soc.* **1993,** *115,* 7978. (b) Whitman, C. P.; Hajipour, G.; Watson, R. J.; Johnson, W. H.; Bembenek, M. E.; Stolowich, N. J. *J. Am. Chem. Soc.* **1992,** *114,* 10104.
69. Stivers, J. T.; Abeygunawardana, C.; Mildvan, A. S.; Hajipour, G.; Whitman, C. P. *Biochemistry* **1996,** *35,* 814.
70. Reardon, J. E.; Abeles, R. H. *Biochemistry* **1986,** *25,* 5609.
71. Peisach, D.; Chipman, D. M.; Van Ophem, P. W.; Manning, J. M.; Ringe, D. *Biochemistry* **1998,** *37,* 4958.
72. Sugio, S.; Petsko, G. A.; Manning, J. M.; Soda, K.; Ringe, D. *Biochemistry* **1995,** *34,* 9661.
73. Jenkins, W. T. *J. Biol. Chem.* **1961,** *236,* 1121.
74. Dunathan, H.; Ayling, J.; Snell, E. *Biochemistry* **1968,** *7,* 4537.
75. Dunathan, H.; Voet, J. *Proc. Natl. Acad. Sci. USA* **1974,** *71,* 3888.
76. Seltzer, S. In *Glutathione,* Part A, Dolphin, D.; Avramovic, O.; Poulson, R., eds., Wiley: New York, 1989, p. 733.
77. Seltzer, S.; Lin, M. *J. Am. Chem. Soc.* **1979,** *101,* 3091.
78. Deigner, P. S.; Law, W. C.; Canada, F. J.; Rando, R. R. *Science* **1989,** *244,* 968.
79. Shi, Y. Q.; Hubacek, I.; Rando, R. R. *Biochemistry* **1993,** *32,* 1257.
80. Jackson, D. Y.; Schultz, P. G., *J. Am. Chem. Soc.* **1991,** *113,* 2319.
81. (a) Ray, W. J., Jr.; Roscelli, G. A. *J. Biol. Chem.* **1964,** *239,* 1228. (b) Ray, W. J., Jr.; Peck, E. *The Enzymes,* vol. 6, 3rd ed., Boyer, P., Ed., Academic Press: New York, 1972, p. 408.
82. Blattler, W. A.; Knowles, J. R. *Biochemistry* **1980,** *19,* 738.
83. Cullis, P. M.; Misra, R. *J. Am. Chem. Soc.* **1991,** *113,* 9679.

CHAPTER 10

Eliminations and Additions

I. *ANTI* ELIMINATIONS AND ADDITIONS

A. Dehydratases and Hydratases

Dehydratases and *hydratases* are enzymes that eliminate and add, respectively, the elements of water—a proton and a hydroxyl group—to a double bond (Scheme 10.1). The olefinic product geometry and the mode of elimination provide information about the mechanism of the elimination. There are typically three fundamental nonenzymatic mechanisms for dehydration each of which uses only bases and acids for catalysis: carbanion (E1cB), concerted (E2), and carbocation (E1) mechanisms (Scheme 10.2). However, Gerlt, Kozarich, Kenyon, and Gassman[1] have presented cogent arguments against enzyme-catalyzed E1cB mechanisms via free enolates, and provide a rationalization for how enzymes can catalyze β-elimination reactions, even though the pK_a values of the α-proton adjacent to carbonyls and carboxylates are so high by the formation of enols (see Chapter 1, Section II.C, including Schemes 1.12–1.14). You may want to reread that discussion. Also discussed in Chapter 1 (in the same section) is the observation of Schwab and co-workers[2] regarding the stereochemistry of enzyme-catalyzed dehydration reactions, namely, that, in general, enzyme-catalyzed β-dehydrations, where the abstracted proton is α to a carboxylic acid, are almost universally *anti*, whereas those where the proton abstracted is α to an aldehyde, ketone, or a thioester, are invariably *syn*. These three

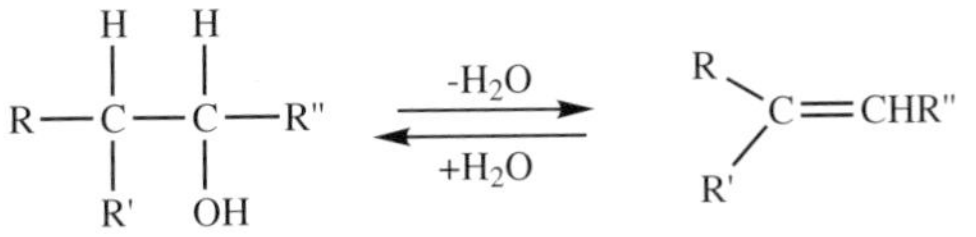

$$R-\overset{\overset{\displaystyle H}{|}}{\underset{\underset{\displaystyle R'}{|}}{C}}-\overset{\overset{\displaystyle H}{|}}{\underset{\underset{\displaystyle OH}{|}}{C}}-R'' \underset{+H_2O}{\overset{-H_2O}{\rightleftharpoons}} \overset{R}{\underset{R'}{>}}C=CHR''$$

SCHEME 10.1   Reactions catalyzed by dehydratases and hydratases.

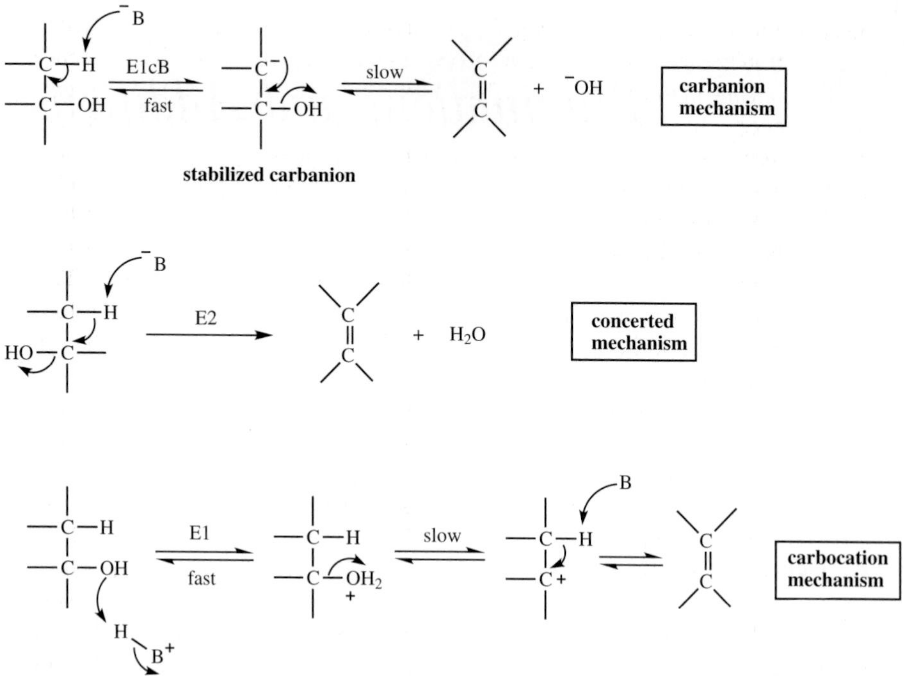

stabilized carbanion

SCHEME 10.2   Three general mechanisms for dehydration.

acid/base mechanisms are not the only possibilities; in this chapter we will see how enzymes use a variety of cofactors, sometimes in unusual ways, to assist some of these reactions. But first, consider a more traditional cofactor, namely, a metal ion.

1. *Metal Ion–Dependent Enzymes*

Enolase catalyzes the reversible elimination of water from 2-phospho-D-glyceric acid (2-PGA, **10.1**) to give phosphoenolpyruvate (PEP, **10.2**), a reaction important in glycolysis (Scheme 10.3; **10.1** is a Fischer projection). Two equivalents of a

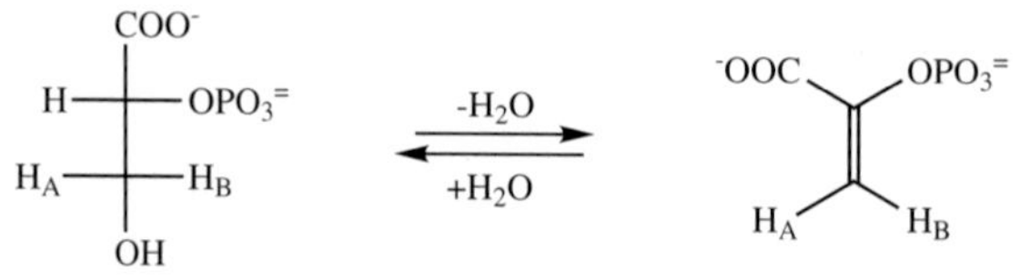

10.1   **10.2**

SCHEME 10.3   Reaction catalyzed by enolase.

divalent metal ion (preferably Mg^{2+}) per subunit are required for full activity.[3] The reaction has an enzymatic equilibrium constant of about 1.[4] All of the evidence points to an *anti* elimination mechanism in which the reaction proceeds in a step-wise manner. The proton NMR of PEP consists of two multiplets (the ^{31}P also splits the protons), one at 5.33 ppm (H_A in **10.2**) and the other at 5.15 ppm (H_B in **10.2**). These assignments were confirmed using ^{13}C-H coupling constants with [1-^{13}C]PEP. When enolase dehydrates $(3R)$-3-[^2H]2-PGA, the isolated 3-[^2H]PEP product has an NMR spectrum with only one resonance (split into a doublet by the ^{31}P), centered at 5.14 ppm, consistent with the deuterium residing at H_A.[5] This indicates that the deuterium in $(3R)$-3-[^2H]2-PGA ends up *trans* to the phosphate group in the product; therefore, the elimination of hydroxide at C-3 had to be *anti* to the proton lost, that is, *anti* elimination (Scheme 10.4).

In the back reaction from PEP catalyzed by enolase, the product 2-PGA has the stereochemistry $(2R)$, so the proton must add to the *si* face of the vinyl double bond of PEP at C-2; therefore the hydroxyl group must add to the *re* face at C-3 for the addition to be *anti* (Scheme 10.5). Isotopic exchange experiments by Dinovo and Boer support a stepwise *anti* mechanism.[6] The enzyme-catalyzed rate of C-2 ^2H exchange of [2-^2H]PGA with H_2O is fast, and there is no deuterium isotope effect on the formation of PEP. As the data in Table 10.1 show, the slow step in the over-all rate of the enolase-catalyzed reaction is the release of product (PEP). The fast step is the removal of the C-2 proton of PGA, and the intermediate step is the loss

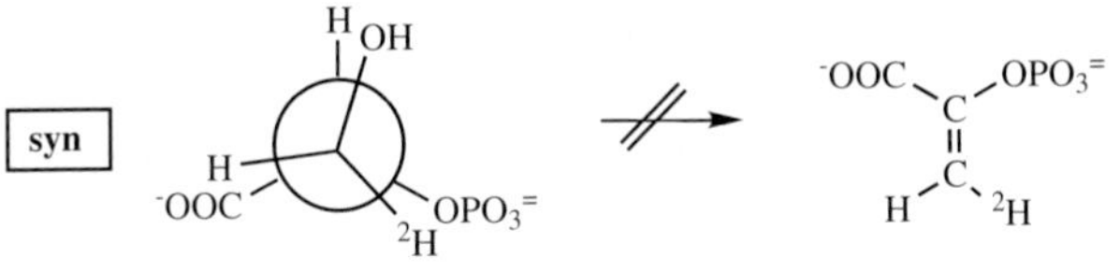

SCHEME 10.4   *Anti* versus *syn* elimination of water from 2-phosphoglycerate (PGA).

SCHEME 10.5   Stereochemistry of water addition to phosphoenolpyruvate catalyzed by enolase.

TABLE 10.1   Relative Rates of Exchange in the
Enolase-Catalyzed Reaction

	k_{rel}
^{14}PGA $\rightleftharpoons$ ^{14}PEP	1.0
[3-^{18}O]PGA $\rightleftharpoons$ H$_2$^{18}O	1.3
[2-^{2}H]PGA $\rightleftharpoons$ ^{2}H$_2$O	1.9

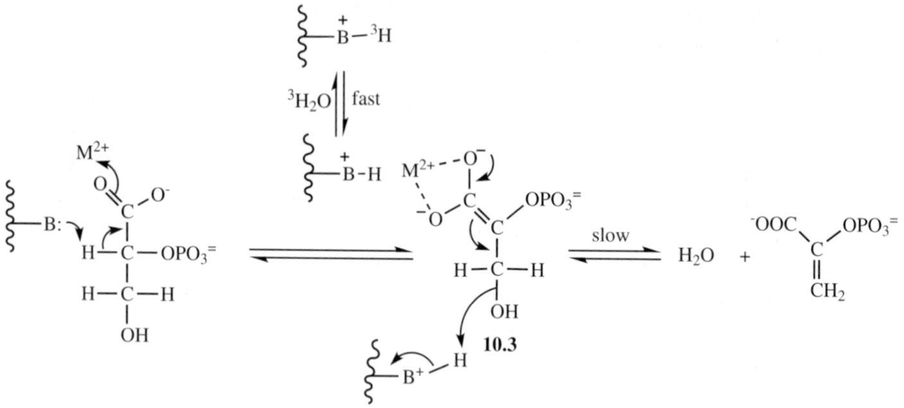

SCHEME 10.6   E1cB mechanism for enolase.

of hydroxide from the PGA. The initial deprotonation is therefore faster than both the loss of hydroxide ion and product formation. This predicts that if 2-PGA and enolase are added to ^{3}H$_2$O, and the PEP formed is trapped so it cannot go back to PGA (by adding ADP and pyruvate kinase), then ^{3}H should be incorporated into "unreacted" 2-PGA as a result of fast initial proton exchange. That, in fact, is what is observed, which supports an E1cB-like mechanism, presumably via the metal ion-stabilized enolate, **10.3** (Scheme 10.6). In this generalized mechanism, a base removes the C-2 proton of 2-PGA with concomitant metal ion coordination to the carboxylate carbonyl, giving a metal ion-stabilized *aci*-carboxylate intermediate (**10.3**), which could undergo an acid-catalyzed elimination of water to PEP. Cleland and co-workers synthesized four analogues of 2-PGA (**10.4–10.7**) which, on

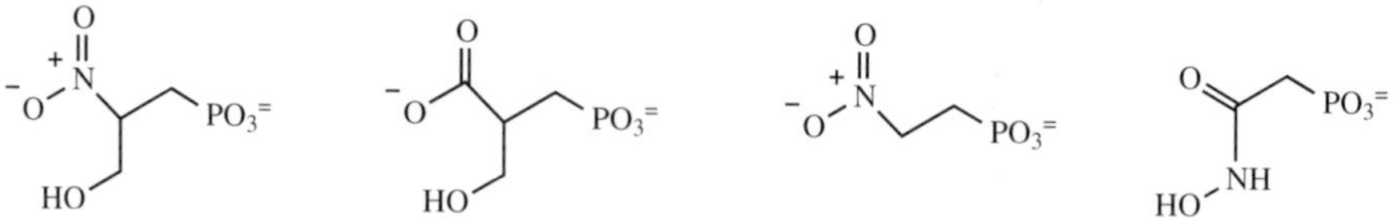

10.4   **10.5**   **10.6**   **10.7**

enolization, would give products that mimic **10.3**.[7] Compounds **10.4** and **10.7** are particularly potent, slow-binding inhibitors of enolase; the K_i value for **10.4** is 6 nM and that for **10.7** is 15 pM! This finding supports the intermediacy of **10.3** in the enolase-catalyzed reaction.

A large amount of crystal structure information is available that provides more specific details about the yeast enzyme. As mentioned, there are two divalent metal ions per subunit; the metal ion that binds with higher-affinity coordinates to the protein through the carboxylate side chains of Asp-246, Asp-320, and Glu-295.[8] When the metal ion is Mg^{2+}, three water molecules complete the octahedral coordination shell.[9] The affinity of the second, weaker-binding metal ion is considerably enhanced by substrate (or inhibitor) binding. Both of the metal ions are believed to stabilize the intermediate by coordination to the *aci*-carboxylate form (**10.3**).

Because Glu-168, Glu-211, and Lys-345 are highly conserved residues in all known enolases, they are probably also involved in catalysis. To test that hypothesis, Reed and co-workers prepared three mutants: E168Q, E211Q, and K345A.[10] All three mutants are very poor catalysts for the enolase reaction. The K345A mutant does not catalyze the exchange of the C-2 proton of 2-PGA in D_2O, whereas the other two mutants do. This result supports K345 as the active-site base responsible for deprotonation of the C-2 proton of 2-PGA. The E211Q and E168Q mutants catalyze exchange of the C-2 proton much faster than dehydration to product, further supporting the formation of an intermediate (**10.3**) in the reaction.

Wild-type enolase catalyzes the hydroxide ion–mediated conversion of (Z)-3-chloro-2-PEP (**10.8**) to the enol of tartronate semialdehyde phosphate (**10.10**), a reaction that mimics the addition of hydroxide to C-3 of PEP in the normal reverse reaction, except for the elimination of chloride ion (Scheme 10.7; note the similarity of **10.9** to **10.3** in Scheme 10.6). All three of the mutants are depressed in catalyzing this reaction, but the activities vary in the order K345A > E168Q >> E211Q. It was concluded, then, that because of the low activity of the E211Q mutant, E211 is important to the second step of the normal enolase reaction, namely, the protonation of the C-3 hydroxyl leaving group. Glu-168 may somehow contribute to both steps of the reaction.

The crystal structure at 1.8 Å resolution of yeast enolase complexed with the

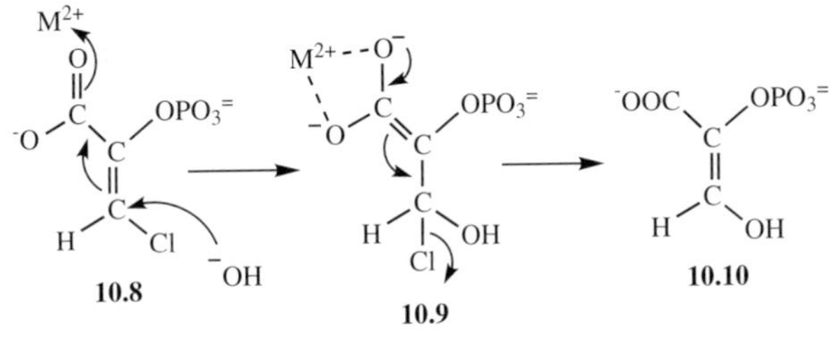

SCHEME 10.7   Enolase-catalyzed hydrolysis of (Z)-3-chloro-2-PEP.

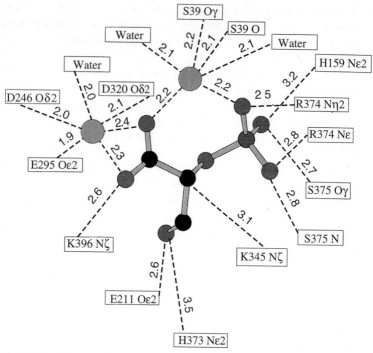

FIGURE 10.1   Schematic of the yeast enolase active site showing the coordination of the residues and the substrate to the two Mg^{2+} ions. The dashed lines from the 2-PGA to amino acids represent possible hydrogen bonds. The dashed lines from the Mg^{2+} ions indicate their coordination. Interatomic distances in angstroms are given on the dashed lines. [Reprinted with permission from Larsen, T. M.; Wedekind, J. E.; Rayment, I.; Reed, G. H. *Biochemistry* **1996**, *35*, 4349. Copyright 1996 American Chemical Society.]

equilibrium mixture of 2-PGA and PEP has provided the most detail of the catalytic machinery of this enzyme (Figure 10.1).[11] Both Mg^{2+} ions are complexed to the substrate and the product. The higher-affinity Mg^{2+} ion coordinates to the carboxylate side chains of Asp-246, Asp-320, and Glu-295, as observed earlier, and also to both carboxylate oxygens of the substrate or product, as well as to a water molecule. The Mg^{2+} ion could stabilize the proposed *aci*-carboxylate (**10.3**) transition state/intermediate. One oxygen of the substrate or product also coordinates to the second, lower-affinity Mg^{2+} ion. The Lys-345 is correctly positioned to serve as the base for removal of the C-2 proton of 2-PGA, and Glu-211 interacts with the C-3 hydroxyl group; they both are oriented to permit the requisite *anti* stereochemistry of the elimination reaction.

These crystallographic results seem to support the mechanism in Scheme 10.6. Also supported is the mechanism by which enolase catalyzes the deprotonation of C-2 of 2-PGA by Lys-345, which should be energetically very unfavorable because the pK_a of that proton is at least 30. As discussed in Chapter 1 (Section II.C), there are two important mechanisms by which an enzyme can facilitate the deprotonation of weak carbon acids, (1) by increasing the basicity of the deprotonating base and (2) by simultaneous protonation or Lewis acid coordination of the adjacent carbonyl group while the proton is being removed.[12] Lys-345 is in a hydrophobic pocket of the active site, which tends to lower its pK_a, increases the free base form of Lys-345, and makes it a more efficient base. The Mg^{2+} ions are Lewis acids that coordinate to the substrate carboxylate, thereby lowering the pK_a of the proton at C-2. In addition to these acidifying effects on the C-2 proton, Glu-211 coordinates to the C-3 hydroxyl group, thereby increasing its leaving group ability.

Using the structural sequence databases for proteins from X-ray crystallographic studies, families of enzymes that catalyze a diversity of reactions can be found. The presence of conserved amino acid residues in enzymes that catalyze the same reaction guides the recognition of an enzyme family. This approach has been used to identify a superfamily of enzymes (see Chapter 2, Section III.A) called the *enolase superfamily*, related by their ability to catalyze the abstraction of the α-proton of a carboxylic acid to form an enolic (*aci*-carboxylate) intermediate as the first step in their overall reaction.[13] This superfamily includes enzymes that catalyze racemization (e.g., mandelate racemase), β-elimination of water (e.g., enolase and galactonate dehydratase), β-elimination of ammonia (e.g., β-methylaspartate ammonia lyase), and cycloisomerization (e.g., muconate-lactonizing enzyme) reactions. From an analysis of sequence and structural similarities of these enzymes, it appears that all of their chemical reactions are mediated by similar active sites, which have been modified during evolution to allow similarly formed enolic intermediates to partition to different products in their respective active sites. It has been predicted that a comparison of active-site structures may lead to predictions of catalytic functions of enzymes for which no function has yet been ascribed. These results support the hypothesis that new enzyme activities evolve by modification of an existing enzyme to catalyze similar chemistry, but with differing substrate specificity.[14]

2. NAD⁺-Dependent Enzymes

In Chapter 9 (Section II.D.2) we saw that some enzymes acidify protons for epimerization reactions by oxidizing an adjacent hydroxyl group to the corresponding ketone. The presence of the carbonyl group then lowers the pK_a of the adjacent proton so that the enzyme can more readily remove it and replace it on the opposite face of the molecule. Some dehydratases use a similar mechanism to facilitate dehydration reactions.

For example, the nucleoside diphosphohexose 4,6-dehydratases (also called nucleoside diphosphohexose oxidoreductases) are NAD^+-dependent enzymes that transform the nucleoside diphosphohexoses (**10.11**) into the nucleoside diphospho-4-keto-6-deoxyhexoses (**10.14**) via the corresponding 4-keto-D-glucose (**10.12**) and 4-keto-$\Delta^{5,6}$-glucoseen (**10.13**, Scheme 10.8).[15] These enzymes are essential for the biosynthesis of all 6-deoxyhexoses, which, other than 2-deoxy-D-ribose of DNA, are the most abundant naturally occurring deoxysugars. 6-Deoxyhexoses

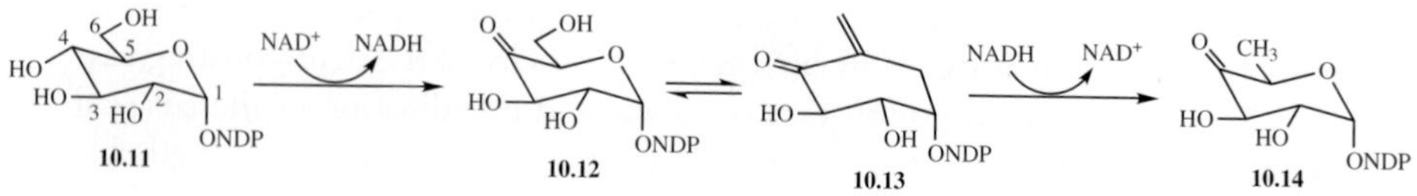

HO OH 6 5 4 O 3 HO 2 1 HO ONDP
10.11

NAD⁺ NADH →

O OH O HO HO ONDP
10.12

⇌

O O HO OH ONDP
10.13

NADH NAD⁺ →

O CH₃ O HO HO ONDP
10.14

SCHEME 10.8 Reactions catalyzed by nucleoside diphosphohexose-4,6-dehydratases (oxido-reductases). NDP stands for nucleoside diphosphate. The sugar positions are numbered.

frequently are constituents of bacterial lipopolysaccharides,[16] occur in macrolide antibiotics, and are the universal precursors for other highly reduced di-and tri-deoxyhexoses.

The nucleoside diphosphohexose 4,6-dehydratases catalyze three discrete chemical steps shown in Scheme 10.8. Because NAD⁺ is regenerated, it acts as a prosthetic group rather than as a cosubstrate, unlike most NAD⁺-dependent enzymes. The mechanism for this series of reactions was established by a variety of isotope experiments carried out originally with the 2′-deoxythymidine diphosphate sugar (dTDP in place of NDP in Scheme 10.8) and the *Escherichia coli* enzyme. Incubation of this enzyme with dTDP-[4-³H]**10.11** gives dTDP-[6-³H]**10.14**.[17] When the reaction is run in ³H₂O or ²H₂O, the product has ³H or ²H, respectively, incorporated at the 5-position. Likewise, incubation of dTDP-[5-³H, 6-¹⁴C]glucose in H₂O gives the product with essentially all of the tritium released.[18] When the enzyme is incubated with [4-³H]NAD⁺ and substrate, the product contains no tritium, indicating that the hydrogen transfer is intramolecular and involves only the *pro-S* hydrogen of NADH. A crossover experiment can be carried out to show that the hydrogen transfer occurs intramolecularly. In a crossover experiment a labeled and unlabeled substrate are added, and you look for the transfer of an atom or group from one molecule to the other. If this crossover occurs, then the reaction is not intramolecular. A mixture of dTDP-**10.11** and dTDP-**10.11**-d_7 gives exclusively dTDP-**10.14** and dTDP-**10.14**-d_6 (the C-5 proton is washed out) with no mono-deuterated or d_5 products. Therefore, the C-4 hydrogen of the unlabeled substrate is only transferred to the C-6 position of the same molecule and the C-4 deuteron of the d_7-labeled substrate is transferred only to C-6 of itself, that is, no crossover occurred. Floss and co-workers showed that the stereochemistry of this reaction involves a displacement of the C-6 hydroxyl group with inversion.[19] Assuming that the transfer is *suprafacial* because it is intramolecular, then the elimination of H₂O from the C-5 and C-6 atoms proceeds in a *syn* relationship, whereas the reduction of the $\Delta^{5,6}$-double bond formally involves an *anti* addition of H⁺ and H⁻. A mechanism that is consistent with all of these results is shown in Scheme 10.9.

The reaction of stereospecifically labeled dTDP-[4-³H]6-deoxyglucose (the substrate except a 6-methyl group in place of the 6-hydroxymethyl group) produces [4β-³H]NADH (the *si* face) bound to the enzyme and without release of the product (because there is no 6-hydroxyl group in the substrate, no 5,6-elimination can occur to give **10.13**).[20] The product (6-deoxy-**10.12**) is not released from the

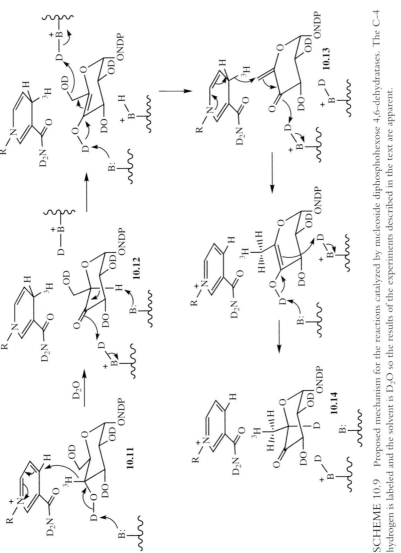

SCHEME 10.9 Proposed mechanism for the reactions catalyzed by nucleoside diphosphohexose 4,6-dehydratases. The C–4 hydrogen is labeled and the solvent is D₂O so the results of the experiments described in the text are apparent.

enzyme in the NADH form, so it is reasonable to conclude that the enzyme in the NADH form has a different conformation from that in the NAD^+ form.[21]

The CDP-D-glucose 4,6-dehydratase from *Yersinia pseudotuberculosis* catalyzes the same reactions with the same stereochemistry as just described.[22] However, this enzyme is a dimeric protein that binds only one NAD^+ per dimer, and, therefore, exhibits half-sites reactivity (see Appendix I, Section VII.C).

You may be wondering how the stereochemistry of these reactions can be determined, given that the C-4 hydrogen is transferred to C-6 to form a methyl group that then has three equivalent protons. How is it possible to determine the stereochemistry of a hydrogen transfer, when the migrating hydrogen ends up in a methyl group? If three different hydrogen isotopes—^1H, ^2H, and ^3H—are used, then a chiral methyl group is generated. Once Cornforth and co-workers[23] and Arigoni and co-workers[24] reported the determination of chiral methyl group stereochemistry, numerous applications of this methodology in enzyme mechanism studies were possible.[25]

Let's take the example of CDP-D-glucose 4,6-dehydratase to see how the stereochemistry of the C-6 methyl group can be determined. The general approach is as follows: Start with the substrate that contains a chirally tritiated methylene group, and label the migrating hydrogen with deuterium, so that the methyl group generated in the product contains a protium, a deuterium, and a tritium atom. Oxidize the carbon in the product adjacent to the chiral methyl group to a carboxylic acid, giving acetic acid containing a chiral methyl group. Then carry out a series of enzyme-catalyzed reactions that converts the acetic acid to a product in which the tritium atom can be stereospecifically removed or retained. If you know the stereospecificity of the enzyme-catalyzed reaction that releases (or does not release) the tritium, the stereochemistry of the tritium in the acetic acid can be determined, from which the stereochemistry of the hydrogen transfer can be deduced. This process should be carried out with both epimers of the labeled substrate as a control for consistency of results.

If (6S)- (**10.15**) and (6R)-CDP-[4-^2H, 6-^3H]D-glucose (**10.17**) are separately incubated with CDP-D-glucose 4,6-dehydratase in H_2O buffer, the CDP-4-keto-6-deoxyglucoses (**10.16** and **10.18,** respectively) produced should have a proton, a tritium (they were at C-6 in the substrate) and a deuteron (transferred from the C-4 position) at the C-6 carbon (Scheme 10.10). Because the transfer of the C-4 deuterium must be the same in both reactions, the stereochemistry of the proton, deuteron, and triton should be enantiomeric for **10.16** and **10.18,** assuming that the transfer is stereospecific. It is not possible to determine the stereochemistry of these products with a polarimeter for two reasons: First, specific rotations of compounds that are chiral by virtue of the presence of one deuterium atom are only about $1-2°$ (usually one to two orders of magnitude smaller than compounds with an alkyl group instead of a deuterium), and, second, the tritium is present in the substrate in trace amounts, so that maybe only one in a billion molecules actually has tritium, whereas essentially all of the molecules have a proton and a deuteron.

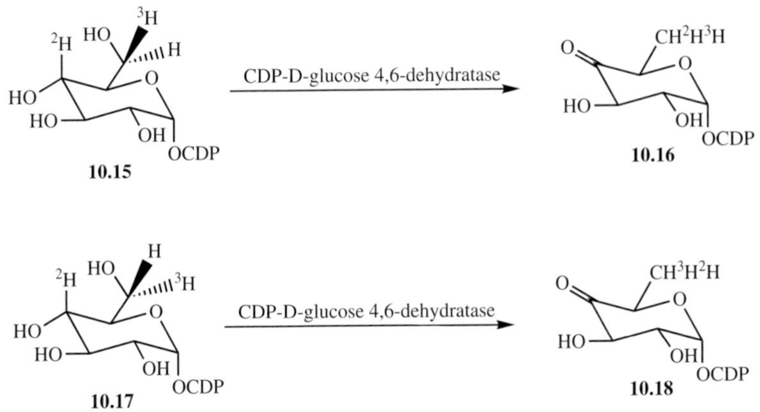

SCHEME 10.10   Transfer of the C-4 hydrogen of (6S)-**10.15** and (6R)-CDP-[4-²H, 6-³H]D-glucose (**10.17**) to the C-6 methyl group in the CDP-4-keto-6-deoxyglucose product.

Therefore, almost all of the molecules have two protons and one deuteron, and are not chiral. So, how is the stereochemistry determined?

First, it is necessary to use a radioactivity measurement, so that only the molecules containing tritium (the chiral ones) are measured. The product of the enzyme reaction is converted into a molecule that can be used as a substrate for a series of known enzyme-catalyzed reactions, each having well-established stereochemical outcomes. A Kuhn–Roth oxidation of sugars[26] converts the CDP-4-keto-6-deoxyglucose product into acetic acid without exchange of the protons on the methyl group (Scheme 10.11). The chiral methyl-containing acetic acid is then converted into fumarate by a series of four enzymatic steps (Scheme 10.12 shows the outcome with the *S*-enantiomer). First, it is converted to acetyl phosphate with acetate kinase and ATP. The acetyl phosphate is converted into acetyl-CoA with phosphotransacetylase and CoASH. Treatment of the acetyl-CoA with glyoxylate and malate synthase makes (2S)-malate with two different stereochemistries—**10.19** (2S,3R) and **10.20** (2S,3S)—regardless of which substrate enantiomer was used in the first step; however, the ratio of these products is opposite for the two enantiomers. The reason for this is that the three isotopic hydrogens are equivalent as a result of free rotation, but there is a kinetic isotope effect in favor of removal of the proton over removal of the deuteron; triton removal requires even more

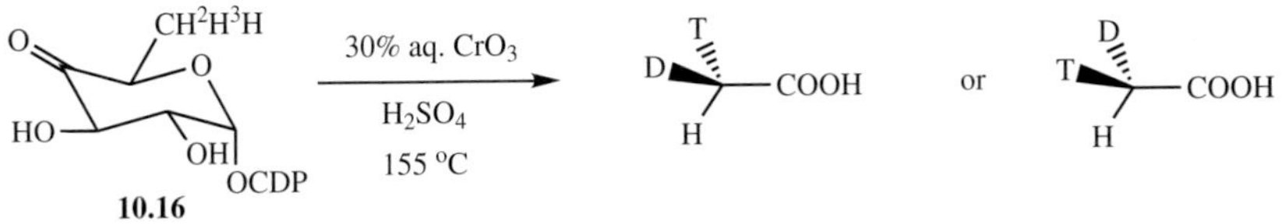

SCHEME 10.11   Kuhn–Roth oxidation of CDP-4-keto-6-deoxyglucose.

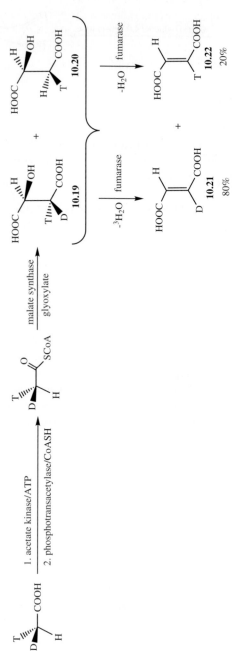

SCHEME 10.12   Enzymatic conversion of chiral methyl-containing acetate into fumarate for determination of the chirality of the methyl group.

energy and is negligible. The ratio of **10.19** and **10.20** is a reflection of both the isotope effect for removal of the proton versus the deuteron, respectively, and the optical purity of the acetyl group. These two compounds cannot be separated, because they are the same molecule (except for a deuterium atom). Fumarase, which is known to catalyze the removal of the *pro-3R* hydrogen from (2S)-malate[27] with *anti* elimination of water to give fumarate, can then be used to determine whether the tritium or the hydrogen is at the (3R)-position of the malate that is generated by malate synthase. As a result of *anti* elimination, epimer **10.19** would lose ^3H_2O and form nonradioactive fumarate (**10.21**), whereas epimer **10.20** would result in loss of H_2O and formation of radioactive fumarate (**10.22**). Lenz and Eggerer established that enzymatically synthesized (R)-[^2H,^3H]acetyl-CoA with malate synthase, glyoxylate, and fumarase releases $21 \pm 2\%$ of the tritium as ^3H_2O and (S)-[^2H,^3H]acetyl-CoA releases $79 \pm 2\%$ of the tritium as ^3H_2O.[28] This corresponds to a kinetic isotope effect (k_H/k_D) of 3.8 ± 0.1 on malate synthase-catalyzed deprotonation of acetyl-CoA.

The result obtained with **10.16** (remember, that is why I went through this whole discussion) is that (6S)-CDP-glucose (**10.15**) retains 71.4% of the tritium in the fumarate (by inference, 28.6% is released as ^3H_2O), which corresponds to (R)-acetate. When (6R)-CDP-glucose (**10.17**) is the substrate, 30.2% of the tritium in the fumarate is retained (by inference, 69.8% is released as ^3H_2O), corresponding to (S)-acetate. Therefore, the CDP-D-glucose 4,6-dehydratase reaction proceeds with inversion of stereochemistry.

3. Iron–Sulfur Clusters in a Nonredox Role

Protein-bound iron–sulfur clusters, discovered in the 1950s (although it was not known at that time that they were iron–sulfur clusters), were thought for years to be involved exclusively in electron transport mechanisms,[29] but more recently it has been found that they also catalyze dehydration (hydration) reactions as hydrolyases.

a. Aconitase

The enzyme aconitase catalyzes the interconversion of two important intermediates in the citric acid (Krebs) cycle, citrate (**10.23**) and isocitrate (**10.25**) via *cis*-aconitate (**10.24**) (Scheme 10.13). At equilibrium the ratio of aconitase substrates is citrate–*cis*-aconitate–isocitrate 0.88:0.04:0.08. Note that this enzyme catalyzes a dehydration reaction (**10.23** to **10.24**) and a hydration reaction (**10.24** to **10.25**), which is therefore an overall isomerization reaction. The enzyme contains a [4Fe–4S]$^{2+(+)}$ iron–sulfur cluster (see the IUB rules[30] for iron–sulfur cluster nomenclature), which participates in catalysis. Before showing how the iron–sulfur cluster is involved, we will first discuss some of the earlier stereochemical questions. Only the (2R,3S) isomer of isocitrate is a substrate. Citrate, however, is a prochiral mole-

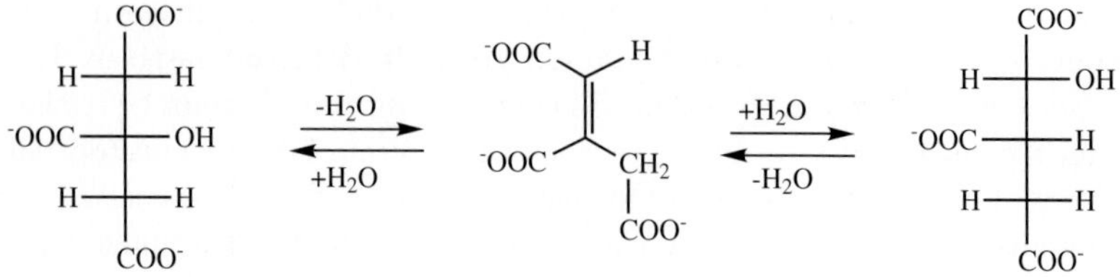

SCHEME 10.13   Aconitase-catalyzed interconversion of citrate (**10.23**) and isocitrate (**10.25**) via *cis*-aconitate (**10.24**).

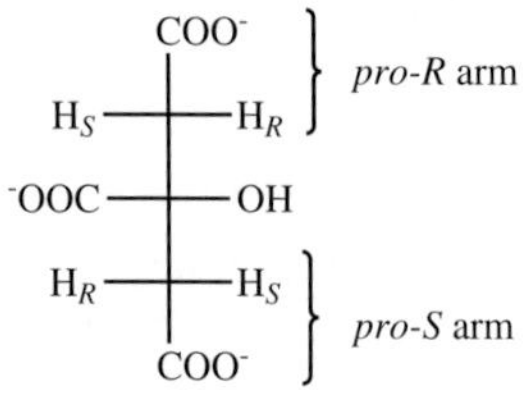

10.26

cule; at C-3 there are two carboxymethyl groups, which the enzyme can distinguish as *pro-R* or *pro-S* arms (**10.26**). The methylene hydrogens of each of the carboxymethyl groups also are prochiral. When aconitase dehydrates citrate to *cis*-aconitate, it specifically removes the *pro-R* hydrogen at the C-2 position of the *pro-R* arm, and of course the hydroxyl group at C-3 is eliminated. Because the product is *cis*-aconitate, this elimination must have been *anti* (Scheme 10.14).[31] For the back reaction (**10.24** to **10.23**), labeling experiments show, as expected from the principle of microscopic reversibility, that it is an *anti* addition, in which the hydroxide attacks the 3-*re*, 2-*si* face at C-3 and the proton adds to the *si-re* face at C-2 (Scheme 10.15). In the second half reaction (**10.24** to **10.25**), the addition of water must occur so that the proton adds to the *re-si* face at C-3 and the hydroxide to the *si-re* face at C-2 in order to produce isocitrate with (2*R*,3*S*) stereochemistry, which is

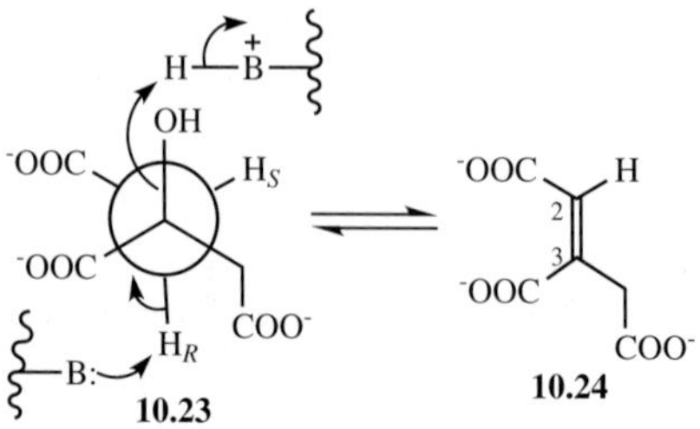

SCHEME 10.14   Stereochemistry of elimination of water from citrate catalyzed by aconitase.

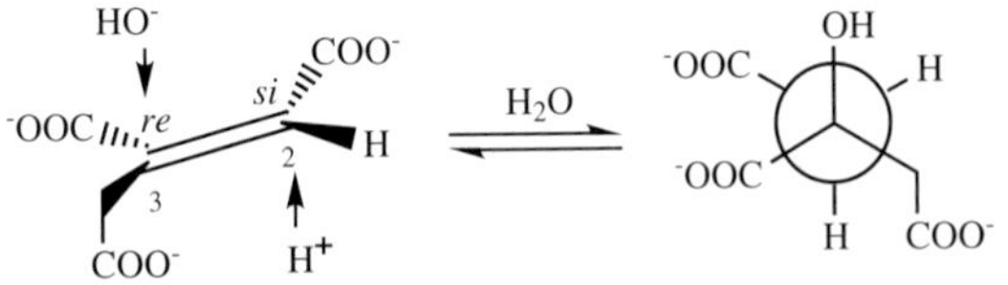

SCHEME 10.15   Stereochemistry of addition of water to *cis*-aconitate to give citrate (back reaction).

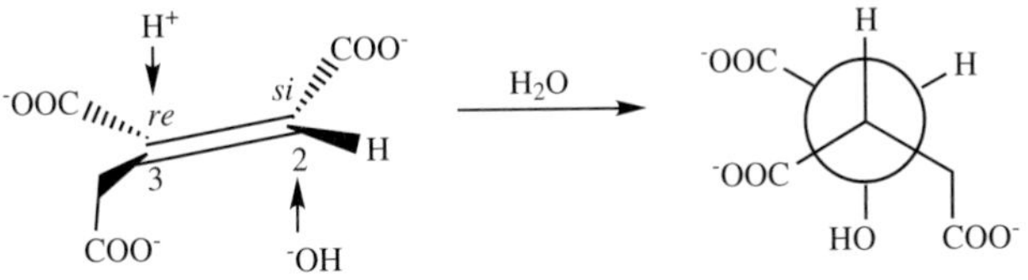

SCHEME 10.16   Stereochemistry of addition of water to *cis*-aconitate to give isocitrate.

again an *anti* addition (Scheme 10.16). Therefore, C-2 is always attacked from one side of the plane by either hydroxide ion or by a proton, and C-3 is always attacked from the opposite side of the plane.

The most intriguing result, however, is that the *pro-R* proton that is removed from C-2 of citrate ends up at C-3 of isocitrate,[32] an example of a proton transfer that occurs without exchange with solvent. If you are following all of this stereochemistry, you are probably really confused about how this is possible, because to account for an *antarafacial* transfer of a proton that is removed from the *si-re* face of citrate and replaced at the *re-si* face of isocitrate, the *cis*-aconitate molecule must flip over in the active site in going from citrate to isocitrate (Scheme 10.17; the proton at C-2 of citrate is shown as *H^+)! A crossover experiment by Rose and O'Connell demonstrated that the proton removed from citrate does not exchange (or exchanges very slowly) with solvent.[33] [2R-³H]Citrate was incubated with aconitase in the presence of the alternative substrate, 2-methyl-*cis*-aconitate (**10.27**); unlabeled *cis*-aconitate was obtained as well as some 2-methyl-[3-³H]isocitrate (**10.28**, Scheme 10.18). This shows that the tritium removed from one substrate molecule (citrate) can be transferred (crossed over) to another substrate molecule (**10.27**). The hydroxyl group, however, exchanges with solvent in every turnover; [2-¹⁸OH]iso-

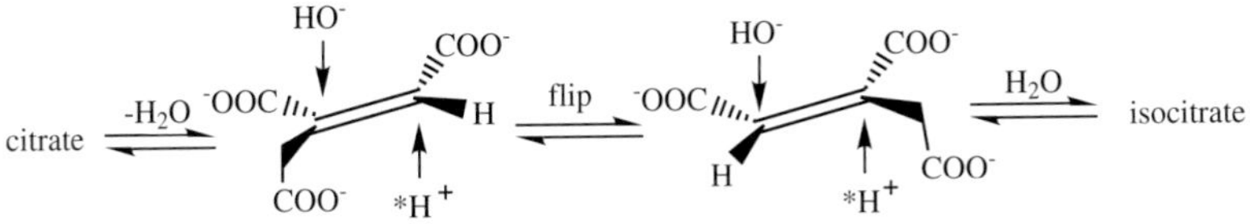

SCHEME 10.17   Overall stereochemistry of the aconitase-catalyzed reaction.

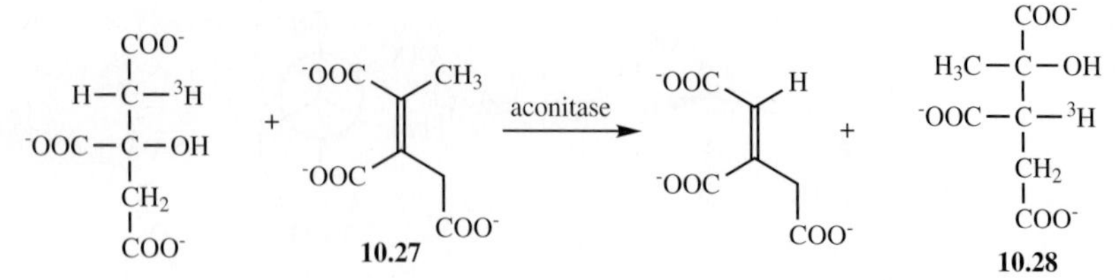

SCHEME 10.18   A crossover experiment with aconitase in which [(2R)-³H]citrate and 2-methyl-*cis*-aconitate (**10.27**) produce unlabeled *cis*-aconitate and 2-methyl-[3-³H]isocitrate (**10.28**).

citrate gives no [3-¹⁸OH]citrate. To rationalize these results, the tritium at C-2 of citrate must be abstracted by the enzyme to give *cis*-aconitate with the tritium tightly bound to an active-site base while the hydroxide exchanges rapidly. The *cis*-aconitate, then, must dissociate from the active site, and 2-methyl-*cis*-aconitate (**10.27**) must bind and accept the bound tritium to make **10.28**! This gives an intermolecular transfer of the proton. Rose suggests that the slow dissociation of the ³H may result from its hydrogen bonding to another basic species. A possible general mechanism is depicted in Scheme 10.19. An alternative to a "flip mechanism" inside the active site is a conformational change in the enzyme that ejects the *cis*-aconitate, but because the enzyme is now in a different conformation, the *cis*-aconitate must return in an orientation that is 180 degrees from what it was when it was released.

The main problem with the mechanism in Scheme 10.19 (and the entire discussion so far) is that no mention has yet been made of the required iron−sulfur cluster. From X-ray crystal structures of aconitase[34] it was obvious that a 4Fe−4S

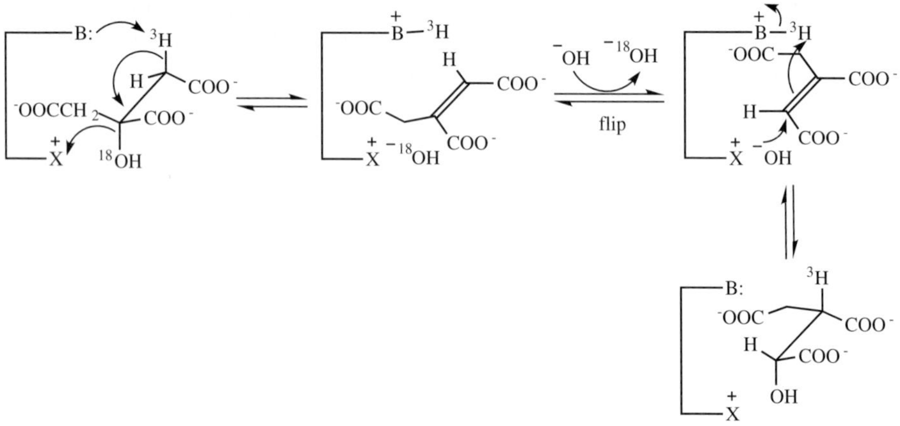

SCHEME 10.19   General mechanism for aconitase.

cluster is bound to the active site (see Chapter 3, **3.147** for this iron–sulfur cluster structure). Ruzicka and Beinert had reported, even before the presence of an iron–sulfur cluster was realized, that Fe^{2+} had to be added to aconitase for activity,[35] although it was not known why. It was later found that one iron is easily lost from the iron–sulfur cluster to give a stable 3Fe–4S cluster (see Chapter 3, **3.146** for this iron–sulfur cluster structure); the enzymatic activity, however, is reduced over 100-fold. Added Fe^{2+} regenerates the 4Fe–4S cluster with return of full activity. This fourth iron atom incorporated, referred to as Fe4, is inserted into the empty corner of the 3Fe–4S cluster. The Fe4 site retains a tetrahedral geometry and acquires a hydroxide ion as its fourth ligand.

Crystal structures at 2.0 Å and 2.7 Å resolution obtained by growing crystals of pig heart aconitase in the presence of either *cis*-aconitate or citrate, respectively, gave a very surprising result: Isocitrate was bound to the enzyme.[36] Given that the equilibrium mixture of substrates is 0.88:0.04:0.08 for citrate–*cis*-aconitate–isocitrate, respectively, and equilibrium would be reached *much* faster than crystallization could occur, it might be expected that citrate would be bound. Possibly the isocitrate–aconitase complex crystallizes most readily, and that shifts the equilibrium. The isocitrate–aconitase structure clearly shows that one oxygen atom of the C-1 carboxyl group and the oxygen atom of the C-2 hydroxyl group are bound to Fe4 of the iron–sulfur cluster (**10.29**) along with a water molecule; because three sulfur atoms are also bound to Fe4, it is a six-coordinate iron. Therefore, it is apparent that the iron–sulfur cluster, in this case, can act as a Lewis acid to activate the hydroxyl group for elimination.

Nitroisocitrate (**10.30**) is isoelectronic with isocitrate, but the acidity of the proton adjacent to the nitro group is much greater than that for the corresponding proton in isocitrate. Nitroisocitrate anion, then, is a reaction intermediate analogue (**10.31**) of the proposed carbanion formed (**10.32**) in going from *cis*-aconitate to isocitrate and is a very tight-binding inhibitor. The crystal structure of beef heart aconitase with **10.31** bound is virtually identical to that with isocitrate bound, indicating that the exchange of nitro for carboxyl is insignificant in the binding

10.29

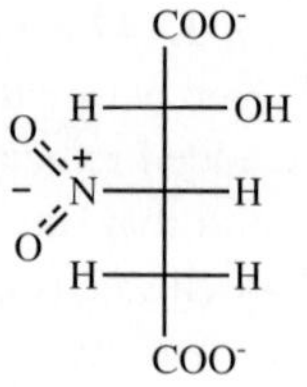

COO⁻ ... structures

10.30 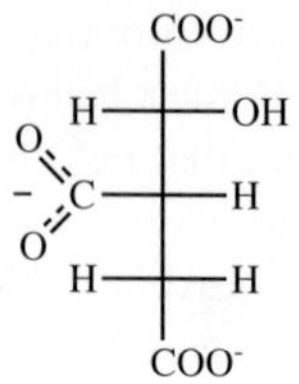   **isocitrate**

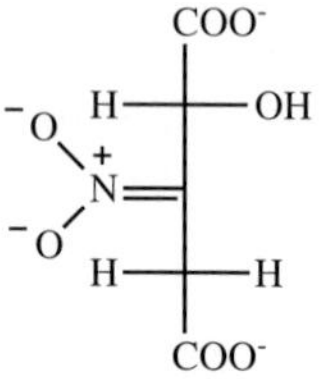

10.31 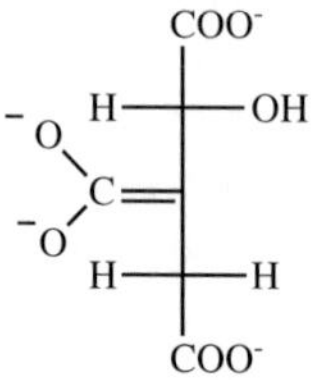   **10.32**

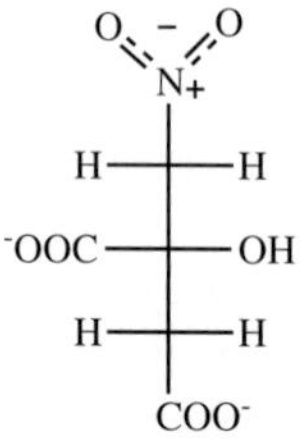

10.33 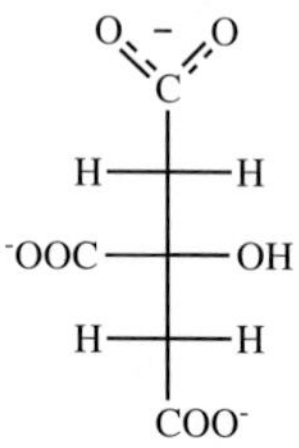   **citrate**

interactions. This similarity also supports a mechanism involving an intermediate carbanion, such as an E1cB mechanism (with the *aci*-carboxylate coordinated to the iron–sulfur cluster). Although the citrate–aconitase complex could not be crystallized, the corresponding nitrocitrate (**10.33**)-aconitase can be.[37] Nitrocitrate binds to Fe4 with one of the oxygens of the C-3 carboxyl group and with the hydroxyl oxygen at C-3. The coordination geometry of these groups to Fe4 is remarkably similar to that for the corresponding C-2 groups of isocitrate. The binding modes for citrate and isocitrate, then, are related by a twofold rotation about the C-2–C-3 bond. Based on these crystal structures, the crystal structure with *trans*-aconitate bound, and spectroscopic data, *cis*-aconitate must bind in two ways, one similar to the way that citrate binds and the other the way that isocitrate binds. The two binding modes are related by a rotation of 180 degrees about the C-2–C-3 bond of the substrate and can account for the required "flip" that must occur to satisfy the stereochemical results (Figure 10.2). The "flip" step is probably not a

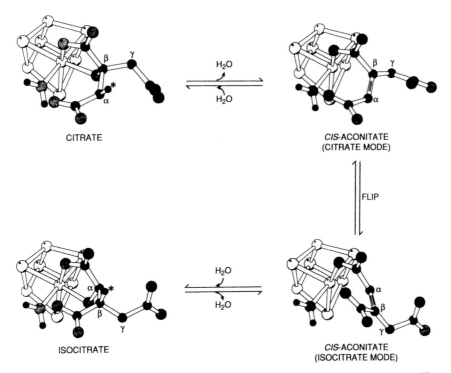

β γ

H₂O

H₂O

CITRATE

CIS-ACONITATE
(CITRATE MODE)

FLIP

H₂O

H₂O

ISOCITRATE

CIS-ACONITATE
(ISOCITRATE MODE)

FIGURE 10.2   Schematic of the interconversion of citrate and isocitrate catalyzed by aconitase. The proton that is removed from citrate and added to *cis*-aconitate to give isocitrate is marked with an asterisk. The gray scale for the atom types is H > C > O > S > Fe. [Reprinted with permission from Lauble, H.; Kennedy, M. C.; Beinert, H.; Stout, C. D. *J. Mol. Biol.* **1994**, *237*, 437.]

single step; one molecule of *cis*-aconitate is more likely replaced by a second molecule in the rotated orientation, perhaps as a result of a conformational change.

A hydrogen bond from His-101 to the isocitrate hydroxyl group suggests that the proton donated to the hydroxide released from citrate and isocitrate to form water comes from this histidine residue. The residue that abstracts the proton from citrate and isocitrate is probably Ser-642. The excess negative charge on the carboxyl oxygens in the *aci*-carboxylate (**10.32**) may be stabilized by a low-barrier hydrogen bond with the water molecule bound to Fe4 of the iron–sulfur cluster.

b. Other Nonredox Iron–Sulfur Cluster-Containing Enzymes

There are a variety of other nonredox enzymes known to contain iron–sulfur clusters that are involved in catalysis; in many cases, the iron–sulfur cluster acts as a Lewis acid.[38] It is interesting that most of these enzymes have counterparts from other sources that either contain no prosthetic group or have different prosthetic

10.34

10.35

groups that catalyze the same reaction. For example, D-serine dehydratase from *Escherichia coli* and L-threonine dehydratases contain PLP as the prosthetic group, but L-serine dehydratase from *Peptostreptococcus asaccharolyticus* has an iron–sulfur cluster instead of PLP.[39] This enzyme functions strictly anaerobically; it is inactivated on exposure to air and is reactivated by Fe^{2+} under anaerobic conditions. A 4Fe–4S cluster is involved in substrate binding, as demonstrated by EPR spectroscopic studies. The mechanism is believed to be similar to that of aconitase. In the case of aconitase the coordination to the iron–sulfur cluster involves both the hydroxyl and α-carboxyl groups of citrate (**10.29**), but L-serine dehydratase coordination of the β-hydroxyl and α-carboxyl groups (**10.34**) or the β-hydroxyl and α-amino groups (**10.35**) would distort the C–H bond to be broken out of an *anti* coplanar orientation with the hydroxyl leaving group. Therefore, it is believed that only the β-hydroxyl group of L-serine coordinates to the iron–sulfur cluster in this enzyme.

4-Hydroxybutyryl-CoA dehydratase is a very unusual enzyme, because it catalyzes a remarkable reversible dehydration of 4-hydroxybutryl-CoA (**10.36**) to crotonyl-CoA (**10.37**) (Scheme 10.20). Note that this is not a simple α,β-elimination of water; the hydroxyl group is at the γ-carbon atom. The enzyme also is unusual because it contains an assortment of cofactors, namely, FAD, nonheme iron, and a $[4Fe–4S]^{2+}$ cluster. With a combination of UV–visible and EPR spectrometries, Buckel and co-workers showed that a flavin semiquinone is generated, but there is no apparent communication between the flavin and the iron–sulfur cluster.[40] A mechanism that uses a one-electron transfer to and from the FAD and confines the iron–sulfur cluster to a role as a Lewis acid is shown in Scheme 10.20, although there is no experimental evidence for the ketyl–radical anion intermediate. There also is no evidence at all for the manner in which I depicted the binding of the iron–sulfur cluster to the substrate; it is just drawn to mimic the structure of other 4Fe–4S cluster-containing enzymes.

4. Heme-Dependent Enzyme in a Nonredox Role

Mansuy and co-workers have shown that cytochrome P450, an enzyme that catalyzes a variety of monooxygenation reactions (see Chapter 4, Section IV.C), also catalyzes a dehydration reaction.[41] Active cytochrome P450-dependent elimination of water from a series of oximes occurs only on the *syn*-isomers (**10.38**), not the

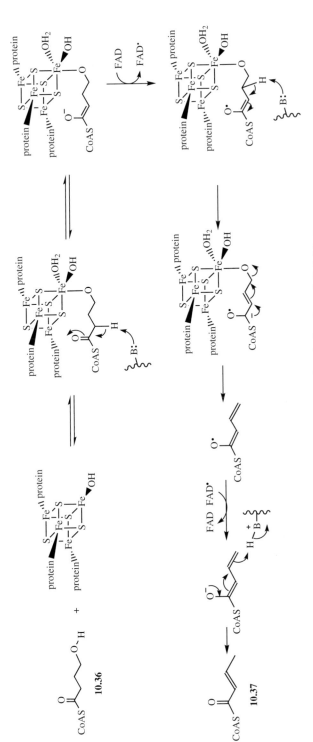

SCHEME 10.20   Proposed mechanism for 4-hydroxybutyryl-CoA dehydratase.

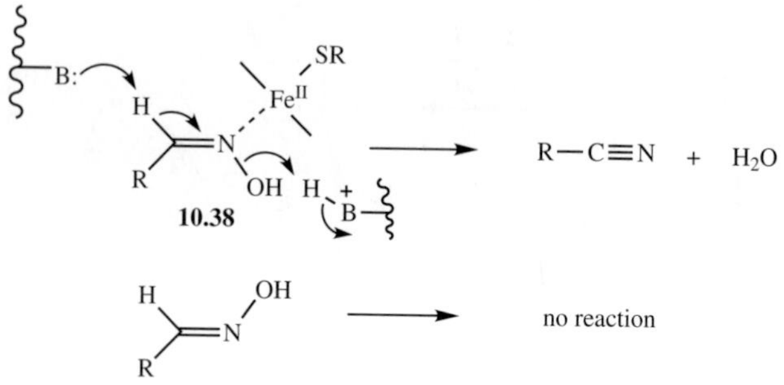

SCHEME 10.21   Dehydration of *syn* oximes by cytochrome P450.

anti-isomers, to give the corresponding nitriles (Scheme 10.21). The characteristic Soret peak of a cytochrome P450 complex was observed during these reactions.

B. Elimination of Phosphate

Chorismate synthase catalyzes the last common step in the shikimic acid pathway, namely, the elimination of phosphate from 5-enolpyruvylshikimate 3-phosphate (EPSP) (**10.39**) to give chorismate (**10.40**) (Scheme 10.22), which is then converted to aromatic amino acids and other aromatic metabolites in bacteria, fungi, and plants.[42] By labeling the hydrogens at C-6, Onderka and Floss showed that the overall stereochemistry of the elimination is 1,4-*trans* with removal of the *pro-R* hydrogen of EPSP at C-6.[43] Therefore, this is an *anti* elimination. Because, according to orbital symmetry arguments, concerted 1,4-eliminations should favor *syn* stereochemistry,[44] this reaction appears to be a stepwise process. A [1,3]-rearrangement of the allylic phosphate of EPSP to give *iso*-EPSP (**10.41**) was ruled out because *iso*-EPSP is not a substrate.[45] An addition/elimination mechanism using an active-site residue as the nucleophile was considered (Scheme 10.23, pathway b).[46] To test this mechanism, two EPSP analogues were synthesized containing fluorine at either the *pro-R* or *pro-S* C-6 position (**10.42**, R_R = F, R_S = H or R_R = H, R_S = F). With the *Neurospora crassa* enzyme neither acted as a substrate

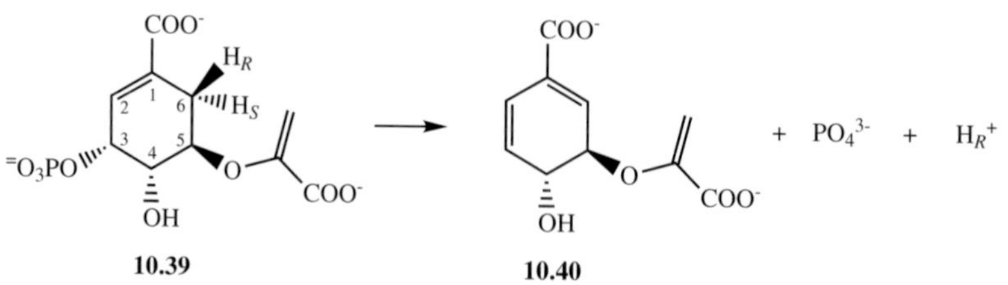

SCHEME 10.22   Reaction catalyzed by chorismate synthase.

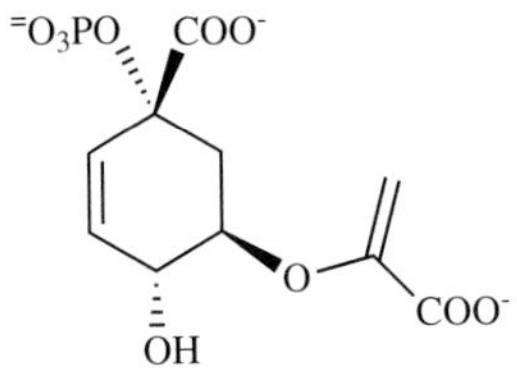

10.41

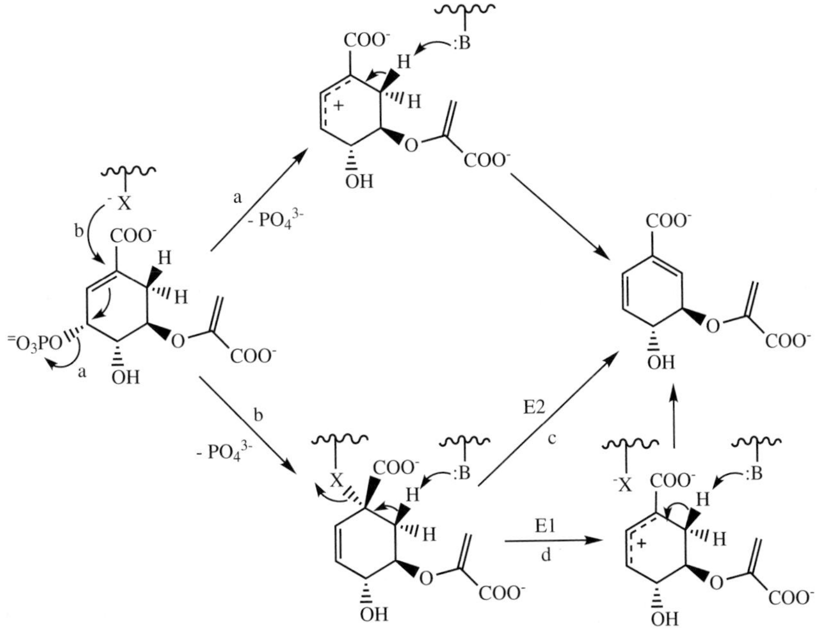

SCHEME 10.23   E1 (pathway a) and addition/elimination (pathway b) mechanisms for chorismate synthase.

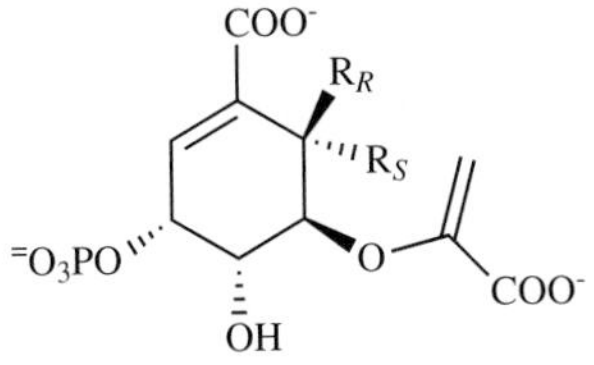

10.42

nor as an inactivator, but both were competitive inhibitors. If the addition–E2 elimination mechanism (pathway b–c) were relevant, then **10.42** (R_R = F) should have inactivated the enzyme because the fluorine could not be abstracted. Analogues **10.42** also provide evidence against the addition–E1 mechanism (pathway b–d). If pathway b–d were important, then treatment of the enzyme with **10.42** should lead to inactivation because the electron-withdrawing effect of the fluorine would prohibit ionization to the electron-deficient cation (pathway d). Therefore, a covalent mechanism seems unlikely.

Two experiments support a direct E1 elimination (Scheme 10.23, pathway a). If an E1 mechanism is important, then analogues **10.42** should prohibit initial ionization, consistent with the fact that they are not substrates. Further evidence to support an E1 elimination mechanism comes from kinetics experiments in which a single turnover of the enzyme did not show a "phosphate burst" (see Chapter 2, Section I.A.1.b.1 for a discussion of "inital bursts").[47] A burst of phosphate ion would occur if that step was fast and prior to the rate-determining step. In the E1 elimination mechanism (pathway a), ionization should be the rate-determining step, and no phosphate burst is expected.

However, one important feature of this enzyme has not yet been mentioned, namely, that it has a requirement for a reduced flavin cofactor, and none of the mechanisms in Scheme 10.23 would require a flavin. The requirement for a flavin is quite unusual, because the overall reaction involves no change in oxidation state in going from EPSP to chorismate; it is simply an elimination of phosphate ion. Single-turnover experiments with a recombinant enzyme from *Escherichia coli* demonstrate that an enzyme-bound flavin species accumulates during turnover.[48] When the *Escherichia coli* enzyme is incubated with the competitive inhibitor (6R)-6-fluoroEPSP (**10.42**, R_R = F and R_S = H), a stable flavin semiquinone signal is observed in the EPR spectrum. It is not known where the electron from the reduced flavin goes, because there are no apparent metals involved. The flavin intermediate also can be observed by stopped-flow absorption spectrophotometry during conversion of EPSP to chorismate. By following the formation and decay of the flavin intermediate, EPSP consumption, chorismate and phosphate formation, and phosphate dissociation, Thorneley and co-workers[49] concluded that the reaction is non-concerted. However, the flavin intermediate forms before the substrate is consumed, so it is not clear if the flavin is really associated with the conversion of substrate to product. Flavin analogues were synthesized to investigate the role of the flavin in catalysis.[50] On substrate binding, the pK_a values of the flavin analogues increase, suggesting that protonation occurs, which induces a more negative redox potential. A two-electron reduction of the flavin is associated with the uptake of one proton, consistent with a reduced flavin being bound to the enzyme.

The mechanism that Thorneley and co-workers proposed[51] is an elimination mechanism initiated by transfer of a hydrogen atom from C-6 of the substrate to the flavin radical. A radical mechanism also had been proposed earlier by Bartlett and co-workers,[52] but it was not clear where the hydrogen atom was going (Scheme 10.24).

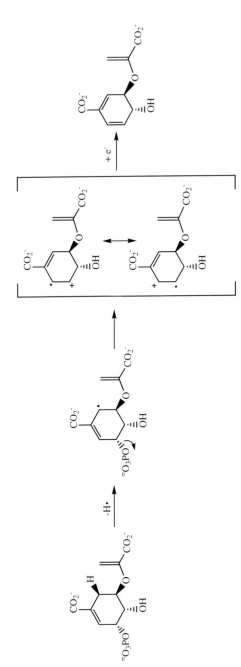

SCHEME 10.24   Radical mechanism proposed for chorismate synthase.

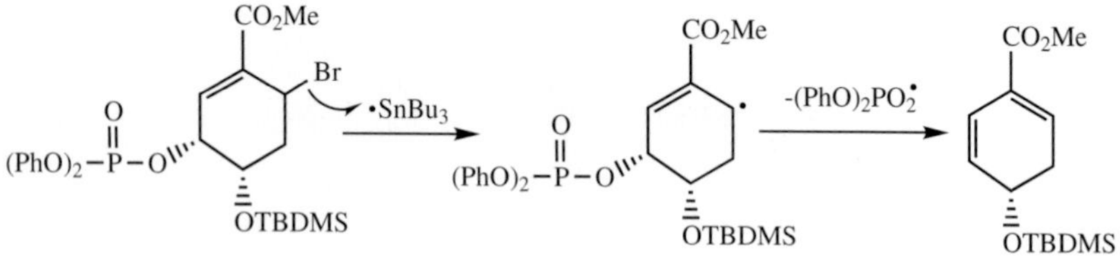

SCHEME 10.25   Chemical model in support of the radical mechanism for chorismate synthase.

Giese and Almstead later reported a chemical model to support such a mechanism (Scheme 10.25).[53]

The two experiments mentioned here that support a direct E1 elimination mechanism (Scheme 10.23, pathway a) also are consistent with the radical mechanism shown in Scheme 10.24, in which case analogues **10.42** would strongly disfavor ionization of the phosphate ion and no phosphate burst would be observed. A later study[54] found that **10.42** (R_R = H and R_S = F) acted as a very poor substrate (one-hundredth that for EPSP) with the enzyme from *Escherichia coli,* but both very low activity or no activity also support the mechanism in Scheme 10.24.

C. Ammonia Lyases: Elimination of Ammonia

In general, the chemical elimination of ammonia from amines is a difficult process because on addition of base to an ammonium salt, one of the ammonium protons is removed, and the leaving group becomes the much stronger base, the amine anion (NH_2^-), rather than ammonia. To get around that problem, the amino group is typically permethylated to give the trimethylammonium ion, which is a stable good leaving group, a process known as a Hofmann elimination.[55] However, in Chapter 2 (Section II.B) we saw that the *anti* elimination of ammonia from L-aspartate was a reaction that enzymes use to generate ammonia that is incorporated into some molecules. The key problem is how to make the β-protons acidic without deprotonating the ammonium group. One approach taken by enzymes is the use of a dehydroalanine prosthetic group.

1. Dehydroalanyl-Dependent Elimination

Histidine ammonia-lyase (HAL) catalyzes the conversion of histidine to urocanic acid (**10.43,** Scheme 10.26),[56] which we saw in Chapter 3 (Section III.A.7) was the substrate for urocanase. One proton is lost from the methylene group of histidine, and the C-5′ proton of the imidazole ring of histidine is washed out by solvent.[57] These results support a carbanionic mechanism. Earlier we found that some enzymes have a pyruvyl group at the active site (Chapter 8, Section V.B) that acts as an electron sink, for example, in decarboxylation reactions. HAL was found to contain another electrophilic species, the dehydroalanyl prosthetic group (**10.44**),

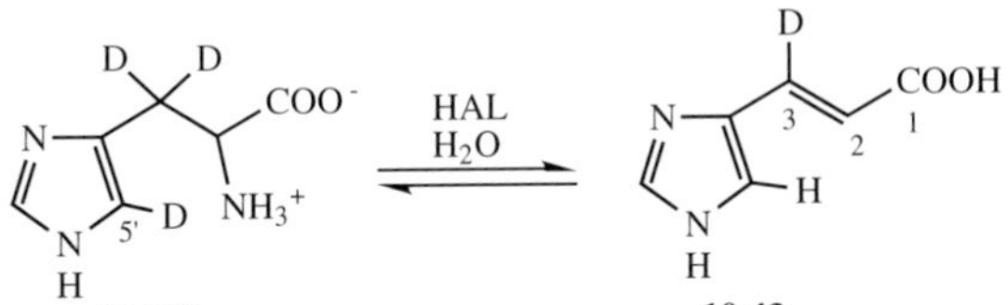

SCHEME 10.26   Reaction catalyzed by histidine ammonia-lyase (HAL).

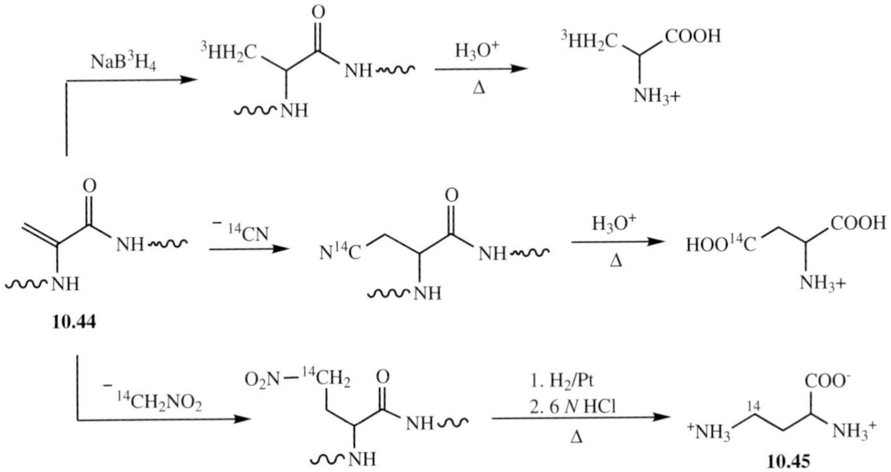

SCHEME 10.27   Reactions to identify the active-site prosthetic group as a dehydroalanyl moiety.

as evidenced by three reactions (Scheme 10.27). Treatment of the enzyme with NaB^{3}H$_4$ or ^{14}CN$^-$, followed by hydrolysis of the inactivated enzyme, gives [^{3}H]alanine or [^{14}C]aspartate, respectively, and treatment of HAL with [^{14}C]nitromethane followed by hydrogenolysis, then hydrolysis, gives [^{14}C]2,4-diaminobutyric acid (**10.45**).[58] Site-specific mutagenesis of the four conserved serine residues in HAL to alanines demonstrates that only Ser-143 is important for activity.[59] Furthermore, the S143T mutant is inactive, whereas the S143C mutant is fully active, indicating that the posttranslational modifying activity can eliminate water from serine or hydrogen sulfide from cysteine, but not water from threonine, to give the dehydroalanyl moiety.[60] The 2.1 Å-resolution crystal structure[60a] revealed that the actual electrophilic species is **10.45a**, which could be formed by cyclization and elimination of the Ala–Ser–Gly at positions 142–144 (Scheme 10.28).

The initial mechanism for HAL was based on a variety of radiolabeling experiments. Treatment of HAL with histidine and [2-^{14}C]urocanate gives [^{14}C]histidine, indicating a reversible reaction. Incubation of HAL with ammonia and urocanate

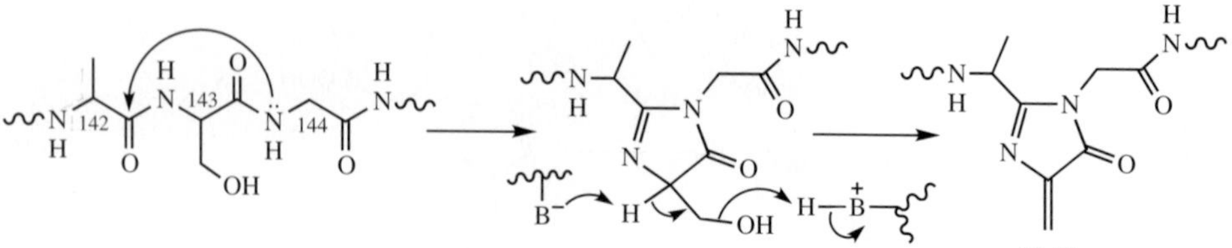

SCHEME 10.28   Posttranslational conversion of the active site Ala-Ser-Gly at positions 142–144 to give a dehydroalanyl-like species.

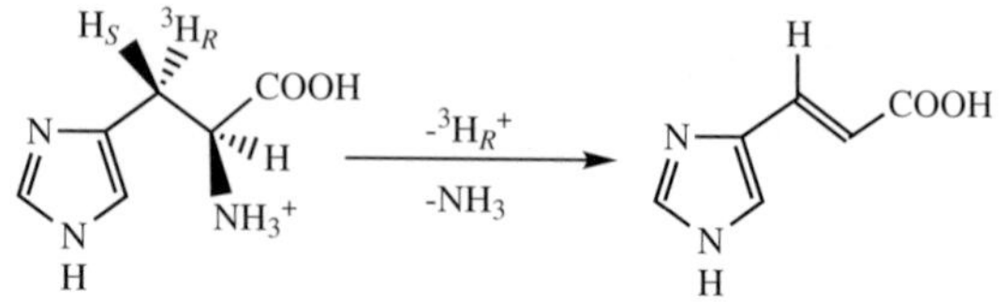

SCHEME 10.29   Stereochemistry of the elimination catalyzed by histidine ammonia-lyase.

gives histidine,[61] again supporting a reversible reaction. When the reaction is run in ^3H_2O, tritium is incorporated into the histidine at the *pro-R* position of the methylene, but not at C-3 of urocanate. Therefore, the proton that is initially exchanged in histidine is the proton that is removed in the elimination; removal of the *pro-R* hydrogen of histidine results in an *anti* elimination (Scheme 10.29). No pyridoxal 5′-phosphate (PLP) was found. Based on these results, the mechanism in Scheme 10.30 was initially proposed. The question still needing an answer in light of this mechanism, however, is how has the enzyme activated the substrate for removal of the proton. Given the low acidity of an allylic proton, this mechanism does not seem to be satisfactory. Rétey[62] has provided a reasonable alternative mechanism that does answer this question. Klee et al.[63] reported earlier that 4′-nitro-L-histidine is a good substrate for HAL. Rétey and co-workers[64] showed that 4-nitro-L-histidine is also a good substrate for their inactive mutants that do not contain a dehydroalanyl prosthetic group (S143A and S143T). Apparently, the nitro group is fulfilling the same role that the dehydroalanyl moiety plays, thereby making the prosthetic group superfluous. The 4′-nitro group would make the C-3 protons much more acidic (Scheme 10.31). The mechanism in Scheme 10.30 uses the dehydroalanyl group as a means of activating the leaving group, but does not use it in an acidifying role for the C-3 proton, so Rétey and co-workers proposed an alternative mechanism (Scheme 10.32). This mechanism is initiated by an electrophilic aromatic substitution mechanism to give **10.46,** which is now activated for deprotonation because of the increased acidity of the protons beta to the iminium ion of the imidazole ring. Intermediate **10.47** has an enamine moiety adjacent to the ammonium leaving group to assist in ejection of the ammonia and produce **10.48,**

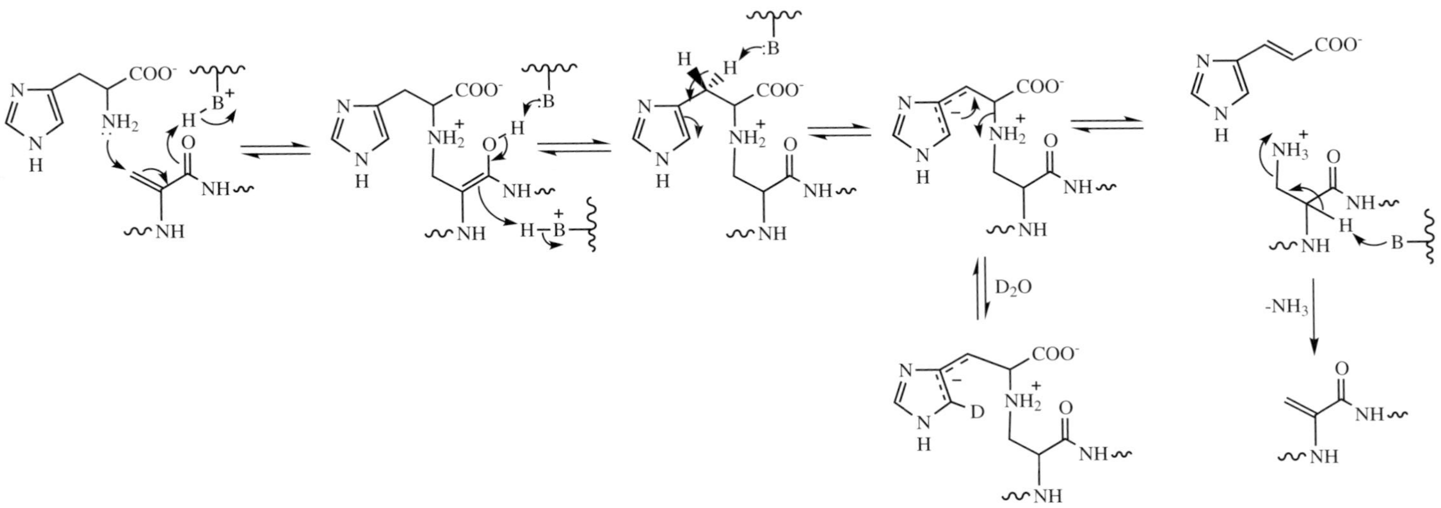

SCHEME 10.30 Original proposed mechanism for histidine ammonia-lyase.

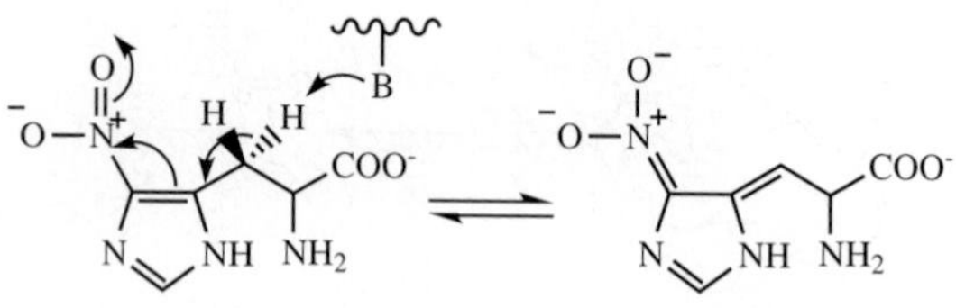

SCHEME 10.31   Activation of C-3 deprotonation by a 2′-nitro group in the histidine ammonia-lyase reaction.

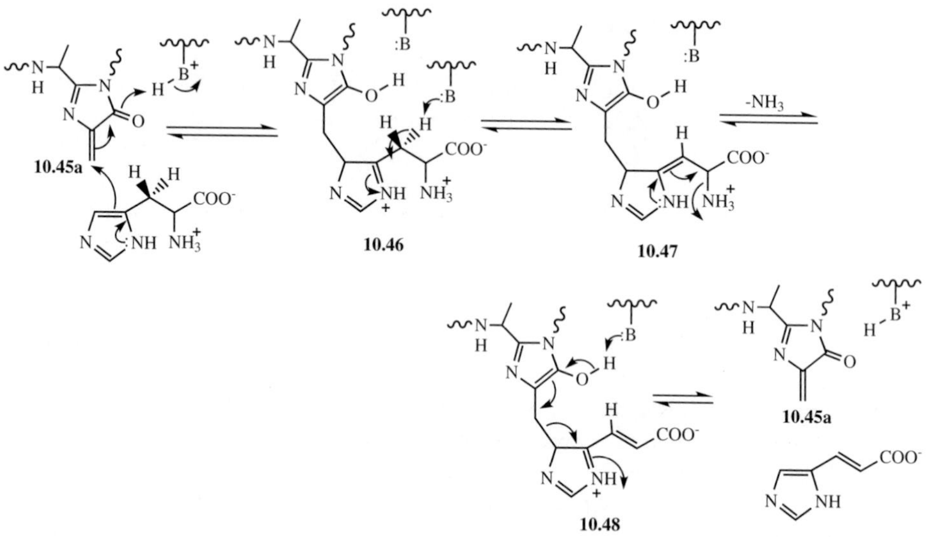

SCHEME 10.32   Proposed alternative (electrophilic aromatic substitution) mechanism for histidine ammonia-lyase.

which can readily break down to the dehydroalanyl prosthetic group (**10.45a**) and urocanate.

But what is the precedence for an imidazole undergoing electrophilic aromatic substitution with the dehydroalanyl double bond? Maybe the best support comes from the enzyme that follows this one, namely, urocanase. As discussed in Chapter 3 (Section III.A.7), it appears that the imidazole of urocanase undergoes electrophilic aromatic substitution with the active-site bound NAD$^+$ cofactor (see Scheme 3.26), another unprecedented reaction.

II. SYN ELIMINATIONS AND ADDITIONS

Some enzyme-catalyzed reactions appear to proceed by *syn* elimination, but generally they are not concerted reactions.[65]

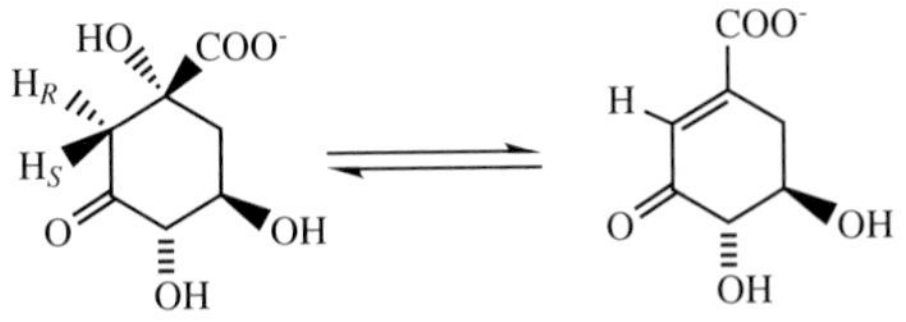

SCHEME 10.33   Reaction catalyzed by 3-dehydroquinate dehydratase (3-dehydroquinase).

A. Schiff Base Mechanism

The dehydration of 3-dehydroquinate (**10.49**) to 3-dehydroshikimate (**10.50**), catalyzed by type I 3-dehydroquinate dehydratase (3-dehydroquinase), proceeds with the loss of the *pro-R* hydrogen, suggesting a *syn* elimination (Scheme 10.33). Nugent and co-workers and Coggins and co-workers found that $NaBH_4$ inactivates the enzyme in the presence of substrate, but not in the absence of substrate.[66] Furthermore, when NaB^3H_4 is used, one tritium is incorporated into the protein, and the reduced substrate can be found attached to Lys-170.[67] These types of experiments may sound familiar to you; we discussed $NaBH_4$ reduction of substrate-bound enzymes in the context of decarboxylases (Chapter 8, Section II.A.1). These are typical experiments done to determine if a Schiff base mechanism is important, and it appears that it is in the case of type I dehydroquinase. Furthermore, the complex of the enzyme bound as a lysine Schiff base to the substrate, as well as the stable $NaBH_4$-reduced enzyme-substrate complex, were detected by Abell and coworkers using electroospray mass spectrometry.[68] These results all point to the Schiff base mechanism shown in Scheme 10.34. This is an E1cB-type mechanism with stereo-specific *syn* elimination; it is not an E2 elimination. However, type II dehydro-

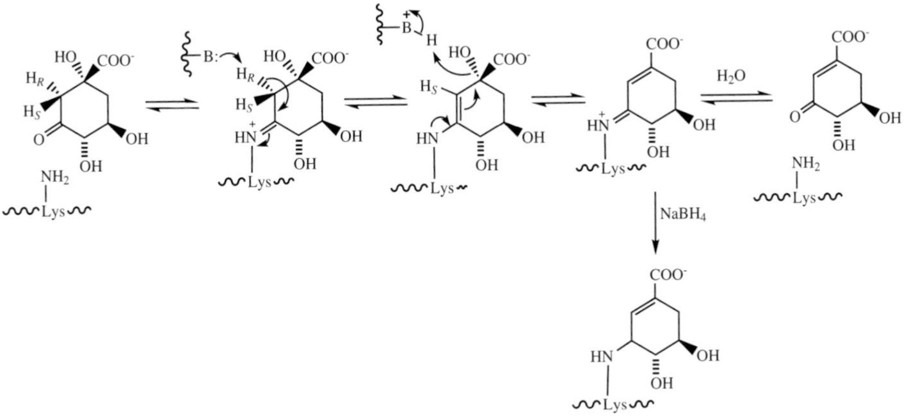

SCHEME 10.34   Proposed mechanism for 3-dehydroquinate dehydratase (3-dehydroquinase).

quinase catalyzes this reaction without Schiff base formation and without metal catalysis,[68a] and the reaction proceeds by *anti*-elimination with loss of the 2-*pro-S* proton.[68b]

B. Pyridoxal 5′-Phosphate-Dependent Eliminations

Already we have seen that pyridoxal 5′-phosphate (PLP) is a coenzyme involved in decarboxylation reactions (Chapter 8, Section V.A), racemization reactions (Chapter 9, Section II.B), and transamination reactions (Chapter 9, Section IV.C). PLP also is important in a variety of enzymes that catalyze both β- (**A**) and γ-elimination reactions (**B**) (Scheme 10.35). Some of the enzymes that catalyze a β-elimination can substitute another group for the leaving group, thereby producing a new amino acid, a reaction known as a β-replacement reaction (**A**); the corresponding reaction with a γ-substituted amino acid is a γ-replacement reaction (**B**) (Scheme 10.36). The replacement reactions are really examples of substitution reactions and thus could be discussed in Chapter 6, but they proceed by an elimination/addition mechanism that uses PLP, so they are discussed here.

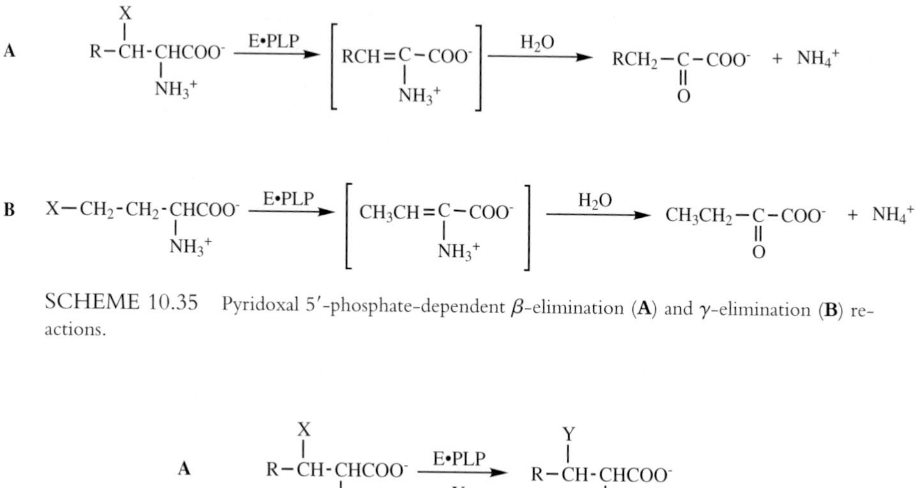

SCHEME 10.35   Pyridoxal 5′-phosphate-dependent β-elimination (**A**) and γ-elimination (**B**) reactions.

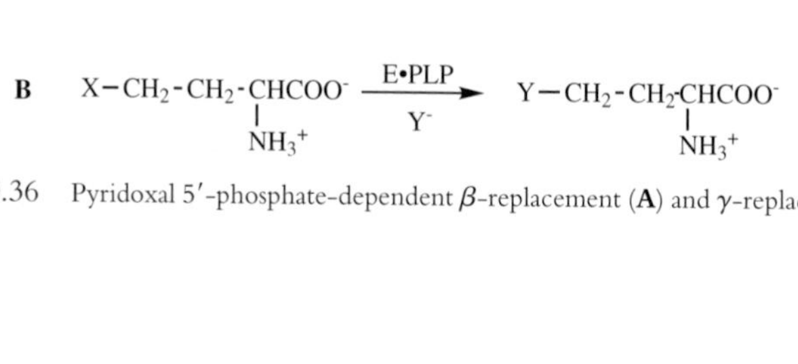

SCHEME 10.36   Pyridoxal 5′-phosphate-dependent β-replacement (**A**) and γ-replacement (**B**) reactions.

1. β-Elimination and β-Replacement Reactions

The overall mechanism for PLP-dependent β-elimination reactions is shown in Scheme 10.37, and the mechanism for a β-replacement reaction in Scheme 10.38. I have stereospecifically labeled two of the protons so you can understand the following stereochemical discussion. Note that β-elimination leads to an α-keto acid and ammonia, whereas β-replacement produces a different amino acid. Some enzymes catalyze only elimination, some only replacement, and some catalyze both. If the replacement nucleophile is omitted, either β-elimination occurs or addition of water (or another nucleophile in the solution) results. There is evidence for each of the intermediates drawn in Scheme 10.37.[69] Much of the discussion centers around two enzymes, tryptophan synthase and tryptophanase; tryptophanase catalyzes the same reaction as the reverse reaction of tryptophan synthase (the X in **10.51** is 3-indolyl). Interestingly, although these two enzymes catalyze very similar reactions on the same substrate, there are no common structural features in the two enzymes that can be detected by antibodies[70] or by comparison of the sequences of tryptophanase[71] and the β₂ subunit of tryptophan synthase.[72] Lane and Kirschner have detected intermediate **10.52** in stopped-flow experiments[73] and in the absorbance and fluorescence spectra[74] during the binding of L-tryptophan to tryptophan synthase. The *quinonoid intermediate* **10.53** also has been detected by transient absorption spectroscopy (470–500 nm) of tryptophan synthase with L-tryptophan and by stopped-flow kinetic studies of tryptophanase.[75] Evidence for the aminoacrylate intermediate **10.54** comes from the treatment of tryptophan synthase with serine and sodium borohydride; the reduced alanine-PLP adduct, derived from reduction of **10.54,** was obtained.[76] Two possible mechanisms for the conversion of **10.54** to pyruvate (**10.56**) are shown in Scheme 10.37. Pathway a involves lysine-induced stereospecific protonation of the aminoacrylate double bond, giving **10.55,** which breaks down to the imine of pyruvate; hydrolysis produces pyruvate. In pathway b the oxygen of the PLP ring assists in the protonation of the aminoacrylate double bond to give **10.57,** which undergoes breakdown and hydrolysis to pyruvate.

A variety of labeling experiments with tryptophan synthase and tryptophanase support the stereochemistry shown in Schemes 10.37 and 10.38.[77] Tryptophan synthase catalyzes the condensation of L-serine with indole to give L-tryptophan and water (Scheme 10.39). The enzyme exists as an $\alpha_2\beta_2$ tetramer (four subunits; two each identical) in which the subunits are separable. Only the β subunits contain the PLP; it is the β subunits that catalyze the β-elimination part of the reaction, and the α subunits are needed for the β-replacement (the α subunits are not functional without the β subunits). Starting with [2-³H]serine and the $\alpha_2\beta_2$ enzyme, the tritium is rapidly exchanged out. The β subunits catalyze the conversion of serine to the aminoacrylate-PLP intermediate (**10.58**, Scheme 10.40). In the presence of the α subunits, indole undergoes a Michael addition to the double bond of **10.58;** following deprotonation at C-3′ of the indole ring and protonation at the α-position, the enzyme releases Trp. The aminoacrylate-PLP intermediate (**10.58**) was ob-

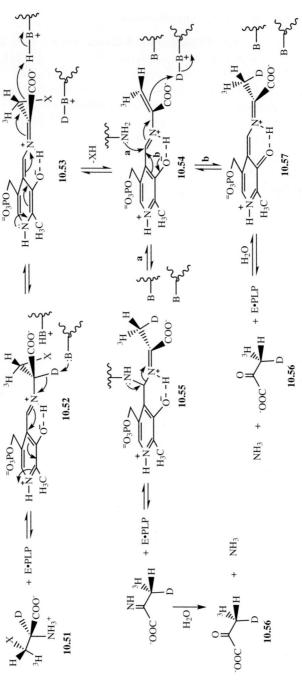

SCHEME 10.37   Proposed mechanisms for PLP–dependent β–elimination reactions.

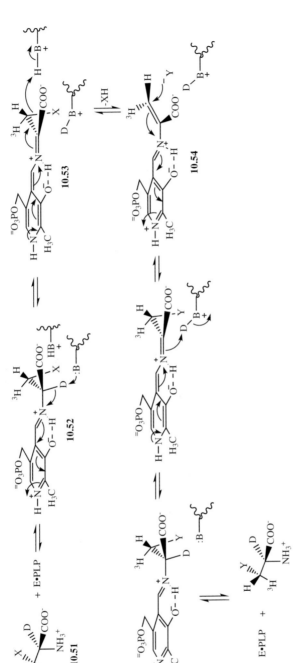

SCHEME 10.38 Proposed mechanisms for PLP-dependent β-replacement reactions.

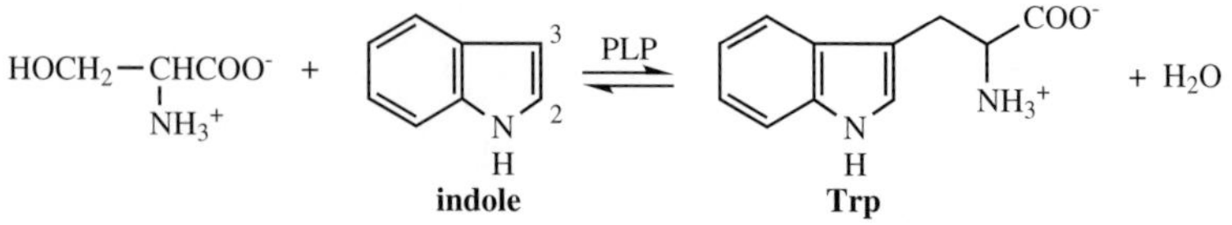

$$HOCH_2-\underset{\underset{NH_3^+}{|}}{CH}COO^- \; + \; \text{indole} \quad \underset{\rightleftharpoons}{\overset{PLP}{\quad\quad}} \quad Trp \; + \; H_2O$$

SCHEME 10.39   Reaction catalyzed by tryptophan synthase.

served by Schaefer, Anderson, and co-workers by solid-state NMR spectrometry during the tryptophan synthase-catalyzed conversion of serine to pyruvate (Scheme 10.40 in the absence of α subunits).[78] Using $(2S,3S)$-[3-^{2}H, 3-^{3}H]serine and the β subunits of tryptophan synthase, Floss and co-workers showed that the pyruvate recovered has a chiral methyl group (see Section I.A.2 for chiral methyl analysis) with R stereochemistry.[79] However, when the same experiment is run in D_2O, the pyruvate is still chiral and has R stereochemistry. For that to happen, a proton must be added to C-3, and the only source for the proton, because the reaction is carried out in D_2O, is the one at the C-2 position of serine. Therefore, this is an intramolecular transfer of the proton from C-2 of serine to C-3 of pyruvate without solvent exchange (as depicted in Scheme 10.37). The C-2 proton of serine replaces the hydroxyl group with retention of configuration. This is therefore a *suprafacial* transfer, mediated by a single base, probably a histidine,[80] and is a *syn* elimination from the *si* face.

In the case of tryptophanase (Scheme 10.41), there is a significant intramolecular transfer of the proton from C-2 of tryptophan to C-3 of indole, the leaving group. The tryptophanase reaction is the exact reverse of the tryptophan synthase reaction (compare the reverse of Scheme 10.40 with Scheme 10.41). Note that an indole anion is *not* the leaving group in this reaction because it would be a very strong base and therefore a very poor (nonexistent) leaving group. By transferring the proton from C-2 to C-3' of the indole ring (**10.59**), tryptophanase converts the indole into a good leaving group (**10.60**) because the pair of electrons in the C-3—C-3' bond can be delocalized into the iminium ion to give [3-^{3}H]indole and the amino-acrylate-PLP intermediate (**10.58**). This suggests that a single base is involved in a *suprafacial* [1,3]-hydrogen shift, and the elimination is *syn*. The lysine-bound amino-acrylate-PLP intermediate (**10.61**) was detected by Phillips using stopped-flow spectrophotometry.[81] When [α-^{2}H]tryptophan is used as the substrate, there is a kinetic isotope effect of 3.6 on the reaction, indicating that the cleavage of the C_α–H bond is the rate-determining step. Floss and co-workers showed that when the enzyme reaction is run with stereospecifically C-3 tritiated tryptophan in D_2O, pyruvate is produced with a chiral methyl group (again with retention of configuration). This suggests that a polyprotic base is involved, and that protonation of the amino-acrylate is enzyme catalyzed, which leads to stereospecific deuteration at C-3 of pyruvate.[82]

A similar stereochemical picture emerges from β-replacement reaction studies. Again, retention of configuration occurs (the incoming nucleophile replaces the leaving group in the same stereochemical position), reactions occur on the *si*-face

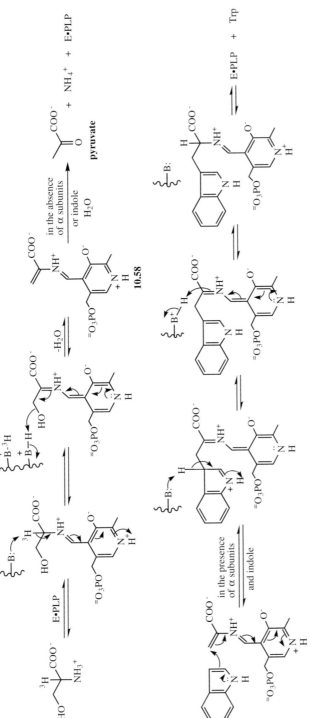

SCHEME 10.40　Proposed mechanism for tryptophan synthase in the absence and presence of α subunits.

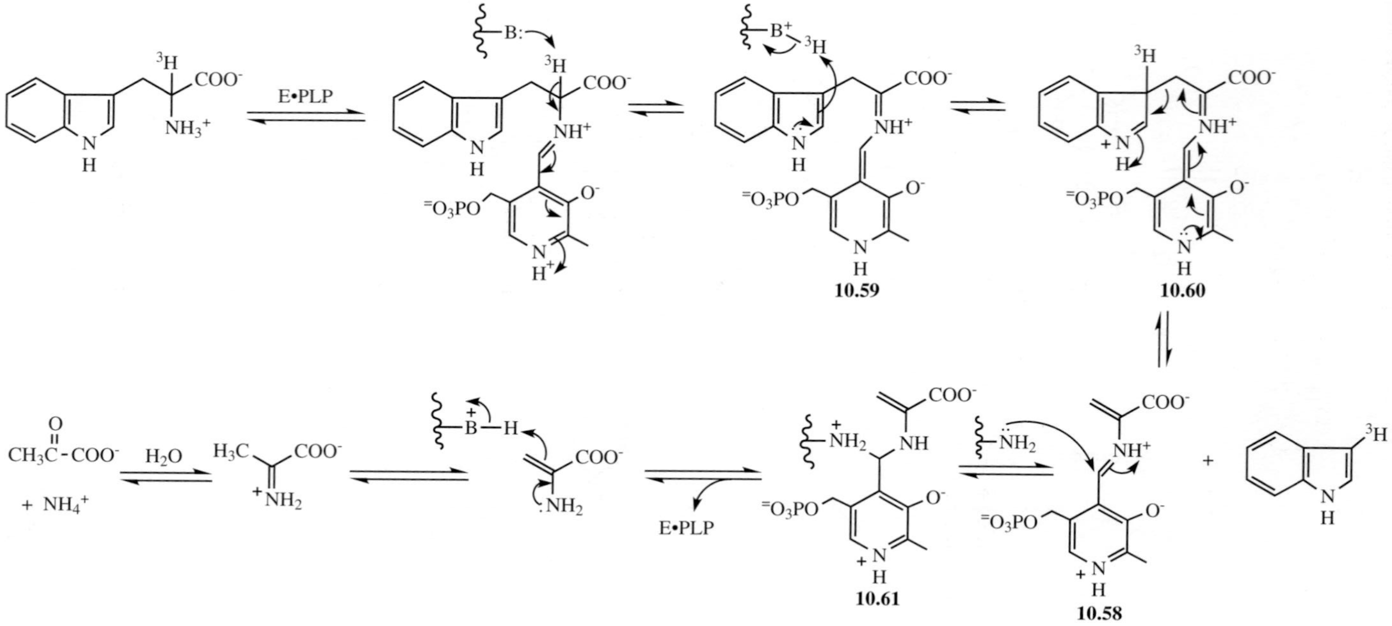

SCHEME 10.41 Proposed mechanism for the reaction catalyzed by tryptophanase.

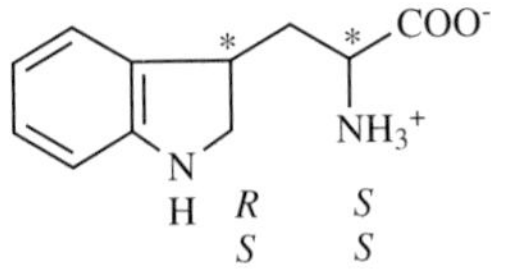

10.62

with the *re*-face inaccessible to external reagents, and a single base appears to be involved, although in some cases it is a monoprotic base (complete retention of the C-2 proton) and in others a polyprotic base (at least some solvent exchange).

To investigate the stereochemical differences between tryptophanase and tryptophan synthase, Phillips and co-workers synthesized the (2S)-diastereomers of 2,3-dihydro-L-tryptophan (**10.62**) and showed them to have opposite potency as inhibitors of tryptophanase and tryptophan synthase. The conclusion reached is that the two enzymes have opposite stereochemistry (Scheme 10.42).[83]

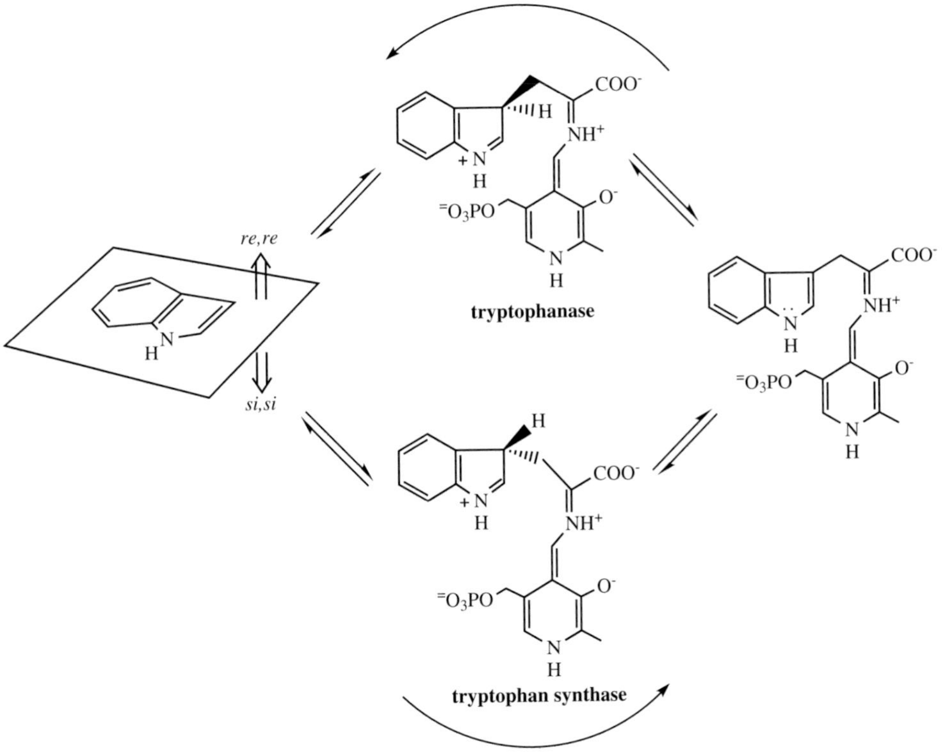

SCHEME 10.42   Proposed difference in the stereochemistry of the reactions catalyzed by tryptophanase and tryptophan synthase.

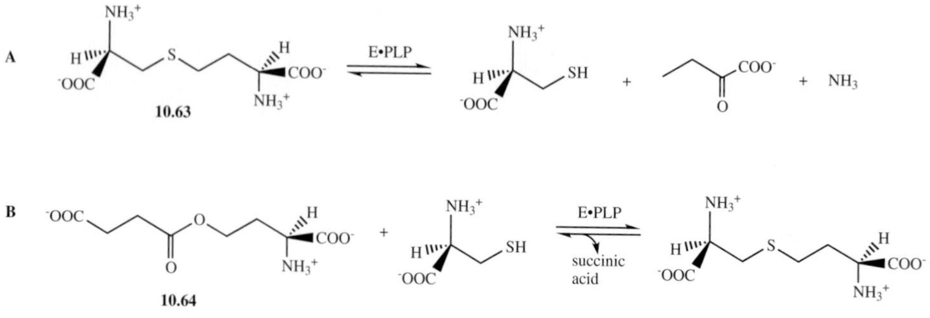

A

B

SCHEME 10.43   Reactions catalyzed by γ-cystathionase (**A**) and cystathionine γ-synthase (**B**).

2. γ-Elimination and γ-Replacement Reactions

Just as the discussion of β-elimination and β-replacement centered around two enzymes that catalyze reactions that are the reverse of each other, a similar situation
arises here. γ-Cystathionase, which catalyzes the γ-elimination of cystathionine
(**10.63**) to α-ketobutyric acid and cysteine (Scheme 10.43A), and cystathionine γ-
synthase, which catalyzes the synthesis of cystathionine by γ-replacement of the
succinate group of O-succinyl-L-homoserine (**10.64**, Scheme 10.43B) with cysteine, are the most studied examples of these reactions. There generally is some
internal return, so Posner and Flavin believe that a single polyprotic base is involved
in a *suprafacial* process.[84] Interestingly, internal hydrogen transfer only occurs for
unlabeled substrates in D_2O, not for deuterated substrates in H_2O; a deuterium isotope effect favors transfer of the proton over the deuteron in both cases.

The mechanism for PLP-dependent γ-elimination reactions is shown in Scheme
10.44 and for γ-replacement reactions in Scheme 10.45. Note that γ-elimination
gives α-ketobutyric acid and ammonia, and γ-replacement gives a different γ-
amino acid. The exchanging proton in Schemes 10.44 and 10.45 is depicted with
the symbol BH_x. The C-2 proton is removed from **10.65** to give **10.66**. This makes
the C-3 protons acidic, so the same base removes one proton stereospecifically; by
NMR spectrometry it is possible to show that the *pro-R* proton is the one that is
rapidly exchanged, yielding **10.67**.[85] Ejection of the γ-leaving group generates the
terminal double bond (**10.68**), which is protonated by one of the protons on the
polyprotic base to give **10.69** (note that any of the protons can be donated, so I
have symbolized that as $H_{x,a,b}$; this is not meant to infer that there are 3 protons on
the base). Because the β-hydrogen is rapidly exchanged, H_x is donated to the α,β-
double bond to give **10.70** followed by hydrolysis, as in the case of β-elimination
(Scheme 10.37), to give α-ketobutyric acid, ammonia, and the PLP. Alternatively,
as in the case of β-elimination, the active-site lysine residue could attack **10.69** to
initiate proton donation to C-3 (see pathway b in Scheme 10.37).

γ-Cystathionase also uses homoserine as a substrate. With homoserine in D_2O,
both the α-proton (H_a) and the 3-*pro-R* proton (H_b) exchange with solvent. The

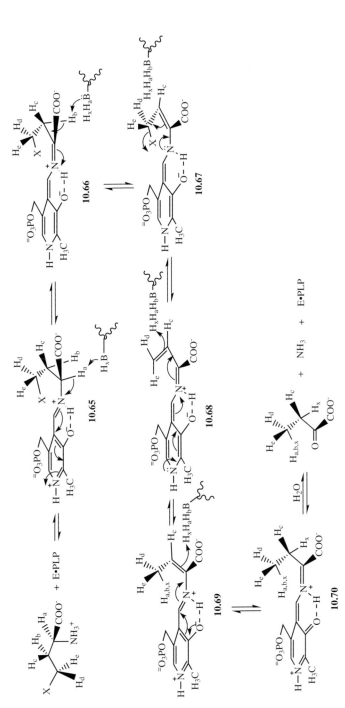

SCHEME 10.44 Proposed mechanism for the reaction catalyzed by PLP-dependent γ-elimination enzymes. H_x represents solvent protons. $H_xH_bH_aB$ implies that one or more of these protons is attached to the base B.

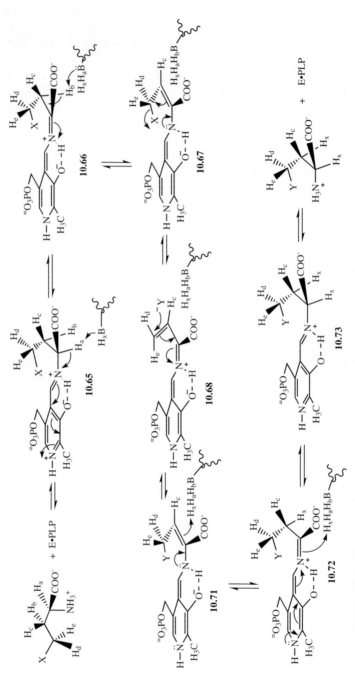

SCHEME 10.45   Proposed mechanism for the reaction catalyzed by PLP-dependent γ-replacement enzymes. H_x represents solvent protons. $H_xH_bH_aB$ implies that one or more of these protons is attached to the base B.

α-ketobutyric acid product has deuterium stereospecifically incorporated at the 3-*pro-S* position (H_x), which is the same proton as the 3-*pro-R* proton of the substrate (note that the R,S priorities change in going from homoserine to α-ketobutyric acid).[86]

γ-Replacement (Scheme 10.45) also occurs with retention of configuration at the γ-position. The β,γ-elimination in that reaction (**10.66–10.68**) is *syn*, because the *pro-R* proton (H_b) is removed at C-3 and (4S)-O-succinyl-[4-^{2}H]-homoserine gives the Z-alkenylglycine intermediate (**10.71**).[87] Both protonations occur at the *si*-face to give **10.72** and **10.73** in a net suprafacial process.

In Chapter 2 (Section II.A.2.a) the concept of mechanism-based inactivation was developed, and it was indicated that these kinds of inactivators can be quite useful in the elucidation of enzyme mechanisms. Two mechanism-based inactivators of γ-cystathionase have been used to show that removal of the protons at C-2 and C-3 of alternative substrates (mechanism-based inactivators) lead to β- and γ-eliminations. Abeles and Walsh showed that [2-^{14}C]propargylglycine (**10.74**) inactivates γ-cystathionase with incorporation of 2 moles of radioactivity per mole of tetrameric enzyme, indicating half-sites reactivity (see Appendix I, Section VII.C), and covalent attachment to the enzyme.[88] Incubation with [α-^{2}H]propargylglycine leads to inactivation with a kinetic isotope effect of 2.2, comparable to the isotope effect on conversion of homoserine to product. Acid hydrolysis of the enzyme inactivated with **10.74** releases the radioactivity from the enzyme as [2-^{14}C]2-amino-4-ketopentanoic acid (**10.75**), indicating that the α-proton is replaced prior to the release of **10.75**.[89] A mechanism that is consistent with these results is shown in Scheme 10.46; **10.77** represents the inactivated enzyme. Note how the mechanism-based inactivator acts as a substrate until the enzyme converts it into the activated form (**10.76**), which reacts with the active-site nucleophile X. Although the identity of X is not known, on the basis of chemical model studies of the rate of hydrolysis of various vinyl–X compounds, the hydrolysis rate for release of **10.75** coincides with X being either a cysteine or tyrosine. Acid-catalyzed hydrolysis of **10.77** gives **10.75**.

The other mechanism-based inactivator of γ-cystathionase that provided mechanism information about the enzyme is β,β,β-trifluoroalanine (**10.78**), which

10.74

10.75

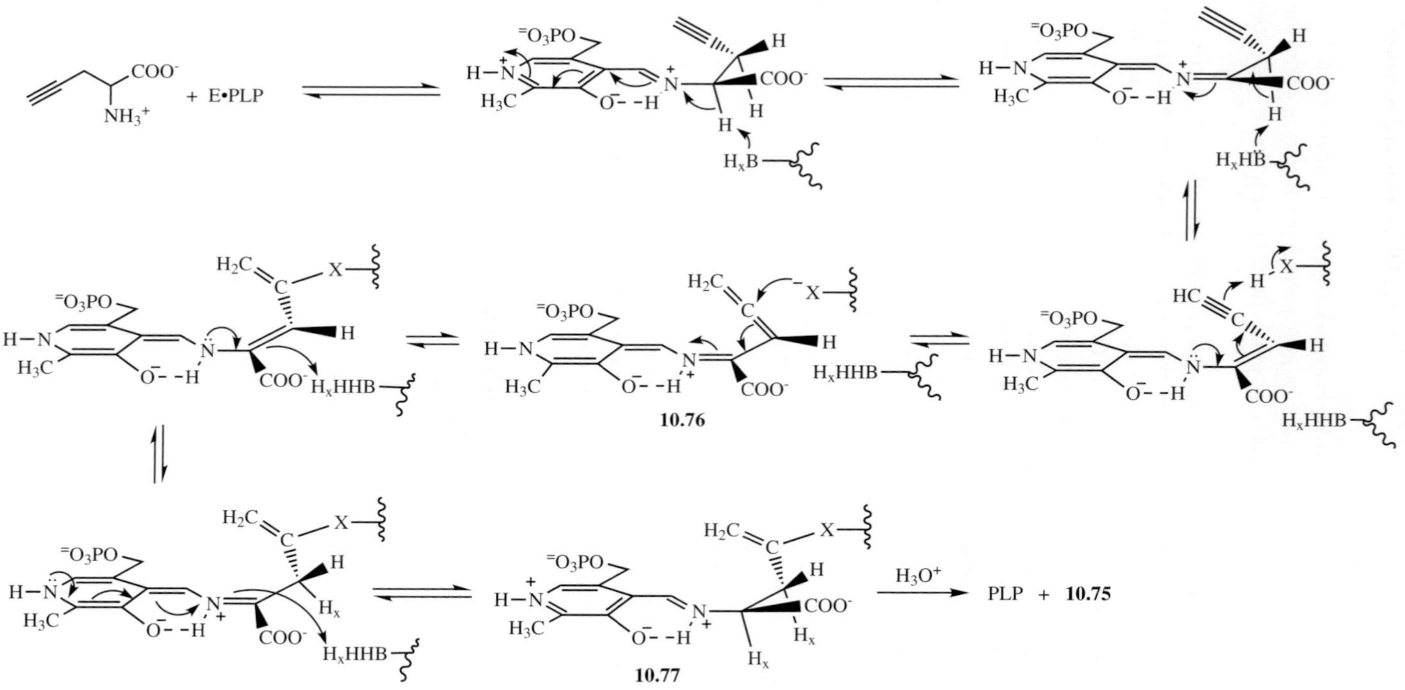

SCHEME 10.46 Mechanism-based inactivation of γ-cystathionase by propargylglycine.

$$F_3C \diagdown \diagup COO^-$$
$$| \atop NH_3^+$$

10.78

inactivates a series of PLP-dependent enzymes, including γ-cystathionase, tryptophan synthase, and tryptophanase; β-fluoroalanine is only a substrate.[90] This suggests that the additional fluorines activate an intermediate for covalent reaction with the enzyme. Inactivation of γ-cystathionase with $[1-^{14}C]$**10.78** results in the covalent attachment of 2 moles of radioactivity per enzyme tetramer, just as propargylglycine did. The rate of inactivation, the rate of ^{14}C incorporation into the enzyme, and the rate of release of fluoride ions from the inactivator are comparable.[91] Approximately 3 equivalents of F^- are released per mole of inactivator incorporated into the enzyme (Scheme 10.47). This, plus the fact that no difluoropyruvate is formed, supports a mechanism in which every turnover of inactivator leads to inactivation, or, in other words, the partition ratio (see Chapter 2, Section II.A.2.a) is zero. On inactivation, the absorption spectrum changes with a λ_{max} developing at 519 nm, unlike propargylglycine inactivation, which does not lead to a spectral change. This long wavelength absorption suggests the formation of a highly delocalized carbanion (**10.79**). Denaturation of the inactivated enzyme results in the release of all of the radioactivity as $^{14}CO_2$ because of the instability of **10.80**. If the denaturation is done in ^3H_2O, tritium becomes incorporated into the inactive enzyme (**10.81**). Hydrolysis of the tritiated modified enzyme produces $[^3H]$glycine. Initial evidence for the identity of X comes from chemical model studies of the rates of base hydrolysis and hydroxylaminolysis of a thioester, an ester, a phenyl ester, and an amide; these rates were then compared to the corresponding rates of release of $[^3H]$glycine from the inactivated enzyme under the same conditions. From those experiments it was surmised that X is a lysine residue. Abeles and co-workers carried out further studies,[92] in which the enzyme was inactivated with β,β,β-trifluoroalanine in ^3H_2O to incorporate a tritium, the inactivated enzyme was cyanogen bromide cleaved to a radioactive peptide, which was then treated with trypsin, and the resultant labeled peptide was sequenced. The heptapeptide obtained contained a modified lysine residue, the same lysine residue shown in a noninactivated control to be the one attached to the PLP; X, then, is the active-site lysine residue. The identity of X was later confirmed by refinement of the 2.3 Å resolution crystal structure of a related enzyme, β-cystathionase, inactivated with β,β,β-trifluoroalanine.[93] In that case, the crystal structure clearly showed **10.79**, in which X is Lys-210, the lysine residue that is bound to PLP in the native enzyme.

As noted, only half-sites reactivity occurs with both propargylglycine and β,β,β-trifluoroalanine, leaving two PLP-containing subunits free within the enzyme tetramer. Even though the inactivated enzyme has no catalytic activity to convert cystathionine to its products, Silverman and Abeles found that when γ-cystathionase

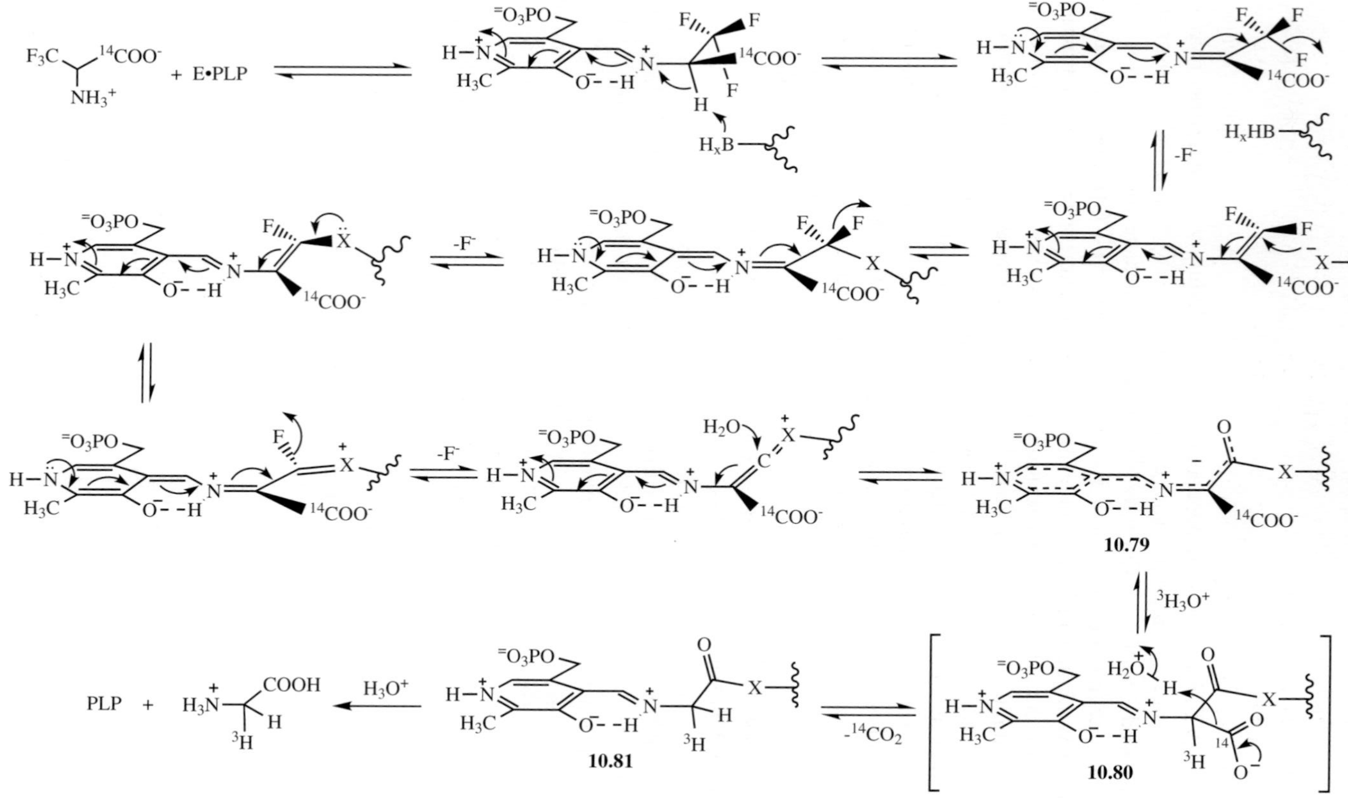

SCHEME 10.47 Mechanism-based inactivation of γ-cystathionase by β,β,β-trifluoroalanine.

is first inactivated with propargylglycine, 2 equivalents of $[^{14}C]\beta,\beta,\beta$-trifluoroalanine can still be incorporated into the inactive enzyme. However, if the converse experiment is done, first inactivating the enzyme with β,β,β-trifluoroalanine, then treating it with $[^{14}C]$propargylglycine, no radioactivity is incorporated. That result could be explained by different conformational changes produced by propargylglycine and β,β,β-trifluoroalanine, which affect the ability of the two additional molecules of inactivator to react with the two empty subunits. These results also can be rationalized by a mechanism in which all four inactivator molecules go into only two subunits. This may be possible if propargylglycine binds to a site far enough away from the PLP that, after inactivation, cystathionine cannot bind to the active site, but the smaller β,β,β-trifluoroalanine can. However, initial inactivation by β,β,β-trifluoroalanine blocks the binding of both cystathionine and propargylglycine because it binds to the active-site lysine residue.

According to evolutionary considerations, iron−sulfur cluster-containing enzymes that catalyze elimination reactions are primative catalysts, whereas PLP-dependent ones are more sophisticated.[94] Anaerobic bacteria, which contain iron−sulfur clusters, for example, are more primitive than the aerobic ones that use PLP.

C. Pyridoxamine 5′-Phosphate and Iron−Sulfur Cluster-Dependent Reactions

In the discussion in Section I.A.2 we saw that nucleoside diphosphohexose 4,6-dehydratases catalyze the dehydration and reduction of nucleoside diphosphohexoses to the nucleoside diphospho-4-keto-6-deoxyhexoses. These compounds are just intermediates in the biosynthesis of 3,6-dideoxyhexoses, which were discussed in Chapter 3 (Section III.D.2). The first step in the the deoxygenation of CDP-4-keto-6-deoxy-D-glucose (**10.82**) to CDP-4-keto-3,6-dideoxy-D-glucose (**10.83**) is a *syn*-dehydration reaction catalyzed by the enzyme CDP-6-deoxy-L-*threo*-D-*glycero*-4-hexulose 3-dehydratase, also referred to as the E_1 enzyme (Scheme 10.48). This enzyme is coupled to a second enzyme, CDP-6-deoxy-L-*threo*-D-*glycero*-4-hexulose 3-dehydratase reductase, or E_3, which we discussed in Chapter 3 (Section III.D.2). The unusual aspect of the *syn*-dehydration is that it uses the cofactor pyridoxamine 5′-phosphate (PMP) and also contains a redox active 2Fe−2S cluster.

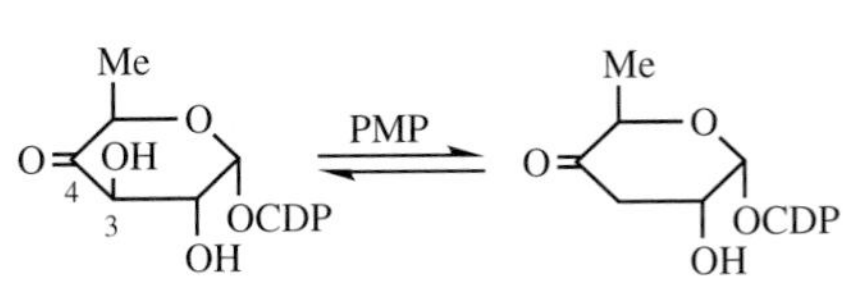

SCHEME 10.48   The first step in the the deoxygenation of CDP-4-keto-6-deoxy-D-glucose (**10.82**) to CDP-4-keto-3,6-dideoxy-D-glucose (**10.83**) by CDP-6-deoxy-L-*threo*-D-*glycero*-4-hexulose 3-dehydratase.

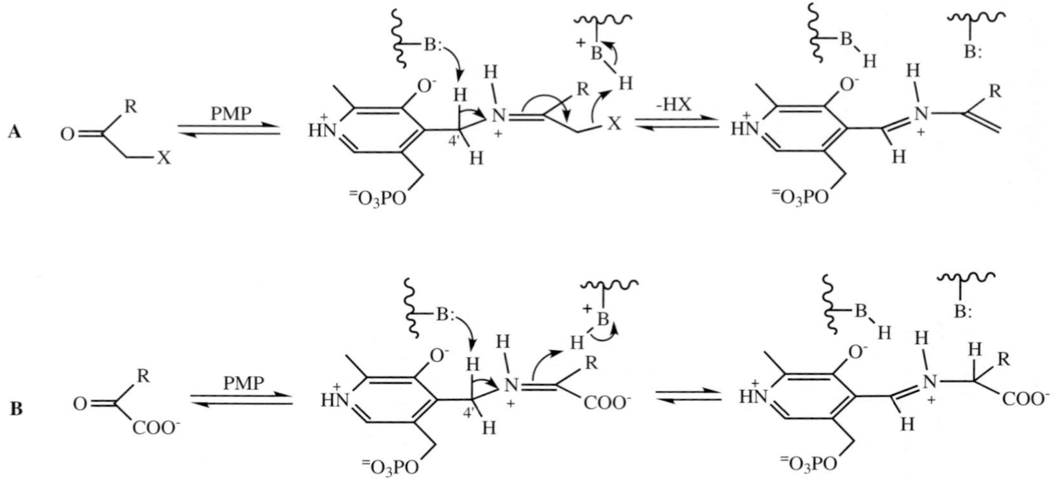

SCHEME 10.49   Comparison of a PMP-dependent elimination reaction (**A**) with the corresponding tautomerization reaction (**B**).

We have already discussed PMP in a completely different context (Chapter 9, Section IV.C.1), namely, pyridoxal 5′-phosphate (PLP)-dependent aminotransferases. These enzymes convert amino acids in the first half reaction to keto acids with concomitant conversion of PLP to PMP (see Scheme 9.38); the PMP is then converted back to PLP with a second keto acid substrate (see Scheme 9.39). The conversion of PMP back to PLP involves the reaction of the amino group of PMP with the ketone carbonyl of the α-keto acid substrate, which activates the PMP methylene proton for base removal and tautomerization.

Except for the fact that PMP was not previously known to be used in a dehydration reaction, it is quite reasonable for PMP to catalyze an elimination reaction (Scheme 10.49A, X is a leaving group) as well as a tautomerization reaction (Scheme 10.49B). Early results with E_1 were consistent with a Schiff base mechanism. Rubenstein and Strominger showed that if the substrate is incubated with E_1 in $H_2^{18}O$, the recovered substrate has ^{18}O incorporated at the 4-keto carbonyl.[95] The hydroxyl at C-3 also is exchanged for ^{18}OH, supporting the reversibility of the elimination reaction. If [*methylene*-^{3}H]PMP is used, the tritium is released into the medium, supporting the proposed further acidification of that proton when it forms a Schiff base with the substrate carbonyl. When $(4'S)$- and $(4'R)$-$[4'-^3H]$PMP are the cofactors, Liu and co-workers demonstrated that the stereochemistry is *pro-S* specific[96] (as is the case with all PLP enzymes that act on the methylene protons). Reprotonation at C-4′ in the reverse reaction also is *pro-S* specific, suggesting that the *si*-face of the cofactor–substrate complex is exposed to solvent (again, as is the case with other PLP-dependent enzymes). The mechanism in Scheme 10.50 is consistent with these results. Formation of the Schiff base between **10.82** and PMP gives **10.84**. Removal of the *pro-S* proton initiates the elimination of the C-3 hydroxyl group, assisted by an active-site acid, producing **10.85**. In $H_2^{18}O$, the reverse reaction with **10.85** incorporates the ^{18}O at C-3 of **10.82**; hydrolysis of the Schiff base would incorporate the ^{18}O into the C-4 position. The active-site base responsible for removal of the proton from the PMP was identified by Liu and co-workers using chemical studies and by site-directed mutagenesis.[97] The enzyme is rapidly inactivated by diethyl pyrocarbonate, a reagent that is highly selective for modification of histidine residues in proteins.[98] In the presence of the substrate, diethyl

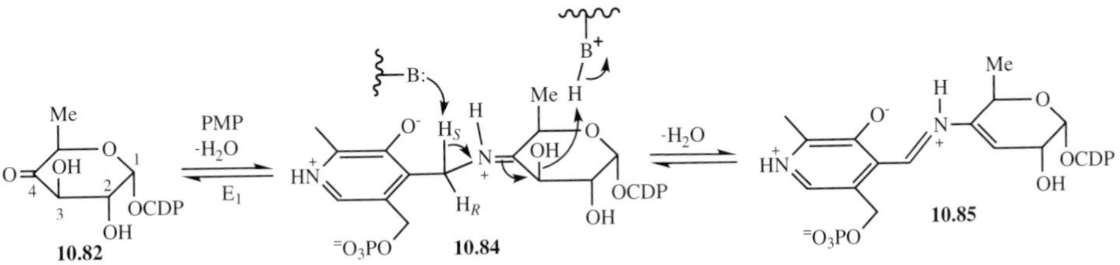

SCHEME 10.50   Proposed mechanism for the dehydration catalyzed by CDP-6-deoxy-L-*threo*-D-*glycero*-4-hexulose 3-dehydratase (E_1).

pyrocarbonate does not cause inactivation, suggesting that the inactivation arises from histidine modification in the active site. By comparing the primary sequence of E_1 with other PLP/PMP enzymes (i.e., aminotransferases), Liu and co-workers found that several invariant residues in aminotransferases also are found in E_1, which may account for the similarity in activity of E_1 and of aminotransferases. Sequence alignment of E_1 with aminotransferases indicates that the highly conserved lysine residue of the aminotransferases is replaced by a histidine (His-220) in E_1. To test whether His-220 is important in catalysis of E_1, the H220N mutant was constructed. This mutant has almost no catalytic activity, and the exchange of the C-4′ proton of PMP is severely impaired; the base in Scheme 10.50, then would be His-220. By analogy to other aminotransferase sequences, the proton donor in Scheme 10.50 would be Arg-403. Liu and co-workers suggest that the evolutionary way to convert an aminotransferase scaffold into a dehydratase may simply be to switch an active-site lysine to a histidine residue. Dunathan[99] hypothesized that the entire family of PLP/PMP enzymes, regardless of their catalytic diversity, has evolved from a common progenitor, and maybe he is right.

We are still left with two problems: first, **10.85** is not the 3,6-dideoxyhexose product, and E_1 does not give that product without E_3 (or some other electron-transfer protein), and second, what is the function of the 2Fe−2S cluster? These questions are answered, or at least discussed, in Chapter 3 (Section III.D.2).

D. A Catalytic Antibody-Catalyzed *Syn* Elimination

Lerner and co-workers raised a catalytic antibody that catalyzes the *syn*-elimination of an alkyl fluoride to give a *cis* alkene (Scheme 10.51), a reaction that would re-

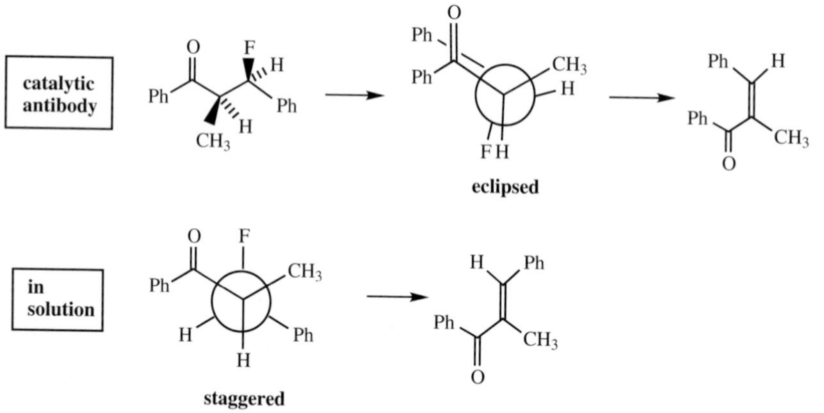

SCHEME 10.51   A *syn* elimination reaction catalyzed by a catalytic antibody compared to the reaction in solution.

quire an eclipsed transition state, resulting in a large amount of torsional energy.[100] In solution, only the normal *anti* elimination occurs to give the *trans* alkene.

To continue in the study of reactions that involve carbanion intermediates, we next turn our attention to enolate chemistry, enzyme-catalyzed aldol- and Claisen-type condensation reactions.

NOTE THAT PROBLEMS AND SOLUTIONS RELEVANT TO EACH CHAPTER CAN BE FOUND IN APPENDIX II.

REFERENCES

1. (a) Gerlt, J. A.; Kozarich, J. W.; Kenyon, G. L.; Gassman, P. G. *J. Am. Chem. Soc.* **1991**, *113*, 9667. (b) Gerlt, J. A.; Gassman, P. G. *J. Am. Chem. Soc.* **1992**, *114*, 5928.
2. Schwab, J. M.; Klassen, J. B.; Habib, A. *J. Chem. Soc. Chem. Commun.* **1986**, 357.
3. Faller, L. D.; Baroudy, B. M.; Johnson, A. M.; Ewall, R. X. *Biochemistry* **1977**, *16*, 3864.
4. Burbaum, J. J.; Knowles, J. R. *Biochemistry* **1989**, *28*, 9306.
5. Cohn, M.; Pearson, J. E.; O'Connell, E.; Rose, I. A. *J. Am. Chem. Soc.* **1970**, *92*, 4095.
6. Dinovo, E. C.; Boer, P. D. *J. Biol. Chem.* **1971**, *246*, 4586.
7. Anderson, V. E.; Weiss, P. M.; Cleland, W. W. *Biochemistry* **1984**, *23*, 2779.
8. Lebioda, L.; Stec, B. *J. Am. Chem. Soc.* **1989**, *111*, 8511.
9. Wedekind, J. E.; Reed, G. H.; Rayment, I. *Biochemistry* **1995**, *34*, 4325.
10. Poyner, R. R.; Laughlin, L. T.; Sowa, G. A.; Reed, G. H. *Biochemistry* **1996**, *35*, 1692.
11. Larsen, T. M.; Wedekind, J. E.; Rayment, I.; Reed, G. H. *Biochemistry* **1996**, *35*, 4349.
12. (a) Gerlt, J. A.; Kozarich, J. W.; Kenyon, G. L.; Gassman, P. G. *J. Am. Chem. Soc.* **1991**, *113*, 9667. (b) Gerlt, J. A.; Gassman, P. G. *J. Am. Chem. Soc.* **1992**, *114*, 5928.
13. Babbitt, P. C.; Hasson, M. S.; Wedekind, J. E.; Palmer, D. R. J.; Barrett, W. C.; Reed G. H.; Rayment, I.; Ringe, D.; Kenyon, G. L.; Gerlt, J. A. *Biochemistry* **1996**, *36*, 16489.
14. Petsko, G. A.; Kenyon, G. L.; Gerlt, J. A. Ringe, D.; Kozarich, J. W. *Trends Biochem. Sci.,* **1993**, *18*, 372.
15. (a) Glaser, L.; Zarkowsky, R. In *The Enzymes*, vol. 5, 3rd ed.; Boyer, P., Ed., Academic Press: New York, p. 465. (b) Liu, H.-w.; Thorson, J. S. *Annu. Rev. Microbiol.* **1994**, *48*, 223. (c) Frey, P. A. In *Pyridine Nucleotide Coenzymes*, Part B, Dolphin, D.; Poulson, R.; Avramovic, O., Eds., Wiley: New York, 1987, p. 461.
16. Lindberg, B. *Adv. Carbohydr. Chem. Biochem.* **1990**, *48*, 279.
17. (a) Melo, A.; Elliott, W. H.; Glaser, L. *J. Biol. Chem.* **1968**, *243*, 1467. (b) Gabriel, O.; Lindquist, L. C. *J. Biol. Chem.* **1968**, *243*, 1479.
18. Herman, K.; Lehman, J. *Eur. J. Biochem.* **1968**, *3*, 369.
19. Snipes, C. E.; Brillinger, G.-U.; Sellers, L.; Mascaro, L.; Floss, H. G. *J. Biol. Chem.* **1977**, *252*, 8113.
20. Wang, S.-F.; Gabriel, O. *J. Biol. Chem.* **1970**, *245*, 8.
21. (a) Zarkowsky, H.; Glaser, L. *J. Biol. Chem.* **1969**, *244*, 4750. (b) Zarkowsky, H.; Lipkin, E.; Glaser, L. *Biochem. Biophys. Res. Commun.* **1970**, *38*, 787.
22. Yu, Y.; Russell, R. N.; Thorson, J. S.; Liu, L. D.; Liu, H. W. *J. Biol. Chem.* **1992**, *267*, 5868.
23. (a) Cornforth, J. W.; Redmond, J. W.; Eggerer, H.; Buckel, W.; Gutschow, C. *Nature* **1969**, *221*, 1212. (b) Cornforth, J. W.; Redmond, J. W.; Eggerer, H.; Buckel, W.; Gutschow, C. *Eur. J. Biochem.* **1970**, *14*, 1.
24. Lüthy, J.; Rétey, J.; Arigoni, D. *Nature* **1969**, *221*, 1213.

25. Floss, H. G.; Tsai, M.-D. *Adv. Enzymol.* **1979,** *50,* 243.

26. (a) Kuhn, R.; Roth, H. *Chem. Ber.* **1933,** *66B,* 1274. (b) Maciak, G. In *Methods in Carbohydrate Chemistry,* vol. 1, Whistler, R. L.; Wolfrom, M. L., Eds., Academic Press: Orlando, 1962, p. 461.

27. Gawron, O.; Glaid, A. J. III; Fondy, T. P. *J. Am. Chem. Soc.* **1961,** *83,* 3634.

28. Lenz, H.; Eggerer, H. *Eur. J. Biochem.* **1976,** *65,* 237.

29. (a) Davenport, H. E.; Hill, R.; Whatley, F. R. *Proc. R. Soc. London Ser. B* **1952,** *139,* 346. (b) Arnon, D. I.; Whatley, F. R.; Allen, M. B. *Nature* **1957,** *180,* 182. (c) San Pietro, A.; Lang, H. M. *J. Biol. Chem.* **1958,** *231,* 211.

30. Nomenclature Committee of the International Union of Biochemistry *J. Biol. Chem.* **1992,** *267,* 665.

31. Alworth, W. L. *Stereochemistry and Its Application in Biochemistry,* Wiley: New York, 1972.

32. (a) Speyer, J. F.; Dickman, S. R. *J. Biol. Chem.* **1956,** *220,* 193. (b) Englard, S. *J. Biol. Chem.* **1960,** *235,* 1510.

33. Rose, I. A.; O'Connell, E. *J. Biol. Chem.* **1967,** *242,* 1870.

34. Beinert, H.; Kennedy, M. C.; Stout, C. D. *Chem. Rev.* **1996,** *96,* 2335.

35. Ruzicka, F. J.; Beinert, H. *J. Biol. Chem.* **1978,** *253,* 2514.

36. Lauble, H.; Kennedy, M. C.; Beinert, H.; Stout, C. D. *Biochemistry* **1992,** *31,* 2735.

37. Lauble, H.; Kennedy, M. C.; Beinert, H.; Stout, C. D. *J. Mol. Biol.* **1994,** *237,* 437.

38. Flint, D. H.; Allen, R. M. *Chem. Rev.* **1996,** *96,* 2315.

39. (a) Grabowski, R.; Buckel, W. *Eur. J. Biochem.* **1992,** *199,* 89. (b) Emptage, M. A. In *Metal Clusters in Proteins,* Que, L., Ed., Am. Chem. Soc.: Washington, DC, 1988, p. 343.

40. Müh, U.; Cinkaya, I.; Albracht, S. P. J.; Buckel, W. *Biochemistry* **1996,** *35,* 11710.

41. Boucher, J.-L.; Delaforge, M.; Mansuy, D. *Biochemistry* **1994,** *33,* 7811.

42. (a) Haslam, E. *Shikimic Acid Metabolism and Metabolites,* Wiley: New York, 1993. (b) Dewick, P. M. *Nat. Prod. Rep.,* **1988,** *5,* 73.

43. Onderka, D. K.; Floss, H. G. *J. Am. Chem. Soc.* **1969,** *91,* 5894.

44. Hill, R. K.; Bock, M. G. *J. Am. Chem. Soc.,* **1978,** *100,* 637.

45. Bartlett, P. A.; Maitra, U.; Chouinard, P. M. *J. Am. Chem. Soc.* **1986,** *108,* 8068.

46. Balasubramanian, S.; Davies, G. M.; Coggins, J. R.; Abell, C. *J. Am. Chem. Soc.* **1991,** *113,* 8945.

47. Hawkes, T. R.; Lewis, T.; Coggins, J. R.; Mousdale, D. M.; Lowe, D. J.; Thorneley, R. N. F. *Biochem. J.,* **1990,** *265,* 899.

48. Ramjee, M. N.; Coggins, J. R.; Hawkes, T. R.; Lowe, D. J.; Thorneley, R. N. F. *J. Am. Chem. Soc.,* **1991,** *113,* 8566.

49. Bornemann, S.; Lowe, D. J.; Thorneley, R. N. F. *Biochemistry* **1996,** *35,* 9907.

50. Macheroux, P.; Bornemann, S.; Ghisla, S.; Thorneley, R. N. F. *J. Biol. Chem.* **1996,** *271,* 25850.

51. Ramjee, M. N.; Balasubramanian, S.; Abell, C.; Coggins, J. R.; Davies, G. M.; Hawkes, T. R.; Lowe, D. J.; Thorneley, R. N. F. *J. Am. Chem. Soc.,* **1992,** *114,* 3151.

52. Bartlett, P. A.; McLaren, K. L.; Alberg, D. G.; Fassler, A.; Nyfeler, R.; Lauhon, C. T.; Grissom, C. B. *Proc. Soc. Chem. Ind. Pesticides Group Meeting,* BCPC Monograph **1989,** *42,* 155.

53. Giese, B.; Almstead, N. G. *Tetrahedron Lett.,* **1994,** *35,* 1677.

54. Bornemann, S.; Ramjee, M. K.; Balasubramanian, S.; Abell, C.; Coggins, J. R.; Lowe, D. J.; Thorneley, N. F. *J. Biol. Chem.* **1995,** *270,* 22811.

55. Cope, A. C.; Trumbull, E. R. *Org. React.* **1960,** *11,* 317.

56. Hanson, K. R.; Havir, E. A. In *The Enzymes,* vol. 7, 3rd ed., Boyer, P. D., Ed., Academic Press: New York, 1972, pp. 75–166.

57. (a) Furuta, T.; Takahaski, H.; Kasuya, Y. *J. Am. Chem. Soc.* **1990,** *112,* 3633. (b) Furuta, T.; Takahashi, H.; Shibasaki, H.; Kasuya, Y. *J. Biol. Chem.* **1992,** *267,* 12600.

58. (a) Givot, I. L.; Smith, T. A.; Abeles, R. H. *J. Biol. Chem.* **1969,** *244,* 6341. (b) Wickner, R. B. *J. Biol. Chem.* **1969,** *244,* 6550. (c) Hanson, K. R.; Havir, E. A. *Arch. Biochem. Biophys.* **1970,** *141,* 1.

59. Langer, M.; Reck, G.; Reed, J.; Rétey, J. *Biochemistry* **1994,** *33,* 6462.

60. Langer, M.; Lieber, A.; Rétey, J. *Biochemistry* **1994,** *33,* 14034.

60a. Schwede, T. F.; Rétey, J.; Schulz, G. E. *Biochemistry* **1999**, *38*, 5355.

61. Williams, V. R.; Hiroms, J. M. *Biochim. Biophys. Acta* **1967**, *139*, 214.

62. Rétey, J. *Naturwissenschaften* **1996**, *83*, 439.

63. Klee, C. B.; Kirk, K. L.; Cohen, L. A. *Biochem. Biophys. Res. Commun.* **1979**, *87*, 343.

64. Langer, M.; Pauling, A.; Rétey, J. *Angew. Chem. Int. Ed. Engl.* **1995**, *34*, 1464.

65. Hanson, K. R.; Rose, I. A., *Acc. Chem. Res.* **1975**, *8*, 1.

66. (a) Butler, J. R.; Alworth, W. L.; Nugent, M. J. *J. Am. Chem. Soc.* **1974**, *96*, 1617. (b) Kleanthous, C.; Reilly, M.; Copper, A.; Kelly, S.; Price, N. C.; Coggins, J. R. *J. Biol. Chem.* **1991**, *266*, 10893.

67. S. Chaudhuri, S.; Duncan, K.; Graham, L. D.; Coggins, J. R. *Biochem. J.* **1991**, *275*, 1.

68. Shneier, A.; Kleanthous, C.; Deka, R.; Coggins, J. R.; Abell, C. *J. Am. Chem. Soc.* **1991**, *113*, 9416.

68a. Harris, J. M.; Watkins, W. J.; Hawkins, A. R.; Coggins, J. R.; Abell, C. *J. Chem. Soc. Perkin Trans. 1* **1996**, 2371 and Dewick, P. M. *Nat. Products Rep.* **1998**, *15*, 17.

68b. Shneier, A.; Harris, J.; Kleanthous, C.; Coggins, J. R.; Hawkins, A. R.; Abell, C. *Bioorg. Med. Chem. Lett.* **1993**, *3*, 1399 and Harris, J.; Kleanthous, C.; Coggins, J. R.; Hawkins, A. R.; Abell, C. *J. Chem. Soc. Chem. Commun.* **1993**, 1080.

69. Miles, E. W. In *Vitamin B$_6$ Pyridoxal Phosphate*, Part B, Dolphin, D.; Poulson, R.; Avramovic, O., Eds., Wiley: New York, 1986, p. 253.

70. Chaffotte, A. F.; Zakin, M. M.; Goldberg, M. E. *Biochem. Biophys. Res. Commun.* **1980**, *92*, 381.

71. Deeley, M. C.; Yanofsky, C. *J. Bacteriol.* **1981**, *147*, 787.

72. Crawford, I. P.; Nichols, B. P.; Yanofsky, C. *J. Mol. Biol.* **1980**, *142*, 489.

73. Lane, A. N.; Kirschner, K. *Eur. J. Biochem.* **1981**, *120*, 379.

74. (a) Miles, E. W.; Hatanaka, M.; Crawford, I. P. *Biochemistry* **1968**, *7*, 2742. (b) Lane, A. N.; Kirschner, K. *Eur. J. Biochem.* **1983**, *129*, 561. (c) Lane, A. N.; Kirschner, K. *Eur. J. Biochem.* **1983**, *129*, 571.

75. June, D. S.; Suelter, C. H.; Dye, J. L. *Biochemistry* **1981**, *20*, 2714.

76. Miles, E. W.; Houck, D. R.; Floss, H. G. *J. Biol. Chem.* **1982**, *257*, 14203.

77. Palcic, M. M.; Floss, H. G. In *Vitamin B$_6$ Pyridoxal Phosphate*, Part A, Dolphin, D.; Poulson, R.; Avramovic, O., Eds.; Wiley: New York, 1986, p. 25.

78. McDowell, L. M.; Lee, M.; Schaefer, J.; Anderson, K. S. *J. Am. Chem. Soc.* **1995**, *117*, 12352.

79. Tsai, M.-D.; Schleicher, E.; Potts, R.; Skye, G. E.; Floss, H. G. *J. Biol. Chem.* **1978**, *253*, 5344.

80. (a) Miles, E. W.; Kumagai, H. *J. Biol. Chem.* **1974**, *249*, 2483. (b) Miles, E. W.; McPhie, P. *J. Biol. Chem.* **1974**, *249*, 2492.

81. (a) Phillips, R.S. *J. Am. Chem,. Soc.* **1989**, *111*, 727. (b) Phillips, R.S. *Biochemistry* **1991**, *30*, 5927.

82. (a) Schleicher, E.; Mascaro, K.; Potts, R.; Mann, D. R.; Floss, H. G. *J. Am. Chem. Soc.* **1976**, *98*, 1043. (b) Vederas, J. C.; Schleicher, E.; Tsai, M.-D.; Floss, H. G. *J. Biol. Chem.* **1978**, *253*, 5350.

83. Phillips, R. S.; Miles, E. W.; Cohen, L. A. *J. Biol. Chem.* **1985**, *260*, 14665.

84. Posner, B. I.; Flavin, M. *J. Biol. Chem.* **1972**, *247*, 6412.

85. (a) Hansen, P. E.; Feeney, J.; Roberts, G. C.K. *J. Magn. Res.* **1975**, *17*, 249. (b) Fuganti, C.; Coggiola, D. *Experientia* **1977**, *33*, 847.

86. Washtien, W.; Cooper, A. J. Abeles, R. H. *Biochemistry* **1977**, *16*, 460.

87. (a) Chang, M. N. T.; Walsh, C. *J. Am. Chem. Soc.* **1980**, *102*, 2499. (b) Chang, M. N. T.; Walsh, C. *J. Am. Chem. Soc.* **1980**, *102*, 7368. (c) Chang, M. N. T.; Walsh, C. *J. Am. Chem. Soc.* **1981**, *103*, 4921.

88. Abeles, R. H.; Walsh, C. T. *J. Am. Chem. Soc.* **1973**, *95*, 6124.

89. Washtien, W.; Abeles, R. H. *Biochemistry* **1977**, *16*, 2485.

90. Silverman, R. B.; Abeles, R. H. *Biochemistry* **1976**, *15*, 4718.

91. Silverman, R. B.; Abeles, R. H. *Biochemistry* **1977**, *16*, 5515.

92. Fearon, C. W.; Rodkey, J. A.; Abeles, R. H. *Biochemistry* **1982**, *21*, 3790.

93. Clausen, T.; Huber, R.; Laber, B.; Pohlenz, H.-D.; Messerschmidt, A. *J. Mol. Biol.* **1996**, *262*, 202.

94. Grabowski, R.; Hofmeister, A. E. M.; Buckel, W. *TIBS,* **1993,** *18,* 297.

95. Rubenstein, P. A.; Strominger, J. L. *J. Biol. Chem.* **1974,** *249,* 3776.

96. Weigel, T. M.; Miller, V. P.; Liu, H.-w. *Biochemistry* **1992,** *31,* 2140.

97. Lei, Y.; Ploux, O.; Liu, H.-w. *Biochemistry* **1995,** *34,* 4643.

98. Lundblad, R. L. *Techniques in Protein Modification,* CRC Press: Boca Raton, 1995.

99. Dunathan, H. C. *Adv. Enzymol. Rel. Areas Mol Biol.* **1971,** *35,* 79.

100. Cravatt, B. F.; Ashley, J. A.; Janda, K. D.; Boger, D. L.; Lerner, R. A. *J. Am. Chem. Soc.* **1994,** *116,* 6013.

Aldol and Claisen Reactions and Retroreactions

I. ALDOL REACTIONS

A. General Chemical Considerations

One of the most useful chemical methods for C–C bond formation is the *aldol reaction* (or condensation),[1] in which the α-carbon of one aldehyde or ketone (**11.1**) adds to the carbonyl carbon of another aldehyde or ketone to give a β-hydroxy aldehyde or ketone (**11.2**), an aldol (Scheme 11.1). When two different aldehydes or ketones are used in solution, and both have α-protons that can be removed, a mixture of all four possible condensation reactions generally results. Furthermore, each reaction can produce all of the possible isomers derived from that reaction. If only one of the compounds has an α-carbon, then only that one can be the donor, but the corresponding enolate can add to the carbonyl group of either compound and with both stereochemistries; the compound with the more reactive carbonyl (aldehydes are generally more reactive than ketones), is the preferential acceptor molecule. Unlike aldol reactions in solution, enzyme-catalyzed aldol reactions generally give only one condensation product stereospecifically.

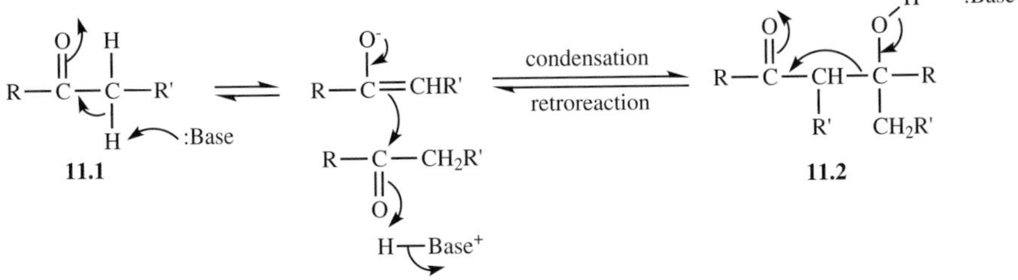

SCHEME 11.1   Generalized aldol reaction.

B. Aldolases

By the early part of the 20th century it was known that there were enzymes capable of catalyzing the reversible formation of hexoses from their three-carbon components, presumably by an aldol-type reaction.[2] Originally, they were referred to as *zymohexases,* but enzymes that catalyze an aldol reaction are now called *aldolases.* These enzymes can be found in all organisms, and are mainly involved in the metabolism of carbohydrates, although they also metabolize some amino acids and hydroxy acids. The majority of the aldolases catalyze the reversible stereospecific addition of a ketone donor to an aldehyde acceptor. As noted, when two aldehydes or ketones are mixed in basic media, a mixture of all four possible aldol reactions is found. Enzymatically, however, there can be two different adjacent binding sites to which only the two different substrate molecules bind, so that a stereospecific condensation with two different molecules is possible. If the active-site base is oriented so that only the donor molecule's α-carbon is nearby, then only one of the possible four different aldol reactions will occur, and only one stereochemistry will be obtained. Stereochemical studies with aldolases show that the electrophilic carbonyl component always adds to the enol carbon (or, as shown later, the enamine) from which the proton was abstracted, with retention of configuration (i.e., the C–C bond replaces the C–H bond that is broken).

Two different classifications of aldolases depend on the type of mechanism that the enzyme catalyzes. *Type I aldolases* have an essential active-site lysine residue that is involved in Schiff base formation with the donor substrate (Scheme 11.2, pathway a). *Type II aldolases* require a metal ion, generally Zn^{2+}, as a cofactor for activity (Scheme 11.2, pathway b). The type I aldolases are primarily found in animals and in higher plants, and type II aldolases are found in bacteria and fungi. Both types of enzymes are specific for the donor molecules utilized as substrates but are less specific about the structures that will act as acceptor substrates.

We have already seen an example of an enzyme-catalyzed aldol-like reaction, namely, the thiamin diphosphate (TDP)-dependent acetolactate synthase reaction (Chapter 8, Section IV.C.1.a). In that case, the TDP generates a stabilized carbanion that reacts with a second molecule of pyruvate, a modification of the aldol reaction. That enzyme, though, is generally not considered an aldolase.

1. Fructose 1,6-Diphosphate Aldolase

Fructose 1,6-diphosphate aldolase catalyzes the reversible condensation of dihydroxyacetone phosphate (**11.3**, DHAP) with D-glyceraldehyde 3-phosphate (**11.4**, G3P) to give D-fructose 1,6-diphosphate (**11.5**, FDP) with $(3S,4R)$ stereochemistry (Scheme 11.3 gives the reaction in two stereoviews, Fischer projections and line drawings).[3] The equilibrium constant is $10^4\,M^{-1}$ in favor of formation of **11.5**. A related enzyme, tagatose 1,6-diphosphate aldolase, catalyzes the same reaction

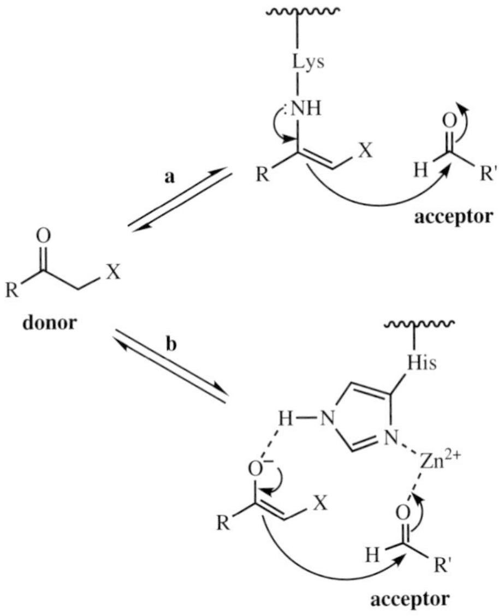

SCHEME 11.2 General mechanisms for type I (pathway a) and type II (pathway b) aldolases.

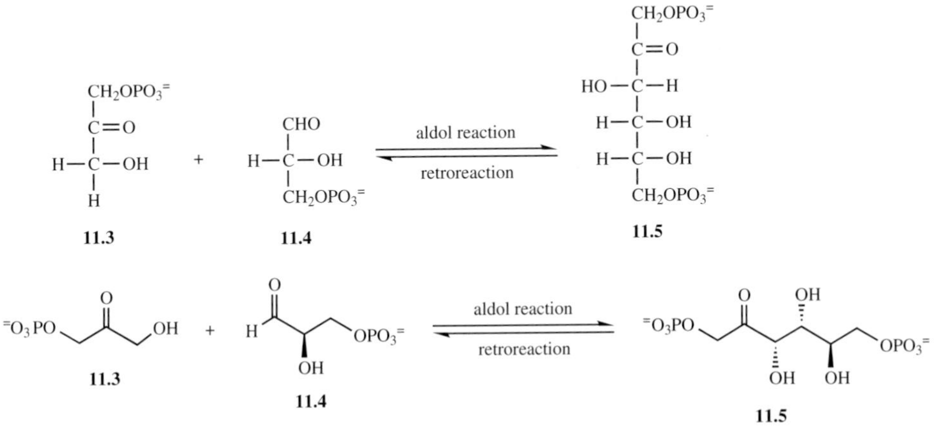

SCHEME 11.3 Reaction catalyzed by fructose 1,6-diphosphate aldolase.

with the same substrates, but produces tagatose 1,6-diphosphate (**11.6**), which is **11.5** except with (3S,4S) stereochemistry. X-ray crystal structures of FDP aldolases from rabbit muscle at 2.7 Å resolution,[4] human muscle at 3 Å resolution,[5] and *Drosophila melanogaster* at 2.5 Å resolution[6] are similar. Both type I and type II FDP aldolases are known.[7] The sequences of the type I enzymes have a high degree of homology (>50%), and the active-site sequence is conserved throughout the evo-

$$CH_2OPO_3^=$$
$$|$$
$$C=O$$
$$|$$
$$HO-C-H$$
$$|$$
$$HO-C-H$$
$$|$$
$$H-C-OH$$
$$|$$
$$CH_2OPO_3^=$$

11.6

$$CH_2OPO_3^=$$
$$|$$
$$C=O$$
$$|$$
$$D-C-OH$$
$$|$$
$$H$$

11.7

11.7

lutionary chain.[8] The rabbit muscle enzyme (type I) and the *Escherichia coli* enzyme (type II) have only a small degree of primary sequence homology, yet they have the same substrate specificity.[9]

The mechanism of FDP aldolase has been elucidated by a variety of kinetic and isotopic experiments. When the rabbit muscle enzyme is incubated with DHAP in D_2O, only one deuterium is incorporated, and it is found at the *pro-3S* position (**11.7**).[10] Therefore, enolization of DHAP can occur in the absence of the second substrate, G3P.

If FDP is incubated with [14C]G3P and enzyme, and the rate of incorporation of [14C] into the FDP is measured, and then the same is done with [14C]DHAP, the time courses for incorporation of [14C] into FDP are different.[11] The rate of incorporation into FDP is slower starting from [14C]DHAP than from [14C]G3P (Figure 11.1). This indicates an ordered product release (see Appendix I, Section VI.A.2), in which G3P is released faster than DHAP (for the direction in which FDP is converted to G3P and DHAP), so that the G3P can exchange with [14C]G3P and bind back to the enzyme before the DHAP has been released from the last turnover (Scheme 11.4). According to the principle of microscopic reversibility, then, because DHAP is released last in the retroreaction, it must bind first in the forward reaction (formation of FDP from G3P and DHAP). This sequence is consistent with the results, which showed that DHAP exchanges with solvent in the absence of G3P.

Type I aldolases utilize an active-site lysine residue to catalyze their reactions. Evidence for Schiff base formation can be obtained by incubation of [14C]DHAP

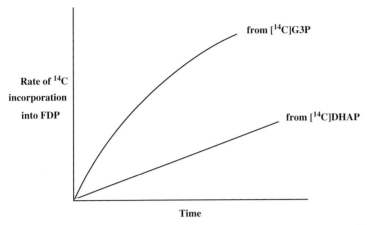

FIGURE 11.1   Rate of incorporation of ^{14}C into fructose 1,6-diphosphate (FDR) from [^{14}C]glyceraldehyde 3-phosphate (G3P) and [^{14}C]dihydroxyacetone phosphate (DHAP) catalyzed by fructose 1,6-diphosphate aldolase.

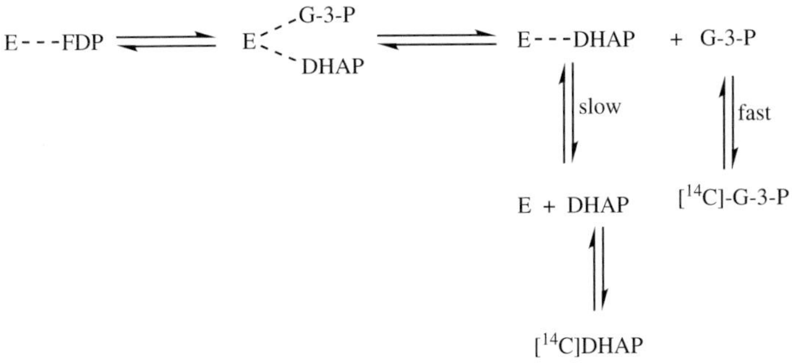

SCHEME 11.4   Ordered reaction of fructose 1,6-diphosphate aldolase.

and FDP aldolase followed by treatment with NaBH$_4$; this procedure produces inactive enzyme with 1 equivalent of ^{14}C bound.[12] After acid hydrolysis of the labeled enzyme, **11.8** (Scheme 11.5) can be isolated. Similarly, if [^{14}C]FDP is incubated with the enzyme followed by NaBH$_4$ reduction, inactivation results with incorporation of radioactivity at a lysine residue. However, if the enzyme is incubated with G3P and reduced with NaBH$_4$, no inactivation or radioactive incorporation is observed. As discussed earlier, DHAP binds first; without the DHAP, the G3P does not bind to the enzyme. Also consistent with a Schiff base mechanism is the release of [18]O from FDP or DHAP labeled in the carbonyl with an [18]O atom following incubation with aldolase; the corresponding experiment with just [^{18}O]G3P results in no loss of [18]O. These types of experiments were mentioned before, when we discussed evidence for a Schiff base mechanism in the reaction catalyzed by acetoacetate decarboxylase (Chapter 8, Section II.A.1). In that reaction a Schiff

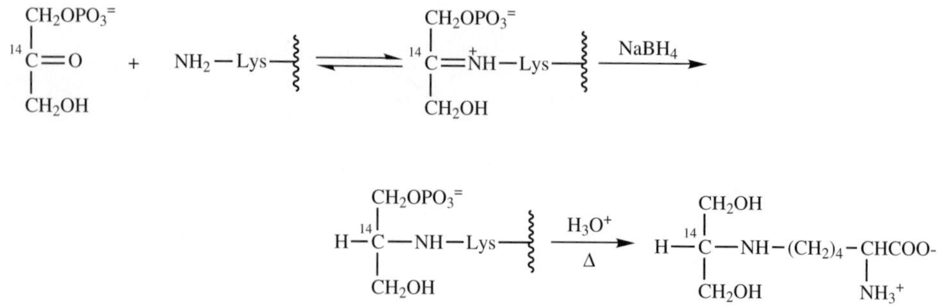

$$
\overset{14}{C}=O \quad + \quad NH_2-Lys-\\
\text{(with } CH_2OPO_3^= \text{ above and } CH_2OH \text{ below)}
$$

SCHEME 11.5   Evidence for the involvement of an active-site lysine residue in the reaction catalyzed by fructose 1,6-diphosphate aldolase.

base mechanism was utilized to activate the substrate for decarboxylation and give a stabilized carbanion. Because the stabilization of electrons following a decarboxylation is equivalent to stabilization of electrons after a deprotonation, a Schiff base mechanism would be a reasonable approach to produce a stabilized carbanion by deprotonation as well as to produce a stabilized carbanion by decarboxylation. That is what is done with the type I aldolases (Scheme 11.6). Following Schiff base formation with the active-site lysine residue, the *pro-3S* proton of DHAP is removed, and the resultant enamine attacks the *si*-face of G3P to give the (3S,4R) stereochemistry (retention of configuration at C-3). For rabbit muscle aldolase, Lubini and Christen found that Lys-229 forms the Schiff base with DHAP.[13]

    If the reaction conditions are changed, the diastereoselectivity of aldolases can change. Under kinetically controlled conditions, Whitesides and co-workers found that D-G3P is accepted as a substrate for FDP aldolase with a 20:1 preference over L-G3P.[14] Under thermodynamically controlled conditions, differentiation of enantiomers by FDP aldolase also occurs. With modified aldehyde substrates (**11.9**), when the aldol product can cyclize to form pyranoses, Durrwachter and Wong found that the product with the least 1,3-diaxial interactions predominates after equilibration because of the reversibility of the reaction (Scheme 11.7).[15]

    Several antibodies raised by Lerner, Barbas, and co-workers catalyze aldolase-type reactions via an enamine mechanism with a Lys residue at the active site.[16] Two of the antibodies are broad specificity abzymes that catalyze ketone–ketone, ketone–aldehyde, aldehyde–ketone, and aldehyde–aldehyde aldol reactions; these catalysts should be useful synthetic tools for stereospecific aldol reactions.[17]

C. Porphobilinogen Synthase
(5-Aminolevulinate Dehydratase)

Porphobilinogen synthase (PBGS), a metalloenzyme that catalyzes the asymmetric condensation of two molecules of 5-aminolevulinic acid (**11.10**) to give porphobilinogen (**11.11,** Scheme 11.8), is the first common step in the biosynthesis of

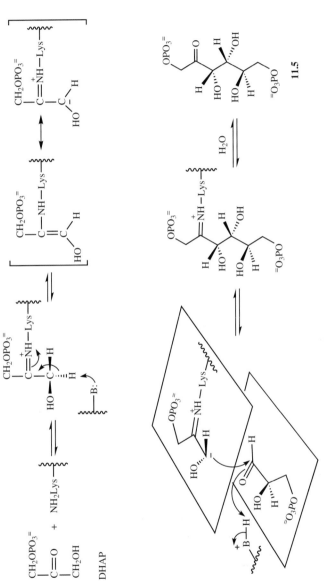

SCHEME 11.6   Overall proposed mechanism and stereochemistry for fructose 1,6–diphosphate aldolase.

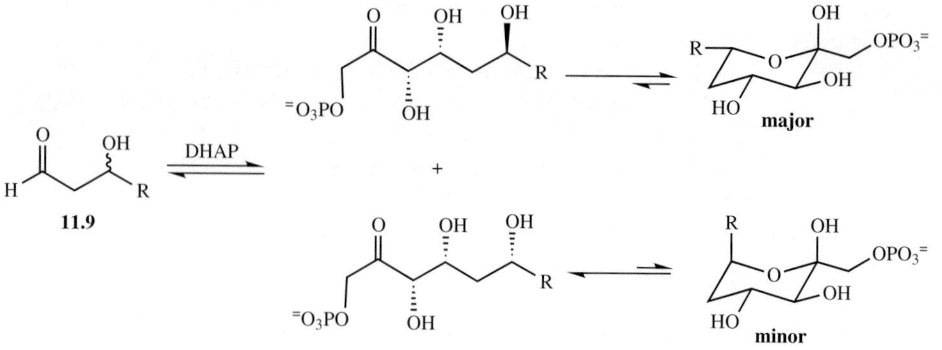

SCHEME 11.7 Substituted pyranose syntheses catalyzed by fructose 1,6-diphosphate aldolase.

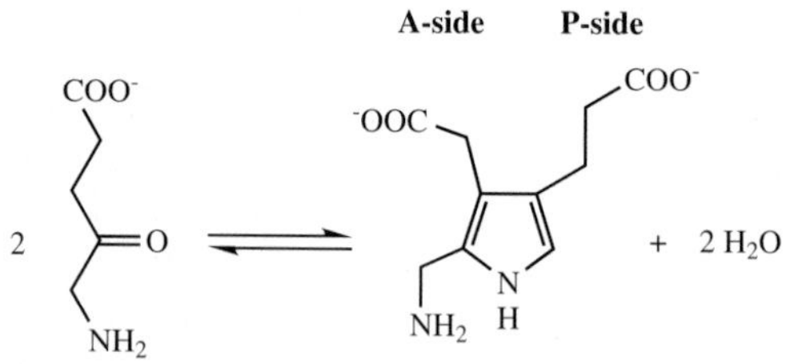

SCHEME 11.8 Reaction catalyzed by porphobilinogen synthase.

tetrapyrroles, such as heme and vitamin B_{12}. Although the two substrate molecules are the same molecule, they react in different ways and are differentiated once bound to the active site. One of the 5-aminolevulinic acid molecules is referred to as the A-side **11.10** (for the one that produces the acetyl half of the porphobilinogen), and the other is called the P-side **11.10** (for the one that produces the propionyl half of **11.11**).[18]

A comparison of the sequences of porphobilinogen synthases from a wide variety of sources indicates a highly conserved protein, which suggests that the mechanism is likely to be the same for the enzyme from all organisms. Mammalian PBGS contains two Zn^{2+} ions at each active site; bacterial and plant PBGS contain a third metal ion, Mg^{2+}, used as an allosteric activator (see Appendix I, Section VII.B). Plant and some bacterial PBGS use Mg^{2+} in place of one or both of the Zn^{2+} ions. The Zn^{2+}-utilizing forms of the enzyme also require a thiol reducing agent, presumably to prevent formation of a disulfide (cystine) linkage, which would prohibit zinc binding. Once the enzyme is activated, it can be inactivated with sulfhydryl reagents. Inactivation by iodoacetate leads to the incorporation of one inactivator molecule (i.e., one sulfhydryl) per monomer, and inactivation by iodoacetamide also leads to the incorporation of one inactivator molecule per monomer.[19] However, the two reagents label two different cysteine residues! Tryptic digestion gives two different labeled peptides.

The P-side **11.10** binds first and forms a Schiff base with Lys-252 of the mammalian enzyme.[20] Jaffe and co-workers have used ^{13}C- and ^{15}N NMR spectroscopy to determine the tautomeric structure, the stereochemistry, and the protonation state of the P-side Schiff base intermediate.[21] All of the ^{13}C NMR data support an imine with E-stereochemistry, and the ^{15}N NMR results indicate a deprotonated amino group. Other evidence for the formation of a Schiff base is that incubation of 5-[^{14}C]aminolevulinate with the enzyme and $NaBH_4$ produces an inactive covalent radioactive protein.

In the case of mammalian PBGS, a catalytically essential Zn^{2+} ion must bind prior to the A-side **11.10**;[22] this ion is known as Zn_A.[23] Four Zn_A ions are required for catalysis;[24] each Zn_A contains five ligands, of which only one is a cysteine (Cys-223), one is water; two more come from three possibilities, namely, a Tyr, a His, and an Asp residue; and the fifth ligand is the substrate. Mammalian PBGS also contains four Zn_B ions, which are tetrahedrally coordinated to four cysteine residues (Cys-119, -122, -124, -132). 5-Chlorolevulinic acid (**11.10** with a chlorine in place of the amino group) is a reactive substrate analogue shown by Seehra and Jordan to inactivate mammalian PBGS.[25] This inactivator labels Cys-223, which then causes a loss in the Zn_A binding and implicates that residue as the one cysteine that binds to Zn_A.[26] Because Zn_A binding is lost under these circumstances, PBGS cannot catalyze the formation of the first bond between the P-side **11.10**-Schiff base and the A-side **11.10**. Zn_A most likely coordinates to the A-side **11.10** carbonyl either to promote the reaction with the P-side **11.10** amino group (Scheme 11.9) or to induce enolization at C-3 of the A-side **11.10** (Scheme 11.10). In Scheme 11.9 the first step after Schiff base formation between Lys-252 and the P-side **11.10** is Schiff base formation between the P-side **11.10** amino group and the carbonyl of the Zn_A-bound A-side **11.10** (i.e., **11.12**) to give **11.13**. That promotes removal of the C-3 proton of the A-side **11.10** to give an enamine (**11.14**), which should readily undergo an aldol reaction with the P-side **11.10** iminium ion and make the 5-membered ring (**11.15**). Base-catalyzed elimination of Lys-252, followed by stereospecific *pro-R* deprotonation[27] gives **11.11**. The alternative mechanism (Scheme 11.10) shows the A-side base removing the C-3 proton from the A-side **11.10** to give the enolate (**11.16**), which can undergo an aldol reaction to give **11.17**. Schiff base formation between the P-side **11.10** amino group and the A-side **11.10** carbonyl gives **11.15,** the same intermediate in Scheme 11.9. This scheme depicts imine formation followed by aldol condensation, and Scheme 11.10 is aldol condensation followed by imine formation. Both mechanisms have precedents in synthetic pyrrole chemistry.

Photoinactivation of PBGS with methylene blue and treatment with diethyl pyrocarbonate inactivate the enzyme with loss of one histidine residue. To determine if active-site histidine residues are involved in the tautomerization steps, Jaffe and co-workers made three *Escherichia coli* mutants, H128A, H126A, and the double

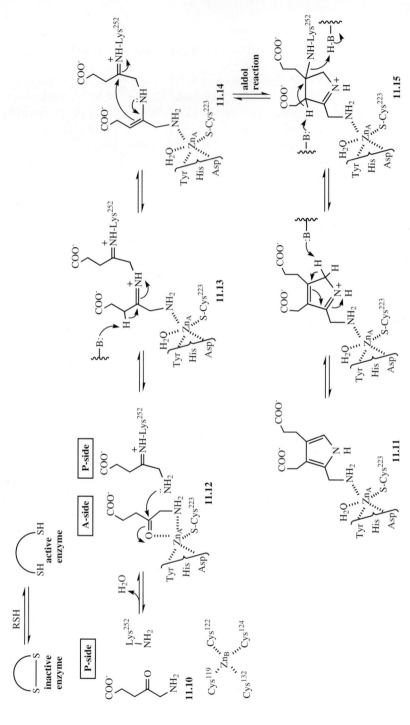

SCHEME 11.9   Proposed mechanism for the reaction catalyzed by porphobilinogen synthase.

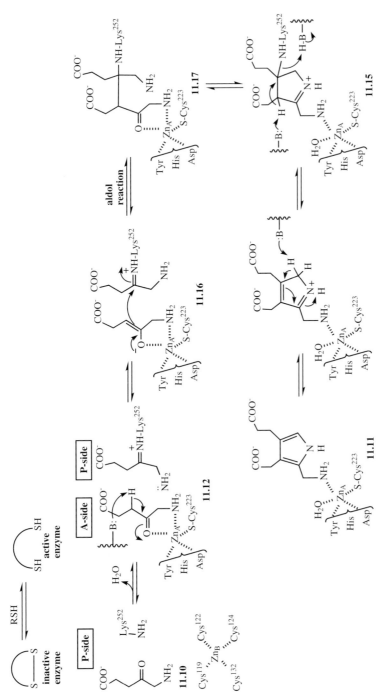

SCHEME 11.10 Alternative proposed mechanism for the reaction catalyzed by porphobilinogen synthase.

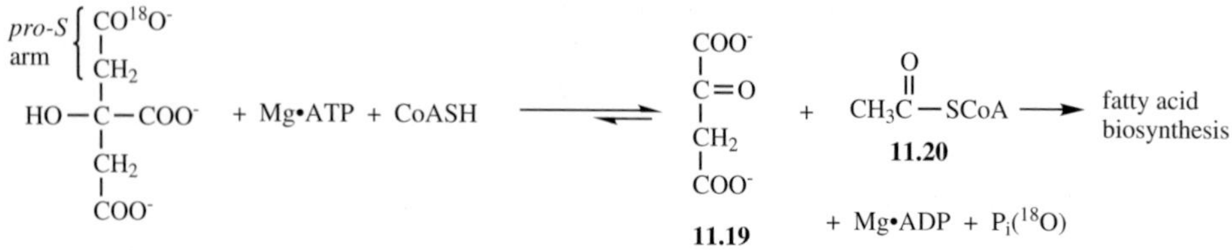

$$\text{pro-S arm} \left\{ \begin{array}{l} CO^{18}O^- \\ | \\ CH_2 \\ | \\ HO-C-COO^- \\ | \\ CH_2 \\ | \\ COO^- \end{array} \right. + Mg{\cdot}ATP + CoASH \rightleftharpoons \begin{array}{l} COO^- \\ | \\ C=O \\ | \\ CH_2 \\ | \\ COO^- \end{array} + CH_3\overset{O}{\overset{||}{C}}-SCoA \longrightarrow \begin{array}{l}\text{fatty acid}\\\text{biosynthesis}\end{array}$$

11.18   **11.20**

11.19 $+ Mg{\cdot}ADP + P_i(^{18}O)$

SCHEME 11.11   Reaction catalyzed by ATP citrate-lyase.

mutant, H126,128A.[28] All were similar in activity to the wild-type enzyme, so H126 and H128 are not involved in the proton transfers. Possibly cysteine residues are involved.

D. ATP Citrate-Lyase

ATP citrate-lyase, one of the principal enzymes in both fatty acid and cholesterol biosynthetic pathways, catalyzes the reaction of citrate (**11.18**) and CoASH to give oxaloacetate (**11.19**) and acetyl-CoA (**11.20**) with the concomitant conversion of ATP into ADP and inorganic phosphate (Scheme 11.11).[29] This reaction is the main supplier of acetyl-CoA for biosynthetic reactions. In Chapter 2 (Section III.C) we saw that acetyl-CoA also is produced by the enzyme acetyl-CoA synthetase from acetate, ATP, and CoASH. Isotope exchange[30] and steady-state kinetic studies[31] support a double displacement mechanism with a phosphoryl enzyme intermediate. As shown in Scheme 11.11, if the *pro-S* carboxylate in citrate is labeled with ^{18}O, the ^{18}O ends up in the inorganic phosphate, supporting phosphorylation of that carboxylate by ATP. In fact, (S)-citrate 1-phosphate (**11.21**) can be substituted for citrate and ATP in the reaction to make acetyl-CoA.[32] Likewise, citryl-CoA (**11.22**) is converted to products in the absence of citrate, ATP, and CoASH.[33] The reaction appears to involve more than two intermediates, because there is an exchange between [^{14}C]ADP and ATP in the absence of all other substrates. That is consistent with the formation of a phosphoryl enzyme (Scheme 11.12). A phosphoryl enzyme has been isolated using [^{32}P]ATP and rapid gel filtration (Scheme 11.13).[34] By ^{31}P NMR spectroscopy Bridger and co-workers showed that the

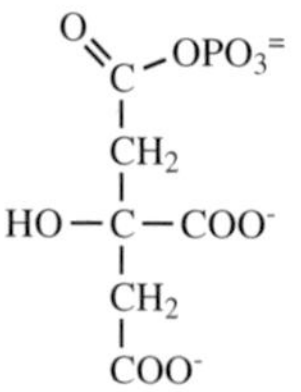

$$\begin{array}{l} \overset{O}{\overset{\diagdown}{C}}-OPO_3^= \\ | \\ CH_2 \\ | \\ HO-C-COO^- \\ | \\ CH_2 \\ | \\ COO^- \end{array}$$

11.21

$$\begin{array}{l} \overset{O}{\overset{\diagdown}{C}}-SCoA \\ | \\ CH_2 \\ | \\ HO-C-COO^- \\ | \\ CH_2 \\ | \\ COO^- \end{array}$$

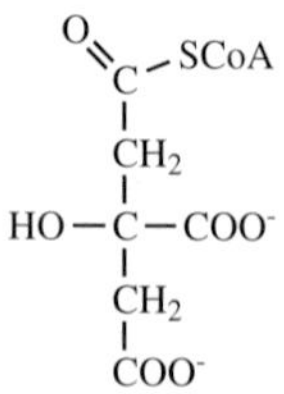

11.22

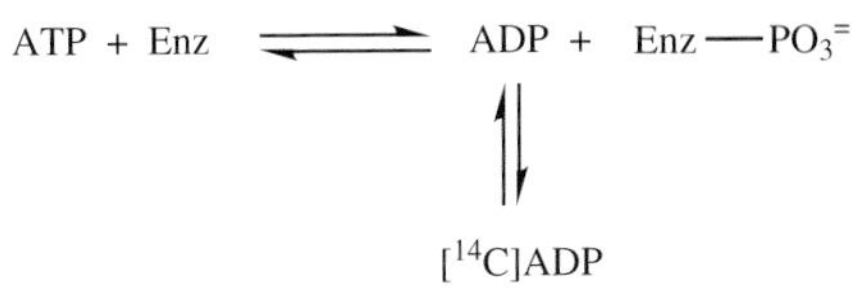

$$ATP + Enz \; \xrightleftharpoons \; ADP + \; Enz\!-\!PO_3^=$$

$$\Updownarrow$$

$$[^{14}C]ADP$$

SCHEME 11.12   ATP phosphorylation of ATP citrate-lyase and exchange with $[^{14}C]$ADP.

$$[\gamma\text{-}^{32}P]ATP + Enz \; \xrightleftharpoons \; [^{32}P]\text{-phosphoenzyme} + ADP$$

SCHEME 11.13   ATP phosphorylation of ATP citrate-lyase.

phosphorylated residue is a histidine; treatment of the phosphoryl enzyme with citrate and CoASH causes the resonance in the spectrum to disappear.[35] When $[^{14}C]$citrate is added to the phosphoryl enzyme, the $[^{14}C]$ appears to become attached to the enzyme (as detected by gel filtration, which indicates that the ^{14}C comigrates with the protein), and the $[^{32}P]PO_4^{3-}$ is released, which supports the formation of a citryl enzyme intermediate. Addition of CoASH to the citryl enzyme gives acetyl-CoA and oxaloacetate. A mechanism consistent with all of these results is shown in Scheme 11.14. There also is a $[^{3}H]$CoASH exchange with acetyl-CoA, suggesting the possible formation of an acetyl enzyme intermediate as well, which could arise from an alternative pathway from a citryl enzyme (Scheme 11.15).[36] However, Wells believes that the evidence cited for the formation of a covalently bound citryl enzyme does not necessarily support that conclusion.[37] When the enzyme is incubated with ATP and $[^{14}C]$citrate in the absence of CoASH, two distinct types of citrate-containing enzyme complexes can be isolated. Early in the enzyme reaction, a highly unstable complex can be isolated by gel filtration, but not by acid precipitation or SDS denaturation, having a half-life of 36 s at 25 °C; this suggests that the complex formed may be a noncovalent one. The addition of CoASH causes a rapid reaction to give **11.22,** as earlier observed. A citryl enzyme adduct could not be isolated under these conditions, which could mean that the citrate-containing enzyme complex is simply citryl phosphate bound tightly, but noncovalently, to the enzyme. The released phosphate that was earlier observed when citrate was added to the phosphoryl enzyme could arise from release of citryl phosphate followed by its (nonenzymatic) hydrolysis. If that is so, then the rate of release of citryl phosphate from the enzyme should be faster than the rate of release of phosphate ion from the phosphoryl enzyme when citrate is added. In fact, the release of citryl phosphate occurs about 15 times faster than the appearance of phosphate ion in solution. A second $[^{14}C]$citrate-containing enzyme complex that forms more slowly than this unstable complex can be isolated by acid precipitation and can be shown to be a nonspecific citrylated enzyme, which does not react with CoASH. A similar covalent complex was formed when the protein cytochrome c

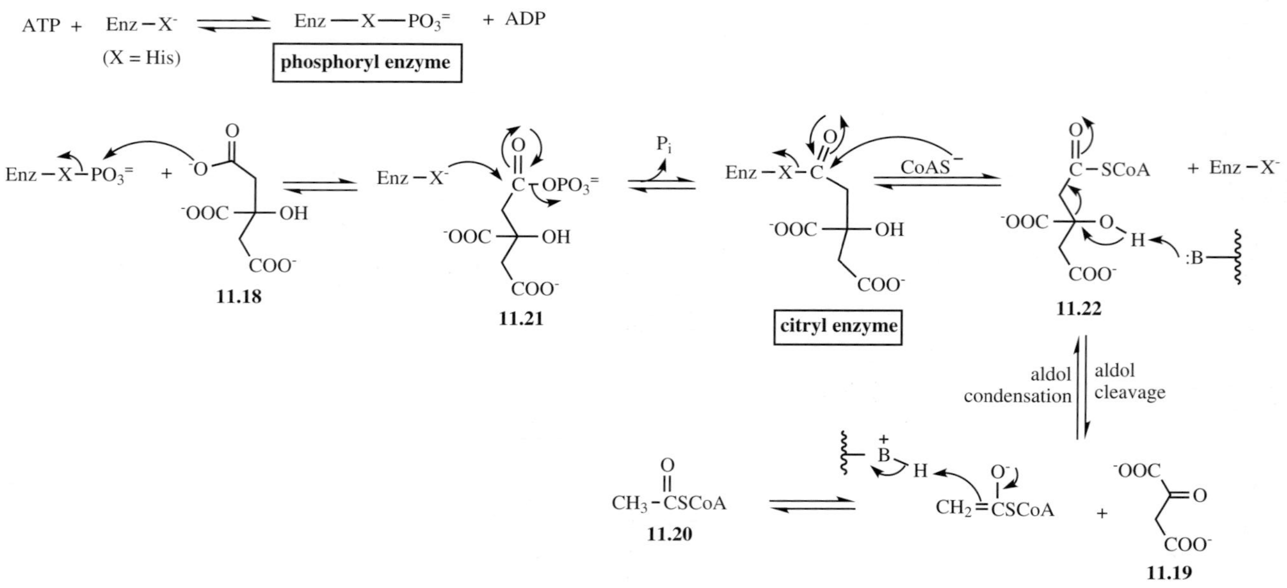

SCHEME 11.14 Mechanism proposed for the reaction catalyzed by ATP citrate-lyase.

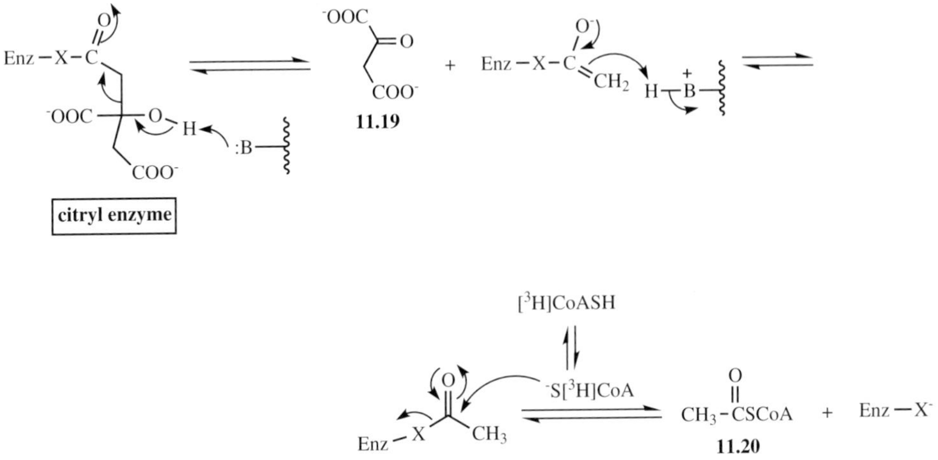

SCHEME 11.15 Possible modification of the mechanism proposed for the reaction catalyzed by ATP citrate-lyase.

was added to the incubation mixture, confirming the random nature of this citrylating activity. This indicates that a covalent citrylated enzyme may not be relevant to the catalytic mechanism and that CoASH may react directly with citryl phosphate (**11.21**) to give citryl-CoA (**11.22**) (see Scheme 11.14).

E. Transketolase

Transketolase catalyzes an important step in the pentose phosphate pathway during glucose metabolism, namely, the reversible transfer of a two-carbon ketol fragment from the C_5 ketose, D-xylulose 5-phosphate (**11.23**, Scheme 11.16), to a C_5 aldose, D-ribose 5-phosphate (**11.24**), to give the C_7 ketose, D-sedoheptulose 7-phosphate (**11.25**) and the C_3 aldose, glyceraldehyde 3-phosphate (**11.26**). Thiamin

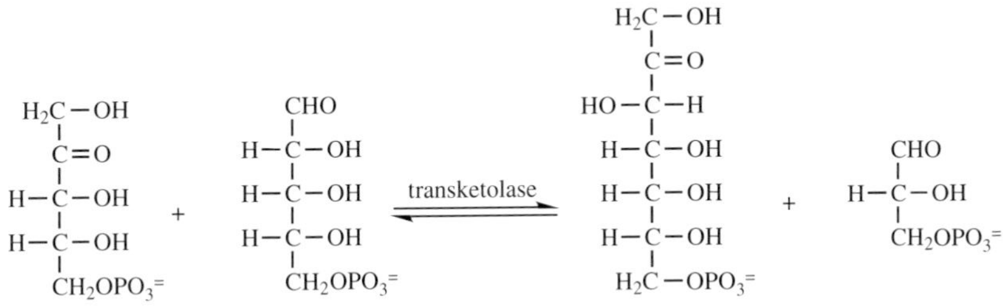

SCHEME 11.16 Reaction catalyzed by transketolase.

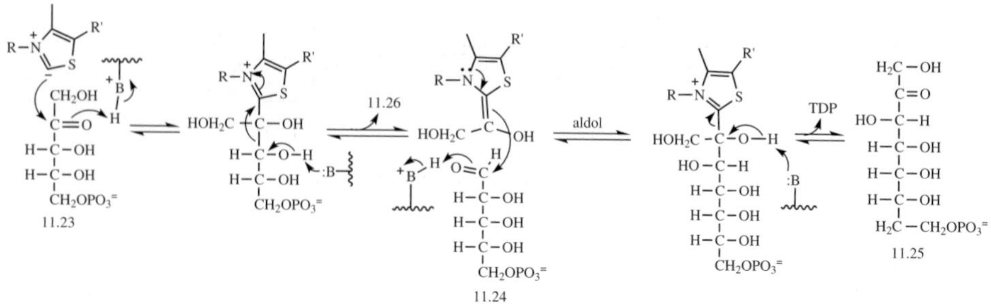

SCHEME 11.17   Mechanism proposed for transketolase.

diphosphate (TDP) and Mg^{2+} are required for activity, although other divalent cations, such as Ca^{2+}, Mn^{2+}, and Co^{2+} can replace Mg^{2+}.[38] As in the case of enzymes that catalyze α-keto acid decarboxylation (Chapter 8, Section IV), the TDP provides an electron sink for carbanion stabilization; this newly generated carbanion can, then, undergo an aldol condensation with **11.24** (Scheme 11.17). The crystal structure of the yeast enzyme at 2.5 Å resolution shows that the active-site thiamin diphosphate is located at the interface between the two identical subunits of the active dimer.[39] The divalent cation binds to the two oxygen anions of the diphosphate group of TDP, apparently assisting in holding the TDP in the proper orientation for reaction. The thiazolium ring binds to residues from both subunits. The mechanism for deprotonation of thiamin diphosphate was discussed in Chapter 8 (Section IV.B).

The enzymes from yeast[40] and *Escherichia coli*[41] are relatively nonspecific for the aldehyde substrate and have been used as synthetic catalysts. Condensation of hydroxypyruvate (**11.27**) as the ketol donor with a variety of 2-hydroxyaldehyde acceptors (**11.28**) produces a series of trihydroxylated compounds having the D-*threo* configuration (**11.29**) (Scheme 11.18).

F. Dehydroquinate Synthase

Aromatic amino acids in plants and microorganisms are biosynthesized by enzymes of the shikimate pathway.[42] Dehydroquinate synthase, the second enzyme in this

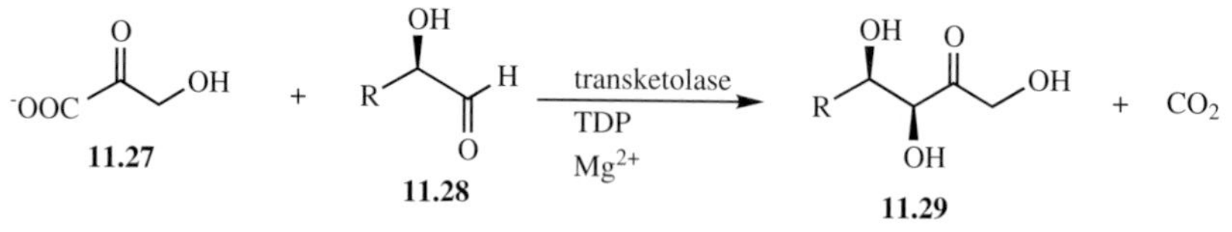

SCHEME 11.18   Synthesis of substituted D-*threo*-trihydroxylated ketones catalyzed by transketolase.

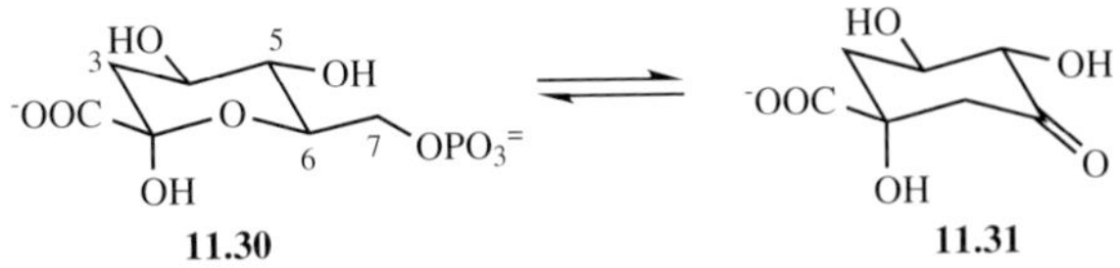

SCHEME 11.19   Reaction catalyzed by dehydroquinate synthase.

pathway, catalyzes an unusual reaction, the conversion of 3-deoxy-D-*arabino*-hep-
tulosonate 7-phosphate (**11.30**) to dehydroquinate (**11.31**) (Scheme 11.19). The
enzyme requires NAD$^+$, even though the product is in the same oxidation state
as the substrate. On the basis of the overall stereochemical course of the reaction,[43]
primary tritium isotope effects,[44] and the use of substrate analogues,[45] the mechanism, originally proposed by Sprinson and co-workers,[46] was supported (Scheme
11.20). The function of the NAD$^+$, then is to oxidize **11.30** to **11.32** so that the
C-6 proton becomes acidic enough for deprotonation and elimination of the phosphate group, giving **11.33**. Once that has been accomplished, the ketone can be
reduced by the NADH that was just generated in the last step to yield **11.34,** which

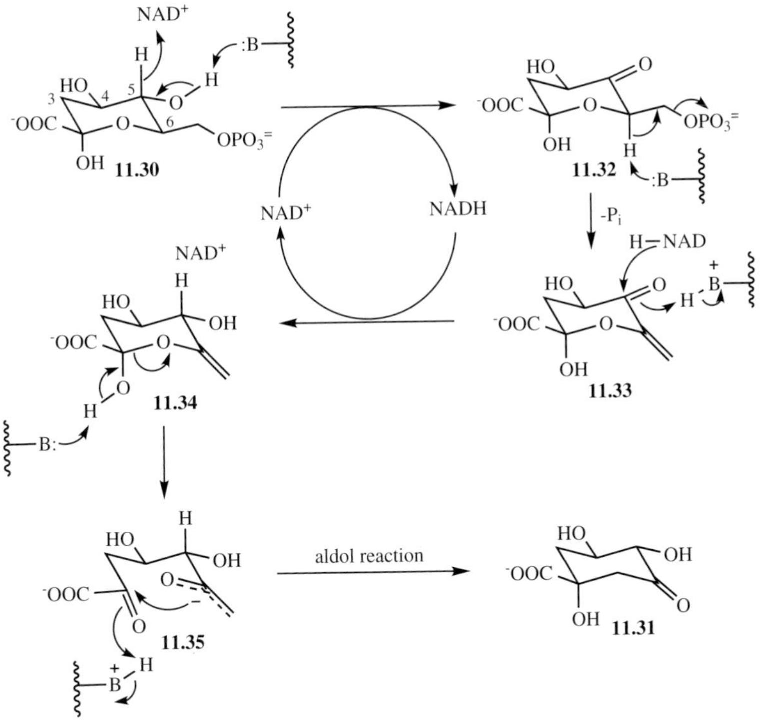

SCHEME 11.20   Originally proposed mechanism for dehydroquinate synthase.

is now activated for hemiketal cleavage of the sugar ring to generate an enolate intermediate (**11.35**). This can undergo intramolecular aldol condensation with protonation, forming **11.31**. The most unusual aspect of this reaction is the complexity of the steps; a single protein appears to be catalyzing five different reactions, namely, oxidation, elimination, reduction, ring cleavage, and aldol reactions. Very remarkable, but can an enzyme do that? Bartlett and Satake[47] did not think so; they therefore synthesized a protected form of **11.34,** deprotected it photochemically, and observed only **11.31** as the product, and **11.34**. Therefore, chemical generation of **11.34** rapidly produces **11.31** and suggests that the enzymatic product of dehydroquinate synthase may actually be **11.34,** which spontaneously rearranges to **11.31** nonenzymatically. That would eliminate the ring cleavage and aldol reactions from the list of reactions that dehydroquinate synthase had to catalyze.

Knowles and co-workers thought that even the three remaining reactions (oxidation, elimination, and reduction) were too numerous for one enzyme, and carried out experiments to lower the number of these reactions to two. Elimination of inorganic phosphate occurs with *syn* stereochemistry, suggesting that an E1cB mechanism is involved.[48] Modified substrate analogues containing phosphonate groups instead of phosphates were synthesized to determine how the elimination step occurs.[49] Substitution by a phosphonate prevents elimination from occurring, but would still permit C-6 deprotonation. Furthermore, cyclohexane rings were used instead of pyranose rings to block the ring cleavage step. Incubation of phosphonate **11.36** in D_2O in the presence of dehydroquinate synthase exchanges the C-6 proton with solvent, but under the same conditions **11.37** does not exchange the C-6 proton with solvent, even though it does undergo C-5 hydroxyl oxidation to **11.38**. Furthermore, whereas the *E*-vinyl analogue **11.39** binds poorly to the enzyme, is not oxidized at C-5, and exhibits no C-6 proton exchange, the corresponding *Z*-vinyl analogue **11.40** is oxidized as well as **11.36** and undergoes C-6 proton exchange at the same rate as **11.36.** All of these results are consistent with the hypothesis that the phosphate group of the substrate, not an active-site base, is responsible for C-6 deprotonation. Whereas a six-membered transition state containing the phosphonate, as would occur with **11.36,** is energetically favorable, the corresponding five-membered transition state generated by **11.37** is highly energetically unfavorable (because of ring strain). Likewise, phosphonate **11.39** is not capable of C-6 deprotonation, but **11.40** is set up very well for deprotonation.

There would be several advantages to having the phosphate group act as the base for C-6 proton removal: (1) The enzyme would utilize the strongest base at physiological pH, namely, the dianionic phosphate ester; (2) the enzyme could get around the issue of steric congestion caused by attempting to remove the tertiary proton (C-6), which also is 1,3-diaxial to the C-2 hydroxyl group; (3) the problem of an active-site base approaching the anionic phosphate group nearby could be avoided; (4) the elimination would be *syn,* as observed; and, most importantly, (5) this hypothesis could explain how a small monomeric protein could catalyze such a complex series of reactions. Furthermore, because the protonated phosphate group is a better leaving group than the corresponding dianion, intramolecular abstraction of C-6 by the phosphate also activates it as the leaving group for elimination in the subsequent step. With a nonenzymatic elimination step, the former complex series of reactions is reduced to a simple oxidation and reduction, easily catalyzed by $NAD^+/NADH$. Following enzyme-catalyzed oxidation of the C-5 hydroxyl group of substrate to **11.32** (Scheme 11.20), a nonenzymatic intramolecular (phosphate dianion-dependent) elimination could occur to give **11.33.** Enzyme-catalyzed reduction by the NADH generated in the oxidation step would give **11.34,** which, as Bartlett and Satake suggest, would be the product of the enzyme reaction that is released into solution and converted nonenzymatically to **11.31.**

Therefore, dihydroquinate synthase does *not* catalyze an aldol reaction; rather, this step is a nonenzymatic reaction. So why did I just have you spend your time reading about a purported enzyme-catalyzed aldol condensation that is not enzyme-catalyzed? First, because often mechanisms turn out to be different from what they were originally proposed to be. Second, because enzyme structures are highly evolved to allow for catalysis of precise chemistry; it is unlikely that an enzyme will be able to catalyze multiple varied mechanisms. If you isolate an enzyme that appears to be catalyzing more than one step in a reaction, it may be that you have not actually isolated a single protein or, as in the unusual case of dihydroquinate synthase, steps may be involved that are not catalyzed by enzymes.

II. CLAISEN REACTIONS

A. General

The *Claisen reaction* is very similar to the aldol reaction, except that the acceptor carbonyl-containing molecule is an ester (or thioester) instead of an aldehyde or ketone, and the product is a β-keto ester (thioester) (**11.41,** Scheme 11.21). In

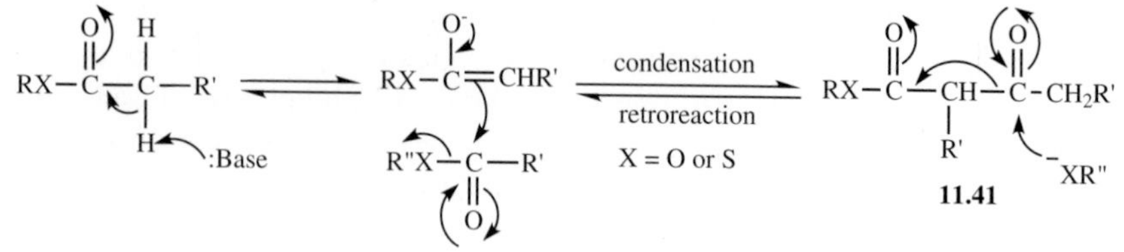

SCHEME 11.21   Generalized Claisen reaction.

general, the α-hydrogens of thioesters are more acidic than those of oxygen esters (the pK_a difference is about 2, so it is a factor of 100 in enolate formation), and therefore carbanion formation of thioesters is lower in energy than that of esters. This phenomenon can be rationalized in two ways. If resonance is considered, then the resonance structure for an ester (**11.42**) is more important than that for the thioester (**11.43**) because there is little C–S π-bond character. In terms of inductive effects, oxygen is more electronegative than sulfur, so **11.45** is lower energy (more favorable) than **11.44**; this enhances the acidity of the α-protons. Because of the advantages of thioesters over esters, few intermediates in metabolism are oxygen esters; generally, intermediates are either phosphate esters or thioesters, such as CoA esters. In contrast to the enzymes that catalyze aldol reactions, which generate the new C–C bond with retention of configuration, those that catalyze Claisen reactions produce the new C–C bond with inversion of stereochemistry relative to the α-proton that was removed.

B. Thiolase

Thiolase catalyzes the reversible reaction of two acyl-CoA thioesters to give the corresponding β-keto CoA thioester (**11.46,** Scheme 11.22 shows the retroreac-

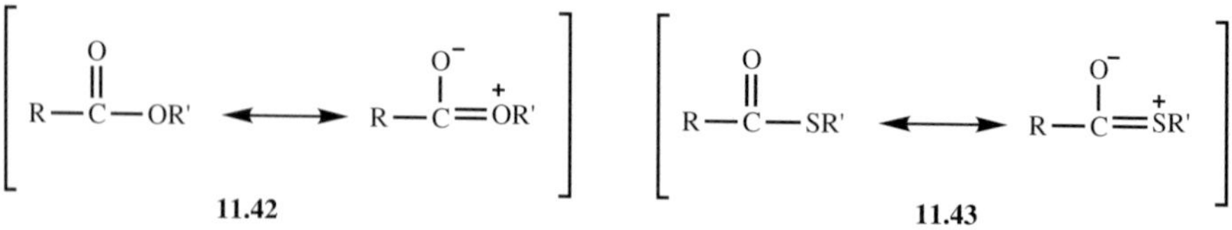

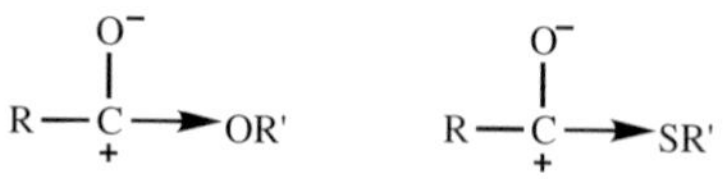

SCHEME 11.22   Reaction catalyzed by thiolase.

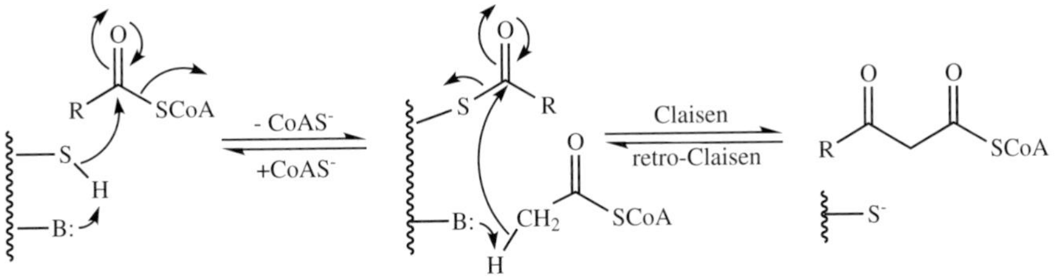

SCHEME 11.23   Mechanism proposed for thiolase.

tion). Iodoacetamide inactivates thiolase and becomes attached to a cysteine residue at the active site.[50] Incubation of the enzyme with [^{14}C]acetyl-CoA gives the acetylated enzyme; acetylation occurs at the same cysteine residue modified by iodoacetamide.[51] A mechanism that accounts for those observations is shown in Scheme 11.23.[52]

Another mechanism also was considered. Sodium borohydride reduction of thiolase in the presence of the substrate acetoacetyl-CoA or acetyl-CoA results in a slow inactivation.[53] Reduction of a thiolimidate intermediate (**11.47**), formed by the reaction of an active-site lysine with the CoA ester, was suggested as a possible cause. However, the reaction of an amine, such as lysine, with a thioester should give an amide (with ejection of the thiolate), not a thiolimidate. Leinhart[54] argued that selective acid catalysis by the enzyme might facilitate the formation of the thiolimidate rather than the amide; the intramolecular version of this proposed reaction is known.[55] To test this possibility with thiolase, several experiments were carried out by Izbicka and Gilbert.[56] Following various chemical modifications of the active-site lysine residue, only partial inactivation is observed. When the enzyme-catalyzed reaction is carried out in H$_2^{18}$O, no ^{18}O is incorporated into the product. Therefore, the thiolimidate is not an intermediate in the thiolase reaction, and the mechanism in Scheme 11.23 is more reasonable.

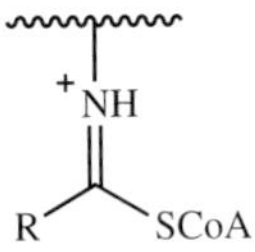

11.47

C. Kynureninase

Kynureninase is a pyridoxal 5'-phosphate (PLP)-dependent enzyme that catalyzes the hydrolytic cleavage of L-kynurenine (**11.48**) to anthranilic acid (**11.49**) and L-alanine (Scheme 11.24). In the presence of benzaldehyde, Phillips and Dua showed that an aldol product, (2S, 4R)-2-amino-4-hydroxy-4-phenylbutanoic acid (**11.50**), is obtained.[57] A mechanism for the conversion of **11.48** to **11.49** and L-alanine is shown in Scheme 11.25. Reaction of kynurenine with the active-site PLP gives the external aldimine **11.51**. Tautomerization, via the L-kynurenine quinonoid intermediate (**11.52**), gives the kynurenine ketimine (**11.53**), which undergoes attack by water, forming the (S)-*gem*-diolate (**11.54**). The *retro*-Claisen reaction releases anthranilic acid (**11.49**) and produces enamine **11.55.** Protonation gives the pyruvate ketimine (**11.56**), and tautomerization of **11.56,** via the alanine quinonoid intermediate **11.57,** produces the L-alanine aldimine **11.58;** L-alanine is released upon recapture of the PLP by the active-site lysine residue. Only the (4R)-isomer of dihydro-L-kynurenine (**11.59**) is a substrate for kynureninase, undergoing a retroaldol cleavage reaction; the (4S)-isomer is a potent competitive inhibitor. These results suggest that the hydration step occurs on the *re*-face of the carbonyl to give the (S)-*gem*-diolate intermediate (**11.54**). Based on this hypothesis, a series of transition-state analogue inhibitors was synthesized containing a sulfone in place of the carbonyl moiety; one analogue, (S)-(2-aminophenyl)-L-cysteine-(S,S)-dioxide (**11.60**) has a K_i value (70 nM) which is 300-fold lower than the K_m for L-kynurenine, and supports the (S)-*gem*-diolate hypothesis.[58] A series of kinetic isotope effect studies by Phillips and co-workers resulted in the conclusion that the rate-determining step is C-4' deprotonation of **11.56** to give **11.57.**[59] With rapid-scanning stopped-flow spectrophotometry and rapid chemical quench methods, both the quinonoid (**11.52**) and the pyruvate ketimine (**11.56**) intermediates can be detected when L-kynurenine is the substrate.[60] When L-alanine is the substrate (i.e., the reverse reaction), three intermediates can be detected, first **11.58,** then

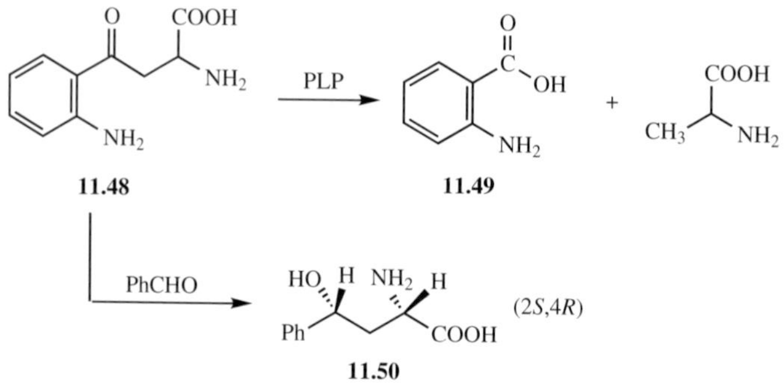

SCHEME 11.24   Reaction catalyzed by kynureninase.

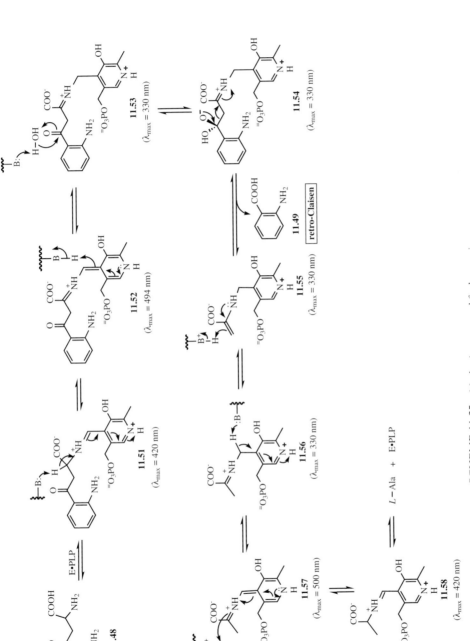

SCHEME 11.25 Mechanism proposed for kynureninase.

11.59

11.60

11.57, and then **11.56.** The rates of formation and decay of these intermediates are consistent with the suggestion that they are catalytically competent species.

Aldol and Claisen reactions are common ways of making C–C bonds both in solution and by enzymes. In the next chapter we explore other ways of making C–C bonds by formylations, hydroxymethylations, and methylations.

NOTE THAT PROBLEMS AND SOLUTIONS RELEVANT TO EACH CHAPTER CAN BE FOUND IN APPENDIX II.

REFERENCES

1. (a) Mukaiyama, T. *Org. React.* **1982,** *28,* 203. (b) Evans, D. A.; Nelson, J. V.; Taber, T. R. *Top. Stereochem.* **1982,** *13,* 1. (c) Heathcock, C. H.; Kim, B. M.; Williams, S. F.; Masamune, S.; Rathke, M. W.; Weipert, P.; Paterson, I. In *Comprehensive Organic Synthesis,* vol. 2, Trost, B. M., Ed., Pergamon: Oxford, 1991, pp. 133–319.
2. Meyerhof, O.; Lohmann, K. *Biochem. Z.* **1934,** *271,* 89.
3. Gijsen, H. J. M.; Qiao, L.; Fitz, W.; Wong, C.-H. *Chem. Rev.* **1996,** *96,* 443.
4. Sygusch, J.; Beaudy, D.; Allaire, M. *Proc. Natl. Acad. Sci. USA* **1987,** *84,* 7846.
5. Gamlin, S. J.; Copper, B.; Millar, J. R.; Davies, G. J.; Littlechild, J. A.; Watson, H. C. *FEBS Lett.* **1990,** *262,* 282.
6. Hester, G.; Brenner-Holzach, O.; Rossi, F. A.; Struck-Donatz, M.; Winterhalter, K.; Smit, J. D. G.; Pointek, K. *FEBS Lett.* **1991,** *292,* 237.
7. Wong, C.-H.; Whitesides, G. M. *Enzymes in Synthetic Organic Chemistry* , Pergamon Press: Oxford, 1994, pp. 195–251.
8. (a) Sugimoto, S.; Nosoh, Y. *Biochim. Biophys. Acta* **1971,** *235,* 210. (b) Hill, H. A. O.; Lobb, R. R.; Sharp, S. L.; Stokes, A. M.; Harris, J. I.; Jack, R. S. *Biochem. J.* **1976,** *153,* 551. (c) Sygusch, J.; Beaudy, D.; Allaire, M. *Proc. Natl. Acad. Sci. USA* **1987,** *84,* 7846.
9. Henderson, I.; Garcia-Junceda, E.; Liu, K. K.-C.; Chen, Y.-L.; Shen, G.-J.; Wong, C.-H. *Bioorg. Med. Chem.* **1994,** *2,* 837.
10. Alworth, W. L. *Stereochemistry and Its Application in Biochemistry,* Wiley: New York, 1972, p. 260.
11. Rose, I. A.; O'Connell, E. L.; Mehler, A. H. *J. Biol. Chem.* **1965,** *240,* 1758.

12. Speck, J.; Rowley, P.; Horecker, B. *J. Am. Chem. Soc.* **1963,** *85,* 1012.

13. Lubini, D. G. E.; Christen, P. *Proc. Natl. Acad. Sci. USA* **1979,** *76,* 2527.

14. Bednarski, M. D.; Simon, E. S.; Bischofbergen, N.; Fessner, W.-D.; Kim, M.; Lees, W.; Saito, T.; Waldmann, H.; Whitesides, G. M. *J. Am. Chem. Soc.* **1989,** *111,* 627.

15. Durrwachter, J. R.; Wong, C.-H. *J. Org. Chem.* **1988,** *53,* 4175.

16. (a) Wagner, J.; Lerner, R. A.; Barbas III, C. F. *Science* **1995,** *270,* 1797. (b) Reymond, J.-L.; Lerner, R. A. *J. Am. Chem. Soc.* **1995,** *117,* 9383. (c) Reymond, J.-L.; Chen, Y. *Tetrahedron Lett.* **1995,** *36,* 2575.

17. Hoffmann, T.; Zhong, G.; List, B.; Shabat, D.; Anderson, J.; Gramatikova, S.; Lerner, R. A.; Barbas, C. F. III *J. Am. Chem. Soc.* **1998,** *120,* 2768.

18. Jaffe, E. K. *J. Bioenerget. Biomembranes* **1995,** *27,* 169.

19. Barnard, G. F.; Itoh, R.; Hohberger, L. H.; Shemin, D. *J. Biol. Chem.* **1977,** *252,* 8965.

20. (a) Nandi, D. L. *Biosciences* **1978,** *33,* 799. (b) Gibbs, P. N.; Jordan, P. M. *Biochem. J.* **1986,** *236,* 447.

21. (a) Jaffe, E. K.; Markham, G. D.; Rajagopalan, J. S. *Biochemistry* **1990,** *29,* 8345. (b) Jaffe, E. K.; Markham, G. D. *Biochemsitry* **1988,** *27,* 4475. (c) Jaffe, E.K.; Markham, G. D. *Biochemistry* **1987,** *26,* 4258, 8030.

22. Jaffe, E. K.; Hanes, D. *J. Biol. Chem.* **1986,** *261,* 9348.

23. Dent, A. J.; Beyersmann, D.; Block, C.; Hasnain, S.S. *Biochemistry* **1990,** *29,* 7822.

24. (a) Bevan, D. R.; Bodlaender, P.; Shemin, D. *J. Biol. Chem.* **1980,** *255,* 2030. (b) Jaffe, E. K.; Salowe, S. P.; Chen, N. T.; DeHaven, P. A. *J. Biol. Chem.* **1984,** *259,* 5032.

25. Seehra, J. S.; Jordan, P. M. *Eur. J. Biochem.* **1981,** *113,* 435.

26. Jaffe, E. K.; Abrams, W. R.; Kaempfen, K. X.; Harris, K. A. *Biochemistry* **1992,** *31,* 2113.

27. Chaudhry, A. G.; Jordan, P. M. *Biochem. Soc. Trans.* **1976,** *4,* 760.

28. Mitchell, L. W.; Volin, M.; Jaffe, E. K. *J. Biol. Chem.* **1995,** *270,* 24054.

29. (a) Dimroth, P.; Eggerer, H. *Eur. J. Biochem.* **1975,** *53,* 227. (b) Srere, P. A. *Adv. Enzymol. Rel. Areas Mol. Biol.* **1975,** *43,* 57.

30. Plowman, K. M.; Cleland, W. W. *J. Biol. Chem.* **1967,** *242,* 4239.

31. Houston, B.; Nimmo, H. G. *Biochem. J.* **1984,** *224,* 437.

32. Walsh, Jr. C. T., Spector, L. B. *J. Biol. Chem.* **1969,** *244,* 4366.

33. (a) Eggerer, H.; Remberger, U. *Biochem. Z.* **1963,** *339,* 62. (b) Srere, P. A.; Bhaduri, A. *J. Biol. Chem.* **1964,** *239,* 714.

34. (a) Inoue, H,; Suzuki, F.; Tanioka, H.; Takeda, Y. *Biochem. Biophys. Res. Commun.* **1967,** *26,* 602. (b) Inoue, H.; Suzuki, F.; Tanioka, H.; Takeda, Y. *J. Biochem.* **1968,** *63,* 89.

35. Williams, S. P.; Sykes, B. D.; Bridger, W. A. *Biochemistry* **1985,** *24,* 5527.

36. Inoue, H.; Tsunemi, T.; Suzuki, F.; Takeda, Y. *J. Biochem.* **1969,** *65,* 889.

37. Wells, T. N. C. *Eur. J. Biochem.* **1991,** *199,* 163.

38. Heinrich, P. C.; Steffen, H.; Janser, P.; Wiss, O. *Eur. J. Biochem.* **1972,** *30,* 533.

39. Lindqvist, Y.; Schneider, G.; Ermler, U.; Sundstrom, M. *EMBO J.* **1992,** *11,* 2373.

40. Kobori, Y.; Myles, D. C.; Whitesides, G. M. *J. Org. Chem.* **1992,** *57,* 5899.

41. Hobbs, G. R.; Lilly, M. D.; Turner, N. J.; Ward, J. M.; Willets, A. J.; Woodley, J. M. *J. Chem. Soc. Perkin Trans. 1* **1993,** 165.

42. (a) Haslam, E. *The Shikimate Pathway,* Wiley: New York, 1974. (b) Weiss, U.; Edwards, J. M. *The Biosynthesis of Aromatic Compounds,* Wiley: New York, 1980.

43. Rotenberg, S. L.; Sprinson, D. B. *Proc. Natl. Acad. Sci. USA* **1970,** *67,* 1669.

44. (a) Le Maréchal, P.; Azerad, R. *Biochimie* **1976,** *58,* 1123. (b) Rotenberg, S. L.; Sprinson, D. B. *J. Biol. Chem.* **1978,** *253,* 2210.

45. (a) Le Maréchal, P.; Azerad, R. *Biochimie* **1976,** *58,* 1145. (b) Le Maréchal, P.; Froussios, C.; Level, M.; Azerad, R. *Biochem. Biophys. Res. Commun.* **1980,** *92,* 1104. (c) Widlanski, T. S.; Bender, S. L.; Knowles, J. R. *J. Am. Chem. Soc.* **1987,** *109,* 1873.

46. (a) Srinivasan, P. R.; Rothschild, J.; Sprinson, D. B. *J. Biol. Chem.* **1963,** *238,* 3176. (b) Maitra, U. S.; Sprinson, D. B. *J. Biol. Chem.* **1978,** *253,* 5426.

47. Bartlett, P. A.; Satake, K. *J. Am. Chem. Soc.* **1988,** *110,* 1628.

48. Widlanski, T. S.; Bender, S. L.; Knowles, J. R. *J. Am. Chem. Soc.* **1987,** *109,* 1873.

49. (a) Widlanski, T. S.; Bender, S. L.; Knowles, J. R. *J. Am. Chem. Soc.* **1989,** *111,* 2299. (b) Bender, S. L.; Widlanski, T. S.; Knowles, J. R. *Biochemistry* **1989,** *28,* 7560. (c) Widlanski, T. S.; Bender, S. L.; Knowles, J. R. *Biochemistry* **1989,** *28,* 7572.

50. Quandt, L.; Huth, W. *Biochim. Biophys. Acta* **1985,** *829,* 103.

51. Willadsen, P.; Eggerer, H. *Eur. J. Biochem* **1975,** *54,* 253.

52. Salam, W. H.; Boxham, D. P. *Biochim. Biophys. Acta* **1986,** *873,* 321.

53. (a) Kornblatt, J. A.; Rudney, H. *J. Biol. Chem.* **1971,** *246,* 4417. (b) Holland, P. C.; Clark, M. G.; Boxham, D. P. *Biochemistry* **1973,** *12,* 3309.

54. Leinhart, G. E. *J. Am. Chem. Soc.* **1968,** *90,* 3781.

55. Martin, R. B.; Parcell, A. *J. Am. Chem. Soc.* **1961,** *83,* 4830.

56. Izbicka, E.; Gilbert, H. F. *Arch. Biochem. Biophys.* **1989,** *272,* 476.

57. Phillips, R. S.; Dua, R. K. *J. Am. Chem. Soc.* **1991,** *113,* 7385.

58. Dua, R. K.; Taylor, E. W.; Phillips, R. S. *J. Am. Chem. Soc.* **1993,** *115,* 1264.

59. Koushik, S. V.; Moore III, J. A.; Sundararaju, B.; Phillips, R. S. *Biochemistry* **1998,** *37,* 1376.

60. Phillips, R. S. ; Sundararaju, B.; Koushik, S. V. *Biochemistry* **1998,** *37,* 8783.

CHAPTER 12

Formylations, Hydroxymethylations, and Methylations

I. TETRAHYDROFOLATE-DEPENDENT ENZYMES: THE TRANSFER OF ONE-CARBON UNITS

A. Formation of the Active Coenzyme

The most important way for one carbon-containing groups to be added to molecules involves enzymes that utilize a coenzyme that we have not yet discussed, namely, tetrahydrofolate (**12.1**). Unfortunately for chemists, many biochemists use the abbreviation THF for this molecule, an abbreviation that, to chemists, stands for the solvent tetrahydrofuran. So be aware that these enzyme-catalyzed reactions are *not* carried out in THF solvent! Because the chemistry of **12.1** occurs at the N^5 and N^{10} nitrogen atoms, this coenzyme will be abbreviated as **12.2**; the remainder of the molecule is important for binding, but is not involved in the chemistry (although it does play a role in modulating the redox potential of the folate).

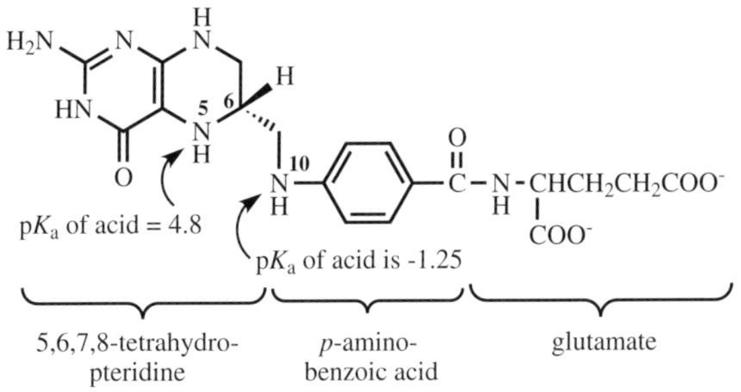

pKa of acid = 4.8

pKa of acid is -1.25

5,6,7,8-tetrahydro-pteridine   *p*-amino-benzoic acid   glutamate

12.1

12.2

Tetrahydrofolate (**12.1**) is a monoglutamate, but the coenzyme can contain an oligomer of as many as twelve glutamate residues, depending on the enzyme. The polyglutamate derivatives, with the glutamate moieties linked via γ-glutamyl linkages, are the major intracellular forms of folate and are the physiological substrates.[1] Tetrahydrofolate derivatives are named as polyglutamate derivatives of tetrahydropteroate (abbreviated H_4Pte; pteroate is the general name for the folate ring class), which has no glutamate residues. Tetrahydrofolate, having one glutamyl residue, is abbreviated H_4PteGlu, so the pentaglutamate form is written as H_4PteGlu$_5$.[2]

Tetrahydrofolate is derived from folic acid (**12.3**), a vitamin for humans. Two of the C—N double bonds in folic acid are reduced by the NADPH-dependent enzyme (see Chapter 3, Section III.A for a discussion of NADPH) dihydrofolate reductase (Scheme 12.1).[3]

Actually, tetrahydrofolate is not the full coenzyme; the complete coenzyme form contains an additional carbon atom between the N^5 and N^{10} positions, which is transferred to other molecules. That additional carbon atom is derived from the methylene group of L-serine in a reaction catalyzed by the PLP-dependent enzyme (see Chapters 8.V.A and 9.II.B) serine hydroxymethyltransferase.[4] The reaction catalyzed by this PLP-dependent enzyme is one that we have not yet discussed, namely, α-cleavage (Scheme 12.2; the carbon atom that will be transferred to the tetrahydrofolate is marked with an asterisk). The hydroxymethyl group is held perpendicular to the plane of the PLP aromatic system (**12.4**), as in the case of the group cleaved during all of the other PLP-dependent reactions, so that deprotonation and a retroaldol reaction can occur readily. The serine methylene is converted into formaldehyde as a result of this retroaldol reaction, without removal of the α-proton of serine or loss of the hydroxyl oxygen.

In the back reaction, the enzyme abstracts only the *pro-S* hydrogen of glycine and the formaldehyde is introduced with retention of configuration at the α-

12.3

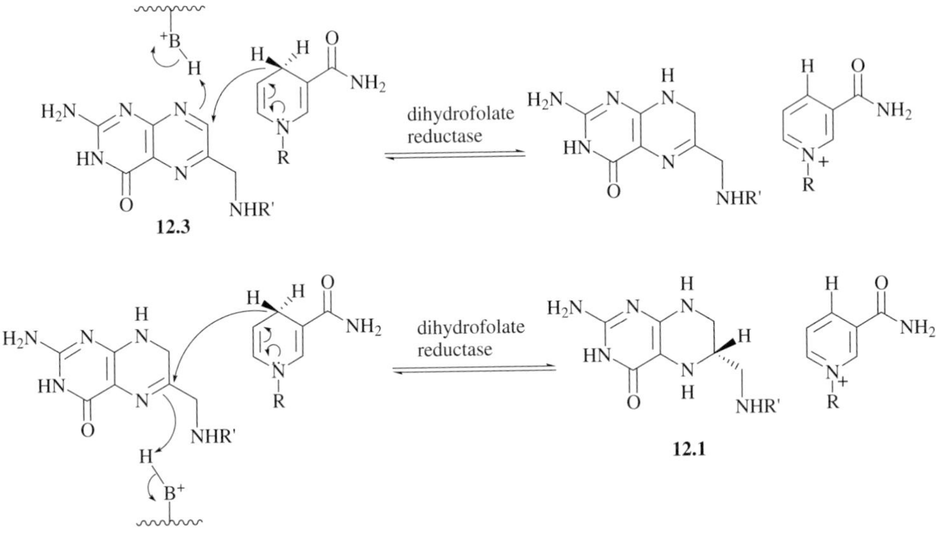

SCHEME 12.1   Reactions catalyzed by dihydrofolate reductase.

carbon.[5] Because of the potential toxicity of released formaldehyde (it is highly electrophilic and reacts readily with amino groups), serine hydroxymethyltransferase does not catalyze the degradation of serine (to an appreciable extent) until *after* the acceptor for formaldehyde, namely, tetrahydrofolate, is already bound at an adjacent site.[6] This is an example of an ordered sequential reaction (see Appendix I, Section VI.A.2).

Once the tetrahydrofolate binds, the degradation of serine is triggered, and the formaldehyde generated reacts directly with the tetrahydrofolate to give, initially, the carbinolamine **12.5**, as shown in Scheme 12.3. Kallen and Jencks showed that the more basic nitrogen of tetrahydrofolate is the N^5 nitrogen,[7] which attacks the formaldehyde first;[8] the N^{10} nitrogen is attached to the aromatic ring *para* to a carbonyl group, and therefore this nitrogen has amide-like character, and is not basic. Stereochemical studies by Floss, Matthews, and co-workers of the transfer of the methylene carbon of serine to tetrahydrofolate demonstrate that $(3R)$-$[3-^3H]$serine is converted to $(11R)$-$[11-^3H]$methylenetetrahydrofolate.[9] Carbinolamine **12.5** is dehydrated to N^5-methylenetetrahydrofolate (**12.6**), which is in equilibrium with N^5,N^{10}-methylenetetrahydrofolate (**12.7**) and N^{10}-methylenetetrahydrofolate (**12.8**). The equilibrium in solution strongly favors the cyclic form (**12.7**) ($K_{eq} = 3.2 \times 10^4$ at pH 7.2).

Let's consider an elimination/addition mechanism for the formation of **12.6**–**12.8** (Scheme 12.4), and see why it is not relevant. In this mechanism both the α-proton and the serine hydroxyl group are lost. Two results argue against the mechanism in Scheme 12.4. When α-methylserine is used as the substrate, Schirch and

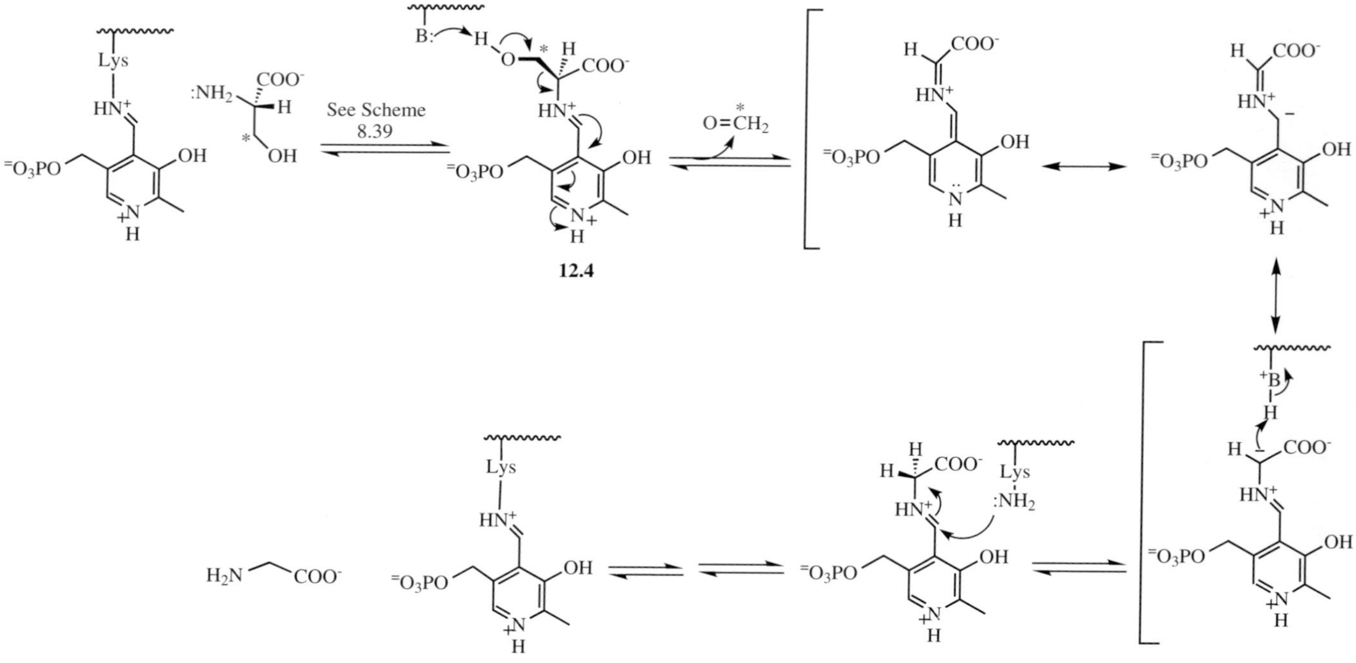

SCHEME 12.2 Serine hydroxymethyltransferase-catalyzed formation of formaldehyde via a proposed α-cleavage mechanism. The asterisk indicates the carbon that becomes the one-carbon unit transferred in tetrahydrofolate–dependent enzymes.

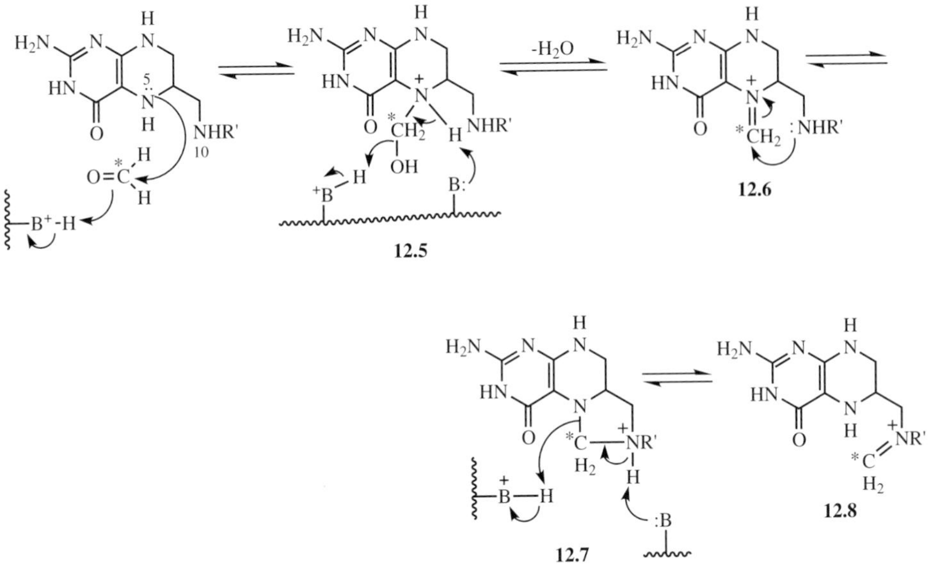

12.5

12.6

12.7

12.8

SCHEME 12.3 Serine hydroxymethyltransferase-catalyzed reaction of formaldehyde and tetrahydrofolate to give methylenetetrahydrofolate.

Mason found that formaldehyde is still generated, indicating that removal of the α-proton is not important.[10] Serine hydroxymethyltransferase also can use threonine as the substrate. Akhtar and co-workers showed that when [^{18}O-hydroxyl]threonine is the substrate, the acetaldehyde produced (with threonine, acetaldehyde, not formaldehyde, is generated) retains all of the ^{18}O, supporting a mechanism in which the hydroxyl group is not lost.[11] Both of these results negate the mechanism in Scheme 12.4. The fact that the ^{18}O is retained also prohibits a mechanism in which the formaldehyde generated forms a Schiff base with an active-site lysine residue.

B. Transfer of One-Carbon Units

The three coenzyme forms of methylenetetrahydrofolate (12.6–12.8) are responsible for the transfer of one-carbon units in three different oxidation states: formate (transferred as a formyl group), formaldehyde (transferred as a hydroxymethyl group), and methanol (transferred as a methyl group). Methylenetetrahydrofolate is in the correct oxidation state to transfer a hydroxymethyl group. What if the cell needs a one-carbon unit at the formate oxidation state? This is one oxidation state higher than the formaldehyde oxidation state. A NADP$^+$-dependent dehydrogenase (N^5,N^{10}-methylenetetrahydrofolate dehydrogenase) is involved in the oxidation of methylenetetrahydrofolate to give N^5,N^{10}-methenyltetrahydrofolate (12.9,

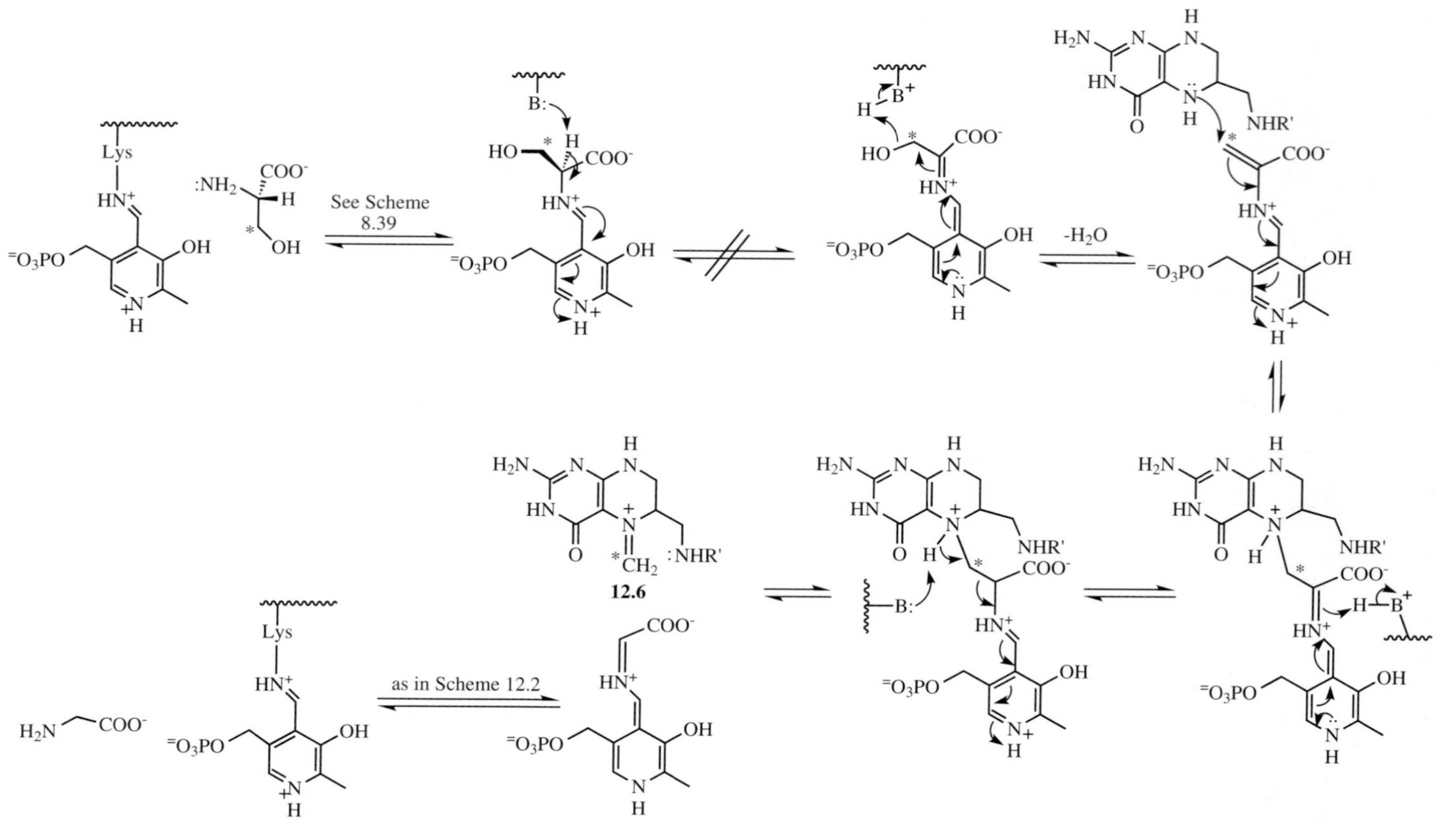

SCHEME 12.4 An alternative incorrect elimination/addition mechanism for serine hydroxymethyltransferase.

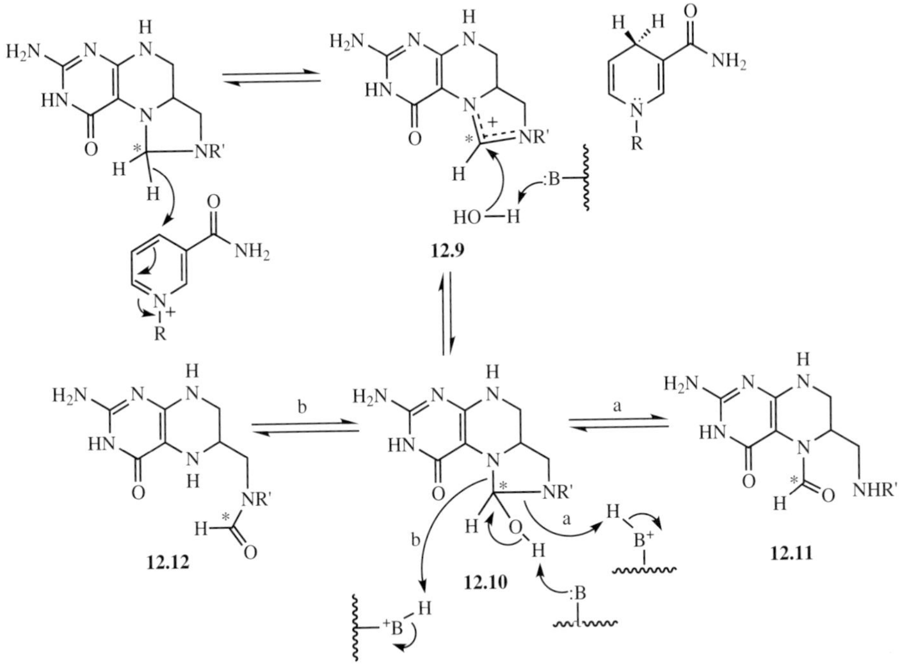

SCHEME 12.5 Oxidation of N^5,N^{10}-methylenetetrahydrofolate to N^5,N^{10}-methenyltetrahydrofolate catalyzed by N^5,N^{10}-methylenetetrahydrofolate dehydrogenase and hydrolysis of N^5,N^{10}-methenyltetrahydrofolate by N^5,N^{10}-methenyltetrahydrofolate cyclohydrolase.

Scheme 12.5). Hydrolysis of **12.9**, catalyzed by the enzyme N^5,N^{10}-methenyltetrahydrofolate cyclohydrolase, gives the carbinolamine **12.10**, which can break down to give either N^5-formyltetrahydrofolate (**12.11**, pathway a) or N^{10}-formyltetrahydrofolate (**12.12**, pathway b). N^5,N^{10}-methenyltetrahydrofolate (**12.9**) also can be hydrolyzed by serine hydroxymethyltransferase.[12] Some enzymes utilize **12.11**, and others use **12.12** as their coenzyme.

Another enzyme that catalyzes the formation of N^{10}-formyltetrahydrofolate is N^{10}-formyltetrahydrofolate synthetase, which uses formate ion and ATP as the formylating agent for tetrahydrofolate (Scheme 12.6).

If the cell needs a one-carbon unit at the methanol oxidation state, then N^5,N^{10}-methylenetetrahydrofolate is reduced by N^5,N^{10}-methylenetetrahydrofolate reductase to N^5-methyltetrahydrofolate (**12.13**). This enzyme requires not only NADPH

$$\text{HCOO}^- + \text{ATP} + \textbf{12.1} \; \underset{\text{Mg}^{++}}{\rightleftharpoons} \; \text{ADP} + \text{P}_i + \textbf{12.12}$$

SCHEME 12.6 Reaction catalyzed by N^{10}-formyltetrahydrofolate synthetase.

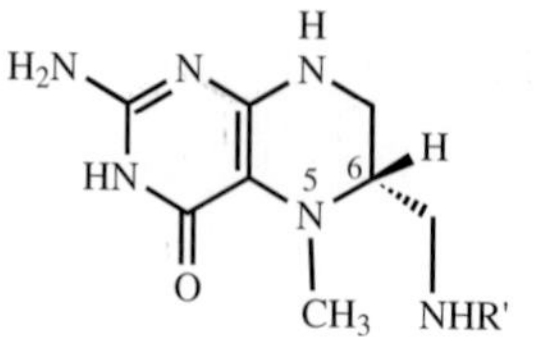

12.13

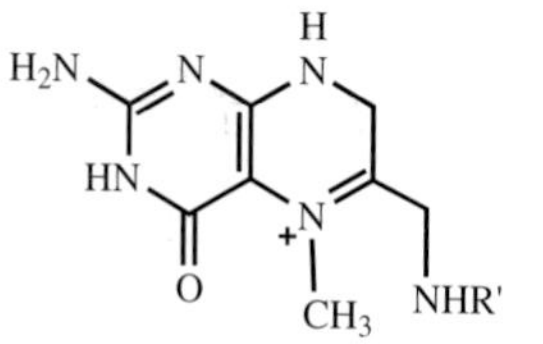

12.14

as a coenzyme, but also flavin adenine dinucleotide.[13] It is generally thought that the NADPH, either directly or via FADH$^-$, reduces N^5-methylenetetrahydrofolate (12.6) or one of its tautomeric forms to 12.13.[14] Matthews and Haywood found that if this enzyme reaction is run in reverse with [6-^{3}H]-N^5-methyltetrahydrofolate as substrate, no tritium is released (thus excluding tautomerization of 12.6 via 12.14), and no tritium is transferred from C-6 to the methyl group in the forward direction (excluding a [1,3]-hydride shift).[15] The enzyme can reduce quinonoid dihydrofolate derivatives that lack an N^5-substituent (12.15) to the corresponding tetrahydrofolate derivatives with k_{cat} and K_m values comparable to those for methylenetetrahydrofolate.[16] This supports a mechanism in which electrons are transferred to the pterin ring rather than to the exocyclic methylene group (Scheme 12.7). Further support for this mechanism ostensibly comes from running the reaction in D$_2$O and showing that the third hydrogen on the methyl group is derived from solvent with little exchange and with an isotope effect (k_H/K_D) of 10.[17] However, these results also can be rationalized by a hydride transfer from NADPH to FAD, followed by hydride transfer from the FADH$^-$ to the *exo*-methylene group

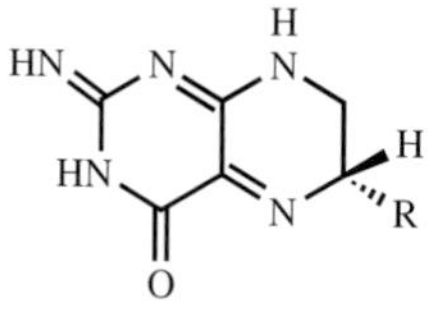

12.15

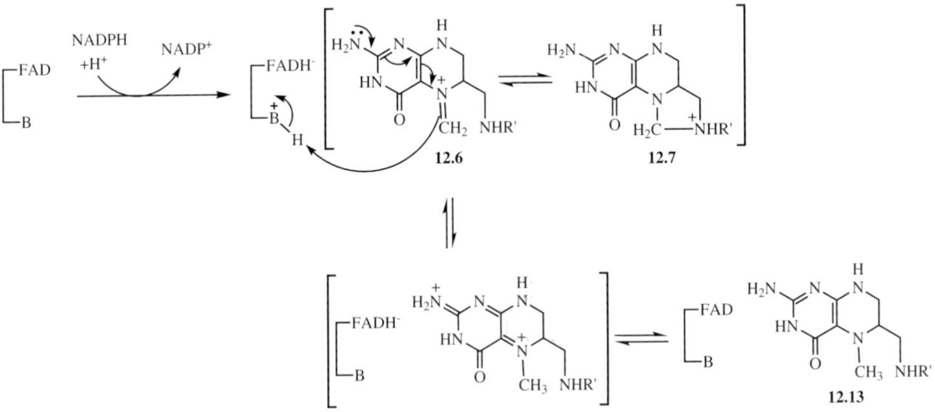

SCHEME 12.7   Proposed mechanism for the reduction of N^5,N^{10}-methylenetetrahydrofolate by N^5,N^{10}-methylenetetrahydrofolate reductase.

of N^5-methylenetetrahydrofolate (**12.6**) (Scheme 12.8). In D_2O, the hydrogen at the N^5-position of $FADH^-$ would exchange with deuterium, and a deuteride would be transferred to the methylene group, thereby accounting for the incorporation of deuterium from solvent into the methyl group of N^5-methyltetrahydrofolate. To test the mechanism in Scheme 12.8, Sumner and Matthews reconstituted pig liver methylenetetrahydrofolate reductase with a reduced 5-deazaflavin (see Chapter 3; **3.62,** Scheme 3.42), so that the hydrogen at the 5-position of the flavin could not

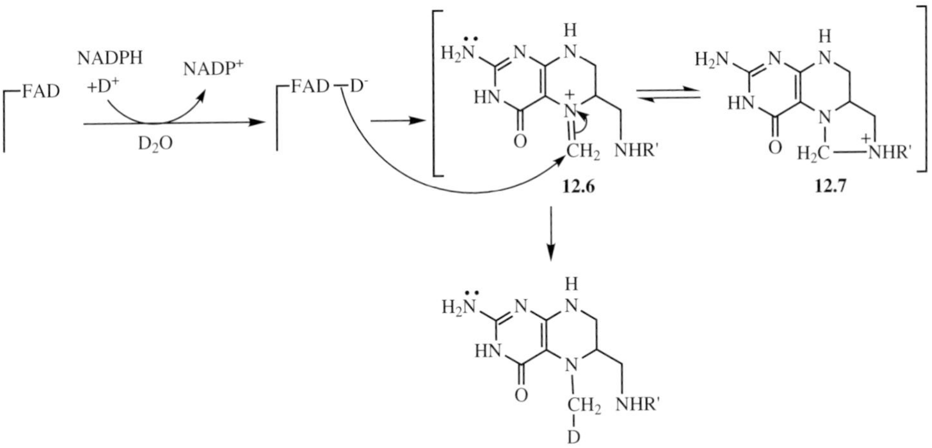

SCHEME 12.8   Proposed alternative hydride mechanism for N^5,N^{10}-methylenetetrahydrofolate reductase.

exchange with solvent.[18] Methylenetetrahydrofolate reductase reconstituted with 8-hydroxy-[5-³H]5-deazaFADH$_2$ retained over half of its activity, and the tritium was quantitatively transferred to methylenetetrahydrofolate with incorporation of the tritium into the methyl group of N^5-methyltetrahydrofolate. This is consistent with a direct hydride transfer from the flavin to the *exo*-methylene of N^5-methyl-enetetrahydrofolate (Scheme 12.8). Because 5-deazaflavins do not undergo one-electron transfer reactions,[19] this experiment also excludes one-electron transfer mechanisms.

1. Transfer at the Formate Oxidation State

An example of an enzyme that catalyzes the transfer of a one-carbon unit at the for-mate oxidation state (i.e., as a formyl group) is glycinamide ribonucleotide (GAR) transformylase. This is the third enzyme in the *de novo* biosynthetic pathway to purines, and it catalyzes the conversion of glycinamide ribonucleotide (**12.16**) to *N*-formylglycinamide ribonucleotide (**12.18**, FGAR), presumably via the tetra-hedral intermediate **12.17** (Scheme 12.9). The formyl group is transferred from 10-formyltetrahydrofolate, which is converted to tetrahydrofolate. High-resolution X-ray crystal structures of the *Escherichia coli* enzyme[20] revealed several amino acid residues that could be involved in binding and/or catalysis. Benkovic and co-workers showed by site-directed mutagenesis of each of the conserved polar resi-dues within 6 Å of the active site of the *Escherichia coli* enzyme (individually) that none of these residues is essential for catalysis.[21] These residues may alter the pK_a values of the attacking and leaving amino groups in **12.17** to facilitate the break-down of the tetrahedral intermediate.

2. Transfer at the Formaldehyde Oxidation State

The enzyme used to exemplify a transfer of the one-carbon unit at the formalde-hyde oxidation state (equivalent to transfer of a hydroxymethyl group) is unique

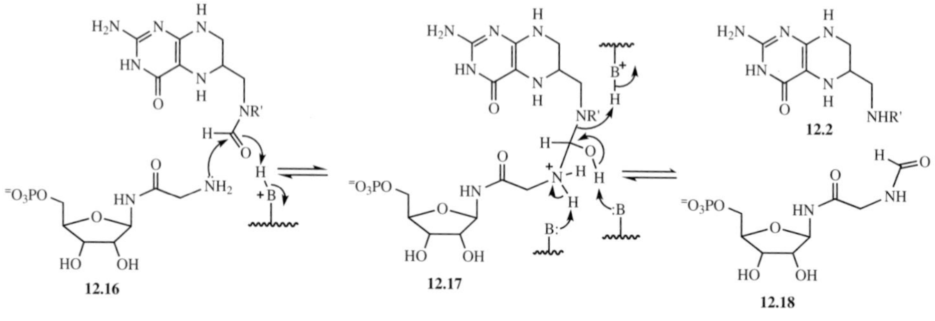

SCHEME 12.9   Proposed mechanism for glycinamide ribonucleotide transformylase.

among enzymes that utilize folate coenzymes because it not only catalyzes the transfer of the one-carbon unit, but also catalyzes the reduction of that unit; the methylenetetrahydrofolate is utilized for both operations. The enzyme is thymidylate synthase,[22] which catalyzes the conversion of 2'-deoxyuridine monophosphate (**12.19**; note the numbering of C-5 and C-6) to 2'-deoxythymidine monophosphate (**12.20**, thymidylate) with concomitant conversion of N^5,N^{10}-methylenetetrahydrofolate to 7,8-dihydrofolate (Scheme 12.10). The product, then, appears to have been methylated rather than hydroxymethylated. This corresponds to a reduction of the hydroxymethyl group to a methyl group and the oxidation of tetrahydrofolate to dihydrofolate. The cell, then, requires the enzyme dihydrofolate reductase (and serine hydroxymethyltransferase) (see Schemes 12.1 and 12.2) to return the consumed coenzyme back to the active N^5,N^{10}-methylenetetrahydrofolate form. In protozoa, interestingly, thymidylate synthase and dihydrofolate reductase exist on a single polypeptide chain; the thymidylate synthase is located on the carboxy terminus and the dihydrofolate reductase is found on the amino terminus![23] Thymidylate synthase is a very important enzyme because it catalyzes the last step in the *de novo* biosynthesis of thymidylate, which is an essential component of DNA. If this enzyme is inhibited, DNA synthesis is blocked, and these inhibitors are used clinically as antitumor and antimicrobial agents.[24]

Early results obtained by Lomax and Greenberg with thymidylate synthase demonstrated the exchange of the C-5 hydrogen of 2'-deoxyuridine monophosphate with solvent in the presence of methyleneTHF.[25] The rate of tritium release is about 80% the rate of dihydrofolate production, indicating an inverse secondary isotope effect, consistent with the rehybridization of the C-5 position from sp^2 to sp^3 during catalysis. Furthermore, kinetic studies by Bruice and Santi with [6-^{3}H]5-fluoro-2'-deoxyuridine monophosphate[26] also revealed an inverse secondary isotope effect at C-6, supporting rehybridization (sp^2 to sp^3) at this position as well. Another curious observation early on is the stoichiometric transfer of a tritium from N^5,N^{10}-methylene-[6-^{3}H]tetrahydrofolate into the methyl group of thymidylate with $(V/K)_H/(V/K)_T = 5.2$ (Scheme 12.11).[27]

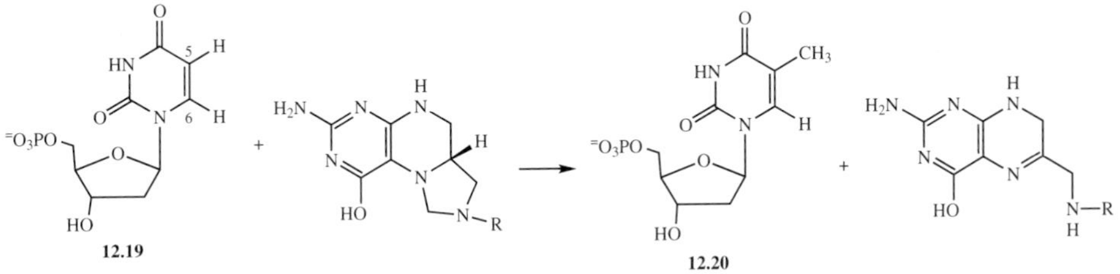

SCHEME 12.10   Reaction catalyzed by thymidylate synthase.

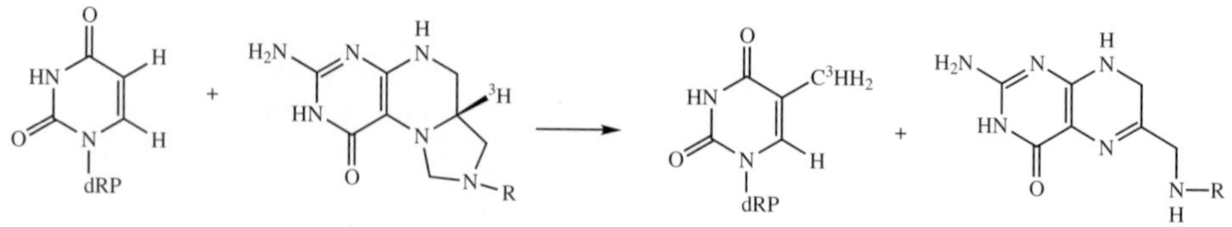

SCHEME 12.11   Transfer of the C-6 hydrogen of N^5,N^{10}-methylenetetrahydrofolate to the methyl group of thymidylate catalyzed by thymidylate synthase.

In an attempt to understand what chemistry would be reasonable to account for some of the early enzymatic observations, Santi and co-workers carried out chemical model studies of the reaction.[28] Incubation of [5-^{2}H]1,3-dimethyluracil (**12.21**) in the presence of hydroxide ions results in release of the deuterium via Michael addition to C-6 (Scheme 12.12). Note that in **12.22** both C-5 and C-6 are sp^3, consistent with the secondary isotope effect results.[29] Thiols, such as mercaptoethylamine, glutathione, and cysteine are much more effective than hydroxide ion at catalyzing the exchange of the C-5 hydrogen, and the same mechanism as shown in Scheme 12.12 (except with the thiolate as the nucleophile) is applicable.[30] Probably the most significant aid to the elucidation of the mechanism of thymidylate synthase was the determination of the structure of the complex formed (**12.24**) when the mechanism-based inactivator, 5-fluoro-2′-deoxyuridylate (**12.23**), inactivates thymidylate synthase (Scheme 12.13).[31] This clearly shows the result of an attack at C-6 by an active-site cysteine residue, leading to attachment of the enolate at C-5 to the methylene of methyleneTHF. Based on these results, it would be reasonable to draw the initial steps in the catalytic mechanism as shown in Scheme 12.14, where **12.25** mimics the structure of the inactivated complex **12.24.** In this mechanism the enolate does not attack the cyclic form of methylene-H$_4$PteGlu, but, instead, the much more reactive Mannich base form of N^5-methylene-H$_4$PteGlu,[32] which is too unstable to be detected.[33] Because the N^5-nitrogen is more basic than the N^{10}-nitrogen, elimination of the N^{10}-containing group would be more facile, and the N^5-iminium ion would be generated. Also, reduction of methylene-H$_4$PteGlu with KBH$_4$ gives N^5-methyl-H$_4$PteGlu, presumably via N^5-

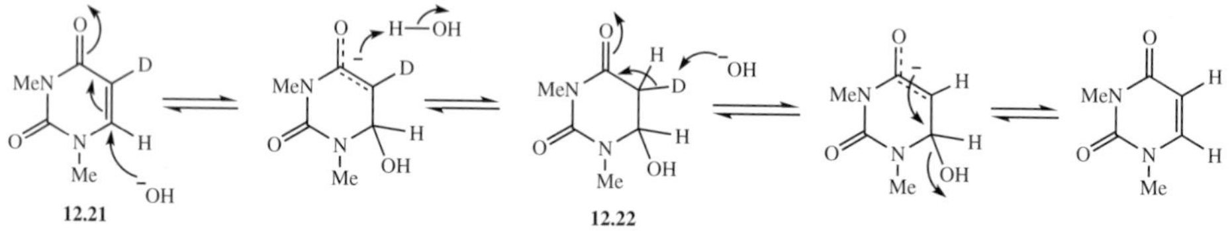

SCHEME 12.12   Chemical model study for thymidylate synthase-catalyzed exchange of the C-5 hydrogen of 2′-deoxyuridine-5′-monophosphate.

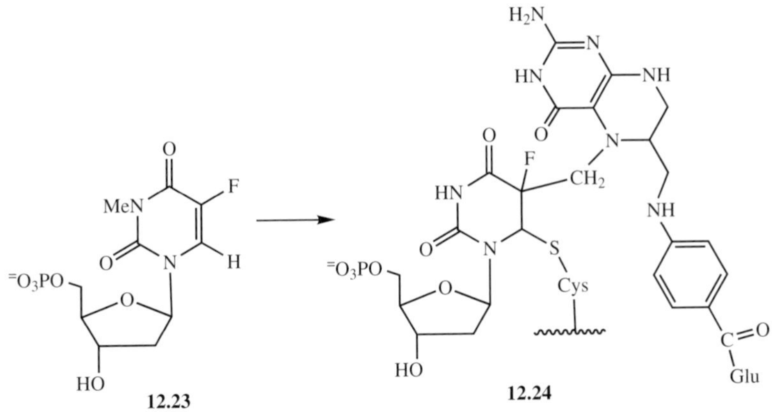

SCHEME 12.13   Inactivation of thymidylate synthase by 5-fluoro-2'-deoxyuridylate.

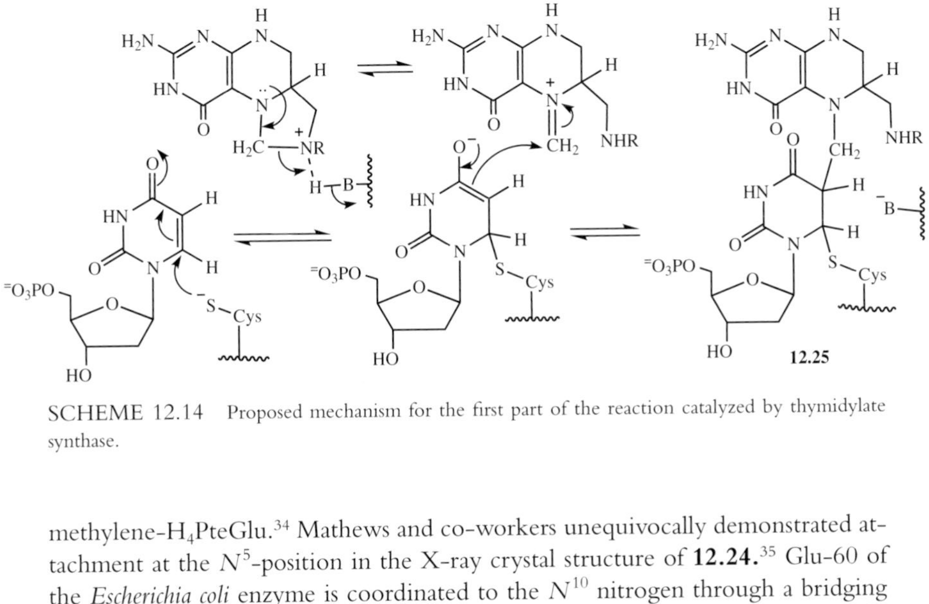

SCHEME 12.14   Proposed mechanism for the first part of the reaction catalyzed by thymidylate synthase.

methylene-H_4PteGlu.[34] Mathews and co-workers unequivocally demonstrated attachment at the N^5-position in the X-ray crystal structure of **12.24**.[35] Glu-60 of the *Escherichia coli* enzyme is coordinated to the N^{10} nitrogen through a bridging water molecule, as predicted for assistance in elimination of that group to give N^5-methylene-H_4PteGlu. The crystal structure also shows that Cys-198 is the active-site cysteine attached to C-6 of the inactivator (and presumably to the substrate). Intermediate **12.25** has been isolated by rapid quench of wild-type thymidylate synthase[36] and of Glu-60 mutants.[37]

The first hypothesis for the reaction from intermediate **12.25** (actually, this was proposed long before **12.25** was conceived), based on the suggestion by Friedkin,[38]

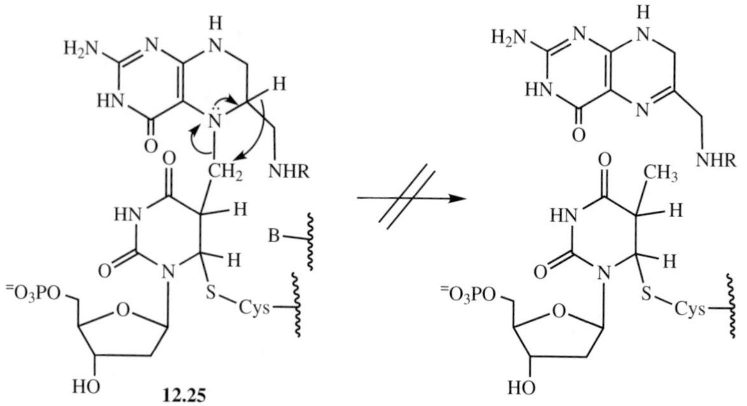

SCHEME 12.15   Highly unlikely [1,3]-hydride shift mechanism for reduction of the substrate catalyzed by thymidylate synthase.

was a [1,3]-sigmatropic hydride shift of the C-6 hydrogen to the N^5-methylene to account for the incorporation of the C-6 hydrogen into the thymidylate methyl group (Scheme 12.15). There are several problems with this proposed mechanism. One is that a [1,3]-*suprafacial* sigmatropic hydrogen shift is forbidden according to the Woodward–Hoffmann rules for pericyclic reactions.[39] If this mechanism were appropriate, then the inactivated complex (**12.24**) would have undergone the same [1,3]-sigmatropic rearrangement, and **12.24** would not have been isolated. The fact that **12.24** can be isolated suggests that the next step in the mechanism should be deprotonation of the C-5 proton of **12.25**; in the inactivated complex there is a fluorine at C-5, which cannot be removed, and that would account for the stability of **12.24**. Therefore, a more reasonable alternative to the [1,3]-sigmatropic hydride shift would be an elimination/hydride addition mechanism via *exo*-methylene **12.26** (Scheme 12.16). Precedence for the formation of the exocyclic methylene comes from a variety of chemical model studies by Pogolotti and Santi with esters and reactive ethers of 5-hydroxymethyluracil (**12.27**, Scheme 12.17).[40] When **12.27** is deuterated in the methylene group, there is a normal secondary deuterium isotope effect (k_H/k_D) of 1.43 (indicative of a change in hybridization at that carbon from sp^3 to sp^2), and when it is C-6 deuterated, there is an inverse secondary deuterium isotope effect of 0.89, as expected for a change in hybridization at C-6 from sp^2 to sp^3. Schultz and co-workers obtained further evidence for the formation of *exo*-methylene **12.26** using the W82Y mutant of thymidylate synthase in the presence of 2-mercaptoethanol as a trap; **12.28** was isolated from the reaction (Scheme 12.18).[41]

Because of the electrophilicity of the *exo*-methylene group, it is expected that proper orbital alignment of the C-6 hydrogen of the coenzyme with the β-position

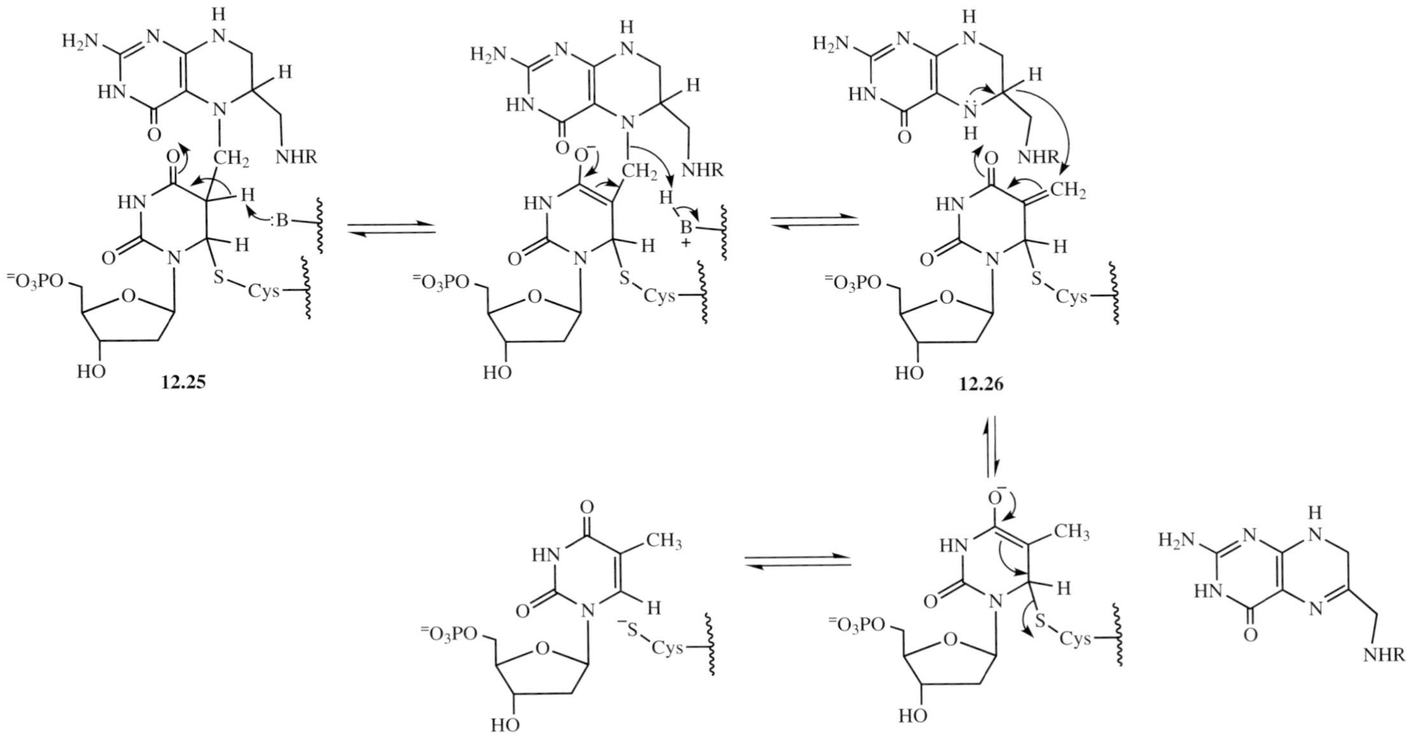

SCHEME 12.16 Proposed mechanism for the second part of the reaction catalyzed by thymidylate synthase.

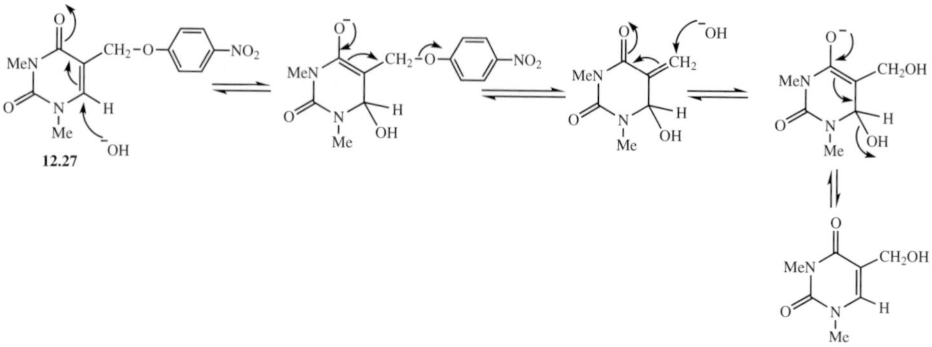

SCHEME 12.17   Model study for the formation of the C-5 *exo*-methylene intermediate proposed in the reaction catalyzed by thymidylate synthase.

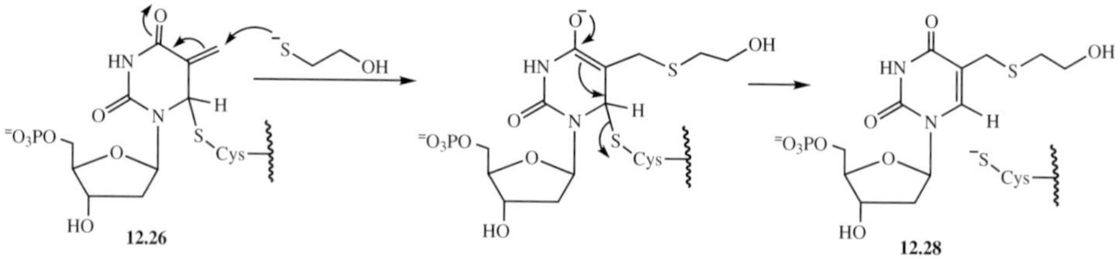

SCHEME 12.18   Trapping of the proposed C-5 *exo*-methylene intermediate during catalytic turnover of thymidylate synthase.

of the *exo*-methylene should be sufficient to catalyze the hydride transfer (see Scheme 12.16). An alternative to hydride transfer is two single-electron transfers via the enzyme-bound substrate radical anion **12.29** and radical **12.30** (Scheme 12.19).[42] However, there is no evidence to support the formation of radical intermediates in the enzyme-catalyzed reaction.

3. Transfer at the Methanol Oxidation State

Finally, one-carbon units can be transferred to acceptors as methyl groups. Methionine synthase (also called methyltetrahydrofolate-homocysteine methyltransferase), which catalyzes the transfer of a methyl group from N^5-methyl-H_4PteGlu to homocysteine to make methionine, exists in different organisms in two forms.[43] One form of the enzyme probably transfers the methyl group directly from N^5-methyl-H_4PteGlu to homocysteine, and the other transfers the methyl group first to an unusual cobalt-containing complex, a cobalamin (see later), to give one of only two currently known organometallic coenzymes, namely, methylcobalamin,

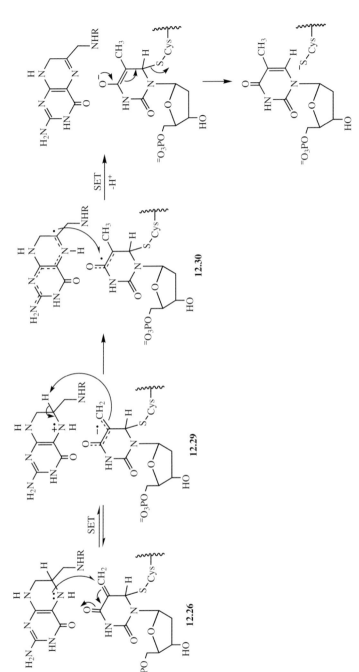

SCHEME 12.19 Alternative proposed electron transfer mechanism for the reduction of the *exo*-methylene intermediate in the re-action catalyzed by thymidylate synthase.

which then transfers the methyl group to homocysteine. There is evidence from model studies[44] that the methanogenic cofactor F430 (a nickel corphin)—and from model studies[45] and a variety of spectroscopic investigations[46] that carbon monoxide dehydrogenase/acetyl-CoA synthase—may operate via methyl–nickel intermediates, which would be the third organometallic complex. The cobalamin-independent enzyme has a strict requirement for methyl-$H_4PteGlu_n$ in the polyglutamate form in addition to magnesium and phosphate ions,[47] whereas the cobalamin-dependent enzyme can utilize either the polyglutamate or monoglutamate form of methyl-$H_4PteGlu_n$. The latter enzyme also requires S-adenosylmethionine and a reducing system for activity (see Section I.B.3.b).

a. Cobalamin-Independent Methionine Synthase

The enzyme cobalamin-independent methionine synthase catalyzes the direct methylation of homocysteine (**12.31**) by N^5-methyl-$H_4PteGlu_n$ to give methionine (**12.32**, Scheme 12.20). It appears to be a simple reaction, possibly an S_N2 reaction; however, the methyl group is at a tertiary amine position, which is sterically hindered, and the leaving group for N^5-methyl-$H_4PteGlu_n$, namely tetrahydrofolate anion, would be a very poor leaving group because of its high pK_a. Furthermore, the pK_a of the thiol of homocysteine is about 10, so under physiological pH it would be protonated and therefore would not be a very good nucleophile. One way to catalyze this reaction would be to lower the pK_a of the leaving group by protonation of N^5-methyl-$H_4PteGlu_n$ to the corresponding ammonium cation; the pK_a for the protonated N^5-amine of N^5-methyl-$H_4PteGlu_n$ is 4.82,[48] which would make it a reasonable leaving group. As a chemical model for this reaction, Pandit and co-workers treated the sodium salt of homocysteine with the dimethylammonium pterin analogue **12.33** (Scheme 12.21) at room temperature, and methionine was produced.[49] It is not clear, however, if this reaction proceeds by a S_N2 or radical mechanism. As a means of activating homocysteine as a nucleophile, Matthews and co-workers[50] showed that one equivalent of zinc per subunit is required for catalysis. Homocysteine coordinates to this zinc with release of the thiol proton, thereby binding at the active site as the corresponding zinc thiolate.

b. Cobalamin-Dependent Methionine Synthase

The more unusual version of this enzyme requires a coenzyme that we have not yet discussed, namely, methylcobalamin (**12.34**); the cyclic ligand is called a corrin

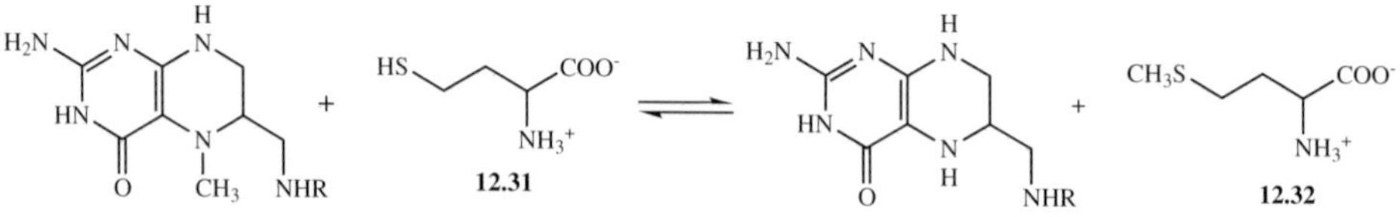

SCHEME 12.20   Reaction catalyzed by the cobalamin-independent methionine synthase.

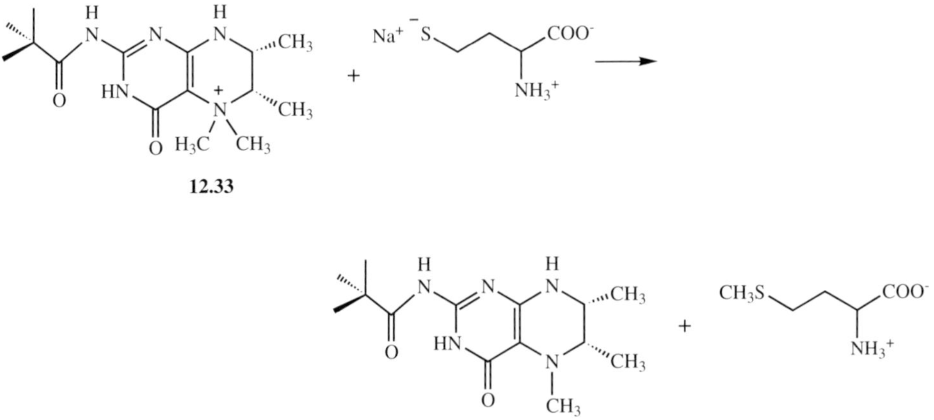

12.33

SCHEME 12.21   Model study for the reaction catalyzed by the cobalamin-independent methionine synthase.

ring, a relative of the porphyrin ring in heme. This coenzyme is generated by N^5-methyl-H_4PteGlu methylation of cob(I)alamin (i.e., cobalamin in the Co^{1+} oxidation state, the reduced form; the cobalt in methylcobalmin is considered to be in the +3 oxidation state). Once the methyl group is transferred to the cobalamin, it is then transferred to homocysteine, producing cob(I)alamin for another cycle; Scheme 12.22 depicts this development in general form (**A**) and in the Cleland notation[51] (**B**) to indicate that it is an ordered sequential mechanism (see Appendix I, Section VI.A.4) in which both substrates bind to the methylcobalamin form

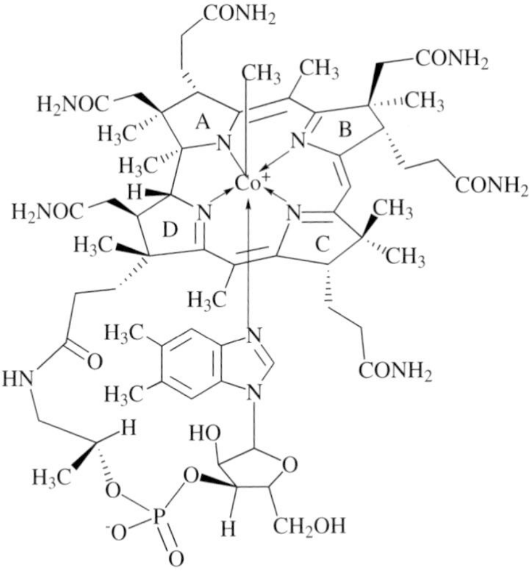

12.34

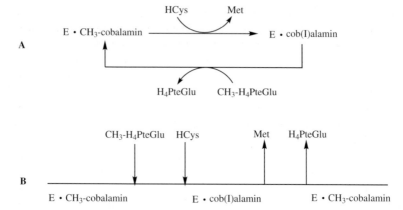

SCHEME 12.22 (**A**) Reaction catalyzed by cobalamin-dependent methionine synthase (**B**) Cleland diagram for the reaction catalyzed by cobalamin-dependent methionine synthase.

of the enzyme prior to methyl transfer. This way, when the homocysteine de-methylates methylcobalamin, the cob(I)alamin is generated in the ternary complex with methyl-H$_4$PteGlu already bound. This minimizes oxidation of cob(I)alamin and loss of enzyme activity (see later). If each of the two methyl transfer steps proceeds by S$_N$2 reactions, then two inversions—which is the same as retention—would result. In fact, the overall transfer of the methyl from N^5-methyl-H$_4$PteGlu to give methionine proceeds with retention of configuration of the methyl group,[52] and cob(I)alamin is a kinetically competent intermediate.[53] Cob(I)alamin is one of the most powerful nucleophiles known,[54] and it reacts with alkyl halides by S$_N$2 reactions, but the reaction rate depends on the leaving group.[55] As in the case of the cobalamin-independent enzyme discussed earlier, the N^5-methyl-H$_4$PteGlu must be activated for nucleophilic attack (this time by the cob(I)alamin); protonation of the N^5 nitrogen, as already discussed, is the most likely way the enzyme could accomplish this. Model studies by Pandit and co-workers and Pratt and co-workers for the methylation of cob(I)alamin, similar to that described earlier for the cobalamin-independent enzyme, demonstrate the feasibility of this reaction (Scheme 12.23).[56] A chemical model study by Hogenkamp and co-workers dem-

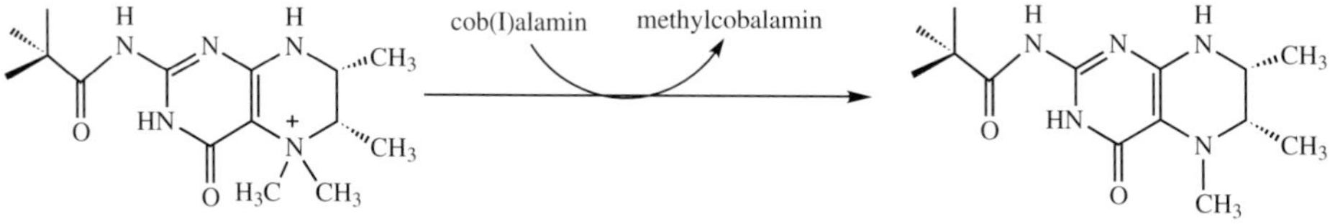

12.33

SCHEME 12.23 Model study for the methylation of cob(I)alamin during the reaction catalyzed by the cobalamin-dependent methionine synthase.

onstrates the feasibility of the transfer of the methyl group from methylcobalamin to a thiolate.[57] As in the case of the cobalamin-independent enzyme discussed earlier, one equivalent of zinc is required for enzyme catalysis; addition of homocysteine is accompanied by the release of one equivalent of protons/mole of enzyme.[58]

The crystal structure of a fragment of the enzyme containing the α/β domain (where the cobalamin resides) reveals that the lower axial 5,6-dimethylbenzimidazole ligand is replaced by a histidine residue (His-759).[59] Furthermore, a water molecule and two hydrogen-bonded residues, Asp-757, and Ser-810, which, along with His-759, are referred to as the *ligand triad,* appear to position and secure His-759 as the lower axial ligand to the cobalt (Figure 12.1). The ligand triad also may be responsible for facilitation of changes in the coordination geometry during conversion of methylcobalamin (six-coordinate) to cob(I)alamin (four-coordinate). The H759G mutant is inactive, but the D757N or D757E and S810A mutants exhibit some enzymatic activity.[60] In the resting state, the methyl group of methylcobalamin is buried in the protein, but on reduction of the cobalt, a conformational change, possibly as a result of the change in coordination geometry, may expose the methyl group for transfer to homocysteine.

During the *in vitro* reaction, cob(I)alamin is slowly oxidized to cob(II)alamin, an inactive form of the prosthetic group. To reactivate the enzyme, *S*-adenosylmethionine (a methylating agent) and a reducing system, in addition to homocysteine and N^5-methyl-H$_4$PteGlu, are needed. The reducing system for cob(II)-alamin can be a dithiol or a reduced flavin.[61] Why should *S*-adenosylmethionine (SAM) be used as the methylating agent instead of N^5-methyl-H$_4$PteGlu for the

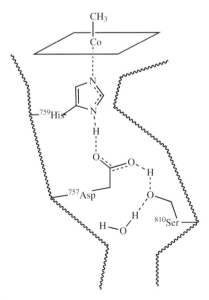

FIGURE 12.1   Ligand triad for cobalamin-dependent methionine synthase.

reactivation process? SAM is a much more reactive methylating agent than N^5-methyl-H_4PteGlu, and it is thought that the highly exergonic reaction with SAM may drive the very endergonic reduction of cob(II)alamin to cob(I)alamin in a coupled electron transfer reaction.

II. S-ADENOSYLMETHIONINE-DEPENDENT ENZYMES: THE TRANSFER OF METHYL GROUPS

In general, neither N^5-methyl-H_4PteGlu nor methylcobalamin is used for methylation of endogenous molecules; rather, S-adenosylmethionine (SAM) (**12.36**), biosynthesized from methionine and ATP (**12.35**) in a reaction catalyzed by methionine adenosyltransferase, is the more important methylating agent (Scheme 12.24). The triphosphate that is released is hydrolyzed to diphosphate and inorganic phosphate to drive this reaction. Methionine adenosyltransferase catalyzes one of only two enzymatic reactions in which a molecule attacks ATP at the C-5′ carbon (the other is in the biosynthesis of coenzyme B_{12}; see Chapter 13, Section III.B.1). In humans, methionine is an essential amino acid, so SAM comes directly from methionine.

SAM is a reactive methylating agent, the endogenous form of methyl iodide. Each of the methyl transfers is catalyzed by a specific methyltransferase, such as catechol-O-methyltransferase, phenol-O-methyltransferase, phenylethanolamine-N-methyltransferase, and nonspecific amine N-methyltransferases and thiol S-methyltransferases. These methyltransferases are responsible for catalyzing the transfer of the methyl group of SAM to a variety of molecules, genericized in Scheme 12.25 as RXH, to give $RXCH_3$ and S-adenosylhomocysteine (**12.37**). With the use of SAM containing a chiral methyl group ($[^1H^2H^3H]SAM$; see Chapter 10, Section I.A.2), Floss, Coward, and co-workers showed that the transmethylation catalyzed by the SAM-dependent catechol-O-methyltransferase oc-

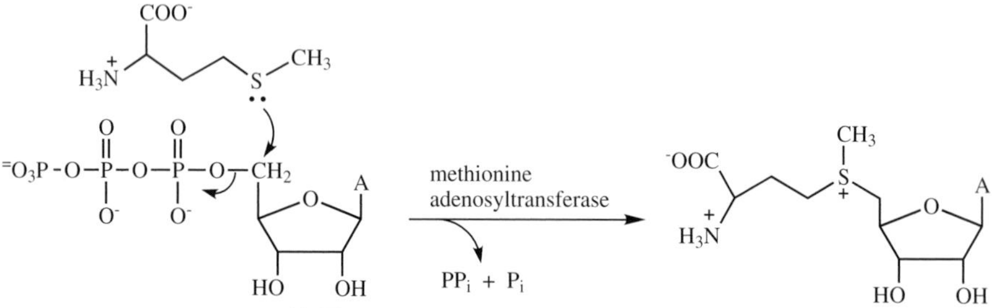

SCHEME 12.24 Proposed mechanism for the synthesis of S-adenosylmethionine catalyzed by methionine adenosyltransferase.

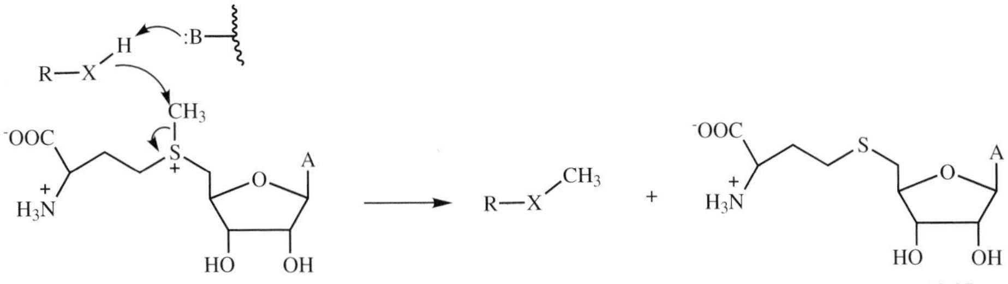

SCHEME 12.25   Generalized reaction catalyzed by *S*-adenosylmethionine-dependent methyltransferases.

curred with inversion of configuration of the methyl group, supporting an S_N2 mechanism.[62] *N*-Methylation occurs at DNA and tRNA bases and at protein side chains; *O*-methylation also occurs at protein carboxylates and at polysaccharides. Even a carbon atom can be methylated by SAM, provided that a carbanion can be generated. Floss and co-workers again used chiral methyl-labeled SAM to show that the methyl group in indolmycin (**12.39**), which is derived from a SAM-dependent methylation of indolylpyruvate (**12.38**), is incorporated with inversion of stereochemistry, as expected for a S_N2 reaction (Scheme 12.26).[63]

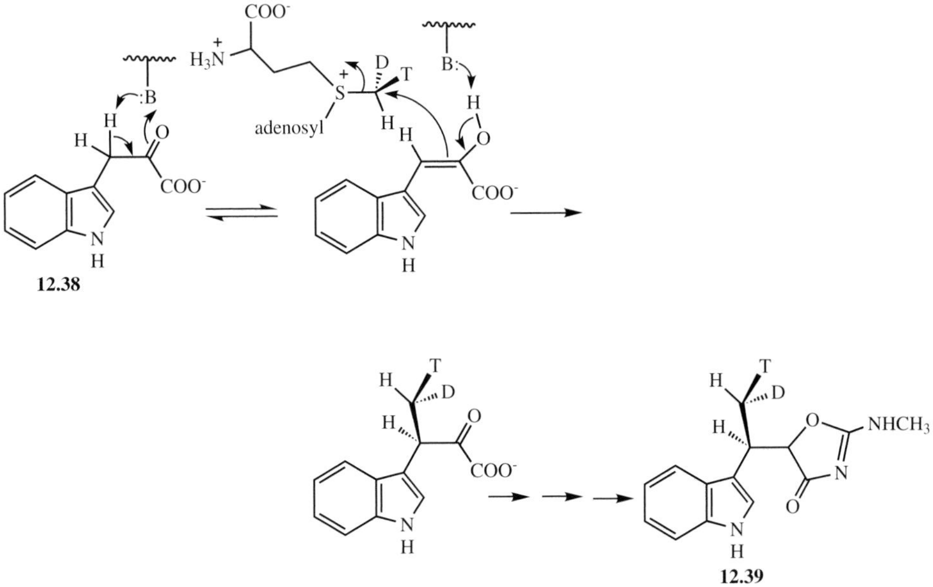

SCHEME 12.26   Stereochemistry of methylation of indolylpyruvate in the biosynthesis of indolmycin.

NOTE THAT PROBLEMS AND SOLUTIONS RELEVANT TO EACH CHAPTER CAN BE FOUND IN APPENDIX II.

REFERENCES

1. Atkinson, I.; Garrow, T.; Brenner, A.; Shane, B. *Meth. Enzymol.* **1997**, *281,* 134.
2. Schirch, V. *Meth. Enzymol.* **1997**, *281,* 146.
3. Bertino, J. R.; Booth, B. A.; Bieber, A. L.; Cashmore, A.; Sartorelli, A. C. *J. Biol. Chem.* **1964**, *239,* 479.
4. (a) Schirch, L. G. In *Folates and Pterins,* vol. 1, Blakely, R. L.; Benkovic, S. J., Eds., Wiley: New York, 1984, p. 399. (b) Matthews, R. G.; Drummond, J. T. *Chem. Rev.* **1990**, *90,* 1275.
5. (a) Besmer, P.; Arigoni, D. *Chimia* **1968**, *22,* 494. (b) Jordan, P. M.; Akhtar, M. *Biochem. J.* **1970**, *116,* 277.
6. (a) Benkovic, S. J.; Bullard, W. P. In *Progress in Bioorganic Chemistry,* vol. 2, Kaiser, E. T.; Kedzy, F., Eds., Wiley: New York, 1973, p. 133. (b) Jordan, P. M.; Akhtar, M. *Biochem. J.* **1970**, *116,* 277.
7. Kallen, R. G.; Jencks, W. P. *J. Biol. Chem.* **1966**, *241,* 5845.
8. Kallen, R. G.; Jencks, W. P. *J. Biol. Chem.* **1966**, *241,* 5851.
9. Vanoni, M. A.; Lee, S.; Floss, H. G.; Matthews, R. G. *J. Am. Chem. Soc.* **1990**, *112,* 3987.
10. Schirch, L. G.; Mason, M. *J. Biol. Chem.* **1963**, *238,* 1032.
11. Jordan, P. M.; El-Obeid, H. A.; Corina, D. L.; Akhtar, M. *J. Chem. Soc. Chem. Commun.* **1976,** 73.
12. Stover, P.; Schirch, V. *Biochemistry* **1992**, *31,* 2155.
13. Daubner, S. C.; Matthews, R. G. *J. Biol. Chem.* **1982**, *257,* 140.
14. Matthews, R. G.; Drummond, J. T. *Chem. Rev.* **1990**, *90,* 1275.
15. Matthews, R. G.; Haywood, B. J. *Biochemistry* **1979**, *18,* 4845.
16. Matthews, R. G.; Kaufman, S. *J. Biol. Chem.* **1980**, *255,* 6014.
17. Matthews, R. G. *Biochemistry* **1982**, *21,* 4165.
18. Sumner, J. S.; Matthews, R. G. *J. Am. Chem. Soc.* **1992**, *114,* 6949.
19. (a) Bruice, T. C. *Prog. Bioorg. Chem.* **1976**, *4,* 1. (b) Hemmerich, P.; Massey, V.; Fenner, H. *FEBS Lett.* **1977**, *84,* 5. (c) Walsh, C. *Acc. Chem. Res.* **1986**, *19,* 216.
20. (a) Chen, P.; Schulze-Gahmen, U.; Stura, E. A.; Inglese J.; Johnson, D. L.; Marloewski, A.; Benkovic, S. J.; Wilson, I. A. *J. Mol. Biol.* **1992**, *227,* 283. (b) Almassy, R. J.; Janson, C. A.; Kan, C. C.; Hostomska, Z. *Proc. Natl. Acad. Sci. USA* **1992**, *89,* 6114. (c) Klein, C.; Chen, P.; Arevalo, J. H.; Stura, E. A.; Marolewski, A.; Warren, M. S.; Benkovic, S. J.; Wilson, I. A. *J. Mol. Biol.* **1995**, *249,* 153.
21. Warren, M. S.; Marolewski, A. E.; Benkovic, S. J. *Biochemistry* **1996**, *35,* 8855.
22. Carreras, C. W.; Santi, D. V. *Annu. Rev. Biochem.* **1995**, *64,* 721.
23. (a) Ivanetich, K. M.; Santi, D. V. *FASEB J.* **1990**, *4,* 1591. (b) Liang, P.-H.; Anderson, K. S. *Biochemistry* **1998**, *37,* 12195, 12206.
24. (a) Costi, M. P. *Med. Res. Rev.* **1998**, *18,* 21. (b) Peters, F. J.; Kuiper, C. M.; van Triest B.; Backus, H.; van der Wilt, C. L.; van Laar, J. A.; Jansen, G.; van Groeningen, C. J.; Pinedo, H. M. *Adv. Exp. Med. Biol.* **1998**, *431,* 699.
25. Lomax, M. I. S.; Greenberg, G. R. *J. Biol. Chem.* **1967**, *242,* 109 and 1302.
26. Bruice, T. W.; Santi, D. V. *Biochemistry* **1982**, *21,* 6703.
27. (a) Pastore, E.; Friedkin, M. *J. Biol. Chem.* **1962**, *237,* 3802. (b) Blakley, R. L.; Ramasastri, B. V.; McDougall, B. M. *J. Biol. Chem.* **1963**, *238,* 3075.
28. Pogolotti, A. L. Jr.; Santi, D. V. In *Bioorganic Chemistry,* vol. 1, van Tamelen, E. E., Ed., Academic Press: New York, 1977; p. 277.
29. Santi, D. V.; Brewer, C. F.; Farber, D. *J. Heterocycl. Chem.* **1970**, *7,* 903.
30. (a) Kalman, T. I. *Biochemistry* **1971**, *10,* 2567. (b) Wataya, Y.; Hayatsu, H.; Kawazoe, Y. *J. Biochem.* **1973**, *73,* 871.

31. (a) Santi, D. V.; McHenry, C. S.; Sommer, H. *Biochemistry* **1974,** *13,* 471. (b) Pogolotti, A. L. Jr.; Ivanetich, K. M.; Sommer, H.; Santi, D. V. *Biochem. Biophys. Res. Commun.* **1976,** *70,* 972. (c) Danenberg, P. V.; Heidelberger, C. *Biochemistry* **1976,** *15,* 1331. (d) Bellisario, R. L.; Maley, G. F.; Galivan, J.; Maley, F. *Proc. Natl. Acad. Sci. USA* **1976,** *73,* 1848.

32. Jencks, W. P. *Prog. Phys. Org. Chem.* **1964,** *2,* 63.

33. Kallen, R. G.; Jencks, W. P. *J. Biol. Chem.* **1966,** *241,* 5851.

34. Gupta, V. S.; Huennekens, F. M. *Arch. Biochem. Biophys.* **1967,** *120,* 712.

35. Matthews, D. A.; Villafranca, J. E.; Janson, C. A.; Smith, W. W.; Welsh, K.; Freer, S. *J. Mol. Biol.* **1990,** *214,* 937.

36. Moore, M. A.; Ahmed, F.; Dunlap, R. B. *Biochemistry* **1986,** *25,* 3311.

37. Huang, W.; Santi, D. V. *J. Biol. Chem.* **1994,** *269,* 31327.

38. Friedkin, M. In *The Kinetics of Cellular Proliferation,* Stohlman, F., Jr., Ed., Grune & Stratton: New York, 1959, p. 97.

39. (a) Woodward, R. B.; Hoffmann, R. *The Conservation of Orbital Symmetry,* Academic Press: New York, 1970, p. 120. (b) Spangler, C. W. *Chem. Rev.* **1976,** *76,* 187.

40. Pogolotti, A. L.; Santi, D. V. *Biochemistry* **1974,** *13,* 456.

41. Barrett, J. E.; Maltby, D. A.; Santi, D. V.; Schultz, P. G. *J. Am. Chem. Soc.* **1998,** *120,* 449.

42. Slieker, L. J.; Benkovic, S. J. *J. Am. Chem. Soc.* **1984,** *106,* 1833.

43. Matthews, R. G.; Drummond, J. T. *Chem. Rev.* **1990,** *90,* 1275.

44. Lin, S.-K.; Jaun, B. *Helv. Chim. Acta* **1991,** *74,* 1725.

45. Ram, M. S.; Riordan, C. G.; Yap, G. P. A.; LiableSands, L.; Rheingold, A. L.; Marchaj, A.; Norton, J. R. *J. Am. Chem. Soc.* **1997,** *119,* 1648.

46. (a) Ragsdale, S. W.; Kumar, M. *Chem. Rev.* **1996,** *96,* 2515. (b) Ragsdale, S. W.; Riordan, C. G. *J. Bioinorg. Chem.* **1996,** *1,* 489.

47. Whitfield, C. D.; Steers, E. J., Jr.; Weissbach, H. *J. Biol. Chem.* **1970,** *245,* 390.

48. Kallen, R. G.; Jencks, W. P. *J. Biol. Chem.* **1966,** *241,* 5845.

49. Hilhorst, E.; Chen, T. B. R. A.; Iskander, A. S.; Pandit, U. K. *Tetrahedron* **1994,** *50,* 7837.

50. (a) González, J. C.; Peariso, K.; Penner-Hahn, J. E.; Matthews, R. G. *Biochemistry* **1996,** *35,* 12228. (b) Matthews, R. G.; Goulding, C. W. *Current Opin. Chem. Biol.* **1997,** *1,* 332.

51. Cleland, W. W. *The Enzymes,* vol. 2, Boyer, P. D., Ed., Academic Press: New York, 1970, p. 1.

52. Zydowsky, T. M.; Courtney, L. F.; Frasca, V.; Kobayashi, K.; Shimizu, H.; Yuen, L.-D.; Matthews, R. G.; Benkovic, S. J.; Floss, H. G. *J. Am. Chem. Soc.* **1986,** *108,* 3152.

53. Banerjee, R. V.; Frasca, V.; Ballou, D. P.; Matthews, R. G. *Biochemistry* **1990,** *29,* 11101.

54. Brown, K. L. In *B₁₂,* Dolphin, D., Ed., Wiley: New York, 1982, p. 245.

55. Schrauzer, G. N.; Deutsch, E. *J. Am. Chem. Soc.* **1969,** *91,* 3341.

56. Hilhorst, E.; Iskander, A. S.; Chem, T. B. R. A.; Pandit, U. K. *Tetrahedron* **1994,** *50,* 8863. See also Pratt, J. M.; Norris, P. R.; Hamza, M. S. A.; Bolton, R. *J. Chem. Soc. Chem. Commun.* **1994,** 1333.

57. Hogenkamp, H. P. C.; Bratt, G. T.; Sun, S.-z. *Biochemistry* **1985,** *24,* 6428.

58. (a) Goulding, C. W.; Matthews, R. G. *Biochemistry* **1997,** *36,* 15749. (b) Matthews, R. G.; Goulding, C. W. *Current Opin. Chem. Biol.* **1997,** *1,* 332.

59. Drennan, C. L.; Huang, S.; Drummond, J. T.; Matthews, R. G.; Ludwig, M. L. *Science* **1994,** *266,* 1669.

60. Jarrett, J. T.; Amaratunga, M.; Drennan, C. L.; Scholten, J. D.; Sanols, R. H.; Ludwig, M. L.; Matthews, R. G. *Biochemistry* **1996,** *35,* 2464.

61. (a) Foster, M. A.; Dilworth, M. J.; Woods, D. D. *Nature* **1964,** *201,* 39. (b) Taylor, R. T.; Weissbach, H. *J. Biol. Chem.* **1967,** *242,* 1502.

62. Woodard, R. W.; Tsai, M. D.; Floss, H. G.; Crooks, P. A.; Coward, J. K. *J. Biol. Chem.* **1980,** *255,* 9124.

63. Mascaro, J.; Horhammer, R.; Eisenstein, S.; Sellers, L.; Mascaro, K.; Floss, H. *J. Am. Chem. Soc.* **1977,** *99,* 273.

CHAPTER 13

Rearrangements

I. PERICYCLIC REACTIONS

A *pericyclic reaction* is one in which bonding changes occur via reorganization of electrons within a loop of interacting orbitals. These reactions are concerted; that is, all bond making and bond breaking occur simultaneously with no intermediates formed in the process. However, at any point along the reaction coordinate, the extent of bond breakage and bond formation need not be the same. So, although bond cleavage and formation occur simultaneously, the process does not have to be synchronous. The basis for understanding why pericyclic reactions work was provided by the Nobel Prize-winning contributions of Woodward and Hoffmann.[1] Pericyclic reactions are generally divided into five categories: sigmatropic, cycloaddition, electrocyclic, cheletropic, and group transfer reactions. There are very few enzyme-catalyzed examples of these reactions, unless catalytic antibodies are included.

A. Sigmatropic Rearrangements

1. Claisen Rearrangement, a Type of [3,3] Sigmatropic Rearrangement

The general form of a *Claisen rearrangement* is shown in Scheme 13.1. This is a thermally allowed [3,3] sigmatropic reaction. The only example of this rearrangement

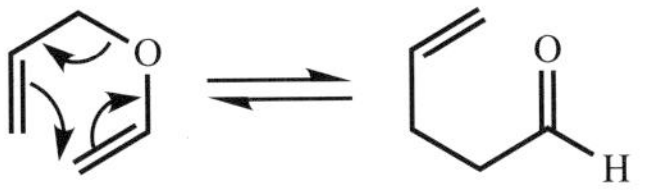

SCHEME 13.1   General form of the Claisen rearrangement.

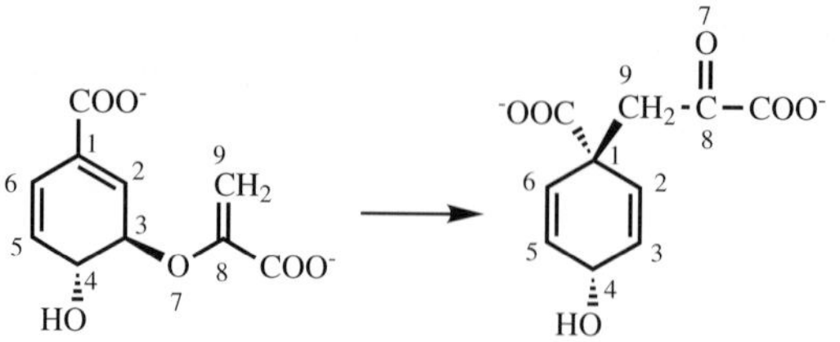

SCHEME 13.2   Chorismate mutase-catalyzed conversion of chorismate to prephenate.

in primary metabolism is found in chorismate mutase, which catalyzes the conversion of chorismate (**13.1**) to prephenate (**13.2,** Scheme 13.2), an important step in the biosynthesis of tyrosine and phenylalanine in bacteria, fungi, and higher plants.[2] At pH 7.5 and 37 °C the corresponding nonenzymatic rearrangement occurs with a $k = 2.6 \times 10^{-5}$ s^{-1}; the rate constant with chorismate mutase under the same conditions is 50 s^{-1}, indicating a rate enhancement by the enzyme of 2×10^{6}.[3] Catalytic antibodies also were raised in the Schultz and Hilvert groups[4] to catalyze the conversion of chorismate to prephenate using a hapten (**13.3**) containing a potent *endo*-oxabicyclo[3.3.1]nonene transition-state inhibitor (**13.4**) designed by Bartlett and co-workers.[5]

An NMR study of chorismate indicated that the predominant conformer in solution has the substituents on the ring in pseudodiequatorial orientations (**13.5**),[6] but the reactive conformation for a Claisen rearrangement would be the one with

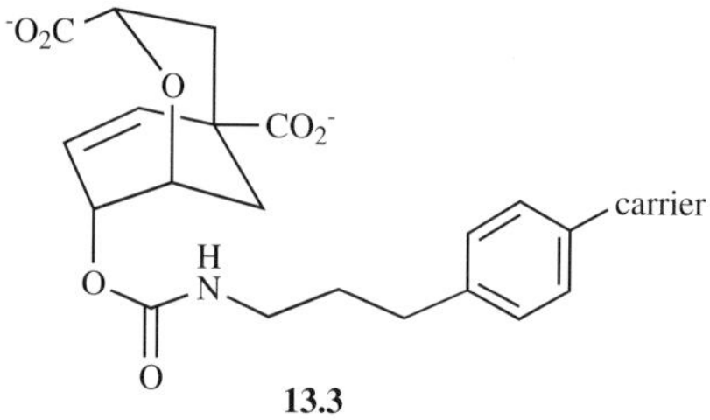

13.3

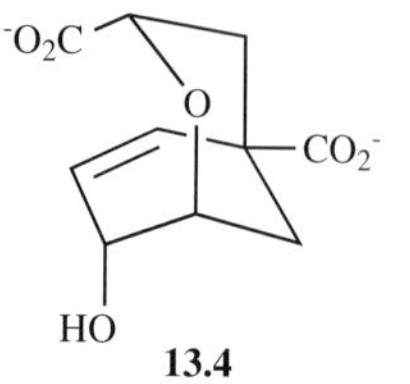

13.4

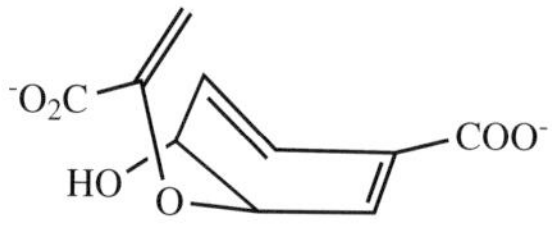

13.5

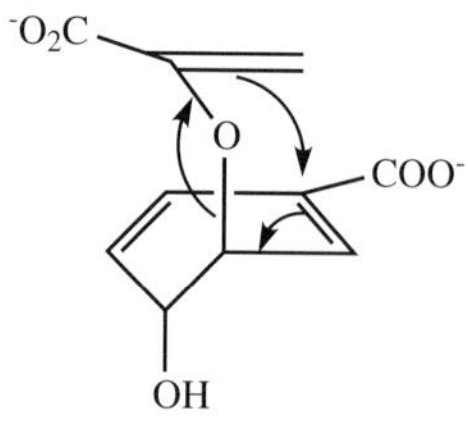

13.6

these groups diaxial (**13.6**). Substrate labeling[7,8] and kinetic isotope effect studies[9] show that both the enzymatic and nonenzymatic reactions proceed via chairlike transition states and that the C^3-O^7 bond cleavage precedes the C^1-C^9 bond formation (see Scheme 13.2 for the numbering system employed). Andrews and co-workers found in nonenzymatic systems that the chair conformation transition state is favored over the boat transition state by about 2 kcal/mol.[10] One of the clever experiments designed by Knowles and co-workers to determine the transition-state structure for the chorismate mutase-catalyzed reaction utilizes (E)- and (Z)-[9-^{3}H]chorismate (**13.7** in Scheme 13.3 is the Z-isomer). These compounds were chemoenzymatically synthesized, incubated with chorismate mutase, and the pre-phenate was analyzed to determine the position of the tritium. If the Z-isomer proceeds via a chairlike transition state (**A**), then the resulting prephenate would have the tritium in the *pro-S* position (**13.8**), but if the reaction proceeds via a boat-like transition state (note that this reaction still starts from **13.7**, except with the enolpyruvyl group flipped over), the tritium in the prephenate would be at the *pro-R* position (**13.9**). How do you know the stereochemistry of the tritium in the prephenate? An enzymatic analysis can be performed by the route shown in Scheme 13.4. The chorismate mutase reaction is run in the presence of phenyl-pyruvate tautomerase (see Chapter 9, Section IV.A) at pH <6 so that the prephe-nate that forms spontaneously undergoes decarboxylative elimination of water to give phenylpyruvate, which is then enolized by the phenylpyruvate tautomerase, known to remove the *pro-R* proton from phenylpyruvate stereospecifically.[11] On incubation of Z-[9-^{3}H]chorismate under these conditions, only 20% of the tritium

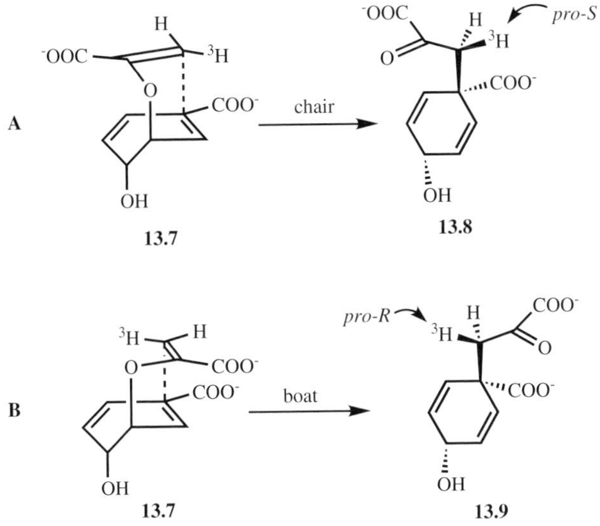

SCHEME 13.3   Stereochemical outcome if chorismate mutase proceeds via chair and boat transition states, respectively, during reaction with (Z)-[9-^{3}H]chorismate.

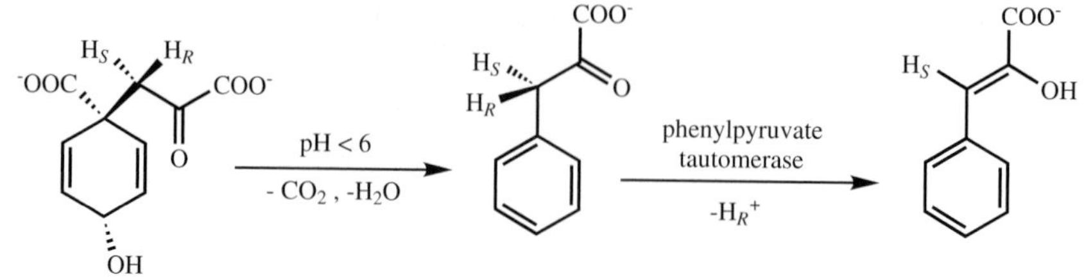

SCHEME 13.4   Chemoenzymatic degradation of the prephenate formed from the chorismate mutase-catalyzed conversion of (Z)-[9-^{3}H]chorismate to determine the position of the tritium.

is released, but when the corresponding E-isomer is used, 67% of the tritium is released, supporting a chairlike transition state (theoretically, no tritium should have been released from the Z-isomer and 100% released from the E-isomer, but biochemical results are rarely that clearcut; that is why it is *essential,* when doing studies with enantiomeric compounds, to carry out parallel experiments with both enantiomers).

On the basis of NMR and reaction rate studies, Copley and Knowles showed that the diaxial conformer of chorismate (the conformer needed for the Claisen rearrangement) binds to chorismate mutase preferentially[12] and that the diaxial conformer exists in 10−40% of the equilibrium mixture in solution; therefore, the enzyme does not have to catalyze a conformational isomerization.

The nonenzymatic rearrangement of chorismate shows a secondary tritium isotope effect at C-3 (the site of bond breaking), but none at C-9 (the site of bond making). This suggests either a stepwise mechanism or a very unsymmetrical transition state. No isotope effect is observed at either C-3 or C-9 in the enzyme-catalyzed reaction, but the rate-determining step may precede this step, in which case no isotope effect would be observed. To determine if the chorismate mutase-catalyzed reaction proceeds by a stepwise mechanism, experiments were carried out by Knowles and co-workers to test five alternative (stepwise) mechanisms (Figure 13.1).[13] A secondary tritium isotope effect (k_H/k_T) on C-4 was found to be 0.96 (a slight inverse isotope effect). This rules out pathways 1−3 because they proceed with a change in hybridization from sp^3 to sp^2, which would have exhibited a normal k_H/k_T isotope effect of 1.10−1.25. Compound **13.10** is a substrate for chorismate mutase, so pathway 5 can be eliminated. Because **13.11** also is a substrate,[14] pathways 1, 2, and 5, again, are not reasonable. The FTIR of the complex between prephenate and chorismate mutase shows one band at 1714 cm^{-1}.[15] With the corresponding ^{13}C-labeled compound, it can be shown that this band corresponds to the ketone carbonyl of enzyme-bound prephenate. The frequency of this bond does not change on binding, indicating that hydrogen bond donation to the carbonyl oxygen to activate the carbonyl is not occurring (otherwise, the carbonyl frequency would not be that of a ketone carbonyl). Therefore, chorismate mutase does not use electrophilic catalysis in its rearrangement of chorismate, and it can be concluded that this reaction is pericyclic; the rate acceleration apparently derives from selective binding of the appropriate conformer of the substrate.

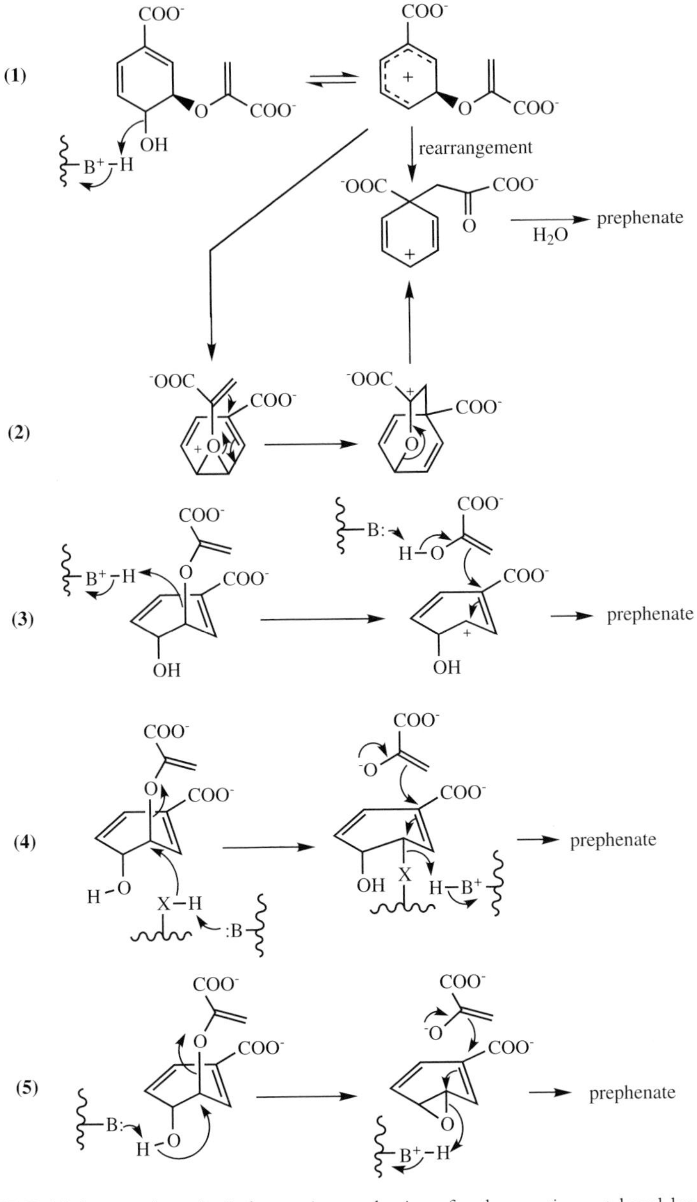

FIGURE 13.1   Five hypothetical stepwise mechanisms for the reaction catalyzed by chorismate mutase.

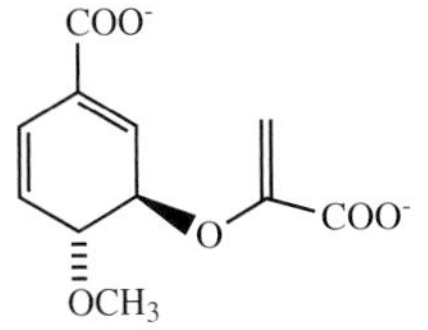

13.10

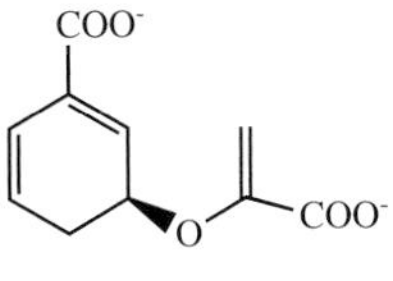

13.11

Several research groups have solved the crystal structures of chorismate mutase from *Bacillus subtilis*,[16] *Escherichia coli*,[17] and catalytic antibody 1F7[18] with the *endo*-oxabicyclo[3.3.1]nonene transition-state inhibitor **13.4** bound to them. In the *Bacillus subtilis* structure, it is apparent that there are no residues capable of proton transfer to the ether oxygen. The general conclusion is that chorismate mutase stabilizes the chairlike transition-state geometry by a variety of electrostatic and hydrogen-bonding interactions. This is consistent with the finding that the nonenzymatic rearrangement of chorismate occurs more readily in hydrogen-bonding solvents.[19] The active-site residues may stabilize developing charge on the enol ether oxygen and the cyclohexadiene ring in a polar transition state.

To test the hypotheses related to the binding interactions in the crystal structures, Schultz and co-workers generated a series of 16 mutants of the *Bacillus subtilis* enzyme[20] and 13 mutants of the *Escherichia coli* enzyme.[21] The results with the mutants indicate that neither general acid/base catalysis (pathways 1–3 and 5) nor nucleophilic catalysis (pathway 4) plays a significant role in the reaction catalyzed by chorismate mutase. Hydrogen bonds to the enolpyruvyl oxygen and carboxylate groups seem to be critical for catalysis. Hydrogen bonds to the C-4 hydroxyl group appear to be involved in orienting that group. The primary function of the enzyme therefore appears to be to stabilize a chairlike transition state via hydrogen bonding and electrostatic interactions, thus supporting a pericyclic mechanism.

2. Oxy-Cope Rearrangement, Another Type of [3,3] Sigmatropic Rearrangement

Cope (**A**) and oxy-Cope (**B**) rearrangements are depicted in Scheme 13.5. Although no example of this type of reaction has yet been discovered in nature, Braisted and Schultz have generated an antibody from hapten **13.12** that catalyzes

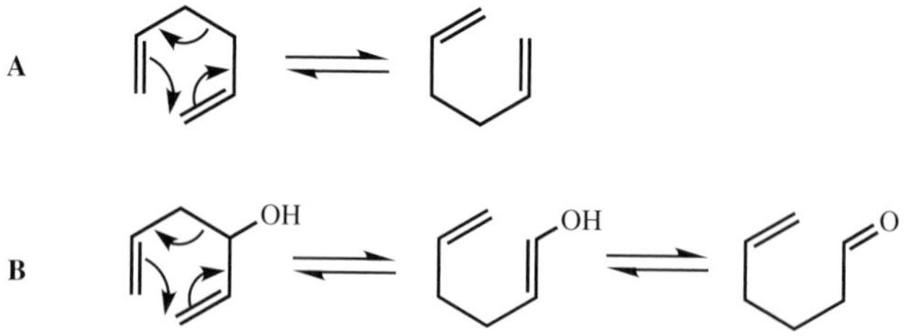

SCHEME 13.5   General form of Cope (**A**) and oxy-Cope (**B**) reactions.

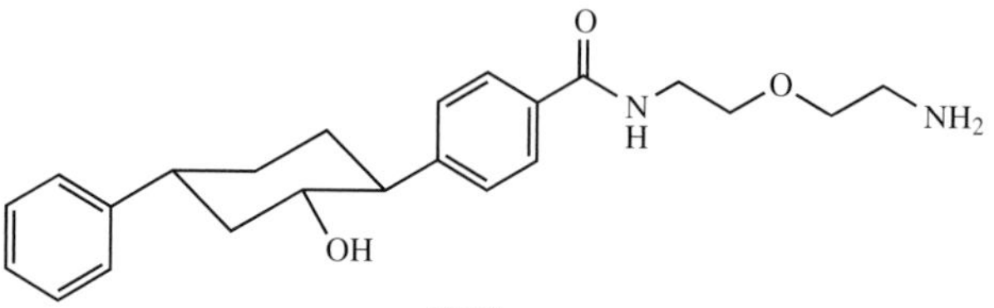

13.12

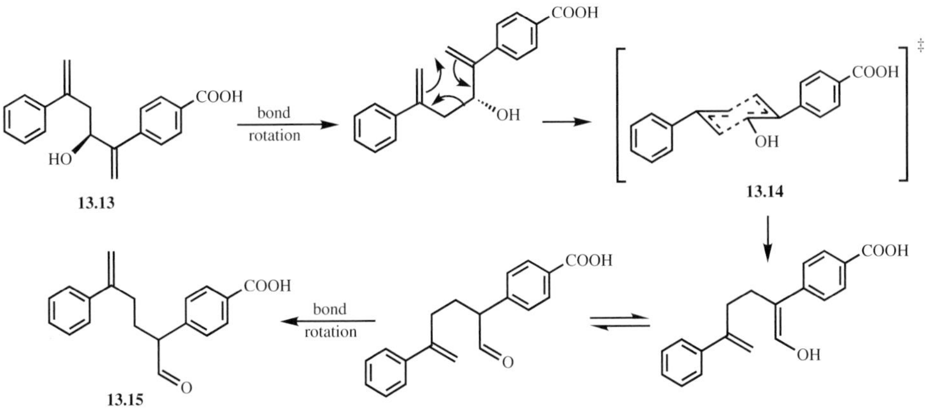

SCHEME 13.6   Oxy-Cope rearrangement catalyzed by an antibody.

the oxy-Cope conversion of **13.13** to **13.15** (Scheme 13.6).[22] NMR studies indicate that although the free substrate adopts an extended conformation in solution, the substrate bound to the antibody has a cyclic conformation, consistent with a chairlike geometry (**13.14**).[23] When both termini of the double bonds are dideuterated, the k_{cat}^{H}/k_{cat}^{D} exhibits a secondary kinetic isotope effect of 0.61, indicating that the rearrangement step is at least partially, if not wholly, rate determining. The magnitude of this inverse secondary isotope effect suggests a significant degree of bond formation between the ends of the two alkene bonds in the transition state. The 2.6 Å resolution crystal structure with the hapten bound is consistent with these NMR and isotope effect results.[24]

3. [2,3] Sigmatropic Rearrangement

Cyclohexanone oxygenase (see Chapter 4, Section II.B.3) catalyzes a [2,3] sigmatropic rearrangement on unsaturated selenides, such as **13.16,** yielding **13.17** (Scheme 13.7).[25] This is the same product as is obtained from the nonenzymatic ozonolysis of **13.16.**

An antibody also was generated to catalyze a [2,3] sigmatropic reaction, namely, an N-oxide elimination reaction (Scheme 13.8).[26]

B. Cycloaddition Reactions

1. Enzyme-Catalyzed Cycloaddition Reactions

The most common of the cycloaddition reactions is the [4+2] cycloaddition or *Diels–Alder reaction* (Scheme 13.9),[27] named for the two Nobel Prize-winning (1950) chemists who first recognized and developed this reaction. Unlike the [3,3] sigmatropic rearrangement, note that the [4+2] cycloaddition reaction proceeds via a

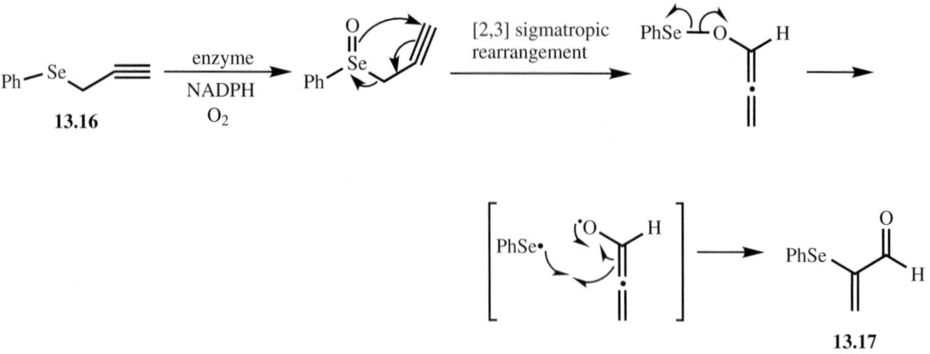

SCHEME 13.7   [2,3] Sigmatropic rearrangement catalyzed by cyclohexanone oxygenase.

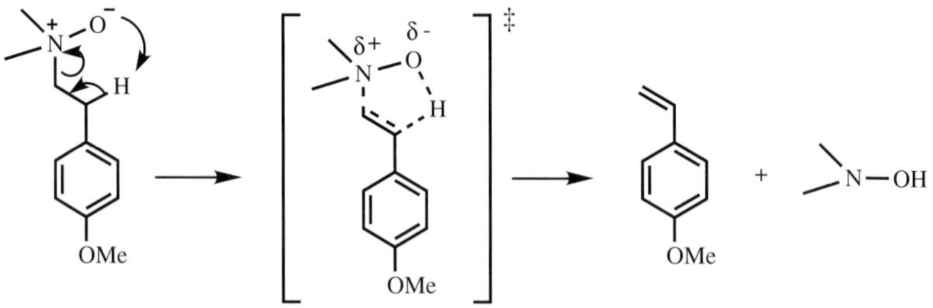

SCHEME 13.8   [2,3] Sigmatropic rearrangement catalyzed by an antibody.

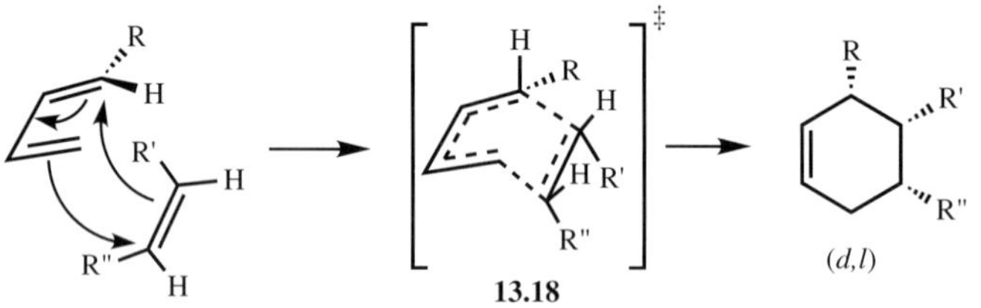

SCHEME 13.9   [4+2] Cycloaddition (Diels-Alder) reaction.

boatlike transition state (**13.18**). The favored stereochemistry derives from an *endo* cyclization, in which the larger side of the dienophile adds under the plane of the diene, as shown in Scheme 13.9.

A series of labeling experiments by Oikawa, Ichihara, and co-workers in the biosynthesis of solanapyrones (**13.19**), two phytotoxins produced by the pathogenic fungus *Alternaria solani,* provided the first direct evidence for an intramolecular Diels–Alder reaction in nature with demonstrated diene and dienophile precursors (Scheme 13.10).[28] Interestingly, the *exo–endo* ratio of products (**13.19a–13.19b**)

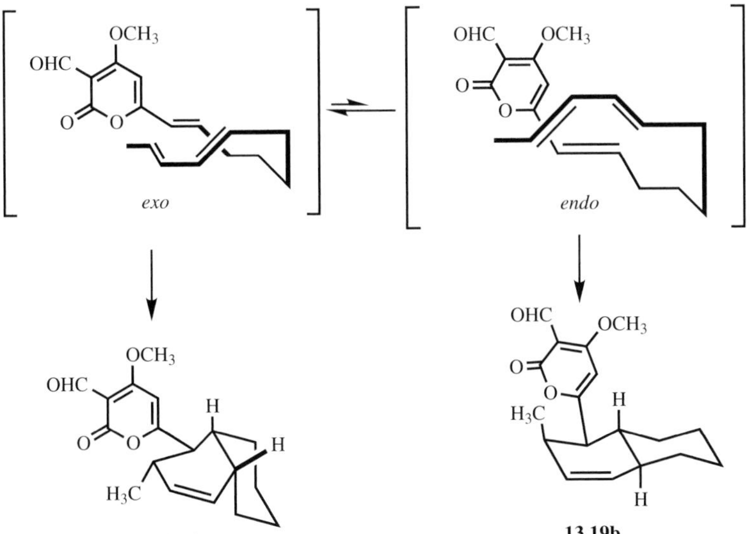

SCHEME 13.10   An intramolecular Diels–Alder reaction catalyzed by *Alternaria solani.*

is 53:47; in aqueous solution the corresponding nonenzymatic Diels–Alder reaction gives an *exo–endo* ratio of 3:97. This latter ratio is typical for nonenzymatic Diels–Alder reactions in water.[29]

   Initially, researchers thought that the enzyme DNA photolyase catalyzed a [2+2] cycloreversion reaction, but the evidence to date suggests a stepwise radical mechanism instead (see Section III.A).

2. Antibody-Catalyzed Cycloaddition Reactions

An antibody (39-A11) that catalyzes the Diels–Alder reaction in Scheme 13.11[30] was generated from hapten **13.20,** which mimics the boatlike transition state of the Diels–Alder reaction. The crystal structure at 2.4 Å resolution of this antibody with **13.20** bound demonstrates that the antibody probably binds the diene and dienophile in a reactive orientation that reduces both translational and rotational degrees of freedom.[31] Hydrogen-bonding and π-stacking interactions with the maleimide ring orient the dienophile, and the diene is bound in a hydrophobic pocket close to the dienophile. The hapten binds to the antibody 1000 times greater than the product because of conformational constraints imposed by the [2.2.2]bicyclic framework, which lock the cyclohexene ring of the hapten into a boatlike geometry distinct from that of the product, and lead to better binding interactions with the active-site residues. A single mutation (S91V) is largely responsible for the increase in both affinity and catalytic activity.

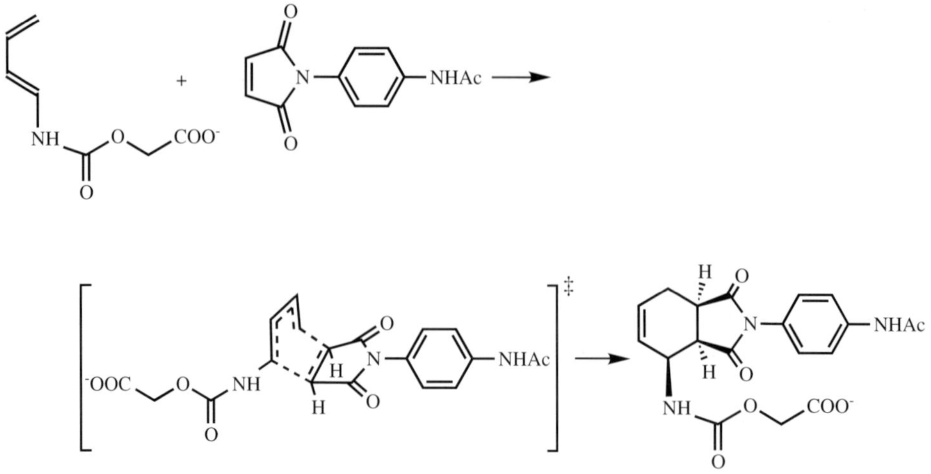

SCHEME 13.11   An antibody-catalyzed Diels−Alder reaction.

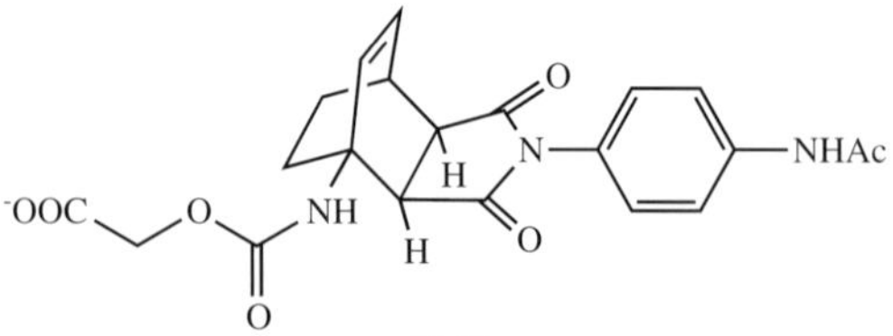

13.20

A catalytic antibody raised by Janda, Lerner, and co-workers catalyzes a favored *endo* Diels−Alder reaction (*endo* Diels−Alderase 7D4), and another related antibody was generated to catalyze the disfavored *exo* pathway (*exo* Diels−Alderase 22C8), yielding exclusively the *cis* (**13.21**) and *trans* (**13.22**) adducts, respectively, in >98% enantiomeric purity (Scheme 13.12; this shows the result of the nonenzymatic re-

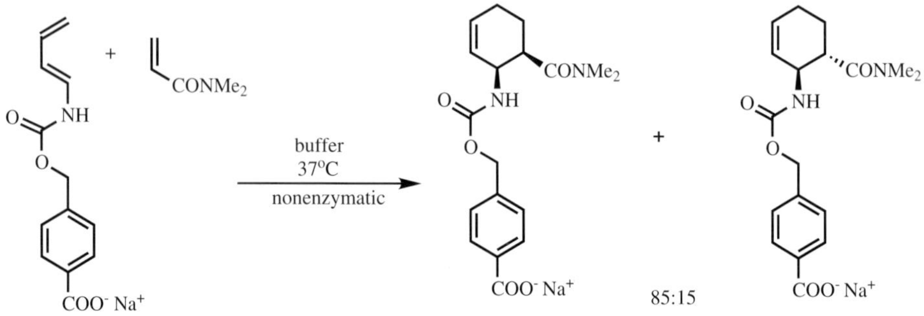

SCHEME 13.12   Nonenzymatic Diels−Alder reaction to mimic an antibody-catalyzed reaction.

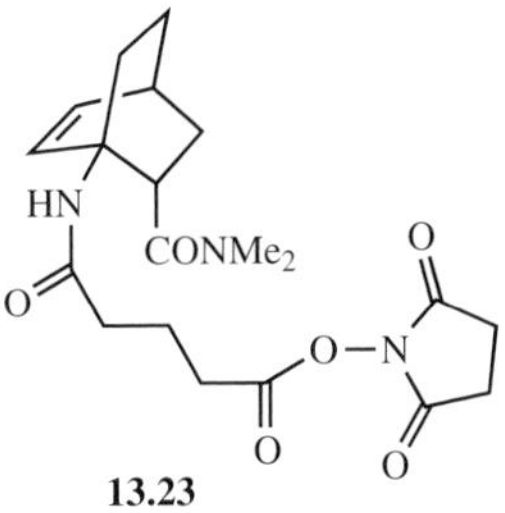

13.23 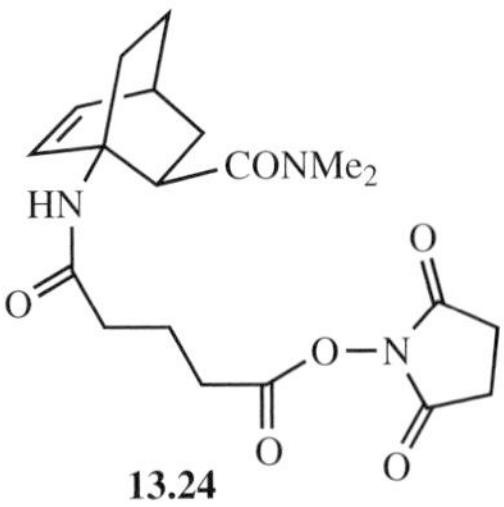   **13.24**

action).[32] The hapten originally used to generate the antibody for exclusive *endo* cyclization is **13.23,** and the one to generate the antibody for exclusive *exo* cyclization is **13.24.** To achieve this high degree of stereoselectivity, the antibody must control both the conformation of the individual reactants and their relation to each other. In the case of the *exo* reaction, the binding energy to the catalytic antibody must be used to reroute the reaction along the higher-energy (*exo*) pathway. It was determined that ~20 kcal/mol of binding energy is the maximum that the antibody can use to reroute a reaction.

To show that the best haptens for reactions that proceed through highly ordered transition states need not be conformationally restricted analogues, Janda and coworkers synthesized the flexible ferrocene hapten **13.25** (conjugated to keyhole limpet hemocyanin), which generates an antibody (13G5) that catalyzes the *exo* Diels–Alder reaction shown in Scheme 13.12 and gives the *ortho* product (the one favored in nonenzymatic Diels–Alder reactions) with high regio-, diastereo-, and enantioselectivity; only one out of eight possible isomers is produced.[33] The *ortho* transition state is predicted to be 2–4 kcal/mol lower in energy than the corresponding *meta* transition state with the *endo* transition states being about 1.9 kcal/mol lower in energy than the corresponding *exo* transition states. In the nonenzymatic reaction, however, no enantioselectivity is observed. A crystal structure at 1.95 Å resolution was obtained with the more stable ferrocene inhibitor **13.26** bound.[34] Three active-site residues appear to be responsible for catalysis and product

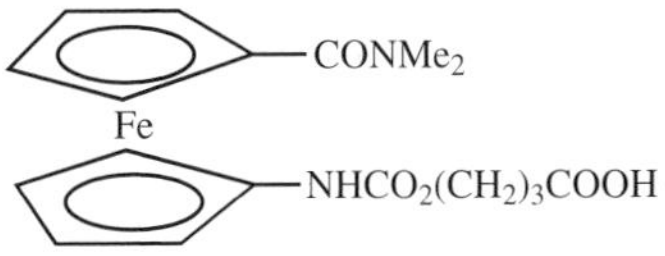

13.25

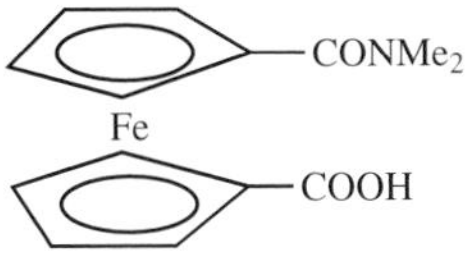

13.26

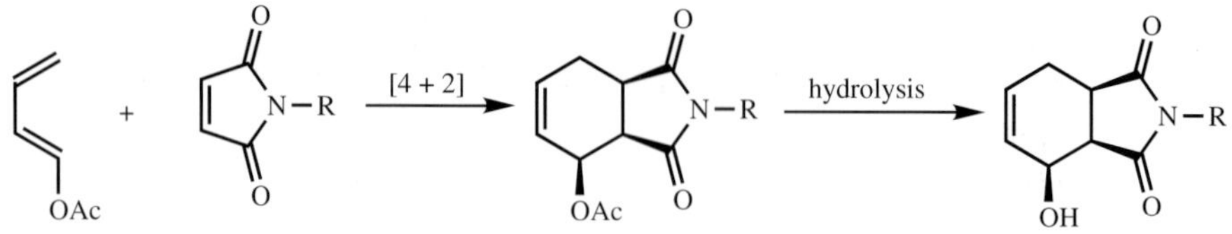

SCHEME 13.13 An antibody-catalyzed Diels–Alder and hydrolysis reactions.

structure control. Tyrosine-L36 (i.e., Tyr-36 on the light chain) acts as a Lewis acid to activate the dienophile, and asparagine-L91 and aspartic acid-H50 (Asp-50 on the heavy chain) hydrogen bond to the carboxylate side chain that substitutes for the carbamate diene substrate. This hydrogen bonding results in the rate acceleration and also the pronounced stereoselectivity.

Antibodies raised by Suckling and co-workers to the bovine serum albumin conjugate of the adduct of acetoxybutadiene with N-(4-carboxybutanoyl)maleimide (**13.29**, R = 4-carboxybutyl) catalyze the Diels–Alder cycloaddition of acetoxybutadiene (**13.27**) with N-ethyl- and N-benzylmaleimide (**13.28**, R = Et or Bn, respectively; Scheme 13.13).[35] The product of this abzyme reaction, however, was not the expected Diels–Alder acetate (**13.29**) but instead the Diels–Alder alcohol (**13.30**). This result suggests that the antibody catalyzes two consecutive reactions, a [4+2] cycloaddition followed by an ester hydrolysis.

II. REARRANGEMENTS THAT PROCEED VIA CARBENIUM ION INTERMEDIATES

A. Acyloin Rearrangement

An *acyloin*-type rearrangement is an acid-catalyzed reorganization of α-hydroxy ketones (**13.31**), also known as acyloins (Scheme 13.14). These acid-catalyzed [1,2]-alkyl migration reactions also can involve other types of molecules, such as glycols (pinacol rearrangement), α-diketones (benzil–benzilic acid rearrangement), and alcohols (Meerwein rearrangements).[36]

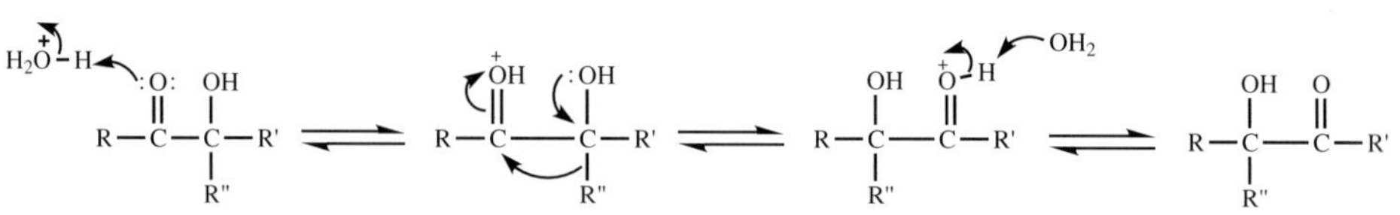

SCHEME 13.14 An acid-catalyzed acyloin-type rearrangement.

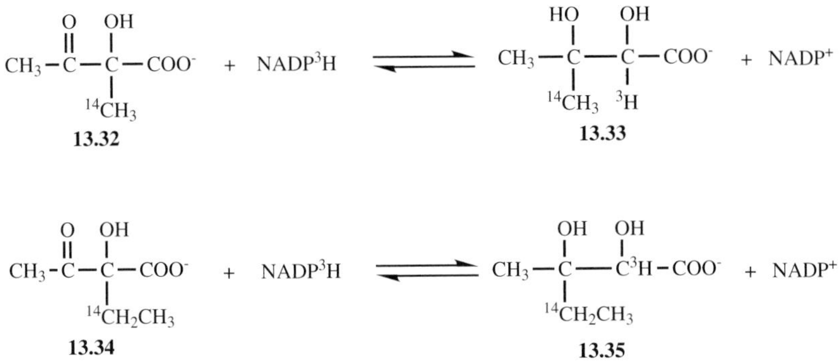

$$CH_3-\overset{\overset{O}{\|}}{C}-\overset{\overset{OH}{|}}{\underset{\underset{^{14}CH_3}{|}}{C}}-COO^- \; + \; NADP^3H \; \rightleftharpoons \; CH_3-\overset{\overset{HO}{|}}{\underset{\underset{^{14}CH_3}{|}}{C}}-\overset{\overset{OH}{|}}{\underset{\underset{^3H}{|}}{C}}-COO^- \; + \; NADP^+$$

13.32   **13.33**

$$CH_3-\overset{\overset{O}{\|}}{C}-\overset{\overset{OH}{|}}{\underset{\underset{^{14}CH_2CH_3}{|}}{C}}-COO^- \; + \; NADP^3H \; \rightleftharpoons \; CH_3-\overset{\overset{OH}{|}}{\underset{\underset{^{14}CH_2CH_3}{|}}{C}}-\overset{\overset{OH}{|}}{C^3H}-COO^- \; + \; NADP^+$$

13.34   **13.35**

SCHEME 13.15   Reactions catalyzed by acetohydroxy acid isomeroreductase.

An enzymatic example of an acyloin rearrangement is the reaction catalyzed by acetohydroxy acid isomeroreductase in the biosynthesis of valine and isoleucine; the same enzyme catalyzes transformations of both acetolactate (**13.32**) and α-aceto-α-hydroxybutyrate (**13.34**) to give α,β-dihydroxyisovalerate (**13.33**) and α,β-dihydroxy-β-methylvalerate (**13.35**), respectively (Scheme 13.15). (4S)-[^{3}H]NADPH introduces a tritium into C-2 of the dihydroxy product. A [1,2] migration can be demonstrated by labeling the methyl or ethyl groups of **13.32** and **13.34**, respectively; these alkyl groups migrate to the C-3 position in **13.33** and **13.35**, respectively.[37] β-Hydroxy-α-ketoisovalerate (**13.36**) was synthesized by Arfin and Umbarger and shown to be a substrate for the enzyme, yielding **13.33**.[38] At lower pH, **13.36** also can undergo a reverse isomerization back to acetolactate (**13.32**); the isomerization, however, still requires the presence of NADPH (and Mg^{2+}). Therefore, **13.36** is a kinetically competent intermediate in the enzyme-catalyzed reaction. Both isomerization and reduction appear to be catalyzed by the same enzyme because both of these activities occur at a constant rate ratio throughout enzyme purification. Further evidence for two catalytic activities is the observation that [^{14}C]**13.32** is converted without dilution of its specific radioactivity to [^{14}C]**13.33**, even in the presence of excess exogenous **13.36**. Therefore, the **13.36** that is generated remains tightly bound to the enzyme without exchange, and then becomes reduced. This is not what would be expected if two different enzymes were involved. In fact, the full-length cDNA clone encoding acetohydroxy acid isomeroreductase from spinach was isolated and sequenced by Douce and co-workers, who

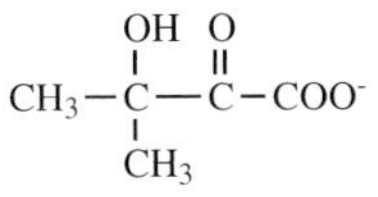

$$CH_3-\overset{\overset{OH}{|}}{\underset{\underset{CH_3}{|}}{C}}-\overset{\overset{O}{\|}}{C}-COO^-$$

13.36

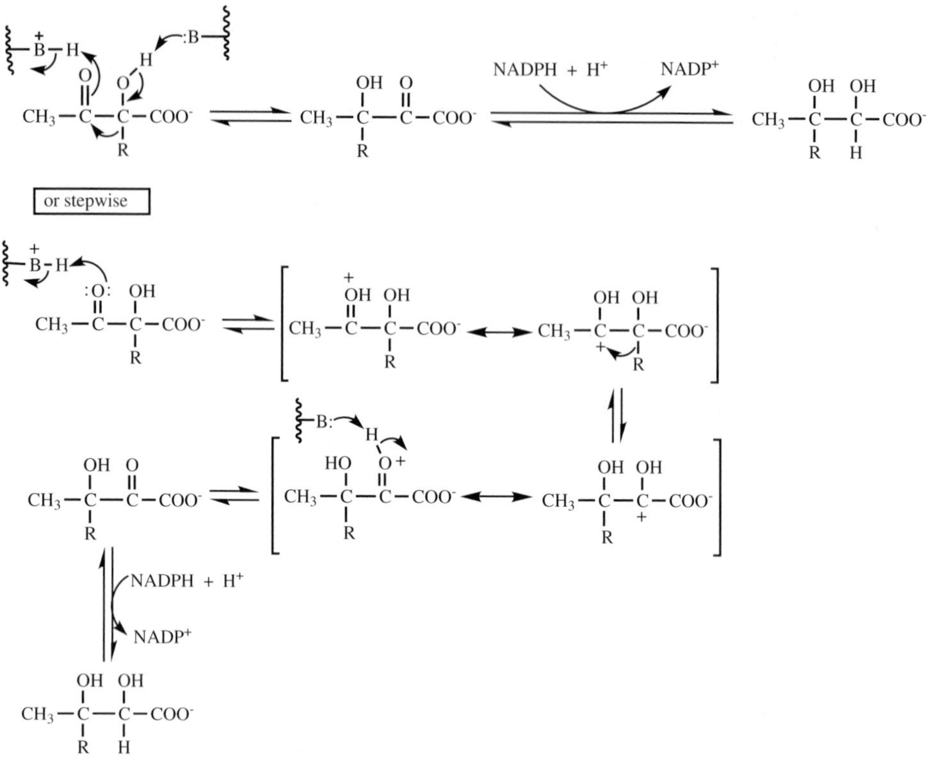

SCHEME 13.16   Proposed acyloin-type mechanism for acetohydroxy acid isomeroreductase.

showed that one protein with both activities is encoded by a single gene.[39] An acyl-oin rearrangement mechanism followed by NADPH reduction is consistent with these experiments (Scheme 13.16). In all eukaryotic and prokaryotic acetohydroxy acid isomeroreductase, there is a lysine residue seven residues upstream of the conserved four amino acid sequence SHGF; therefore, Lys-Xaa-Xaa-Xaa-Xaa-Xaa-Xaa-Xaa-Ser-His-Gly-Phe may be in the active site, in which case the active-site acidic group in Scheme 13.16 would be Lys-NH$_3^+$ and the basic group would be the imidazole of His.

B. Cyclizations

1. Sterol Biosynthesis

In mammals and fungi, squalene (**13.37**) is the precursor to the primary sterol product, lanosterol (**13.38**) (Scheme 13.17), which is converted by oxidative demethy-

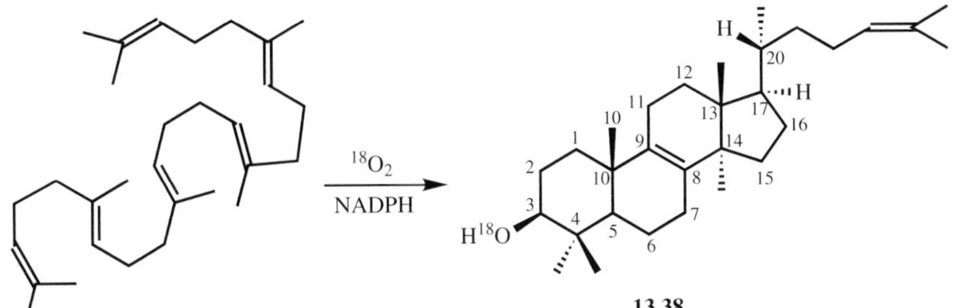

SCHEME 13.17   Conversion of squalene to lanosterol.

lations, isomerizations, and saturations to cholesterol and other steroid hormones. The C-3β-hydroxyl group of lanosterol comes from molecular oxygen, not from water. Subsequently, a squalene monooxygenase, squalene 2,3-epoxidase, which catalyzes the epoxidation of the 2,3-double bond of squalene, was isolated and found to require NADPH, O$_2$, flavin, and nonheme iron. Squalene 2,3-epoxide (**13.39**) is then converted into lanosterol (**13.38**) by the enzyme 2,3-oxidosqualene-lanosterol cyclase via the hypothetical intermediate protosterol (**13.40;** shown enzyme bound) (Scheme 13.18 depicts the initially proposed electrophilic addition mechanism).[40] In this reaction squalene 2,3-epoxide (**13.39**), which has no stereogenic centers, is converted directly into protosterol (**13.40;** originally thought to be enzyme bound), which has seven stereogenic centers. Thus there are 2^7 (128) possible isomers of protosterol, but apparently only one isomer is produced, although it has not been isolated. Isotopic labeling experiments have been used to

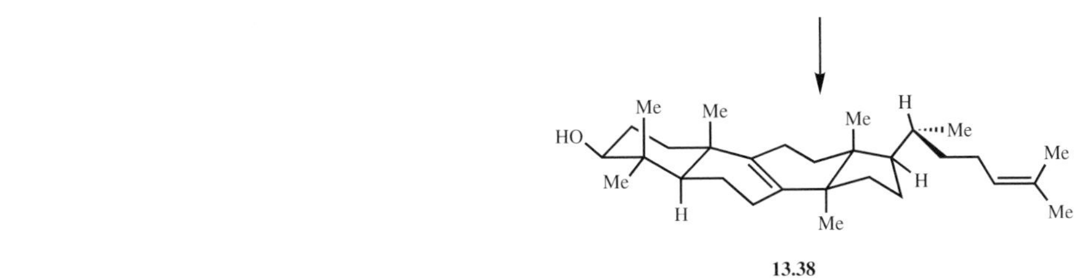

SCHEME 13.18   Initial mechanism proposed for 2,3-oxidosqualene-lanosterol cyclase.

demonstrate that the four migrations shown (H–13 to C–17; H–17 to C–20; methyl at C–8 to C–14; methyl at C–14 to C–13) occur intramolecularly, and the 9β-proton is lost to the medium.[41]

In the conversion of squalene 2,3-epoxide to lanosterol, an intermediate with a 17α side chain had originally been assumed (as shown in Scheme 13.18). To get lanosterol with the natural R-stereochemistry at C–20, assuming 17α stereochemistry of the side chain, the H–17 to C–20 hydride shift would require a 120-degree rotation of the side chain; the more entropically favored 60-degree rotation in the opposite direction gives the unnatural S-isomer. To assure the control of a 120-degree rotation, a covalent enzyme intermediate (**13.40**) was proposed. However, the corresponding squalene epoxide ether (**13.41**) was synthesized, and it was found that squalene epoxidase/cyclase converted it into **13.43** having C–17 with the β-configuration, not the formerly presumed α-configuration (Scheme 13.19).[42] This suggests that a C–20 cation intermediate, having a C–17β side chain (**13.42**), is produced and that there is no need to invoke a covalent enzyme intermediate for control.

Further support for the structure of protosterol was provided with an enzymatic experiment and a chemical model study.[43] In the enzymatic work, (20E)-

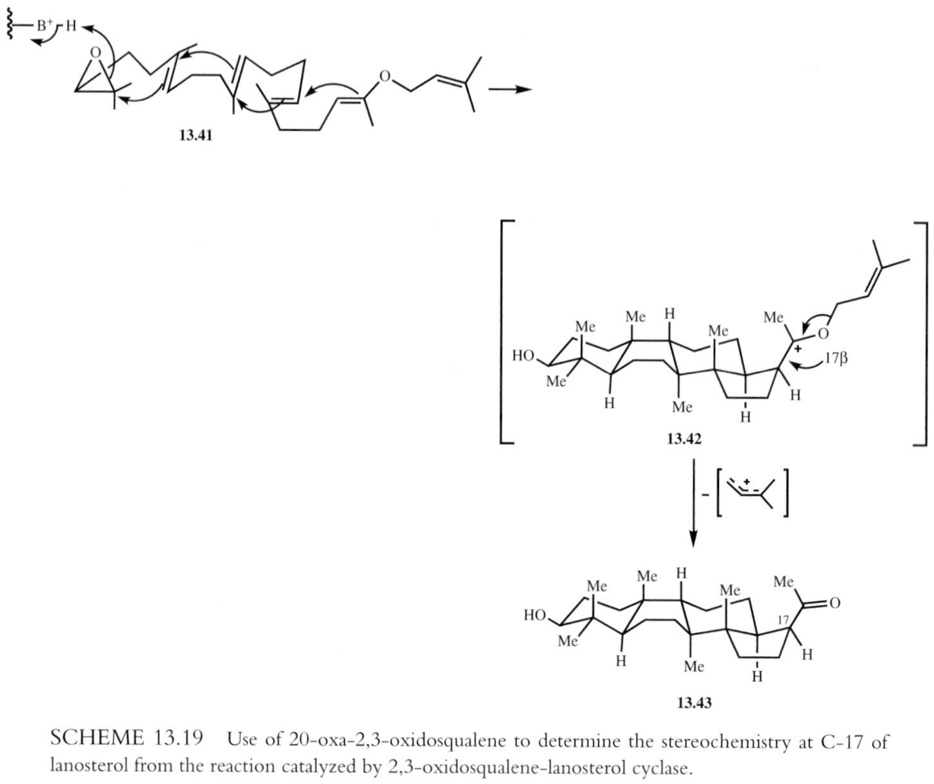

SCHEME 13.19   Use of 20-oxa-2,3-oxidosqualene to determine the stereochemistry at C-17 of lanosterol from the reaction catalyzed by 2,3-oxidosqualene-lanosterol cyclase.

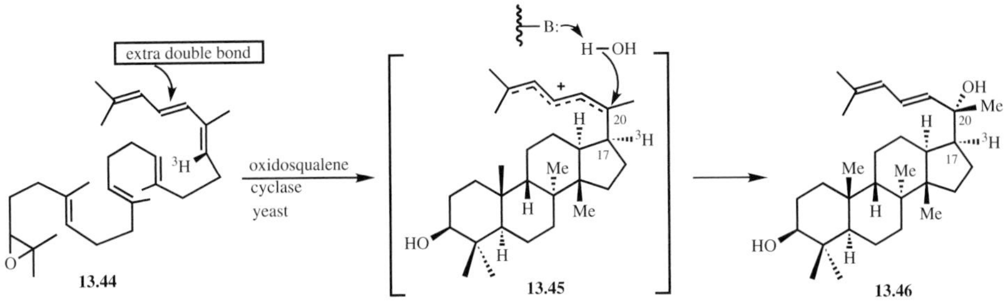

SCHEME 13.20   Use of (20E)-20,21-dehydro-2,3-oxidosqualene to determine the stereochemistry at C-17 of lanosterol from the reaction catalyzed by 2,3-oxidosqualene-lanosterol cyclase.

20,21-dehydro-2,3-oxidosqualene (**13.44**) was converted into the protosterol derivative **13.46** (having C-17β stereochemistry) by 2,3-oxidosqualene-lanosterol cyclase (Scheme 13.20). Apparently the stabilized intermediate carbocation (**13.45**) reacts with water faster than it undergoes the [1,2]-hydride shift from C-17 to C-20 and faster than rotation of the C^{17}–C^{20} bond. It is interesting that addition of water occurs only at C-20 and not at the other end of the C-17 side chain; possibly there is a binding pocket that shields attack by water at that site or maybe the 17β orientation of the side chain, which hinders rotation about the C^{17}–C^{20} bond because of a steric interaction with the *cis*-14β methyl group, favors stereospecific attachment of the hydroxyl at C-20.

The chemical model for the stereospecificity of the oxidosqualene cyclase-catalyzed reaction and for the importance of the 17β configuration at C-17 involves the treatment of **13.47** (having a 17β side chain) with the Lewis acid boron trifluoride at −90 °C for just 3 minutes. Under these conditions a 90% yield of **13.48** is obtained stereoselectively with overall retention of configuration in the replacement of the C-20 hydroxyl group by hydrogen (Scheme 13.21). However, the same reaction with the corresponding 17α-isomer is nonstereoselective at C-20, giving a 1:1 mixture of C-20 epimers of **13.48**.[44]

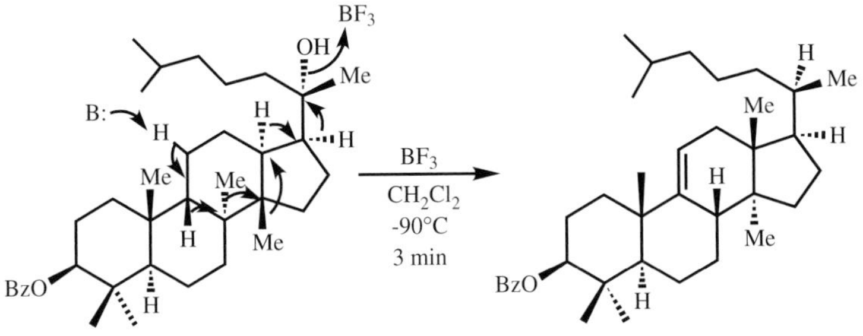

SCHEME 13.21   Chemical model for the conversion of protosterol to lanosterol.

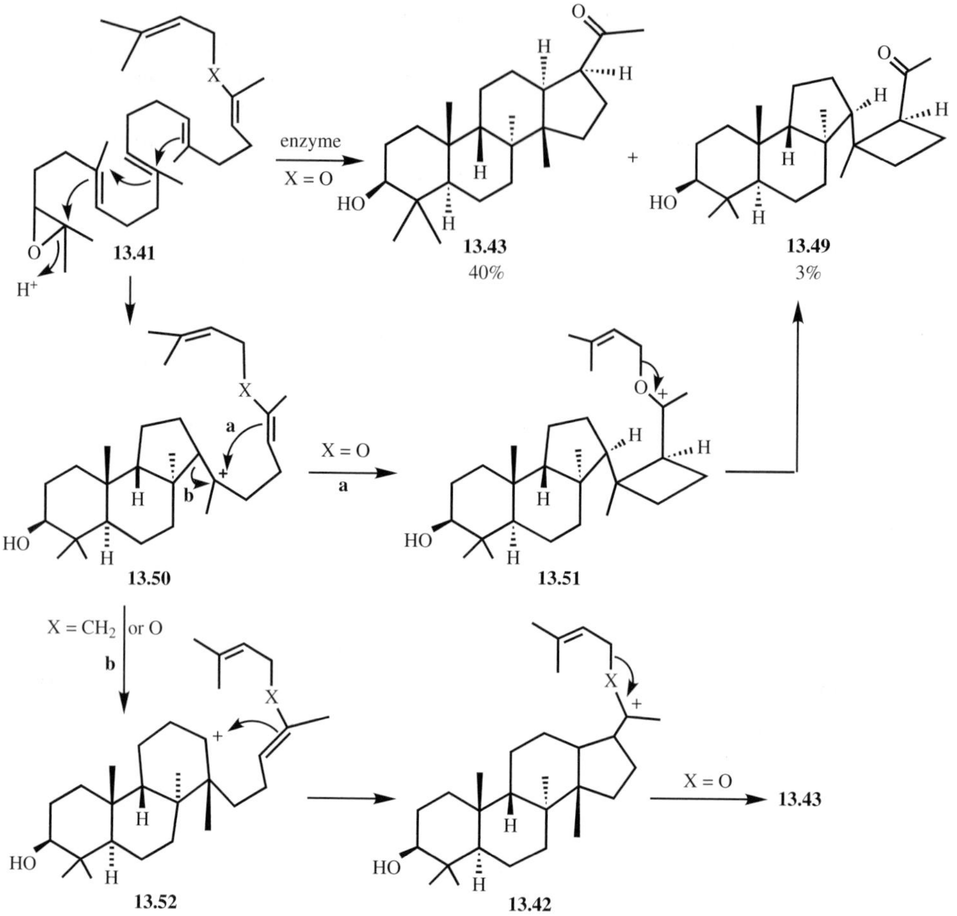

SCHEME 13.22   Mechanism proposed for the formation of the minor product isolated in the 2,3-oxidosqualene cyclase-catalyzed reaction with 20-oxa-2,3-oxidosqualene.

Evidence that the conversion of (S)-2,3-oxidosqualene to lanosterol (via protosterol), catalyzed by oxidosqualene cyclase, is not concerted was obtained by demonstrating that a minor product (13.49), characterized by X-ray crystallography, is obtained when 13.41 is used as the substrate (Scheme 13.22).[45] The formation of 13.49 is not easily rationalized as the product of a concerted reaction. The key intermediate is 13.50, obtained by a favored Markovnikov addition (i.e., via the more stable carbocation) to give the five-membered C-ring, rather than the unfavored anti-Markovnikov cyclization (i.e., via the less stable carbocation) that produces the expected six-membered C-ring of steroids (note that in the usually drawn mechanism for the conversion of 13.39 to 13.40 shown in Scheme 13.18, cycliza-

tion to make the C^8-C^{14} bond of the C-ring occurs by an anti-Markovnikov reaction). When X = O, closure of **13.50** to the 4-membered ring of **13.51** (pathway a) is a minor pathway, but is still more favorable than when X = CH_2 (pathway b). This is because an oxocarbenium ion (**13.51**) is formed, but a carbenium
ion (**13.51**, CH_2 instead of O), which is less stable, would have been formed if
X = CH_2 instead of O. Because a four-membered ring is generated, even the formation of an oxocarbenium ion is barely stable enough to allow this step to occur,
which accounts for the low yield of **13.49**. Cyclization of **13.41** to **13.50** (instead
of directly to **13.52**) is favored not only because it is the Markovnikov addition,
but because there also is a large steric factor in favor of closure to **13.50** rather than
directly to **13.52**: direct formation of **13.52** entails severe nonbonded steric repulsions between the two methyl groups on the carbons being joined and their nearest
neighbors; ring closure to **13.50** involves relatively minor steric repulsions.

Experiments were done to determine if the initial closure to **13.50** followed by
ring expansion to **13.52** is reasonable. When the enzyme reaction is carried out
with 1,1'-bisnor-2,3-oxidosqualene (**13.41**, but with no methyls on the epoxide
ring), no reaction occurs. This supports formation of a carbocation at the 1-position
followed by alkene addition, not a concerted opening of the protonated epoxide
by the alkene. The 20-thia analogue (**13.41**, X = S) also is completely inert to cyclization. The larger size of the sulfur does not fit into the space for the D-ring/side
chain, and this misfit prevents epoxide activation. This argues in favor of the requirement of a correctly folded form of the substrate, prior to initiation of the cyclization. Appropriate substrate binding may induce correct folding of the substrate,
which positions the electrophilic groups for epoxide activation, then cyclization via
13.50 and **13.52**.

The V_{max}/K_m values for 2,3-oxidosqualene (**13.53**, R = CH_3), 6-desmethyl-
2,3-oxidosqualene (**13.53**, R = H), and 6-chloro-6-desmethyl-2,3-oxidosqualene
(**13.53**, R = Cl) were determined to be 138, 9.4, and 21.9 pmol $\mu g^{-1} h^{-1} \mu M^{-1}$,
respectively, which correlates with the carbocation-stabilizing ability of the group

13.53

at C-6 (CH$_3$ > Cl > H).[46] This suggests that the nucleophilicity of the 6,7–double bond influences the rate of oxirane cleavage, which is consistent with a concerted cleavage of the oxirane and formation of the A-ring. However, this conclusion is contrary to the conclusion made from the preceding experiment using **13.41** without the oxirane methyl groups. Further experiments are needed to establish **13.50** as a viable intermediate. Site-directed mutagenesis indicates that Asp–456 may be the active-site residue that protonates the oxirane.

2. Squalene Biosynthesis

The precursor to sterol formation, as just discussed, is squalene (**13.37**), which is synthesized from two molecules of farnesyl diphosphate (**13.54**) via presqualene diphosphate (**13.55**) in a reaction catalyzed by squalene synthase (Scheme 13.23).[47] Poulter and co-workers have proposed an interesting rearrangement mechanism for the conversion of **13.55** to **13.37** via carbocation intermediates **13.56** and **13.57** (Scheme 13.24).[48] Zhang and Poulter found that in the absence of NADPH, **13.54** is converted to **13.55** followed by a slow enzyme-catalyzed hydrolysis of **13.55** to a

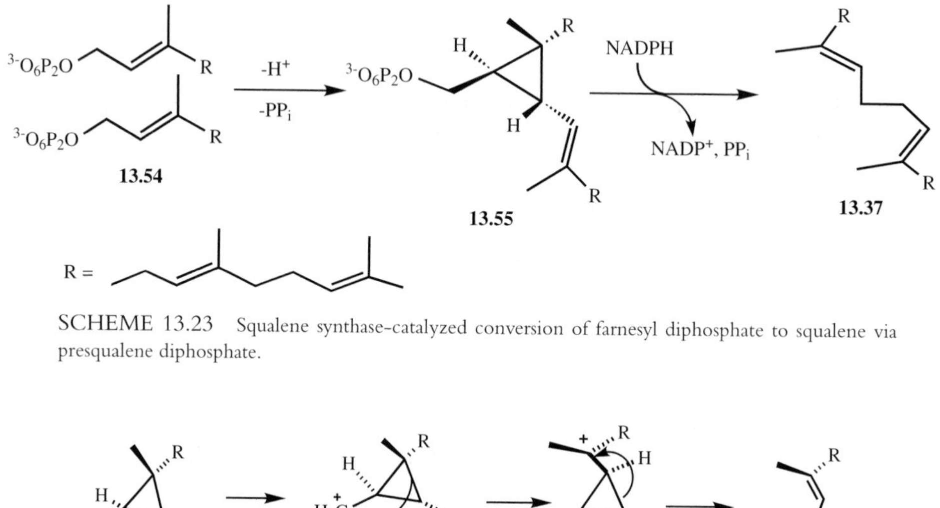

SCHEME 13.23   Squalene synthase-catalyzed conversion of farnesyl diphosphate to squalene via presqualene diphosphate.

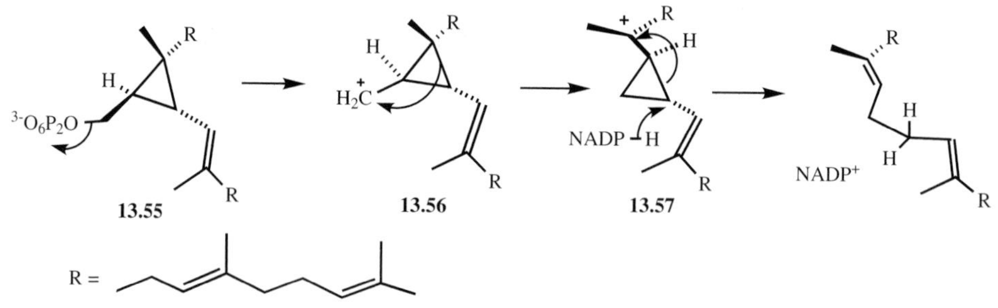

SCHEME 13.24   Mechanism proposed for the conversion of presqualene diphosphate to squalene by squalene synthase.

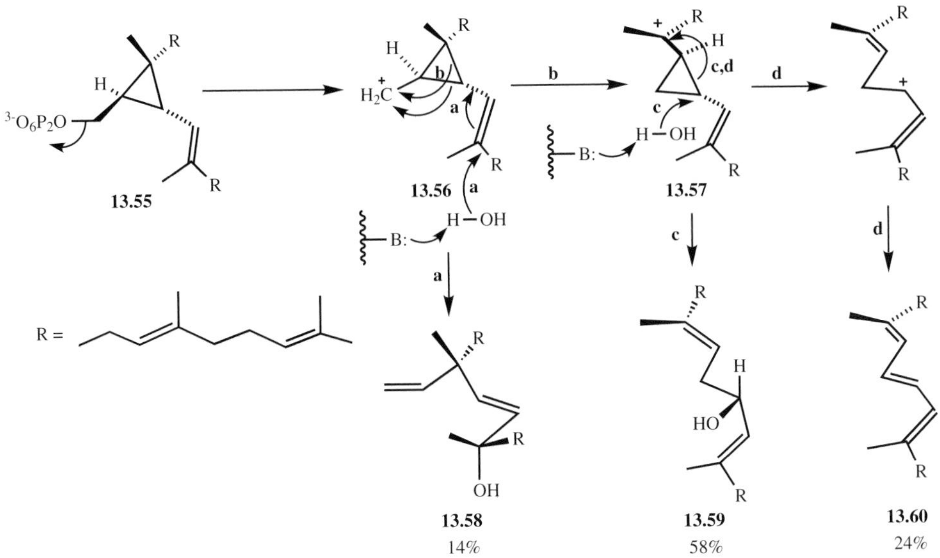

SCHEME 13.25   Mechanisms proposed for the squalene synthase-catalyzed hydrolysis of presqualene diphosphate to several different products in the absence of NADPH.

mixture of two C_{30} alcohols (**13.58** and **13.59**), via pathways a and b−c, respectively, from **13.56**, and to (Z)-dehydrosqualene (**13.60**), via pathway b−d, supporting the formation of **13.56** and **13.57** (Scheme 13.25).[49] The nucleophilic water, presumably, resides in the space normally taken up by NADPH. Further support for intermediate **13.57** was gathered by substituting an unreactive, reduced analogue of NADPH (dihydroNADPH, **13.61**) for NADPH itself. Under these conditions **13.62** is obtained, which presumably is derived from **13.57** (Scheme 13.26).

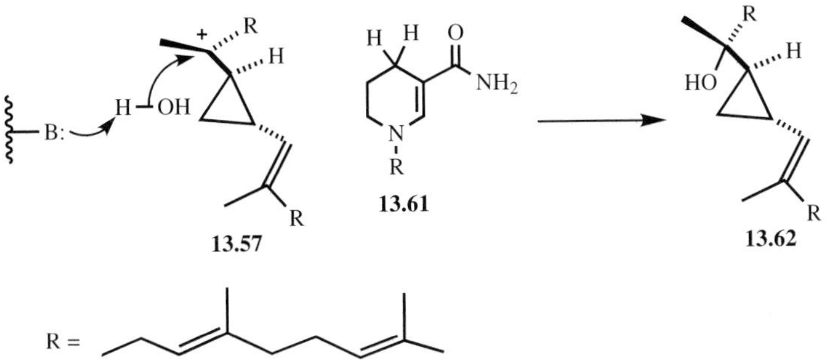

SCHEME 13.26   Use of dihydro-NADPH to provide evidence for the formation of an intermediate in the reaction catalyzed by squalene synthase.

III. REARRANGEMENTS THAT PROCEED VIA RADICAL INTERMEDIATES

A. DNA Photolyase, Formally a [2+2] Cycloreversion Reaction

Exposure of DNA to ultraviolet light induces two major types of damage, the formation of cyclobutane pyrimidine dimers (**13.63**) and (6−4) photoproducts (**13.64**) (Scheme 13.27). These lesions block replication and transcription and therefore are carcinogenic, mutagenic, and cytotoxic.[50] Depletion of the stratospheric ozone layer leads to a higher degree of ultraviolet light reaching the surface of the Earth, and DNA damage is increasing in the biosphere.[51] Cells protect themselves against the effects of this (200−300 nm) UV-induced DNA damage by either excision repair or by *photoreactivation,* a process that involves exposure to blue light (350−450 nm). Photoreactivation repair, which is catalyzed by enzymes known as photolyases, has been observed in bacteria, archea, and eukaryotes, but not in humans.[52] To differentiate photoreactivation of the two major types of UV damage, the enzymes that catalyze repair of cyclobutane pyrimidine dimers (CPD) are referred to as DNA photolyases (although, to differentiate them from the other class of photolyases, they have been referred to as CPD photolyases[53]), and the enzymes that catalyze photoreactivation of the (6−4) photoproducts are called (6−4) photolyases.[54]

Photolyases are unique enzymes in that they use visible light as a second substrate. DNA photolyase contains two chromophoric cofactors, reduced flavin adenine dinucleotide (**13.65**)[55] and either N^5, N^{10}-methenyltetrahydropteroylpoly-

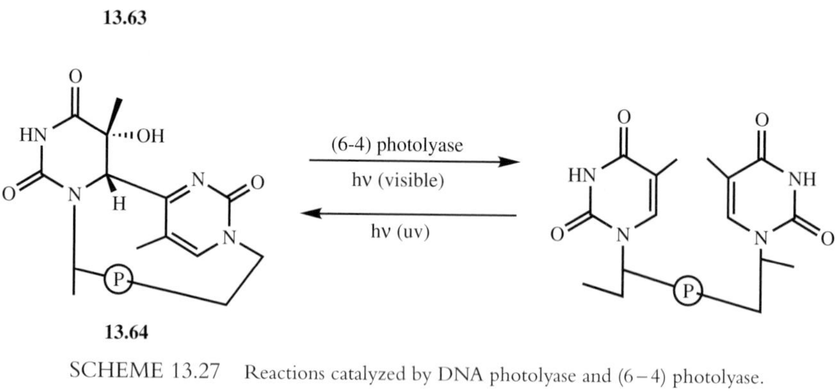

13.63

13.64

SCHEME 13.27   Reactions catalyzed by DNA photolyase and (6−4) photolyase.

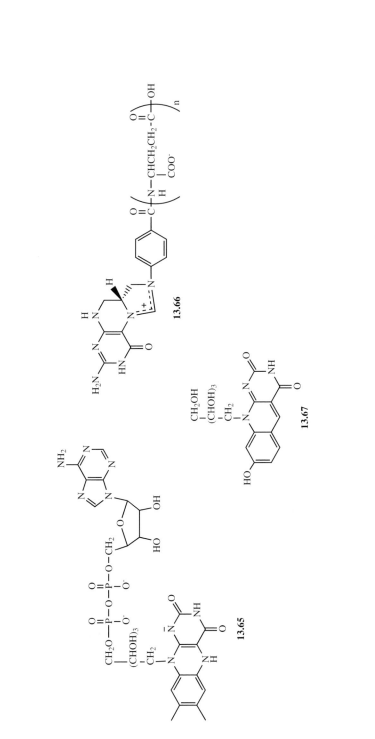

13.66

13.67

13.65

glutamate with $3-6$ glutamate residues (**13.66,** $n = 3-6$)[56] or 8-hydroxy-7,8-didemethyl-5-deazariboflavin (**13.67**).[57] Although the reduced flavin cofactor is found in the blue neutral semiquinone form on isolation of the enzyme,[58] the active form in photolyases is the dihydro flavin anion form.[59] Sancar and co-workers showed that excitation of the semiquinone form at 366 nm causes a transfer of an electron from Trp-306 to the flavin to generate the reduced flavin anion.[60] The second cofactor (either **13.66** or **13.67**) acts as a photoantenna to absorb the blue light photon and transmit the excitation energy to the FADH$^-$. Both of these second cofactors are unique in two ways from the roles that they play in DNA photolyase compared to their functions in other enzymes. First, they act only as photoantennae, in contrast to their usual functions in one-carbon transfer (Chapter 12, Section I.B.) and redox reactions (Chapter 3, Section III.B), respectively. Second, these cofactors complete the catalytic cycle unchanged,[61] unlike their behavior in other enzyme systems where they act as cosubstrates and must be converted back to their catalytic form in a second set of enzyme-catalyzed reactions.

With this background, we are ready to consider a mechanism Sancar proposed for DNA photolyase (Scheme 13.28),[62] which is based on the original proposal by Rupert and co-workers.[63] Following binding of the substrate (**13.63**) to DNA photolyase, a photon of visible light is absorbed by **13.66** (or **13.67**) to give an excited state **13.66***. This excitation energy is transferred to the reduced flavin (**13.65***), which then transfers an electron to the substrate to give the radical anion **13.68** and the flavin semiquinone radical. Intermediate **13.68** decomposes by a concerted, but nonsynchronous pathway, possibly via **13.69** to **13.70,** which donates an electron back to the flavin semiquinone to regenerate reduced flavin (**13.65**) and the repaired DNA (**13.71**). Note that there is no net gain or loss of electrons, so this is not a redox reaction. The observation of electron spin polarization in the EPR spectrum at 4 K with the fully reduced DNA photolyase·substrate complex is evidence for the formation of radical pairs, such as the neutral flavin semiquinone radical and one of the substrate radical anions (**13.68–13.70**).[64] Fluorescence-quenching experiments established that the reduced flavin is converted to the singlet state during catalysis.[65] Direct evidence for singlet−singlet energy transfer from **13.66** to singlet FADH$^-$ (^{1}FADH$^-$) was obtained by Lipman and Jorns in fluorescence quantum yield studies, which show that the efficiency of this energy transfer process (E_{ET}) is 0.92 (92% efficient).[66] The rate and efficiency of electron transfer from ^{1}FADH$^-$ to the substrate was investigated by Sancar and co-workers using time-resolved fluorescence, absorbance, and EPR spectroscopy.[67] The rate of electron transfer was determined for the enzyme from *Escherichia coli*[68] and *A. nidulans*[69] to be 5.5 $\times$ 10^9 s^{-1} at a quantum yield of 0.88 for the *Escherichia coli* enzyme and 6.5 $\times$ 10^9 s^{-1} at a quantum yield of 0.92 for the *A. nadulans* enzyme. From the crystal structure of the *Escherichia coli* enzyme at 2.3 Å resolution, it appears that electron transfer between the flavin and the substrate occurs over van der Waals contact distance,[70] which could account for the efficiency of electron transfer. Calculations by Sancar and co-workers of the free-energy changes for single-electron transfer in either

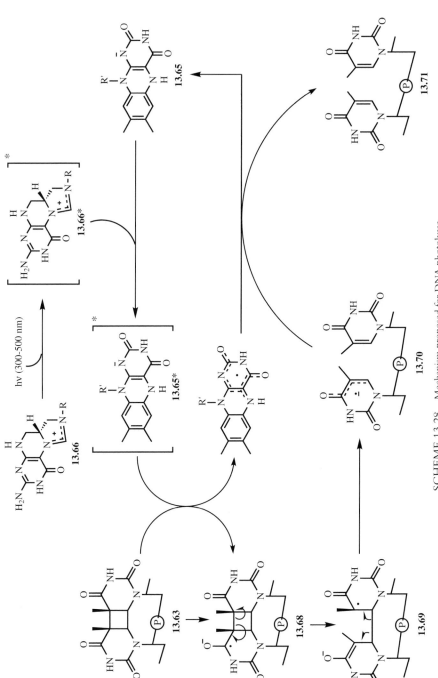

SCHEME 13.28 Mechanism proposed for DNA photolyase.

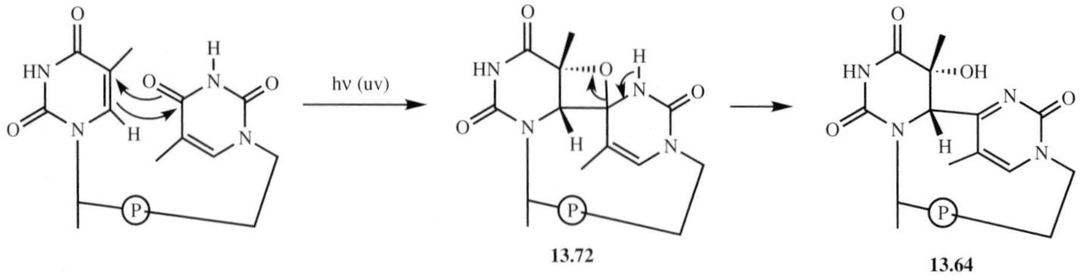

SCHEME 13.29   Proposed mechanism for the formation of the (6−4) photoproduct.

direction by relevant excited states of enzyme-bound flavin and substrate dimer, demonstrate that, on thermodynamic grounds, electron transfer from ^{1}FADH$^-$ to the substrate dimer is the only reasonable mechanism for the photolyase.[71] This could account for why only the reduced anionic ^{1}FADH$^-$ form of the enzyme is active. Photosensitized splitting of dimers in solution is favored by nonpolar solvents,[72] and the flavin binding site in *Escherichia coli* photolyase is hydrophobic,[73] presumably to accommodate the reaction.

The (6−4) photoproduct (**13.64**) could arise from a [2+2]-type cycloaddition between the C^5–C^6 double bond of one thymine and the C^4 carbonyl of another thymine to give the oxetane **13.72**, which would be unstable and would fragment to **13.64** (Scheme 13.29).[74] Although little is known of the mechanism for the (6−4) photolyase, a pathway related to that for DNA photolyase is reasonable (Scheme 13.30).[75] A chemical model study by Prakash and Falvey supports this oxetane-photoinduced mechanism.[76] Because CPD photolyase and (6−4) photolyase are related in evolution,[77] it also is reasonable that their mechanisms could be related as well.

B. Coenzyme B$_{12}$-Dependent and Related Rearrangements

1. General Information

In Chapter 12 (Section I.B.3.b) we talked about reactions catalyzed by methylcobalamin, a naturally occurring organometallic coenzyme. The other major organometallic coenzyme in nature is coenzyme B$_{12}$ (**13.73a**; adenosylcobalamin or AdoCbl),[77a] derived from the largest vitamin, namely, vitamin B$_{12}$ (**13.73b**).[78] In 1948 Folkers and co-workers[79] and Smith and Parker[80] independently reported that a red crystalline compound, designated vitamin B$_{12}$, from liver extracts was the clinically active material in the raw liver treatment for pernicious anemia. In 1958 Barker and others[81] recognized the existence of a catalytically active form of vitamin B$_{12}$, the coenzyme form, during studies on fermentation of L-glutamate by *Clostridium tetanomorphum*. In 1961 Lenhert and Hodgkin reported the crystal structure of coenzyme B$_{12}$.[82] Nobel laureate (1964) Dorothy Hodgkin was the head

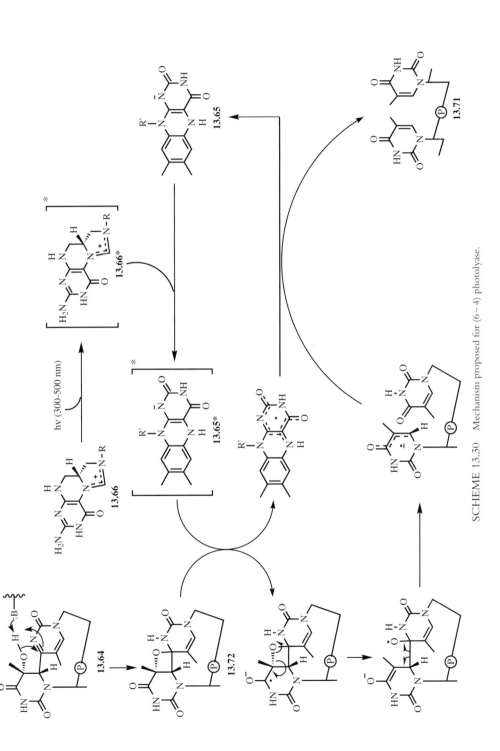

SCHEME 13.30 Mechanism proposed for (6−4) photolyase.

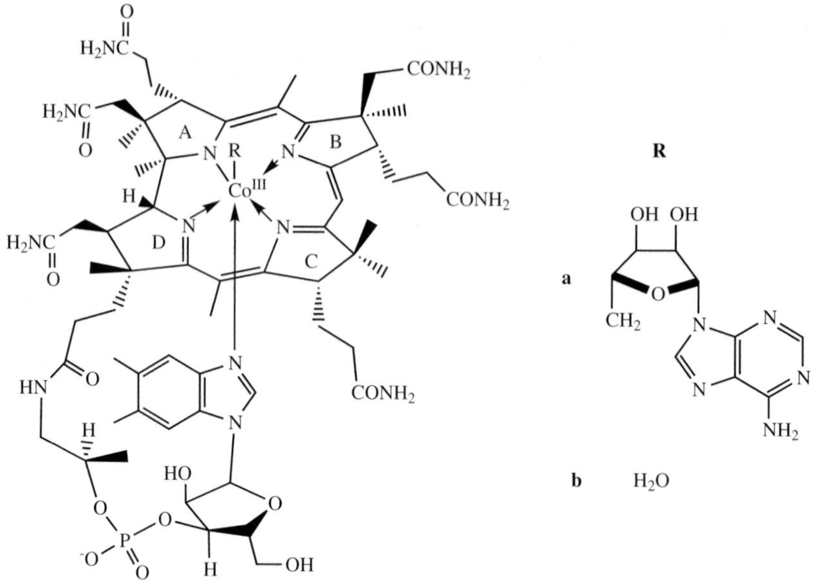

13.73

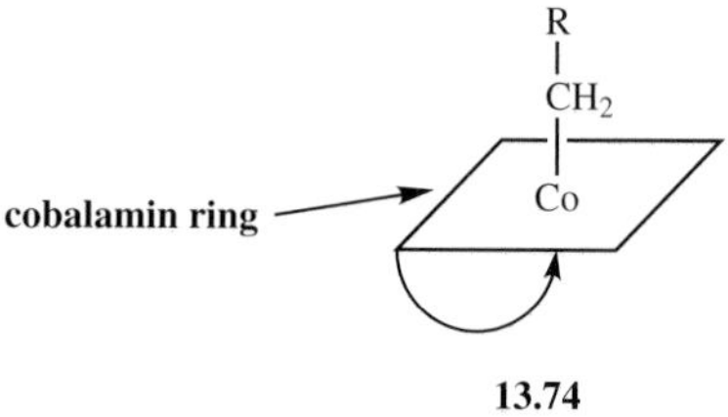

13.74

of the group that determined the crystal structure of vitamin B_{12} five years earlier.[83] Because of the complex structure of coenzyme B_{12}, it is generally abbreviated by a structure such as **13.74** (RCH_2 = 5′-deoxyadenosyl), which will be adapted in this section, unless a specific point needs to be made regarding the more precise structure. Vitamin B_{12} is converted into coenzyme B_{12} in three steps (Scheme 13.31): Two NADH/FAD-dependent enzymes catalyze single-electron reductions to convert the Co(III) first to Co(II) (cob(III)alamin reductase); then the Co(II) to Co(I) (**13.75**) (cob(II)alamin reductase), and a third enzyme (adenosylating enzyme), which requires ATP and a divalent metal ion, converts the cob(I)alamin to coenzyme B_{12}.[84] This last reaction with ATP is one of only two enzyme-catalyzed reactions that occurs at the 5′-position of ATP (the other was in the conversion of me-

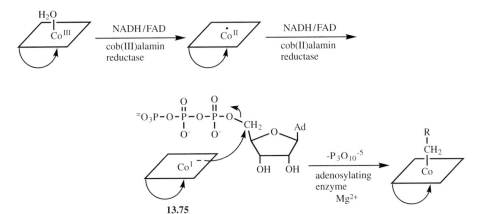

SCHEME 13.31   Biosynthesis of coenzyme B_{12}.

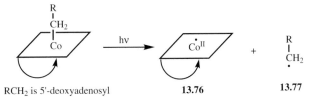

RCH$_2$ is 5'-deoxyadenosyl   **13.76**   **13.77**

SCHEME 13.32   Light sensitivity of the Co–C bond of coenzyme B_{12}.

thionine to *S*-adenosylmethionine (Chapter 12, Section II). Cob(I)alamin (also referred to as B_{12s}) is one of the most powerful nucleophiles known;[85] the absolute reactivities of Co(I) nucleophiles are up to 10^7 times greater than those of iodide ion. As in the case of the formation of methylcobalamin, cob(I)alamin becomes alkylated via an S_N2-like reaction.[86]

The cobalt–carbon bond is very labile, especially to light, and it undergoes rapid homolytic cleavage to the radical species, cob(II)alamin (**13.76**; also referred to as B_{12r}) and the 5'-deoxyadenosyl radical (**13.77**) (Scheme 13.32). Therefore, all of the work with 5'-deoxyadenosylcobalamin or with enzymes that require this coenzyme must be done in the dark!

2. Rearrangements

Coenzyme B_{12} has been reported as a required cofactor for 10 different enzymes (Table 13.1). At first sight, it appears that these reactions have little in common. However, all of these reactions involve a net substrate rearrangement in which a hydrogen exchanges places with an alkyl, acyl, or electronegative group on an

TABLE 13.1   Coenzyme B_{12}-Dependent Enzyme-Catalyzed Reactions

Enzyme	Reaction catalyzed
CARBON SKELETAL REARRANGEMENTS	

Methylmalonyl-CoA mutase

$$HOOCCH_2CH_2\text{-}COSCoA \rightleftharpoons HOOC\underset{CH_3}{-}CH\text{-}COSCoA$$

2-Methyleneglutarate mutase

$$HOOCCH_2CH_2\text{-}\underset{\underset{CH_2}{\|}}{C}\text{-}COOH \rightleftharpoons HOOC\underset{CH_3}{-}CH\text{-}\underset{\underset{CH_2}{\|}}{C}\text{-}COOH$$

Glutamate mutase

$$HOOCCH_2CH_2\text{-}\underset{\underset{NH_2}{|}}{CH}\text{-}COOH \rightleftharpoons HOOC\underset{CH_3}{-}CH\text{-}\underset{\underset{NH_2}{|}}{CH}\text{-}COOH$$

Isobutyryl-CoA mutase

$$CH_3CH_2CH_2\text{-}COSCoA \rightleftharpoons H_3C\underset{CH_3}{-}CH\text{-}COSCoA$$

ELIMINATIONS

Diol dehydratase

$$R\text{-}\underset{\underset{OH}{|}}{CH}\text{-}CH_2OH \rightleftharpoons RCH_2CHO \quad R = CH_3 \text{ or } H$$

Glycerol dehydratase

$$HOCH_2\text{-}\underset{\underset{OH}{|}}{CH}\text{-}CH_2OH \rightleftharpoons HOCH_2\text{-}CH_2CHO$$

Ethanolamine ammonia lyase

$$\underset{\underset{NH_2}{|}}{CH_2}\text{-}CH_2OH \rightleftharpoons CH_3CHO$$

ISOMERIZATIONS

L-β-Lysine-5,6-aminomutase

$$H_2\underset{\underset{NH_2}{|}}{C}\text{-}CH_2\text{-}CH_2\text{-}\underset{\underset{NH_2}{|}}{CH}CH_2\text{-}COOH \rightleftharpoons H_3C\text{-}\underset{\underset{NH_2}{|}}{CH}\text{-}CH_2\text{-}\underset{\underset{NH_2}{|}}{CH}CH_2\text{-}COOH$$

D-Ornithine-4,5-aminomutase

$$H_2\underset{\underset{NH_2}{|}}{C}\text{-}CH_2\text{-}CH_2\text{-}\underset{\underset{NH_2}{|}}{CH}\text{-}COOH \rightleftharpoons H_3C\text{-}\underset{\underset{NH_2}{|}}{CH}\text{-}CH_2\text{-}\underset{\underset{NH_2}{|}}{CH}\text{-}COOH$$

REDUCTION

Ribonucleotide reductase

adjacent carbon atom (Scheme 13.33). See Figure 13.2 for three of the examples, glutamate mutase (**A**), a carbon skeletal rearrangement of L–glutamate to β–methyl-aspartate; diol dehydratase (**B;** think of an aldehyde as equivalent to a hydrate),

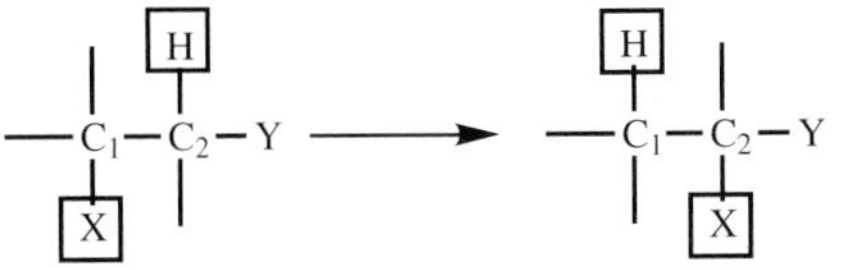

SCHEME 13.33   General form of coenzyme B_{12}-dependent rearrangements.

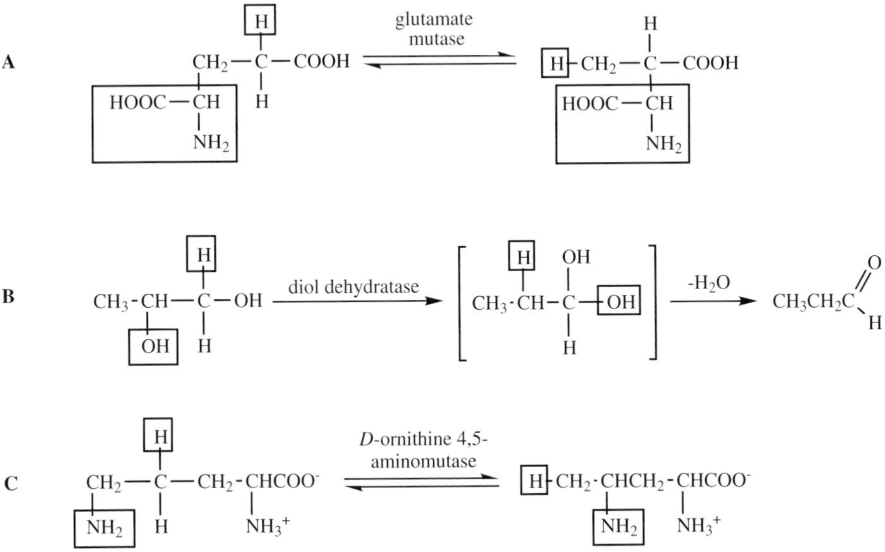

FIGURE 13.2   Three examples of coenzyme B_{12}-dependent rearrangements showing how the hydrogen and an adjacent group appear to change places.

an "elimination" of propylene glycol to propanal (and ethylene glycol to acetaldehyde); and D-ornithine 4,5-aminomutase (**C**), an isomerization of D-ornithine to (2R,4S)-2,4-diaminovaleric acid.[87]

a. Mechanism(s) for Diol Dehydratase and Ethanolamine Ammonia-Lyase

Much of the early mechanistic work[88] was carried out on diol dehydratase (called *dioldehydrase* in the earlier literature) and ethanolamine ammonia-lyase, so we will start there, then take a look at different potential mechanisms for these and other coenzyme B_{12}-dependent rearrangements.

There is no incorporation of solvent hydrogens at any stage in the diol dehydratase-catalyzed reaction.[89] Thus there can be no elimination of water; otherwise, an enol would form that must be protonated. Diol dehydratase converts (1R,2R)-[1-^{2}H]-[1-^{14}C]propanediol (**13.78**) to (2S)-[2-^{2}H]-[1-^{14}C]propionaldehyde (**13.79**)

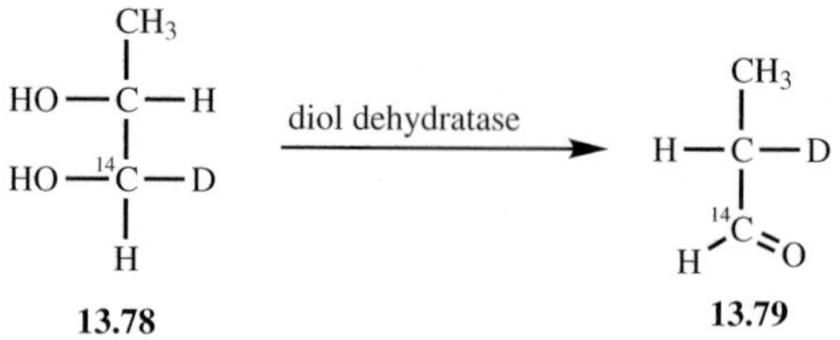

13.78   **13.79**

SCHEME 13.34   Stereospecific conversion of $(1R,2R)$-$[1-^2H]$-$[1-^{14}C]$propanediol to $(2S)$-$[2-^2H]$-$[1-^{14}C]$propionaldehyde catalyzed by diol dehydratase.

(Scheme 13.34; these structures, as well as the ones in Schemes 13.35-13.37 are Fischer projections),[90] indicating a stereospecific [1,2] migration of the *pro-R* hydrogen with inversion of configuration.[91] When the same experiment is carried out with the epimer, $(1R,2S)$-$[1-^2H]$-$[1-^{14}C]$propanediol (**13.80**), $[1-^2H]$-$[1-^{14}C]$propionaldehyde (**13.81**) is obtained, indicating that the *pro-S* hydrogen has migrated with this epimer (Scheme 13.35)! Therefore, the hydrogen that migrates from C-1 to C-2 depends on the stereochemistry at C-2.

The intermediate hydrate has to eliminate a molecule of water. Is this a random process of elimination? When $(2S)$-$[1-^{18}O]$propanediol (**13.82**) is incubated with diol dehydratase, 88% of the ^{18}O is retained in the product aldehyde (**13.83**), indicating that after the *pro-S* hydrogen migrates (as shown in the preceding experiment), the *pro-R* hydroxyl group is eliminated (Scheme 13.36A).[92] But when the $(2R)$-enantiomer (**13.84**) is used (the *pro-R* hydrogen migrates), loss of the *pro-R* hydroxyl group (this time, the *pro-R* hydroxyl is ^{18}O labeled) again is the one that is eliminated, and only 8% of the ^{18}O is retained in the product (Scheme 13.36B). Therefore the *pro-R* hydroxyl group is eliminated regardless of the enantiomer of the diol used. Overall, then, the C-1 hydrogen and the C-2 hydroxyl group migrate from opposite sides, giving inversion of configuration at both C-1 and C-2, and the *pro-R* hydroxyl group is always eliminated.

To test for intermolecular hydrogen atom transfer, a crossover experiment between $(1S, 2S)$-$[1-^3H]$propanediol (**13.85**) and ethylene glycol (the enzyme uses either glycol) was carried out by Frey and Abeles; tritium was found at C-2 of both

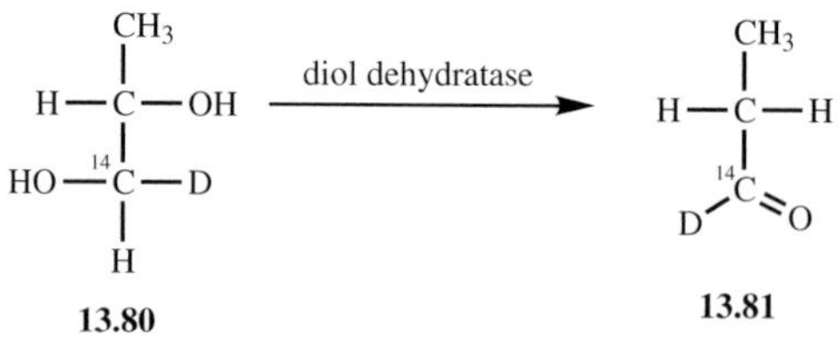

13.80   **13.81**

SCHEME 13.35   Stereospecific conversion of $(1R,2S)$-$[1-^2H]$-$[1-^{14}C]$propanediol to $[1-^2H]$-$[1-^{14}C]$propionaldehyde catalyzed by diol dehydratase.

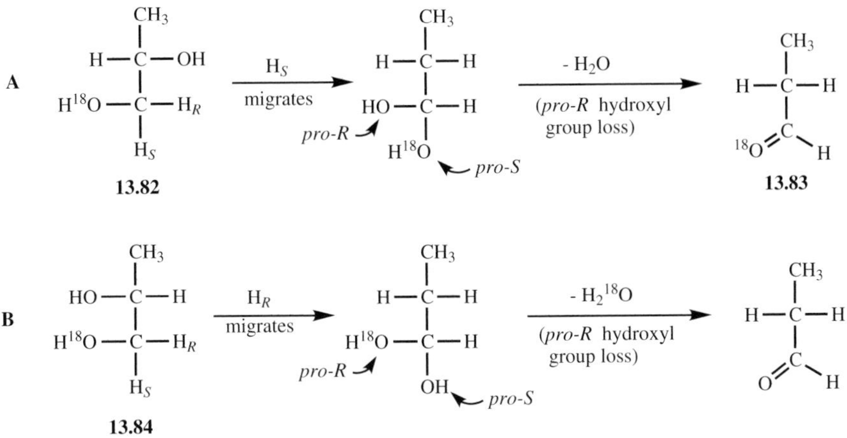

SCHEME 13.36   Diol dehydratase-catalyzed conversion of (2S)-[1-^{18}O]propanediol to [^{18}O]pro-pionaldehyde (**A**) and of (2R)-[1-^{18}O]propanediol to propionaldehyde (**B**).

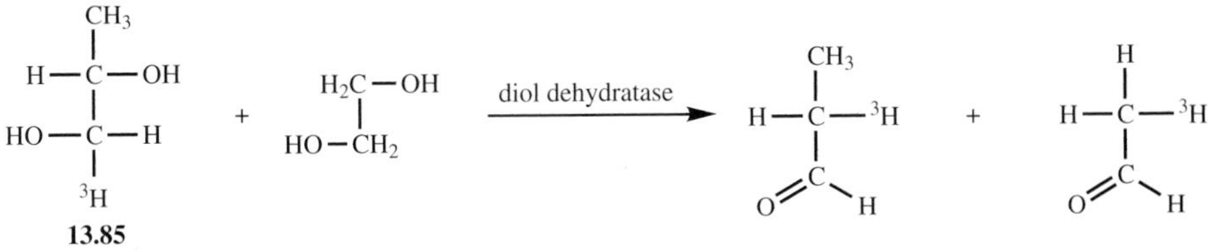

SCHEME 13.37   Crossover experiment to show that diol dehydratase catalyzes an intermolecular transfer of a hydrogen from C-1 to C-2.

the propionaldehyde and the acetaldehyde products (Scheme 13.37).[93] This indicates that an intermolecular hydrogen transfer occurs in which tritium from one substrate molecule can end up in the product generated by a second substrate molecule. This experiment also provides evidence for the fact that the proton removed in the first turnover may not necessarily be the one that is transferred to the product molecule of that turnover (a protium was removed from ethylene glycol and a tritium replace it). When either (R)- or (S)-[1,1-^{2}H$_2$]propanediol is used as the substrate, the $k_H/k_D = 10-12$, a deuterium isotope effect larger than the theoretical maximum of about 7.[94] To rationalize these results, an intermediate has to form that contains the migrating hydrogen and also has other hydrogens to equilibrate with it. Coenzyme B$_{12}$ was suspected of being involved in the formation of the intermediate.

When diol dehydratase is incubated with [1-^{3}H]propanediol, the reaction quenched periodically, then the coenzyme isolated (in the dark), there is a time-dependent incorporation of tritium into the recovered coenzyme B$_{12}$ (Figure 13.3).[95]

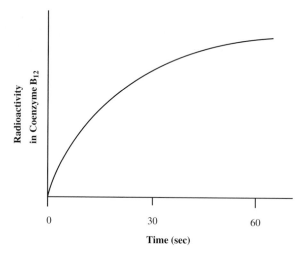

FIGURE 13.3   Time course for incorporation of tritium from [1-³H]propanediol into the cobalamin of diol dehydratase.

Degradation of the isolated radioactive coenzyme B$_{12}$ by photolysis either aerobically, which cleaves the Co—C bond and oxidizes the 5′-methylene group to an aldehyde (**13.86**) with loss of one proton, or anaerobically, which cleaves the Co—C bond, but retains both protons on the 5′-methylene group (**13.87**) can be used to determine the incorporation of tritium (Scheme 13.38). In this case, all of

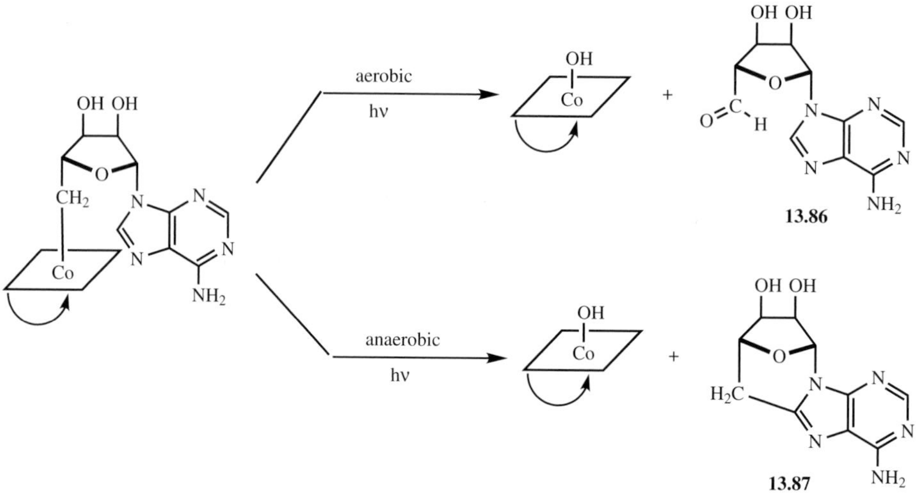

SCHEME 13.38   Aerobic and anaerobic photolytic degradation of coenzyme B$_{12}$ to locate the position of the tritium incorporated from [1-³H]propanediol in a reaction catalyzed by diol dehydratase.

the radioactivity initially in the coenzyme B_{12} is found in **13.87,** but only half of the radioactivity initially in coenzyme B_{12} is in the **13.86;** the other half is released as tritiated water. The tritium is therefore incorporated equally and randomly into the 5′-methylene group of the coenzyme. No radioactivity is found in the released aquocobalamin. The isolated [³H]coenzyme B_{12} can then be reconstituted into *apo*-diol dehydratase and incubated with unlabeled propanediol; the product is [2-³H]propionaldehyde. *All* of the tritium that was in the coenzyme B_{12} is transferred to the product. Therefore, coenzyme B_{12} is simply a hydrogen transfer agent, taking a hydrogen from substrate and giving it to an intermediate to make product. The incorporation of the hydrogen atom into the 5′-methylene group of the coenzyme is random, and the atom is transferred out of the coenzyme randomly. The final proof of concept comes from the use of chemically synthesized $(R,S-)$-[5′-³H]coenzyme B_{12} (randomly incorporated tritium) reconstituted into *apo*-diol dehydratase; all of the tritium is transferred to the propionaldehyde product, demonstrating that the enzyme does not distinguish between the two prochiral hydrogens at the 5′-methylene position. That may seem unusual, because enzymes are supposed to catalyze stereospecific reactions. One possible rationale for this behavior would be if the Co−C bond is somehow cleaved and the resultant 5′-adenosyl methylene radical abstracts the substrate hydrogen to make a 5′-methyl group (i.e., 5′-deoxyadenosine; **13.88**); in this way the three hydrogens of the newly-formed methyl group would equilibrate just by bond rotation.

Abeles and co-workers found that when they used pseudosubstrates such as glycolaldehyde[96] and 2-chloroacetaldehyde,[97] which produce coenzyme cleavage but no rearrangement, diol dehydratase becomes inactivated concomitant with formation of 5′-deoxyadenosine. With other pseudosubstrates that produce Co−C bond cleavage reversibly but without rearrangement, 5′-deoxyadenosine is produced on enzyme denaturation. A similar observation was made with ethanolamine ammonia-lyase.[98] EPR spectroscopic experiments by Abeles and co-workers and by Babior and co-workers established the formation of Co(II) as well as of other radicals (presumably substrate, product, and 5′-deoxyadenosyl radicals) during catalysis[99] (Scheme 13.39).

Based on these astonishing results, the mechanism shown in Scheme 13.40 can be imagined. Following homolytic cleavage of the Co−C bond of the coenzyme, the 5′-deoxyadenosyl radical abstracts a hydrogen atom from the substrate to give

13.88

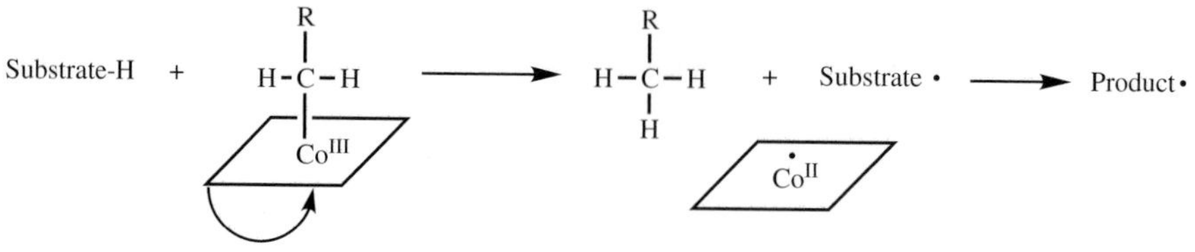

SCHEME 13.39   Formation of 5′-deoxyadenosine, cob(II)alamin, and substrate radicals during co-enzyme B_{12}-dependent reactions.

5′-deoxyadenosine (**13.88**) and substrate radical **13.89.** This radical rearranges to product radical **13.90,** which abstracts a hydrogen atom from 5′-deoxyadenosine (the hydrogens at this point have equilibrated, so any one can be abstracted at random) to give the product hydrate (**13.91**) and 5′-deoxyadenosyl radical, which recombines with the cob(II)alamin to give the coenzyme back. It is not clear how radical **13.89** rearranges to product radical **13.90,** or whether the substrate radical combines with the cob(II)alamin to give **13.92** (dashed box), which then rearranges on the cobalt to **13.93** prior to homolytic cleavage to **13.90.** I did my doctoral thesis research on development of a chemical model for the rearrangement of diol dehydratase− and ethanolamine ammonia-lyase−dependent enzyme reactions.[100] This model provides support for the formation and rearrangement of **13.92** to **13.93** via a Co(III)−olefin π-complex (Scheme 13.41). To simplify matters, a cobaloxime complex (a much smaller cobalt ligand), rather than a cobalamin complex, was synthesized with a $[2-^{13}C]$acetoxyethyl axial ligand (**13.94**). Incubation of **13.94** in methanol gave two products in a 1:1 ratio, **13.96** and **13.97.** These products can be rationalized as occurring via the Co(III)−olefin π-complex **13.95.** As applied to either the diol dehydratase− or ethanolamine ammonia-lyase−catalyzed reaction, once the cob(III)alamin−alkyl complex forms (e.g., **13.92**), an active-site acid would donate a proton to the β-hydroxyl (or amino) group to activate it as a leaving group. The corresponding cob(III)alamin−olefin π-complex would form, followed by addition to the adjacent carbon by the water (ammonia) molecule that was just released on ionization, yielding **13.93** (for diol dehydratase).

However, as much as I would like everyone to think that my doctoral thesis research provided the long-sought-after answer to this question, I must admit that there also is known chemistry to support a mechanism that does not require reformation of a Co−C bond and proceeds completely via radical rearrangements.[101] The reaction is the conversion of ethylene glycol (and related compounds) to acetaldehyde (exactly the reaction catalyzed by diol dehydratase), catalyzed by ferrous ion and hydrogen peroxide, known as the *Fenton reaction* (Scheme 13.42). This is the same mechanism as is shown in Scheme 13.40 for the enzyme-catalyzed reactions, except for formation of the hydrate, although it is not known if a hydrate forms in the Fenton reaction. MO calculations by Golding and co-workers indicate that protonation of the X group catalyzes this reaction.[102]

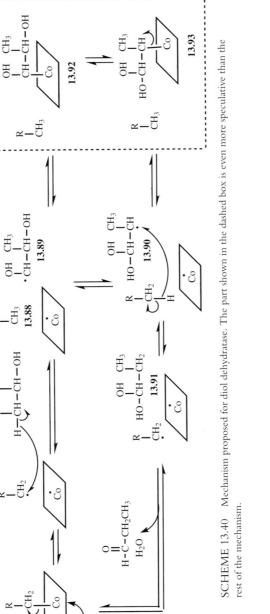

SCHEME 13.40 Mechanism proposed for diol dehydratase. The part shown in the dashed box is even more speculative than the rest of the mechanism.

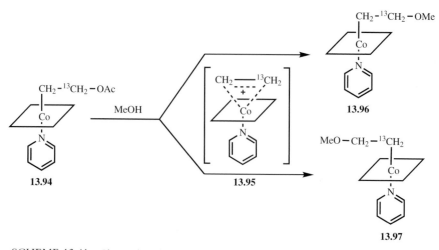

SCHEME 13.41   Chemical model study for a proposed diol dehydratase-catalyzed rearrangement involving a Co(III)–olefin π-complex. The trapezoid represents a cobaloxime ligand.

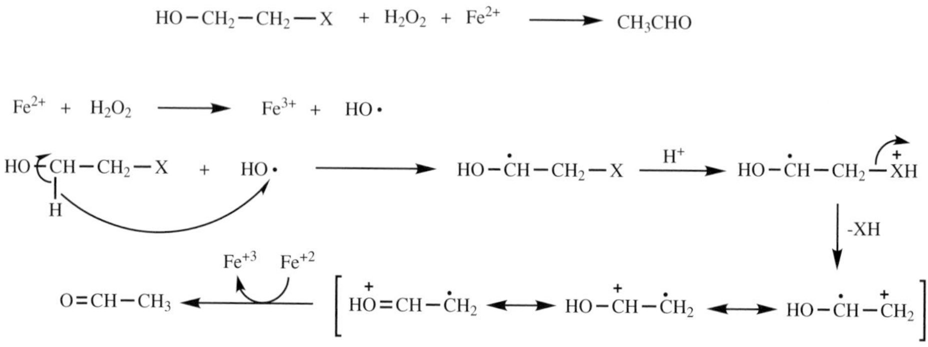

$$HO-CH_2-CH_2-X \quad + \quad H_2O_2 \quad + \quad Fe^{2+} \quad \longrightarrow \quad CH_3CHO$$

$$Fe^{2+} \quad + \quad H_2O_2 \quad \longrightarrow \quad Fe^{3+} \quad + \quad HO\cdot$$

SCHEME 13.42   The Fenton reaction as a model for a proposed diol dehydratase-catalyzed free radical rearrangement.

A closer chemical model for the diol dehydratase reaction by Golding and co-workers involves the photolytic or thermolytic cleavage of a cobaloxime complex (**13.98**) to generate radical **13.99**, which can undergo a [1,5] hydrogen atom abstraction to give **13.100**. This radical presumably rearranges, as in the case of diol dehydratase, to give the aldehyde (**13.101**) (Scheme 13.43).[103] The cobaloxime complex was used just to generate the desired starting radical (**13.99**); it is not meant to imply the involvement of a cob(III)alamin–substrate intermediate.

Grissom and co-workers have taken an interesting approach to the identification of radical pair formation in enzyme-catalyzed reactions.[104] The rate or product distribution of reactions that involve geminate radical pair intermediates can be al-

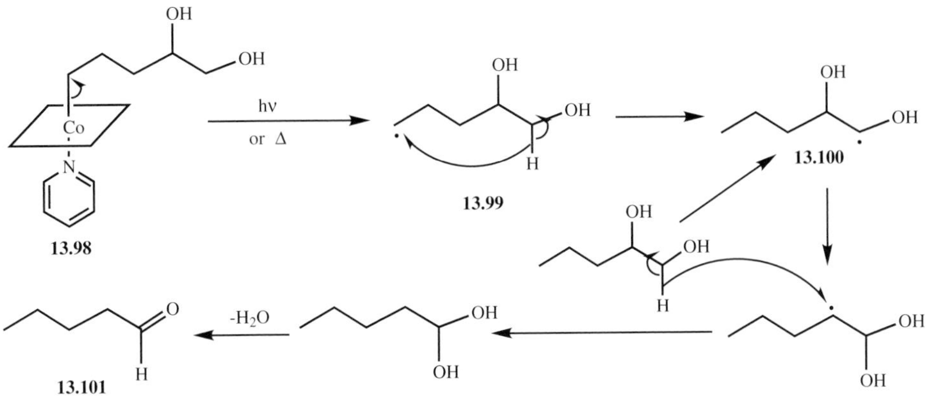

SCHEME 13.43   Chemical model study for a proposed diol dehydratase-catalyzed free radical rearrangement.

tered by an external magnetic field that increases or decreases intersystem crossing rates between the singlet and triplet states. If radical character exists on both the substrate and the enzyme (or cofactor) during the course of the reaction, then the E·S complex constitutes a geminate radical pair. If intersystem crossing occurs in the singlet E·S complex to produce a triplet E·S complex, the probability of nonproductive radical recombination and dissociation of S from the E·S complex is decreased. This would result in an increase in V_{max}/K_m (if it occurs before the first irreversible step). Therefore, if an external magnetic field is applied during an enzyme-catalyzed reaction that proceeds via a radical pair, a change in V_{max}/K_m should be induced. In fact, the V_{max}/K_m for ethanolamine ammonia-lyase was decreased by 25% by a static magnetic field. This effect is caused by a magnetic field-induced change in intersystem crossing rates between the singlet and triplet states in the cob(II)alamin:5′-deoxyadenosyl radical spin correlated radical pair. Using rapid-scanning stopped-flow spectrophotometry as a function of magnetic field, Harkins and Grissom showed that one of the magnetic field-dependent steps is nonproductive recombination of the enzyme-bound 5′-deoxyadenosyl radical and cob(II)alamin,[105] which supports the initial Co−C bond homolysis step.

b. Mechanism(s) for Methylmalonyl-CoA Mutase, Glutamate Mutase, and 2-Methyleneglutarate Mutase

Coenzyme B_{12}-dependent carbon skeletal rearrangements also could proceed by a mechanism similar to that shown in Scheme 13.40 for the "elimination"-type rearrangements. As in the case of diol dehydratase, the intermolecular hydrogen transfer in the carbon skeletal enzyme-catalyzed rearrangements proceeds without exchange of solvent.[106] The hydrogen from the coenzyme is removed at a rate comparable to that of its appearance in the product with a tritium isotope effect of 13.5–18 (deuterium isotope effect of about 6–7.4).[107] These results exclude the intermediacy of a protein-based radical during catalysis and the incorporation of the migrating hydrogen into the protein. Whereas the hydrogen is transferred with inversion of stereochemistry for glutamate mutase[108] and 2-methylglutarate mutase,[109] it is transferred with retention of configuration with methylmalonyl-CoA mutase.[110] By using (2S,3S)-[3-^{2}H, 3-^{3}H]glutamate as the substrate and converting the methylaspartate from the glutamate mutase reaction into chiral acetate (see Chapter 10, Section I.A.2 for a discussion of chiral methyl groups), Hartrampf and Buckel showed that racemic acetate was formed.[111] This is consistent with the formation of a methylene radical that can freely rotate. EPR spectroscopic investigations of the reactions catalyzed by methylmalonyl-CoA mutase,[112] glutamate mutase,[113] and 2-methyleneglutarate mutase[114] confirm the substrate-induced generation of cob(II)alamin and organic radical species during catalysis.

It probably seems pretty amazing that reactive species such as free radicals could be viable intermediates in enzyme-catalyzed reactions. On the basis of the crystal structure of *P. shermanii* methylmalonyl-CoA mutase at 2.0 Å resolution with[115] and without[116] substrates bound, Evans and co-workers have demonstrated that binding of the substrate causes a major conformational change by which the active site is closed to create a deeply buried active-site environment that shields intermediates from solvent. Furthermore, site-directed mutagenesis (Y89F) shows that the loss of the hydroxyl group of Tyr-89 affects both the stability of the radical intermediates and decreases the rate of interconversion of the substrate- and product-derived radicals.[117]

On the basis of UV–visible and EPR spectroscopic studies, Banerjee and co-workers showed that on coenzyme B_{12} binding to methylmalonyl-CoA mutase, the bottom axial dimethylbenzimidazole ligand is displaced by an active-site histidine residue.[118] The crystal structure of bacterial methylmalonyl-CoA mutase reveals that His-610 is coordinated to the cobalt via an unusually long bond length of 2.45 Å. Initially, it was thought that this ligand exchange may assist in the labilization of the Co–C bond, but this notion has been questioned by Banerjee.[119] Whatever the effect(s) may be, they must be very significant, because there is a rate acceleration of about 10^{12} between the rate of thermolysis of the Co–C bond of the coenzyme in solution (the bond dissociation energy is about 31 kcal/mol[120]) versus the k_{cat} for the enzyme-catalyzed reaction,[121] or about 15.5 kcal/mol of destabilization energy during the reaction.[122] A chemical model study by Waddington and Finke involving the thermolysis of neopentylcobalamin showed that the bulky neopentyl ligand can account for a factor of 10^6 of the 10^{12} rate acceleration observed with the enzyme.[123]

Banerjee and co-workers used UV–visible stopped-flow spectrophotometry to show that the rate of Co–C bond homolysis in methylmalonyl-CoA mutase in the presence of protiated methylmalonyl-CoA is more than 21 times greater than the rate with [CD$_3$]methylmalonyl-CoA.[124] This suggests that Co–C bond cleavage is coupled to hydrogen atom abstraction from the substrate, supporting a concerted

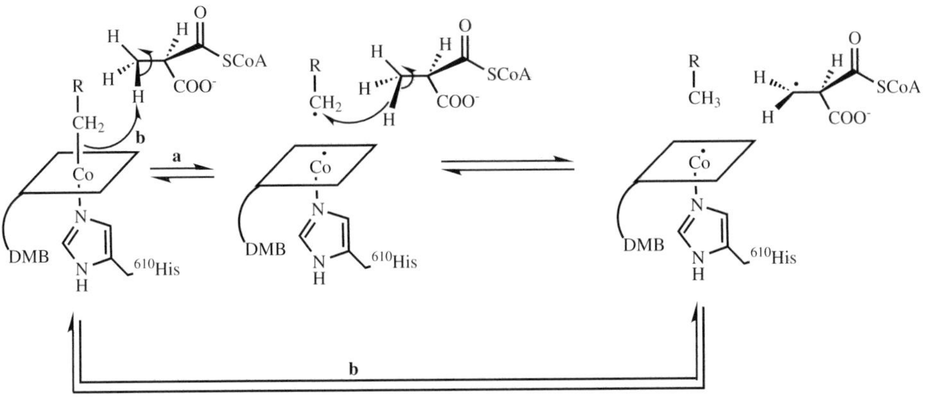

SCHEME 13.44   Stepwise (a) versus concerted (b) mechanisms for the methylmalonyl-CoA mutase-catalyzed generation of 5′-deoxyadenosine, cob(II)alamin, and substrate radical.

reaction (pathway b, Scheme 13.44). The same conclusion was reached by Marsh and Ballou based on an isotope effect of 28 on Co–C bond homolysis, when L-glutamate and deuterated L-glutamate are substrates for glutamate mutase.[125] EPR spectra of (S)-glutamate labeled with deuterium or ^{13}C at C-2, C-3, or C-4 all contain a similar signal. With the aid of spectral simulations and a comparison of the hyperfine couplings with and without the isotopic label, Buckel and co-workers identified the principal contributor to the spectra as the 4-glutamyl radical, supporting a hydrogen atom abstraction of the substrate.[126]

Once the methyl radical is generated, there are six pathways that can be contemplated for the rearrangement mechanism (Figure 13.4 shows the hypothetical mechanisms for the methylmalonyl-CoA mutase-catalyzed conversion of methyl-malonyl-CoA radical (13.102) to the succinyl-CoA radical (13.103), but similar mechanisms could be drawn for the other carbon skeletal rearrangements). Pathway a involves the transfer of the electron from 13.102 to cob(II)alamin to give cob(I)alamin and the corresponding carbocation (13.104), which rearranges to carbocation 13.105, accepts an electron back from cob(I)alamin to give cob(II)-alamin and 13.103. The opposite electron transfer, namely, from cob(II)alamin to 13.102 (pathway b) gives cob(III)alamin and carbanion 13.106. After rearrangement (13.107) and transfer back of the electron, 13.103 is obtained. Pathway c is a concerted radical rearrangement mechanism, whereas pathway d proceeds via the cyclopropanoxy radical 13.108. In pathway e, radical 13.102 undergoes radical fragmentation to acrylic acid (13.109) and 2-formyl-CoA radical (13.110), which come together at the other end of the olefin and reform 13.103. Finally, pathway f involves initial combination of 13.102 with cob(II)alamin to give 13.111, which rearranges to the organocobalamin complex 13.112 before Co–C bond homolysis to 13.103. *Ab initio* calculations indicate that pathway e for 2-methyleneglutarate mutase has a much higher activation energy than other pathways.[126a]

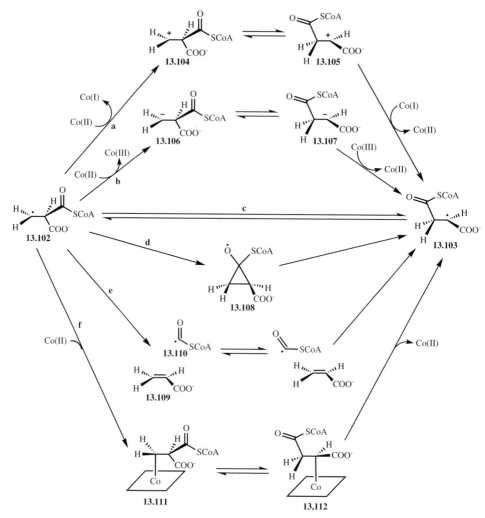

FIGURE 13.4   Six possible mechanisms for the conversion of methylmalonyl-CoA radical to succinyl-CoA radical catalyzed by methylmalonyl-CoA mutase.

None of these mechanisms has direct enzymological support, so we must rely on chemical model studies to determine what chemistry is reasonable. Unfortunately, there is no consensus, and, in fact, depending on what year you look at, you can find the same authors supporting different mechanisms. Instead of going through the numerous model studies, I will just mention a few of the studies and which mechanism they support. For example, some model studies have argued against radical intermediates [127] or against carbanion intermediates. [128] Some models support radical intermediates [129] or radical intermediates that proceed via a cyclopropane intermediate,[130] or a radical intermediate that proceeds via a cyclopropane but is still bound to the cobalamin,[131] or protein-bound radical intermediates.[132] Others support rearrangements via an initial radical intermediate followed by electron transfer to a carbanion (pathway b).[133] Models also support an organocobalt complex[134] and reactions via carbanion intermediates.[135] The fragmentation mechanism[136] (pathway e) is supported by the finding of synergistic inhibition of glutamate mutase by glycine and acrylate.[137]

c. Mechanism for Ribonucleotide Reductase

Ribonucleotide reductases[138] catalyze the conversion of ribonucleotides to deoxyribonucleotides, which are required for DNA biosynthesis and repair. There are four classes of ribonucleotide reductases, based on the mechanisms they use for radical generation and on their structural differences, but only one class (class II) utilizes coenzyme B_{12}, so this is the class that will be emphasized here. After a discussion of that enzyme class, the other ribonucleotide reductases will be discussed briefly.

Ribonucleotide reductase from *Lactobacillus leichmannii* is the coenzyme B_{12}-dependent enzyme for which the mechanism is most well understood. Some early results represent a diversion from the results with the coenzyme B_{12}-dependent enzymes discussed previously (III.B.2.a and b). First, there is an isotope effect on the cleavage of the C^3–H bond of [$3'$-^{3}H]UTP with release of 0.01−0.1% of tritium, and no tritium from [$3'$-^{3}H]UTP is found in the adenosylcobalamin. No tritium is found in dATP when [$3'$-^{3}H]UTP and ATP are mixed, indicating that the intermediate is not released, and there are no crossover products. Tritium incorporated at C-3 of the substrate is found at C-3 of the product. In addition to nucleotide reduction, the enzyme catalyzes two other reactions: One is an exchange of tritium from [$5'$-^{3}H]AdoCbl with solvent in the presence of an allosteric effector and a reductant but in the absence of substrate,[139] and the other is the conversion of AdoCbl to $5'$-deoxyadenosine and cob(II)alamin.[140]

Stubbe and co-workers demonstrated that the function of the adenosylcobalamin is to generate a thiyl radical at the active site.[141] EPR experiments, using rapid freeze quench techniques to trap intermediates, with wild-type ribonucleotide reductase and ribonucleotide reductase in which all its cysteine residues are replaced with [3-^{2}H$_2$]cysteine, indicate that a thiyl radical interacting with cob(II)alamin is generated. Simulations of these spectra are consistent with this conclusion.[142] When ribonucleotide reductase is mixed with dGTP (an allosteric effector), the rate of formation of cob(II)alamin is identical to the rate of formation of $5'$-deoxyadenosine (from the coenzyme), suggesting that Co−C bond cleavage is concerted with hydrogen atom abstraction from a cysteine residue to form the thiyl radical.

Stubbe and co-workers investigated by mutagenesis experiments the roles of five cysteine residues postulated to be required for catalysis.[143] Various single and double mutants (C731S, C736S, C119S, C419S, C408S, and C305S) were prepared. Mutants C731S and C736S catalyze the formation of dCTP from CTP in the presence of an artificial reductant (dithiolthreitol), but could not catalyze multiple turnovers in the presence of the *in vivo* reducing system (thioredoxin). This experiment indicates that their function may be to transfer reducing equivalents from thioredoxin (which is reduced by NADPH, catalyzed by thioredoxin reductase[144]) into the active-site disulfide of ribonucleotide reductase. Neither C119S nor C419S catalyzes the formation of dCTP, but they are capable of catalyzing the formation of cytosine and of destroying the cofactor, two reactions that require the active-site cysteines in the class I ribonucleotide reductases. This experiment suggests that C119 and C419 are active-site cysteines, but not directly involved in initiation of the substrate reduction; however, they become oxidized concomitant with substrate reduction. Mutant C409S is unable to catalyze production of dCTP or cytosine release or other reactions unique to this enzyme, namely, exchange of tritium from [5′-³H]AdoCbl and the formation of 5′-deoxyadenosine and cob(II)alamin under anaerobic conditions. This result suggests that C408 is the active-site cysteine that is oxidized to a thiyl radical as a result of homolysis of the Co−C bond of AdoCbl. A mechanism consistent with these data is shown in Scheme 13.45. The first step is concerted homolysis of AdoCbl and abstraction of the hydrogen atom from C408. This active-site C408 radical abstracts the C-3 hydrogen atom (H_a) from the substrate to give a substrate C-3 radical (**13.113**). Because the bond dissociation energy for S−H is 88−91 kcal/mol,[145] the corresponding thiyl radical is thermodynamically capable of abstraction of a hydrogen of a deoxyribose sugar, given that the bond dissociation energy for secondary alcohols is about 91 kcal/mol.[146] Protonation of the C-2 hydroxyl group by either C119 or C419 and deprotonation of the C-3 hydroxyl group by an active-site base assists in the loss of water to the α-keto radical **13.114.** Transfer of an electron from the thiolate ion and a proton from the other cysteine residue (either C419 or C119) gives a ketone and a disulfide radical anion at the active site (**13.115**). Electron transfer from this disulfide radical anion to the carbonyl at C-3 gives **13.116,** which can abstract the hydrogen atom from C408. Because this hydrogen has not exchanged with solvent, the same hydrogen (H_a) that was initially removed from C-3 of the substrate by the C408 radical is replaced at C-3 of the product (**13.117**). Note the similarity of this mechanism up to **13.115** with the Fenton mechanism for conversion of ethylene glycol to acetaldehyde (Scheme 13.42), which utilizes hydroxyl radical to initiate the reaction instead of a thiyl radical as shown in Scheme 13.45.

Cysteine residues C731 and C736 are proposed to shuttle electrons into the active-site disulfide from the *in vivo* protein reductant thioredoxin, thereby reducing the active-site disulfide generated during turnover and setting up the active site for the next turnover (Scheme 13.46).

Evidence for the formation of substrate radicals during catalysis is derived from experiments by Stubbe and co-workers with mechanism-based inactivators, such as

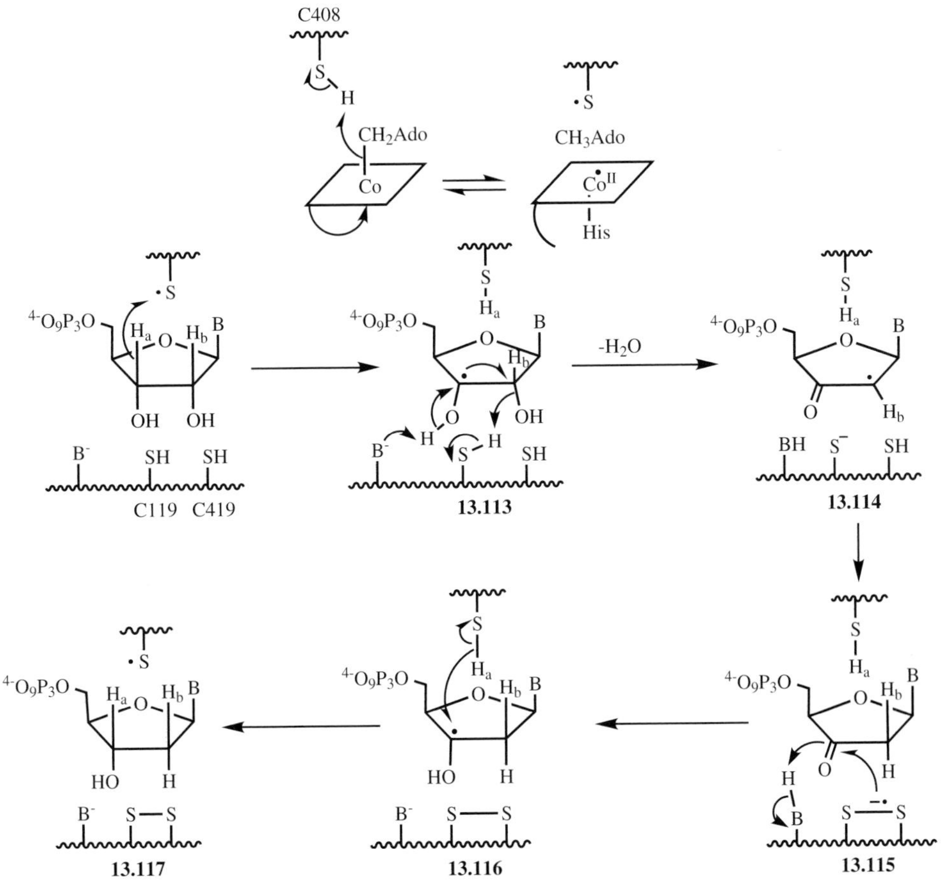

SCHEME 13.45   Mechanism proposed for coenzyme B_{12}-dependent ribonucleotide reductase.

the 5′-triphosphates of 2′-(fluoromethylene)-2′-deoxycytidine,[147] 2′-methylene-2′-deoxycytidine, 2′-spirocyclopropyl-2′-deoxyctyidine, and 2′-difluoro-2′-deoxycytidine.[148] EPR spectroscopy during inactivation shows the initial formation of a thiyl radical-cob(II)alamin couple followed by the spectrum of cob(II)alamin interacting with a nucleotide-based radical.

Unlike the other coenzyme B_{12}-dependent enzymes, the cobalamin in ribonucleotide reductase serves the function of initiating the radical reaction by abstraction of a hydrogen atom from the active-site cysteine residue, rather than serving to transfer a hydrogen atom. The other members of the family of ribonucleotide reductases appear to proceed by the same mechanism as that shown for class II ribonucleotide reductase,[149] except that they use different cofactors for abstraction of

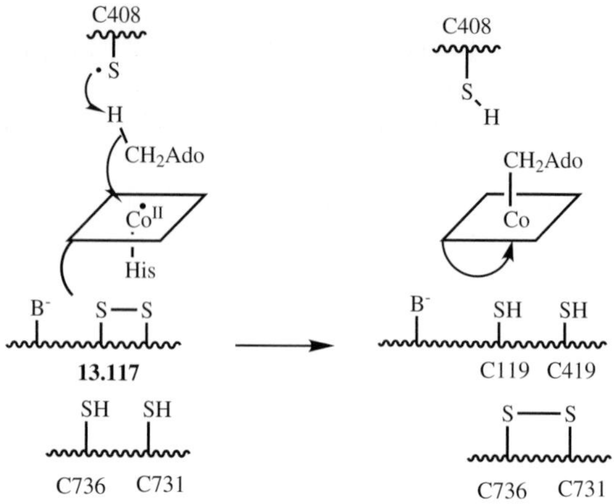

SCHEME 13.46   Mechanism proposed for reducing and reestablishing the active site of coenzyme B_{12}-dependent ribonucleotide reductase.

the active-site cysteine hydrogen atom (Figure 13.5). Class I ribonucleotide reductases contain a μ-oxo-bridged diferric cluster adjacent to a tyrosine residue (**13.118**; this is the complex after hydrogen atom abstraction). In the presence of O_2 and a reductant the diferric cluster abstracts a hydrogen atom from the tyrosine residue (Tyr-122 in *Escherichia coli*) to form a diferric–tyrosyl radical,[150] which abstracts a hydrogen atom from Cys-439 at the active site. Actually, there are two class

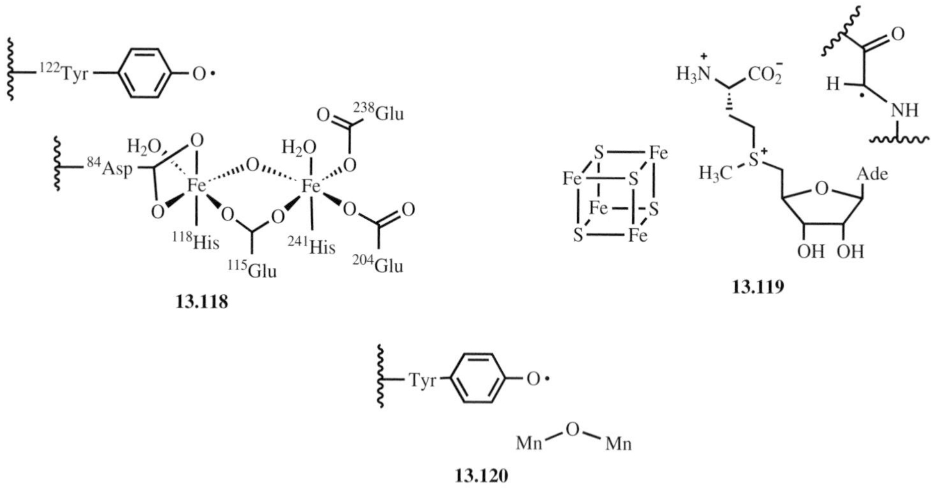

FIGURE 13.5   Cofactors for class I (**13.118**), class III (**13.119**), and class IV (**13.120**) ribonucleotide reductases.

I ribonucleotide reductases, which differ in their reductant sources[151] and in their mechanism of expression.[152] This class has therefore been divided into class Ia and class Ib. Class III ribonucleotide reductases are found in *Escherichia coli* grown under anaerobic conditions.[153] This class of enzymes utilizes a glycine radical (Gly-681) as the abstractor of the substrate hydrogen atom and requires an [4Fe−4S] cluster and (*S*)-adenosylmethionine (**13.119**) for the generation of this glycine radical.[154] The mechanism for glycine radical formation is not known, but it may be related to the mechanism of hydrogen atom abstraction from the substrate by lysine 2,3-aminomutase, discussed in the next section. The class IV ribonucleotide reductases have a putative dimanganese cluster and a protein radical (**13.120**) that may be a tyrosine radical.[155]

d. Mechanism for a Related Rearrangement Not Dependent on Cobalamin

A reaction that appears to be related to the coenzyme B_{12}-dependent rearrangements is the conversion of L-α-lysine (**13.121**) to L-β-lysine (**13.122**), catalyzed by lysine 2,3-aminomutase (Scheme 13.47).[156] This is the first step in lysine metabolism in *Clostridia,* and the enzyme requires an unusual mixture of cofactors, namely, pyridoxal 5′-phosphate (PLP), (*S*)-adenosylmethionine (SAM),[157] and an iron−sulfur cluster;[158] a reducing agent also is needed. The reaction involves the transfer of the 3-*pro-R* hydrogen of L-α-lysine to the 2-*pro-R* position of β-lysine and the migration of the 2-amino group of L-α-lysine to C-3 of β-lysine;[159] the migrating hydrogen does not exchange with solvent. This is definitely a unique transformation for a PLP-dependent enzyme.

Frey and co-workers have reported a series of elegant studies to elucidate the mechanism of this fascinating enzyme. Moss and Frey found that when (*S*)-[5′-^{3}H]adenosylmethionine is used, the tritium ends up in both the L-α-lysine and L-β-lysine in the equilibrium mixture amounts.[160] When L-α-[3-^{2}H$_2$]lysine is the substrate, β-[2,3-^{2}H$_2$]lysine is formed, indicating that the hydrogen from C-3 in α-lysine becomes C-2 in β-lysine.[161] One equivalent each of methionine and 5′-deoxyadenosine are produced.[162] When L-α-[3-^{3}H]lysine is the substrate, 1−6% of the tritium is incorporated into the SAM, consistent with exchange of the SAM protons with the 3-*pro-R* proton of L-α-lysine.[163] This suggests that the SAM acts as a hydrogen carrier, just as coenzyme B_{12} does with coenzyme B_{12}-dependent reactions. However, unlike the Co−C bond of cobalamin, the C−S bond of SAM is not a labile bond. So how does SAM get activated? In the presence of a reducing

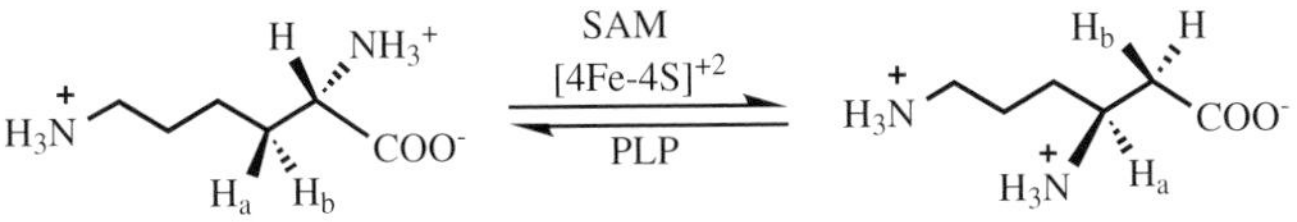

SCHEME 13.47   Reaction catalyzed by lysine 2,3-aminomutase.

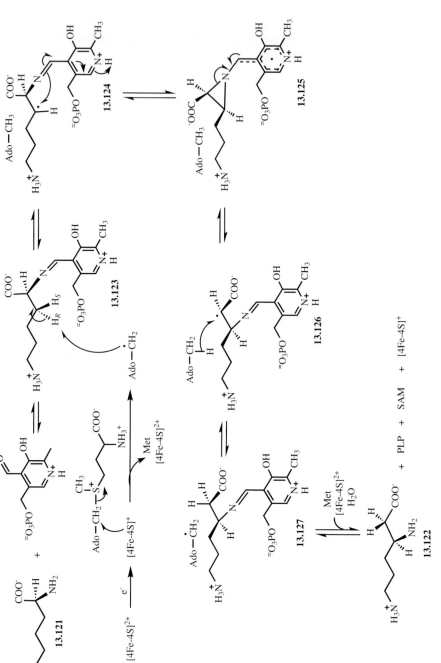

SCHEME 13.48 Mechanism proposed for lysine 2,3-aminomutase.

agent (dithionite) and SAM, the EPR-silent $[4Fe-4S]^{2+}$ cluster at the active site of the enzyme is reduced to the paramagnetic $[4Fe-4S]^{1+}$ cluster, and the EPR spectrum shows an interaction between this reduced iron–sulfur cluster and SAM.[164] Apparently the reduced iron–sulfur cluster reduces SAM to methionine and 5′-deoxyadenosyl radical.

A mechanism that is consistent with these results is shown in Scheme 13.48. Schiff base formation of **13.121** with PLP gives **13.123.** One-electron reduction of the iron–sulfur cluster and then transfer of an electron from the $[4Fe-4S]^{1+}$ to SAM releases methionine and generates the 5′-deoxyadenosyl radical, which abstracts the 3-*pro-R* hydrogen atom from **13.123,** giving radical **13.124.** The unique step (and a very clever proposal) in this mechanism is the radical cyclization to give the aziridine-containing PLP-stabilized radical intermediate **13.125.** Cleavage of the aziridine ring at the other C–N bond completes the C-2 to C-3 migration of the amino group (**13.126**). Abstraction of one of the 5′-deoxyadenosine hydrogen atoms with the product (**13.127**) regenerates the Schiff base of PLP and regenerates the 5′-deoxyadenosyl radical, which reacts with methionine and $[4Fe-4S]^{2+}$ to release **13.122** and reform SAM.

Radical cycloadditions to carbonyls are well-known reactions,[165] but when this type of cyclization step (**13.124** to **13.125**) was proposed for lysine 2,3-aminomutase, there was no chemical precedent for the utilization of PLP as a means of radical stabilization. Consequently, Han and Frey devised a chemical model reaction to test the chemistry (Scheme 13.49).[166] To make the reaction simpler, benzaldehyde was used in place of PLP, and tributyltin radical was the simulated 5′-deoxyadenosyl radical.

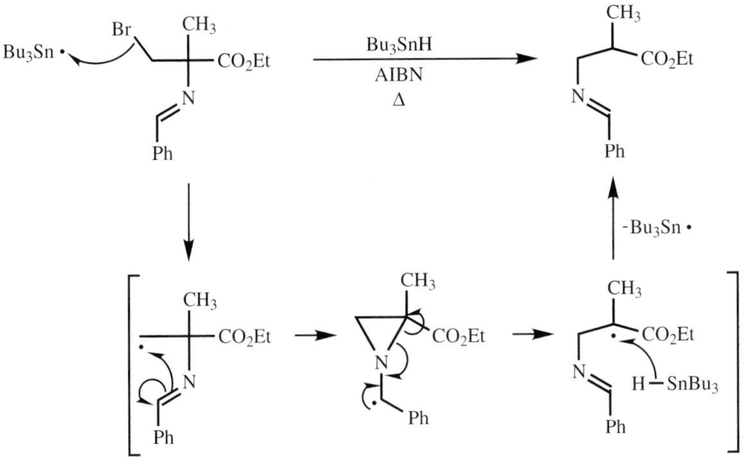

SCHEME 13.49   Chemical model study to test the proposed rearrangement mechanism for lysine 2,3-aminomutase.

$$\text{H}_3\overset{+}{\text{N}} \diagup\diagdown \text{S} \diagup\diagdown \overset{\text{COO}^-}{\underset{\text{NH}_2}{\text{ }}}\text{H}$$

13.128

A series of EPR spectroscopic studies provided support for most of the proposed radical intermediates. Purified lysine 2,3-aminomutase was shown by EPR spectroscopy to contain low levels of a stable radical identified as an iron−sulfur cluster, [4Fe−4S].[167] Incubation of the enzyme with lysine and SAM results in the formation of an EPR signal for an organic radical (g = 2.001) with multiple hyperfine transitions.[168] The signals changed over the time course of the reaction, suggesting the presence of more than one radical species. Using [2-^{2}H]Lys and [2-^{13}C]Lys, a π-radical was detected at steady state having its electron in a p-orbital on C-2 of β-lysine, thereby supporting the formation of a product-related radical (such as **13.126**).[169] Presumably, the product radical is observed and not the substrate radical (**13.124**) because of the relative stabilities of these two radicals: **13.126** is stabilized by the carboxylate group, so its concentration should be higher than that of **13.124**. The rate of formation of this radical is identical to the turnover number of the reaction; therefore, it is a kinetically competent intermediate.[170]

To change the relative stabilities of the substrate and product radicals, Frey and co-workers synthesized an alternative substrate that was designed to lead to a more stabilized substrate radical.[171] 4-Thialysine (**13.128**) has a K_m the same as that for lysine and a V_{max} about 3% of that for lysine; however, the initial substrate-derived radical (**13.129**) is sulfur stabilized, so it is expected to be much more stable than the corresponding product-derived radical (**13.130**) which is only carboxylate stabilized (Scheme 13.50). The EPR spectra of lysine 2,3-aminomutase incubated with **13.128** and with ^{13}C- and ^{2}H-labeled **13.128** contain one signal at g = 2.003 (on rapid freeze quenching), corresponding exclusively to the substrate radical **13.129**. Furthermore, malonic acid semialdehyde (**13.131**) is a product of this reaction, consistent with the mechanism shown in Scheme 13.50. Electron spin echo envelope modulation (ESEEM) spectroscopy with PLP and [4'-^{2}H]PLP establishes that the PLP is in a Schiff base with the β-nitrogen of β-lysine when the α-radical is generated, supporting the idea that PLP is involved during the radical rearrangement.[172] The only two radical intermediates not yet identified are the 5'-deoxyadenosyl radical and the aziridine-containing radical **13.125** (Scheme 13.48). Both of these radicals are expected to be much higher in energy than the substrate and product radicals: 5'-Deoxyadenosyl radical is an unstabilized primary radical, and **13.125** is a radical adjacent to a highly strained ring. It is highly unlikely that they will be detected.

IV. EPILOGUE

The purpose of this text has been to summarize how enzymes catalyze reactions and the organic pathways that they utilize. New enzymes are discovered all the

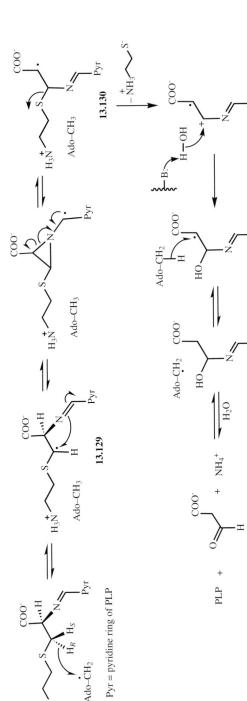

SCHEME 13.50   Lysine 2,3-aminomutase-catalyzed rearrangement of 4-thialysine to generate a more stable substrate radical.

time, but rarely are new mechanisms uncovered. With the foundation you have received from this text, you should be able to devise reasonable mechanisms, or at least be able to design experiments to uncover reasonable mechanisms, for these enzymes. However, keep in mind that although cofactors generally catalyze the same kind of chemistry from enzyme to enzyme, there are cases where unusual chemistry is recruited for specific needs; for example, just look in the previous section for a new utilization of PLP. In all cases, though, reasonable organic chemistry is employed. Do not expect an enzyme to catalyze chemistry that has no theoretical foundation. Enzymes are efficient organic chemists; they are not magicians.

Note that problems and solutions relevant to each chapter can be found in Appendix II.

REFERENCES

1. (a) Woodward, R. B.; Hoffmann, R. *The Conservation of Orbital Symmetry,* Academic Press: New York, 1970. (b) Woodward, R. B.; Hoffmann, R. *Angew. Chem. Int. Ed. Engl.* **1969,** *8,* 781.
2. Weiss, U.; Edwards, J. M. *The Biosynthesis of Aromatic Amino Compounds.* Wiley: New York, 1980, pp. 134−184.
3. Gill, G. *Quart. Rev. Chem. Soc.* **1968,** *22,* 338.
4. (a) Jackson, D. Y.; Liang, M. N.; Bartlett, P. A.; Schultz, P. G. *Angew. Chem. Int. Ed. Engl.* **1992,** *31,* 182. (b) Jackson, D. Y.; Jacobs, J. W.; Sugasawara, R.; Reich, S. H.; Bartlett, P. A.; Schultz, P. G. *J. Am. Chem. Soc.* **1988,** *110,* 4841. (c) Hilvert, D.; Carpenter, S. H.; Nared, K. D.; Auditor, M.-T. M. *Proc. Natl. Acad. Sci. USA* **1988,** *85,* 4953. (d) Hilvert, D.; Nared, K. D. *J. Am. Chem. Soc.* **1988,** *110,* 5593.
5. Bartlett, P. A.; Johnson, C. R. *J. Am. Chem. Soc.* **1985,** *107,* 7792.
6. Edwards, J. M.; Jackman, L. M. *Aust. J. Chem.* **1965,** *18,* 1227.
7. Sogo, S. G.; Widlanski, T. S.; Hoare, J. H.; Grimshaw, C. E.; Berechtold, G. A.; Knowles, J. R. *J. Am. Chem. Soc.* **1984,** *106,* 2701.
8. Copley, S. D.; Knowles, J. R. *J. Am. Chem. Soc.* **1985,** *107,* 5306.
9. Addadi, L.; Jaffe, E. K.; Knowles, J. R. *Biochemistry* **1983,** *22,* 4494.
10. (a) Andrews, P. R.; Smith, G. D.; Young, I. G. *Biochemistry* **1973,** *12,* 3492. (b) Andrews, P. R.; Haddon, R. C. *Aust. J. Chem.* **1979,** *32,* 1921.
11. Rétey, J.; Bartl, K.; Ripp, E.; Hull, W. *Eur. J. Biochem.* **1977,** *72,* 251.
12. Copley, S. D.; Knowles, J. R. *J. Am. Chem. Soc.* **1987,** *109,* 5008.
13. Guilford, W. J.; Copley, S. D.; Knowles, J. R. *J. Am. Chem. Soc.* **1987,** *109,* 5013.
14. Pawlak, J. L.; Padykula, R. E.; Kronis, J. D.; Aleksejczyk, R. A.; Berchtold, G. A. *J. Am. Chem. Soc.* **1989,** *111,* 3374.
15. Gray, J. V.; Knowles, J. R. *Biochemistry* **1994,** *33,* 9953.
16. (a) Chook, Y.-M.; Ke, H.; Lipscomb, W. N. *Proc. Natl. Acad. Sci. USA,* **1993,** *90,* 8600. (b) Chook, Y.-M.; Gray, J. V.; Ke, H.; Lipscomb, W. N. *J. Mol. Biol* **1994,** *240,* 476.
17. (a) Lee, A. Y.; Karplus, P. A.; Ganem, B.; Clardy, J. *J. Am. Chem. Soc.* **1995,** *117,* 3627. (b) Lee, A. Y.; Stewart, J. D.; Clardy, J.; Ganem, B. *Chem. Biol.* **1995,** *2,* 195.
18. Haynes, M. R.; Stura, E. A.; Hilvert, D.; Wilson, I. A. *Science* **1994,** *263,* 646.
19. Gajewski, J. J.; Jurayi, J.; Kimbrough, D. R.; Gande, M. E.; Ganem, B.; Carpenter, B. K. *J. Am. Chem. Soc.* **1987,** *109,* 1170.
20. Cload, S. T.; Liu, D. R.; Pastor, R. M.; Schultz, P. G. *J. Am. Chem. Soc.* **1996,** *118,* 1787.

21. Liu, D. R.; Cload, S. T.; Pastor, R. M.; Schultz, P. G. *J. Am. Chem. Soc.* **1996**, *118*, 1789.

22. Braisted, A. C.; Schultz, P. G. *J. Am. Chem. Soc.* **1994**, *116*, 2211.

23. Driggers, E. M.; Cho, H. S.; Liu, C. W.; Katzka, C. P.; Braisted, A. C.; Ulrich, H. D.; Wemmer, D. E.; Schultz, P. G. *J. Am. Chem. Soc.* **1998**, *120*, 1945.

24. Ulrich, H. D.; Mundorff, E.; Santarsiero, B. D.; Driggers, E. M.; Stevens, R. C.; Schultz, P. G. *Nature* **1997**, *389*, 271.

25. Latham, Jr., J. A.; Branchaud, B. P.; Chen, Y.-C.J.; Walsh, C. *J. Chem. Soc. Chem. Commun.* **1986**, 528.

26. Yoon, S. S.; Oei, Y.; Sweet, E.; Schultz, P. G. *J. Am. Chem. Soc.* **1996**, *118*, 11686.

27. (a) March, J. *Advanced Organic Chemistry,* 4th ed., Wiley: New York, 1992, p. 839. (b) Wassermann, A. *Diels−Alder Reactions,* Elsevier: Amsterdam, 1965.

28. (a) Oikawa, H.; Suzuki, Y.; Naya, A.; Katayama, K.; Ichihara, A. *J. Am. Chem. Soc.* **1994**, *116*, 3605. (b) Oikawa, H.; Katayama, K.; Suzuki, Y.; Ichihara, A. *J. Chem. Soc. Chem. Commun.* **1995**, 1321.

29. (a) Breslow, R.; Guo, T. *J. Am. Chem. Soc.* **1988**, *110*, 5613. (b) Blokzijl, W.; Blandamer, M. J.; Engberts, J. B. F. N. *J. Am. Chem. Soc.* **1991**, *113*, 4241.

30. Braisted, A. C.; Schultz, P. G. *J. Am. Chem. Soc.* **1990**, *112*, 7430.

31. Romesberg, F. E.; Spiller, B.; Schultz, P. G.; Stevens, R. C. *Science* **1998**, *279*, 1929.

32. Gouverneur, V. E.; Houk, K. N.; dePasculal-Teresa, B.; Beno, B.; Janda, K. D.; Lerner, R. A. *Science* **1993**, *262*, 204.

33. Yli-Kauhaluoma, J. T.; Ashley, J. A.; Lo, C.-H.; Tucker, L.; Wolfe, M. M.; Janda, K. D. *J. Am. Chem. Soc.* **1995**, *117*, 7041.

34. Heine, A.; Stura, E. A.; Yli-Kauhaluoma, J. T.; Gao, C.; Deng, Q.; Beno, B. R.; Houk, K. N.; Janda, K. D.; Wilson, I. A. *Science* **1998**, *279*, 1934.

35. Suckling, C. J.; Tedford, M. C.; Bence, L. M.; Irvine, J. I.; Stimson, W. H. *J. Chem. Soc. Perkin Trans. 1* **1993**, 1925.

36. March, J. *Advanced Organic Chemistry,* 4th ed., Wiley: New York, 1992.

37. (a) Strassman, M.; Thomas, A.; Weinhouse, S. *J. Am. Chem. Soc.* **1953**, *75*, 5135. (b) Strassman, M.; Thomas, A.; Weinhouse, S. *J. Am. Chem. Soc.* **1955**, 77, 1261. (c) Adelberg, E. *J. Biol. Chem.* **1955**, *216*, 431.

38. Arfin, S.; Umbarger, H. *J. Biol. Chem.* **1969**, *244*, 1118.

39. Dumas, R.; Lebrun, M.; Douce, R. *Biochem. J.* **1991**, *277*, 469.

40. Johnson, W. S. *Bioorg. Chem.* **1976**, *5*, 51.

41. Mulheirn, L. J.; Ramm, P. J. *Chem. Soc. Rev.* **1972**, *1*, 259.

42. Corey, E. J.; Virgil, S. C. *J. Am. Chem. Soc.* **1991**, *113*, 4025.

43. Corey, E. J.; Virgil, S. C.; Sarshar, S. *J. Am. Chem. Soc.* **1991**, *113*, 8171.

44. Corey, E. J.; Virgil, S. C. *J. Am. Chem. Soc.* **1990**, *112*, 6429.

45. Corey, E. J.; Virgil, S. C.; Cheng, H.; Baker, C. H.; Matsuda, S. P. T.; Singh, V.; Sarshar, S. *J. Am. Chem. Soc.* **1995**, *117*, 11819.

46. Corey, E. J.; Cheng, H.; Baker, C. H.; Matsuda, S. P. T.; Li, D.; Song, X. *J. Am. Chem. Soc.* **1997**, *119*, 1277.

47. (a) Poulter, C. D.; Rilling, H. C. *Biosynthesis of Isoprenoid Compounds,* vol. 1, Porter, J. W.; Spurgeon, S. L., Eds., Wiley: New York, 1981, pp. 413−441. (b) Poulter, C. D. *Acc. Chem. Res.* **1990**, *23*, 70.

48. Jarstfer, M. B.; Flagg, B. S. J.; Rogers, D. H.; Poulter, C. D. *J. Am. Chem. Soc.* **1996**, *118*, 13089.

49. Zhang, D.; Poulter, C. D. *J. Am. Chem. Soc.* **1995**, *117*, 1641.

50. Harm, W. *Biological Effects of Ultraviolet Radiation;* Cambridge University Press: Cambridge, 1980.

51. Stolarski, R.; Bojkov, R.; Bishop, L.; Zerefos, C.; Staehelin, J.; Zawodny, J. *Science* **1992**, *256*, 342.

52. Li, Y. F.; Kim. S.-T.; Sancar, A. *Proc. Natl. Acad. Sci. USA* **1993**, *90*, 4389.

53. Sancar, A. *Science* **1996**, *272*, 48.

54. Zhao, X.; Liu, J.; Hsu, D. S.; Zhao, S.; Taylor, J.-S.; Sancar, A. *J. Biol. Chem.* **1997**, *272*, 32580.

55. (a) Iwatsuki, N.; Joe, C. D.; Werbin, H. *Biochemistry* **1980,** *19,* 1172. (b) Sancar, A.; Sancar, G. B. *J. Mol. Biol.* **1984,** *172,* 223.

56. Johnson, J. L.; Hamm-Alvarez, S.; Payne, G.; Sancar, G. B.; Rajagopalan, K. V.; Sancar, A. *Proc. Natl. Acad. Sci. USA* **1988,** *85,* 2046.

57. Eker, A. P. M.; Dekker, R. H.; Berends, W. *Photochem. Photobiol.* **1981,** *33,* 65.

58. Jorns, M. S.; Sancar, G. B.; Sancar, A. *Biochemistry* **1984,** *23,* 2673.

59. Hartman, R. F.; Rose, S. D. *J. Am. Chem. Soc.* **1992,** *114,* 3559.

60. (a) Heelis, P. F.; Payne, G. P.; Sancar, A. *Biochemistry* **1987,** *26,* 4634. (b) Li, Y. F.; Heelis, P. F.; Sancar, A. *Biochemistry* **1991,** *30,* 6322.

61. (a) Hamm-Alverez, S. F.; Sancar, A.; Rajagolpalan, K. V. *J. Biol. Chem.* **1989,** *264,* 9649. (b) Hamm-Alverez, S. F.; Sancar, A.; Rajagolpalan, K. V. *J. Biol. Chem.* **1990,** *265,* 18656.

62. Sancar, A. *Biochemistry* **1994,** *33,* 2.

63. (a) Rupert, C. S.; Goodgal, S. H.; Herriott, R. M. *J. Gen. Physiol.* **1958,** *41,* 451. (b) Rupert, C. S. *J. Gen. Physiol.* **1962,** *45,* 703. (c) Rupert, C. S. *J. Gen. Physiol.* **1962,** *45,* 725.

64. Rustandi, R. R.; Jorns, M. S. *Biochemistry* **1995,** *34,* 2284.

65. Jordan, S. P.; Jorns, M. S. *Biochemistry* **1988,** *27,* 8915.

66. Lipman, R. S. A.; Jorns, M. S. *Biochemistry* **1992,** *31,* 786.

67. Kim, S.-T.; Heelis, P. F.; Sancar, A. *Biochemistry* **1992,** *31,* 11244.

68. Kim, S.-T.; Heelis, P. F.; Okamura, T.; Hirata, Y.; Mataga, N.; Sancar, A. *Biochemistry* **1991,** *30,* 11262.

69. Kim, S.-T.; Sancar, A.; Essenmacher, C.; Babcock, G. *J. Am. Chem. Soc.* **1992,** *114,* 442.

70. Park, H.-W.; Kim, S.-T.; Sancar, A.; Deisenhofer, J. *Science* **1995,** *268,* 1866.

71. Heelis, P. F.; Deeble, D. J.; Kim, S.-T.; Sancar, A. *Int. J. Radiat. Biol.* **1992,** *62,* 137.

72. Hartzfeld, D. G.; Rose, S. D. *J. Am. Chem. Soc.* **1993,** *115,* 850.

73. Jorns, M. S.; Wang, B. Y.; Jordan, S. P.; Chanderkar, L. P. *Biochemistry* **1990,** *29,* 552. (b) Chanderkar, L. P.; Jorns, M. S. *Biochemistry* **1991,** *30,* 745.

74. (a) Taylor, J.-S.; Cohrs, M. P. *J. Am. Chem. Soc.* **1987,** *109,* 2834. (b) Mitchell, D. L.; Nairn, R. S. *Photochem. Photobiol.* **1989,** *49,* 805.

75. (a) Zhao, X.; Liu, J.; Hsu, D. S.; Zhao, S.; Taylor, J.-S.; Sancar, A. *J. Biol. Chem.* **1997,** *272,* 32580. (b) Kim, S.-T.; Malhotra, K.; Smith, C. A.; Taylor, J.-S.; Sancar, A. *J. Biol. Chem.* **1994,** *264,* 8535.

76. Prakash, G.; Falvey, D. E. *J. Am. Chem. Soc.* **1995,** *117,* 11375.

77. Todo, T.; Ryo, H.; Yamamoto, K.; Toh, H.; Inui, T.; Ayaki, H.; Nomura, T.; Ikenaga, M. *Science* **1996,** *272,* 109.

77a. Banerjee, R., Ed., *Chemistry & Biochemistry of B$_{12}$;* John Wiley & Sons: New York, 1999.

78. (a) Dolphin, D. H. *B$_{12}$,* vols. 1, 2, Wiley: New York, 1982. (b) Abeles, R. H.; Dolphin, D. H. *Acc. Chem. Res.* **1976,** *9,* 114.

79. Rickes, E. L.; Brink, N. G.; Koniuszy, F. R.; Wood, T. R.; Folkers, K. *Science* **1948,** *107,* 396.

80. Smith, E. L.; Parker, L. F. J. *Biochem. J.* **1948,** *43,* viii.

81. Barker, H. A.; Weissbach, H.; Smyth, R. D. *Proc. Natl. Acad. Sci. USA* **1958,** *44,* 1093.

82. Lenhert, P. G.; Hodgkin, D. C. *Nature* **1961,** *192,* 937.

83. Hodgkin, D. C.; Kamper, J.; Mackay, M.; Pickworth, J.; Trueblood, K. N.; White, J. G. *Nature* **1956,** *178,* 64.

84. Huennekens, F. M.; Vitols, K. S.; Fujii, K.; Jacobsen, D. W. In *B$_{12}$;* vol. 1 Dolphin, D., Ed., Wiley: New York, 1982, p. 145.

85. Schrauzer, G. N.; Deutsch, E. *J. Am. Chem. Soc.* **1969,** *91,* 3341.

86. Jensen, F. R.; Madan, V.; Buchanan, D. H. *J. Am. Chem. Soc.* **1970,** *92,* 1414.

87. Somack, R.; Costilow, R. N. *Biochemistry* **1973,** *12,* 2597.

88. (a) Abeles, R. H. In *The Enzymes,* vol. 5, 3rd ed.; Boyer, P., Ed., Academic Press: New York, 1972, p. 481. (b) Abeles, R. H.; Dolphin, D. *Acc. Chem. Res.* **1976,** *9,* 114. (c) Babior, B. M. *Acc. Chem. Res.* **1975,** *8,* 376.

89. Abeles, R. H.; Lee, H. A., Jr. *Brookhaven Symp. Biol.* **1962,** *15,* 310.

90. Frey, P. A.; Karabatsos, G. L.; Abeles, R. H. *Biochem. Biophys. Res. Commun.* **1965,** *18,* 551.

91. (a) Zagalak, B.; Frey, P. A.; Karabatsos, G. L.; Abeles, R. H. *J. Biol. Chem.* **1966,** *241,* 3028. (b) Rétey, J.; Umani-Ronchi, A.; Arigoni, D. *Experientia* **1966,** *22,* 72.

92. Rétey, J.; Umani-Ronchi, A.; Seibl, J.; Arigoni, D. *Experientia* **1966,** *22,* 502.

93. Frey, P. A.; Abeles, R. H. *J. Biol. Chem.* **1966,** *241,* 2732.

94. Kresge, A. J. In *Isotope Effects on Enzyme-Catalyzed Reactions;* Cleland, W. W.; O'Leary, M. H.; Northrup, D. B., Eds., University Park Press: Baltimore; 1977, p. 37.

95. Frey, P. A.; Eisenberg, M. K.; Abeles, R. H. *J. Biol. Chem.* **1967,** *242,* 5369.

96. Wagner, O. W.; Lee, H. A., Jr.; Frey, P. A.; Abeles, R. H. *J. Biol. Chem.* **1966,** *241,* 1751.

97. Findlay, T. H.; Valinsky, J.; Sato, K.; Abeles, R. H. *J. Biol. Chem.* **1972,** *247,* 4197.

98. Babior, B. M.; Carty, T. J.; Abeles, R. H. *J. Biol. Chem.* **1974,** *249,* 1689.

99. (a) Findlay, T. H.; Valinsky, J.; Mildvan, A. S.; Abeles, R. H. *J. Biol. Chem.* **1973,** *248,* 1285. (b) Valinsky, J. E.; Abeles, R. H.; Fee, J. A. *J. Am. Chem. Soc.* **1974,** *96,* 4708. (c) Babior, B. M.; Moss, T. H.; Gould, D. C. *J. Biol. Chem.* **1972,** *247,* 4389.

100. (a) Silverman, R. B. Ph.D. Dissertation; Harvard University, 1974. (b) Silverman, R. B.; Dolphin, D. *J. Am. Chem. Soc.* **1976,** *98,* 4626. (c) Silverman, R. B.; Dolphin, D. *J. Am. Chem. Soc.* **1976,** *98,* 4633.

101. (a) Walling, C. *Acc. Chem. Res.* **1975,** *8,* 125. (b) Walling, C.; Johnson, R. A. *J. Am. Chem. Soc.* **1975,** *97,* 2405. (c) Golding, B. T.; Radom, L. *J. Chem. Soc. Chem. Commun.* **1973,** 939.

102. Golding, B. T.; Sell, C. S.; Sellars, P. J. *J. Chem. Soc. Chem. Commun.* **1976,** 773.

103. Anderson, R. J.; Ashwell, S; Dixon, R. M.; Golding, B. T. *J. Chem. Soc. Chem. Commun.* **1990,** 70.

104. (a) Grissom, C. B. *Chem. Rev.* **1995,** *95,* 3. (b) Harkins, T. T.; Grissom, C. B., *Science* **1994,** *263,* 958.

105. Harkins, T. T.; Grissom, C. B. *J. Am. Chem. Soc.* **1995,** *117,* 566.

106. (a) Marsh, E. N. G. *Biochemistry* **1995,** *34,* 7542. (b) Meier, T. W.; Thomä, N. H.; Leadlay, P. F. *Biochemistry* **1996,** *35,* 11791.

107. (a) Marsh, E. N. G. *Biochemistry* **1995,** *34,* 7542. (b) Hartzoulakis, B.; Gani, D. *Proc. Ind. Acad. Sci. (Chem. Sci.)* **1994,** *106,* 1165.

108. Sprecher, M.; Switzer, R. L.; Sprinson, D. B. *J. Biol. Chem.* **1966,** *241,* 864.

109. Hartrampf, G.; Buckel, W. *Eur. J. Biochem.* **1986,** *156,* 301.

110. (a) Sprecher, M.; Clark, M. S.; Sprinson, D. B. *J. Biol. Chem.* **1966,** *241,* 872. (b) Rétey, J. In *B₁₂,* vol. 2, Dolphin, D., Ed., Wiley: New York, 1982, p. 357.

111. Hartrampf, G.; Buckel, W. *FEBS Lett.* **1984,** *171,* 73.

112. (a) Zhao, Y.; Such, P.; Rétey, J. *Angew. Chem. Int. Ed. Engl.* **1992,** *31,* 215. (b) Padmakumar, R.; Banerjee, R. *J. Biol. Chem.* **1995,** *270,* 9295. (c) Keep, N. H.; Smith, G. A.; Evans, M. C. W.; Diakun, G. P.; Leadlay, P. F. *Biochem. J.* **1993,** *295,* 387. (d) Zhao, Y.; Abend, A.; Kunz, M.; Such, P.; Rétey, J. *Eur. J. Biochem.* **1994,** *225,* 891.

113. Zelder, O.; Beatrix, B.; Leutbecher, U.; Buckel, W. *Eur. J. Biochem.* **1994,** *226,* 577.

114. (a) Michel, C.; Albracht, S. P. J.; Buckel, W. *Eur. J. Biochem.* **1992,** *205,* 767. (b) Beatrix, B.; Zelder, O.; Kroll, F.; Orlygsson, G.; Golding, B. T.; Buckel, W. *Angew. Chem. Int. Ed. Engl.* **1995,** *34,* 2398.

115. Mancia, F.; Keep, N. H.; Nakagawa, A.; Leadlay, P. F.; McSweeney, S.; Rasmussen, B.; Bösecke, P.; Diat, O.; Evans, P. R. *Structure* **1996,** *4,* 339.

116. Mancia, F.; Evans, P. R. *Structure* **1998,** *6,* 711.

117. Thomä, N. H.; Meier, T. W.; Evans, P. R.; Leadlay, P. F. *Biochemistry* **1998,** *37,* 14386.

118. Padmakumar, R.; Taoka, S.; Padmakumar, R.; Banerjee, R. *J. Am. Chem. Soc.* **1995,** *117,* 7033.

119. Banerjee, R. *Chem. Biol.* **1997,** *4,* 175.

120. Finke, R. G.; Hay, B. P. *Inorg. Chem.* **1984,** *23,* 3041.

121. Hay, B. P.; Finke, R. G. *J. Am. Chem. Soc.* **1987,** *109,* 8012.

122. Garr, C. D.; Sirovatka, J. M.; Finke, R. G. *Inorg. Chem.* **1996,** *35,* 5912.

123. Waddington, M. D.; Finke, R. G. *J. Am. Chem. Soc.* **1993,** *115,* 4629.

124. Padmakumar, R.; Padmakumar, R.; Banerjee, R. *Biochemistry* **1997,** *36,* 3713.

125. Marsh, E. N. G.; Ballou, D. P. *Biochemistry* **1998**, *37*, 11864.

126. Bothe, H.; Darley, D. J.; Albracht, S. P. J.; Gerfen, G. J.; Golding, B. T.; Buckel, W. *Biochemistry* **1998**, *37*, 4105.

126a. Smith, D. M.; Golding, B. T.; Radom, L. *J. Am. Chem. Soc.* **1999**, *121*, 1037.

127. (a) Dowd, P.; Shapiro, M. *Tetrahedron* **1984**, *40*, 3063. (b) Dowd, P.; Shapiro, M.; Kang, J. *Tetrahedron* **1984**, *40*, 3069. (c) Choi, G.; Choi, S.-C.; Galan, A.; Wilk, B.; Dowd, P. *Proc. Natl. Acad. Sci. USA* **1990**, *87*, 3174. (d) He, M.; Dowd, P. *J. Am. Chem. Soc.* **1996**, *118*, 711. (e) He, M.; Dowd, P. *J. Am. Chem. Soc.* **1998**, *120*, 1133.

128. (a) Dowd, P.; Wilk, B.; Wilk, B. *J. Am. Chem. Soc.* **1992**, *114*, 7949. (b) Choi, G.; Choi, S.-C.; Galan, A.; Wilk, B.; Dowd, P. *Proc. Natl. Acad. Sci. USA* **1990**, *87*, 3174. (c) Wollowitz, S.; Halpern, J. *J. Am. Chem. Soc.* **1984**, *106*, 8319.

129. Dowd, P.; Wilk, B.; Wilk, B. *J. Am. Chem. Soc.* **1992**, *114*, 7949.

130. (a) Golding, B. T.; Mwesigye-Kibende, S. *J. Chem. Soc. Chem. Commun.* **1983**, 1103. (b) Davies, A. G.; Golding, B. T.; Hay-Motherwell, R. S. Mwesigye-Kibende, S.; Rao, D. N. R.; Symons, M. C. R. *J. Chem. Soc. Chem. Commun.* **1988**, 378. (c) Wollowitz, S.; Halpern, J. *J. Am. Chem. Soc.* **1984**, *106*, 8319. (d) Wollowitz, S.; Halpern, J. *J. Am. Chem. Soc.* **1988**, *110*, 3112.

131. Bury, A.; Ashcroft, M. R.; Johnson, M. D. *J. Am. Chem. Soc.* **1978**, *100*, 3217.

132. Ashwell, S.; Davies, A. G.; Golding, B. T.; Hay-Motherwell, R.; Mwesigye-Kibende, S. *J. Chem. Soc. Chem. Commun.* **1989**, 1483.

133. (c) Scott, A. I.; Karuso, P.; Williams, H. J.; Lally, J.; Robinson, J.; Nayar, G. P. *J. Am. Chem. Soc.* **1994**, *116*, 777.

134. (a) Dowd, P.; Shapiro, M. *Tetrahedron* **1984**, *40*, 3063. (b) Dowd, P.; Shapiro, M.; Kang, J. *Tetrahedron* **1984**, *40*, 3069. (c) He, M.; Dowd, P. *J. Am. Chem. Soc.* **1996**, *118*, 711. (d) He, M.; Dowd, P. *J. Am. Chem. Soc.* **1998**, *120*, 1133.

135. (a) Scott, A. I.; Kang, J.; Dalton, D.; Chung, S. K. *J. Am. Chem. Soc.* **1978**, *100*, 3603. (b) Grate, J. H.; Grate, J. W.; Schrauzer, G. N. *J. Am. Chem. Soc.* **1982**, *104*, 1588. (c) He, M.; Dowd, P. *J. Am. Chem. Soc.* **1998**, *120*, 1133.

136. Buckel, W.; Golding, B. T. *Chem. Soc. Rev.* **1996**, 329.

137. (a) Beatrix, B.; Zelder, O.; Kroll, F.; Orlygsson, G.; Golding, B. T.; Buckel, W. *Angew. Chem. Int. Ed. Engl.* **1995**, *34*, 2398. (b) Edwards, C.; Golding, B.T.; Kroll, F.; Beatrix, B.; Broker, G.; Buckel, W. *J. Am. Chem. Soc.* **1996**, *118*, 4192.

138. (a) Stubbe, J.; van der Donk, W. A. *Chem. Rev.* **1998**, *98*, 705. (b) Stubbe, J.; van der Donk, W. A. *Chem. Biol.* **1995**, *2*, 793. (c) Jordan, A.; Reichard, P. *Annu. Rev. Biochem.* **1998**, *67*, 71. (d) Stubbe, J. *Adv. Enzymol. Relat. Areas Mol. Biol.* **1990**, *63*, 349. (d) Fontecave, M.; Nordlund, P.; Eklund, H.; Reichard, P. *Adv. Enzymol. Relat. Areas Mol. Biol.* **1992**, *65*, 147.

139. (a) Beck, W. S.; Abeles, R. H.; Robinson, W. G. *Biochem. Biophys. Res. Commun.* **1966**, *25*, 421. (b) Hogenkamp, H. P. C.; Ghambeer, R. K.; Brownson, C.; Blakley, R. L.; Vitols, E. *J. Biol. Chem.* **1968**, *243*, 799.

140. (a) Hamilton, J. A.; Yamada, R.; Blakley, R. L.; Hogenkamp, H. P. C.; Looney, F. D.; Winfield, M. E. *Biochemistry* **1971**, *10*, 347. (b) Yamada, R.; Tamao. Y.; Blakley, R. L. *Biochemistry* **1971**, *10*, 3959.

141. Licht, S.; Gerfen, G. J.; Stubbe, J. A. *Science* **1996**, *271*, 477.

142. Gerfen, G. J.; Licht, S. L.; Willems, J.-P.; Hoffman, B. M.; Stubbe, J. *J. Am. Chem. Soc.* **1996**, *118*, 8192.

143. Booker, S.; Licht, S.; Broderick, J.; Stubbe, J. *Biochemistry* **1994**, *33*, 12676.

144. Holmgren, A.; Björnstedt, M.; *Methods Enzymol.* **1995**, *252*, 199.

145. (a) Benson, S. W. *Chem. Rev.* **1978**, *78*, 23. (b) McMillen, D. F.; Golden, D. M. *Annu. Rev. Phys. Chem.* **1982**, *33*, 493.

146. Schöneich, C.; Bonifacic, M.; Dillinger, U.; Asmus, K.-D. In *Sulfur-Centered Reactive Intermediates in Chemistry and Biology,* Chatgilialoglu, C.; Asmus, K.-D., Eds., Plenum Press: New York, 1990, p. 367.

147. van der Donk, W. A.; Gerfen, G. J.; Stubbe, J. *J. Am. Chem. Soc.* **1998,** *120,* 4252.
148. (a) Silva, D. J.; Stubbe, J.; Samano, V.; Robins, M. J. *Biochemistry* **1998,** *37,* 5528. (b) Stubbe, J.; van der Donk, W. A. *Chem. Rev.* **1998,** *98,* 705.
149. (a) Stubbe, J. *J. Biol. Chem.* **1990,** *265,* 5329. (b) Eliasson, R.; Reichard, P.; Mulliez, E.; Ollagnier, S.; Fontecave, M.; Liepinsh, E.; Otting, G. *Biochem. Biophys. Res. Commun.* **1995,** *214,* 28.
150. (a) Reichard, P.; Ehrenberg, A. *Science* **1983,** *221,* 514. (b) Fontecave, M.; Nordlund, P.; Eklund, H.; Reichard, P. *Adv. Enzymol. Relat. Areas Mol. Biol.* **1992,** *65,* 147. (c) Bollinger, J. M., Jr.; Edmonson, D. E.; Huynh, B. H.; Filley, J.; Norton, J. R.; Stubbe, J. *Science* **1991,** *253,* 292. (d) Ravi, N.; Bollinger, J. M., Jr.; Huynh, B. H.; Edmondson, D. E.; Stubbe, J. *J. Am. Chem. Soc.* **1994,** *116,* 8007. (e) Bollinger, J. M., Jr.; Tong. W. H.; Ravi, N.; Huynh, B. H.; Edmondson, D. E.; Stubbe, J. *J. Am. Chem. Soc.* **1994,** *116,* 8015. (f) Bollinger, J. M., Jr.; Tong. W. H.; Ravi, N.; Huynh, B. H.; Edmondson, D. E.; Stubbe, J. *J. Am. Chem. Soc.* **1994,** *116,* 8024.
151. Jordan, A.; Åslund, F.; Pontis, E.; Reichard, P.; Holmgren, A. *J. Biol. Chem.* **1997,** *272,* 18044.
152. Jordan, A.; Aragakki, E.; Gibert, I.; Barbé, J. *Mol. Microbiol.* **1996,** *19,* 777.
153. Fontecave, M.; Eliasson, R.; Reichard, P. *Proc. Natl. Acad. Sci. USA* **1989,** *86,* 2147.
154. (a) Sun, X.; Ollagnier, S.; Schmidt, P. P.; Atta, M.; Mulliez, E.; Lepape, L.; Eliasson, R.; Gräslund, A.; Fontecave, M.; Reichard, P.; Sjöberg, B.-M. *J. Biol. Chem.* **1996,** *271,* 6827. (b) Ollagnier, S.; Mulliez, E.; Gaillard, J.; Eliasson, R.; Fontecave, M.; Reichard, P. *J. Biol. Chem.* **1996,** *271,* 9410.
155. Griepenburg, U.; Lassmann, G.; Auling, G. *Free Rad. Res.* **1996,** *26,* 473.
156. Frey, P. A. *Chem. Rev.* **1990,** *90,* 1343.
157. Petrovich, R. M.; Ruzicka, F. J.; Reed, G. H.; Frey, P. A. *J. Biol. Chem.* **1991,** *266,* 7656.
158. Petrovich, R. M.; Ruzicka, F. J.; Reed, G. H.; Frey, P. A. *Biochemistry* **1992,** *31,* 10774.
159. (a) Chirpich, T. P.; Zappia, V.; Costilow, R. N.; Barker, H. A. *J. Biol. Chem.* **1970,** *245,* 1778. (b) Aberhart, D. J.; Gould, S. J.; Lin, H.-J.; Thiruvengadam, T. K.; Weiller, B. H. *J. Am. Chem. Soc.* **1983,** *105,* 5461.
160. Moss, M. L.; Frey, P. A. *J. Biol. Chem.* **1987,** *262,* 14859.
161. Baraniak, J.; Moss, M. L.; Frey, P. A. *J. Biol. Chem.* **1989,** *264,* 1357.
162. Moss, M. L.; Frey, P. A. *J. Biol. Chem.* **1990,** *265,* 18112.
163. Kilgore and Alberhart, *J. C. S. Perkin Trans. 1* **1991,** 79.
164. Lieder, K. W.; Booker, S.; Ruzicka, F. J.; Beinert, H.; Reed, G. H.; Frey, P. A. *Biochemistry* **1998,** *37,* 2578.
165. Dowd, P.; Zhang, W. *Chem. Rev.* **1993,** *93,* 2091.
166. Han, O.; Frey, P. A. *J. Am. Chem. Soc.* **1990,** *112,* 8982.
167. Petrovich, R. M.; Ruzicka, F. J.; Reed, G. H.; Frey, P. A. *Biochemistry* **1992,** *31,* 10774.
168. Ballinger, M. D.; Reed, G. H.; Frey, P. A. *Biochemistry* **1992,** *31,* 949.
169. Ballinger, M. D.; Frey, P. A.; Reed, G. H. *Biochemistry* **1992,** *31,* 10782.
170. Chang, C. H.; Ballinger, M. D.; Reed, G. H.; Frey, P. A. *Biochemistry* **1996,** *35,* 11081.
171. Wu, W.; Lieder, K. W.; Reed, G. H.; Frey, P. A. *Biochemistry* **1995,** *34,* 10532.
172. Ballinger, M. D.; Frey, P. A.; Reed, G. H.; LoBrutto, R. *Biochemistry* **1995,** *34,* 10086.

Enzyme Kinetics

I. SUBSTRATE KINETICS

A. Michaelis–Menten Equation

1. Derivation of the Michaelis–Menten Equation

The simplest form of an enzyme-catalyzed reaction is shown in Scheme A.1, where k_1 is a second-order rate constant, and k_{-1} and k_2 are first-order rate constants. K_s is the dissociation constant for the E·S complex, which equals k_{-1}/k_1. We will only be concerned with *steady-state kinetics,* that is, when the rate of E·S formation equals the rate of E·S breakdown. The *pre-steady state* is the initial period during which intermediates form until the steady state is reached.

The rates of enzyme-catalyzed reactions show a characteristic dependence on substrate concentration (Figure A.1). At low substrate concentration the initial rate is proportional to both $[E]_0$ and $[S]$ (a second-order reaction). As the substrate concentration increases, it becomes easier for the enzyme to find the substrate, until when the $[S] >> [E]_0$, all of the enzyme active sites are occupied with bound substrate (or product). At this point the enzyme is said to be *saturated* (i.e., all of the

$$\text{E} + \text{S} \underset{k_{-1}}{\overset{\overset{\textstyle K_s}{\overset{\textstyle k_1}{\rightleftharpoons}}}{}} \text{E·S} \underset{}{\overset{k_2}{\rightleftharpoons}} \text{E·P} \rightleftharpoons \text{E} + \text{P}$$

E = free enzyme   S = substrate   P = product

E·S = enzyme-substrate complex (also called the *Michaelis complex*)

SCHEME A.1   Simplest form of an enzyme-catalyzed reaction.

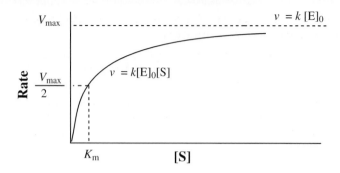

[E]$_0$ is the total enzyme concentration

[E] is free enzyme

[E] = [E]$_0$ - [E·S]

FIGURE A.1   Dependence of the rate of an enzyme-catalyzed reaction on the substrate concentration.

active sites have substrate or product bound), and further increases in the substrate concentration will not increase the rate of the enzyme reaction. This is the maximal rate or V_{max}, which is attained really only at infinite substrate concentration, so it is never fully achieved. At this point, the rate is zeroth order with respect to [S] and is dependent only on the enzyme concentration. The actual rate, then, depends on how much of the total amount of enzyme is in the E·S complex [Eq. (1)]. The rate of formation

$$\nu = k_2[E\cdot S] \tag{1}$$

of E·S is given in Eq. (2), and the rate of decomposition of E·S back to E + S and to products is given in Eq. (3). During the period that the initial rate is measured,

$$\frac{+d\,[E\cdot S]}{dt} = k_1([E]_0 - [E\cdot S])[S] \tag{2}$$

$$\frac{-d\,[E\cdot S]}{dt} = k_{-1}[E\cdot S] + k_2[E\cdot S] \tag{3}$$

[S], ([E]$_0$ − [E·S]), and [E]$_0$ remain approximately constant. Therefore, [E·S] can be assumed to remain constant (i.e., it remains in a steady state). If this steady-state assumption of George E. Briggs and J. B. S. Haldane[1] is made, then the rate of formation of E·S [Eq. (2)] equals the rate of its decomposition [Eq. (3), as shown in Eq. (4)].

$$k_1([E]_0 - [E\cdot S])\,[S] = k_{-1}[E\cdot S] + k_2\,[E\cdot S] \tag{4}$$

Rearranging Eq. (4) gives Eq. (5). The K_m is called the *Michaelis–Menten constant,* which is given in the units of concentration.

$$\frac{([E]_0 - [E \cdot S])[S]}{[E \cdot S]} = \frac{k_{-1} + k_2}{k_1} = K_m \tag{5}$$

The K_m is the concentration of the substate required to produce a rate of $V_{max}/2$. Under certain conditions (see I.A.2), this is a *dissociation* constant for the E·S complex. Therefore, the smaller the K_m, the stronger the interaction between E and S.

Solving Eq. (5) for [E·S] yields Eq. (6).

$$[E \cdot S] = \frac{[E]_0 [S]}{K_m + [S]} \tag{6}$$

Substituting Eq. (6) into Eq. (1) gives Eq. (7).

$$\nu = \frac{k_2 [E]_0 [S]}{K_m + [S]} \tag{7}$$

When $k_2 \ll k_{-1}$, then k_2 is called k_{cat}, and Eq. (7) becomes Eq. (8), which is the *Michaelis–Menten equation,* named after Leonor Michaelis and Maud Menten.[2]

$$\nu = \frac{k_{cat} [E]_0 [S]}{K_m + [S]} \tag{8}$$

The term k_{cat} is the rate constant for conversion of the E·S complex to product. It is a measure of the rate at which the enzyme catalyzes the reaction. Because the k_{cat} can be represented by Eq. (9), the maximum rate divided by the total enzyme concentration, we can substitute this equality for k_{cat} into Eq. (8), which then can be reduced to Eq. (10). This is an

$$k_{cat} = \frac{V_{max}}{[E]_0} \tag{9}$$

$$\nu = \frac{V_{max} [S]}{K_m + [S]} \tag{10}$$

alternative form of the Michaelis–Menten equation. At high [S], all of the enzyme is in the E·S form, and K_m is negligible relative to [S]. Consequently, Eq. (10) becomes Eq. (11), indicating that the rate is at its maximum,

$$\nu = \frac{V_{max} [\cancel{S}]}{[\cancel{S}]} = V_{max} \tag{11}$$

which as we saw earlier (Figure A.1) is independent of [S]. When $[S] = K_m$, Eq. (10) can be rewritten as Eq. (12),

$$\nu = \frac{V_{max} K_m}{K_m + K_m} = \frac{V_{max}}{2} \tag{12}$$

which shows that the rate is at half the maximum value. At low [S], where [S] $<<$ K_m, [S] is negligible relative to K_m, so Eq. (10) becomes Eq. (13).

$$v = \frac{V_{max} [S]}{K_m} \tag{13}$$

From this equation it can be seen that at low [S], the rate is proportional to [S], where V_{max}/K_m is the proportionality constant.

2. Difference between K_m and K_s

As noted, when k_{-1} and k_1 are large relative to k_2, then k_2 is referred to as k_{cat}. If k_2 is negligible, then Eq. (5) reduces to Eq. (14), which is the K_s (the dissociation constant for E·S). Only under these special conditions—that is, $k_2 << k_{-1}$— is $K_m = K_s$.

$$K_m = \frac{k_{-1} + k_2}{k_1} \tag{5}$$

$$K_m \approx \frac{k_{-1}}{k_1} \tag{14}$$

When k_2 is not $<< k_{-1}$, then Eq. (5) holds. Substitution of $K_s = k_{-1}/k_1$ (actually, we are substituting $k_{-1} = k_1 K_s$), into Eq. (5) gives Eq. (15).

$$K_m = K_s + \frac{k_2}{k_1} \tag{15}$$

In this case, $K_m > K_s$. When intermediates occur after the E·S complex, as shown in Scheme A.2, K_m and k_{cat} are a combination of various rate and equilibrium constants, but K_m is always less than or equal to K_s.

3. k_{cat}

The term k_{cat} represents the maximum number of substrate molecules converted to product molecules per active site per unit time, the number of times the enzyme "turns over" substrate to product per unit time (called the *turnover number*). Values for k_{cat} on the order of 10^3 s^{-1} are typical.

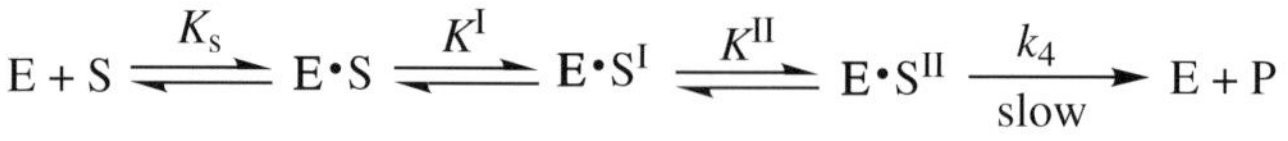

$$E + S \underset{K_s}{\rightleftharpoons} E·S \underset{K^I}{\rightleftharpoons} E·S^I \underset{K^{II}}{\rightleftharpoons} E·S^{II} \xrightarrow[\text{slow}]{k_4} E + P$$

SCHEME A.2   Form of an enzyme-catalyzed reaction when there are intermediates between the E·S and E·P complexes.

4. k_{cat}/K_m

The term k_{cat}/K_m is called the *specificity constant*. At low [S] the Michaelis–Menten equation [Eq. (8)] reduces to Eq. (16). Therefore,

$$v = \frac{k_{cat}\,[E]_0\,[S]}{K_m + [S]} \tag{8}$$

$$v = \frac{k_{cat}\,[E]_0\,[S]}{K_m} \tag{16}$$

k_{cat}/K_m is a second-order proportionality constant for the reaction of free E+S to give product (at low [S], $[E]_0$ is approximately the same as [E]). The value of the term k_{cat}/K_m allows you to rank an enzyme according to how good it is with different substrates. It contains information about how fast the reaction of a given substrate would be when bound to the enzyme (k_{cat}) and how much of the substrate is required to reach V_{max}. The value of k_{cat}/K_m can approach the rate of diffusion (about $10^9\ M^{-1}\ s^{-1}$), which is when a reaction occurs with every collision between molecules.

Substituting k_2 for k_{cat} and $(k_{-1} + k_2)/k_1$ for K_m [Eq. (5)], then it can be seen in Eq. (17) that when $k_2 \gg k_{-1}$, then $k_{cat}/K_m = k_1$, the rate of encounter between E and S.

$$k_{cat}/K_m = \frac{k_2}{(k_{-1} + k_2)/k_1} = \frac{k_1 k_2}{k_{-1} + k_2} = \frac{k_1 k_2}{k_2} = k_1 \tag{17}$$

B. Graphical Representations

Because the V_{max} is attained only at infinite substrate concentration, it can never really be achieved. The difficulty in obtaining V_{max} experimentally and the curved nature of the v versus [S] plot (Figure A.1) led to the transformation of the Michaelis–Menten equation into linear forms. This allows for more accurate determination of K_m and V_{max} values. We will consider three graphical representations of the kinetic data:

$$\frac{1}{v} \text{ versus } \frac{1}{[S]} \qquad \text{Lineweaver–Burk plot}$$

$$v \text{ versus } \frac{v}{[S]} \qquad \text{Eadie–Hofstee plot}$$

$$\frac{S}{v} \text{ versus } [S] \qquad \text{Hanes–Woolf plot}$$

The most commonly used is the *Lineweaver–Burk plot*,[3] but this is the least accurate for linear regression analysis if there is significant experimental error in the

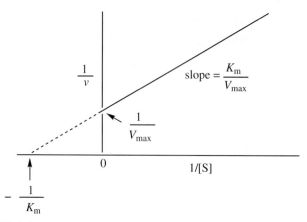

FIGURE A.2   Lineweaver–Burk plot for determination of substrate kinetic constants.

initial rate measurements. Taking the reciprocal of the Michaelis–Menten equation [Eq. (10)] gives Eq. (18). A plot of $1/\nu$ versus $1/[S]$ gives a straight line ($a = b + mx$), and this is called a Lineweaver–Burk plot (Figure A.2). The interception of the x-axis is $-1/K_m$, and the interception of the y-axis is $1/V_{max}$. The slope

$$\nu = \frac{V_{max}\,[S]}{K_m + [S]} \tag{10}$$

$$\frac{1}{\nu} = \frac{1}{V_{max}} + \frac{K_m}{V_{max}\,[S]} \tag{18}$$

defines K_m/V_{max}. As the $[S]$ increases, the rate approaches V_{max}. The Lineweaver–Burk plot, however, has the disadvantages of compressing the data points at high concentrations into a small region and emphasizing the points at low concentrations. Also, small errors in the determination of ν are magnified when the reciprocals are taken. Because of these problems, very careful choice of substrate concentrations and a large number of data points are required to increase the accuracy.

Rearranging the Michaelis–Menten equation [Eq. (10)] a different way, that is, by multiplying the Lineweaver–Burk equation [Eq. (18)] by $V_{max} \cdot \nu$ gives Eq. (19) or (20). A plot of ν versus $\nu/[S]$, an *Eadie–Hofstee plot*[4] (Figure A.3), gives a straight line with the intercept at the x-axis equal to

$$V_{max} = K_m \frac{\nu}{[S]} + \nu \tag{19}$$

$$\nu = V_{max} - K_m \frac{\nu}{[S]} \tag{20}$$

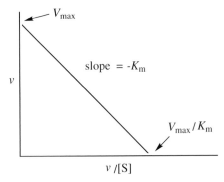

FIGURE A.3   Eadie–Hofstee plot for determination of substrate kinetic constants.

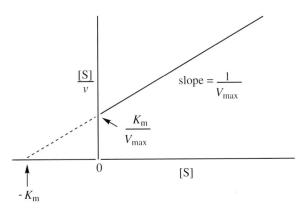

FIGURE A.4   Hanes–Woolf plot for determination of substrate kinetic constants.

V_{max}/K_m, the y-axis intercept equal to V_{max}, and the slope $= -K_m$. The Eadie–Hofstee plot does not compress the higher substrate concentration values. However, the fact that v appears on both the x- and y-axis means that errors in v affect both axes and cause deviations toward or away from the origin. Therefore, instead of making poor data look better, as in the case of the Lineweaver–Burk plot, it tends to make good data look worse. Because of the exaggerated deviations from linearity, the data must be excellent to obtain a straight line with this kind of a plot. If you want to detect small deviations from Michaelis–Menten kinetics, this is the plot to use.

Multiplication of the Lineweaver–Burk equation [Eq. (18)] by [S] gives a third transformation [Eq. (21)]. A plot of [S]/v versus S, a *Hanes–Woolf plot*[5] (Figure A.4), gives a straight line where the x-axis intercept corresponds to $-K_m$, the y-axis intercept is K_m/V_{max}, and the slope is $1/V_{max}$. This is a popular way to represent the kinetic data graphically; over a wide range of [S] values, the errors in $[S]/\nu$ provide a fair reflection of the errors in ν, and therefore, this should be the preferred plot for most purposes.

$$\frac{[S]}{\nu} = \frac{K_m}{V_{max}} + [S]\frac{1}{V_{max}} \tag{21}$$

II. KINETICS OF ENZYME INHIBITION

A. Reversible Enzyme Inhibition

A *reversible enzyme inhibitor* is a molecule that binds reversibly to the enzyme in such a way as to slow down or prevent enzyme turnover. There are naturally occurring inhibitors that control metabolism and synthetic inhibitors that are used as drugs or for agricultural purposes (e.g., insecticides, fungicides, and weed-killers). There are three simple types of inhibition: competitive, noncompetitive/mixed, and uncompetitive inhibition.

1. Competitive Inhibition

Competitive inhibitors (I) bind to the active site of the free enzyme and prevent the substrate from binding (and vice versa). Therefore, I and S compete for the active site (Scheme A.3). The term K_i is the dissociation constant for the E·I complex. The rate of the reaction depends on [I], K_i, [S], and K_m. By solving the equilibrium and rate equations, Eq. (22) can be derived (but not here).

$$\nu = \frac{V_{max}[S]}{K_m\left(1 + \dfrac{[I]}{K_i}\right) + [S]} \tag{22}$$

Eq. (22) is identical to the Michaelis–Menten equation [Eq. (10)] except that the presence of the inhibitor results in K_m being multiplied by the factor $[1 + ([I]/K_i)]$.

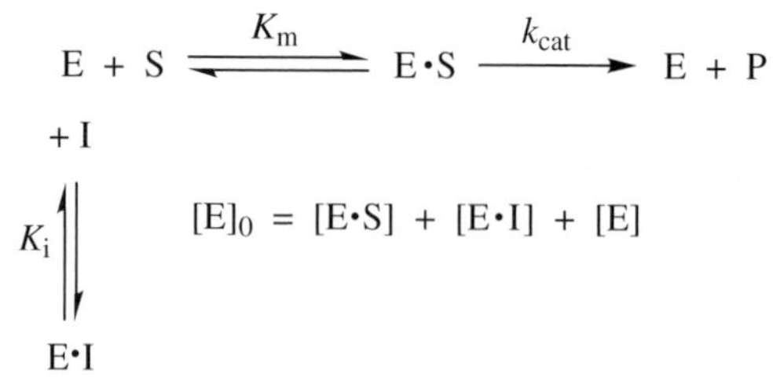

$$E + S \xrightleftharpoons{K_m} E \cdot S \xrightarrow{k_{cat}} E + P$$

$$+ I$$

$$K_i \updownarrow \qquad [E]_0 = [E \cdot S] + [E \cdot I] + [E]$$

$$E \cdot I$$

SCHEME A.3   Competitive inhibition.

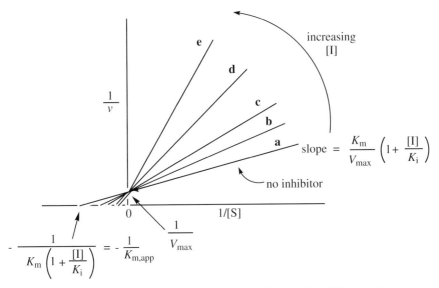

FIGURE A.5   Lineweaver–Burk plot for determination of competitive inhibitor kinetic constants.

Consequently, the Lineweaver–Burk equation becomes Eq. (23).

$$\frac{1}{v} = \frac{K_m}{V_{max}}\left(\frac{1}{[S]}\right)\left(1 + \frac{[I]}{K_i}\right) + \frac{1}{V_{max}} \tag{23}$$

This equation is plotted graphically in Figure A.5. The rate is determined as a function of [S] in the absence of inhibitor (line **a**), then the experiment is repeated except at each [S] a constant amount of inhibitor is added (line **b**). The experiment is repeated with more inhibitor added at each [S] (lines **c, d,** and **e**). As more I is added, the slope increases, but there is no effect on V_{max}. As 1/[S] approaches 0 (infinite [S]), all of the inhibitor is displaced by the substrate, and the enzyme is present only as the E·S complex, so the same V_{max} is observed regardless of whether I is present or not.

At low [S], the enzyme is predominantly in the E form. The competitive inhibitor can bind to E, and this decreases the rate (there are then fewer E molecules available to bind to S). Because the rate is proportional to V_{max}/K_m, the slopes of the lines are affected. Increasing [I] causes the K_m/V_{max} to increase. This is because it now requires more substrate to give the same V_{max} than would be obtained in the absence of the inhibitor, thereby making it appear that the K_m is larger (i.e., the apparent K_m or $K_{m,app}$ is larger), and the V_{max} is smaller. A larger K_m/V_{max} means a steeper slope (**a** < **b** < **c** < **d**). At each [I] the apparent K_m ($K_{m,app}$) can be calculated from each x-axis intercept using Eq. (24).

$$-\frac{1}{K_m\left(1 + \dfrac{[I]}{K_i}\right)} = -\frac{1}{K_{m,app}} \tag{24}$$

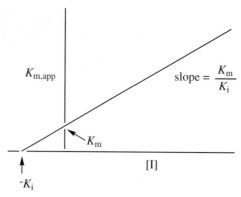

FIGURE A.6   Replot of data from Lineweaver–Burk plots for determination of K_i values for competitive inhibitors.

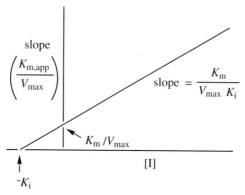

FIGURE A.7   Alternative replot of data from Lineweaver–Burk plots for determination of K_i values for competitive inhibitors.

To obtain the K_i values, the data from the Lineweaver–Burk plot (Figure A.5) are replotted as $K_{m,app}$ versus [I] (Figure A.6) or as $K_{m,app}/V_{max}$ (the slope) versus [I] (Figure A.7).

The other graphical representation we will consider for competitive inhibition is the Dixon plot[6] (Figure A.8), a plot of Eq. (25). The [I] is plotted against $1/v$, varying the amount of substrate. The negative K_i value can be read directly from the plot at the intersection of the lines.

$$\frac{1}{v} = \frac{K_m}{V_{max}[S]K_i} [I] + \frac{1}{V_{max}} \left(1 + \frac{K_m}{[S]} \right) \tag{25}$$

To determine what kind of inhibition has occurred, the data from the Dixon plot must be replotted in a *Cornish-Bowden plot*[7] (Figure A.9), that is, S/v versus [I]. Parallel lines indicate competitive inhibition.

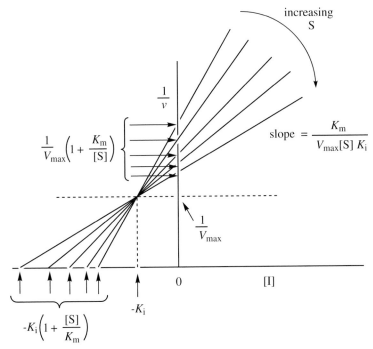

FIGURE A.8   Dixon plot for determination of competitive inhibitor kinetic constants.

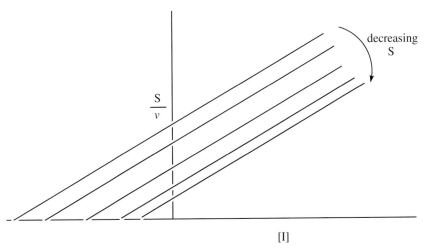

FIGURE A.9   Cornish-Bowden replot of the data from the Dixon plot for determination of the type of inhibition (in this case, competitive).

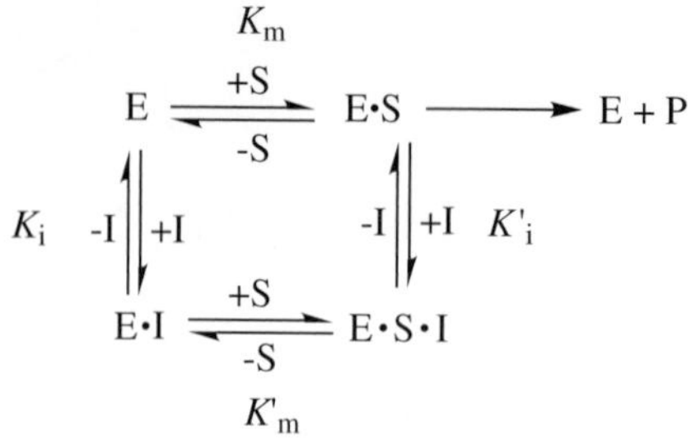

$$K_m$$

$$
\text{E} \underset{-\text{S}}{\overset{+\text{S}}{\rightleftharpoons}} \text{E·S} \longrightarrow \text{E} + \text{P}
$$

$$
K_i \quad -\text{I} \Big\uparrow \Big\downarrow +\text{I} \qquad -\text{I} \Big\uparrow \Big\downarrow +\text{I} \quad K'_i
$$

$$
\text{E·I} \underset{-\text{S}}{\overset{+\text{S}}{\rightleftharpoons}} \text{E·S·I}
$$

$$K'_m$$

SCHEME A.4   Noncompetitive inhibition.

2. Noncompetitive Inhibition

When the inhibitor binds to both E and E·S, *noncompetitive inhibition* results. This is depicted in Scheme A.4. It is rare for single-substrate reactions to exhibit this inhibition, but it is common in multiple-substrate systems. So that the ternary complex, E·S·I, can form, the inhibitor must bind at a site other than the active site. If you assume that the dissociation of S from E·S is the same as from E·S·I (i.e., $K_m = K'_m$), then the rate is described by Eq. (26), derived from the Michaelis–Menten equation.

$$
v = \frac{V_{max}[S] \Big/ \left(1 + \dfrac{[I]}{K_i}\right)}{K_m + [S]} \tag{26}
$$

This is pure noncompetitive inhibition. The Lineweaver–Burk transformation of this equation is shown in Eq. (27);

$$
\frac{1}{v} = \left[\frac{K_m}{V_{max}}\left(\frac{1}{[S]}\right) + \frac{1}{V_{max}}\right]\left(1 + \frac{[I]}{K_i}\right) \tag{27}
$$

Figure A.10 gives a graphical representation. As is apparent from Figure A.10, noncompetitive inhibitors affect the V_{max} (note that the interception on the y-axis varies) but do not affect the K_m (all of the lines intersect at the same point on the x-axis). The K_i value can be obtained by a replot of the data from Figure A.10, either as a plot of the slope versus [I] (Figure A.11A) or $1/V_{max,app}$ versus [I] (Figure A.11B).

Equation 27 can be rearranged to Eq. (28). A plot of $1/v$ versus [I], a Dixon plot, gives a straight line (Figure A.12) from which the K_i can be read directly at the $-$x-axis intercept.

$$
\frac{1}{v} = \frac{\left(1 + \dfrac{K_m}{[S]}\right)}{V_{max}K_i}[I] + \frac{1}{V_{max}}\left(1 + \frac{K_m}{[S]}\right) \tag{28}
$$

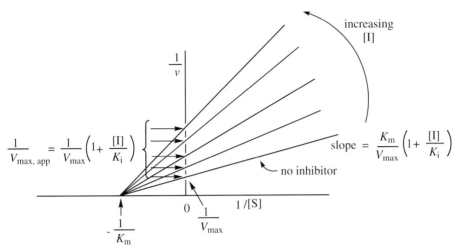

FIGURE A.10 Lineweaver–Burk plot for determination of noncompetitive inhibitor kinetic constants.

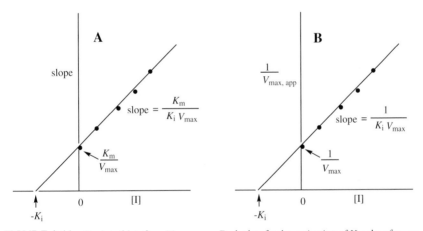

FIGURE A.11 Replot of data from Lineweaver–Burk plots for determination of K_i values for non-competitive inhibitors. **A,** plot of the slope of lines from the Lineweaver–Burk plot versus inhibitor concentration; **B,** plot of $1/V_{max,app}$ from the Lineweaver–Burk plot versus inhibitor concentration.

To determine what kind of inhibition has occurred, the data from the Dixon plot must be replotted in a Cornish-Bowden plot (Figure A.13), that is, S/ν versus [I]. Lines that intersect on the $-$x-axis indicate a noncompetitive inhibitor.

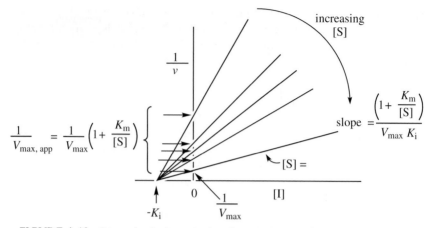

FIGURE A.12   Dixon plot for determination of noncompetitive inhibitor kinetic constants.

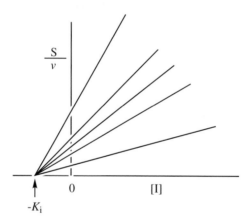

FIGURE A.13   Cornish-Bowden replot of the data from the Dixon plot for determination of the type of inhibition (in this case, noncompetitive).

3. Mixed Inhibition

It is more common that the dissociation constant of S from the E·S complex (K_m) in Scheme A.4 is different from that of S from the E·S·I complex (K'_m). When E·I has a lower affinity for S than does E, the E·S·I complex is nonproductive. In this case, the inhibition is a mixture of partial competitive and pure noncompetitive inhibition, called *mixed inhibition*. As long as the inhibitor is present, some of the enzyme will always be in the nonproductive E·S·I form, even at infinite substrate concentration. In effect, this lowers the concentration of free enzyme. Therefore, the V_{max} will be less than that of the free enzyme. Also, because a portion of the enzyme available for substrate binding will exist in the lower affinity form, E·I, the K_m will be greater than that for the free enzyme. The reciprocal form of the rate equation is shown in Eq. (29) (note that in the Cleland nomenclature K_i would be K_{is} and K'_i would be K_{ii}).

$$\frac{1}{v} = \frac{K_m}{V_{max}} \left(\frac{1}{[S]}\right)\left(1 + \frac{[I]}{K_i}\right) + \frac{1}{V_{max}}\left(1 + \frac{[I]}{K'_i}\right) \tag{29}$$

With this type of inhibition, the interception of the lines in a Lineweaver–Burk plot does not occur on the x-axis; it may occur either above the x-axis (Figure A.14A) or below the x-axis (Figure A.14B).

To obtain the K_i values, the data from the Lineweaver–Burk plots are replotted as $K_m/V_{max} (1 + ([I]/K_i))$ (the slope) versus [I] (Figure A.15A) or as $1/V_{max,app}$ (the y-axis intercept) versus [I] (Figure A.15B). The x-axis intercept for the slope versus [I] plot is $-K_i$; the x-axis intercept for the $1/V_{max,app}$ versus [I] plot is $-K'_i$.

Dixon plots, $1/v$ versus [I], for most mixed inhibitors are curved; however, when the E·S·I complex is not catalytically active, the plot is linear (Figure A.16). Eq. (30) is the equation from which the Dixon plot is derived.

$$\frac{1}{v} = \frac{\left(1 + \dfrac{K'_m}{[S]}\right)}{V_{max}K'_i} [I] + \frac{1}{V_{max}}\left(1 + \frac{K_m}{[S]}\right) \tag{30}$$

For a Dixon plot in which the lines intersect above the x-axis, the corresponding Cornish-Bowden plot lines intersect below the x-axis (Figure A.17) and vice versa.

4. Uncompetitive Inhibition

When the inhibitor binds to the E·S complex, but not to free enzyme, thereby producing an inactive E·S·I complex, *uncompetitive inhibition* results (Scheme A.5). This is rare in single-substrate reactions, but common in multiple-substrate systems. Often an inhibitor of a two-substrate enzyme that is competitive against one of the substrates is found to give uncompetitive inhibition when the other substrate is varied. The inhibitor combines at the active site but only prevents the binding of one of the substrates. The form of the Michaelis–Menten equation that describes this behavior is given in Eq. (31).

$$v = \frac{\dfrac{V_{max}[S]}{(1 + [I]/K_i)}}{\dfrac{K_m}{\left(1 + \dfrac{[I]}{K_i}\right)} + [S]} \tag{31}$$

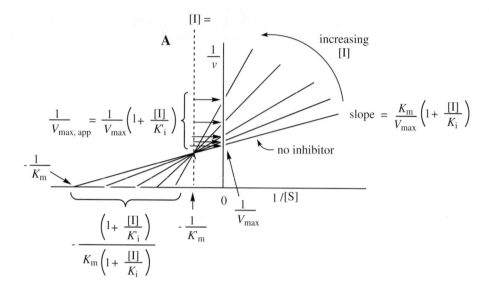

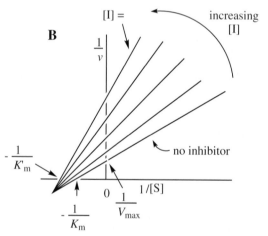

FIGURE A.14   Two possible Lineweaver–Burk plots for determination of mixed inhibitor kinetic constants. **(A)**, intersection of lines above the x-axis; **(B)**, intersection of lines below the x-axis.

The Lineweaver–Burk transformation of this equation is shown in Eq. (32),

$$\frac{1}{v} = \frac{K_m}{V_{max}}\left(\frac{1}{[S]}\right) + \frac{1}{V_{max}}\left(1 + \frac{[I]}{K_i}\right) \tag{32}$$

which is depicted graphically in Figure A.18. Because an uncompetitive inhibitor does not bind to free enzyme, the inhibitor has no effect on the V_{max}/K_m, and the slopes are independent of inhibitor concentration. The K_i values can be obtained

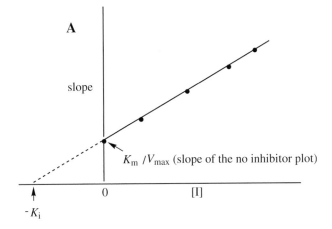

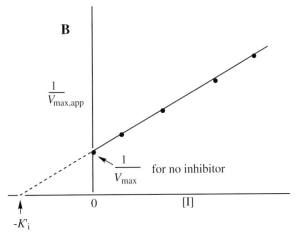

FIGURE A.15   Replot of data from Lineweaver–Burk plots for determination of K_i values for mixed inhibitors. **(A),** plot of the slope of lines from the Lineweaver–Burk plot versus inhibitor concentration; **(B),** plot of $1/V_{\text{max,app}}$ from the Lineweaver–Burk plot versus inhibitor concentration.

from replots of either $1/V_{\text{max,app}}$ (the y-axis intercepts) versus [I] (Figure A.19A) or $1/K_{\text{m,app}}$ (the x-axis intercepts) versus [I] (Figure A.19B).

Equation 33 is the equation for a Dixon plot of uncompetitive inhibition, which is represented graphically in Figure A.20.

$$\frac{1}{v} = \frac{1}{V_{\text{max}}K_i}[I] + \frac{1}{V_{\text{max}}}\left(1 + \frac{K_m}{[S]}\right) \tag{33}$$

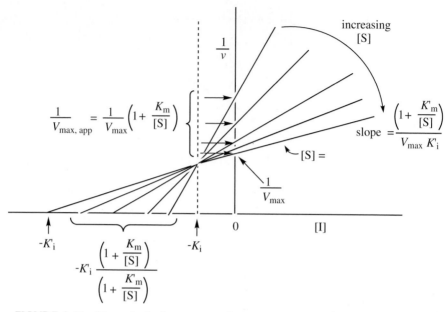

FIGURE A.16   Dixon plot for determination of mixed inhibitor kinetic constants when the E·S·I complex is not catalytically active.

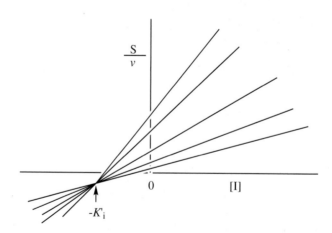

FIGURE A.17   Cornish-Bowden replot of the data from the Dixon plot for determination of the type of inhibition (in this case, mixed, when the lines of the Dixon plot intersect above the x-axis).

Because the slope expression does not contain a [S] term, the lines are parallel for all substrate concentrations. The Cornish-Bowden replot of these data is shown in Figure A.21.

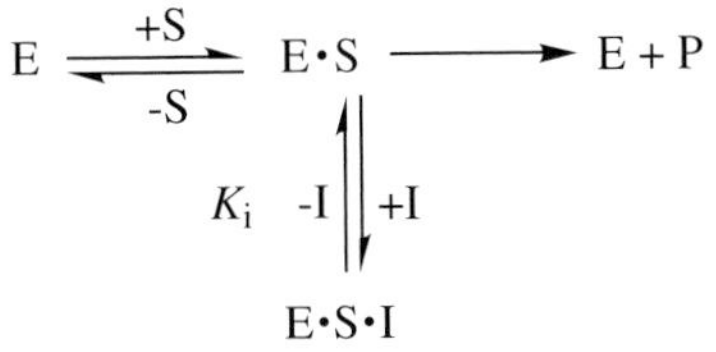

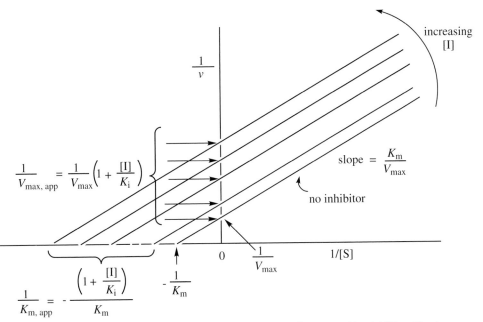

SCHEME A.5   Uncompetitive inhibition.

FIGURE A.18   Lineweaver–Burk plot for determination of uncompetitive inhibitor kinetic constants.

5. Slow-Binding and Slow, Tight-Binding Inhibition

Some reversible inhibitors attain the equilibrium between enzyme, inhibitor, and E·I complex slowly (with a steady-state time scale of seconds to minutes instead of milliseconds, as for the classical competitive inhibitor). These inhibitors are called *slow-binding inhibitors*. In some cases the ratio of total inhibitor to total enzyme must be high, as in the case of the classical competitive inhibitors, but in other cases, the attainment of the equilibrium of E, I, and the E·I complex occurs when the [I] is approximately the same as the [E], in which case the inhibitors are called *slow, tight-binding inhibitors*. The lifetimes of enzyme complexes with slow-binding and slow, tight-binding inhibitor exhibit slow off rates (k_{off}), but the on rates (k_{on}) may be

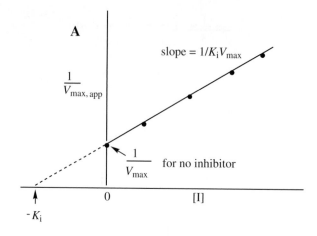

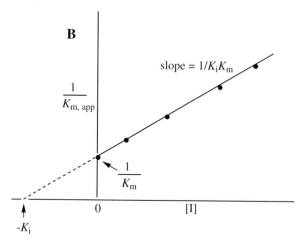

FIGURE A.19   Replot of data from Lineweaver–Burk plots for determination of K_i values for un-competitive inhibitors. **(A)**, plot of $1/V_{max,app}$ from the Lineweaver–Burk plot versus inhibitor concentration; **(B)**, plot of $1/K_{m,app}$ from the Lineweaver–Burk plot versus inhibitor concentration.

fast or slow. As the rate of release of the inhibitor from the E·I complex becomes vanishing slow, the inhibition approaches irreversible (see II.B). Inhibition by both of these types of inhibitors cannot be described by Michaelis–Menten kinetics.

The principal cause for slow-binding inhibition is believed to be a slow isomer-ization of the initially formed E·I complex, as the result of a conformational change, to another E·I complex (E·I*), in which the enzyme is in a tighter com-plex with I than it was in the E·I complex (Scheme A.6).[8] In this case the overall

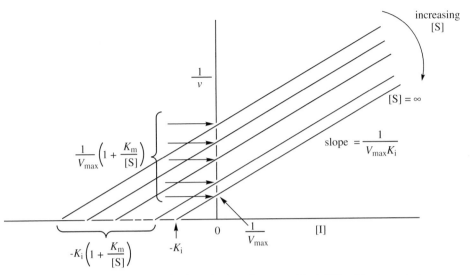

FIGURE A.20   Dixon plot for determination of uncompetitive inhibitor kinetic constants.

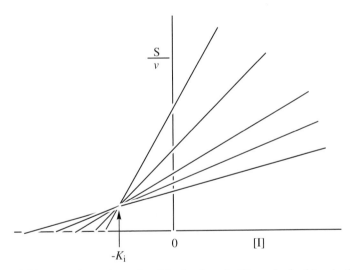

FIGURE A.21   Cornish-Bowden replot of the data from the Dixon plot for determination of the type of inhibition (in this case, uncompetitive).

$$E + I \underset{k_{-1}}{\overset{k_1}{\rightleftharpoons}} E \cdot I \underset{\underset{\text{slow}}{k_{-2}}}{\overset{k_2}{\rightleftharpoons}} E \cdot I^*$$

SCHEME A.6   Slow-binding and slow, tight-binding inhibition.

dissociation constant (K_i^*) is defined by Eq. (34).

$$K_i^* = \frac{(E)(I)}{(E \cdot I) + (E \cdot I^*)} = \frac{K_i k_{-2}}{k_2 + k_{-2}} \tag{34}$$

In addition to a slow conversion of $E \cdot I$ to $E \cdot I^*$, the reverse isomerization rate (k_{-2}) must be less than the forward isomerization rate (k_2).

Progress curves for the attainment of a slow-binding or slow, tight-binding inhibitor are described by the general integrated Eq. (35),

$$P = \nu_s t + \frac{(\nu_o - \nu_s)(1 - e^{-kt})}{k} \tag{35}$$

where ν_o, ν_s, and k represent, respectively, the initial rate, the final steady-state rate, and the apparent first-order rate constant for establishment of the equilibrium between $E \cdot I$ and $E \cdot I^*$. The initial rate is obtained from Eq. (36),

$$\nu_o = \frac{V_{max}[S]}{K_m(1 + [I]/K_i) + [S]} \tag{36}$$

where V_{max} is the maximum rate, $[S]$ is the concentration of the substrate for which I is an inhibitory analogue, K_m is the Michaelis constant for S, and K_i is the dissociation constant for the $E \cdot I$ complex. Note that Eq. (36) is the same as the rate equation for a competitive reversible inhibitor (Eq. (22)). The final steady-state rate is obtained from Eq. (37),

$$\nu_s = \frac{V_{max}[S]}{K_m(1 + [I]/K_i^*) + [S]} \tag{37}$$

where K_i^* is the overall inhibition constant as defined in Eq. (34). The apparent first-order rate constant (k) for the interconversion of $E \cdot I$ and $E \cdot I^*$ in the presence of substrate S is expressed in equation 38. Figure A.22 gives an example of a progress curve for a slow-binding inhibitor. For each curve with inhibitor present, there is an initial burst followed by a slower steady-state rate.

$$k = k_{-2} + k_2 \left(\frac{[I]/K_i}{1 + ([S]/K_m) + ([I]/K_i)} \right) \tag{38}$$

B. Irreversible Enzyme Inhibition

The two types of irreversible inhibition that will be discussed are *affinity labeling* and *mechanism-based inactivation*.

1. Affinity Labeling

These inactivators are reactive compounds that initially form a reversible $E \cdot I$ complex, then an active-site nucleophile reacts with the enzyme-bound inactivator to

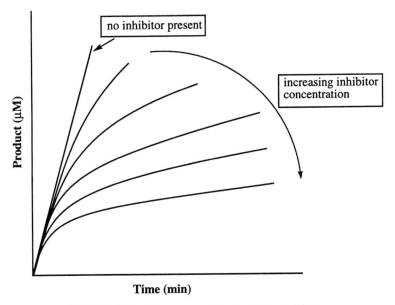

FIGURE A.22   Progress curves for slow-binding inhibitors.

$$E + I \; \underset{}{\overset{K_I}{\rightleftharpoons}} \; E \cdot I \; \xrightarrow{k_{inact}} \; E\text{-}I$$

SCHEME A.7   Affinity labeling inactivation.

form a covalent adduct (Scheme A.7). Loss of enzyme activity is time dependent, a measure of the rate of the reaction of the enzyme with the inactivator. To determine the kinetic constants, K_I and k_{inact}, a plot of the log of the enzyme activity versus time is constructed (Figure A.23). From this plot, the half-lives for inactivation $(t_{1/2})$ at each inactivator concentration (**a–e**) can be determined. A replot of the half-lives versus the inverse of the inactivator concentration (Figure A.24), known as a Kitz and Wilson replot,[9] is then made to determine the kinetic constants for inactivation.

The term K_I represents the concentration of the inactivator that gives half-maximal inactivation, just like K_m represents the concentration of substrate that gives half-maximal rate. Likewise, k_{inact} is the maximal rate constant for inactivation.

2. Mechanism-Based Inactivation

These inactivators are unreactive compounds that have a structural similarity to the substrate or product for the target enzyme and are converted by the target enzyme

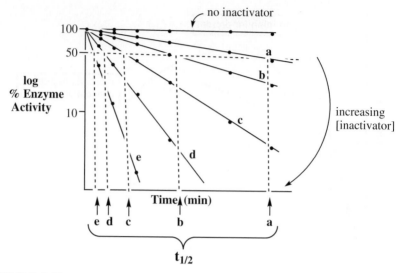

FIGURE A.23   Time-dependent inactivation by affinity labeling agents or mechanism-based inactivators.

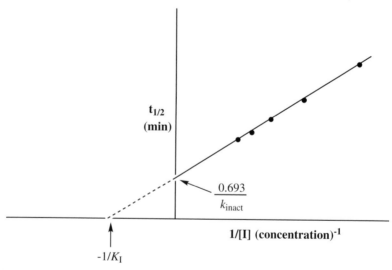

FIGURE A.24   Kitz and Wilson replot of the data from Figure A.22 for determination of the kinetic constants of inactivation by affinity labeling agents or mechanism-based inactivators.

into species that inactivate the enzyme prior to its release from the active site. Because of this additional step of converting the inactivator into the active form, an
additional step(s) has to be added to the kinetic expression (Scheme A.8). The ki-

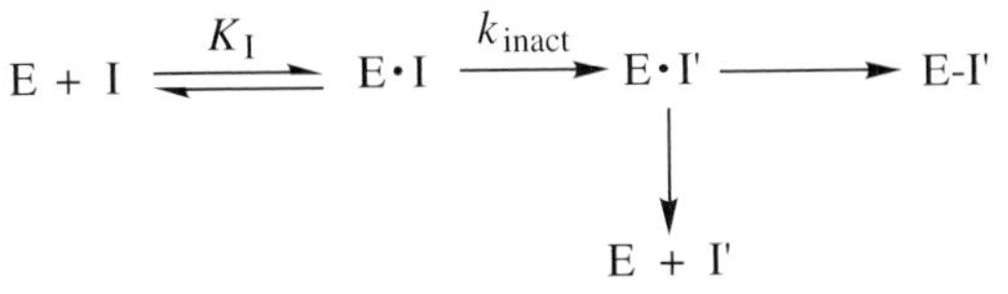

$$E + I \underset{}{\overset{K_I}{\rightleftharpoons}} E \cdot I \xrightarrow{k_{inact}} E \cdot I' \longrightarrow E\text{-}I'$$

$$\downarrow$$

$$E + I'$$

SCHEME A.8   Mechanism-based inactivation.

netic constants (K_I and k_{inact}) are determined the same way as for affinity labeling agents (Figures A.23 and A.24).

III. SUBSTRATE INHIBITION

Occasionally, the dependence of rate on substrate concentration does not give a hyperbolic curve, as in Figure A.1; instead, the rate may reach a maximum, then diminish as [S] increases further (Figure A.25). This is known as *substrate inhibition* (Figure A.25A demonstrates complete substrate inhibition, and Figure A.25B shows partial substrate inhibition), depicted in Scheme A.9. At high [S], an inactive E·S·S complex forms, which decreases the rate. A possible physical representation for this phenomenon is shown in Figure A.26. One substrate molecule is bound to the active site in A, and two substrate molecules are bound in B.

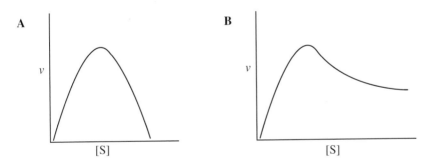

FIGURE A.25   Substrate inhibition.

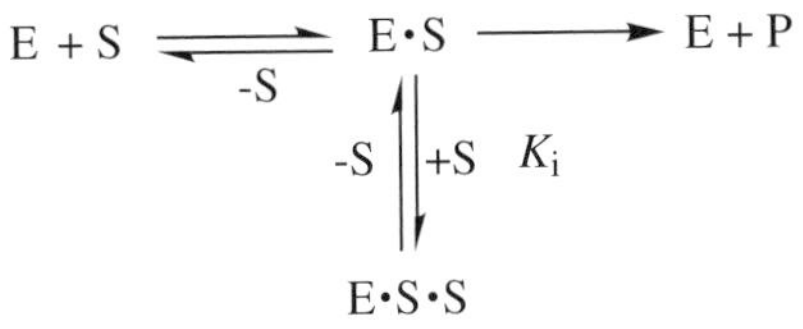

$$E + S \underset{-S}{\overset{}{\rightleftharpoons}} E \cdot S \longrightarrow E + P$$

$$-S \updownarrow +S \quad K_i$$

$$E \cdot S \cdot S$$

SCHEME A.9   Substrate inhibition.

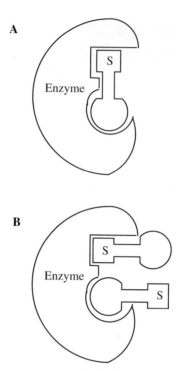

FIGURE A.26   Possible representation of substrate inhibition.**(A)**, E·S complex; **(B)**, inactive E·S·S complex.

The rate of an enzyme reaction that shows substrate inhibition is described by Eq. (39).

$$v = \frac{V_{max}[S]}{K'_m + [S] + K_i[S]^2} \tag{39}$$

K_i, in this case, is the equilibrium constant for formation of E·S·S from E·S and S. K'_m is a modified Michaelis constant. The reciprocal of Eq. (39) gives Eq. (40), which is analogous to the Lineweaver–Burk equation and is depicted graphically in Figure A.27.

$$\frac{1}{v} = \frac{K'_m}{V_{max}[S]} + \frac{1}{V_{max}} + \frac{K_i[S]}{V_{max}} \tag{40}$$

At low [S] (large 1/[S]), the term $K_i[S]/V_{max}$ becomes negligible, and Eq. (40) reduces to the Lineweaver–Burk equation [Eq. (18)], which gives a straight line (the linear part of Figure A.27).

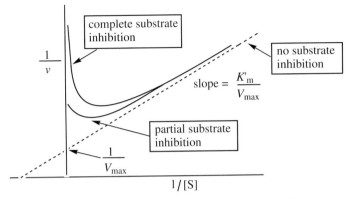

FIGURE A.27   Graphical representation of substrate inhibition at high substrate concentrations.

The two principal reasons for the failure of the Michaelis–Menten equation are substrate inhibition, as described earlier, and *substrate activation*, that is, an E·S·S complex that is *more* active than E·S.

IV. NONPRODUCTIVE BINDING

When a substrate binds at the active site in an alternative unreactive mode in competition with its productive mode of binding, it is called *nonproductive binding* (Scheme A.10). The effect is to lower both k_{cat} and K_m. The k_{cat} is lowered because, at saturation, only a fraction of the enzyme molecules has substrate bound productively; therefore, fewer enzyme molecules are active. The K_m is lower than the K_s because the existence of additional binding modes apparently leads to tighter binding. Eq. (41) describes this behavior.

$$\nu = \frac{[E]_0 \, [S] \, k_2}{K_s + [S] \, (1 + K_s/K'_s)} \tag{41}$$

By a comparison of Eq. (41) with the Michaelis–Menten equation [Eq. (8)], k_{cat} and K_m can be redefined under these conditions as Eq. (42) and Eq. (43), respectively.

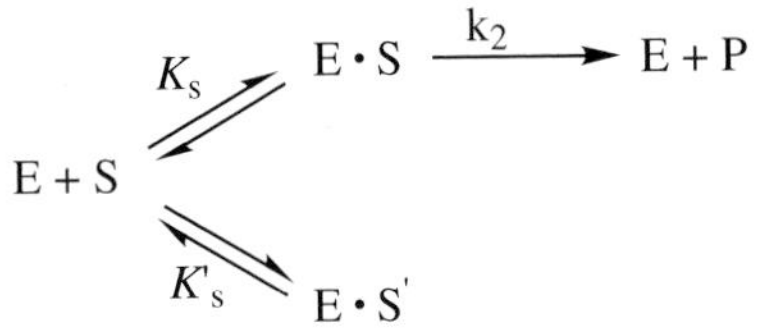

$$E + S \underset{K'_s}{\overset{K_s}{\rightleftharpoons}} \quad \begin{array}{c} E \cdot S \xrightarrow{k_2} E + P \\ \\ E \cdot S' \end{array}$$

SCHEME A.10   Nonproductive binding.

Dividing k_{cat} by K_m gives Eq. (44). Therefore,

$$k_{cat} = \frac{k_2}{1 + K_s/K'_s} \tag{42}$$

$$K_m = \frac{K_s}{1 + K_s/K'_s} \tag{43}$$

$$k_{cat}/K_m = k_2/K_s \tag{44}$$

k_{cat}/K_m is unaffected by the presence of the additional binding mode, because k_{cat} and K_m are altered in a compensating manner. For example, if the nonproductive site binds the substrate 1000 times more strongly than the productive site, K_m will be 1000 times lower than K_s, but because only 1 molecule in 1000 is productively bound, k_{cat} is 1000 times lower than k_2.

V. COMPETING SUBSTRATES

Suppose two different substrates compete for the active site of an enzyme (Scheme A.11). An alternative way to calculate reaction rates besides making the Michaelis–Menten assumptions is from Eq. (45). The consumption of substrate A is described by Eq. (46) and consumption of substrate B is described by Eq. (47).

$$\nu = \frac{k_{cat}}{K_m} [E] [S] \tag{45}$$

$$\frac{-d\,[A]}{dt} = \nu_A = \left(\frac{k_{cat}}{K_m}\right)_A [E][A] \tag{46}$$

$$\frac{-d\,[B]}{dt} = \nu_B = \left(\frac{k_{cat}}{K_m}\right)_B [E][B] \tag{47}$$

The ratio of the rate with substrate A to substrate B is given by Eq. (48).

$$\frac{\nu_A}{\nu_B} = \frac{(k_{cat}/K_m)_A\,[A]}{(k_{cat}/K_m)_B\,[B]} \tag{48}$$

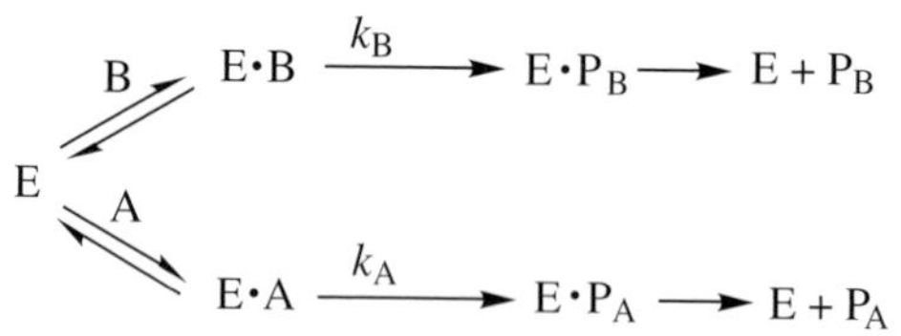

SCHEME A.11   Competing substrates.

Because the reaction specificity is determined by the ratios of k_{cat}/K_m, not just by K_m, nonproductive binding, which we have already learned has no effect on k_{cat}/K_m, does not affect substrate specificity.

VI. MULTISUBSTRATE SYSTEMS

For the most part, the preceding discussions dealt with single-substrate enzymes. The equations become *much* more complex and difficult to solve when multiple substrates are involved. Therefore, if you are interested in these systems, I recommend you read an advanced text on enzyme kinetics (see General References at the end of this appendix). Here, however, I have described, in a nonmathematical way, some of the standard multisubstrate pathways, so you can get, at least, a physical understanding of their meaning.

A. Sequential Mechanisms

1. Random Sequential Mechanism

In this mechanism there is no obligatory order of combination of the substrates and enzyme or release of products (Scheme A.12).

2. Ordered Sequential Mechanism

In this mechanism there is an obligatory order for how substrates combine with the enzyme and how products dissociate (Scheme A.13). If substrate A has a lower K_m for the enzyme than does substrate B, A will bind first. This may cause a conformational change in the enzyme that increases the affinity of the enzyme for substrate B, producing the ternary complex E·A·B.

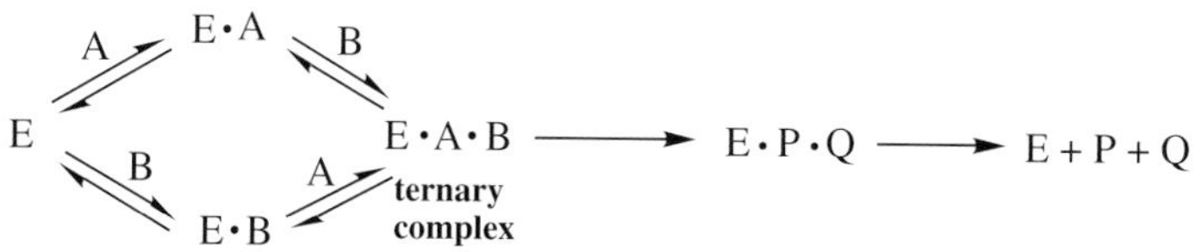

SCHEME A.12   Random sequential mechanism.

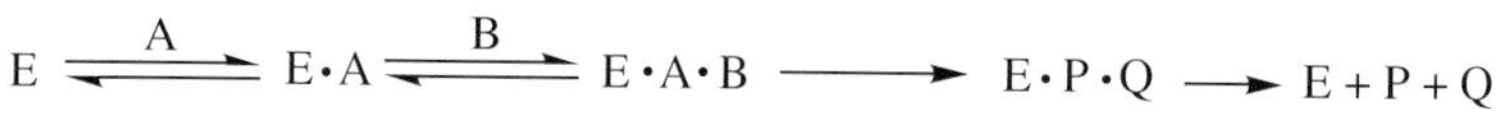

SCHEME A.13   Ordered sequential mechanism.

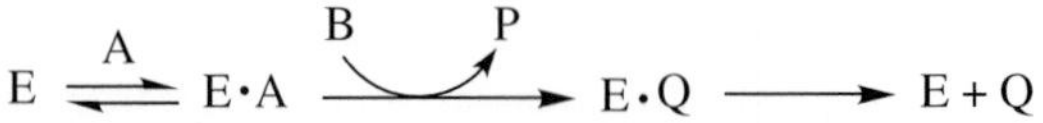

$$E \overset{A}{\rightleftharpoons} E{\cdot}A \overset{B \qquad P}{\curvearrowright} E{\cdot}Q \longrightarrow E + Q$$

SCHEME A.14   Theorell-Chance sequential mechanism.

$$E + A \rightleftharpoons E{\cdot}A \rightleftharpoons F{\cdot}P \rightleftharpoons F + P$$

$$F + B \rightleftharpoons F{\cdot}B \rightleftharpoons E{\cdot}Q \rightleftharpoons E + Q$$

SCHEME A.15   Ping-Pong mechanism; E is the free enzyme, A if the first substrate, F is the modified enzyme, P is the first product, B is the second substrate, Q is the second product.

3. Theorell–Chance Sequential Mechanism

This is a type of ordered mechanism in which the ternary complex does not accumulate (Scheme A.14; P is one product and Q is the second product).

4. Ping Pong Mechanism

Reactions in which one or more products are released before all the substrates are added are called *Ping Pong reactions* (Scheme A.15). When two substrates are involved and two products are formed, this is referred to as a Ping Pong Bi-Bi mechanism, such as that shown in Scheme A.15. If one substrate is cleaved into two products, it would be a Ping Pong Uni-Bi mechanism, and so forth. An example of a Ping Pong mechanism is phosphoglycerate mutase. The enzyme is phosphorylated by one substrate, and then the phosphoryl group is transferred to a second substrate. This reaction is depicted in Figure A.28 in the shorthand notation for enzyme mechanisms that was described by Cleland (except using $A\text{-}PO_3^{2-}$ as the phosphorylated substrate instead of A, $E\text{-}PO_3^{2-}$ as the phosphorylated enzyme instead of F, and $B\text{-}PO_3^{2-}$ as the phosphorylated product instead of P).[10] The horizontal line represents the enzyme. Substrate $A\text{-}PO_3^{2-}$ binds to the enzyme (downward vertical arrow), forming an $E{\cdot}A\text{-}PO_3^{2-}$ complex, which forms phosphorylated en-

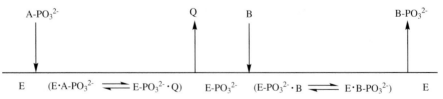

FIGURE A.28   Modified Cleland shorthand notation for the Ping Pong mechanism of phosphoglycerate mutase.

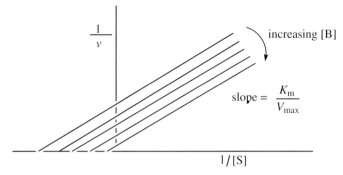

FIGURE A.29   Lineweaver–Burk plot for a Ping Pong mechanism.

zyme (E-PO$_3^{2-}$) and the bound dephosphorylated substrate Q (E-PO$_3^{2-}$·Q). Q is released from the enzyme (upward vertical arrow), leaving phosphorylated enzyme (E-PO$_3^{2-}$). Then the second substrate B binds to the enzyme (downward vertical arrow) to give the E-PO$_3^{2-}$·B complex. The phosphate is transferred to B (E·B-PO$_3^{2-}$), and B-PO$_3^{2-}$ is released from the enzyme (upward vertical arrow), leaving free enzyme (E).

A plot of $1/v$ versus $1/[S]$ for a Ping Pong reaction gives parallel lines (Figure A.29). As the [B] increases, V_{max} increases, as does the K_m for A-PO$_3^{2-}$, but V_{max}/K_m remains constant.

VII. ALLOSTERISM AND COOPERATIVITY

A. General

The terms *allosterism* and *cooperativity* generally apply to multisubunit enzymes. Binding (and catalytic) events at one active site can influence binding (and catalytic) events at another active site in a multimeric protein. Generally, the active sites are located on different subunits. The binding of the *effector molecule* (the one causing the effect) causes a conformational change in the protein in such a way that signals the other subunits that it has been bound. If the effector acts at another site and is not the substrate, the effect is called *allosteric* and *heterotropic*. If the effector is the substrate, the effect is termed *cooperative* and *homotropic*.

Positive cooperativity means that binding of the substrate at one site makes it easier for another substrate molecule to bind to the active site. *Negative cooperativity* means binding at one site causes a decrease in affinity of the substrate for the active site. Cooperative enzymes show sigmoidal kinetics, not Michaelis–Menten-like kinetics (Figure A.30).

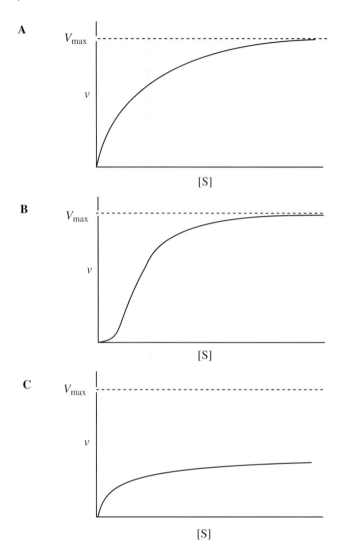

FIGURE A.30   Rate versus substrate concentration plots for enzyme-catalyzed reactions that exhibit **(A)**, no cooperativity; **(B)**, positive cooperativity; **(C)**, negative cooperativity.

B. Monod–Wyman–Changeux (MWC) Concerted Model[11]

This is one of the models that is used to rationalize the effects of cooperativity. It proposes that the protein is in a dynamic equilibrium between a state that has a low affinity for the substrate (called the T state, the tense state) and much less of a state

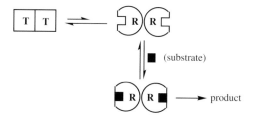

FIGURE A.31   Monod–Wyman–Changeux (MWC) concerted model for cooperativity.

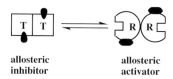

**allosteric
inhibitor**

**allosteric
activator**

FIGURE A.32   Allosteric inhibitor and activator effects on the Monod–Wyman–Changeux (MWC) concerted model for cooperativity.

that has a high affinity for the substrate (the R state, the relaxed state). According to this model, in a multisubunit protein *all* subunits of an enzyme molecule are either in the T conformation or the R conformation. The T and R states coexist in the absence of substrate (Figure A.31). When the substrate is added, it binds to the R state and drives the equilibrium toward R.

Allosteric (or heterotropic) inhibitors bind specifically to the T state and make it harder for the substrate to switch the enzyme into the R state (Figure A.32).

Allosteric (or heterotropic) activators bind specifically to the R state and pull more of the enzyme into the more active R state, thereby activating the enzyme. A substrate or effector that binds preferentially to the R state increases the concentration of the R state, giving positive cooperativity. However, the MWC model cannot account for negative cooperativity (this is rare), because binding of the first ligand can only stabilize the R state and cannot increase the proportion of the T state.

C. Koshland–Némethy–Filmer (KNF) Sequential Model[12]

This model states that the progress from the T state to the R state is a sequential process, not an either T or R process (Figure A.33). The conformation of each

FIGURE A.33   Koshland–Némethy–Filmer (KNF) sequential model for cooperativity.

subunit changes in turn as the substrate binds. In the KNF model the protein is completely in the T state in the absence of substrate. On substrate binding, a conformational change is induced in only that subunit. This change may be transmitted to neighboring vacant subunits. The KNF model, then, can account for negative cooperativity. This would occur when the binding of a ligand to one site causes in that subunit a conformational change that is transmitted to an adjacent subunit, and this conformational change blocks binding in that subunit. *Half-of-the-sites* (or half-sites) *reactivity* is an example of this, by which an enzyme containing $2n$ sites reacts at only n of them.

GENERAL REFERENCES

Segel, I. H. *Enzyme Kinetics,* Wiley: New York, 1975.
Cornish-Bowden, A. *Fundamentals of Enzyme Kinetics,* Portland Press: London, 1995.
Silverman, R. B. *Meth. Enzymol.* **1995,** *249,* 240 (for mechanism-based inactivation).

REFERENCES

1. Briggs, G. E.; Haldane, J. B. S. *Biochem. J.* **1925,** *19,* 338.
2. Michaelis, L.; Menten, M. L. *Biochem. Z.* **1913,** *49,* 333.
3. Lineweaver, H.; Burk, D. *J. Am. Chem. Soc.* **1934,** *56,* 658.
4. (a) Eadie, G. S. *J. Biol. Chem.* **1942,** *146,* 85. (b) Hofstee, B. H. J. *Nature* **1959,** *184,* 1296.
5. (a) Woolf, B., cited by Haldane, J. B. S.; Stern, K. G. *Allgemeine Chemie der Enzyme,* Steinkopf: Dresden, 1932, pp. 119–120. (b) Hanes, C. S. *Biochem. J.* **1932,** *26,* 1406.
6. Dixon, M. *Biochem. J.* **1953,** *55,* 170.
7. Cornish-Bowden, A. A. *Biochem. J.* **1974,** *137,* 143.
8. (a) Morrison, J. F.; Walsh, C. T. *Adv. Enzymol. Rel. Areas Mol. Biol.* **1988,** *61,* 201. (b) Schloss, J. V. *Acc. Chem. Res.* **1988,** *21,* 348.
9. Kitz, R.; Wilson, I. B. *J. Biol. Chem.* **1962,** *237,* 3245.
10. Cleland, W. W. *The Enzymes,* vol. 2, Boyer, P. D., Ed., Academic Press: New York, 1970, p. 1.
11. Monod, J.; Wyman, J.; Changeux, J.-P. *J. Mol. Biol.* **1965,** *12,* 88.
12. Koshland, Jr., D. E.; Némethy, G.; Filmer, D. *Biochemistry* **1966,** *5,* 365.

Problems and Solutions

I. PROBLEMS

A. Chapter 1

P1-1. a. Which amino acids have side chains that are cationic or anionic at pH 4, 7, and 10? Draw them.
b. Draw the nucleophilic amino acids.

P1-2. Describe how the concept of orbital steering applies to the following reaction.

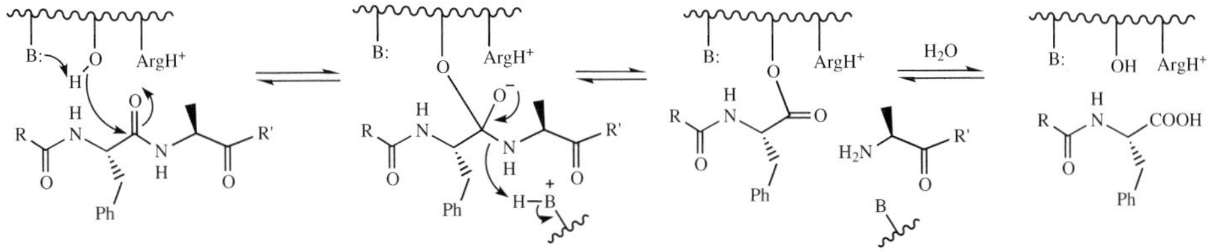

P1-3. What is the maximum k_{cat} for a substrate whose K_m is 150 μM?

P1-4. The following reaction is catalyzed by an enzyme at pH 7 and 25 °C, whereas nonenzymatically, this reaction does not occur under these conditions. Explain how the enzyme can easily catalyze this reaction.

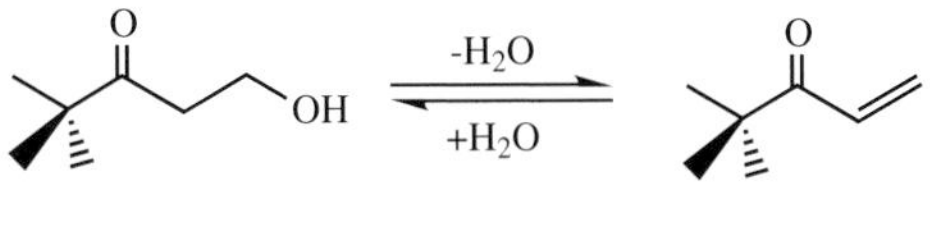

P1-5. Porphobilinogen synthase is a zinc-dependent enzyme (contains two zinc ions) that catalyzes the condensation of two molecules of 5-aminolevulinic acid

(**P1.1**) to give porphobilinogen (**P1.2**). A condensed mechanism is shown next. Indicate with an arrow every place possible where the enzyme catalyzes this reaction, and name the catalytic mechanism.

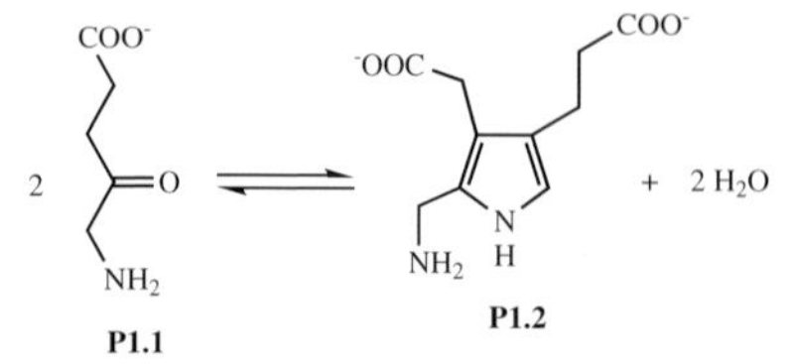

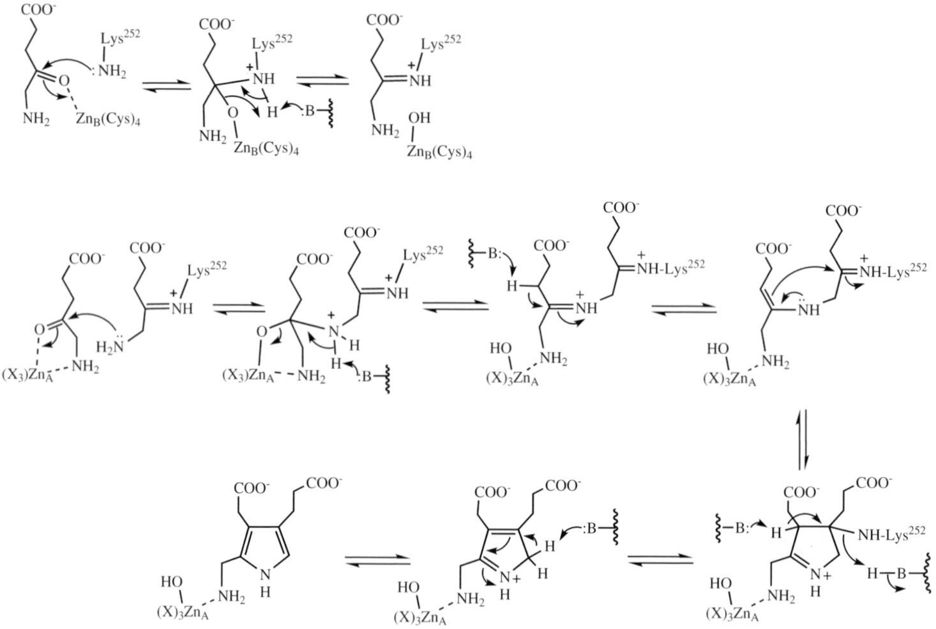

P1-6. β-Glycosidases catalyze the hydrolysis of β-oligosaccharides via enzyme-catalyzed protonation of the leaving group. Construct a chemical model that may catalyze a similar reaction.

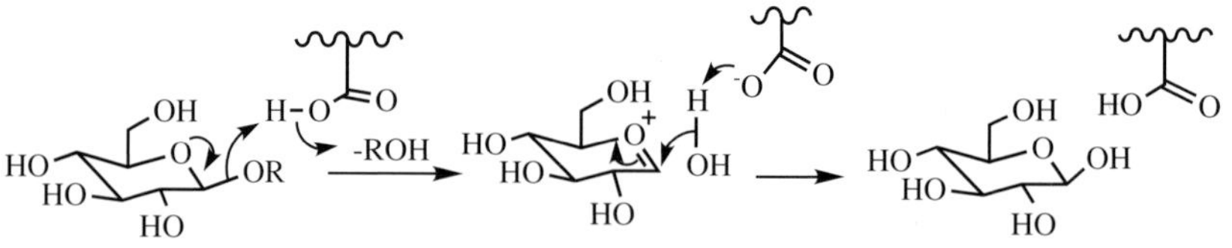

P1-7. Indicate what interaction is relevant at each letter.

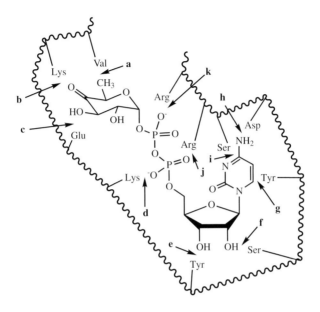

P1-8. Draw mechanisms for enzyme-catalyzed enolization of a ketone utilizing (a) an electrostatic interaction and (b) a low-barrier hydrogen bond.

P1-9. Transition-state stabilization accounts for a large portion of the stabilization energy of an enzyme-catalyzed reaction. Haloalkane dehalogenase catalyzes an S_N2 reaction of an active-site carboxylate with an alkyl halide followed by hydrolysis to the alcohol. What transition-state stabilization processes could the enzyme utilize? (Draw the transition state for guidance.)

P1-10. What effect do you suppose the residue on the right-hand side of each pair of active-site residues has on the pK_a of the residue on the left-hand side?

a. Lys^{162} Lys^{163}
    | |
    NH_3^+ NH_3^+

b. Lys^{81} Asp^{82}
    | |
    NH_3^+ COO^-

c. Glu^{245} Glu^{246}
    | |
    COO^- COO^-

P1-11. The (S,S)-isomer of deuterated compound **P1.3** is a substrate for a dephosphorylase that gives an *anti* elimination to the Z-isomer **P1.4**. The corresponding (R,S)-isomer is not a substrate. How do you explain that?

(S,S)-**P1.3**   **P1.4**

B. Chapter 2

P2-1. Why does an affinity labeling agent exhibit time-dependent inhibition, even in the absence of substrate?

P2-2. An esterase was isolated, and experiments were carried out to determine its mechanism of action. Explain the significance of each experiment, and deduce a mechanism.

a.

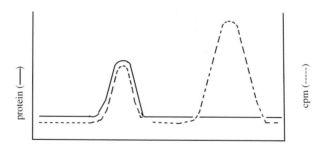

b.

$$RCO_2{}^{14}CH_3 \xrightarrow[\text{0}^\circ,\ \text{pH 5}]{\text{esterase}} \xrightarrow[\text{filtration}]{\text{gel}}$$

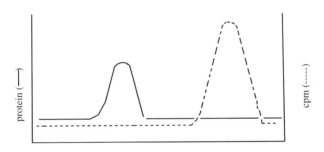

c.

$$^{14}RCO_2CH_3 \xrightarrow[\text{0}^\circ,\ \text{pH 9}]{\text{esterase}} \xrightarrow[\text{filtration}]{\text{gel}}$$

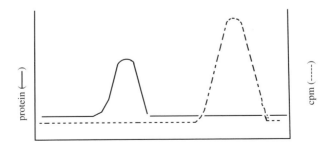

d. The protein fractions from part a were treated with NH_2OH. Two radioactive products were isolated in the ratio 95:5; the major product was determined to be $^{14}RCONHOH$. Only 95% of the enzyme activity was recovered. What was the other product? How do you account for the loss of 5% of the enzyme activity?

P2-3. Consider a new broad-spectrum phosphatase that catalyzes the hydrolysis of any phosphate ester that binds. When the reaction is carried out at 37 °C with its best substrates, no covalent catalysis is observed. What experiments would you devise, if you suspected that covalent catalysis is occurring?

P2-4. Acetate kinase catalyzes the conversion of acetate to acetyl phosphate. ATP and Mg^{2+} are required, and ADP and inorganic phosphate are products. The acetyl phosphate is then transformed into acetyl coenzyme A with coenzyme A, catalyzed by phosphotransacetylase.

$$CH_3COO^- + ATP \cdot Mg^{2+} \longrightarrow CH_3\overset{O}{\overset{\|}{C}}OPO_3^{2-} \xrightarrow{CoASH} CH_3\overset{O}{\overset{\|}{C}}SCoA + P_i$$

with ADP leaving in the first step.

(As we'll see later, CoASH is used to help transport carboxylic acids and to activate them.) What experiments would you do to test the mechanism and reaction sequence?

P2-5. Explain the following result (a mechanism might help): A sugar 6-phosphate undergoes a reversible hydrolysis to sugar plus P_i catalyzed by sugar phosphatase. The reaction was carried out in $H_2^{18}O$ and incubated with the enzyme and substrate for a long time. At various time intervals aliquots were removed, and the sugar 6-phosphate was isolated (how?) and analyzed for ^{18}O content (how was this done?). As time progressed, the specific activity of ^{18}O in the sugar 6-phosphate *increased*.

P2-6. The biosynthetic sequence for the production of pyrimidines (precursors of DNA) begins with carbamoyl phosphate (**P2.1**). If carbamoyl phosphate synthetase is incubated with $H^{14}CO_3^-$ plus other required cosubstrates and/or cofactors, the carbamoyl phosphate produced contains ^{14}C. If the enzyme is incubated with $^{15}NH_4Cl$ as the nitrogen source, no ^{15}N is found in the carbamoyl phosphate.

$$H^{14}CO_3^- \xrightleftharpoons[\text{synthetase}]{\text{carbamoyl phosphate}} H_2N^{14}CO_2PO_3^{2-}$$

P2.1

a. What cosubstrates and/or cofactors are involved?

b. What are all the products?

c. Draw detailed mechanism.

d. Draw a compound for irreversible inactivation of this enzyme.

P2-7. Draw a mechanism for the following enzymatic reaction:

$$Asp + Gln + Mg \cdot ATP \rightleftharpoons Asn + Glu + Mg \cdot AMP + PP_i$$

a. If $AspCO^{18}O^-$ produces $[^{18}O]AMP$

b. If $AspCO^{18}O^-$ produces $[^{18}O]PP_i$

c. If you incubate the enzyme with $[^{32}P]PP_i$ in the absence of Gln, you get $[^{32}P]ATP$. What does this indicate? Explain.

P2-8. Selenophosphate synthetase from *Escherichia coli* catalyzes the formation of the reactive selenium donor, selenophosphate $(SePO_3H^{2-})$, which is required for the biosynthesis of selenocysteinyl-tRNA^{SEC}, the precursor of specific seleno-cysteine residues in selenoproteins. In the absence of selenide, the enzyme catalyzes a slow conversion of ATP to AMP and two molecules of inorganic phosphate.

$$ATP + HSe^- + H_2O \longrightarrow SePO_3H^= + P_i + AMP$$

Incubation of the enzyme with $[\gamma\text{-}^{18}O_4]ATP$ in the absence of selenide resulted in complete scrambling of the ^{18}O label in the β,γ-bridge position with the two ^{16}O labels at the β-nonbridge positions.

a. Draw a mechanism that accounts for this observation.

b. What would you expect to observe if $[\beta\text{-}^{18}O_4]ATP$ were used?

c. Draw a mechanism for selenophosphate synthetase.

P2-9. Phosphonoacetaldehyde hydrolase catalyzes the hydrolysis of the C–P bond of phosphonoacetaldehyde. When chiral $[^{17}O,^{18}O]$thiophosphonoacetaldehyde is used as the substrate, the phosphate produced has its stereochemistry retained:

$$^=O_3P \cdot CH_2CHO \xrightarrow{H_2O} HPO_4^= + CH_3CHO$$

Incubation of the enzyme with substrate and sodium borohydride inactivates the enzyme (no inactivation occurs in the absence of substrate). When the reaction is run in D_2O, trideuteriomethylacetaldehyde is produced.

a. Draw a mechanism that is compatible with these results.

Phosphonoacetaldehyde hydrolase is a member of the haloacid superfamily of enzymes, which includes 2-haloalkanoic acid dehalogenase, epoxide hydrolase, 2-phosphoglycolate phosphatase, and phosphoserine phosphatase. Superfamilies contain enzymes that catalyze apparently diverse reactions but have a common reaction step, and the active-site structures reflect the catalytic similarity. This suggests that proteins are constructed in a modular fashion with separate domains performing different catalytic functions.

b. Based on the notion of superfamilies, the mechanism for phosphoserine phosphatase, and your mechanism for phosphonoacetaldehyde hydrolase in part a, predict the mechanism for 2-haloalkanoic acid dehalogenase (see Chapter 6, Section III for related enzymes).

C. Chapter 3

P3-1. The following reaction is catalyzed by a NAD^+-dependent enzyme:

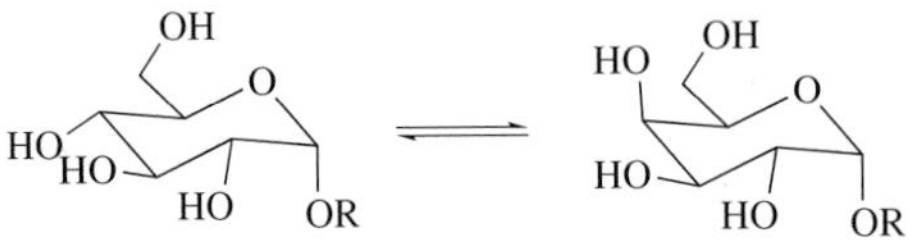

If the reaction is carried out in ^3H_2O, no ^3H is incorporated into the substrate or the product. In $H_2{}^{18}O$, no ^{18}O is incorporated.

a. Draw a mechanism to account for this.

b. Devise two more experiments to test your mechanism, one of which must be directed at determining the structure of an intermediate.

P3-2. D-Amino-acid oxidase (a flavoenzyme) can catalyze the conversion of nitroethane into acetaldehyde and nitrite ion ($NO_2{}^-$). If the reaction is done in $H_2{}^{18}O$ with [^{14}C]nitroethane, the CH_3CHO contains ^{18}O and ^{14}C. If the reaction is carried out with [^{14}C]nitroethane in the presence of NaCN, enzyme inactivation occurs.

a. Write a mechanism compatible with these results.

b. Why is this enzyme oxygen dependent?

P3-3. Write a mechanism for the biosynthesis of NAD^+ from nicotinic acid (niacin; vitamin B_3; **P3.1**).

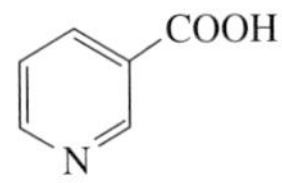

P3.1

P3-4. Succinate (**P3.2**) is converted into fumarate (**P3.3**) by a flavin-dependent enzyme.

a. Draw a mechanism.

b. This is a stereospecific reaction. Design experiments to determine the stereochemistry.

P3-5. a. Pargyline (**P3.4**) is a mechanism-based inactivator of monoamine oxidase. Propose a mechanism for inactivation on the basis of the catalytic mechanism
of monoamine oxidase.

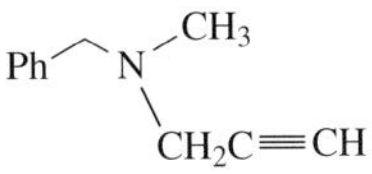

P3.4

b. Describe two experiments to determine the partition ratio.

P3-6. a. Based on the mechanism of general acyl-CoA dehydrogenase, design
a mechanism-based inactivator (not **3.70** in Chapter 3) and show its inactivation
mechanism.

b. Design an affinity labeling agent.

P3-7. Plasma amine oxidase (PAO) was incubated with $[(1R)-^2H, (2R)-^3H]$
and $[(1S)-^2H, (2R)-^3H]$ dopamines. The C-1 protons are indiscriminantly removed
by PAO. The loss of tritium at C-2 is in a ratio of 68:18 for the $[(1R)-^2H, (2R)-$
$^3H]:[(1S)-^2H, (2R)-^3H]$ isomers. Predict whether this goes by *syn* or *anti* removal
of the C-1 and C-2 hydrogens. The favored conformer at the active site is **P3.5.**
Draw a mechanism.

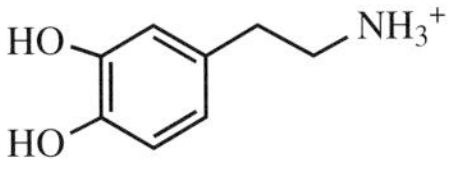

P3.5

P3-8. The mechanism of galactose oxidase was discussed in the text, and it was
noted that this enzyme is inactivated by β-haloethanols. Draw a mechanism for the
inactivation of galactose oxidase by β-haloethanols.

P3-9. Show how pyruvyl- and dehydroalanyl-containing enzymes can be differentiated by reduction with $[^3H]NaBH_4$ followed by hydrolysis in 6 N HCl at
110 °C.

D. Chapter 4

P4-1. Two steps in the catabolism of L-tryptophan are needed to convert **P4.1**
to **P4.2**.

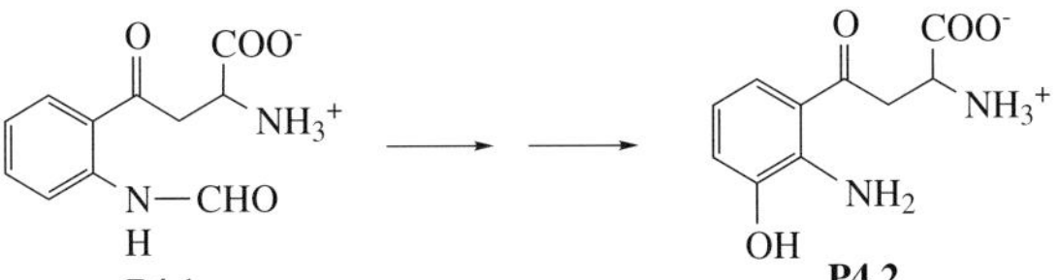

P4.1   **P4.2**

a. Draw two reasonable mechanisms for the first of these two enzyme-catalyzed reactions, one involving hydrolysis and the other oxidative decarboxylation. Also draw the second hydroxylation reaction.

b. Describe experiments to test them.

P4-2. Phenylalanine hydroxylase, a pterin-dependent enzyme, can oxygenate *p*-chlorophenylalanine (**P4.3**).

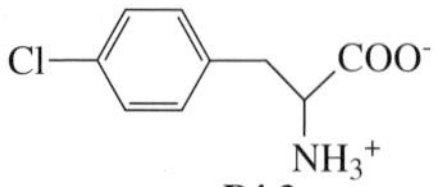

P4.3

a. Draw possible mechanisms. What is the product?

b. Design experiments that *may* differentiate between a chloronium ion intermediate or chloride ion migration in favor of a Cl⁻ elimination mechanism.

P4-3. Compound **P4.4** is a substrate for cyclohexanone oxygenase. Draw a mechanism, and show the stereochemistry.

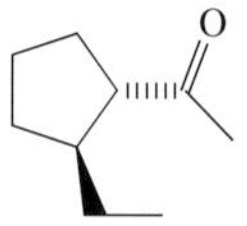

P4.4

P4-4. Draw a mechanism to account for the irreversible inactivation of aromatase by each of the following mechanism-based enzyme inactivators.

a.

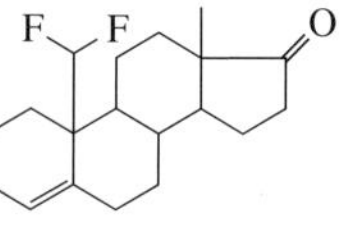

b.

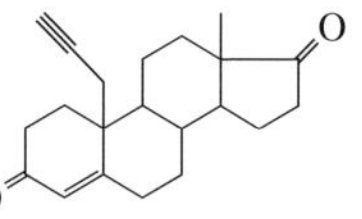

P4-5. Nitric oxide synthase converts L-arginine to L-citrulline and nitric oxide (·NO). It requires NADPH, O_2, and a heme cofactor. The urea carbonyl oxygen of citrulline and the oxygen of NO each come from separate molecules of O_2. Draw a mechanism.

L-Arg → L-citrulline + ·NO

P4-6. Draw mechanisms for the formation of flavin C4a–hydroperoxide that utilize triplet O_2 with a metal ion present and with no metal ion present.

P4-7. 1-Aminocyclopropane 1-carboxylate oxidase (ACC oxidase), a nonheme iron(II)-dependent enzyme, catalyzes the last step in the biosynthesis of the plant-signaling molecule ethylene from 1-aminocyclopropane 1-carboxylate (**P4.5**). The enzyme requires ascorbate as a reducing agent. Hydrogen cyanide and carbon dioxide are found as by-products of the oxidation as well. Draw a reasonable mechanism for this reaction.

P4-8. Two other heme-dependent oxygenases that utilize heteroatomic substrates are horseradish peroxidase and chloroperoxidase. The former enzyme catalyzes the formation of sulfoxides (**P4.7**), an aldehyde (**P4.8**), and a thiol (**P4.9**) from the corresponding sulfide (**P4.6**), but the latter enzyme only gives the sulfoxide. Draw mechanisms to account for these products and give an explanation for why there should be a difference.

P4-9. 2-Methyl-3-hydroxypyridine-5-carboxylic acid oxygenase is a flavoenzyme (FAD) that catalyzes the incorporation of two oxygen atoms into 2-methyl-3-hydroxypyridine-5-carboxylate (**P4.10**) and cleavage of the aromatic ring to give α-(N-acetylaminomethylene) succinic acid (**P4.11**). At first sight, this reaction

appears to be a textbook case of an example of a dioxygenase (see Chapter 5), but it is actually a monooxygenase-catalyzed reaction. Using $^{18}O_2$ and $H_2^{18}O$, it was shown that the amide carbonyl oxygen comes from O_2 and the incipient carboxylate oxygen comes from H_2O. From stopped-flow spectrophotometric experiments, C^{4a}-hydroperoxyflavin and C^{4a}-hydroxyflavin intermediates can be detected, indicating that a typical monooxygenase reaction is functional.

    a. Draw a mechanism that is compatible with these results.

    b. Design an experiment to test this mechanism.

E. Chapter 5

P5-1. The "old wives' tale" about eating carrots to make your eyes healthy has some validity. β-Carotene (**P5.1**), which is orange, is a constituent of carrots that is converted to vitamin A (retinal; **P5.2**), a substrate for an eye enzyme.

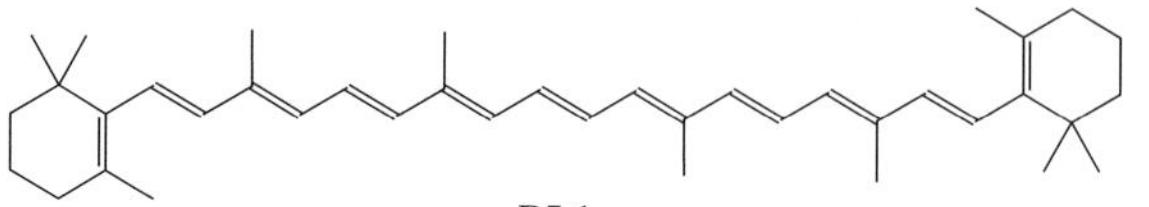

P5.1

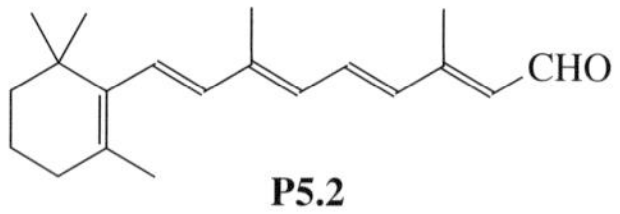

P5.2

    a. Draw a mechanism for the Fe^{2+}-dependent reaction.

    b. Devise experiments to test your mechanism.

P5-2. a. Draw a mechanism for the following enzymatic reaction:

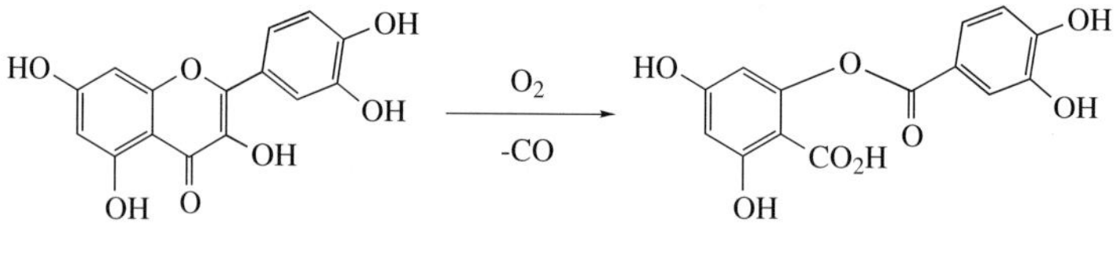

    b. Where do the atoms from O_2 end up?

    c. Devise an experiment to determine the influence of the substituents (i.e., the hydroxyl groups) on the mechanism.

P5-3. The α-ketoglutarate-dependent (nonheme Fe) conversion of L-(α-amino-δ-adipoyl)-L-cysteinyl-D-valine (**P5.3**) to isopenicillin N (**P5.4**) requires two molecules of O_2. No oxygen atoms are inserted into the β-lactam product. Draw a mechanism for this reaction.

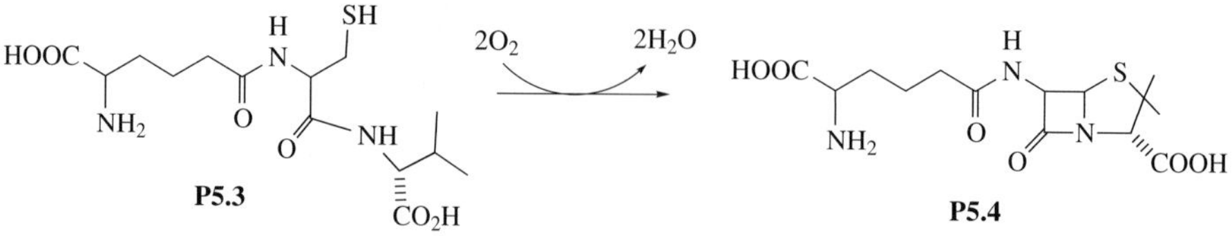

P5.3

P5.4

P5-4. Tryptophan 2,3-dioxygenase is a heme–dependent enzyme that catalyzes the conversion of L-tryptophan (**P5.5**) to N-formylkynurenine (**P5.6**). Both atoms of O_2 are incorporated into the product.

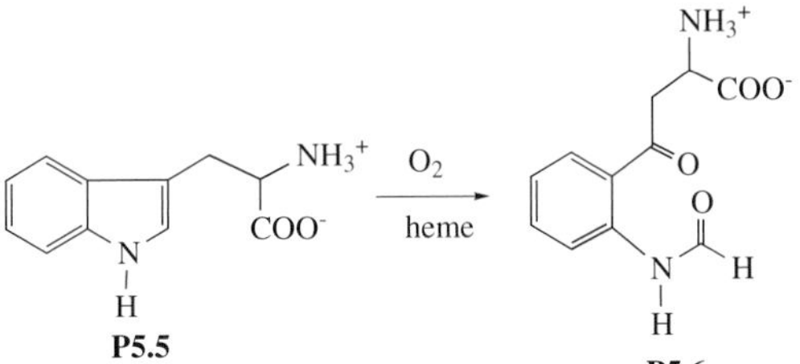

P5.5

P5.6

N-Methyl-L-tryptophan (**P5.7**) and the benzofuran analogue of L-tryptophan (**P5.8**) are not substrates, but are potent competitive inhibitors.

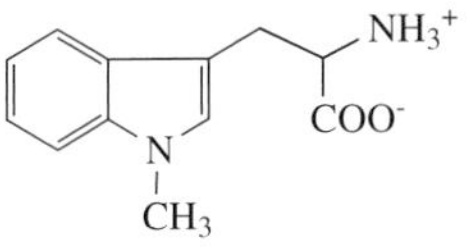

P5.7

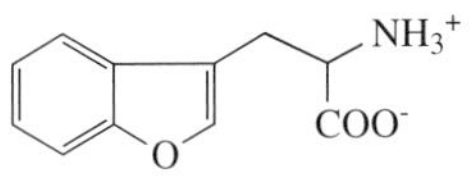

P5.8

The V_{max} is larger with electron-donating substituents than with electron-withdrawing substituents on the ring. Draw two mechanisms, one ionic and one radical, for this reaction, both using a Criegee rearrangement, consistent with these results.

P5-5. Compound **P5.5** was used to provide evidence for a radical intermediate in the conversion

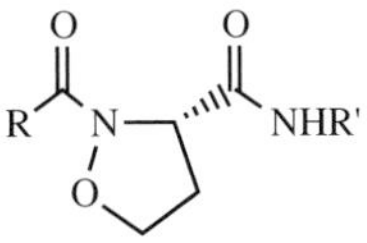

P5.5

of proline-containing peptides (**P5.6**) to 4-hydroxyproline peptides (**P5.7**), catalyzed by proline 4-hydroxylase, a nonheme iron enzyme that requires ascorbate, α-ketoglutarate, and oxygen for activity.

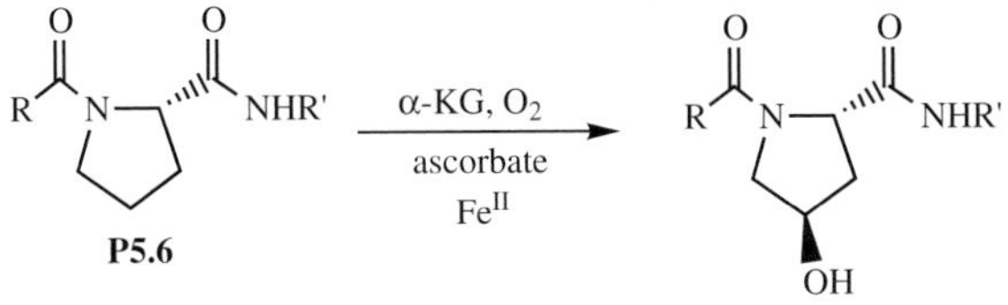

P5.6

P5.7

The product isolated from the enzyme reaction was **P5.8**. Draw a mechanism to show how a radical intermediate is supported by the reaction with **P5.5**.

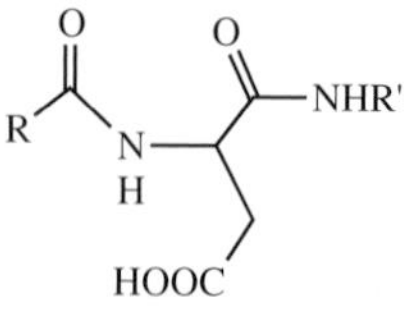

P5.8

P5-6. Mammalian and plant lipoxygenases catalyze the oxidation of unsaturated fatty acids containing a *cis,cis*-1,4-pentadiene moiety, such as linoleic acid (**P5.9**) to the corresponding 1-hydroperoxy-*trans,cis*-2,4-diene (**P5.10**).

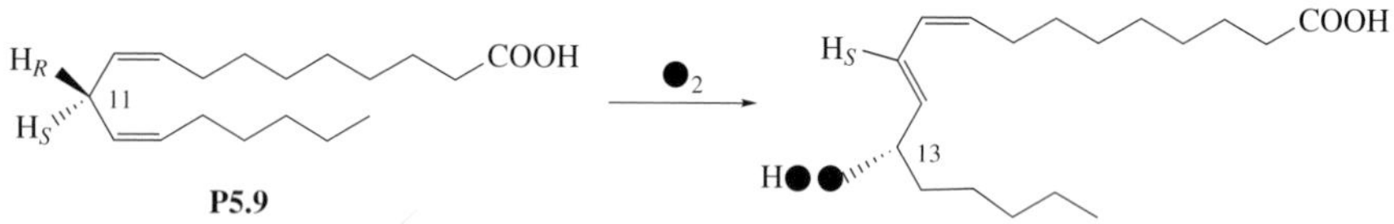

P5.9

P5.10

The substrate for plant lipoxygenase is linoleic acid (9*Z*, 12*Z*-octadecadienoic acid). In the case of the mammalian enzyme, arachidonate is the substrate, and the hydroperoxides are precursors to leukotrienes and lipoxins, which have been implicated as mediators of inflammation. The C-11 *pro-R* hydrogen is removed and the C-13 (*S*)-hydroperoxide is formed. Full enzyme activity occurs when the non-heme iron is in the high-spin Fe(III) state. There is a question as to the role of the Fe(III) in the mechanism. Draw two possible mechanisms, one involving a radical intermediate and the other an organoiron(III) complex. Propose experiments to differentiate these mechanisms.

P5-7. Gentisate 1,2-dioxygenase catalyzes the cleavage of the C-1−C-2 bond of 2,5-dihydroxybenzoate (gentisate, **P5.11**) to give maleylpyruvate (**P5.12**).

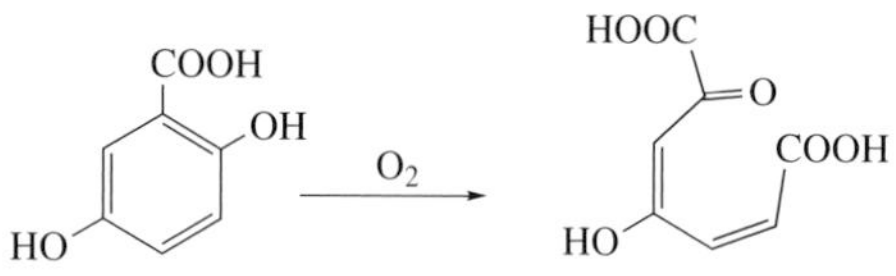

P5.11   **P5.12**

It requires nonheme Fe(II) for activity and incorporates both oxygen atoms of O_2 into the product. ^{17}O-Labeling indicates that the metal coordinates to the carboxylate and the 2-hydroxyl group, but not to the 5-hydroxyl group. Either a hydroxyl or an amino group must be at the C-5 position for the compound to be a substrate. Draw a mechanism using $^{18}O_2$.

F. Chapter 6

P6-1. Draw a mechanism for the prostaglandin endoperoxide reductase reaction, which reduces PGH_1 to $PGF_{1\alpha}$.

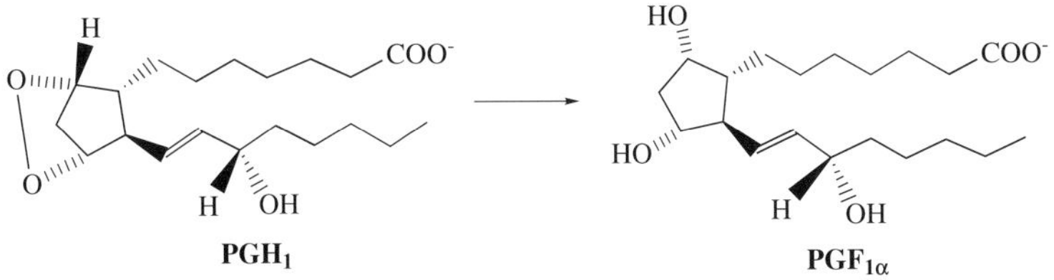

PGH₁ → **PGF₁ₐ**

P6-2. Glutathione S-transferase catalyzes the following reaction:

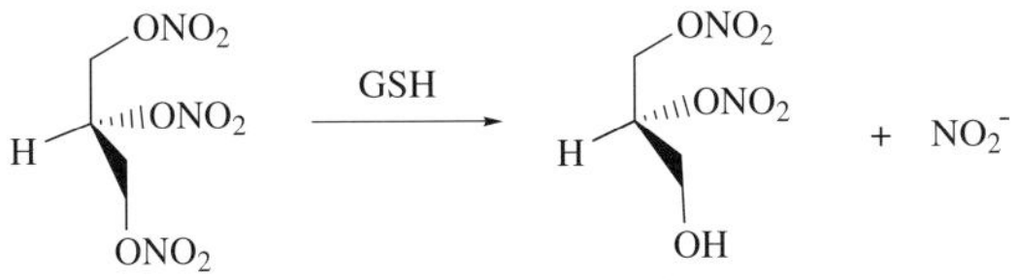

Draw a mechanism.

P6-3. The bromoketone **P6.1** inactivates yeast α-glucosidase and labels Glu-276.

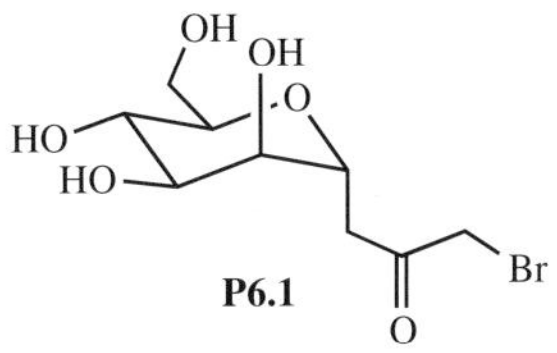

P6.1

a. Devise two analytical methods that could be used to identify this modified active-site residue.

b. How would you know that the modified residue was in the active site?

P6-4. The diterpene cyclase, taxadiene synthase, catalyzes the first step in the proposed biosynthesis of the anticancer agent taxol (**P6.2,** R = N-benzoyl-3-phenylisoserine). [1-$^2H_2$, 18-$^2H_3$]Geranylgeranyl diphosphate (**P6.3**) leads, in the first step, to taxa-4(5),11(12)–diene (**P6.4**). Draw a mechanism for the conversion of **P6.3** to **P6.4**.

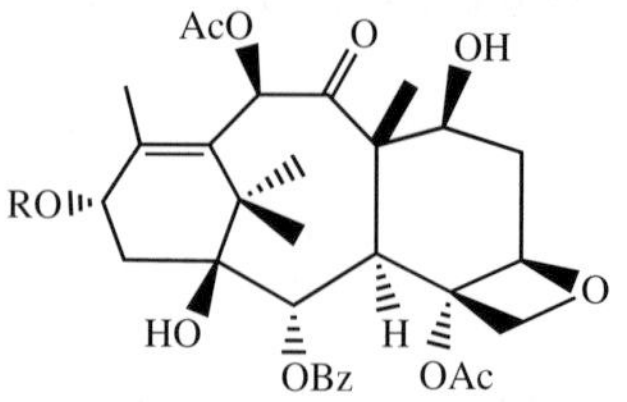

P6.2

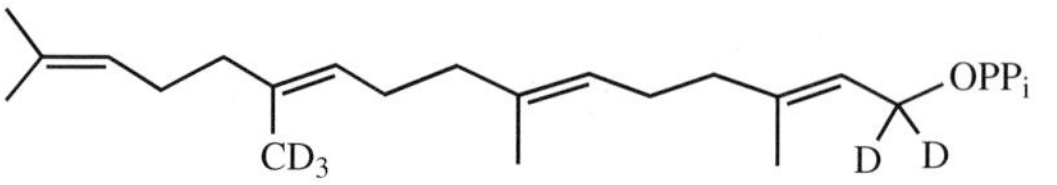

P6.3

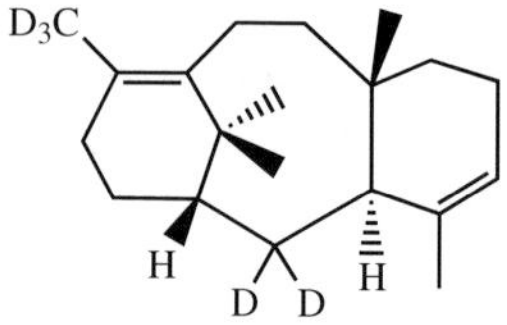

P6.4

P6-5. α-1,3-Fucosyltransferase V catalyzes the transfer of 1-fucose from guanosine 5′-diphospho-β-1-fucose (**P6.5**) to an acceptor sugar (**P6.6**), which is the terminal step in the biosynthesis of Lewis X (**P6.7a**) or sialyl Lewis X (**P6.7b**), a tetrasaccharide ligand involved in inflammatory cell adhesion and metastasis.

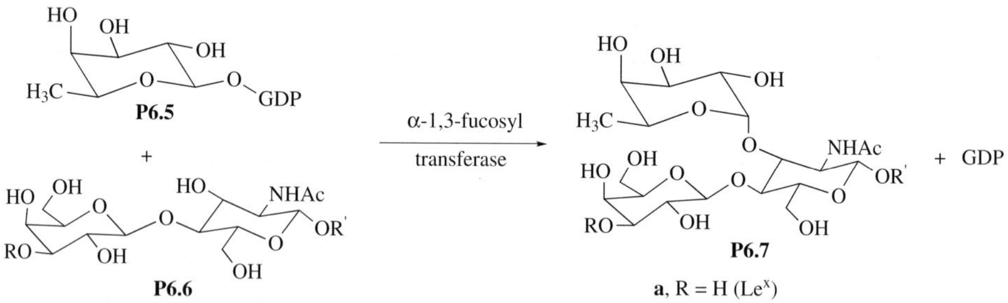

P6.5

+

P6.6

a, R = H

b, R = NeuAcα-2

α-1,3-fucosyl transferase

P6.7

a, R = H (Le^x)

b, R = NeuAcα-2 (sLe^x)

+ GDP

Inhibition of this enzyme may impede inflammation of cancer progression. The reaction occurs with inversion of configuration at the anomeric center of 1-fucose. There is a solvent isotope effect, and a secondary isotope effect is observed with guanosine 5′-diphospho-[1-^{2}H]β-1-fucose (**P6.8**). Furthermore, guanosine 5′-diphospho-2-deoxy-2-fluoro-β-1-fucose (**P6.9**) is a potent inhibitor; it is not an inactivator or slow substrate. Draw a mechanism consistent with these observations.

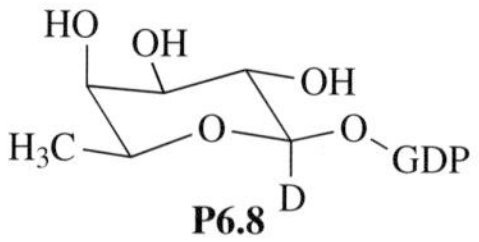

P6.8

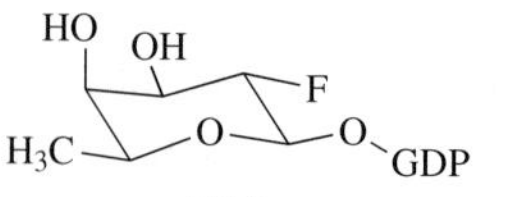

P6.9

G. Chapter 7

P7-1. You have just isolated a new carboxylase that acts on any molecule with the following structure:

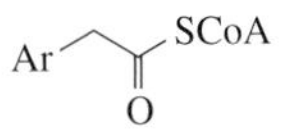

a. Describe experiments you would do to determine the mechanism of action (use substrate analogues).

b. Design two irreversible inactivators, one an affinity labeling agent, the other a mechanism-based inactivator, for this enzyme.

P7-2. A new enzyme was isolated that catalyzes the following reaction:

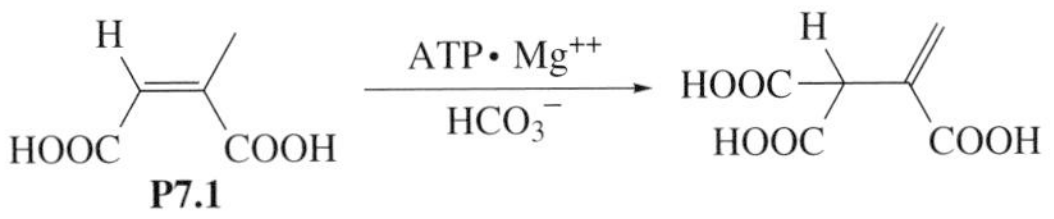

Addition of the carboxylate group occurs at the *re-si* face of **P7.1.** The enzyme does *not* bind to avidin.

a. If $HC^{18}O_3^-$ is used, two atoms of ^{18}O are incorporated into the product and one in P_i. What does this mean?

b. In the absence of substrate, but the presence of HCO_3^- and Mg^{2+}, $[^{14}C]ADP$ exchanges with ATP (i.e., $[^{14}C]ATP$ is formed). What does this mean?

c. Incubation of the enzyme with $H^{14}CO_3^- + ATP$ in the absence of substrate gives ^{14}C-labeled enzyme, which produces product when substrate is added. What does this mean?

d. Draw a mechanism, and show the stereochemistry of the reaction.

P7-3. The following enzyme-catalyzed carboxylation reaction is not inhibited by avidin. In D_2O the C-2 hydrogen exchanges. Draw a mechanism to account for both the proton exchange and carboxylation.

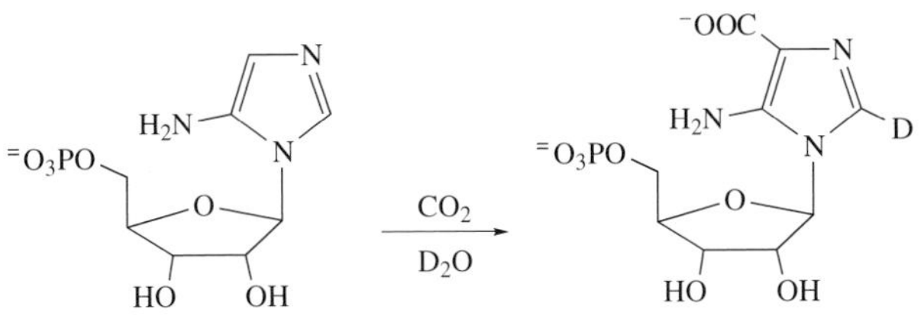

P7-4. Ribulose 1,5-diphosphate carboxylase (Rubisco) catalyzes the carboxylation of ribulose 1,5-diphosphate (**P7.2**) in the presence of CO_2 to give two moles of 3-phosphoglyceric acid (**P7.3**). In the absence of CO_2 and the presence of O_2, one mole of 3-phosphoglyceric acid and one mole of 2-phosphoglycolic acid (**P7.4**) are produced. The following labeling patterns are observed:

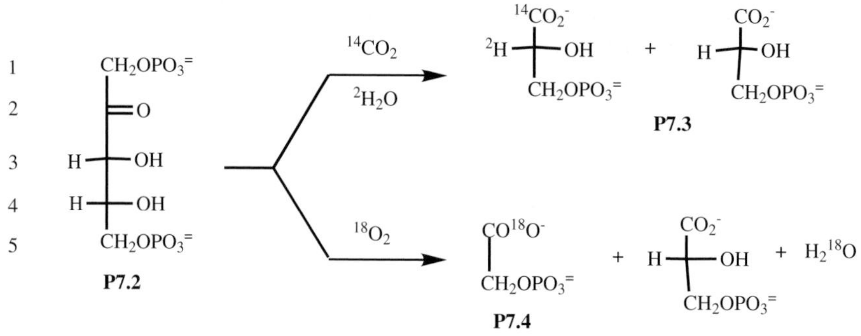

If 3-[^{2}H]ribulose 1,5-diphosphate is incubated with the enzyme with either CO_2 or O_2, the rate decreases by a factor of 3.5.

a. Draw reasonable mechanisms to account for these data that utilize a common intermediate to both pathways (give mechanisms for both the CO_2 and O_2 reactions).

b. Show the C–C bond cleavage step of the carboxylation mechanism as both a general base mechanism and as a covalent catalytic mechanism (presumably, one of these was already done in experiment a).

P7-5. An enzyme was discovered that carboxylates **P7.5**. It requires ATP, Mg^{2+}, and bicarbonate, and is inhibited by avidin. Bicarbonate labeled with three ^{18}O atoms produces **P7.6** and inorganic phosphate with one ^{18}O atom. Draw a possible mechanism for this reaction.

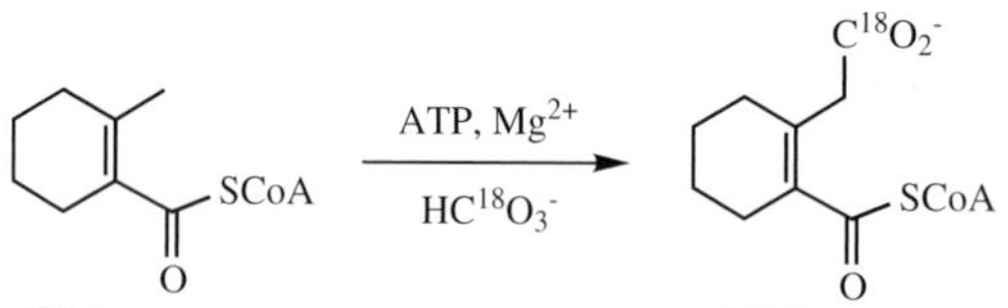

H. Chapter 8

P8-1. L-Aspartate can be decarboxylated to produce β-alanine (**P8.1**) (α-decarboxylation) or α-alanine (**P8.2**) (β-decarboxylation).

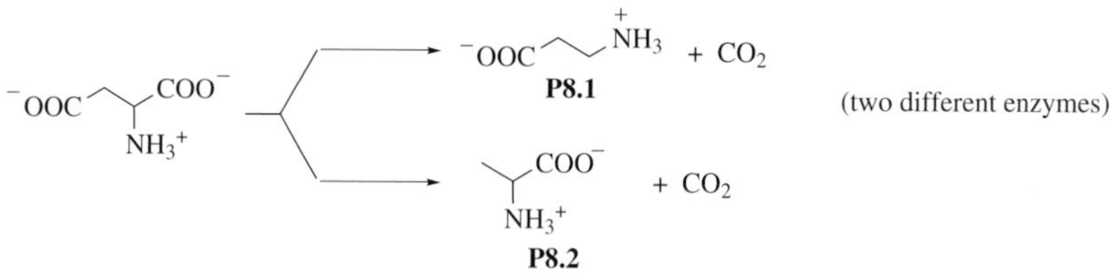

a. Draw mechanisms for each reaction.

b. Explain how an enzyme can control which of these two reactions will occur.

c. Describe two experiments to test each mechanism.

P8-2. Draw mechanisms for the following reactions, and design a useful experiment to test each mechanism.

a.
$$
\begin{array}{c}
\text{COOH} \\
\parallel \text{O} \\
\text{OH} \\
\text{OH} \\
\text{CH}_2\text{OPO}_3^=
\end{array}
\longrightarrow
\begin{array}{c}
\text{CHO} \\
\text{OH} \\
\text{OH} \\
\text{CH}_2\text{OPO}_3^=
\end{array}
$$

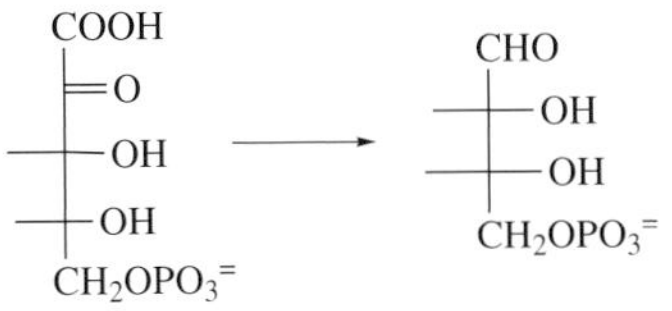

b.
$$
\begin{array}{c}
\text{COO}^- \\
\text{OH} \\
\text{HO} \\
\text{OH} \\
\text{OH} \\
\text{CH}_2\text{OPO}_3^{2-}
\end{array}
\longrightarrow
\begin{array}{c}
\text{CH}_2\text{OH} \\
\parallel \text{O} \\
\text{OH} \\
\text{OH} \\
\text{CH}_2\text{OPO}_3^{2-}
\end{array}
$$

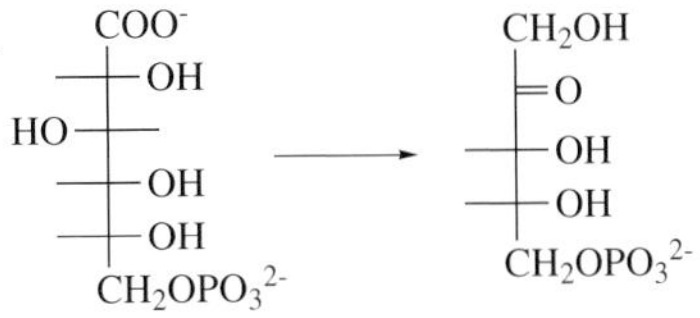

P8-3. *Escherichia coli* has an enzyme that converts pyruvate into acetate in which the added oxygen atom of acetate that replaces the carboxylate group of pyruvate derives from water, and no lipoic acid is involved. Draw a mechanism for this enzyme-catalyzed reaction.

P8-4. Two enzymes are required to catalyze the following sequence of reactions:

$$
\begin{array}{c}
\text{O} \\
\parallel \\
2 \ \text{HCCOOH}
\end{array}
\longrightarrow \longrightarrow \quad \text{HOCH}_2\text{CHO} + 2\text{CO}_2
$$

Draw a mechanism for each reaction, and describe an experiment to test each of the two enzyme mechanisms.

P8-5. Acetolactate synthase catalyzes the conversion of two molecules of pyruvate to carbon dioxide and acetolactate (**P8.1**), a precursor to valine and leucine in plants.

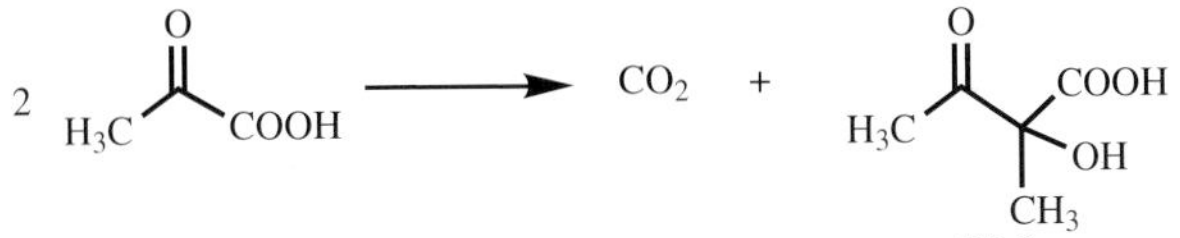

P8.1

a. Draw a mechanism for this reaction.

b. This enzyme exhibits a side oxygenase reaction similar to that observed with Rubisco (Problem 7-4). Draw a mechanism for this reaction.

I. Chapter 9

P9-1. If L-alanine is incubated with alanine aminotransferase in the presence of α-ketoglutarate, pyruvate and L-glutamate are produced.

If the α-ketoglutarate is omitted, and the incubation is done in D_2O, perdeutero L-alanine (L-alanine with all of its hydrogens replaced by deuterium) is isolated.

a. Draw mechanisms for both the substrate reaction with α-ketoglutarate and without α-ketoglutarate in D_2O.

b. Draw a mechanism for the inactivation of this alanine aminotransferase by **P9.1**.

$$HC\equiv CCH_2CHCOO^-$$
$$|$$
$$NH_3^+$$

P9.1

P9-2. Triosephosphate isomerase catalyzes the interconversion of dihydroxy-acetone phosphate (**P9.2**) and glyceraldehyde 3-phosphate (**P9.3**), an exceedingly fast and efficient reaction in which the slow step is product release. A small amount of solvent proton incorporation is observed.

a. Draw a mechanism for this reaction, and devise an experiment to test it.

b. Design an affinity labeling agent for this enzyme.

P9-3. γ-Aminobutyric acid aminotransferase catalyzes the conversion of γ-aminobutyric acid (GABA; **P9.4**) and α-ketoglutarate to succinic semialdehyde (**P9.5**) and L-glutamate.

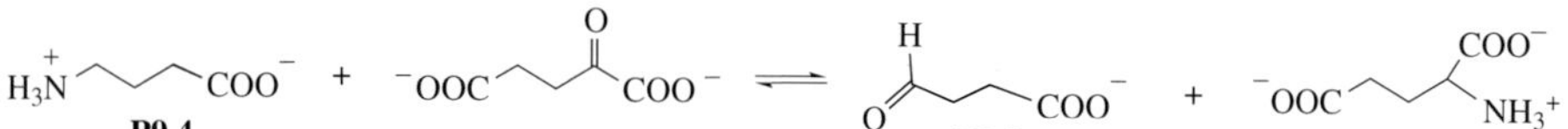

a. Draw a mechanism for this reaction and for recycling of the cofactor.

b. Draw a mechanism for its inactivation by 4-amino-5-hexenoic acid (vinyl-GABA, vigabatrin; **P9.6**) that gives PMP on hydrolysis.

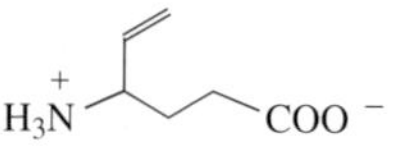

P9.6

P9-4. An enzyme that catalyzes peptide epimerization was discussed in this chapter, and some evidence was presented for a two-base mechanism. What other experiments could be used to determine if a two-base mechanism was applicable?

P9-5. a. Draw a carbanion and a carbocation mechanism for vinylacetyl-CoA isomerase.

b. Describe two experiments to support your mechanism, one of which to differentiate a carbanion from a carbocation mechanism.

P9-6. Draw a mechanism for the following reaction:

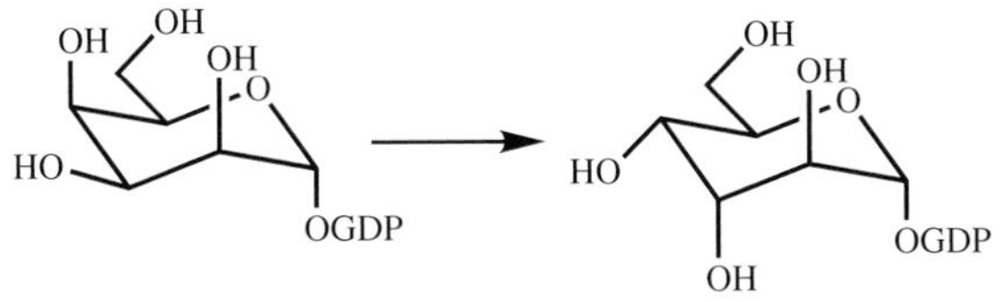

J. Chapter 10

P10-1. a. Draw a mechanism for the following enzymatic reaction:

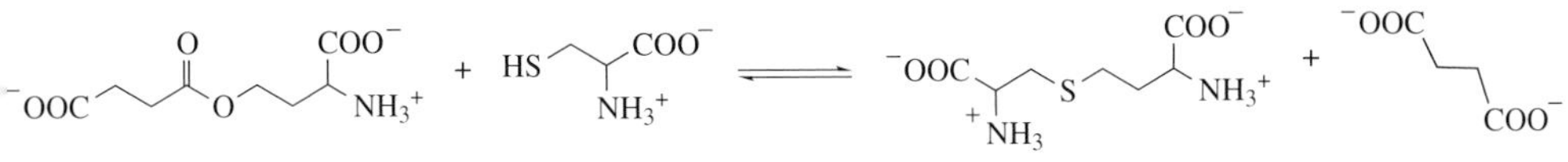

b. Describe an experiment to test the mechanism.

c. Draw a mechanism for the reaction in the absence of cysteine.

d. What would happen if the reaction in part c were carried out in D_2O?

P10-2. An enzyme catalyzes the following reaction:

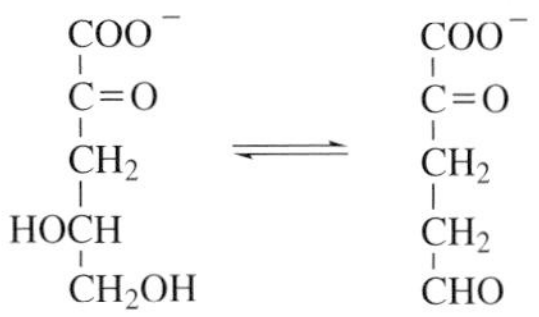

Incubation of the enzyme with substrate and $[^3H]NaBH_4$, followed by gel filtration gives a peak of radioactivity corresponding to the protein peak.

a. Draw a mechanism consistent with this result.

b. Describe two other experiments to test your mechanism.

c. The rate of exchange of the C-3 hydrogens with solvent is twice the rate of loss of the C-4 OH, which is faster than the rate of formation of product. What does this tell you about the mechanism?

P10-3. Rationalize the following stereochemical outcome with a mechanism.

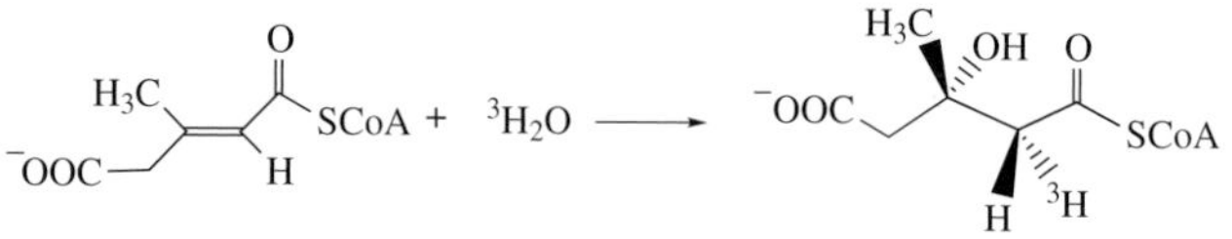

P10-4. An enzyme was isolated that catalyzes the following reaction:

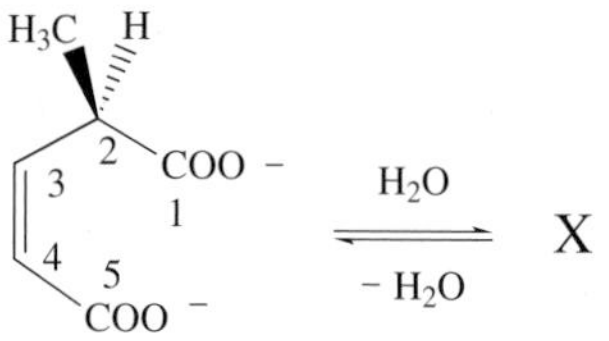

In the foward direction, the (2R, 3S)-isomer of X is formed. In the back reaction the pro-4S hydrogen is removed from X. Deuteration of the pro-4S hydrogen of X has no effect on the rate of dehydration.

a. In the forward direction, which face of the alkene gets hydroxylated?
b. Is this a syn or anti addition (elimination) of H_2O?
c. Draw a mechanism.

P10-5. The PLP-dependent enzyme tryptophanase was shown to be irreversibly inactivated by **P10.1**. Denaturation of the radiolabeled enzyme at low pH produced radioactive **P10.2**.

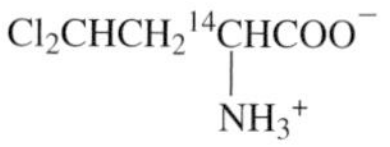

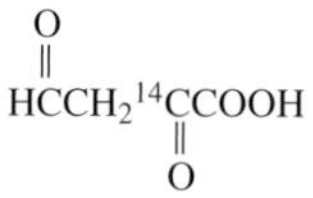

a. Draw a mechanism to account for the mechanism-based inactivation.
b. Design two experiments to test your mechanism.
c. How is **P10.2** produced on denaturation?

P10-6. Anthranilate synthase catalyzes the following reaction:

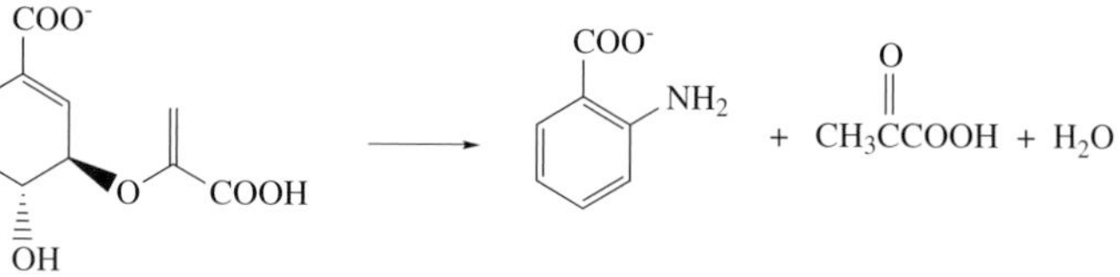

a. Draw a mechanism for this reaction.
b. Describe two experiments to test your mechanism.

P10-7. In the chiral methyl determination methodology, acetyl-CoA condenses with glyoxylate (**P10.3**) in the presence of malate synthase to give (2S)-malate.

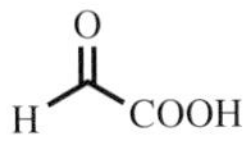

P10.3

Fumarase catalyzes an *anti* elimination of (2S)-malate with removal of the (3R)-hydrogen. If (2S)acetyl CoA is incubated with glyoxylate, malate synthase, and fumarase, 80% of the tritium is released. What is the stereochemistry of the malate synthase reaction (draw it)?

P10-8. If the mechanism for CDP-D-glucose 4,6-dehydratase shown in Scheme 10.9 in the text is correct, and the observed result that (6S)-CDP-glucose (**10.15** in Scheme 10.10) gives (R)-acetate after the enzyme product (**10.16** in Scheme 10.10) is converted to acetate by a Kuhn–Roth oxidation, does a *syn* or *anti* elimination of water occur across the C^5–C^6 bond? Show the stereochemistry.

P10-9. Phenylalanine ammonia-lyase, which catalyzes the degradation of L-phenylalanine to *trans*-cinnamic acid (**P10.4**)—a precursor of lignins, flavonoids, and coumarins in plants—has a dehydroalanyl group at the active site. When the phenyl ring is deuterated, there is a normal secondary kinetic isotope effect, as is observed for nonenzymatic electrophilic aromatic substitution reactions. Draw a reasonable mechanism, and describe some experiments you might carry out to support the mechanism.

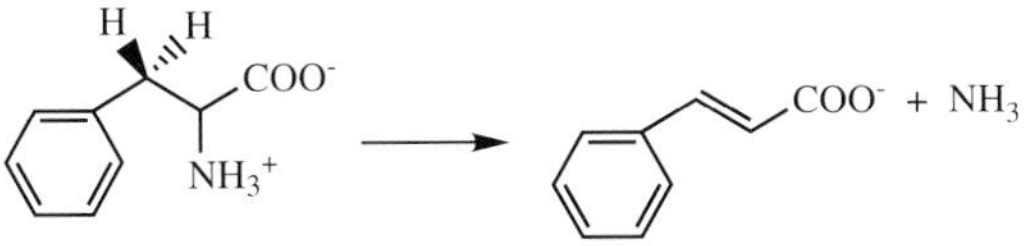

P10.4

P10-10. Porphobilinogen deaminase catalyzes the condensation of 4 units of porphobilinogen (**P10.5**), with the loss of 4 moles of ammonia, to give hydroxy-methylbilane (**P10.6**), the substrate for uroporphyrinogen-III synthase, which catalyzes the cyclization of the tetrapyrrole.

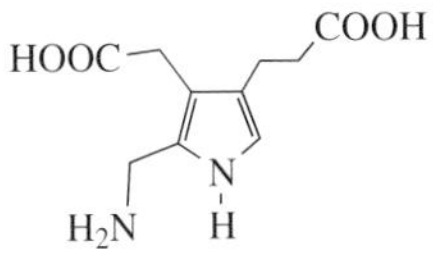

P10.5 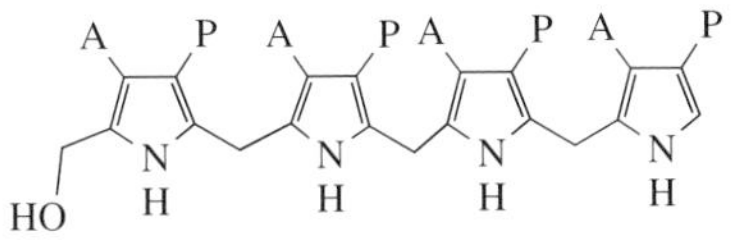   **P10.6**

A = carboxymethyl   P = carboxyethyl

a. Draw three mechanisms for the elimination of ammonia.

b. Propose analogues of **P10.5** that could be tried to test the mechanisms.

K. Chapter 11

P11-1. An enzyme catalyzes the following reaction:

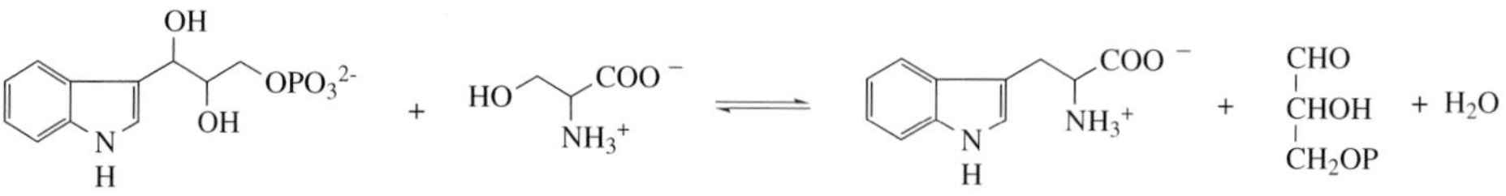

$$\text{indole-OPO}_3{}^{2-}\text{,OH} + \text{HO-CH}_2\text{-CH(NH}_3{}^+)\text{-COO}^- \rightleftharpoons \text{indole-CH}_2\text{-CH(NH}_3{}^+)\text{-COO}^- + \text{CHO-CHOH-CH}_2\text{OP} + H_2O$$

Other reactions catalyzed by this enzyme include the following:

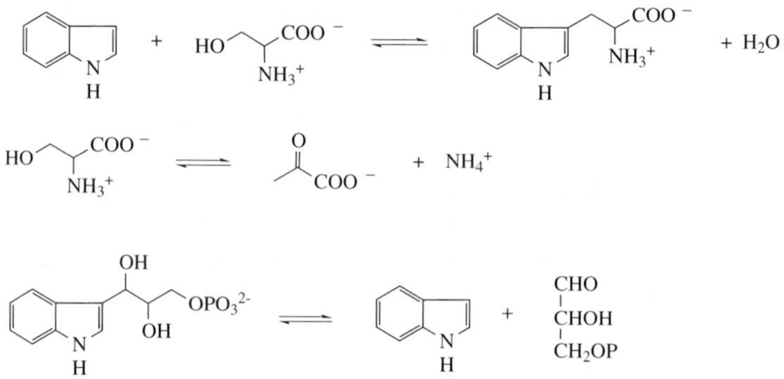

$$\text{indole} + \text{HO-CH}_2\text{-CH(NH}_3{}^+)\text{-COO}^- \rightleftharpoons \text{indole-CH}_2\text{-CH(NH}_3{}^+)\text{-COO}^- + H_2O$$

$$\text{HO-CH}_2\text{-CH(NH}_3{}^+)\text{-COO}^- \rightleftharpoons \text{CH}_3\text{-CO-COO}^- + NH_4{}^+$$

$$\text{indole-CH(OH)-CH(OH)-OPO}_3{}^{2-} \rightleftharpoons \text{indole} + \text{CHO-CHOH-CH}_2\text{OP}$$

a. Write a mechanism compatible with these results.

b. Would you expect the enzyme to be inactivated by β-fluoroalanine? Why?

P11-2. An enzyme catalyzes the conversion of **P11.1** to pyruvate and glyceraldehyde 3-phosphate (**P11.2**). No metal ions are required.

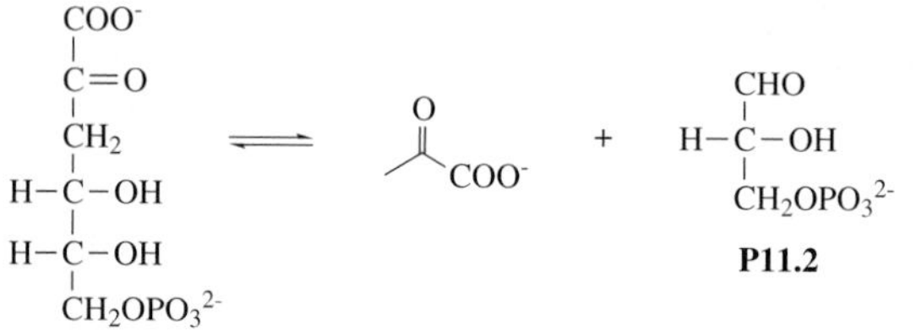

$$
\begin{array}{l}
\text{COO}^- \\
\text{C}=\text{O} \\
\text{CH}_2 \\
\text{H}-\text{C}-\text{OH} \\
\text{H}-\text{C}-\text{OH} \\
\text{CH}_2\text{OPO}_3{}^{2-}
\end{array}
\rightleftharpoons
\quad \text{CH}_3\text{-CO-COO}^- \quad + \quad
\begin{array}{l}
\text{CHO} \\
\text{H}-\text{C}-\text{OH} \\
\text{CH}_2\text{OPO}_3{}^{2-}
\end{array}
$$

P11.1   **P11.2**

a. Draw a mechanism.

b. Bromopyruvate inactivates the enzyme. If the enzyme is treated with [¹⁴C]bromopyruvate, one mole of ¹⁴C/mol enzyme is detected after dialysis. Treatment of the labeled enzyme with hydroxylamine releases the label, but the enzyme remains inactive. Draw a mechanism for this. What is the structure of the labeled product?

P11-3. An enzyme catalyzes the formation of **P11.3** from pyruvate and glyceraldehyde 3-phosphate. No metal ions are required.

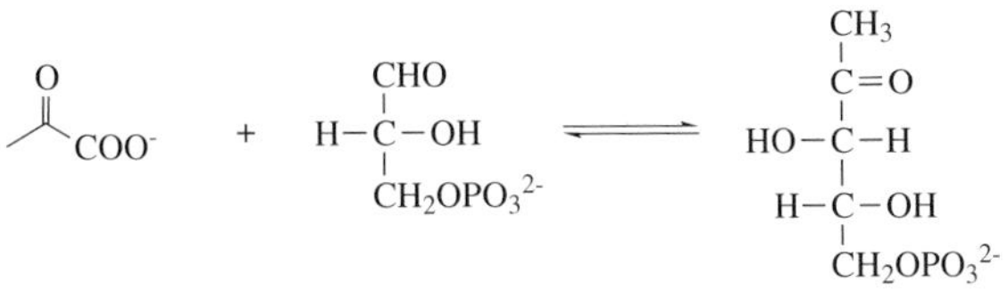

P11.3

a. Draw a mechanism for this hypothetical enzyme.

b. [^{14}C]Bromopyruvate inactivates this enzyme, but if you dialyze the inactive enzyme and assay again under normal conditions with any needed cofactors added, the enzyme is found to contain no radioactive label and to be fully active. Explain.

P11-4. Draw a mechanism for the reaction shown in Scheme 11.18 in the text.

P11-5. Draw a mechanism for the conversion of kynurenine (**11.48** in Scheme 11.24 in the text) and benzaldehyde to give **11.50** (in the text).

P11-6. Tetracenomycin F2 cyclase, the first discrete enzyme for C−C bond formation via an intramolecular aldol condensation-dehydration, catalyzes a key step in the biosynthesis of all aromatic polyketides. It also is an enzyme in the biosynthetic pathway of the anthracycline antibiotic tetracenomycin C (**P11.6**) from **P11.4**. At pH ≥ 8.0, **P11.5** (R = COOH) is the aldol product, but at pH ≤ 6.5 decarboxylation occurs to give a different aldol product (**P11.5**, R = H). Draw a mechanism that accounts for the conversion of **P11.4** to **P11.5**, R = COOH and H.

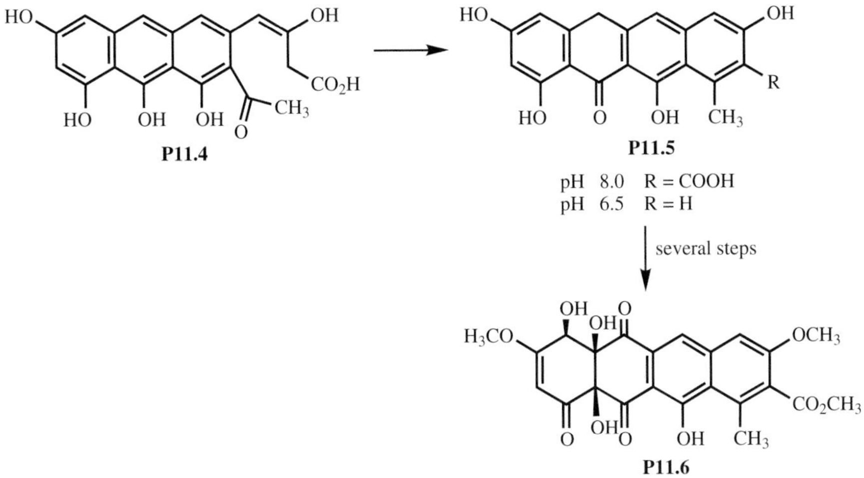

P11.4

P11.5

| pH | 8.0 | R = COOH |
| pH | 6.5 | R = H |

several steps

P11.6

P11-7. Transaldolase catalyzes the conversion of the seven-carbon ketose, se-doheptulose 7-phosphate (**P11.7**), and the three-carbon aldose, glyceraldehyde 3-phosphate (**P11.8**), to the six-carbon ketose, fructose 6-phosphate (**P11.9**), and the four-carbon aldose, erythrose 4-phosphate (**P11.10**). If the C-2 carbonyl oxygen of **P11.7** is labeled with ^{18}O, the ^{18}O is released as $H_2{}^{18}O$. Incubation of the enzyme in the presence of **P11.7** with $NaBH_4$ causes irreversible inactivation. Draw a mechanism consistent with these results.

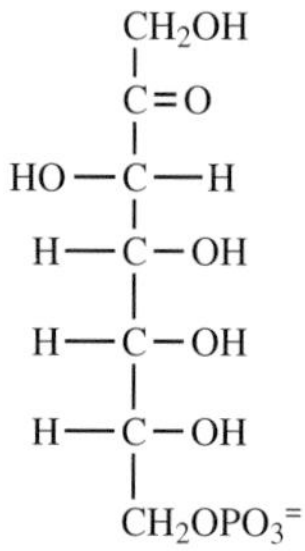

$$CH_2OH$$
$$C=O$$
$$HO-C-H$$
$$H-C-OH$$
$$H-C-OH$$
$$H-C-OH$$
$$CH_2OPO_3^=$$
P11.7

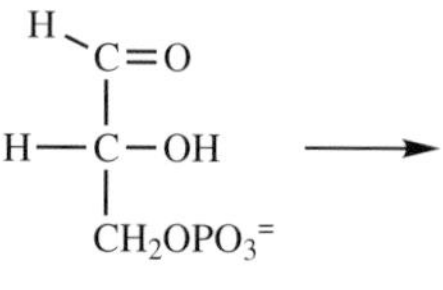

$$H\,C=O$$
$$H-C-OH$$
$$CH_2OPO_3^=$$
P11.8

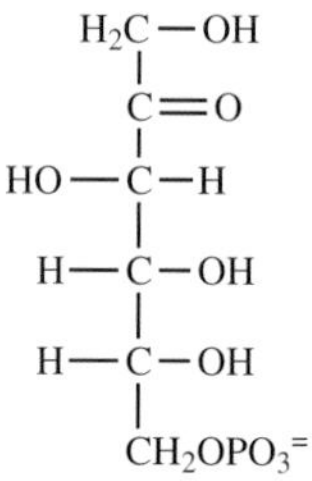

$$H_2C-OH$$
$$C=O$$
$$HO-C-H$$
$$H-C-OH$$
$$H-C-OH$$
$$CH_2OPO_3^=$$
P11.9

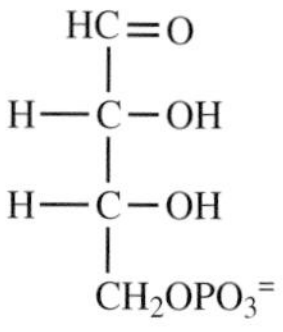

$$HC=O$$
$$H-C-OH$$
$$H-C-OH$$
$$CH_2OPO_3^=$$
P11.10

L. Chapter 12

P12-1. A number of polyamines can be found in semen, which may contribute to its fishy odor. One of these amines, spermidine, is biosynthesized from methionine and ornithine. If the methionine is labeled as follows, the ^{14}C appears in the spermidine as shown:

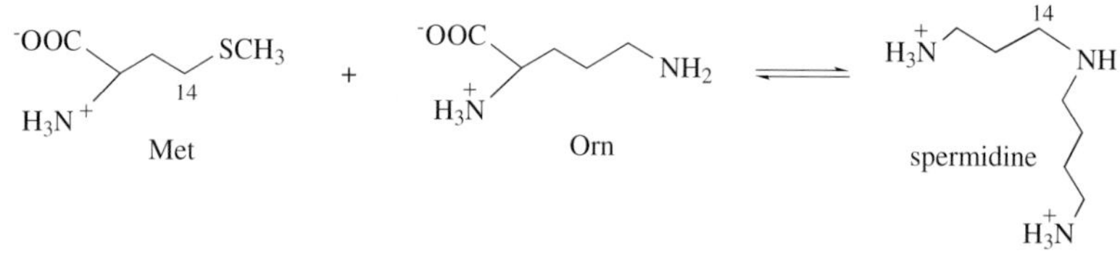

Met   +   Orn   ⇌   spermidine

Write a mechanism to show how spermidine is biosynthesized.

P12-2. The carbon with the asterisk in **P12.1** is derived from serine. Show how serine gives this product.

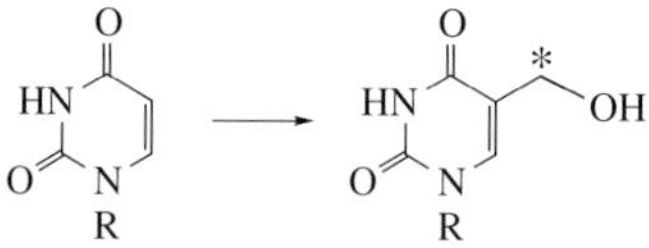

P12.1

P12-3. Draw a mechanism for the formation of N^{10}-formyltetrahydrofolate, catalyzed by N^{10}-formyltetrahydrofolate synthetase (see Scheme 12.6 in the text).

P12-4. How would you make [6-^{3}H]methylenetetrahydrofolate to carry out a study to determine the fate of the tritrium at that position?

P12-5. (3R)-[3-^{3}H]Serine is converted into (11R)-[11-^{3}H]methylenetetrahydrofolate. Draw a mechanism in which tetrahydrofolate reacts at the *re*-face of the generated formaldehyde.

M. Chapter 13

P13-1. a. Draw a mechanism for the following reaction:

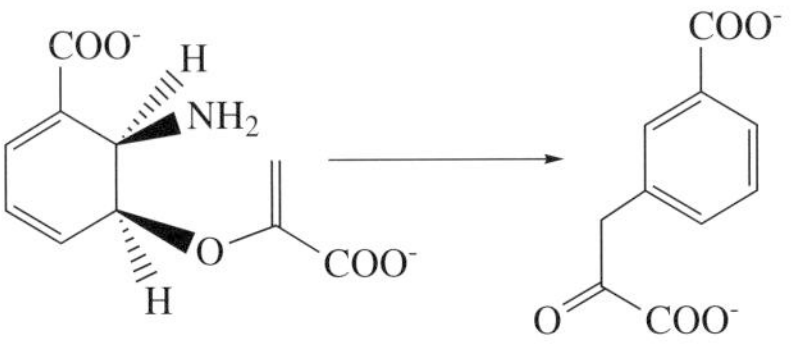

b. Describe two experiments to test the mechanism.

P13-2. The tuberculosis-causing pathogen *Mycobacterium tuberculosis* catalyzes the conversion of oleate (**P13.1**) to 10-methylstearate (**P13.2**) in the presence of *S*-adenosylmethionine (SAM) and a reducing agent.

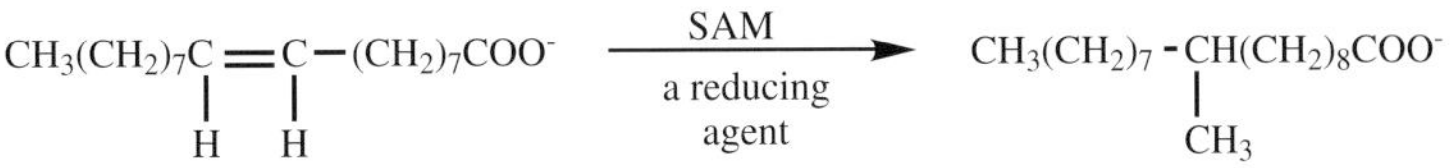

$$CH_3(CH_2)_7C \!=\! C-(CH_2)_7COO^- \xrightarrow[\text{a reducing agent}]{\text{SAM}} CH_3(CH_2)_7-CH(CH_2)_8COO^-$$

with H, H on the double bond carbons (P13.1) and CH$_3$ substituent (P13.2).

P13.1   **P13.2**

When CD$_3$-SAM is used, 10-[CHD$_2$]methyl stearate is produced.

$$CH_3(CH_2)_7CH\!=\!CH(CH_2)_7COO^- + CD_3SAM \longrightarrow CH_3(CH_2)_7-CH(CH_2)_8COO^-$$

with CHD$_2$ substituent.

9,10-Dideuteriostearate gives C$_9$-[^{2}H$_2$]methyl stearate.

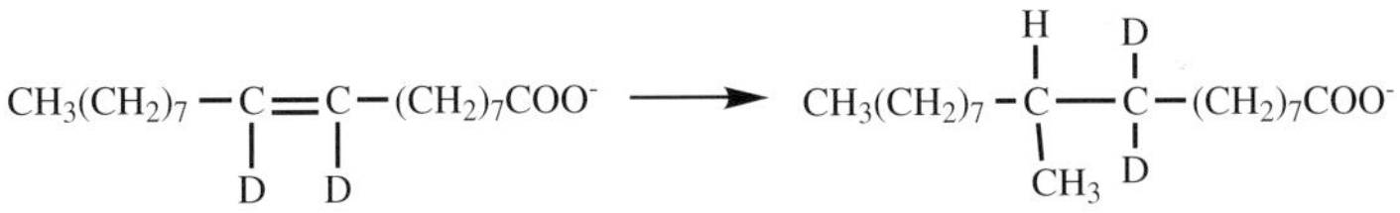

$$CH_3(CH_2)_7-C\!=\!C-(CH_2)_7COO^- \longrightarrow CH_3(CH_2)_7-\overset{H}{\underset{CH_3}{C}}-\overset{D}{\underset{D}{C}}-(CH_2)_7COO^-$$

with D, D on the double bond carbons.

Draw a reasonable mechanism for this reaction.

P13-3. Figure 13.4 in the text shows several potential mechanisms for the rearrangement of methylmalonyl-CoA. Draw mechanisms a, e, and f for the 2-methyleneglutarate mutase-catalyzed reaction.

P13-4. Trichodiene synthase catalyzes the conversion of farnesyl diphosphate (**P13.3**) to trichodiene (**P13.5**), the parent hydrocarbon of the trichothecane antibiotics and mycotoxins. The first step is the rearrangement of the diphosphate group from C–1 to C–3 to give **P13.4.** The hydrogen at C-6 (*H) ends up in trichodiene as follows. Draw a reasonable mechanism.

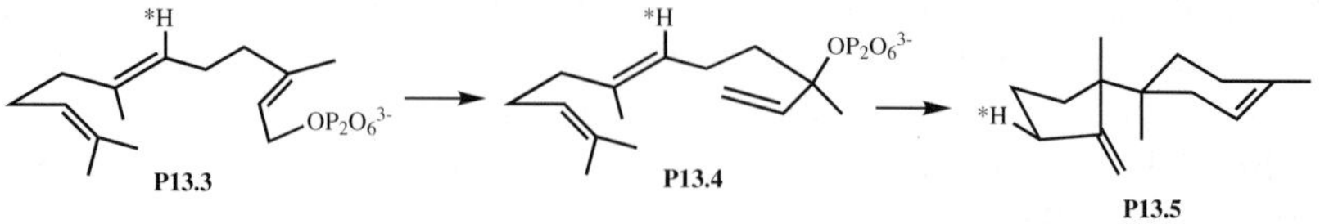

P13.3   P13.4

P13.5

P13-5. Draw a mechanism for the inactivation of ribonucleotide reductase by the mechanism-based inactivator (E)-2′-(fluoromethylene)-2′-deoxycytidine triphosphate (**P13.6**), given that 1 equiv of radioactive inactivator becomes covalently attached to an active-site residue and 1 equiv of F^- and 1 equiv of cytosine are released.

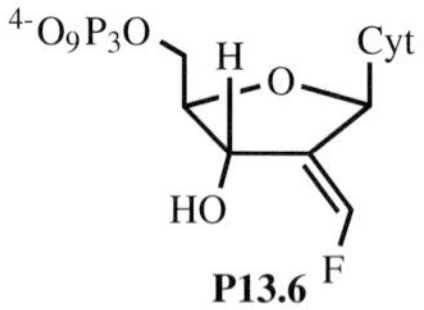

P13.6

P13-6. The following model study reactions for the coenzyme B_{12}-dependent methylmalonyl-CoA mutase rearrangement were carried out. What do you conclude from these results?

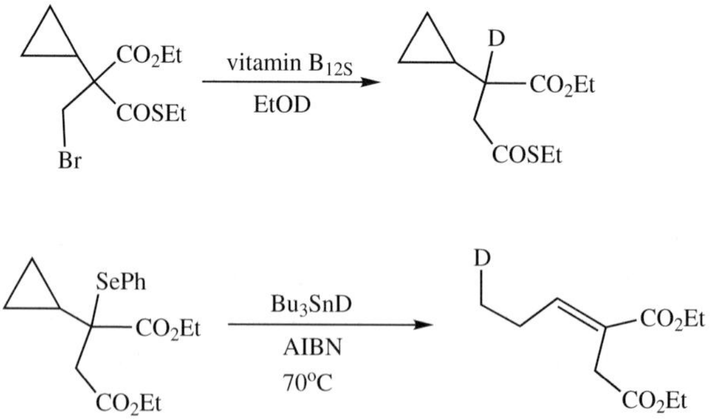

N. Biosynthetic Pathways

Congratulations! You now should be able to determine probable mechanisms for enzymes you have not seen before. Open any biochemistry book to any page that has metabolic pathways, and draw mechanisms for every enzyme in the pathway.

To make it more challenging, jot down the first substrate in the pathway and a substrate five or six enzymes downstream of that enzyme in the metabolic pathway
(without looking at the intermediates). Then see if you can devise an appropriate
(reasonable) metabolic pathway to get from the first substrate to the later one.
Think of this as a synthesis problem, and *work backward* from the end. Starting from
the last product, ask yourself, From what substrate could this have come? The following few problems give you some practice doing this.

   PBP-1. Draw all of the intermediates and coenzymes in the following biosynthetic pathways. Work backward.

a.

b.

c.

The third enzyme in part c is inactivated by $NaBH_4$ in the presence of substrate. If
fumarate is added to the enzyme in $H_2^{18}O$, there is an enzyme-catalyzed exchange
reaction:

1. Write a mechanism for the third enzyme consistent with these results.
2. Describe two other experiments to test the mechanism.

   PBP-2. Draw a mechanism for insertion of the methylene group of glycine into
tetrahydrofolate; in the mechanism, use PLP, reduced lipoic acid, and NAD^+.

   PBP-3. *N*-Formylglycinamidine ribonucleotide (**BP-1**), an intermediate in
purine biosynthesis, is derived from glycine, 5-phosphoribosyl-1-amine (**BP-2**),

and serine in six enzyme reactions. If the glycine is labeled in the methylene group, the *N*-formylglycinamidine ribonucleotide obtained has the label as shown in **BP-2.** If $^{15}NH_4Cl$ is used in any of these enzyme reaction, no ^{15}N is incorporated into the product. It has also been shown that aspartic acid is not a substrate for any of the enzymes involved.

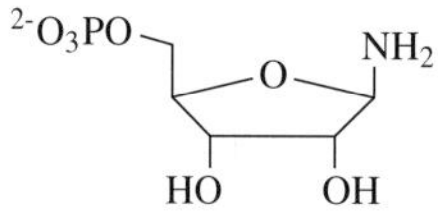

BP-1

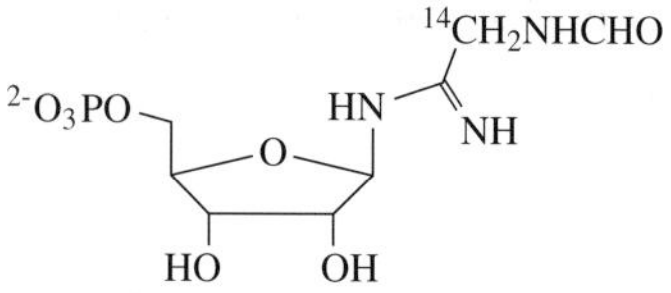

BP-2

Draw mechanisms for these reactions that are compatible with the results.

II. SOLUTIONS TO PROBLEMS

A. Chapter 1

S1-1. a.

	Cationic	*Anionic*
pH 4	Arg (R), Lys (K), His (H)	none
pH 7	Arg, Lys, His (partially)	Asp (D), Glu (E)
pH 10	Arg	Asp, Glu, Cys (C), Tyr (Y) (partially)

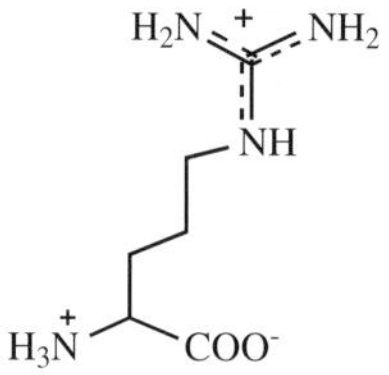

Arg 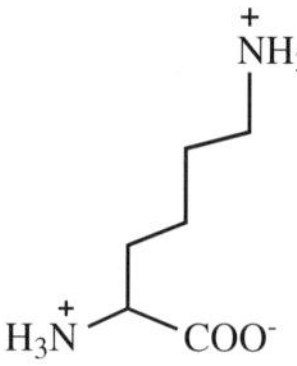   **Lys** 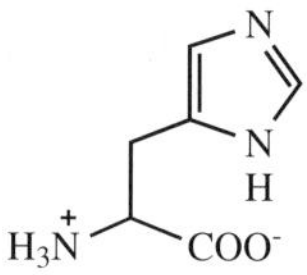   **His**

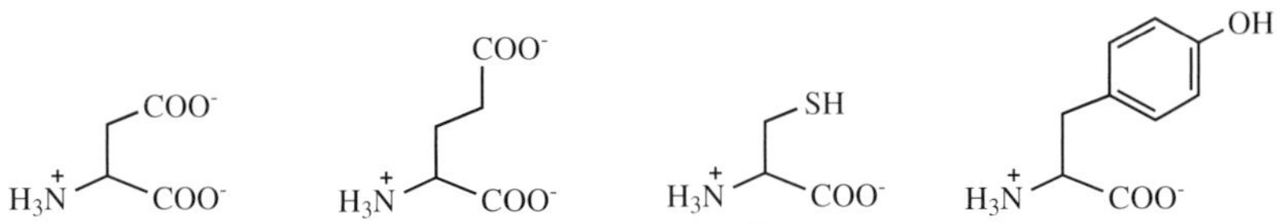

Asp **Glu** **Cys** **Tyr**

b. The nucleophilic amino acids include all of the ones shown above plus the following:

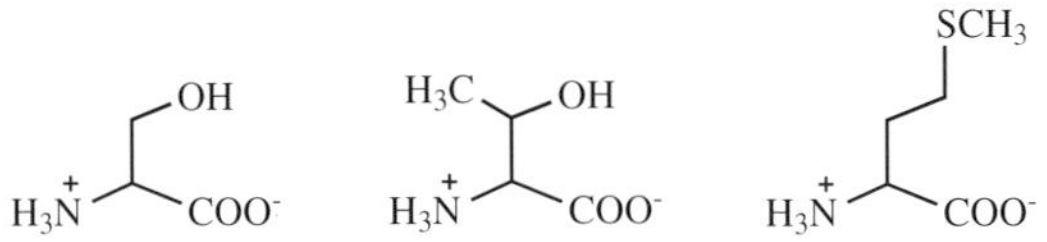

Ser **Thr** **Met**

S1-2. The nonbonded orbitals of the active-site base could be lined up with the proton of the serine residue at the active site, and the attacking orbital of the serine oxygen could be aligned parallel with the amide carbonyl π-orbital for maximum overlap.

S1-3. If diffusion controlled, $k_{cat}/K_m = 10^9 \, M^{-1} \, s^{-1}$; $k_{cat} = K_m \, (10^9 \, M^{-1} \, s^{-1})$. In this case, $k_{cat} = 1.5 \times 10^{-4} \, M \, (10^9 \, M^{-1} \, s^{-1}) = 1.5 \times 10^5 \, s^{-1}$.

S1-4.

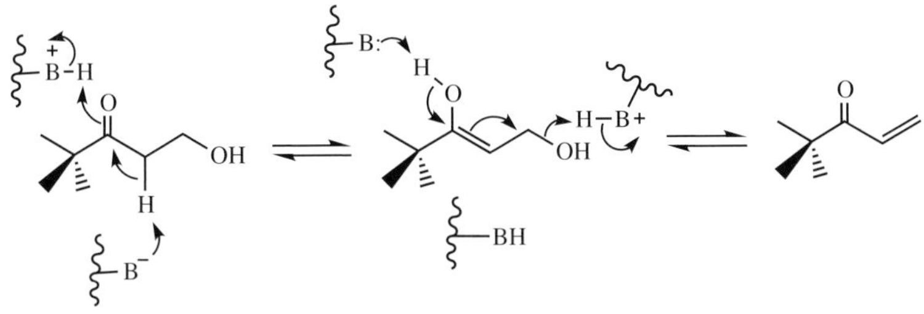

Simultaneous protonation of the ketone carbonyl and deprotonation of the α-proton gives the enol, which can be coupled with protonation of the β-hydroxyl group to effect β-elimination. Because the pK_a values of residues inside an enzyme can be quite different from those in solution, these protonations and deprotonations may be more facile than expected.

S1.5.

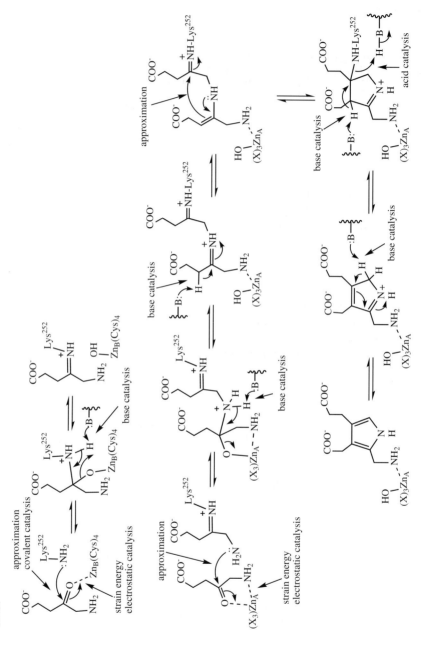

S1.6.

S1.7.
a. hydrophobic interaction
b. ion–dipole interaction
c. hydrogen bonding/ion–dipole
d. electrostatic (ionic) interaction
e. hydrogen bonding
f. hydrogen bonding
g. charge transfer interaction
h. ion–dipole/hydrogen bonding
i. dipole–dipole interaction
j. ion–dipole
k. electrostatic (ionic) interaction
S1.8.

a.

b.

S1.9.

S1.10. a. Having two cationic groups adjacent lowers the pK_a of the amine so it has more tendency to be neutral.

b. The carboxylate stabilizes the positive charge of the amine, leading to a maintenance or increase in pK_a.

c. To avoid the two adjacent negative charges, the pK_a of the carboxylate increases to remain protonated.

S1.11.

B. Chapter 2

S2-1. An affinity labeling agent forms an initial noncovalent E·I complex in a rapidly established equilibrium. The rate-determining step is the covalent reaction with the active-site nucleophile from the E·I complex.

S2-2. a. This experiment shows that the radioactivity is associated with the protein (the large peak of radioactivity is the excess substrate). Therefore, R is attached to enzyme.

b. This is the same experiment as part a and shows that the ester CH_3 is not attached to the enzyme.

c. This experiment shows that incubation at the higher pH results in no R attached to the protein. Therefore the protein−substrate bond is very base labile (pH 9 is not very basic). Structure **S2.1** is a reasonable intermediate (covalent catalysis) in which the X-carbonyl bond is labile to base.

$$\text{Enz}-\text{X}-\overset{\overset{\textstyle O}{\|}}{\text{C}}-\text{R}$$

S2.1

d. The next step in the mechanism is deacylation by NH_2OH (pathway a).

$$\text{Enz}-\text{X}-\overset{\overset{\textstyle O}{\|}}{\text{C}}\text{R} \xrightarrow[a]{NH_2OH} \text{Enz}-\text{X}^- \ + \ \text{HO}-\overset{\text{H}}{\underset{}{\text{N}}}-\overset{\overset{\textstyle O}{\|}}{\text{C}}-\text{R}$$

$$\xrightarrow[b]{H_2O} \text{Enz}-\text{X}^- \ + \ \text{HO}-\overset{\overset{\textstyle O}{\|}}{\text{C}}-\text{R}$$

This accounts for the 95% of the product. It would be reasonable to deduce that the other 5% of the product was the corresponding carboxylate, produced by hydrolysis of the acyl enzyme (pathway b), except that this would not account for the loss of 5% of the enzyme activity. However, if X were a carboxyl group (Asp, Glu), then both formation of the hydroxamate (pathway c) and loss of the enzyme activity (pathway d) could be accounted for.

$$\text{Enz}-\overset{\overset{\textstyle O}{\|}}{\text{C}}-\text{O}-\overset{\overset{\textstyle O}{\|}}{\text{C}}-\text{R} \xrightarrow[95\%]{c} \text{Enz}-\overset{\overset{\textstyle O}{\|}}{\text{C}}-\text{O}^- \ + \ \text{HO}-\overset{\text{H}}{\underset{\text{H}}{\text{N}}}-\overset{\overset{\textstyle O}{\|}}{\text{C}}-\text{R}$$

$$\underset{NH_2OH}{}$$

d | 5%

$$\text{Enz}-\overset{\overset{\textstyle O}{\|}}{\text{C}}-\text{NHOH} \ + \ {}^-\text{OOCR}$$

This could account for
the loss of activity if it
was not hydrolyzed.

S2-3. Many of the experiments discussed for chymotrypsin are applicable, such as the following:

a. Carry out the incubation at 0 °C or at a different pH to see if these conditions stabilize the covalent intermediate.

b. Try a more reactive phosphate (e.g., **S2.2**), so that phosphorylation of the enzyme is faster than dephosphorylation, and the covalent adduct may be isolated.

$$NO_2 - \underset{\textbf{S2.2}}{\bigcirc} - O - PO_3{}^{2-}$$

c. Look for initial burst in kinetics with **2.2** as the substrate (p-nitrophenolate is released).

d. See if V_{max} is the same for $ROPO_3{}^{2-}$, $RSPO_3{}^{2-}$, and $RSePO_3{}^{2-}$. If so, dephosphorylation of the enzyme is probably the slow step.

e. Look for a partial exchange reaction:

$$ROPO_3{}^{2-} + {}^{14}ROH \xrightarrow{\text{phosphatase}} \text{look for } {}^{14}ROPO_3{}^{2-}$$

This would suggest a reversible covalent intermediate:

f. Use $RO^{32}PO_3{}^{2-}$ and attempt to isolate the covalent adduct by phenol quench and gel filtration.

g. Once covalent catalysis is established, the identity of X can be determined by proteolytic digestion of the labeled enzyme and peptide mapping.

S2-4. a. Label the acetate with ^{18}O, and look for ^{18}O in the released P_i.

b. Label ATP with γ-^{32}P, and look for ^{32}P in the P_i.

c. Synthesize **S2.3,** and incubate it with the enzyme in the absence of ATP; look for the product.

$$\underset{\textbf{S2.3}}{\overset{\displaystyle O}{\overset{\displaystyle \|}{CH_3C - OPO_3{}^{=}}}}$$

d. Look for covalent catalysis (as in **S2-3**).
    S2-5.

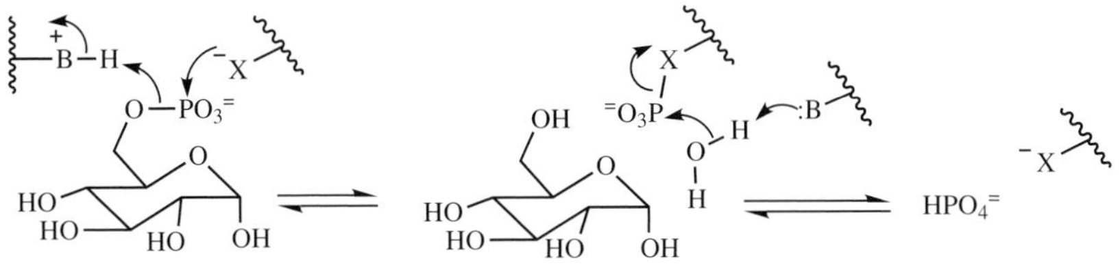

In $H_2{}^{18}O$, the reverse reaction is as follows:

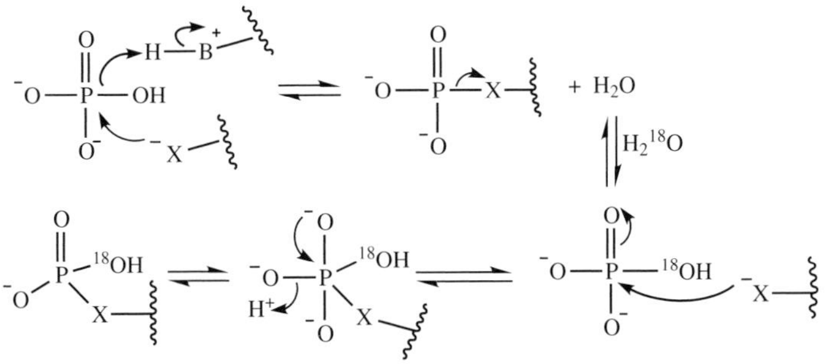

Because of a small ^{18}O isotope effect, $H_2{}^{16}O$ will preferentially leave, so that with time, the P_i will become higher in specific activity of ^{18}O (more ^{18}O atoms incorporated). Because this is all reversible, the increase in ^{18}O specific activity eventually will show up in the substrate. This also could have been drawn as a general acid/base-catalyzed reaction instead of as covalent catalytic. The small molecules can be separated by HPLC (gel filtration only separates protein from *all* small molecules) and analyzed by mass spectrometry.

S2-6. a. Gln, 2ATP, Mg^{2+}
    b. Glu, **P2.1,** 2 ADP

c.

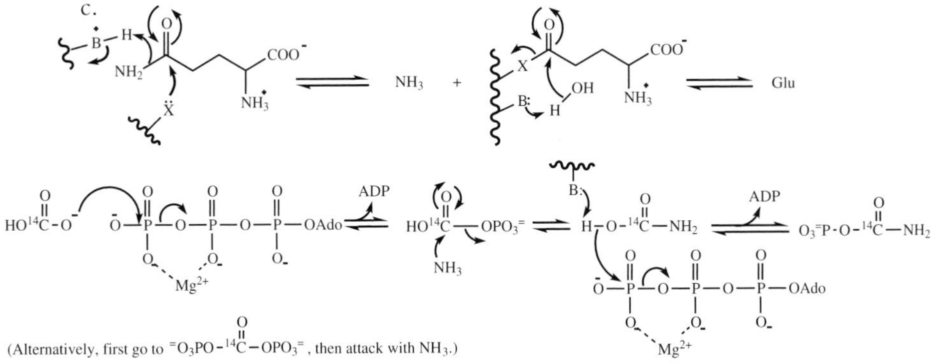

(Alternatively, first go to $^-O_3PO\text{-}^{14}C\text{-}OPO_3^=$, then attack with NH_3.)

d. Inactivate with any of the glutamine inactivators, e.g.,

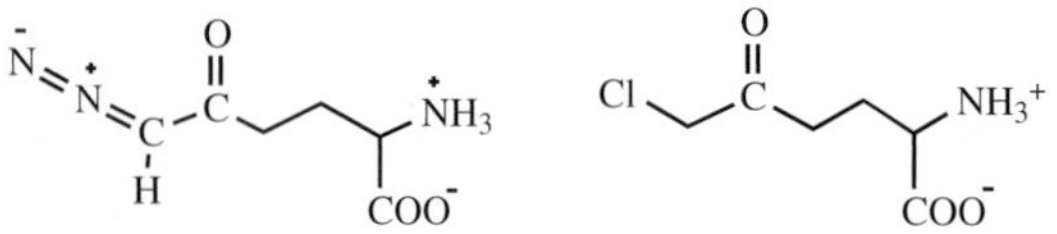

S2-7.

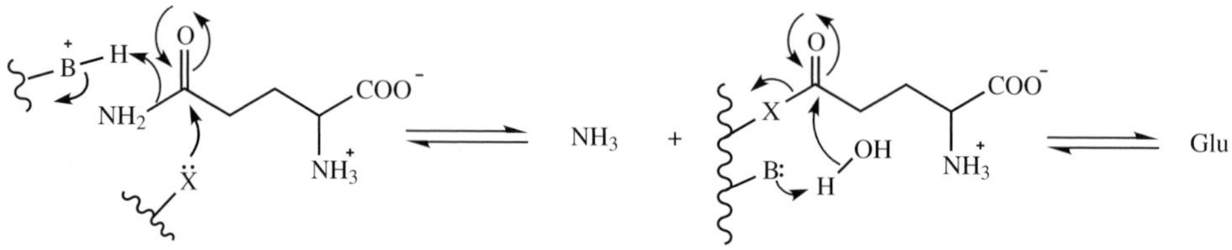

a.

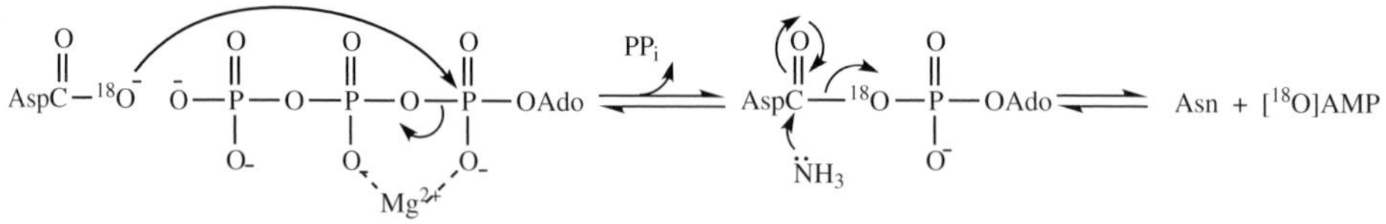

b.

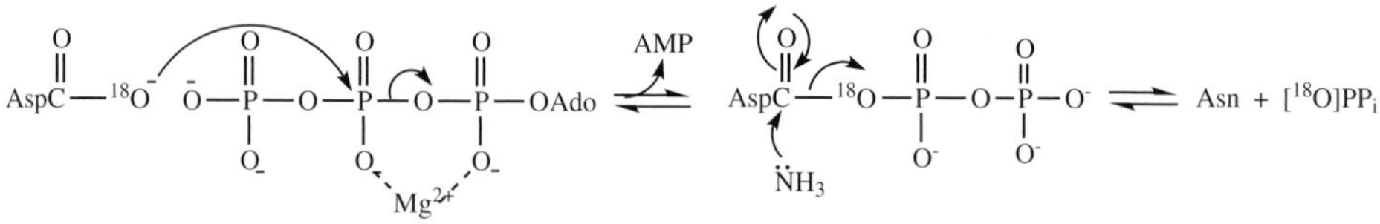

c. This would support mechanism a:

$$\text{Asp} + \text{ATP} \rightleftharpoons \text{AspC(=O)}-\text{OAMP} + PP_i$$

If [^{32}P]PP$_i$ exchanges with PP$_i$, then the reverse reaction would give [^{32}P]ATP. In mechanism b, PP$_i$ is not generated until after Asn is formed.

S2-8. a. Covalent catalysis can rationalize this result.

b. There would be no change.

Attack at the β-position would give a positional isotope exchange if attack occurred at the β-position instead of at the γ-position.

c.

(See Raushel, F. M., and co-workers, *J. Am. Chem. Soc.* **1997,** *119,* 6684.)

S2-9. a. Retention of chirality suggests a double inversion mechanism, which is consistent with covalent catalysis followed by hydrolysis. Inactivation by NaBH$_4$ suggests a Schiff base intermediate involving an active-site lysine residue.

Site-directed mutagenesis of Lys-53 completely inactivates the enzyme.
(See Dunaway-Mariano, D., and co-workers, *Biochemistry* **1998**, *37*, 9305.)

b.

In this case, there is no need for a Schiff base, but the covalent catalysis followed by hydrolysis scaffold is functional. The active-site nucleophiles do not have to be the same for the members of the superfamily.

C. Chapter 3

S3-1. a.

b. 1. Start with ^{14}C-**S3.3,** and look for ^{14}C-**S3.1** (is it reversible?).

2. Synthesize **S3.2,** and incubate with the enzyme to see if **S3.1** or **S3.3** is obtained.

3. Synthesize **S3.4,** and look for either affinity labeling of the enzyme (**S3.6**) or formation of **S3.5**.

4. Incubate **S3.1** with the enzyme in the NADH form to show nothing happens.

S3-2. a. Intermediate **S3.7** can react with water or cyanide.

b. It is an oxidase, which requires O_2 to convert reduced flavin back to oxidized flavin, and thereby, to reactivate the enzyme (not when cyanide is present).

S3-3.

S3-4. a.

b. 1. Synthesize chirally labeled **S3.8.** Removal of H_R and H_S from **S3.8** gives an overall *anti* elimination; removal of H_R and H_R or H_S and H_S gives a *syn* elimination.

S3.8

2. Synthesize stereospecifically labeled succinate, and determine the stereo-chemistry of the products.

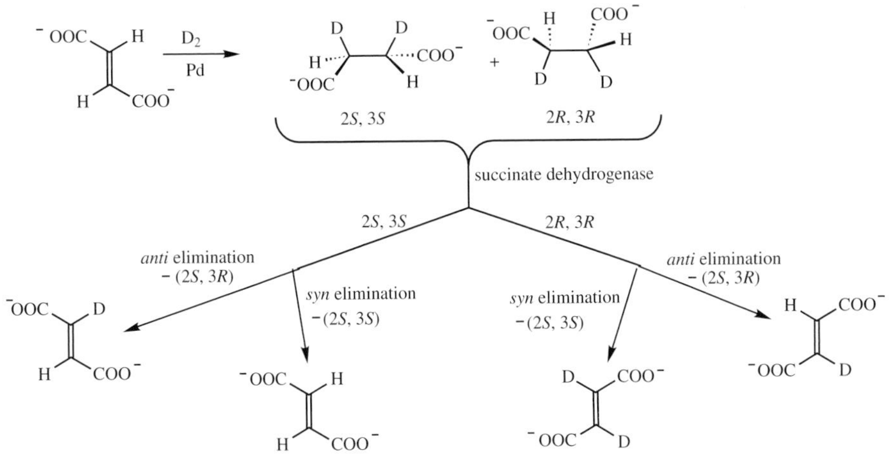

2S, 3S

2R, 3R

succinate dehydrogenase

2S, 3S

2R, 3R

anti elimination
– (2S, 3R)

syn elimination
–(2S, 3S)

syn elimination
–(2S, 3S)

anti elimination
– (2S, 3R)

Anti elimination gives all monodeuterated fumarate; *syn* elimination gives 50% nondeuterated, 50% dideuterated. Take the mass spectrum (experimentally, *anti* elimination occurs). Of course, you should do the complementary experiment with the 2R,3S (same as 2S,3R).

3.

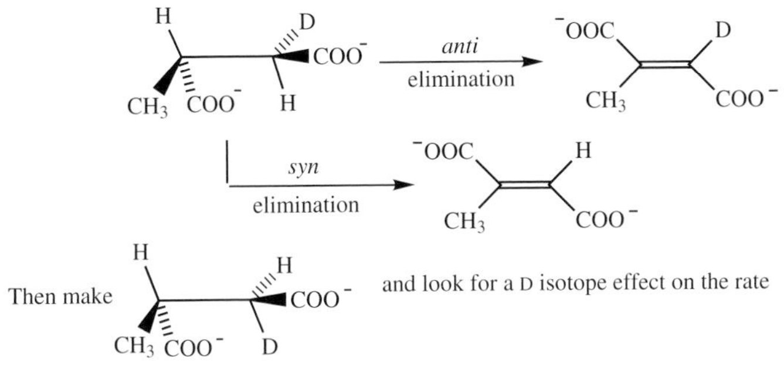

anti
elimination

syn
elimination

Then make

and look for a D isotope effect on the rate

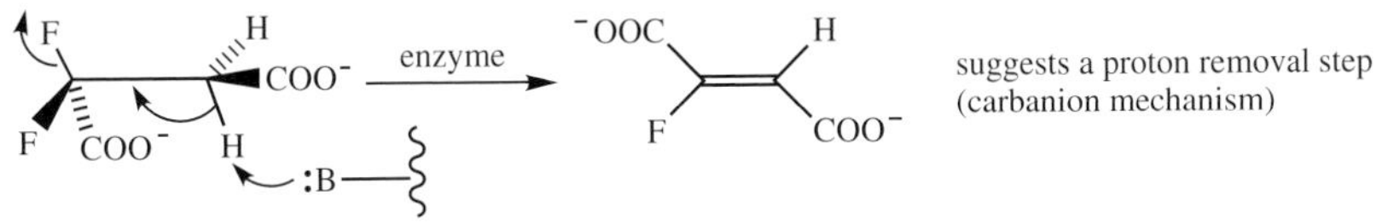

enzyme

suggests a proton removal step
(carbanion mechanism)

S3-5. a.

b. Measure the rate constant for inactivation (k_{inact}) and the catalytic rate constant (k_{cat}). The partition ratio is k_{cat}/k_{inact}. Use ^{14}C-labeled pargyline and compare the amount of radioactive metabolites to the amount of radioactivity bound to the enzyme (presumably equal to the amount of enzyme).

S3-6. There are many correct answers. There are two intermediates to consider—a carbanion and radical.

a.

2.

b.

S3-7.

$$\left[1R\text{-}^2H,\ 2R\text{-}^3H\right]$$ $$\left[1S\text{-}^2H,\ 2R\text{-}^3H\right]$$

Both C-1 protons can be removed, but when there is a deuterium, removal of the proton will be favored. Because almost four times as much ^3H is lost from **S3.9** than **S3.10**, it suggests that *syn* removal of C-1 and C-2 hydrogens is favored.

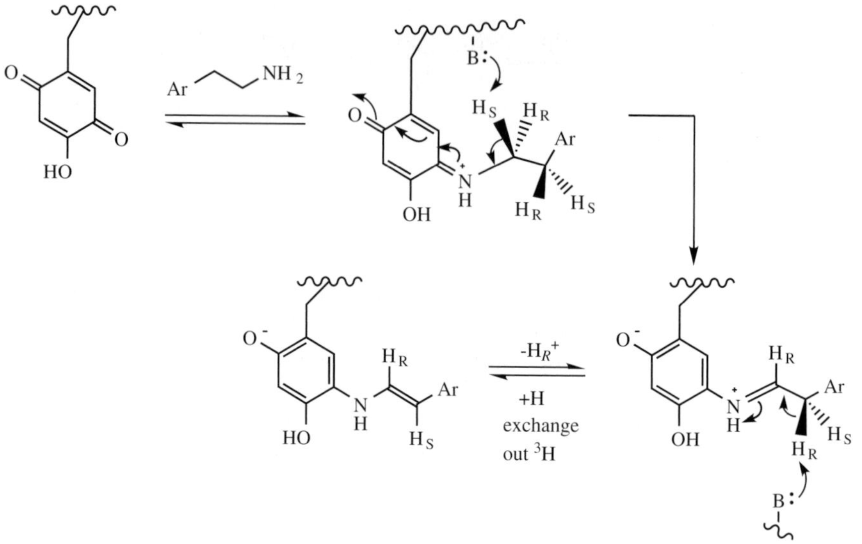
HO
O
O
Ar
NH 2
B:
O
HS
HR
Ar
N
H
OH
HR
HS
O-
HR
Ar
N
H
HO
HS
-HR+
+H
exchange
out 3H
O-
HR
Ar
N
H
OH
HR
HS
B:

S3-8.

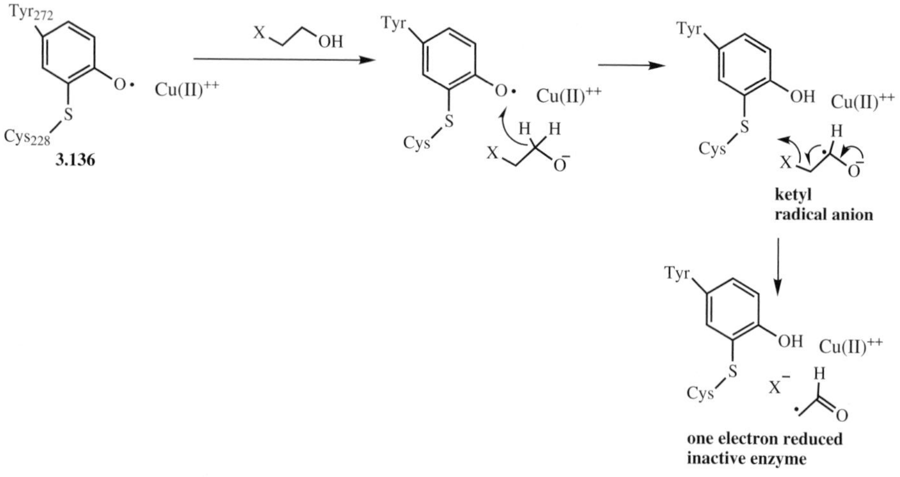
Tyr272
O·
Cu(II)++
S
Cys228
3.136
X
OH
Tyr
O·
Cu(II)++
S
Cys
H
H
X
O−
Tyr
OH
Cu(II)++
S
Cys
H
X
O−
ketyl
radical anion
Tyr
OH
Cu(II)++
S
Cys
X−
H
O
one electron reduced
inactive enzyme

S3-9.

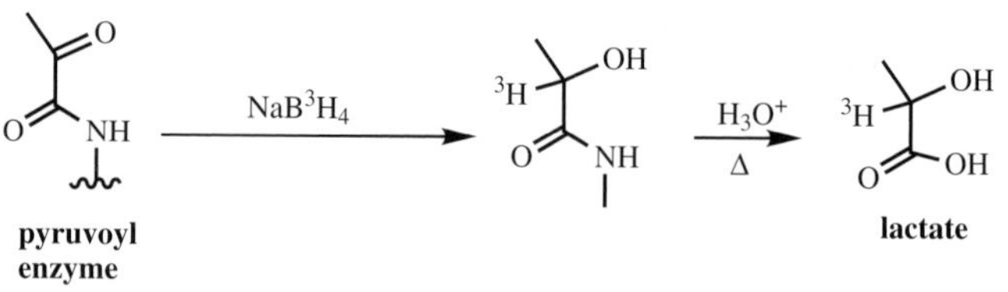
O
O
NH
pyruvoyl
enzyme
NaB3H4
3H
OH
O
NH
H3O+
Δ
3H
OH
O
OH
lactate

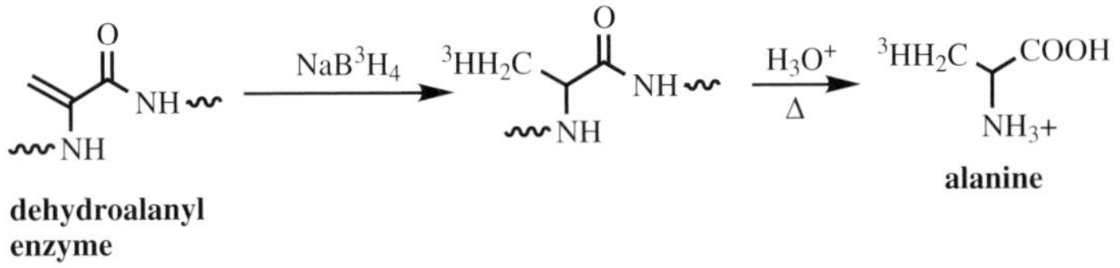

dehydroalanyl enzyme

NaB³H₄

³HH₂C

H₃O⁺ Δ

³HH₂C COOH
NH₃+

alanine

D. Chapter 4

S4-1. It often is useful to think backward (as in any "synthesis") when trying to determine multiple steps in a biosynthetic pathway. The two steps in this case are hydrolysis of the formyl group and monooxygenation of the benzene ring. Aromatic monooxygenation is an electrophilic aromatic substitution reaction, which goes better (chemically) with e⁻-donating substituents on the ring. Therefore, it may be more reasonable to do the monooxygenation *after* the hydrolysis of the formyl group, so there is a more electron-donating NH₂ group instead of a NHCHO group attached to the benzene when monooxygenation occurs.

a.

Hydrolysis

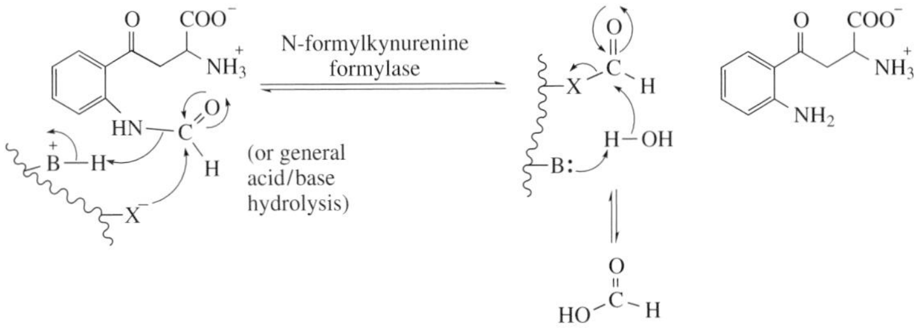

N-formylkynurenine formylase

(or general acid/base hydrolysis)

Hydroxylation

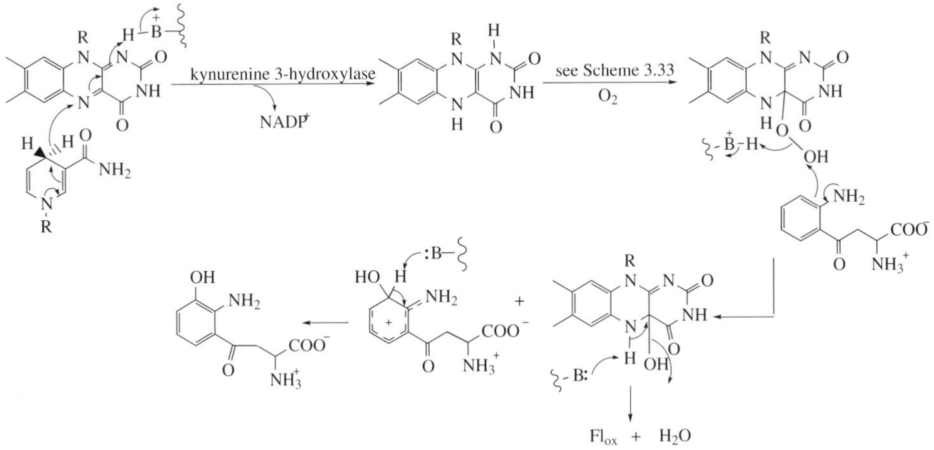

kynurenine 3-hydroxylase

NADP⁺

see Scheme 3.33

O₂

Flox + H₂O

An alternative mechanism for the first enzyme is oxidation−decarboxylation:

b. Experiments to differentiate these possible mechanisms:

For the formylase reaction,
1. Label the substrate with ^{14}C and look for either $[^{14}C]$formate or $^{14}CO_2$.

2. Determine if $NAD(P)^+$ is required. Purify the enzyme, and determine if $NAD(P)^+$ is needed for activity.
3. Look for the incorporation of $H_2^{18}O$ into HCO_2H.
4. Do the usual covalent catalytic experiments described in Chapters 2 and 3.
5. Label the substrate with a tritium in the formyl group, and look for tritium in either formate or NADH.

For the hydroxylase reaction,
1. Demonstrate the requirement of $NAD(P)H$ and FAD for activity.
2. Demonstrate the requirement for O_2, and do an $^{18}O_2$ experiment to see where the ^{18}O ends up.
3. Do experiments discussed for lactate oxidase.

S4-2.

b. Pathway b is unlikely (+ charge next to carbonyl), but either add $^{36}Cl^-$ (radioactive) and look for ^{36}Cl in the product or start with ^{36}Cl-labeled substrate and add a large excess of normal Cl^- to the buffer and look for Cl in the product. If the chloride exchanges, it supports pathway b. If not, however, pathway b may still be correct, but the chloride ion is not released into the medium.

S4-3. This is an example of ketone monooxygenase (Baeyer–Villiger reaction). The flavin hydroperoxide is generated by reduction of the flavin with NADPH and oxygenation as shown in Scheme 3.33.

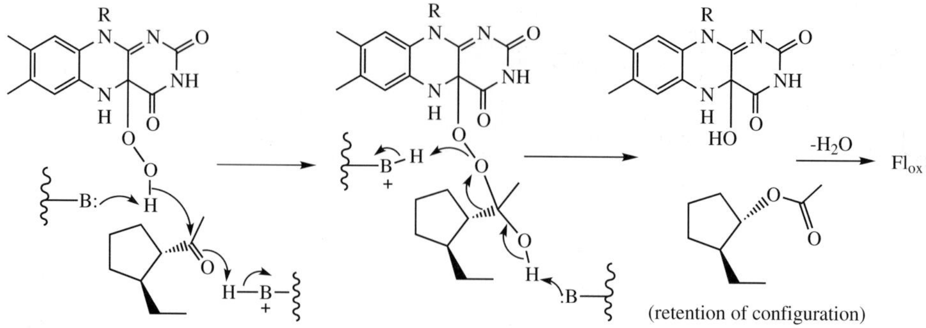

(retention of configuration)

S4-4. a.

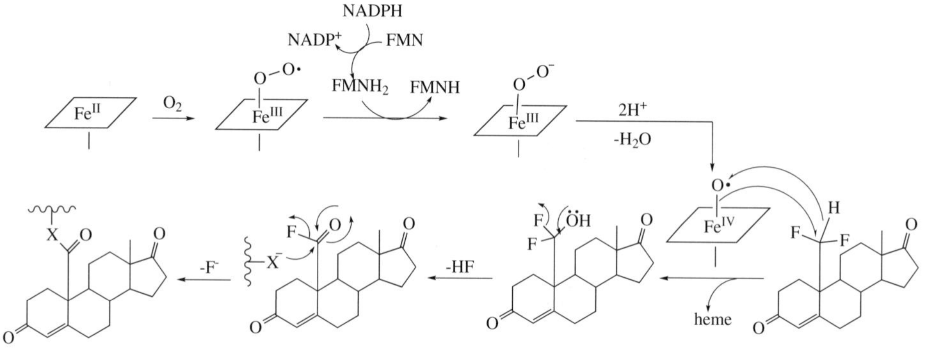

b. Two possibilities are shown next.

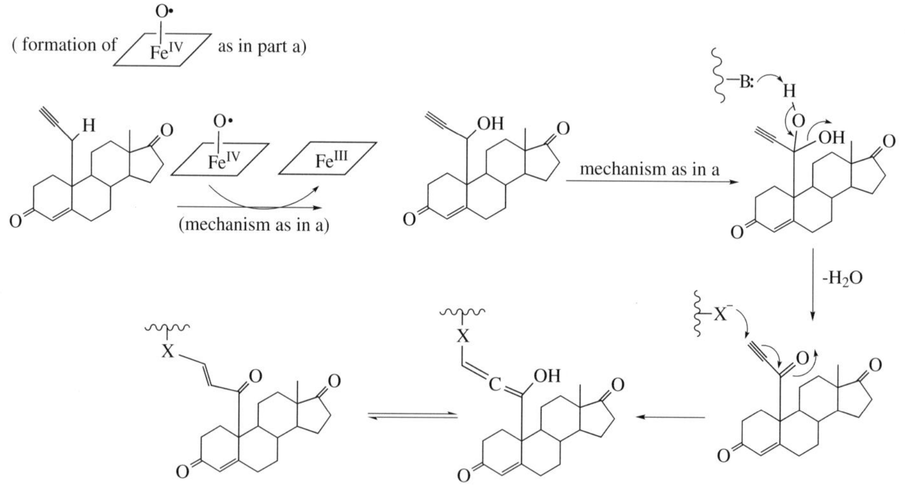

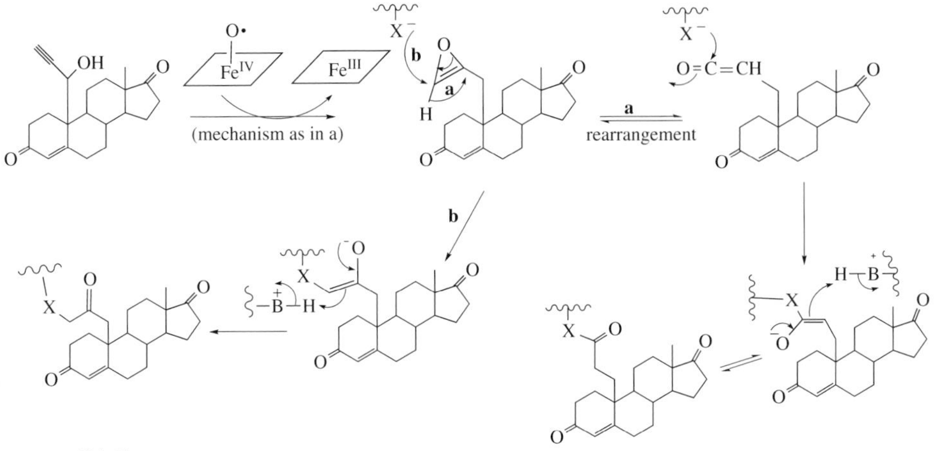

S4-5.

(see problem S4-4)

isolatable intermediate

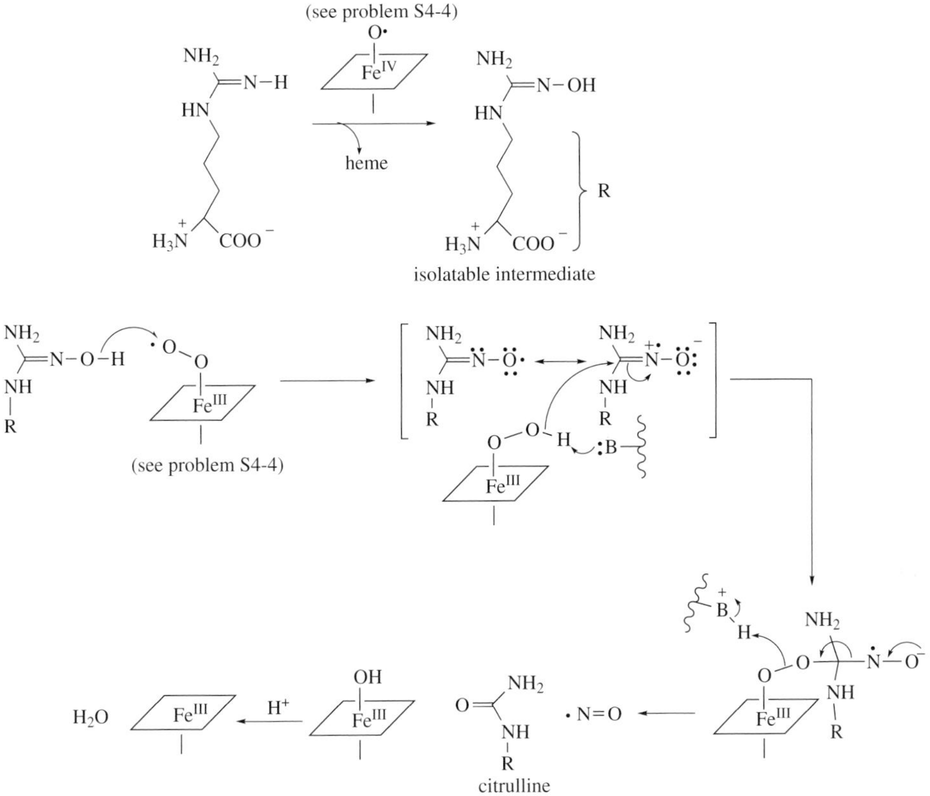

(see problem S4-4)

citrulline

S4-6.

$$M^{2+} \quad \cdot O{-}O\cdot \longrightarrow \left[M{-}O{-}O\cdot \right] \longrightarrow M^{2+} \quad O{=}O$$

$$\text{triplet} \qquad\qquad\qquad\qquad\qquad \text{singlet}$$

S4-7.

$$Fe^{(II)}L_6 \xrightarrow{\ O_2\ } \dot{O}{-}O{-}Fe^{(II)}L_5 \xrightarrow[\substack{H^+ \\ -H_2O}]{\text{ascorbate}} O{=}Fe^{(IV)}L_5$$

(See Schofield and co-workers, *Biochemistry* **1997,** *36,* 15999.)

S4-8.

Horseradish peroxidase partitions between both heteroatom oxygenation (first reaction) and α-hydroxylation (second reaction), but chloroperoxidase undergoes only heteroatom oxygenation. Either oxygen rebound to the sulfur radical cation is much faster than α-deprotonation in chloroperoxidase, the substrate is bound in chloroperoxidase so that the α-proton is not close enough to the Fe(IV)-O⁻ for removal, or the sulfur radical cation is not a real intermediate.

(See Baciocchi, E. and co-workers, *J. Am. Chem. Soc.* **1996**, *118*, 8973.)

S4-9. a.

b. On hydroxylation of the aromatic ring, the methyl group prevents rearomatization, and that may be why hydrolysis occurs. To test this hypothesis, the corresponding desmethyl analogue (no methyl group) can be used as a substrate to see if the catechol (**S4.1**) is obtained instead.

S4.1

This experiment was done, and the ring-cleaved product was still obtained, suggesting that hydrolysis is part of the normal catalytic mechanism.

(See Ballou, D. P., Massey, V., and co-workers, *Biochemistry* **1997,** *36,* 8060, and 13856.)

E. Chapter 5

S5-1. a.

b. Use $^{18}O_2$ and $H_2{}^{18}O$, and see where the labels go. Label the β-carotene with tritium at the double bond that is oxygenated, and look for tritium in the vitamin A. See if ^3H_2O incorporates tritium into the product. Substitute Fe^{2+} with other metals. Add a singlet oxygen scavenger or a superoxide scavenger, and see if either of these species is involved.

    S5-2. a.

b. According to this mechanism, the only OH group needed is the one attached to the center ring. Synthesize analogues with H in place of OH in all positions (one at a time) to determine if this is so.

S5-3.

S5-4.

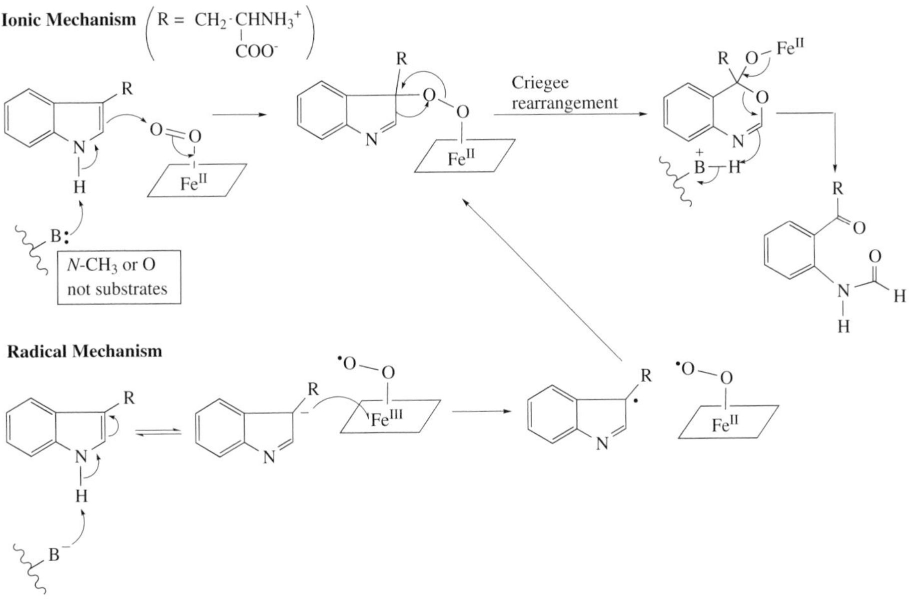

Ionic Mechanism $\left(R = \begin{array}{c} CH_2\cdot CHNH_3^+ \\ | \\ COO^- \end{array} \right)$

Criegee rearrangement

N-CH$_3$ or O not substrates

Radical Mechanism

(See Wiseman, J., and co-workers, *J. Biol. Chem.* **1993,** *268,* 17781.)

S5-5.

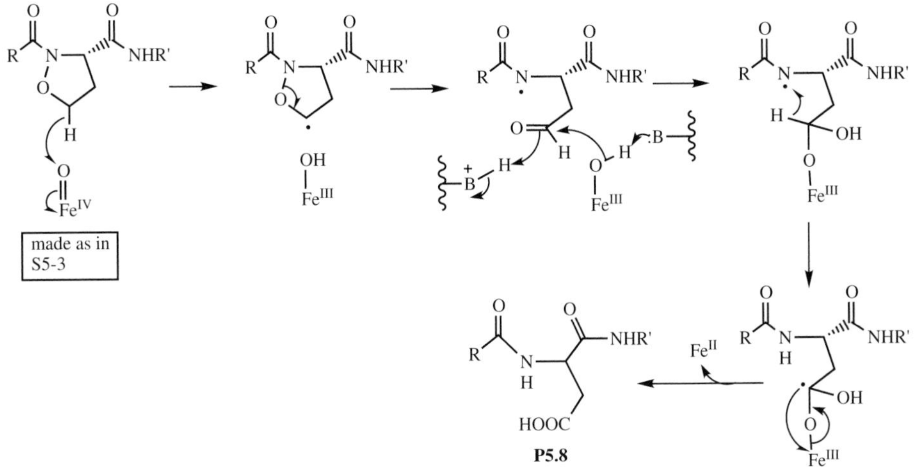

made as in S5-3

P5.8

N−O bond cleavage must be faster than oxygen rebound.
(See Begley, T. P., and co-workers, *J. Am. Chem. Soc.* **1999,** *121,* 587.)

S5-6.

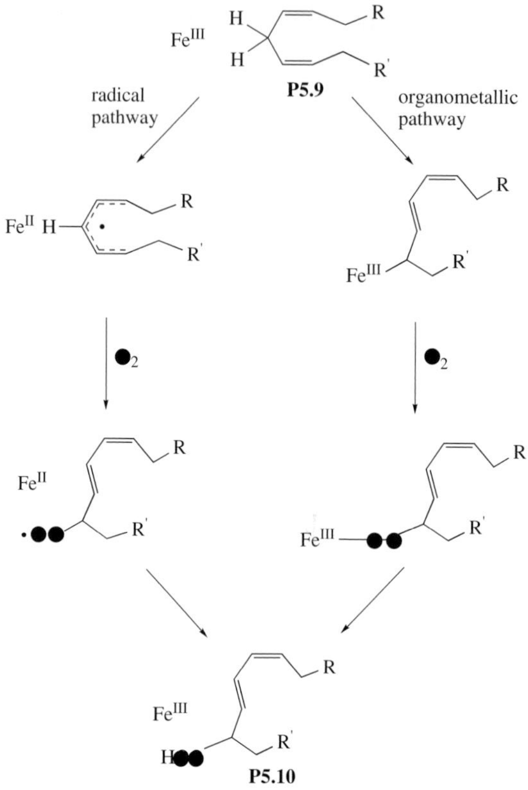

P5.9

P5.10

By both mechanisms there should be a kinetic isotope effect on the cleavage of the C^{11}–H bond in linoleic acid. EPR spectroscopy should give a carbon allylic radical for the radical pathway, but not the organometallic pathway. This experiment has been done, and an allylic radical was observed.

(See Nelson, M. J.; Cowling, R. A.; Seitz, S. P. *Biochemistry* **1994**, *33*, 4966.)

S5-7.

(See Harpel, M. R.; Lipscomb, J. D. *J. Biol. Chem.* **1990**, *265*, 22187.)

F. Chapter 6

S6-1.

S6-2.

S6-3. a. Either MALDI–TOF (matrix-assisted laser desorption/ionization–time of flight) or, preferably (because of the increased sensitivity), electrospray ionization mass spectrometry of the modified enzyme and the native enzyme will give the mass difference as a result of inactivation. Proteolytic digests of the enzyme and of the modified enzyme can be analyzed by comparative HPLC/electrospray mass spectrometry. At least one unique peptide should be produced from the modified enzyme that is not found in the digest of the native enzyme. Tandem mass spectrometric analysis of the unique peptide will give a fragmentation pattern from which the peptide structure can be deduced, presumably showing that the modification occurs at Glu-276.

Alternatively, a radioactively labeled inactivator can be synthesized. After inactivation of the enzyme, dialysis to remove excess inactivator, proteolytic digestion, HPLC separation of the peptides, and scintillation counting to identify where the radioactivity is located, preferably only one of the peptides will have a radioactive label. This peptide can be further purified by HPLC, and Edman degradation carried out to determine the sequence and which residue is modified.

b. To determine if this residue is in the active site, the experiment can be repeated, except in the presence of substrate or a potent reversible inhibitor. Active-site modification should be prevented by the presence of the substrate, and the same spectrum as the native enzyme should be observed.

(See Withers, S. G., and co-workers, *Biochemistry* **1998**, *37*, 3858.)

S6-4. The best approach is to draw the structure of the polyene in the conformation that resembles the product prior to doing the cyclization, so you can see what needs to be done.

(See Croteau, R., and co-workers, *Biochemistry* **1996**, *35*, 2968.)

S6-5. The solvent isotope effect indicates the transfer of a proton in the transition state. The normal secondary deuterium isotope effect with guanosine 5'-
diphospho-[1-^{2}H]β-1-fucose as substrate indicates the conversion of an sp^3 carbon
to sp^2 in the transition state, supporting a S$_N$1 mechanism. Also, potent inhibition by **P6.9,** which has a strong electron-withdrawing group (F) adjacent to the
anomeric carbon, is consistent with the formation of oxocarbenium character in
the transition state, again S$_N$1.

(See Wong, C.-H., and co-workers, *Biochemistry* **1997**, *36*, 823.)

G. Chapter 7

S7-1. a. Do experiments to test a carbanionic mechanism (similar to what was
done with D-amino-acid oxidase):

1. Synthesize a series of aryl analogues (**S7.1**), vary the electronic character
of X, and look for the effect on V_{max} (Hammett analysis).

S7.1

2. Synthesize **S7.2,** and look for the formation of the elimination product (**S7.3**). This is a reactive Michael acceptor, so there may be inactivation.

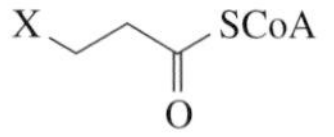

S7.2

X = F, Cl

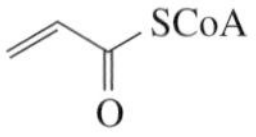

S7.3

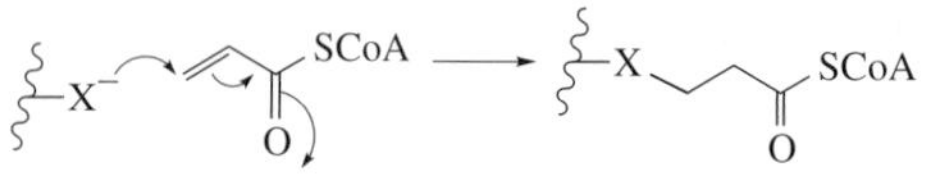

3. See if biotin is involved.

a. Treat the enzyme with avidin to determine if it is a biotin-dependent enzyme.

b. Treat the enzyme with biotinase to see if biotin is involved.

4. See if CO_2 or HCO_3^- is the substrate (initial rate experiments discussed in the text).

b. Affinity labeling agent:

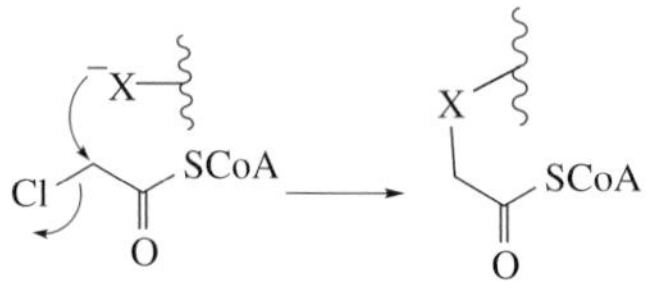

Mechanism-based inactivator:

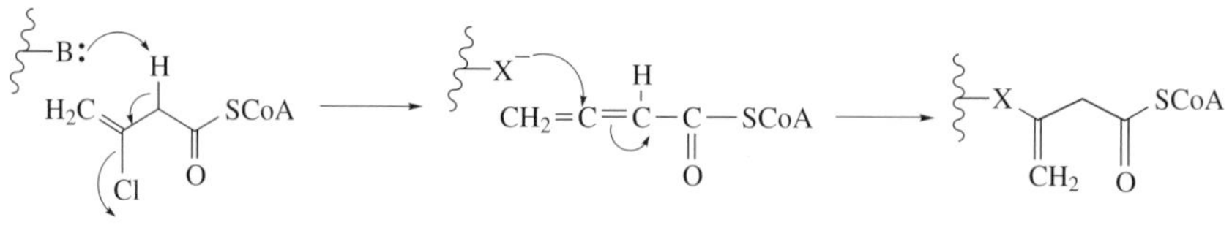

S7-2. a. This suggests that the bicarbonate is phosphorylated (presumably by $ATP \cdot Mg^{2+}$); therefore, HCO_3^-, not CO_2, is involved.

b.

$$ATP + HCO_3^- \rightleftharpoons HCO_3PO_3^{2-} + ADP$$

$$[^{14}C]ADP$$

$[^{14}C]ADP$ exchanges with ADP and the reaction reverses to $[^{14}C]ATP$.

c. The enzyme is carboxylated, which is a viable intermediate. Because the enzyme does not bind to avidin, it does not contain biotin.

d.

top named *re-si* because the right side has higher priority than left

S7-3.

S7-4. a.
Carboxylation

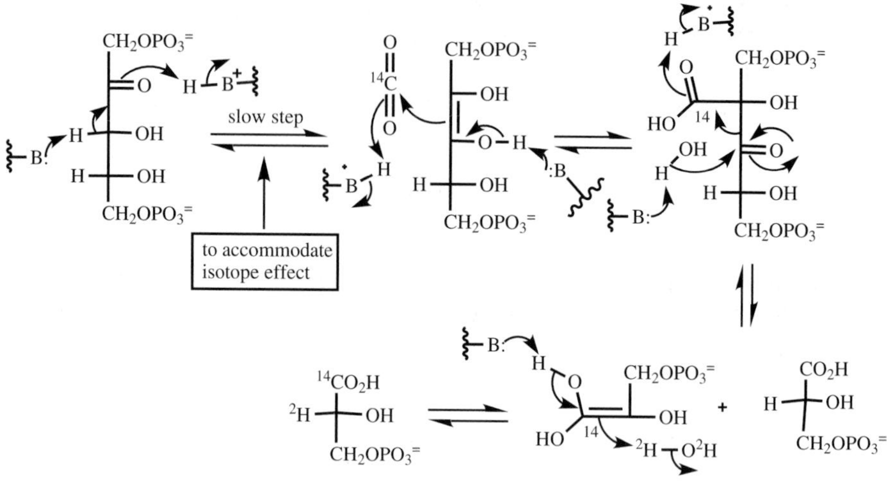

Oxygenation

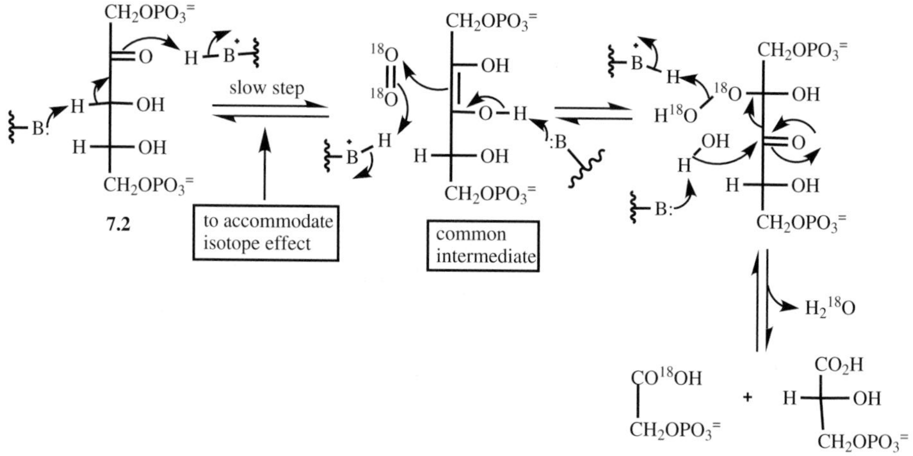

b. The mechanism in part a is a general acid/base mechanism. This is the co-valent catalytic mechanism.

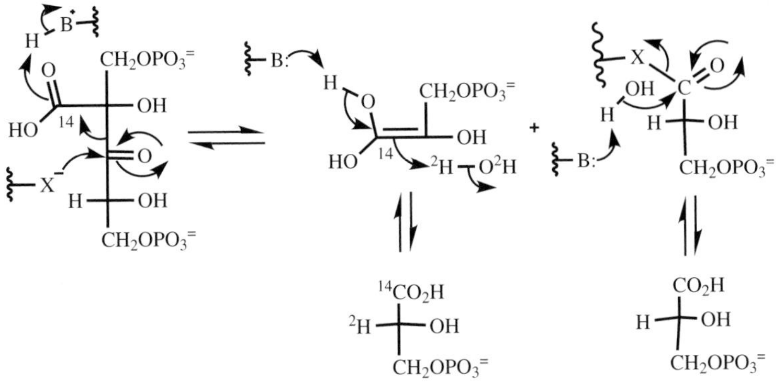

S7-5. Inhibition by avidin indicates a biotin-dependent enzyme. The ^{18}O results support bicarbonate as the substrate. A possible mechanism for this fictitious enzyme is shown here.

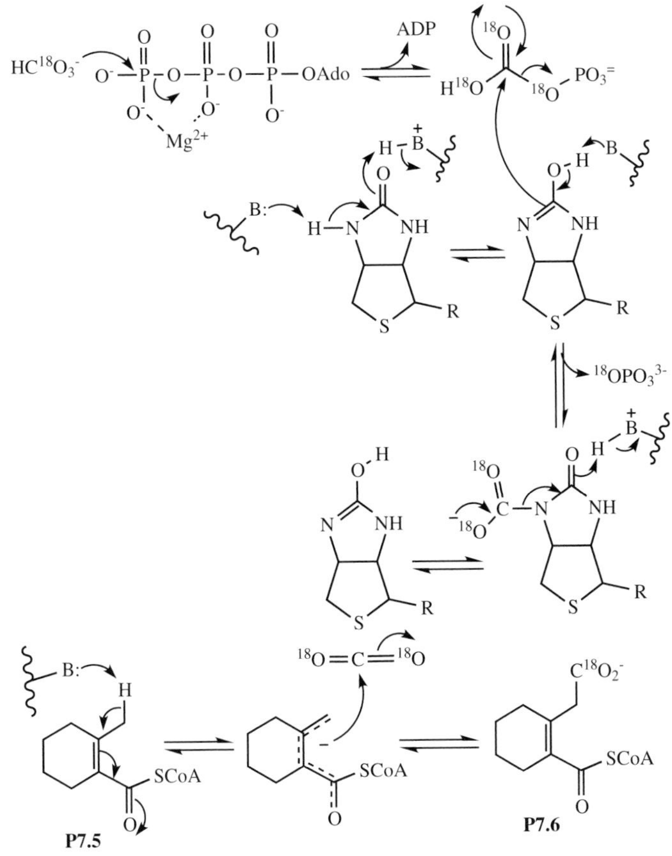

H. Chapter 8

S8-1. a.

b. If the α-carboxylate is held perpendicular to the PLP plane by a positively charged group, α-decarboxylation will occur; if the α-H is perpendicular, β-decarboxylation will occur (Dunathan hypothesis).

c. *α-Decarboxylation*
1. Show that PLP is a required cofactor by UV–visible spectroscopy.
2. Reduce with $NaBH_4$, and show that the enzyme is not active (NaB^3H_4 can be used and the enzyme hydrolyzed to show that PMP is enzyme bound).
3. Carry out the reaction in D_2O, and look for D in the product.
4. Label the α-H, and show it is not released.
5. Label the α-carboxylate, and show that it is lost as CO_2 (but not the β-carboxylate).

β-Decarboxylation
1. Show that the α-H is lost to solvent or is put back at the β-position.
2. Show that PLP is required (as earlier).
3. Label the β-carboxylate, and show it is lost as CO_2 (but not the α-carboxylate).
S8-2. a.

In D_2O, the aldehyde hydrogen would be a deuterium.

b.

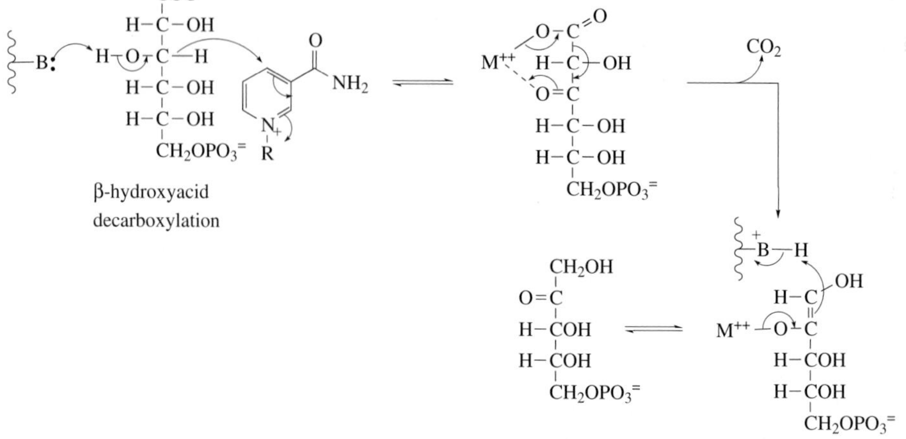

β-hydroxyacid
decarboxylation

Label the C-1 carbon and show the loss of CO_2.
S8-3.

$$CH_3\overset{O}{\overset{\|}{C}}-COOH \longrightarrow CH_3COO^-$$

This is an oxidative decarboxylation.

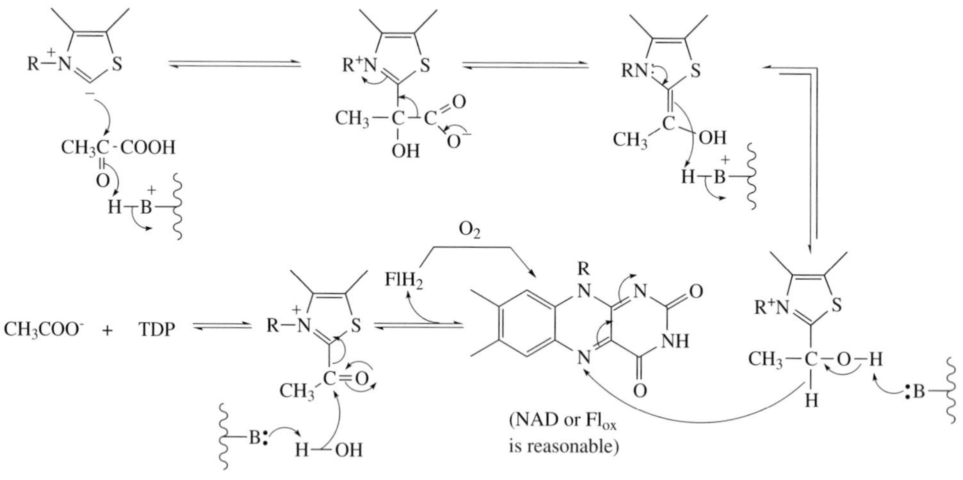

O_2

$CH_3COO^- + TDP \rightleftharpoons$

(NAD or Fl_{ox}
is reasonable)

S8-4.

First enzyme:

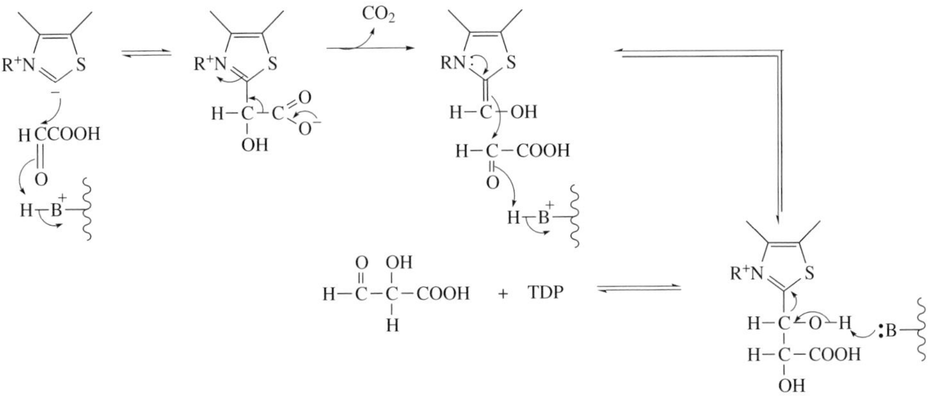

Synthesize [^{14}C]hydroxyethyl TDP, and reconstitute it into the enzyme with substrate; look for ^{14}C in the product.

Second enzyme:

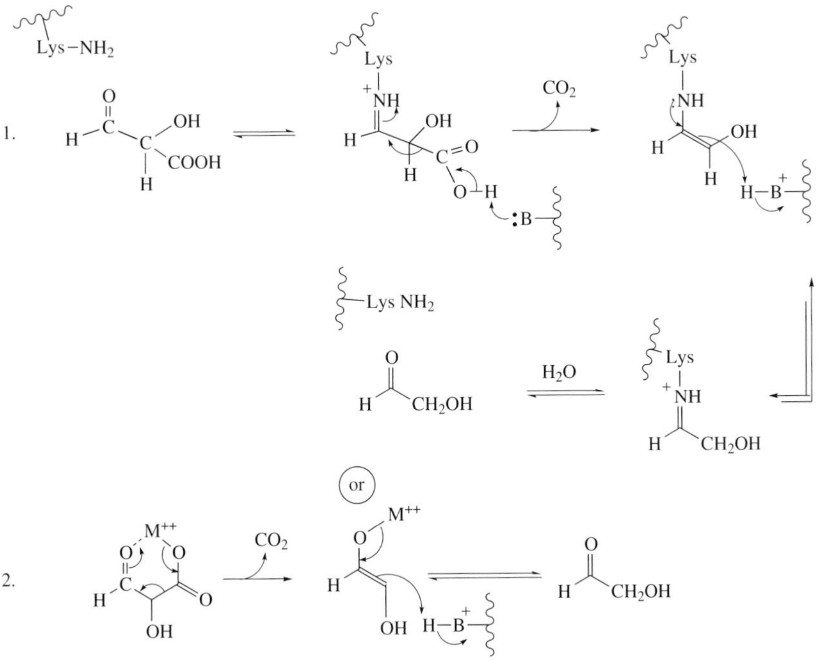

To test the Schiff base mechanism, label the aldehyde with ^{18}O, and demonstrate that it is lost in the product. Add NaBH$_4$ with the substrate and show that inactivation occurs, but not in the absence of substrate. Label the substrate with ^{14}C, add NaBH$_4$, and show that radioactivity is attached to the lysine residue.

In the metal-catalyzed mechanism, the ^{18}O is not lost, no inactivation occurs with NaBH$_4$, and no radioactivity is bound with radioactive substrate and NaBH$_4$.

S8-5. a.

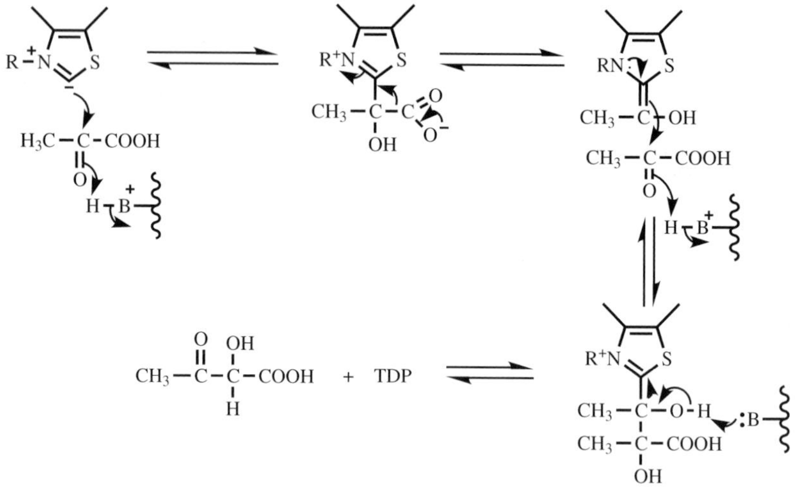

CH$_3$—C—C—COOH + TDP ⇌

b.

CH$_3$—C—O—OH + TDP ⇌

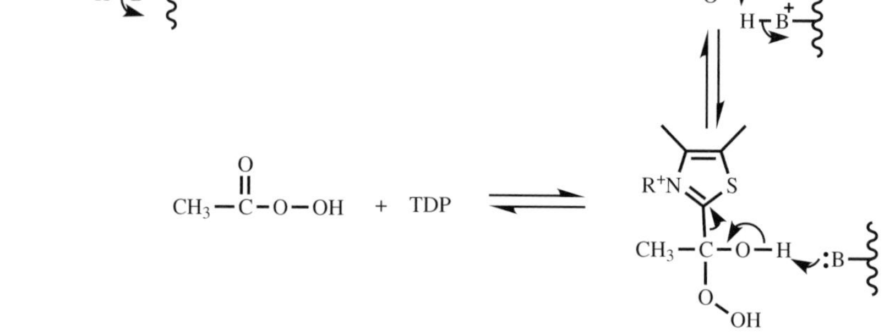

The peracetic acid can then oxygenate other groups to give acetic acid.
(See Abell, L. M.; Schloss, J. V. *Biochemistry* **1991**, *30*, 7883.)

I. Chapter 9

S9-1. a.

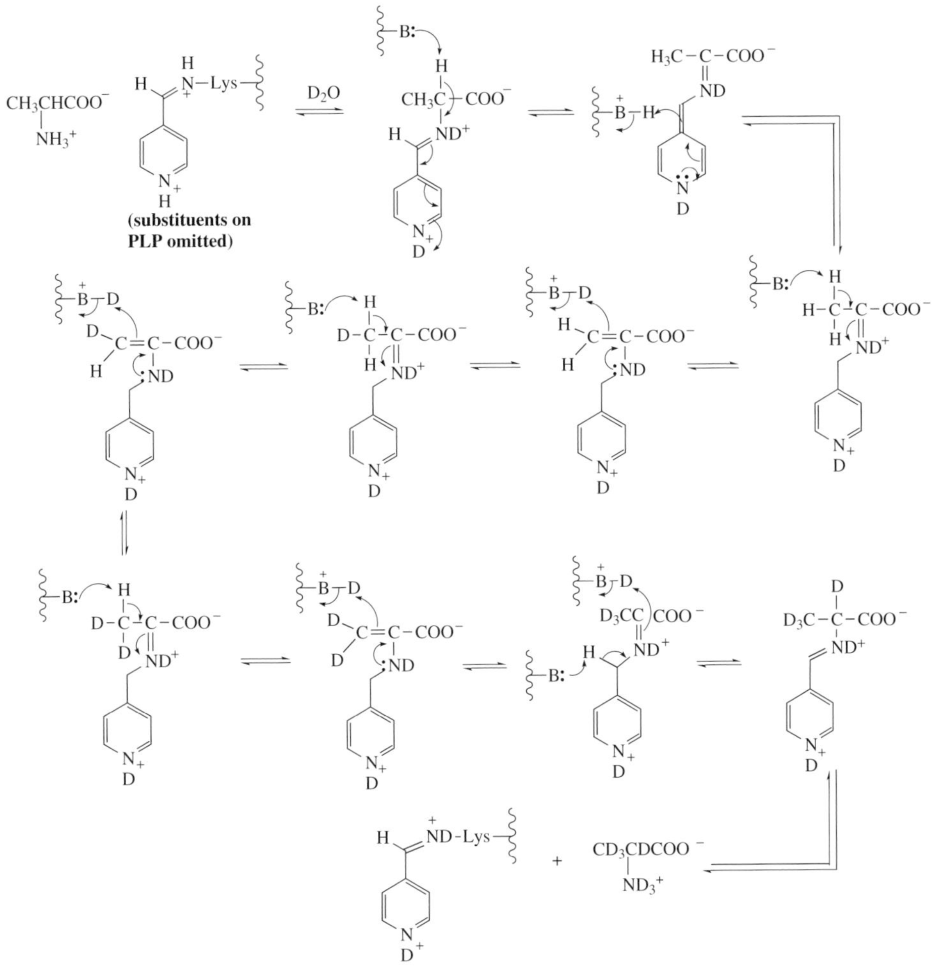

b.

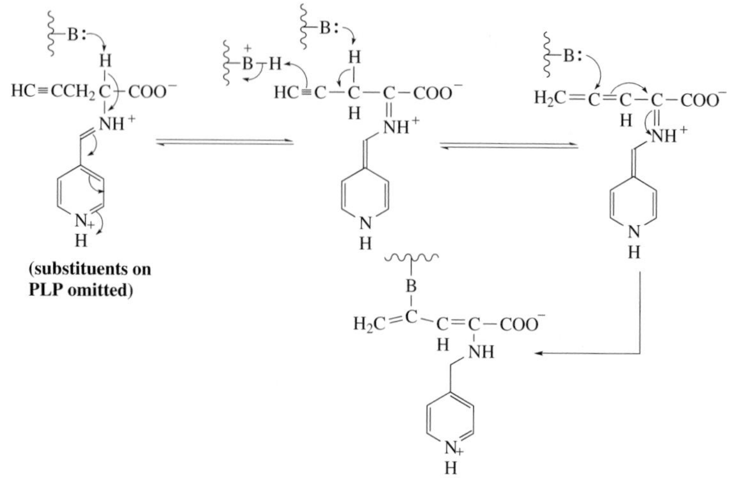

(substituents on
PLP omitted)

S9-2. a.

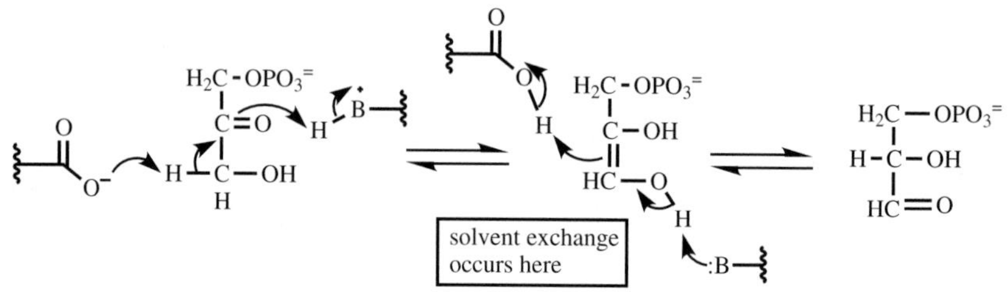

solvent exchange
occurs here

The other mechanism could be a [1,2]-hydride shift, but incorporation of solvent disfavors that. Also, could label the C-3 hydrogen, and see if it is released. Label the carbonyl oxygen, and see if it is released as water (Schiff base mechanism).

b.

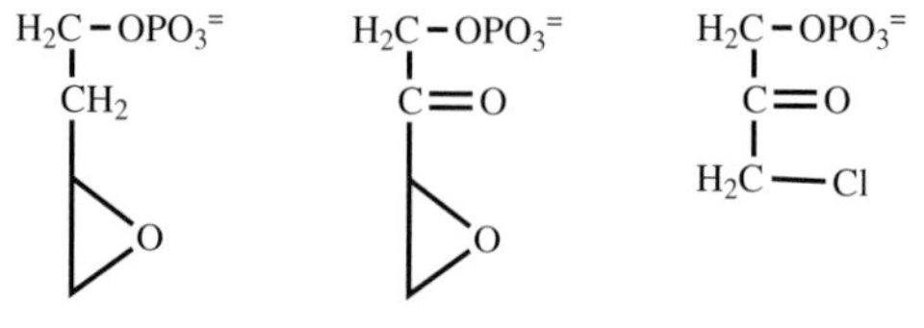

S9-3. a.

b.

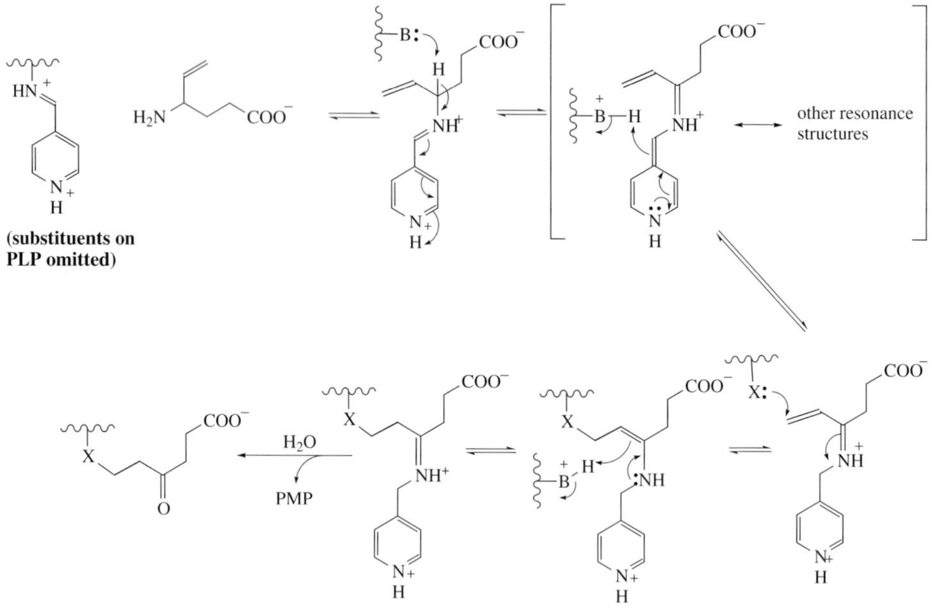
(substituents on
PLP omitted)
other resonance
structures
H2O
PMP

This pathway accounts for only 70% of the inactivation.

(Nanavati, S. M.; Silverman, R. B. *J. Am. Chem. Soc.*, **1991,** *113,* 9341–9349.)

S9-4. As discussed in this chapter, competitive deuterium washout and overshoot experiments could be tried. Also, mutation of active-site groups may provide support for which amino acids are involved in deprotonation and protonation of the substrate.

S9-5. a.

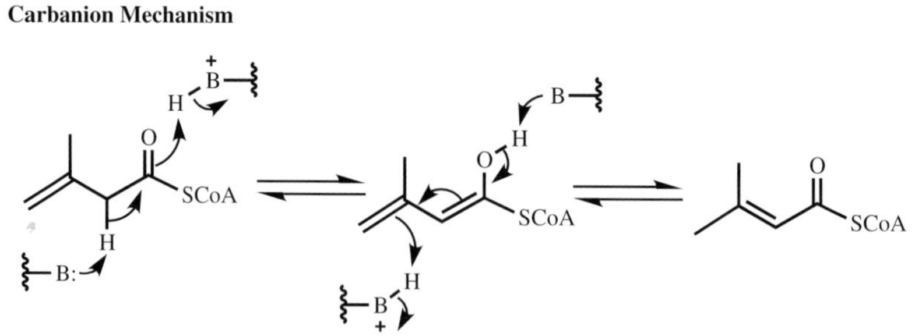
Carbanion Mechanism
SCoA
SCoA
SCoA

Carbocation Mechanism

b. To differentiate a carbanion and carbocation mechanism, carry out the re-action in D_2O, and look for incorporation of deuterium into substrate and product. A carbocation mechanism incorporates deuterium into the product, but not the substrate, whereas a carbanion mechanism exchanges into substrate. Label the C-2 hydrogen, and look for transfer into C-4. If a one-base mechanism is important, suprafacial transfer of the C-2 hydrogen into C-4 should occur.

S9-6. The C–H protons are not acidic enough for an active-site base to remove; oxidation of the C-4 hydroxyl group gives a ketone, which makes the adjacent protons much more acidic.

J. Chapter 10

S10-1. a. This is a γ-replacement.

b. Label the α-hydrogen and the β-hydrogen, and determine if they are lost or not. Treat the enzyme with substrate and NaBH₄, and look for inactivation.

c. This is a γ-elimination.

(substituents on PLP omitted)

d. In D_2O deuterium should be incorporated into the γ-CH_3, unless the α-H removed is put back at the γ-position without exchange of solvent. One deuterium also should be incorporated at the β-position as a result of hydrolysis of the released enamine. If the enamine is enzyme-bound, stereospecific deuteration at the β-position of product will result.

S10-2. a.

b. Label the substrate carbonyl with ^{18}O, and show its loss to solvent. Label the C-4 hydroxyl group with ^{18}O, and show its loss to solvent. Label the hydrogens at C-3, C-4, C-5 (separate experiments), and see if they are lost to the solvent, transferred, or if there is no exchange.

c. These results suggest an ElcB elimination of H_2O (fast exchange of the C-3 hydrogens). Because loss of the C-4 hydroxyl group is slower than the formation of the product, this means that either something else is happening after the hydroxyl group is eliminated or the product release from the enzyme is the slow step.

S10-3. This is a *syn* addition, and therefore probably not concerted.

S10-4. The OH must add to C-3 and the H to C-4 (i.e., Michael-wise addition), otherwise there is no stabilized carbanion. Therefore the structure of X is **S10.1.**

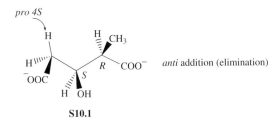

S10.1

a.

b. *Anti* addition (elimination)

c.

S10-5. a.

(substituents on
PLP omitted)

b. 1. Use $[\alpha\text{-}^3H]$ inactivator, and look for the release of ^3H (as ^3H_2O).
   2. Use $[\alpha\text{-}^2H]$ inactivator, and look for an isotope effect.
   3. Do experiments 1 and 2, except with the β-hydrogens.
   4. Use a Cl^- electrode, and monitor the loss of Cl^-.
   5. Look for a change in the UV–visible spectrum from a PLP- to a PMP-like product.

c.

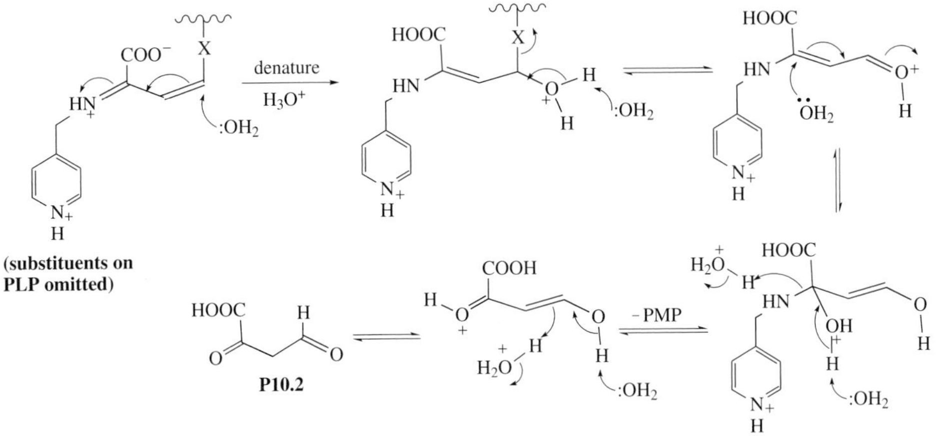

(substituents on PLP omitted)

P10.2

There are several variant mechanisms similar to this mechanism that are reasonable, including initial hydrolysis of the iminium bond.
S10-6.

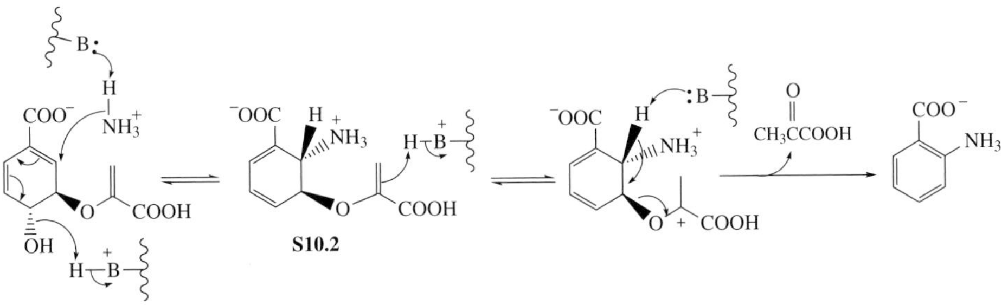

S10.2

Synthesize **S10.2,** and see if it is a substrate. Determine if the NH_4^+ comes from Gln (see Chapter 2, Section II.A) by ^{15}N labeling. Run the reaction in D_2O, and look for a solvent isotope effect.

S10-7.

Therefore, inversion of stereochemistry.

S10-8. *Syn* elimination of water

S10-9.

Labeling experiments, as were done for histidine ammonia-lyase, could be tried. 3-Hydroxyphenylalanine, much more than 4-hydroxyphenylalanine, should accelerate the rate of the electrophilic aromatic substitution mechanism.

Because the function of the dehydroalanyl moiety is to acidify the β-proton for removal, a mutant lacking the dehydroalanyl moiety should be able to utilize 4-nitrophenylalanine better than phenylalanine.

(See Rétey, J. *Naturwissenschaften* **1996**, *83*, 439 and Rétey, J., and co-workers, *Arch. Biochem. Biophys.* **1998**, *359*, 1.)

S10-10. a.

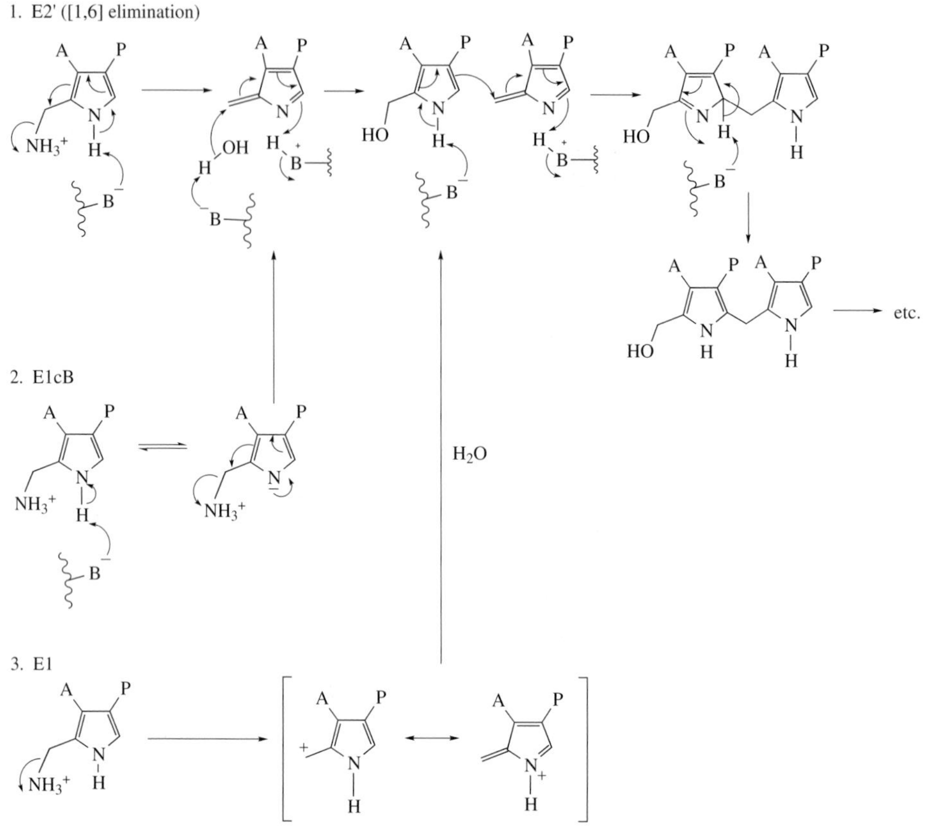
1. E2' ([1,6] elimination)
2. E1cB
3. E1
etc.
H2O

b. To test these mechanisms, the following analogues could be synthesized:

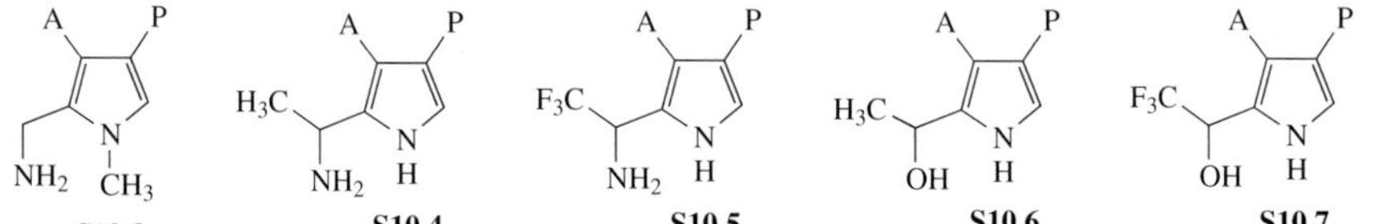

S10.3 **S10.4** **S10.5** **S10.6** **S10.7**

Compound **S10.3** should not be a substrate if mechanisms 2 and 3 are important. Compounds **S10.4** and **S10.6** should be better substrates and **S10.5** and **S10.7** should be worse substrates (or not at all), if mechanism 3 is important because of the electron-donating effect of the methyl substituent and the electron-withdrawing effect of the trifluoromethyl substituent. In fact, **S10.3** is a substrate, **S10.4** and **S10.6** are excellent substrates, and **S10.5** and **S10.7** are not substrates.

(See Pichon, C.; Clemens, K. R.; Jacobson, A. R.; Scott, A. I. *Tetrahedron* **1992,** *48,* 4687.)

K. Chapter 11

S11-1. a.

b. No. Elimination of fluoride would give the same intermediate as the one obtained by elimination of water from serine, so it is a substrate (Silverman, R. B.; Abeles, R. H. *Biochemistry*, **1976**, *15*, 4718–4723).

S11-2. Retro-aldols are catalyzed by a lysine residue at the active site or by metal ions; because metal ions are not required in this example, a Schiff base mechanism must be involved.

a.

b.

This does not explain the NH_2OH experiment, unless $X = CO_2^-$.

inactive enzyme

S11-3. a. Analysis of the product shows the loss of one carbon atom, indicating an α-decarboxylation, which suggests that thiamin diphosphate is involved.

b.

Because TDP is not covalently bound, dialysis will slowly remove the acetyl TDP, and when reassayed (in the presence of TDP) the *apo*-enzyme will become re-activated.

S11-4.

S11-5.

S11-6.

(See Shen, B.; Hutchinson, C. R. *Biochemistry* **1993**, *32*, 11149-54.)

S11-7.

L. Chapter 12

S12-1.

(substituents on
PLP omitted)

(SAM)

(ATP)

S12-2.

S12-3.

S12-4.

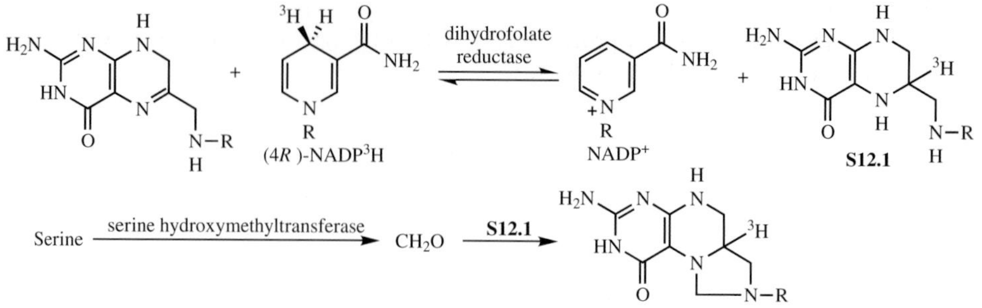

S12-5.

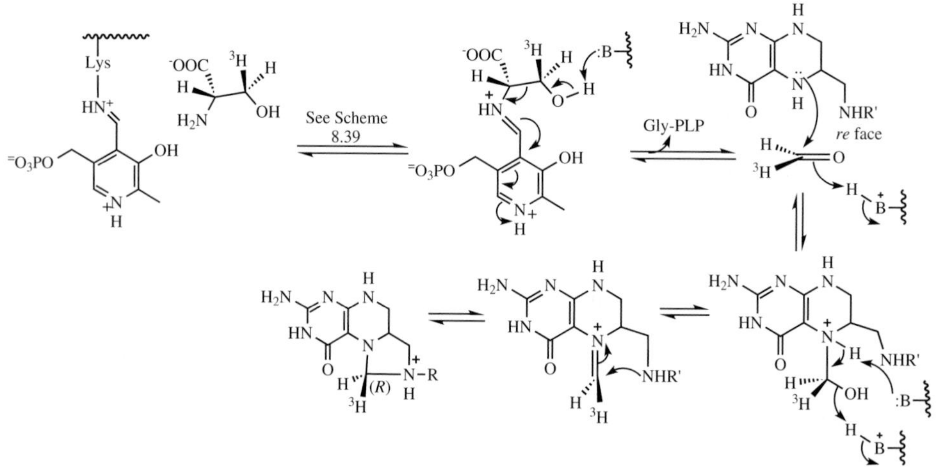

M. Chapter 13

S13-1. a.

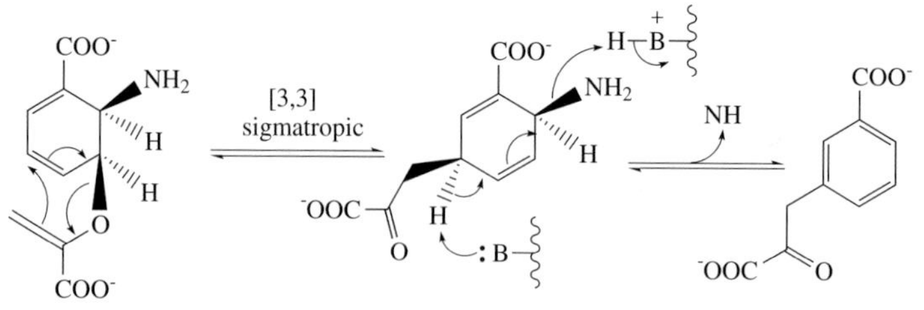

b. Label the ether oxygen with ^{18}O, and show that it becomes the ketone car-bonyl oxygen of the product.

Put a group other than a proton (e.g., a CH_3 or F) at the position where the proton is removed, and look for **S13.1**.

S13.1

S13-2.

S13-3.

S13-4.

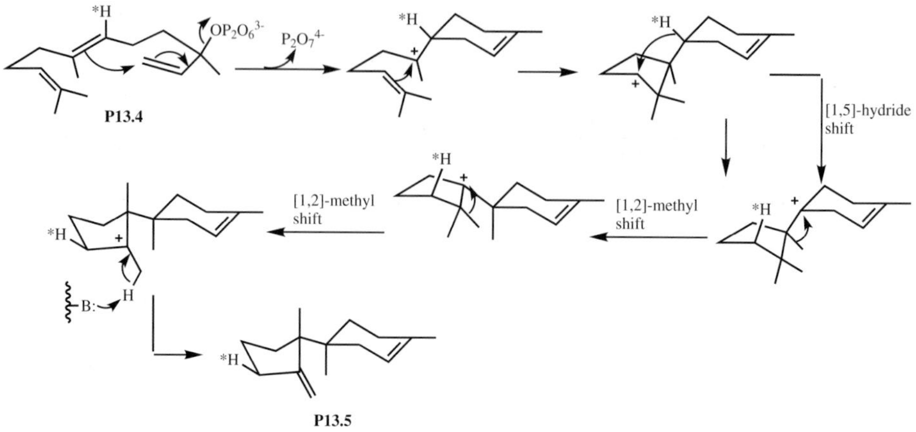
*H
$OP_2O_6^{3-}$
$P_2O_7^{4-}$
P13.4
[1,5]-hydride shift
[1,2]-methyl shift
[1,2]-methyl shift
B:
P13.5

(See Cane, D. E. and co-workers, *Biochemistry* **1997,** *36,* 8332.)

S13-5.

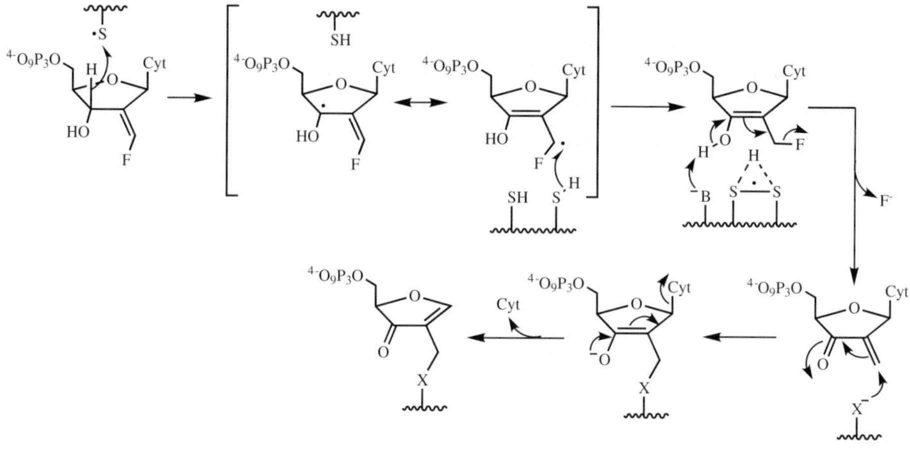
$^{4-}O_9P_3O$
Cyt
HO
F
SH
X
F^-

(See Stubbe, J., and co-workers, *Biochemistry* **1996,** *35,* 8381; *J. Am. Chem. Soc.* **1998,** *120,* 3823.)

S13-6. The second reaction is under radical-generating conditions. Because the first reaction does not undergo cyclopropyl ring cleavage, it suggests that a free-radical intermediate is either very short-lived ($<10^{-7}$ s) or nonexistent.

(See He, M.; Dowd, P. *J. Am. Chem. Soc.* **1996**, *118*, 711.)

N. Biosynthetic Pathways

SBP-1. Try working backward from the final product and forward from the first substrate. The answers in this section are written backward. Always count carbons to get an idea of carboxylation, decarboxylation, and condensations. If there is a choice of two possible routes (like, should the last two steps be dehydrogenase then carboxylase or carboxylase then dehydrogenase), think of the chemistry involved. You should be able to draw mechanisms for every step, but they are not given here.

a.

b.

c. The NaBH$_4$ inactivation experiment suggests imine formation with the substrate. The exchange experiment suggests an initial (in the back reaction) enzyme-catalyzed loss of water.

1.

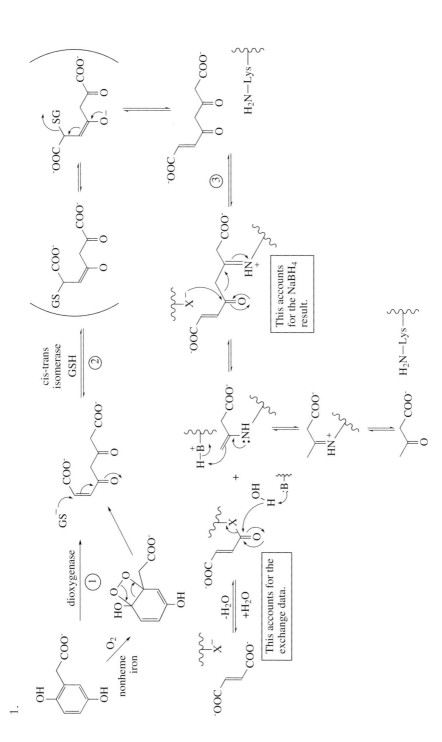

2. Label the ketone oxygen that forms the Schiff base with ^{18}O, and show it is released. Carry out the back reaction with fumarate, quench with NH_2OH, and look for fumarate hydroxamate.

SBP-2.

SBP–3.

The ^{15}N data exclude either NH_4Cl directly or Asp as N donors; therefore Gln is the source.

Index